INTRODUCTION TO FACTS CONTROLLERS

Books in the IEEE Press Series on Power Engineering

Principles of Electric Machines with Power Electronic Applications, Second Edition
M. E. El-Hawary

Pulse Width Modulation for Power Converters: Principles and Practice
D. Grahame Holmes and Thomas Lipo

Analysis of Electric Machinery and Drive Systems, Second Edition
Paul C. Krause, Oleg Wasynczuk, and Scott D. Sudhoff

Risk Assessment for Power Systems: Models, Methods, and Applications
Wenyuan Li

Optimization Principles: Practical Applications to the Operations of Markets of the Electric Power Industry
Narayan S. Rau

Electric Economics: Regulation and Deregulation
Geoffrey Rothwell and Tomas Gomez

Electric Power Systems: Analysis and Control
Fabio Saccomanno

Electrical Insulation for Rotating Machines: Design, Evaluation, Aging, Testing, and Repair
Greg Stone, Edward A. Boulter, Ian Culbert, and Hussein Dhirani

Signal Processing of Power Quality Disturbances
Math H. J. Bollen and Irene Y. H. Gu

Instantaneous Power Theory and Applications to Power Conditioning
Hirofumi Akagi, Edson H. Watanabe and Mauricio Aredes

Maintaining Mission Critical Systems in a 24/7 Environment
Peter M. Curtis

Elements of Tidal-Electric Engineering
Robert H. Clark

Handbook of Large Turbo-Generator Operation and Maintenance, Second Edition
Geoff Klempner and Isidor Kerszenbaum

Introduction to FACTS Controllers: Theory, Modeling, and Applications
Kalyan K. Sen and Mey Ling Sen

INTRODUCTION TO FACTS CONTROLLERS
Theory, Modeling, and Applications

Kalyan K. Sen
Mey Ling Sen

IEEE PRESS SERIES ON POWER ENGINEERING

IEEE

IEEE Press

WILEY

A JOHN WILEY & SONS, INC., PUBLICATION

IEEE Press
445 Hoes Lane
Piscataway, NJ 08855

Published by John Wiley & Sons, Inc., Hoboken, New Jersey
Published simultaneously in Canada.

Library of Congress Cataloging-in-Publication Data is available.

Sen, Kalyan K.
Introduction to FACTS controllers : theory, modeling, and applications / Kalyan K. Sen, Mey Ling Sen.
p. cm.
Includes bibliographical references.
ISBN 978-0-470-47875-2 (cloth)
1. Flexible AC transmission systems. I. Sen, Mey Ling. II. Title.
TK3148.S46 2009
621.319'13—dc22 2009009769

Printed in the United States of America.

10 9 8 7 6 5 4 3 2 1

To our family, friends,
and all our gurus who brought us to this point

CONTENTS

FOREWORD

This book is an outgrowth of two decades of the authors' engineering experience with flexible AC transmission systems (FACTS) design. Today, more than ever before, we are faced with problems of grid congestion and the limitations of power flow in electric power transmission systems. As an integral part of the grid modernization, all alternatives that allow for increased power flow are being explored. FACTS controllers represent the latest and one of the most important technological advances in recent years. It is an emerging field that will require, in the near future, significant research and development. I am convinced that the FACTS controllers will penetrate the high-voltage transmission systems and spread to applications in medium-voltage distribution networks.

Utilities need better modeling techniques for FACTS controllers to use in their system studies. More than anything else, power engineers need to have a basic understanding of the FACTS controllers. When alternatives are considered, the FACTS controllers must be compared from the technical as well as the economical viewpoints.

The book covers a wide area of development and applications of FACTS controllers and is unique in many ways in that it

- Provides the basic theory and the step-by-step evolution to understand FACTS controllers
- Offers modeling techniques that are so essential to electric utilities when conducting their system studies
- Provides computer codes for the FACTS controllers
- Describes a new class of FACTS controllers based on the transformer technology proposed by the authors and named Sen transformer

The topic is of great interest to many utility engineers, large power equipment manufacturers, university professors, and students. The book should benefit all power engineering professionals who want to stay abreast of the evolution of the FACTS technology and associated power electronics. There is no other book that helps a reader to actually "do something" in the field, the way this one does. I believe this book demystifies many of the topics discussed.

In summary, the subject of "FACTS Controllers" cannot be overstressed; it is a very important topic in the electric power industry and electric utilities, particularly in today's environment. Due to the current deregulation trend, the aging of transmission components, the need to reduce costs of operations, the need to ensure reliability of power flow, and the need to achieve efficient transmission utilization, electric utilities will seriously consider all alternatives, including FACTS controllers to improve the power flow along the transmission lines. This is a timely book in the beautiful field of power electronics; it is an advanced application-oriented manual that gives the reader clear instructions on how to model, design, build, evaluate, and install FACTS controllers. It supplements nicely the very few existing books. I realize that this is the first FACTS book that is practical, hands-on, and a true guide for the practicing engineers. I think the authors have an excellent feeling for the prospective readers. The book gives significant amounts of detail in modeling and presentation that will be much appreciated by researchers/engineers in the field. This book is a cornucopia of practical information that is missing in the existing engineering literature.

I have read this book with great satisfaction. It is a work of love, written by two spouses who are former students of Worcester Polytechnic Institute. Kalyan Sen was my Ph.D. student. This book reminds me of the words of Henri Frederic Amiel: "The highest function of the teacher consists not so much in imparting knowledge as in stimulating the pupil in its love and pursuit."

ALEXANDER E. EMANUEL
D.Sc., P.E., Life Fellow, IEEE

Worcester, Massachusetts
January 2009

PREFACE

Both authors have been involved in exploring flexible AC transmission systems (FACTS) controllers since the early 1990s. Kalyan Sen developed VSC-based FACTS models while working at Westinghouse. He has written an award-winning technical committee paper on the modeling of UPFC in the *IEEE Transactions on Power Delivery*. Mey Ling Sen explored an alternate approach to VSC-based FACTS controllers that is cost effective for most utility applications. This effort led to the concept of the Sen transformer (ST). She has modeled all kinds of FACTS controllers in great detail.

Since 2002, Kalyan Sen has traveled around the world as an IEEE Distinguished Lecturer, speaking in more than 30 places. When he gives a presentation on FACTS controllers, his approach is to start from the basics and lead up to the advanced concept of VSC-based FACTS controllers and the ST. His emphasis is based on real-world experience in modeling, simulation, design, and commissioning. He was requested in many places to compile his lecture material in the form of a book. A considerable amount of effort has been put into modeling FACTS controllers. Modeling is the only approach, before any hardware construction, for the verification of the performance of any concept. The book includes some of the major FACTS controllers' models in EMTP.

The book is divided into nine chapters and three appendices. Chapter 1 explains the origin of modern FACTS controllers and guides the reader to the selection process of the right FACTS controller.

Chapter 2 is for anyone who would like to be familiar with the subject and use the book as a reference. It explains various topics of the book in simple engineering terms and corroborates the theory with relevant mathematics. The characteristics of various FACTS controllers, shown from the derived formulae, are verified with detailed simulation in the subsequent chapters.

Chapter 3 gives the "nuts and bolts" of modeling in EMTP and explains the basic differences of modeling various FACTS controllers, such as the voltage regulating transformer (VRT), phase angle regulator (PAR), unified power flow controller (UPFC), and the ST.

Chapter 4 describes the transformer-based FACTS controllers and sets some baselines for comparison with power electronic types in the following chapters.

Chapter 5 describes some early FACTS controllers that use mechanical switches and sets some baselines for comparison in the following chapters.

Chapter 6 describes the heart of the voltage-sourced converter (VSC) using harmonic neutralized (HN) techniques. It also briefly describes other approaches.

Chapter 7 shows intricate modeling and analysis of the basic building block of a VSC—a two-level pole. This chapter may be of interest to those who are interested in switch design. The analysis technique is applicable to other semiconductor switches as well.

Chapter 8 presents the modeling techniques and control implementations of the three basic VSC-based FACTS controllers: shunt-connected static synchronous compensator (STATCOM), series-connected static synchronous series compensator (SSSC), and shunt-series-connected UPFC. A comparison of simulation and field results is presented. The protection strategy of VSC-based FACTS controllers is discussed.

Chapter 9 presents the motive and evolution of the ST and its wide variety of applications.

Appendix A covers the operation of various items, such as (a) three-phase balanced and unbalanced voltage, current, and power; (b) symmetrical components; (c) d–q transformation; and (d) Fourier analysis. The reader will find it useful to see industry techniques and the relevance of theory and applications.

Appendix B presents the power flow control equations in a lossy transmission line and compares the derived results from those in Chapter 2 for lossless lines.

Appendix C presents the sample models that are discussed in this book. However, a more comprehensive set of models is available at the following ftp address:

ftp://ftp.wiley.com/public/sci_tech_med/facts_controllers

Color representations of the figures of the book can be downloaded from the same ftp address.

The Bibliography includes some books and reports and more than 160 technical papers.

KALYAN K. SEN
MEY LING SEN

Pittsburgh, Pennsylvania
January 2009

ACKNOWLEDGMENTS

We both would like to thank our colleagues at Westinghouse where the pioneering development of FACTS controllers took place. We also appreciate our brief time spent in Sweden with our fine colleagues at ABB. We are very grateful for all those who dedicated their time to review the manuscript thoroughly and provide valuable feedback. We thank Prof. Venkata Dinavahi for identifying various references. We appreciate feedback from our readers. Our email addresses are senkk@ieee.org and senml@ieee.org, respectively.

K. K. S.
M. L. S.

Very special thanks go to my two mentors at Westinghouse—Theodore Heinrich and Michael Brennen who are heroes of the power electronics industry. Proper thanks go to Dr. Laszlo Gyugyi who convinced me to change my job from academia to industry, a change that I never regretted.

Throughout my tenure at Westinghouse Science & Technology Center in Pittsburgh, I had a rare opportunity of working with extraordinary people from all over the world. Not only were they a fine class of engineers, they also had hobbies that could be considered as careers in their own rights. During the long hours of commissioning of the world's first UPFC in Kentucky, and a STATCOM in Texas, the team of engineers kept everyone amused with their life stories. I thank the fine colleagues at the Tennessee Valley Authority, American Electric Power, New York Power Authority, Bonneville Power Administration, Western Area Power Administration, and the Electric Power Research Institute.

K. K. S.

NOMENCLATURE

α	Relative phase angle of the shunt-connected compensating voltage
β	Relative phase angle of the series-connected compensating voltage
γ	"Dead" angle of a three-level voltage-sourced converter (VSC) when the output voltage is zero
δ	Power angle (difference of phase angles of the voltages at the two ends of a transmission line)
ε	Least error of the calculated voltage and actually tapped voltage in the Sen transformer
θ	Phase-locked loop (PLL) angle
$^\circ$	degree
ϕ	Power factor angle
ω	Angular frequency
ψ	Angle between the voltages at the modified sending end and the sending end of a line
ζ	Dynamic angle of the series-connected VSC
Ω	Unit of resistance, reactance, and impedance
A	Ampere (unit of current)
BTB-SSSC	Back-to-back SSSC
BTB-STATCOM	Back-to-back STATCOM
BYPBRK	Bypass breaker
C	Capacitance
cos (ϕ)	Power factor
cp	Compensating points in the Sen transformer
CSC	Current-sourced converter
DCLS	DC link switch
$Di_{G(Tfall)}$	Slope of i_G at the midpoint of T_{fall}
$Di_{G(Ttail)}$	Slope of i_G at the midpoint of T_{tail}
E	Compensating voltage behind a tie reactance
ES	Electronic switch
F	Farad (unit of capacitance)
GPFC	Generalized power flow controller
GST	Generalized Sen transformer

H	Henry (unit of inductance)
Hz	Hertz (unit of frequency)
i	Instantaneous current
I	Phasor current
IPFC	Interline power flow controller
I_m	Peak reverse current
I_O	Forward conducting current of a VSC pole
K	Ratio of the GTO currents at the end of T_{fall} to that at the beginning of T_{fall}
L	Inductance
LTC	Load tap changer
MC	Magnetic circuit
MST	Multiline Sen transformer
Mvar	Mega VAR (unit of reactive power)
p	Three-phase instantaneous active power
P	Active power
PAR	Phase angle regulator
PFC	Power flow controller
PST	Phase shifting transformer
q	Three-phase instantaneous quadrature power
Q	Quality factor
Q	Reactive power
Q_a	Reverse recovery charge
r_k	Voltage error at a possible kth operating point in the Sen transformer
R	Resistance
$_r$	Receiving end
S	Apparent power
$_s$	Sending end
$_{s'}$	Modified sending end
SSSC	Static synchronous series compensator
ST	Sen transformer
STATCOM	Static synchronous compensator
SVC	Static var compensator
t	Time
TCR	Thyristor-controlled reactor
T_{fall}	GTO current fall time
THD	Total harmonic distortion
t_{ON}	Turn-on time of a GTO
TSC	Thyristor-switched capacitor
T_{tail}	GTO current tail time
X	Reactance
UPFC	Unified power flow controller
v	Instantaneous voltage
V	Phasor voltage

V	Volt (unit of voltage)
VA	Volt-ampere (unit of apparent power)
VAR	Volt-ampere reactive (unit of reactive power)
v_B	Forward blocking voltage of a GTO
VRT	Voltage regulating transformer
VSC	Voltage-sourced converter
$\mathbf{V}_{\mathbf{s's}}$	Series compensating voltage
W	Watt (unit of active power)

1

APPLICATIONS OF FACTS CONTROLLERS

The locations for electricity generation are based on the presence of energy sources, availability of land for new power plants or substations, need for power in a given area, and availability of a transmission network. Electrical energy is transported from the generating point to the point of use through interconnected transmission lines as shown in Figure 1-1. The flow of electricity takes place freely through the path of least impedance and this natural flow of electricity may cause certain transmission lines to be overloaded or underloaded. The flow of electricity in a particular line of a transmission system can be controlled with the use of a power flow controller (PFC), as shown in Figure 1-2.

The demand for electrical energy around the world increases continuously. The ever-growing need for transporting more electricity can be met either by installing new transmission lines or by using the existing ones in a more efficient way. The construction of new transmission lines is increasingly difficult because of various reasons, such as regulatory, environmental, and public policies, as well as the escalating cost. The free flow of electricity from one particular point to another might not take the shortest path. Any unwanted path along the way causes extra power loss, loop flow of power, and reduced stability with increased voltage variation in the line. In the present environment, the power industry is in constant search for the most economic ways to transfer bulk power along a desired path. Before considering new transmission lines, it is

Introduction to FACTS Controllers. By Kalyan K. Sen and Mey Ling Sen

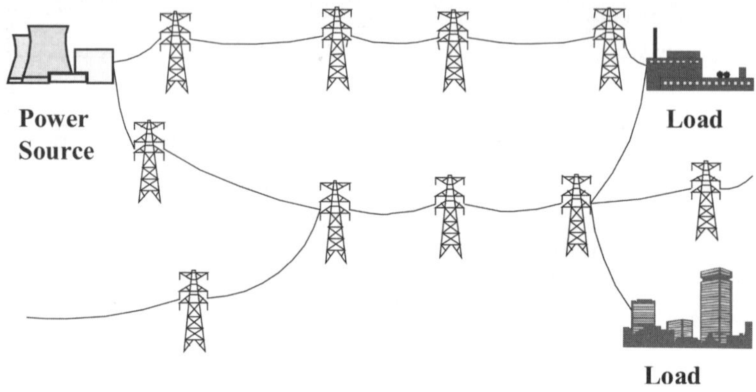

Figure 1-1. Part of a large interconnected transmission system supplying electric power from the generating point to the loads.

desirable to explore other ways to increase the usage of existing transmission lines by increasing their power flow.

The flow of electric power has two components: active power and reactive power. A transmission line consists of electrical conductors that have resistance, inductance, and capacitance. The active power, except for the loss in the resistance of the conductor, reaches from one end of the line to the other. This active power can be converted into lighting, heating, motion force in electric motors, and so on, generating revenue. The inherent inductive and capacitive reactances of the conductor absorb and generate reactive power. This reactive power flow causes an extra loss in the resistance of the conductor.

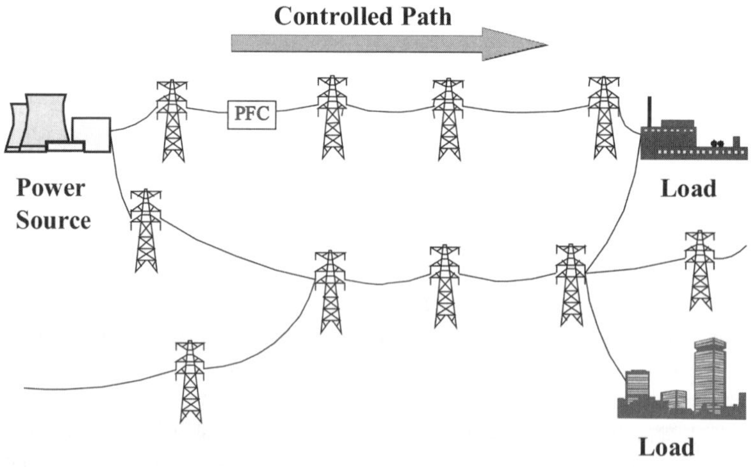

Figure 1-2. Power flow along a controlled path.

The natural or uncompensated power flow through a transmission line in a power network is, in general, not economically optimal. The natural active and reactive power flows (P_{rn} and Q_{rn}) at the receiving end are

$$P_{rn} = \frac{V_s V_r}{X} \sin(\delta_s - \delta_r)$$

and

$$Q_{rn} = \frac{V_s V_r}{X} \left\{ \cos(\delta_s - \delta_r) - \frac{V_r}{V_s} \right\}$$

where the magnitudes of the voltages at the sending and receiving ends are V_s and V_r, the corresponding phase angles are δ_s and δ_r, and the line reactance is X. The power flow control parameters are voltage magnitudes, their phase angles, and line reactance. Any of these parameters can be controlled with the use of the following, now considered conventional, equipment:

- Voltage regulating transformer (VRT), shunt or parallel-connected switched inductor/capacitor, static var compensator (SVC), or static synchronous compensator (STATCOM) for voltage regulation, as shown in Figure 1-3
- Phase angle regulator (PAR) or phase shifting transformer (PST) for phase angle regulation, as shown in Figure 1-4
- Thyristor-controlled series capacitor (TCSC) for series reactance regulation, as shown in Figure 1-5

For more than a century, the transmission line voltage has been regulated with transformers and tap changers. They are referred to in this book as the VRT in the form of a two-winding transformer with isolated windings and an autotransformer with electrical connection between the windings. In both transformers, the line voltage is applied to the primary windings. In the two-winding transformer, the full line voltage is induced in the secondary windings, whereas, in the autotransformer, only a fraction of the line voltage is induced in the secondary windings that are connected to the primary windings to produce the full line voltage. In both cases, the magnitude of the line voltage is regulated. The secondary voltage is varied with the use of load tap changers (LTCs). An LTC can step up/down the voltage without interruption of the load current. Both primary and secondary windings in the two-winding transformer carry the full transmitted power. Both primary and secondary windings in the autotransformer carry only a fraction of the full transmitted power. The indirect way to regulate the line voltage is to connect an inductor or a capacitor in shunt with the transmission line. A shunt-connected inductor absorbs reactive power from the line and lowers the line voltage, whereas a shunt-connected capacitor raises the line voltage with its generated reactive power. The SVC connects fixed capacitors in a step-like manner in shunt with the line through thyristor switches and also connects an inductor in shunt with the line through thyristor switches whose duty cycle can be varied, thereby making it function

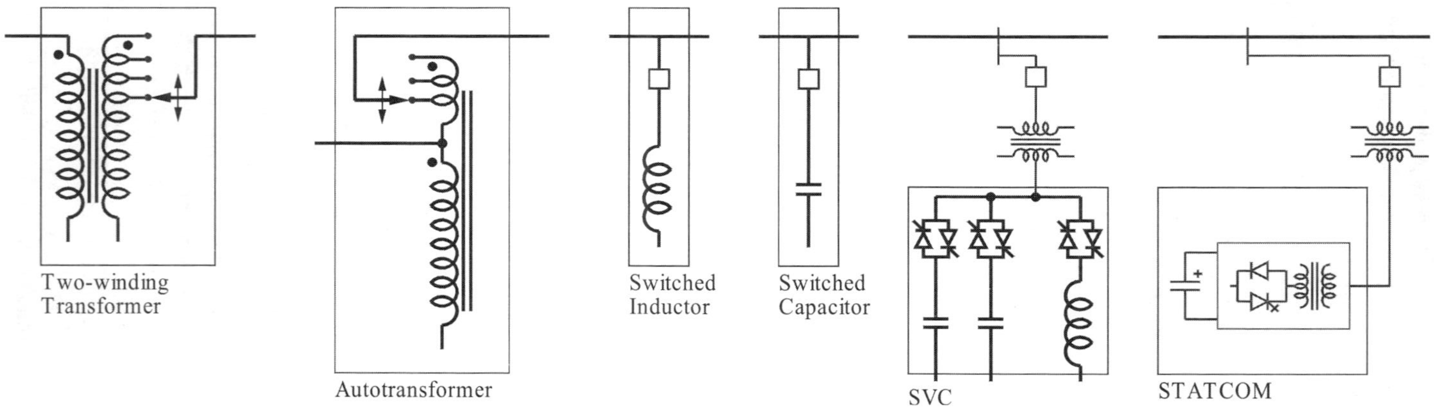

Figure 1-3. Transmission line voltage regulators.

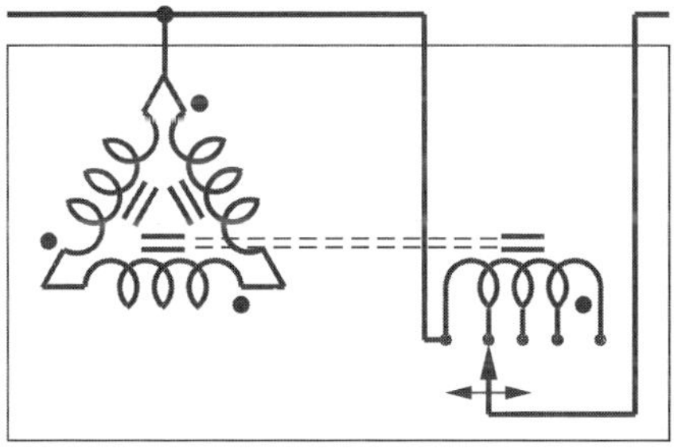

Figure 1-4. Transmission line voltage phase angle regulator.

as a variable inductor. The indirect way to regulate the shunt-connected inductor or capacitor is to use a voltage-sourced converter (VSC)-based STATCOM that connects an electronically generated sinusoidal voltage (with some harmonic components) in shunt with the transmission line through a tie inductor. The same concept has been practiced with the use of a synchronous condenser. The power flow in a transmission line has also been regulated with the use of the PAR. The line voltage is applied to the primary windings and the induced secondary voltage that is varied with the use of LTCs is connected in series with the line. Through the use of the TCSC, a series-connected variable capacitor or a variable inductor can be implemented. As a result, both the magnitude and the phase angle of the line voltage are varied simultaneously.

In a lightly loaded transmission line, the reactive power absorbed by the line inductance becomes much less in comparison to the reactive power generated by the line capacitance. The resulting voltage increase in the line may reach or exceed the allowable limits for the system equipment. In a heavily loaded transmission line, the reactive power needed by the line inductance becomes much more in comparison to the reactive power generated by the line capacitance. The resulting voltage along the line may decrease to a point that is below an acceptable limit. If the voltage along the transmission line is increased to be regulated at its nominal value by using a voltage regulator, the active power flow increases over the natural flow. If the phase angle between the voltages at the two ends of the transmission line is increased by using the PAR, the active power flow also increases. The unintended consequence of increasing active power flow by voltage regulation or phase angle regulation is that the reactive power flow

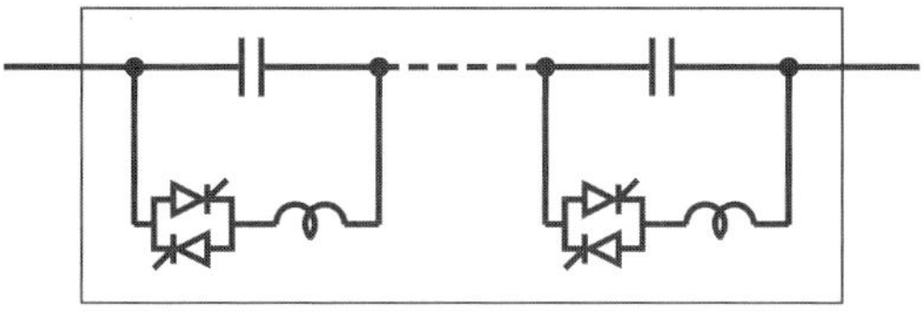

Figure 1-5. Thyristor-controlled series capacitor for transmission line reactance regulation.

in the line is also affected. When the line reactance is regulated, both the active and reactive power flows in the transmission line are varied simultaneously.

If the reactive power along the line is reduced, the freed-up capacity of the line can be used to increase the flow of active power. As a consequence, the generator will be required to supply less reactive power. Furthermore, the efficiencies of the generators and step-up transformers under this condition also increase. In the final analysis, the transmission line needs to be compensated to operate under independent, not simultaneous, control of the active and reactive power flows, so that the line can facilitate the delivery of active power with the greatest value. The active and reactive power flows in a transmission line can be regulated independently by recently developed PFCs that control both the magnitude and phase angle of the transmission line voltage independently.

The magnitude and phase angle of the transmission line voltage can be controlled independently by a shunt-connected compensating voltage, using a shunt–shunt power converter, as shown in Figure 1-6. This concept dates back to the time when rectifiers and inverters were introduced to convert AC power from one voltage and frequency level to another with active power (P_{exch}) transfer through a DC link. The most frequently used topology is an AC-to-DC rectifier followed by a DC-to-AC inverter for variable speed motor drives and, if combined with local energy storage, an uninterruptible AC power supply. To improve the power quality at the rectifier's AC terminal and to accomplish a bidirectional power flow, two DC-to-AC inverters are connected back to back via their DC links, as shown in the figure. This configuration in electric utility applications is known as a back-to-back static synchronous compensator (BTB-STATCOM).

The transfer of power from one line to another can be achieved with the use of a BTB-STATCOM that consists of at least two VSCs, each of which is connected in shunt (parallel) with the transmission line through a coupling transformer. All the VSCs are connected at their shared DC link. The shunt-connected compensating voltage is of variable magnitude and phase angle, and it is also at any phase angle with the prevailing line current. Accordingly, it exchanges active and reactive powers with the line. The exchanged active power flows bidirectionally through the shared link. Each

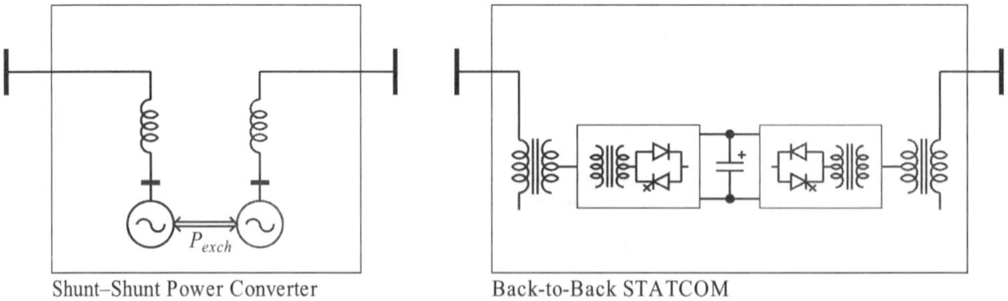

Figure 1-6. Point-to-point transfer of power with local reactive power compensation using a shunt–shunt power converter (BTB-STATCOM).

shunt-connected VSC can also provide independent shunt reactive power compensation at its AC terminal and, thereby, regulates the voltage of the transmission line at the point of compensation. Each shunt-connected VSC is rated for the full line voltage and carries the full line current and, therefore, is rated for its full transmitted power. As an example, in order to increase the power flow in a line from its natural flow of 800 MVA to 1000 MVA, the rating of each of the two units of the shunt–shunt power converter would be 1000 MVA. In certain special cases for point-to-point transfer of power between two isolated networks with different voltages, phase angles, or frequencies, the use of the shunt–shunt power converter remains the preferred topology.

The concept of shunt-connected power conversion can be further extended to include the use of a shared magnetic link in which the compensating voltage is generated from either an electrical machine or a transformer and LTCs. In this case, both active and reactive powers flow through the link. The point-to-point transfer of power from one line to another with different voltages, phase angles, or frequencies can be accomplished with the use of shunt–shunt connected electrical machines. The Sen transformer (ST) can also generate a shunt-connected compensating voltage for the interconnection of two nearby transmission lines with different voltages and phase angles, but of the same frequency.

The shunt–shunt power converter is capable of controlling the power angle (the phase angle between the voltage at the point of compensation and the voltage at the far end of the transmission line) over its full 360° range. The maximum transfer of power along a lossless transmission line (with quality factor $Q = X/R = \infty$) between the sending and receiving ends takes place at the 90° power angle. At the same time, the actual power angle is significantly lower and depends on the line length, system characteristics, and load flows. A transmission line with the natural (uncompensated) power angle in the range of 15° to 20° may have a possible range of compensation of an additional 5° to 10°. Therefore, the shunt–shunt power converter is severely restricted to operate within the first quadrant when used as a PFC. Then the question arises if it is possible to design a new PFC that can control the same amount of transmitted power as the shunt–shunt power converter can and operate within the allowable range of power angles, while the power rating of this new PFC is a fraction of the conventional shunt–shunt power converter. For the same example of increasing the power flow in the line from its natural flow of 800 MVA to 1000 MVA, the rating of each of the two units of the shunt–series power converter might be only 200 MVA, resulting in a saving of 80% of the power rating over a shunt–shunt power converter.

An ideal PFC controls the values of the power flow control parameters (voltage magnitude, its phase angle, and line reactance) to regulate the magnitude and the phase angle of the line voltage independently by adding a series-connected compensating voltage to the original voltage with the use of a shunt–series power converter, as shown in Figure 1-7. The compensating voltage is variable in magnitude and phase angle with respect to the transmission line voltage. This electronic circuit topology was patented by Westinghouse first in 1987 (U.S. patent number 4,651,265 titled "Active Power Conditioner System"). Later this concept was adapted for transmission line applications in a unified power flow controller (UPFC), as shown in the figure. The series-connected VSC is rated for a fraction of the line voltage, but carries the full line

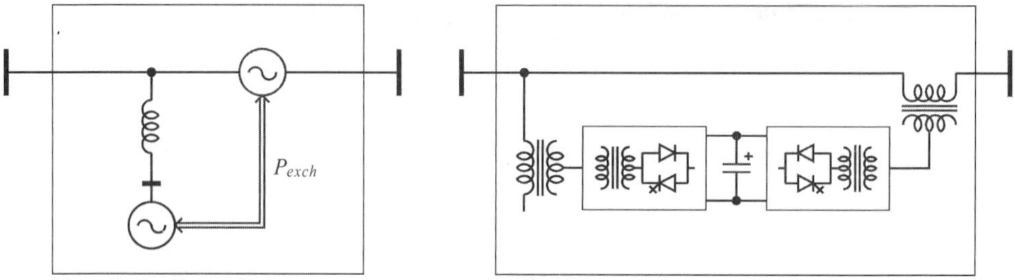

Figure 1-7. Independent active and reactive power flows controller with local reactive power compensation using a shunt–series power converter (UPFC).

current. The shunt-connected VSC is rated for the full line voltage, but carries only a fraction of the line current. Therefore, each VSC carries only a fraction of the full transmitted power.

The shunt–series power converter connects a compensating voltage in series with the line at any phase angle in its full 360° range with respect to the transmission line voltage at the point of compensation. A detailed analysis in Chapter 2, Section 2.1.1, shows that, for example, a series compensating voltage of 0.2 pu modifies the power angle by 11.54°, which may be near the allowable limit. The most important and unique feature of the shunt–series topology is that for a given amount of transmission line power, the series-connected VSC has a large leverage between its own rating and the controlled transmission line power and it needs to be rated for only *a fractional amount of transmitted power,* whereas the shunt-connected VSC in the shunt–shunt topology has no such leverage and it needs to be rated for *the full amount of transmitted power.* Because of this uniqueness, *the shunt–series connection is a preferred topology for a PFC.*

The UPFC consists of two VSCs with a joint DC link capacitor. The two VSCs are connected to the same transmission line through two coupling transformers: one connected in shunt and one connected in series. The series-connected compensating voltage is of variable magnitude and phase angle and it is also at any phase angle with the prevailing line current. Therefore, it exchanges active and reactive powers with the line. The exchanged active power (P_{exch}) flows bidirectionally through the shared link to and from the same transmission line under compensation. Both shunt and series-connected VSCs can also provide independent reactive power compensation at their respective AC terminals. As a special case, when the DC link capacitors of the two VSCs are not connected together, both the shunt-connected VSC (STATCOM) and the series-connected VSC (static synchronous series compensator, SSSC) provide independent reactive power compensation at their respective AC terminals and there is no exchange of active power between them.

In the early 1990s, there was renewed interest to experiment with novel electronic PFCs in power flow control projects due to the availability of semiconductor switches, such as 4500 V, 4000 A-rated gate-turn-off (GTO) thyristors. A new definition, namely flexible alternating current transmission systems (FACTS) was proposed as "alternating current transmission systems incorporating power electronic based and other static controllers to enhance controllability and increased power transfer capability."

In 1998, Westinghouse installed a ±160 MVA-rated flexible AC transmission systems (FACTS) controller (UPFC) at the American Electric Power Inez substation in the state of Kentucky, USA. This UPFC demonstrated for the first time that active and reactive power flows in a transmission line could be regulated independently while maintaining a fixed line voltage at the point of compensation.

In 1998, a patent was granted to General Electric Company, which proposed to implement the independent control of active and reactive power flows such that the compensating voltage was generated using electrical machines (U.S. patent number 5,841,267, titled "Power Flow Control with Rotary Transformers").

The Sens proposed the idea of independent control of active and reactive power flows in a radically low-cost way by using the established technology of regular stationary transformers and LTCs. This implementation is completely different from the original Westinghouse and the GE concepts. The Sens were awarded five U.S. patents (four patents in 2002, all titled "Versatile Power Flow Transformers for Compensating Power Flow in a Transmission Line" and numbered 6,335,613, 6,384,581, 6,396,248, and 6,420,856, and one patent in 2005, titled "Multiline Power Flow Transformer for Compensating Power Flow Among Transmission Lines," numbered 6,841,976).

The VSC-based technology has the capability of providing fast (subcycle) dynamic response for a given transmission line impedance, although in a PFC the dynamic response of at least a few line cycles is necessary to operate safely under contingencies. Most utility applications in the AC system allow regulation of the power flow in the line(s) in a "slow" manner as permitted by the mechanical contacts in the LTCs. If faster response is needed, the mechanical LTCs can be replaced with faster LTC switches. Although the ST, shown in Figure 1-8, provides simultaneous voltage regulation at a point in a transmission line and almost the same independent control of active and reactive power flows as the UPFC, albeit at a reduced dynamic rate, the ST

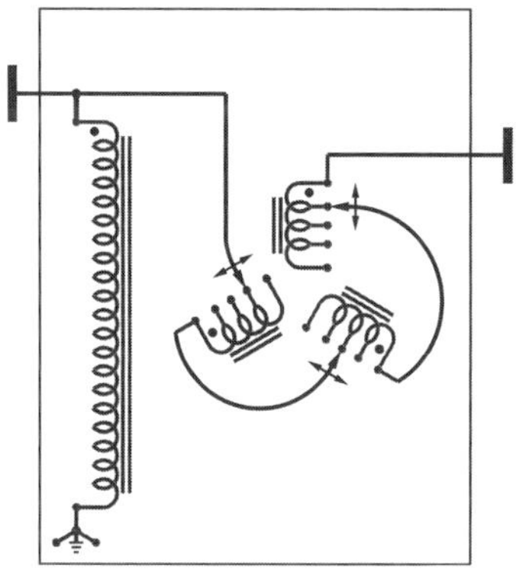

Figure 1-8. Sen transformer.

uses reliable and proven transformer and LTC-based technology that is nearly an order of magnitude more efficient at a fractional cost of the UPFC. The LTC contacts are immersed in the transformer oil and do not need deionized water cooling and heat exchangers to get rid of the semiconductor and snubber losses as in the electronic UPFC. The switching and conduction losses in the two VSCs of the UPFC can be 3–8% of their throughput power depending on its VSC configuration, whereas that in the ST is less than 1%. Therefore, the operating cost of the ST is an order of magnitude less than that of the UPFC.

The compensating voltage in an autotransformer is in phase (0°) or out of phase (180°) with the line voltage and, therefore, regulates the magnitude of the transmission line voltage. The compensating voltage in the PAR is in quadrature (+90° or –90°) with the line voltage and, therefore, regulates the phase angle of the transmission line voltage. The ST creates a series compensating voltage that is variable in magnitude and phase angle and can control the transmission line voltage in both magnitude and phase angle in order to achieve independent control of active and reactive power flows in the line. This compensating voltage may be thought of as two orthogonal compensating voltages of separate autotransformer and PAR. Therefore in the ST, the functions of the autotransformer and the PAR are combined in a single unit that results in a reduced amount of hardware from what is required for separate autotransformer and PAR.

Both the ST and UPFC are suitable for independent control of active and reactive power flows in a single transmission line in which they are installed. However, several transmission lines in close proximity may be connected to a shared voltage bus. Therefore, any change in the power flow in one line will affect the power flows in the other lines as well. Thus, the excessive power from one specific line cannot be transferred directly to another specific line. In a multiline transmission network, it would be advantageous to be able to transfer power from an overloaded to an underloaded line with minimum undesirable impact on the power flows in the other uncompensated lines.

The shared DC link concept can be extended for power exchange between transmission lines with series–series VSCs. The back-to-back static synchronous series compensator (BTB-SSSC), also called interline power flow controller (IPFC), as shown in Figure 1-9, consists of at least two VSCs, each of which is connected in series with a

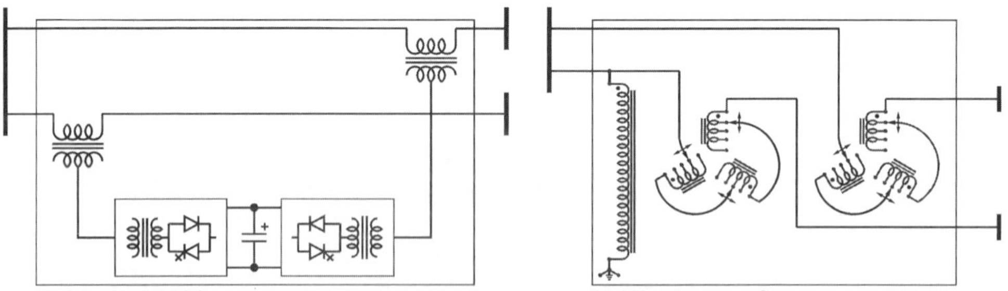

Figure 1-9. Multiline power flow concepts.

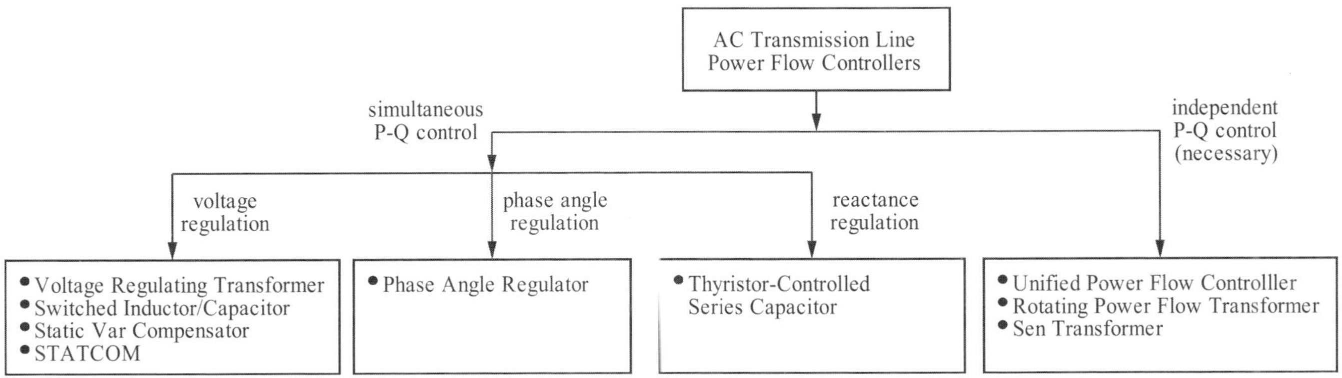

Figure 1-10. Choices for transmission line control equipment.

transmission line. All the VSCs are connected at their shared DC link. The BTB-SSSC transfers active power from one or more transmission lines, referred to as "master" lines, to the others, referred to as "slave" lines, and provides independent series reactive power compensation in each line. A BTB-SSSC selectively controls the active and reactive power flows in each line in a multiline transmission system and provides a power flow management for the transmission system by decreasing the power flow in an overloaded line and increasing the power flow in an underloaded line. The multiline Sen transformer (MST), shown in the figure, provides the same functionality.

In summary, mechanically or electronically switched compensators are used as PFCs, but each of these compensators can control only one of the three power flow control parameters: voltage magnitude, its phase angle, and line reactance. Although the active power flow in the line is regulated, the unwanted reactive power flow is also affected simultaneously, but the optimization of power flow that generates the most revenue can be achieved through independent control of active and reactive power flows in the transmission line. Therefore, the power industry's present need requires the use of PFCs that can independently control the active and reactive power flows in a transmission line, decrease the power flow in an overloaded line, and increase it in an underloaded line, while at the same time keeping the system voltage within the allowable limits. The summary of choices for transmission line power flow control equipment is shown in Figure 1-10 in chronological order of their introduction.

It is well established that the UPFC topology is the most versatile PFC that has been developed. A detailed comparative analysis of the ST and UPFC is given in Chapter 9. The fact is that for a one per-unit (pu) throughput power, the ST requires a less than two pu transformer rating, whereas the "all electronic" UPFC requires a more than four pu transformer rating. Although the ST requires the use of LTCs, the UPFC requires more than eight pu of installed power electronics with its cooling system and heat exchanger. The ST rides through the fault current, but the UPFC requires a protection scheme with an additional electronic bypass switch. In comparison to the UPFC, a 5:1 reduction in equipment cost and a 10:1 improvement in operational cost of the ST are expected. The transformers and LTCs used in the ST are time-tested and proven to be reliable. Hence, the ST is adequate and economically attractive to meet most of the present utility's need for independent control of the active and reactive power flows in the transmission line.

2

POWER FLOW CONTROL CONCEPTS

The theory and implementation of various power flow control methods are described in this chapter. A simple power transmission system with a sending-end voltage, $\mathbf{V_s}$ (i.e., $V_s \angle \delta_s$), and a receiving-end voltage, $\mathbf{V_r}$ (i.e., $V_r \angle \delta_r$), connected by a transmission line reactance (X) and the related phasor diagrams are shown in Figure 2-1. The natural voltage, $\mathbf{V_{Xn}}$ (i.e., $\mathbf{V_s} - \mathbf{V_r}$), across the line reactance (X) is the difference between the sending- and receiving-end voltages. The resulting line current ($\mathbf{I}$) lags the voltage ($\mathbf{V_{Xn}}$) by 90°. The magnitude and phase angle of the voltage with respect to the line current are different at every point along the transmission line. The intermediate line voltages (i.e., $\mathbf{V_1}$, $\mathbf{V_2}$, etc.) are smaller in magnitude than the sending- and receiving-end voltages ($\mathbf{V_s}$ and $\mathbf{V_r}$). The smallest voltage ($\mathbf{V_m}$) is at the midpoint of the transmission line in this illustration. The direct or active and quadrature or reactive components of the line current at the sending end are $\mathbf{I_{ds}}$ and $\mathbf{I_{qs}}$, and the same at the receiving end are $\mathbf{I_{dr}}$ and $\mathbf{I_{qr}}$.

2.1 THEORY

The natural active and reactive power flows P_{sn} and Q_{sn} at the sending end, and P_{rn} and Q_{rn} at the receiving end of the line are derived as follows. Considering the sending-end voltage phasor as reference (i.e., $\delta_s = 0°$), the current through the transmission line is

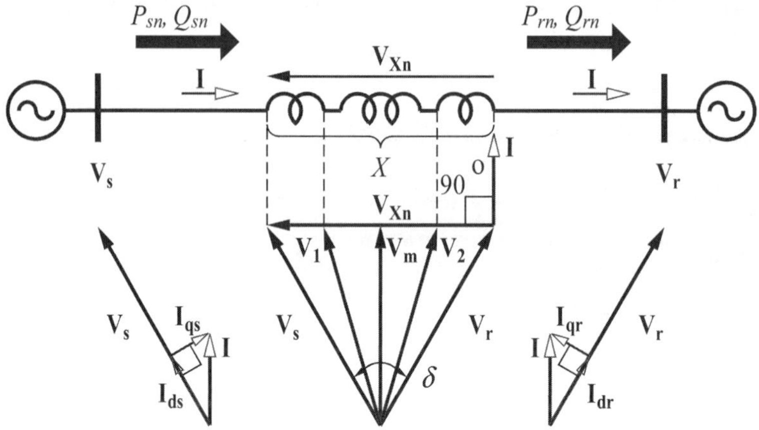

Figure 2-1. Simple power transmission system and the related phasor diagrams.

$$\mathbf{I} = \frac{\mathbf{V_{Xn}}}{jX} = \frac{V_S\angle 0^\circ - V_r\angle -\delta}{jX} = \frac{V_S - V_r\cos\delta + jV_r\sin\delta}{jX}$$

$$= \frac{V_r\sin\delta}{X} + j\frac{V_r\cos\delta - V_S}{X} = I\angle\theta_I$$

or

$$\mathbf{I} = I\cos\theta_I + jI\sin\theta_I \tag{2-1}$$

where

$$I\cos\theta_I = \frac{V_r\sin\delta}{X} \tag{2-2a}$$

and

$$I\sin\theta_I = \frac{V_r\cos\delta - V_S}{X} \tag{2-2b}$$

The natural active power flow at the sending end is

$$P_{sn} = V_S I\cos(\delta_S - \theta_I) = V_S I\cos(-\theta_I) = V_S I\cos\theta_I \tag{2-3}$$

Then equations 2-2a and 2-3 can be combined as

$$P_{sn} = A_n\sin\delta \tag{2-4}$$

where

$$A_n = \frac{V_s V_r}{X} \tag{2-5}$$

The natural reactive power flow at the sending end is

$$Q_{sn} = V_s I \sin(\delta_s - \theta_I) = V_s I \sin(-\theta_I) = -V_s I \sin \theta_I \tag{2-6}$$

Then equations 2-2b and 2-6 can be combined as

$$Q_{sn} = A_n \left(\frac{V_s}{V_r} - \cos \delta \right) \tag{2-7}$$

where A_n is given in equation 2-5.

Equations 2-4 and 2-7 can then be rearranged using equation 2-5 as

$$\frac{-P_{sn}}{Q_{sn} - \frac{V_s^2}{X}} = \tan \delta \tag{2-8}$$

Considering the receiving-end voltage phasor as reference (i.e., $\delta_r = 0°$), the current through the transmission line is

$$\mathbf{I} = \frac{\mathbf{V_{Xn}}}{jX} = \frac{V_s \angle \delta - V_r \angle 0°}{jX} = \frac{V_s \cos \delta + jV_s \sin \delta - V_r}{jX}$$

$$= \frac{V_s \sin \delta}{X} + j \frac{V_r - V_s \cos \delta}{X} = I \angle \theta_I$$

or

$$\mathbf{I} = I \cos \theta_I + jI \sin \theta_I \tag{2-1}$$

where

$$I \cos \theta_I = \frac{V_s \sin \delta}{X} \tag{2-9a}$$

and

$$I \sin \theta_I = \frac{V_r - V_s \cos \delta}{X} \tag{2-9b}$$

The natural active power flow at the receiving end is

$$P_{rn} = V_r I \cos\left(\delta_r - \theta_I\right) = V_r I \cos\left(-\theta_I\right) = V_r I \cos\ \theta_I \tag{2-10}$$

Then equations 2-9a and 2-10 can be combined as

$$P_{rn} = A_n \sin\delta \tag{2-11}$$

where A_n is given in equation 2-5.

The natural reactive power flow at the receiving end is

$$Q_{rn} = V_r I \sin\left(\delta_r - \theta_I\right) = V_r I \sin\left(-\theta_I\right) = -V_r I \sin\ \theta_I \tag{2-12}$$

Then equations 2-9b and 2-12 can be combined as

$$Q_{rn} = A_n \left(\cos\delta - \frac{V_r}{V_s} \right) \tag{2-13}$$

where A_n is given in equation 2-5.

Equations 2-11 and 2-13 can then be rearranged using equation 2-5 as

$$\frac{P_{rn}}{Q_{rn} + \dfrac{V_r^2}{X}} = \tan\delta \tag{2-14}$$

In general terms, the natural active and reactive power flows (P_{sn} and Q_{sn}) at the sending end are

$$P_{sn} = \frac{V_s V_r}{X} \sin\left(\delta_s - \delta_r\right) \tag{2-15a}$$

and

$$Q_{sn} = \frac{V_s V_r}{X} \left\{ \frac{V_s}{V_r} - \cos\left(\delta_s - \delta_r\right) \right\} \tag{2-15b}$$

The natural active and reactive power flows (P_{rn} and Q_{rn}) at the receiving end are

$$P_{rn} = \frac{V_s V_r}{X} \sin\left(\delta_s - \delta_r\right) \tag{2-16a}$$

and

$$Q_{rn} = \frac{V_s V_r}{X} \left\{ \cos\left(\delta_s - \delta_r\right) - \frac{V_r}{V_s} \right\} \tag{2-16b}$$

The active and reactive power flows in the line can be regulated if the magnitude and phase angle of the line voltage are controlled. Alternately, an additional inductor or capacitor in series with the line can control the effective reactance of the transmission line between its sending and receiving ends and the consequence is the regulation of the active and reactive power flows in the line. When one or more power flow control parameters (voltage magnitude, its phase angle, and line reactance) are controlled, the line is said to be "compensated."

Check

Consider the magnitudes (V_s and V_r) of the sending- and receiving-end voltages to be one pu, the phase angle, δ (i.e., $\delta_s - \delta_r$), between them to be 30°, and the line reactance (X) to be 0.5 pu. What is the natural voltage ($\mathbf{V_{Xn}}$) across the line reactance (X)? What is the resulting current? What are the active and reactive powers at the sending end, midpoint, and the receiving end of the transmission line?

Answer

The natural voltage, $\mathbf{V_{Xn}}$ (i.e., $\mathbf{V_s} - \mathbf{V_r}$), is the difference between the sending- and receiving-end voltages, the magnitude of which is 0.5176 pu. The resulting line current, $\mathbf{I}$ (i.e., V_{Xn}/X = 1.0352 pu), lags the voltage ($\mathbf{V_{Xn}}$) by 90°. In the transmission line shown in Figure 2-1, the natural active power flow is the same at the sending end ($P_{sn} = V_s I_{ds}$), at the midpoint ($P_{mn} = V_m I$), and at the receiving end ($P_{rn} = V_r I_{dr}$), and it is one pu from equations 2-4 or 2-11. However, the natural reactive power flow at the sending end is ($Q_{sn} = V_s I_{qs}$) 0.268 pu (inductive) from equation 2-7. At the midpoint, it is (Q_{mn}) zero and at the receiving end, it is ($Q_{rn} = V_r I_{qr}$) 0.268 pu (capacitive) from equation 2-13. Note that there are only active but no reactive current flows at the midpoint of the transmission line. Therefore, there are only active but no reactive power flows at the midpoint of the transmission line.

In this example of a two-bus network, the reactive power absorbed by the "left half" of the transmission line is supplied by the generator and that absorbed by the "right half" of the transmission line is supplied by the load that can be thought of as a back emf of a motor.

Consider the transmission line compensated at its sending end. Assuming that there are no changes in the transmission line's reactance (X) and the receiving-end voltage, $\mathbf{V_r}$ (i.e., $V_r \angle \delta_r$), the power flow in the line can be regulated by controlling the magnitude and phase angle of the sending-end voltage so that it becomes the modified sending-end voltage, $\mathbf{V_{s'}}$ (i.e., $V_{s'} \angle \delta_{s'}$). Controlling one parameter (voltage magnitude or its phase angle) using a power flow controller (PFC) affects both the active and reactive power flows (P_r and Q_r) in the line simultaneously. The key to regulate the active and reactive power flows (P_r and Q_r) so they become a particular pair of values is to control the sending-end voltage to a specific magnitude ($V_{s'}$) and phase angle ($\delta_{s'}$). The modified sending-end voltage $\mathbf{V_{s'}}$ (i.e., $V_{s'} \angle \delta_{s'}$) can be obtained by adding a series-connected compensating voltage $\mathbf{V_{s's}}$ (i.e., $V_{s's} \angle \delta_s + \beta$) to the original sending-end voltage $\mathbf{V_s}$ (i.e., $V_s \angle \delta_s$) as shown in Figure 2-2a. For point-to-point transfer of power between two isolated networks or interconnection of two transmission lines with different voltages or

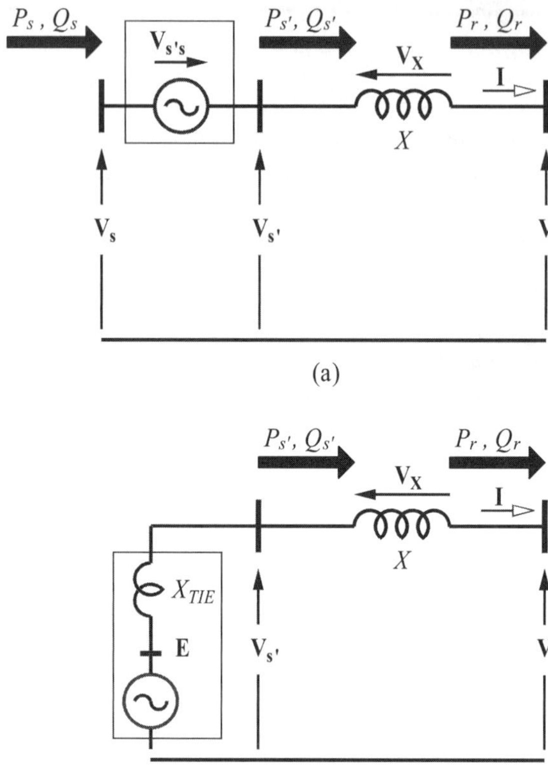

Figure 2-2. Two methods of controlling a transmission line voltage: (a) with a series-connected compensating voltage and (b) with a shunt-connected compensating voltage.

phase angles (or frequencies), the modified sending-end voltage $\mathbf{V}_{s'}$ (i.e., $V_{s'} \angle \delta_{s'}$) can be obtained from a shunt-connected compensating voltage **E** (i.e., $E \angle \delta_{s'} + \alpha$) as shown in Figure 2-2b. Regardless of whether the compensating voltage is series-connected or shunt-connected, it is at any phase angle with respect to its terminal current and, therefore, exchanges both active and reactive powers with the line.

There are three known methods to implement the series and shunt-connected compensating voltages using

1. A voltage-sourced converter (VSC)-based shunt–series and shunt–shunt power converters
2. An electric machine-based shunt–series and shunt–shunt power converters
3. A transformer and load tap changers (LTCs)-based shunt–series and shunt–shunt power converters

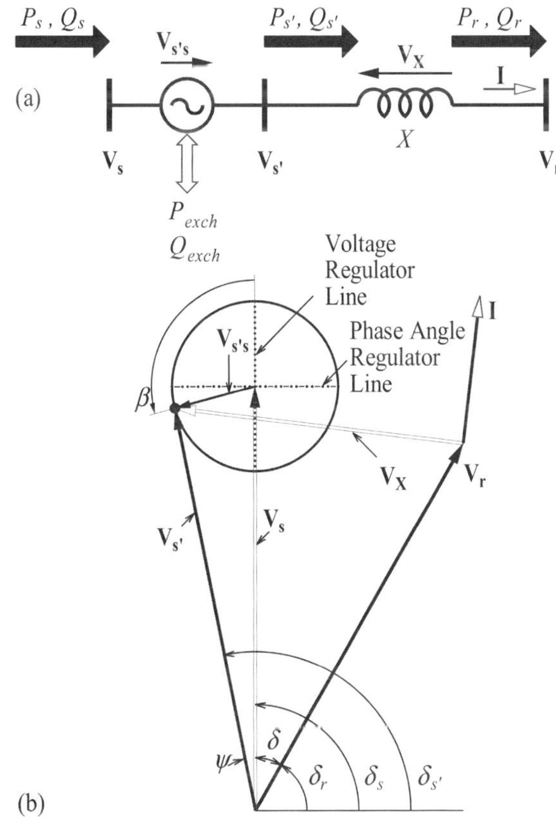

Figure 2-3. (a) Power transmission system with a series-connected compensating voltage ($\mathbf{V}_{s's}$). (b) Phasor diagram.

2.1.1 Series-Connected Compensating Voltage

Consider the simple power transmission system mentioned above with a compensating voltage $\mathbf{V}_{s's}$ (i.e., $V_{s's} \angle \delta_s + \beta$), as shown in Figure 2-3a. When the compensating voltage ($\mathbf{V}_{s's}$) is added in series with the transmission line at the sending-end voltage ($\mathbf{V}_s$), the modified sending-end voltage becomes $\mathbf{V}_{s'}$ (i.e., $\mathbf{V}_s + \mathbf{V}_{s's}$). Figure 2-3b shows the phasor diagram related to a series-connected compensating voltage with a fixed magnitude of 0.2 pu and its entire controllable range of $0° \leq \beta \leq 360°$. The difference voltage ($\mathbf{V}_{s'} - \mathbf{V}_r$) provides the voltage ($\mathbf{V}_X$) across the line reactance (X). As the relative phase angle (β) is varied over its full 360° range, the end of phasor ($\mathbf{V}_{s's}$) moves along a circle with its center located at the end of phasor ($\mathbf{V}_s$). The rotation of phasor ($\mathbf{V}_{s's}$) with an angle (β) modulates both the magnitude and phase angle of the phasor ($\mathbf{V}_X$). In a special case, when $\beta = 0°$ or 180°, the compensating voltage controls the magnitude of the line voltage, as in the case of the voltage regulating transformer

(VRT). In another special case, when $\beta = \pm 90°$, the compensating voltage controls mainly the phase angle of the line voltage, as in the case of the phase angle regulator (PAR).

When a series-connected compensating voltage is used to control the transmission line voltage, the compensating voltage is also at any phase angle with the prevailing line current. In the process, it emulates in series with the line a capacitor that increases the power flow of the line or an inductor that decreases the power flow of the line, and a positive resistor that absorbs active power from the line or a negative resistor that delivers active power to the line. Therefore, the desired compensating voltage is actually an impedance emulator.

Considering the sending-end voltage phasor as reference (i.e., $\delta_s = 0°$), it can be written from Figure 2-3b that

$$V_{s'} \cos \psi = V_s + V_{s's} \cos \beta \tag{2-17a}$$

and

$$V_{s'} \sin \psi = V_{s's} \sin \beta \tag{2-17b}$$

Therefore, the magnitude ($V_{s'}$) and relative phase angle, ψ (i.e., $\delta_{s'} - \delta_s$), of the modified sending-end voltage ($\mathbf{V}_{s'}$) are derived from equations 2-17a and 2-17b as

$$V_{s'} = \sqrt{V_s^2 + V_{s's}^2 + 2V_s V_{s's} \cos \beta} \tag{2-18a}$$

and

$$\psi = \tan^{-1} \frac{V_{s's} \sin \beta}{V_s + V_{s's} \cos \beta} \tag{2-18b}$$

In the transmission system considered above, the magnitude and phase angle of the modified sending-end voltage ($\mathbf{V}_{s'}$) are modulated between $V_s + V_{s's}$ at $\beta = 0°$ and $V_s - V_{s's}$ at $\beta = 180°$ and between $\delta_s + \psi_{max}$ and $\delta_s - \psi_{max}$, respectively, where $V_{s's} = 0.2$ pu and $\psi_{max} = 11.54°$, as shown in Figure 2-4.

2.1.1.1 Power at the Sending End. The equation for the current through the transmission line with series compensation can be derived as

$$\mathbf{I} = \frac{\mathbf{V_X}}{jX} = \frac{\mathbf{V_{s'}} - \mathbf{V_r}}{jX} = \frac{V_s\angle 0° + V_{s's}\angle\beta - V_r\angle -\delta}{jX}$$

$$= \frac{V_s + V_{s's}\cos\beta + jV_{s's}\sin\beta - V_r\cos\delta + jV_r\sin\delta}{jX}$$

$$= \frac{V_r \sin\delta + V_{s's}\sin\beta}{X} + j\frac{V_r\cos\delta - V_s - V_{s's}\cos\beta}{X} = I\angle\theta_I$$

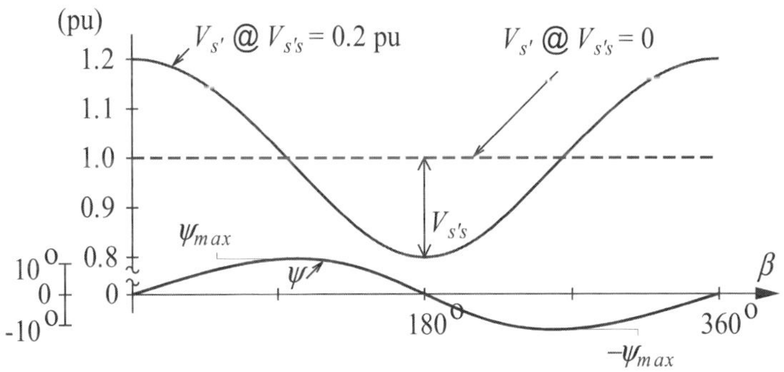

Figure 2-4. Variations of the magnitude ($V_{s'}$) and phase angle (ψ) of the modified sending-end voltage as a function of the relative phase angle (β) of the compensating voltage ($\mathbf{V}_{s's}$) with a fixed magnitude of 0.2 pu.

or

$$\mathbf{I} = I\cos\theta_I + jI\sin\theta_I \tag{2-1}$$

where

$$I\cos\theta_I = \frac{V_r\sin\delta + V_{s's}\sin\beta}{X} \tag{2-19a}$$

and

$$I\sin\theta_I = \frac{V_r\cos\delta - V_s - V_{s's}\cos\beta}{X} \tag{2-19b}$$

The active power flow at the sending end of the transmission line with compensation is

$$P_s = V_s I\cos(\delta_s - \theta_I) = V_s I\cos(-\theta_I) = V_s I\cos\theta_I \tag{2-20}$$

Then equations 2-19a and 2-20 can be combined as

$$P_s = \frac{V_s V_r}{X}\sin\delta + \frac{V_s V_{s's}}{X}\sin\beta$$

or

$$P_s = P_{sn} + A_s\sin\beta \tag{2-21}$$

where P_{sn} is given in equation 2-4 and

$$A_s = \frac{V_s V_{s's}}{X} \tag{2-22}$$

The reactive power flow at the sending end of the transmission line with compensation is

$$Q_S = V_S I \sin(\delta_S - \theta_I) = V_S I \sin(-\theta_I) = -V_S I \sin\theta_I \tag{2-23}$$

Then equations 2-19b and 2-23 can be combined as

$$Q_S = -V_S \frac{V_r \cos\delta - V_S - V_{s's}\cos\beta}{X} = \frac{V_S V_r}{X}\left(\frac{V_S}{V_r} - \cos\delta\right) + \frac{V_S V_{s's}}{X}\cos\beta$$

or

$$Q_S = Q_{Sn} + A_S \cos\beta \tag{2-24}$$

where Q_{sn} and A_s are given in equations 2-7 and 2-22, respectively.

Equations 2-21 and 2-24 can be rearranged as

$$(P_S - P_{Sn})^2 + (Q_S - Q_{Sn})^2 = A_S^2 \tag{2-25}$$

and

$$\frac{P_S - P_{Sn}}{Q_S - Q_{Sn}} = \tan\beta \tag{2-26}$$

Equation 2-25 defines the relationship between P_s and Q_s as a circle centered at (P_{sn}, Q_{sn}) with a radius of A_s. The active and reactive power flows (P_s and Q_s) at the sending end as functions of the relative phase angle (β) are plotted in Figure 2-5a. The active and reactive power flows at the sending end of the line can be regulated within the range defined by the *P–Q* plot in Figure 2-5b by controlling the magnitude ($V_{s's}$) of the compensating voltage ($\mathbf{V}_{s's}$) within its rated value and the relative phase angle (β) between 0° and 360°.

The magnitude ($V_{s's}$) of the compensating voltage from equations 2-22 and 2-25 is

$$V_{s's} = \frac{X}{V_S}\sqrt{(P_S - P_{Sn})^2 + (Q_S - Q_{Sn})^2} \tag{2-27}$$

The relative phase angle (β) of the compensating voltage ($\mathbf{V}_{s's}$) from equation 2-26 is

$$\beta = \tan^{-1}\frac{P_S - P_{Sn}}{Q_S - Q_{Sn}} \tag{2-28}$$

For the desired active and reactive power flows (P_S^*, Q_S^*) within the allowable range, the magnitude ($V_{s's}$) and the relative phase angle (β) of the compensating volt-

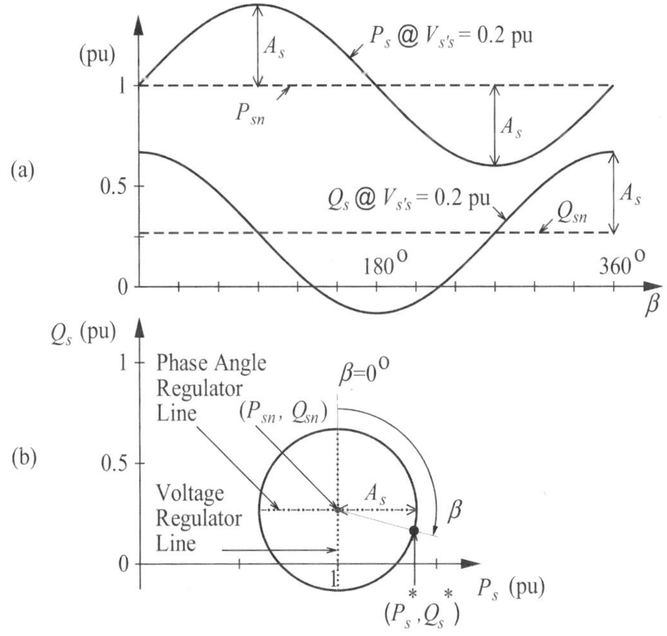

Figure 2-5. (a) Variation of the sending-end active and reactive power flows (P_s and Q_s) as a function of the relative phase angle (β) of the compensating voltage ($\mathbf{V}_{s's}$) with a fixed magnitude of 0.2 pu. (b) Sending end Q_s versus P_s.

age are calculated by substituting $P_s = P_s^*$ and $Q_s = Q_s^*$ in equations 2-27 and 2-28. The relative phase angle (β) is the angle between the voltage regulator line (i.e., $\beta = 0°$ line) and the line that passes through (P_{sn}, Q_{sn}) and (P_s^*, Q_s^*).

For implementation of the VRT, the relative phase angle is set to $\beta = 0°$ or $\beta = 180°$ in equations 2-21 and 2-24. For $\beta = 0°$, the variations of P_s and Q_s are given by

$$P_s = P_{sn} + A_s \sin 0° = P_{sn} \tag{2-29a}$$

and

$$Q_s = Q_{sn} + A_s \cos 0° = Q_{sn} + A_s \tag{2-29b}$$

For $\beta = 180°$, the variations of P_s and Q_s are given by

$$P_s = P_{sn} + A_s \sin 180° = P_{sn} \tag{2-30a}$$

and

$$Q_s = Q_{sn} + A_s \cos 180° = Q_{sn} - A_s \tag{2-30b}$$

The variations of P_s and Q_s are then given by

$$P_s = P_{sn} \tag{2-31a}$$

and

$$Q_s = Q_{sn} \pm A_s \tag{2-31b}$$

for $\beta = 0°$ or 180°.

For implementation of the PAR, the relative phase angle is set to $\beta = 90°$ or $\beta = -90°$ in equations 2-21 and 2-24. For $\beta = 90°$, the variations of P_s and Q_s are given by

$$P_s = P_{sn} + A_s \sin 90° = P_{sn} + A_s \tag{2-32a}$$

and

$$Q_s = Q_{sn} + A_s \cos 90° = Q_{sn} \tag{2-32b}$$

For $\beta = -90°$, the variations of P_s and Q_s are given by

$$P_s = P_{sn} + A_s \sin(-90°) = P_{sn} - A_s \tag{2-33a}$$

and

$$Q_s = Q_{sn} + A_s \cos(-90°) = Q_{sn} \tag{2-33b}$$

The variations of P_s and Q_s are then given by

$$P_s = P_{sn} \pm A_s \tag{2-34a}$$

and

$$Q_s = Q_{sn} \tag{2-34b}$$

for $\beta = \pm 90°$.

In the case of $\beta = 0°$ or $\beta = 180°$, only the reactive power (Q_s), and, in the case of $\beta = \pm 90°$, only the active power (P_s) at the sending end vary with the magnitude ($V_{s's}$) of the compensating voltage ($\mathbf{V}_{s's}$).

2.1.1.2 Power at the Receiving End.

The active power flow at the receiving end of the transmission line with compensation is

$$P_r = V_r I \cos(\delta_r - \theta_I) = V_r I \cos(-\delta - \theta_I) = V_r I \cos(\delta + \theta_I)$$

or

$$P_r = V_r \cos\delta I \cos\theta_I - V_r \sin\delta I \sin\theta_I \tag{2-35}$$

Equations 2-19a, 2-19b, and 2-35 can then be rearranged as

$$P_r = V_r \cos\delta \frac{V_r \sin\delta + V_{s's} \sin\beta}{X} - V_r \sin\delta \frac{V_r \cos\delta - V_s - V_{s's} \cos\beta}{X}$$

$$= \frac{V_r V_{s's} \cos\delta \sin\beta}{X} + \frac{V_r V_s \sin\delta}{X} + \frac{V_r V_{s's} \sin\delta \cos\beta}{X}$$

$$= \frac{V_s V_r}{X} \sin\delta + \frac{V_r V_{s's}}{X} \sin(\delta + \beta)$$

or

$$P_r = P_{rn} + A_r \sin(\delta + \beta) \tag{2-36}$$

where P_{rn} is given in equation 2-11 and

$$A_r = \frac{V_r V_{s's}}{X} \tag{2-37}$$

The reactive power flow at the receiving end of the transmission line with compensation is

$$Q_r = V_r I \sin(\delta_r - \theta_I) = V_r I \sin(-\delta - \theta_I) = -V_r I \sin(\delta + \theta_I)$$

or

$$Q_r = -V_r \sin\delta I \cos\theta_I - V_r \cos\delta I \sin\theta_I \tag{2-38}$$

Equations 2-19a, 2-19b, and 2-38 can then be rearranged as

$$Q_r = -V_r \sin\delta \frac{V_r \sin\delta + V_{s's} \sin\beta}{X} - V_r \cos\delta \frac{V_r \cos\delta - V_s - V_{s's} \cos\beta}{X}$$

$$= -\frac{V_r^2 \sin^2\delta}{X} - \frac{V_r V_{s's} \sin\delta \sin\beta}{X} - \frac{V_r^2 \cos^2\delta}{X} + \frac{V_s V_r \cos\delta}{X} + \frac{V_r V_{s's} \cos\delta \cos\beta}{X}$$

$$= \frac{V_s V_r}{X} \cos\delta - \frac{V_r^2}{X} + \frac{V_r V_{s's}}{X} \cos(\delta + \beta)$$

$$= \frac{V_s V_r}{X}\left(\cos\delta - \frac{V_r}{V_s}\right) + \frac{V_r V_{s's}}{X} \cos(\delta + \beta)$$

or

$$Q_r = Q_{rn} + A_r \cos(\delta + \beta) \tag{2-39}$$

where Q_{rn} and A_r are given in equations 2-13 and 2-37, respectively.

Equations 2-36 and 2-39 can be rearranged as

$$(P_r - P_{rn})^2 + (Q_r - Q_{rn})^2 = A_r^2 \tag{2-40}$$

and

$$\frac{P_r - P_{rn}}{Q_r - Q_{rn}} = \tan(\delta + \beta) \tag{2-41}$$

Equation 2-40 defines the relationship between P_r and Q_r as a circle centered at (P_{rn}, Q_{rn}) with a radius of A_r. The active and reactive power flows (P_r and Q_r) at the receiving end as functions of the relative phase angle (β) are plotted in Figure 2-6a. The active and reactive power flows at the receiving end of the line can be regulated within the range defined by the P–Q plot in Figure 2-6b by controlling the magnitude ($V_{s's}$) of the compensating voltage ($\mathbf{V}_{s's}$) within its rated value and the relative phase angle (β) between 0° and 360°.

The magnitude ($V_{s's}$) of the compensating voltage is derived from equations 2-37 and 2-40 as

$$V_{s's} = \frac{X}{V_r}\sqrt{(P_r - P_{rn})^2 + (Q_r - Q_{rn})^2} \tag{2-42}$$

The relative phase angle (β) of the compensating voltage ($\mathbf{V}_{s's}$) is derived from equation 2-41 as

$$\beta = \tan^{-1}\frac{P_r - P_{rn}}{Q_r - Q_{rn}} - \delta \tag{2-43}$$

For the desired active and reactive power flows (P_r^*, Q_r^*) within the allowable range, the magnitude ($V_{s's}$) and the relative phase angle (β) of the compensating voltage are calculated by substituting $P_r = P_r^*$ and $Q_r = Q_r^*$ in equations 2-42 and 2-43. The relative phase angle (β) is the angle between the voltage regulator line (i.e., $\beta = 0°$ line) and the line that passes through (P_{rn}, Q_{rn}) and (P_r^*, Q_r^*).

For implementation of the VRT, the relative phase angle is set to $\beta = 0°$ or $\beta = 180°$ in equations 2-36 and 2-39. For $\beta = 0°$, the variations of P_r and Q_r are given by

$$P_r = P_{rn} + A_r \sin(\delta + 0°) = P_{rn} + A_r \sin\delta \tag{2-44a}$$

and

$$Q_r = Q_{rn} + A_r \cos(\delta + 0°) = Q_{rn} + A_r \cos\delta \tag{2-44b}$$

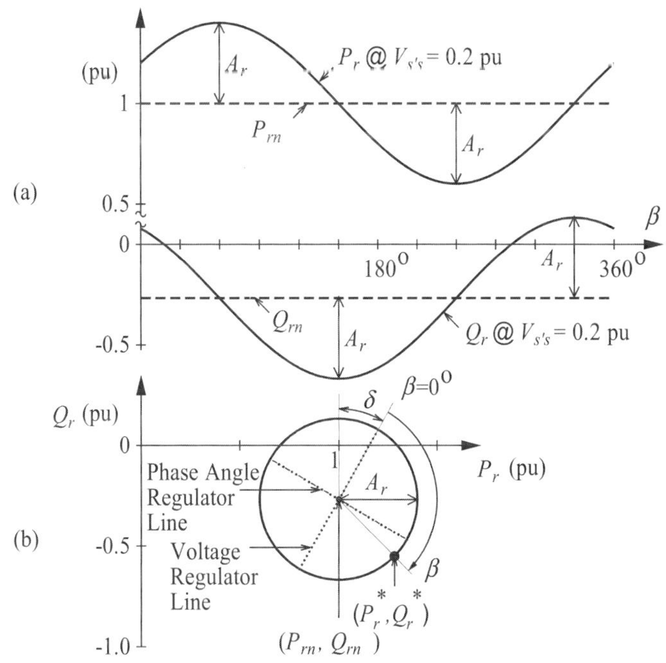

Figure 2-6. (a) Variation of the receiving-end active and reactive power flows (P_r and Q_r) as a function of the relative phase angle (β) of the compensating voltage ($\mathbf{V_{s's}}$) with a fixed magnitude of 0.2 pu. (b) Receiving-end Q_r versus P_r.

For $\beta = 180°$, the variations of P_r and Q_r are given by

$$P_r = P_{rn} + A_r \sin(\delta + 180°) = P_{rn} - A_r \sin \delta \tag{2-45a}$$

and

$$Q_r = Q_{rn} + A_r \cos(\delta + 180°) = Q_{rn} - A_r \cos \delta \tag{2-45b}$$

The variations of P_r and Q_r are then given by

$$P_r = P_{rn} \pm A_r \sin \delta \tag{2-46a}$$

and

$$Q_r = Q_{rn} \pm A_r \cos \delta \tag{2-46b}$$

for $\beta = 0°$ or $\beta = 180°$.

Substituting $\beta = 0°$ or $\beta = 180°$ in equation 2-41 leads to

$$\frac{P_r - P_{rn}}{Q_r - Q_{rn}} = \tan \delta \tag{2-47}$$

Equations 2-14 and 2-47 can then be rearranged as

$$\frac{P_{rn}}{Q_{rn}+\frac{V_r^2}{X}}=\frac{P_r-P_{rn}}{Q_r-Q_{rn}}$$

or

$$P_{rn}\left(Q_r-Q_{rn}\right)=\left(P_r-P_{rn}\right)\left(Q_{rn}+\frac{V_r^2}{X}\right)$$

or

$$Q_r-Q_{rn}=\left(\frac{P_r}{P_{rn}}-1\right)\left(Q_{rn}+\frac{V_r^2}{X}\right)$$

or

$$Q_r=\frac{P_r}{P_{rn}}\left(Q_{rn}+\frac{V_r^2}{X}\right)-\frac{V_r^2}{X} \tag{2-48}$$

For implementation of the PAR, the relative phase angle is set to $\beta = 90°$ or $\beta = -90°$ in equations 2-36 and 2-39. For $\beta = 90°$, the variations of P_r and Q_r are given by

$$P_r=P_{rn}+A_r\sin\left(\delta+90°\right)=P_{rn}+A_r\cos\delta \tag{2-49a}$$

and

$$Q_r=Q_{rn}+A_r\cos\left(\delta+90°\right)=Q_{rn}-A_r\sin\delta \tag{2-49b}$$

For $\beta = -90°$, the variations of P_r and Q_r are given by

$$P_r=P_{rn}+A_r\sin\left(\delta-90°\right)=P_{rn}-A_r\cos\delta \tag{2-50a}$$

and

$$Q_r=Q_{rn}+A_r\cos\left(\delta-90°\right)=Q_{rn}+A_r\sin\delta \tag{2-50b}$$

The variations of P_r and Q_r are then given by

$$P_r=P_{rn}\pm A_r\cos\delta \tag{2-51a}$$

and

$$Q_r=Q_{rn}\mp A_r\sin\delta \tag{2-51b}$$

for $\beta = \pm 90°$.

Substituting $\beta = 90°$ or $\beta = -90°$ in equation 2-41 or rearranging equations 2-51a and 2-51b leads to

$$\frac{P_r - P_{rn}}{Q_r - Q_{rn}} = -\cot\delta \qquad \text{or} \qquad -\frac{Q_r - Q_{rn}}{P_r - P_{rn}} = \tan\delta \tag{2-52}$$

Equations 2-14 and 2-52 can then be rearranged as

$$\frac{P_{rn}}{Q_{rn} + \frac{V_r^2}{X}} = -\frac{Q_r - Q_{rn}}{P_r - P_{rn}} \qquad \text{or} \qquad Q_r = \frac{-P_{rn}P_r + P_{rn}^2}{Q_{rn} + \frac{V_r^2}{X}} + Q_{rn} \tag{2-53}$$

Therefore, in the case of $\beta = 0°$, 180°, or ± 90°, both the active and reactive power flows (P_r and Q_r) at the receiving end vary with the magnitude ($V_{s's}$) of the compensating voltage ($\mathbf{V}_{s's}$). For a given $V_{s's}$ and $\beta = 0°$, 180°, or ± 90°, first calculate P_r from equation 2-36 and then use equation 2-48 for $\beta = 0°$, 180° and equation 2-53 for $\beta = \pm 90°$ to calculate Q_r.

2.1.1.3 *Power at the Modified Sending End.* The active power flow at the modified sending end of the transmission line with compensation is

$$P_{s'} = V_{s'}I\cos(\delta_{s'} - \theta_I) = V_{s'}I\cos(\psi - \theta_I) = V_{s'}I(\cos\psi\cos\theta_I + \sin\psi\sin\theta_I)$$

or

$$P_{s'} = V_{s'}\cos\psi\, I\cos\theta_I + V_{s'}\sin\psi\, I\sin\theta_I \tag{2-54}$$

or

$$P_{s'} = \left(V_s + V_{s's}\cos\beta\right)\frac{V_r\sin\delta + V_{s's}\sin\beta}{X} + V_{s's}\sin\beta\frac{V_r\cos\delta - V_s - V_{s's}\cos\beta}{X}$$

with substitutions from equations 2-17a and 2-17b, and 2-19a and 2-19b, or

$$P_{s'} = \frac{V_sV_r\sin\delta}{X} + \frac{V_{s's}V_r\cos\beta\sin\delta}{X} + \frac{V_{s's}V_r\sin\beta\cos\delta}{X}$$

$$= \frac{V_sV_r}{X}\sin\delta + \frac{V_rV_{s's}}{X}\sin(\delta + \beta)$$

or

$$P_{s'} = P_{sn} + A_r\sin(\delta + \beta) \tag{2-55}$$

where P_{sn} and A_r are given in equations 2-4 and 2-37, respectively.

The reactive power flow at the modified sending end of the transmission line with compensation is

$$Q_{s'} = V_{s'} I \sin(\delta_{s'} - \theta_I) = V_{s'} I \sin(\psi - \theta_I) = V_{s'} I(\sin\psi \cos\theta_I - \cos\psi \sin\theta_I)$$

or

$$Q_{s'} = V_{s'} \sin\psi I \cos\theta_I - V_{s'} \cos\psi \, I \sin\theta_I \tag{2-56}$$

or

$$Q_{s'} = V_{s's} \sin\beta \frac{V_r \sin\delta + V_{s's} \sin\beta}{X} - \left(V_s + V_{s's} \cos\beta\right) \frac{V_r \cos\delta - V_s - V_{s's} \cos\beta}{X}$$

with substitutions from equations 2-17a and 2-17b, and 2-19a and 2-19b, or

$$Q_{s'} = \frac{V_r V_{s's} \sin\beta \sin\delta}{X} + \frac{V_{s's}^2 \sin^2\beta}{X} - \frac{V_s V_r \cos\delta}{X} + \frac{V_s^2}{X} + \frac{V_s V_{s's} \cos\beta}{X}$$
$$- \frac{V_r V_{s's} \cos\beta \cos\delta}{X} + \frac{V_s V_{s's} \cos\beta}{X} + \frac{V_{s's}^2 \cos^2\beta}{X}$$
$$= \frac{V_s V_r}{X}\left(\frac{V_s}{V_r} - \cos\delta\right) + \frac{V_{s's}^2}{X} + \frac{2V_s V_{s's}}{X}\cos\beta - \frac{V_r V_{s's}}{X}\cos(\delta + \beta)$$

or

$$Q_{s'} = Q_{sn} + \frac{V_{s's}^2}{X} + 2A_s \cos\beta - A_r \cos(\delta + \beta) \tag{2-57}$$

where Q_{sn}, A_s, and A_r are given in equations 2-7, 2-22, and 2-37, respectively.

Equations 2-55 and 2-57 can be rearranged as

$$\left(P_{s'} - P_{sn}\right)^2 + \left(Q_{s'} - Q_{sn}\right)^2 = \left(\frac{V_{s's}^2}{X}\right)^2 + A_r^2 + 4A_s^2 \cos^2\beta - 4A_s A_r \cos\beta \cos(\delta + \beta)$$
$$+ 2\left\{2A_s \cos\beta - A_r \cos(\delta + \beta)\right\} \frac{V_{s's}^2}{X}$$

or

$$\left(P_{s'} - P_{sn}\right)^2 + \left(Q_{s'} - Q_{sn}\right)^2 = \frac{1}{X^2}\left[\begin{array}{l} V_{s's}^4 + 2\left\{2V_s \cos\beta - V_r \cos(\delta + \beta)\right\} V_{s's}^3 \\ + V_{s's}^2 \left(V_r^2 + 4V_s^2 \cos^2\beta - 4V_s V_r \cos\beta \cos(\delta + \beta)\right) \end{array}\right] \tag{2-58a}$$

$$\left(P_{s'} - P_{sn}\right)^2 + \left(Q_{s'} - Q_{sn} - \frac{V_{s's}^2}{X}\right)^2 = A_r^2 + 4A_s^2\cos^2\beta - 4A_sA_r\cos\beta\cos(\delta+\beta) \tag{2-58b}$$

and

$$-\frac{P_{s'} - P_{sn}}{Q_{s'} - Q_{sn} - \frac{V_{s's}^2}{X} - 2A_s\cos\beta} = \tan(\delta+\beta) \tag{2-59}$$

Equations 2-58a and 2-58b define the relationship between $P_{s'}$ and $Q_{s'}$ as an ellipse. The active and reactive power flows ($P_{s'}$ and $Q_{s'}$) at the modified sending end as functions of the relative phase angle (β) are plotted in Figure 2-7a. The active and reactive power flows at the modified sending end of the line can be regulated within the range defined by the *P–Q* plot in Figure 2-7b by controlling the magnitude ($V_{s's}$) of the compensating voltage ($\mathbf{V_{s's}}$) within its rated value and the relative phase angle (β) between 0° and 360°.

For the desired power flows ($P_{s'}$* and $Q_{s'}$*) at the modified sending end, the required magnitude ($V_{s's}$) and the relative phase angle (β) of the compensating voltage can be calculated from equations 2-55 and 2-57 following the steps as shown before.

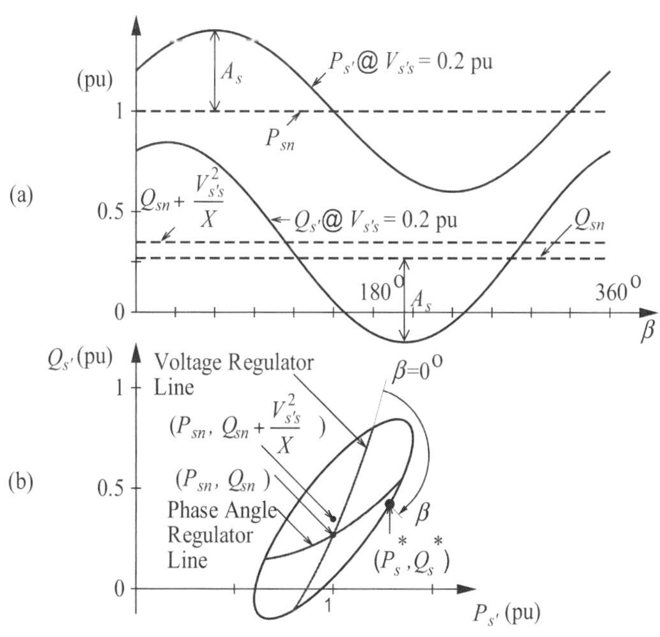

Figure 2-7. (a) Variation of the modified sending-end active and reactive power flows ($P_{s'}$ and $Q_{s'}$) as a function of the relative phase angle (β) of the compensating voltage ($\mathbf{V_{s's}}$) with a fixed magnitude of 0.2 pu. (b) Modified sending-end $Q_{s'}$ versus $P_{s'}$.

However, this is cumbersome and a simpler method is given below. The active and reactive power flows at the sending end are P_s and Q_s. If the desired active and reactive power flows at the modified sending end are P_s^* and Q_s^*, the required exchanged active and reactive powers (P_{exch}^* and Q_{exch}^*) are simply

$$P_{exch}^* = P_S - P_{S'}^* \tag{2-60a}$$

and

$$Q_{exch}^* = Q_S - Q_{S'}^* \tag{2-60b}$$

The required magnitude ($V_{s's}$) and the relative phase angle (β) of the compensating voltage ($\mathbf{V_{s's}}$) can be calculated as shown in the next section.

The active and reactive power flows at any point in the transmission line can also be controlled independently by using a closed-loop control scheme. The automatic power flow control mode of operation of the UPFC for achieving independent control of active and reactive power flows at the modified sending end has been implemented at the world's first UPFC installation. The field results from this ±160 MVA-rated UPFC are shown in Chapter 8. This UPFC demonstrated for the first time that the active and reactive power flows in a transmission line could be regulated independently while maintaining a fixed line voltage at the point of compensation.

The A_s in equation 2-22 and A_r in equation 2-37 are the maximum power flow enhancements and are proportional to the magnitude ($V_{s's}$) of the compensating voltage ($\mathbf{V_{s's}}$). When the magnitude of the compensating voltage increases, so does the maximum power flow enhancement and vice versa. The same applies to the controllable range of the power flow in the *P–Q* plane. When the controllable range extends to all four quadrants, the bidirectional and independent control of active and reactive power flows is achieved.

For implementation of the VRT, the relative phase angle is set to $\beta = 0°$ or $\beta = 180°$ in equations 2-55 and 2-57. For $\beta = 0°$, the variations of $P_{s'}$ and $Q_{s'}$ are given by

$$P_{S'} = P_{Sn} + A_r \sin(\delta + 0°) = P_{Sn} + A_r \sin\delta \tag{2-61a}$$

$$Q_{S'} = Q_{Sn} + \frac{V_{s's}^2}{X} + 2A_S \cos 0° - A_r \cos(\delta + 0°) = Q_{Sn} + \frac{V_{s's}^2}{X} + 2A_S - A_r \cos\delta \tag{2-61b}$$

For $\beta = 180°$, the variations of $P_{s'}$ and $Q_{s'}$ are given by

$$P_{S'} = P_{Sn} + A_r \sin(\delta + 180°) = P_{Sn} - A_r \sin\delta \tag{2-62a}$$

$$Q_{S'} = Q_{Sn} + \frac{V_{s's}^2}{X} + 2A_S \cos 180° - A_r \cos(\delta + 180°) = Q_{Sn} + \frac{V_{s's}^2}{X} - 2A_S + A_r \cos\delta \tag{2-62b}$$

The variations of $P_{s'}$ and $Q_{s'}$ are then given by

$$P_{s'} = P_{sn} \pm A_r \sin\delta \tag{2-63a}$$

and

$$Q_{s'} = Q_{sn} + \frac{V_{s's}^2}{X} \pm 2A_s \mp A_r \cos\delta \tag{2-63b}$$

for $\beta = 0°$ or $\beta = 180°$.

Substituting $\beta = 0°$ or $\beta = 180°$ in equation 2-59 or rearranging equations 2-62a and 2-62b leads to

$$-\frac{P_{s'} - P_{sn}}{Q_{s'} - Q_{sn} - \frac{V_{s's}^2}{X} \mp 2A_s} = \tan\delta \tag{2-64}$$

Equations 2-8 and 2-64 can then be rearranged as

$$\frac{P_{sn}}{Q_{sn} - \frac{V_s^2}{X}} = \frac{P_{s'} - P_{sn}}{Q_{s'} - Q_{sn} - \frac{V_{s's}^2}{X} \mp 2A_s}$$

or

$$P_{sn}\left(Q_{s'} - Q_{sn} - \frac{V_{s's}^2}{X} \mp 2A_s\right) = \left(P_{s'} - P_{sn}\right)\left(Q_{sn} - \frac{V_s^2}{X}\right)$$

or

$$Q_{s'} - Q_{sn} - \frac{V_{s's}^2}{X} \mp 2A_s = \left(\frac{P_{s'}}{P_{sn}} - 1\right)\left(Q_{sn} - \frac{V_s^2}{X}\right)$$

or

$$Q_{s'} = \left(Q_{sn} - \frac{V_s^2}{X}\right)\frac{P_{s'}}{P_{sn}} + \frac{V_s^2}{X} + \frac{V_{s's}^2}{X} \pm 2A_s \tag{2-65}$$

or

$$Q_{s'} = \left(Q_{sn} - \frac{V_s^2}{X}\right)\frac{P_{s'}}{P_{sn}} + \frac{V_s^2}{X} + \frac{V_{s's}^2}{X} \pm 2\frac{V_s V_{s's}}{X}$$

or

$$Q_{s'} = \left(Q_{sn} - \frac{V_s^2}{X}\right)\frac{P_{s'}}{P_{sn}} + \frac{\left(V_s \pm V_{s's}\right)^2}{X} \tag{2-66}$$

with a substitution for A_s from equation 2-22.

For implementation of the PAR, the relative phase angle is set to $\beta = 90°$ or $\beta = -90°$ in equations 2-55 and 2-57. For $\beta = 90°$, the variations of $P_{s'}$ and $Q_{s'}$ are given by

$$P_{s'} = P_{sn} + A_r \sin(\delta + 90°) = P_{sn} + A_r \cos\delta \tag{2-67a}$$

$$Q_{s'} = Q_{sn} + \frac{V^2_{s's}}{X} + 2A_s \cos 90° - A_r \cos(\delta + 90°) = Q_{sn} + \frac{V^2_{s's}}{X} + A_r \sin\delta \tag{2-67b}$$

For $\beta = -90°$, the variations of $P_{s'}$ and $Q_{s'}$ are given by

$$P_{s'} = P_{sn} + A_r \sin(\delta - 90°) = P_{sn} - A_r \cos\delta \tag{2-68a}$$

$$Q_{s'} = Q_{sn} + \frac{V^2_{s's}}{X} + 2A_s \cos(-90°) - A_r \cos(\delta - 90°) = Q_{sn} + \frac{V^2_{s's}}{X} - A_r \sin\delta \tag{2-68b}$$

The variations of $P_{s'}$ and $Q_{s'}$ are then given by

$$P_{s'} = P_{sn} \pm A_r \cos\delta \tag{2-69a}$$

and

$$Q_{s'} = Q_{sn} + \frac{V^2_{s's}}{X} \pm A_r \sin\delta \tag{2-69b}$$

for $\beta = \pm 90°$.

Substituting $\beta = 90°$ or $\beta = -90°$ in equation 2-59 or rearranging equations 2-69a and 2-69b leads to

$$-\frac{P_{s'} - P_{sn}}{Q_{s'} - Q_{sn} - \dfrac{V^2_{s's}}{X}} = -\cot\delta$$

or

$$\frac{Q_{s'} - Q_{sn} - \dfrac{V^2_{s's}}{X}}{P_{s'} - P_{sn}} = \tan\delta \tag{2-70}$$

Equations 2-8 and 2-70 can then be rearranged as

$$-\frac{P_{sn}}{Q_{sn} - \dfrac{V_s^2}{X}} = \frac{Q_{s'} - Q_{sn} - \dfrac{V^2_{s's}}{X}}{P_{s'} - P_{sn}}$$

or

$$Q_{s'} = \frac{-P_{sn}P_{s'} + P_{sn}^2}{Q_{sn} - \frac{V_s^2}{X}} + Q_{sn} + \frac{V_{s's}^2}{X} \tag{2-71}$$

Therefore, in the case of $\beta = 0°$, $180°$, or $\pm 90°$, both the active and reactive power flows ($P_{s'}$ and $Q_{s'}$) at the modified sending end vary with the magnitude ($V_{s's}$) of the compensating voltage ($\mathbf{V_{s's}}$). For a given $V_{s's}$ and $\beta = 0°$, $180°$, or $\pm 90°$, first calculate $P_{s'}$ from equation 2-55 and then use equation 2-67b for $\beta = 0°$, $180°$ and equation 2-71 for $\beta = \pm 90°$ to calculate $Q_{s'}$.

2.1.1.4 Exchanged Power by the Series-Connected Compensating Voltage. The compensating voltage ($\mathbf{V_{s's}}$) is at any phase angle with the prevailing line current (**I**) and, therefore, it exchanges with the line both active and reactive powers (P_{exch} and Q_{exch}), which are defined as

$$P_{exch} = -\mathbf{V_{s's}} \bullet \mathbf{I} = \mathbf{V_{dq}} \bullet \mathbf{I} = V_d I \tag{2-72a}$$

and

$$Q_{exch} = \left|-\mathbf{V_{s's}} \times \mathbf{I}\right| = \left|\mathbf{V_{dq}} \times \mathbf{I}\right| = V_q I \tag{2-72b}$$

where $\mathbf{V_d}$ and $\mathbf{V_q}$ are the respective active or direct and reactive or quadrature components of the compensating voltage with load convention as shown in Figure 2-8a.

Then,

$$\mathbf{V_{dq}} = \mathbf{V_d} + j\mathbf{V_q} \tag{2-73}$$

The exchanged active power (P_{exch}) by the compensating voltage with the transmission line can also be derived from equations 2-21 and 2-55 as

$$P_{exch} = P_s - P_{s'} = A_s \sin\beta - A_r \sin(\delta + \beta) \tag{2-74}$$

or

$$P_{exch} = \frac{V_{s's}}{X}\{V_s \sin\beta - V_r (\sin\delta\cos\beta + \cos\delta\sin\beta)\}$$

$$= \frac{V_{s's}}{X}\{(V_s - V_r \cos\delta)\sin\beta - V_r \sin\delta\cos\beta\}$$

with substitutions for A_s and A_r from equations 2-22 and 2-37 or

$$P_{exch} = \frac{V_{s's}V_r}{X}\left\{\left(\frac{V_s}{V_r} - \cos\delta\right)\sin\beta - \sin\delta\cos\beta\right\} \tag{2-75}$$

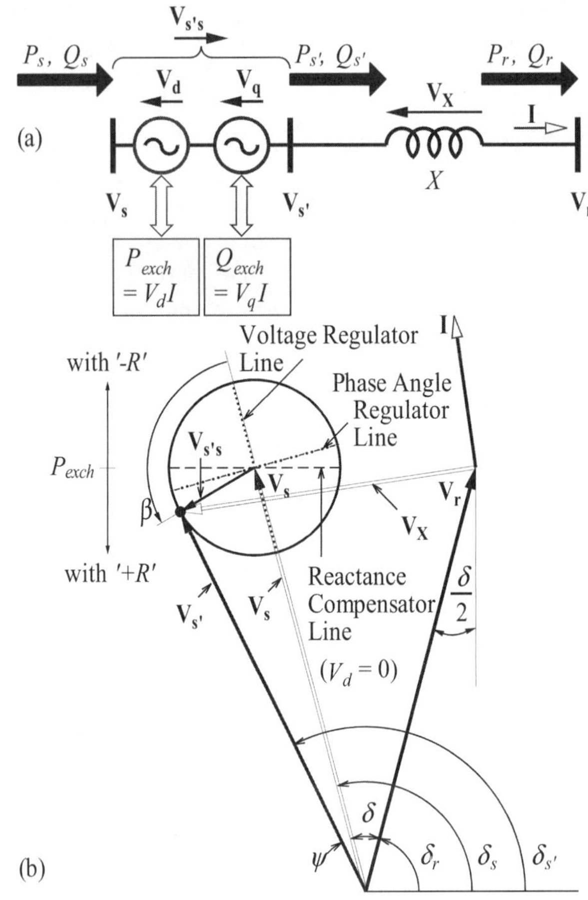

Figure 2-8. (a) Power transmission system with a series-connected compensating voltage ($\mathbf{V}_{s's}$). (b) Phasor diagram.

Equations 2-4, 2-5, and 2-7 can be rearranged as

$$\frac{V_r \sin\delta}{X} = \frac{P_{sn}}{V_s} \tag{2-76a}$$

and

$$\frac{V_r}{X}\left(\frac{V_s}{V_r} - \cos\delta\right) = \frac{Q_{sn}}{V_s} \tag{2-76b}$$

Substituting equations 2-76a and 2-76b into equation 2-75 leads to

$$P_{exch} = \frac{V_{s's}}{V_s}\left(Q_{sn}\sin\beta - P_{sn}\cos\beta\right)$$

or

$$P_{exch} = \frac{V_{s's}}{V_s} A \sin(\beta - \alpha) \tag{2-77}$$

where

$$P_{sn} = A \sin \alpha \tag{2-78a}$$

$$Q_{sn} = A \cos \alpha \tag{2-78b}$$

$$A = \sqrt{P_{sn}^2 + Q_{sn}^2} \tag{2-78c}$$

$$\alpha = \tan^{-1} \frac{P_{sn}}{Q_{sn}} \tag{2-78d}$$

Equations 2-76a and 2-76b can be rearranged as

$$P_{sn}^2 + Q_{sn}^2 = \left(\frac{V_s V_r}{X}\right)^2 \left\{1 + \left(\frac{V_s}{V_r}\right)^2 - 2\frac{V_s}{V_r}\cos\delta\right\}$$

$$= \left(\frac{V_s V_r}{X}\right)^2 \frac{1}{V_r^2}\left(V_r^2 + V_s^2 - 2V_s V_r \cos\delta\right) = \frac{V_s^2}{X^2} V_{Xn}^2 = \left(\frac{V_s V_{Xn}}{X}\right)^2$$

where the natural voltage (V_{Xn}) across the transmission line is

$$V_{Xn} = \sqrt{V_s^2 + V_r^2 - 2V_s V_r \cos\delta} \tag{2-79}$$

Equation 2-78c is then

$$A = \sqrt{P_{sn}^2 + Q_{sn}^2} = \frac{V_s V_{Xn}}{X} \tag{2-80}$$

By substituting A from equation 2-80 and α from equation 2-78d into equation 2-77, it can be written that

$$P_{exch} = A_{exch} \sin\left(\beta - \tan^{-1}\frac{P_{sn}}{Q_{sn}}\right) \tag{2-81}$$

where

$$A_{exch} = \frac{V_{s's} V_{Xn}}{X} \tag{2-82}$$

The exchanged reactive power (Q_{exch}) by the compensating voltage with the transmission line can also be derived from equations 2-24 and 2-57 as

$$Q_{exch} = Q_s - Q_{s'} = A_s \cos\beta - \frac{V_{s's}^2}{X} - 2A_s \cos\beta + A_r \cos(\delta+\beta)$$

$$Q_{exch} = -\frac{V_{s's}^2}{X} - A_s \cos\beta + A_r \cos(\delta+\beta) \tag{2-83}$$

or

$$Q_{exch} = -\frac{V_{s's}^2}{X} + \frac{V_{s's}}{X}\{-V_s \cos\beta + V_r(\cos\delta\cos\beta - \sin\delta\sin\beta)\}$$

with substitutions for A_s and A_r from equations 2-22 and 2-37 or

$$Q_{exch} = -\frac{V_{s's}^2}{X} - \frac{V_{s's}V_r}{X}\left\{\left(\frac{V_s}{V_r} - \cos\delta\right)\cos\beta + \sin\delta\,\sin\beta\right\} \tag{2-84}$$

or

$$Q_{exch} = -\frac{V_{s's}^2}{X} - \frac{V_{s's}}{V_s}(Q_{sn}\cos\beta + P_{sn}\sin\beta)$$

with substitutions from equations 2-76a and 2-76b or

$$Q_{exch} = -\frac{V_{s's}^2}{X} - \frac{V_{s's}}{V_s} A\cos(\beta-\alpha) \tag{2-85}$$

where A and α are given in equations 2-80 and 2-78d, respectively.

By substituting A and α as above, equation 2-85 becomes

$$Q_{exch} = -\frac{V_{s's}^2}{X} - A_{exch}\cos\left(\beta - \tan^{-1}\frac{P_{sn}}{Q_{sn}}\right) \tag{2-86}$$

where A_{exch} is given in equation 2-82.

Equations 2-81 and 2-86 can be rearranged as

$$P_{exch}^2 + \left(Q_{exch} + \frac{V_{s's}^2}{X}\right)^2 = A_{exch}^2 \tag{2-87}$$

and

$$\frac{P_{exch}}{Q_{exch} + \frac{V_{s's}^2}{X}} = -\tan\left(\beta - \tan^{-1}\frac{P_{sn}}{Q_{sn}}\right) \tag{2-88}$$

Equation 2-87 defines the relationship between P_{exch} and Q_{exch} as a circle centered at $(0, -V_{s's}^2/X)$ with a radius of A_{exch}. The exchanged active and reactive powers (P_{exch} and Q_{exch}) as functions of the relative phase angle (β) are plotted in Figure 2-9a. For a given magnitude of the compensating voltage, the exchanged capacitive power (Q_{exch}) is larger than its inductive counterpart due to the fact that the capacitive compensation produces a larger line current. The compensating voltage, being at any phase angle with the prevailing line current, emulates a series capacitor (C) or an inductor (L) and a positive resistor ($+R$) or a negative resistor ($-R$). The exchanged active and reactive powers can be regulated within the range defined by the P–Q plot in Figure 2-9b by controlling the magnitude ($V_{s's}$) of the compensating voltage ($\mathbf{V}_{s's}$) within its rated value and the relative phase angle (β) between 0° and 360°.

The magnitude ($V_{s's}$) of the compensating voltage is expressed by rearranging equations 2-82 and 2-87 as follows:

$$P_{exch}^2 + Q_{exch}^2 + 2Q_{exch}\frac{V_{s's}^2}{X} + \left(\frac{V_{s's}^2}{X}\right)^2 - \left(\frac{V_{s's}V_{Xn}}{X}\right)^2 = 0$$

or

$$\left(V_{s's}^2\right)^2 - \left(V_{Xn}^2 - 2XQ_{exch}\right)V_{s's}^2 + X^2\left(P_{exch}^2 + Q_{exch}^2\right) = 0 \quad (2\text{-}89)$$

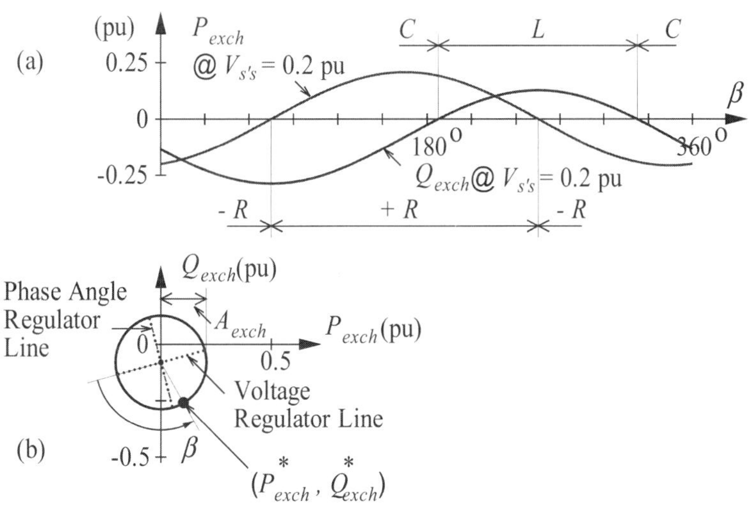

Figure 2-9. (a) Variation of the exchanged active and reactive powers (P_{exch} and Q_{exch}) as a function of the relative phase angle (β) of the compensating voltage ($\mathbf{V}_{s's}$) with a fixed magnitude of 0.2 pu. (b) Exchanged reactive power (Q_{exch}) versus active power (P_{exch}).

Solving the above quadratic equation results in

$$V_{s's}^2 = \frac{\left(V_{Xn}^2 - 2XQ_{exch}\right) \pm \sqrt{\left(V_{Xn}^2 - 2XQ_{exch}\right)^2 - 4X^2\left(P_{exch}^2 + Q_{exch}^2\right)}}{2}$$

or

$$V_{s's}^2 = \left(\frac{V_{Xn}^2}{2} - XQ_{exch}\right) \pm \sqrt{\left(\frac{V_{Xn}^2}{2} - XQ_{exch}\right)^2 - X^2\left(P_{exch}^2 + Q_{exch}^2\right)}$$

or

$$V_{s's} = +\sqrt{\left(\frac{V_{Xn}^2}{2} - XQ_{exch}\right) - \sqrt{\left(\frac{V_{Xn}^2}{2} - XQ_{exch}\right)^2 - X^2\left(P_{exch}^2 + Q_{exch}^2\right)}} \tag{2-90}$$

The relative phase angle (β) of the compensating voltage ($\mathbf{V_{s's}}$) is derived from equation 2-88 as

$$\beta = \tan^{-1}\frac{P_{sn}}{Q_{sn}} - \tan^{-1}\frac{P_{exch}}{Q_{exch} + \dfrac{V_{s's}^2}{X}} \tag{2-91}$$

For the desired exchanged active and reactive powers (P_{exch}^*, Q_{exch}^*) within the allowable control range, the magnitude ($V_{s's}$) and the relative phase angle (β) of the compensating voltage ($\mathbf{V_{s's}}$) are calculated by substituting $P_{exch} = P_{exch}^*$ and $Q_{exch} = Q_{exch}^*$ in equations 2-90 and 2-91, respectively.

For implementation of the VRT, the relative phase angle is set to $\beta = 0°$ or $\beta = 180°$ in equations 2-81 and 2-86. For $\beta = 0°$, the variations of P_{exch} and Q_{exch} are given by

$$P_{exch} = A_{exch}\sin\left(0° - \tan^{-1}\frac{P_{sn}}{Q_{sn}}\right) = -A_{exch}\sin\tan^{-1}\frac{P_{sn}}{Q_{sn}} \tag{2-92a}$$

$$Q_{exch} = -\frac{V_{s's}^2}{X} - A_{exch}\cos\left(0° - \tan^{-1}\frac{P_{sn}}{Q_{sn}}\right) = -\frac{V_{s's}^2}{X} - A_{exch}\cos\tan^{-1}\frac{P_{sn}}{Q_{sn}} \tag{2-92b}$$

For $\beta = 180°$, the variations of P_{exch} and Q_{exch} are given by

$$P_{exch} = A_{exch} \sin\left(180° - \tan^{-1}\frac{P_{sn}}{Q_{sn}}\right) = A_{exch} \sin\tan^{-1}\frac{P_{sn}}{Q_{sn}} \qquad (2\text{-}93a)$$

$$Q_{exch} = -\frac{V_{s's}^2}{X} - A_{exch} \cos\left(180° - \tan^{-1}\frac{P_{sn}}{Q_{sn}}\right) = -\frac{V_{s's}^2}{X} + A_{exch} \cos\tan^{-1}\frac{P_{sn}}{Q_{sn}} \qquad (2\text{-}93b)$$

The variations of P_{exch} and Q_{exch} are then given by

$$P_{exch} = \mp A_{exch} \sin\tan^{-1}\frac{P_{sn}}{Q_{sn}} \qquad (2\text{-}94a)$$

and

$$Q_{exch} = -\frac{V_{s's}^2}{X} \mp A_{exch} \cos\tan^{-1}\frac{P_{sn}}{Q_{sn}} \qquad (2\text{-}94b)$$

for $\beta = 0°$ or $\beta = 180°$.

Substituting $\beta = 0°$ or $\beta = 180°$ in equation 2-91 or rearranging equations 2-94a and 2-94b leads to

$$\frac{P_{exch}}{Q_{exch} + \dfrac{V_{s's}^2}{X}} = \frac{P_{sn}}{Q_{sn}}$$

or

$$Q_{exch} = \frac{Q_{sn}}{P_{sn}} P_{exch} - \frac{V_{s's}^2}{X} \qquad (2\text{-}95)$$

For implementation of the PAR, the relative phase angle is set to $\beta = 90°$ or $\beta = -90°$ in equations 2-81 and 2-86. For $\beta = 90°$, the variations of P_{exch} and Q_{exch} are given by

$$P_{exch} = A_{exch} \sin\left(90° - \tan^{-1}\frac{P_{sn}}{Q_{sn}}\right) = A_{exch} \cos\tan^{-1}\frac{P_{sn}}{Q_{sn}} \qquad (2\text{-}96a)$$

$$Q_{exch} = -\frac{V_{s's}^2}{X} - A_{exch} \cos\left(90° - \tan^{-1}\frac{P_{sn}}{Q_{sn}}\right) = -\frac{V_{s's}^2}{X} - A_{exch} \sin\tan^{-1}\frac{P_{sn}}{Q_{sn}} \qquad (2\text{-}96b)$$

For $\beta = -90°$, the variations of P_{exch} and Q_{exch} are given by

$$P_{exch} = A_{exch} \sin\left(-90° - \tan^{-1}\frac{P_{sn}}{Q_{sn}}\right) = -A_{exch} \cos\tan^{-1}\frac{P_{sn}}{Q_{sn}} \quad (2\text{-}97a)$$

$$Q_{exch} = -\frac{V_{s's}^2}{X} - A_{exch} \cos\left(-90° - \tan^{-1}\frac{P_{sn}}{Q_{sn}}\right) = -\frac{V_{s's}^2}{X} + A_{exch} \sin\tan^{-1}\frac{P_{sn}}{Q_{sn}} \quad (2\text{-}97b)$$

The variations of P_{exch} and Q_{exch} are then given by

$$P_{exch} = \pm A_{exch} \cos\tan^{-1}\frac{P_{sn}}{Q_{sn}} \quad (2\text{-}98a)$$

and

$$Q_{exch} = -\frac{V_{s's}^2}{X} \mp A_{exch} \sin\tan^{-1}\frac{P_{sn}}{Q_{sn}} \quad (2\text{-}98b)$$

for $\beta = \pm 90°$.

Substituting $\beta = 90°$ or $\beta = -90°$ in equation 2-91 or rearranging equations 2-98a and 2-98b leads to

$$\frac{P_{exch}}{Q_{exch} + \dfrac{V_{s's}^2}{X}} = -\frac{Q_{sn}}{P_{sn}}$$

or

$$Q_{exch} = -\frac{P_{sn}}{Q_{sn}} P_{exch} - \frac{V_{s's}^2}{X} \quad (2\text{-}99)$$

Therefore, in the case of $\beta = 0°$, 180°, or ±90°, both exchanged active and reactive powers (P_{exch} and Q_{exch}) vary with the magnitude ($V_{s's}$) of the compensating voltage ($\mathbf{V}_{s's}$). For a given $V_{s's}$ and $\beta = 0°$, 180°, or ±90°, first calculate P_{exch} from equation 2-81 and then use equation 2-94 for $\beta = 0°$, 180° and equation 2-99 for $\beta = \pm 90°$ to calculate Q_{exch}.

A series-connected compensating voltage ($\mathbf{V}_{s's}$) can control the line voltage to a specific magnitude and phase angle and achieve independent control of active and reactive power flows in the transmission line. In the process, the compensating voltage exchanges bidirectional active and reactive powers with the line. To deliver or absorb the active power to and from the line, an active source or a sink is required. To deliver or absorb the reactive power to and from the line, a capacitive or an inductive reactance is required. When only series reactance compensation is used, the exchanged active power is zero.

Figure 2-9 shows that the series reactance compensation can be emulated with a series-connected compensating voltage with $\beta = 75°$ or $\beta = 255°$ in this particular example.

In order to achieve a series-connected compensating voltage, $\mathbf{V}_{s's}$ (i.e., $V_{s's} \angle \delta_s + \beta$), operating in the entire range ($\beta = 0°$ to $360°$), it is necessary to provide a bidirectional path for the exchanged power of the compensating voltage with the line. Figure 2-10 shows a power flow control scheme for this purpose with shunt–series compensating units. The exchanged power by the series-connected compensating voltage ($\mathbf{V}_{s's}$) with the line flows bidirectionally through the shunt-connected voltage, **E** (i.e., $E \angle \delta_s + \alpha$) that is coupled through a shared magnetic link. This is the basic principle of the Sen transformer (ST) that consists of a transformer and LTCs.

2.1.2 Shunt-Connected Compensating Voltage

A shunt-connected compensating voltage can also modify the transmission line's sending-end voltage. In certain special cases for point-to-point transfer of power between two isolated networks or interconnection of two transmission lines with different voltages or phase angles (or frequencies), this scheme is a preferred choice.

2.1.2.1 Power at the Modified Sending End. When the transmission line is compensated, the active and reactive power flows ($P_{s'}$ and $Q_{s'}$) at the modified sending end are

$$P_{s'} = \frac{V_{s'}V_r}{X} \sin\left(\delta_{s'} - \delta_r\right) \tag{2-100a}$$

and

$$Q_{s'} = \frac{V_{s'}V_r}{X}\left\{\frac{V_{s'}}{V_r} - \cos\left(\delta_{s'} - \delta_r\right)\right\} \tag{2-100b}$$

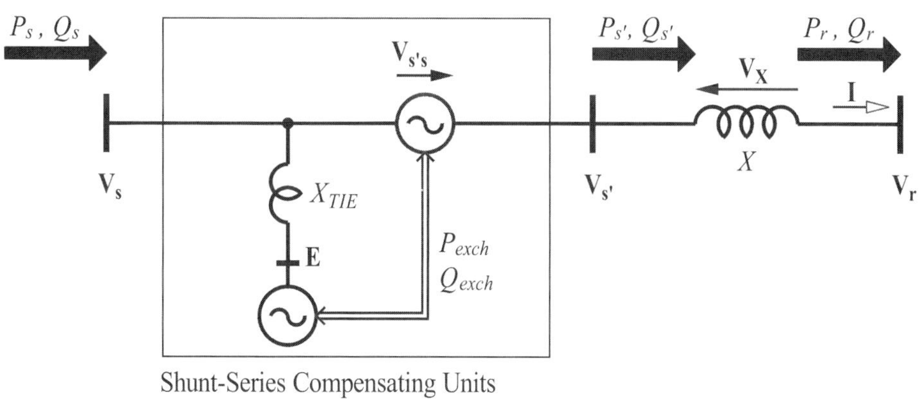

Figure 2-10. Scheme for implementing a series-connected compensating voltage.

Equations 2-100a and 2-100b can be rearranged as

$$P_{s'}^2 + \left(Q_{s'} - \frac{V_{s'}^2}{X}\right)^2 = \left(\frac{V_{s'}V_r}{X}\right)^2 \text{ or } \left(\frac{V_{s'}^2}{X}\right)^2 - \left(\frac{2Q_{s'}}{X} + \frac{V_r^2}{X^2}\right)V_{s'}^2 + P_{s'}^2 + Q_{s'}^2 = 0$$

or

$$\left(V_{s'}^2\right)^2 - \left(V_r^2 + 2XQ_{s'}\right)V_{s'}^2 + X^2\left(P_{s'}^2 + Q_{s'}^2\right) = 0 \qquad (2\text{-}101)$$

Solving the above quadratic equation results in

$$V_{s'}^2 = \frac{\left(V_r^2 + 2XQ_{s'}\right) \pm \sqrt{\left(V_r^2 + 2XQ_{s'}\right)^2 - 4X^2\left(P_{s'}^2 + Q_{s'}^2\right)}}{2}$$

or

$$V_{s'}^2 = \left(\frac{V_r^2}{2} + XQ_{s'}\right) \pm \sqrt{\left(\frac{V_r^2}{2} + XQ_{s'}\right)^2 - X^2\left(P_{s'}^2 + Q_{s'}^2\right)}$$

or

$$V_{s'} = +\sqrt{\left(\frac{V_r^2}{2} + XQ_{s'}\right) \pm \sqrt{\left(\frac{V_r^2}{2} + XQ_{s'}\right)^2 - X^2\left(P_{s'}^2 + Q_{s'}^2\right)}} \qquad (2\text{-}102)$$

Equations 2-100a and 2-100b can be rearranged as

$$\frac{-P_{s'}}{Q_{s'} - \frac{V_{s'}^2}{X}} = \tan\left(\delta_{s'} - \delta_r\right) \qquad (2\text{-}103)$$

Therefore,

$$\delta_{s'} = \tan^{-1}\frac{P_{s'}}{\frac{V_{s'}^2}{X} - Q_{s'}} + \delta_r \qquad (2\text{-}104)$$

For independent control of active and reactive power flows ($P_{s'}$ and $Q_{s'}$), the magnitude ($V_{s'}$) and phase angle ($\delta_{s'}$) of the modified sending-end voltage ($\mathbf{V}_{s'}$) are calculated from equations 2-102 and 2-104, respectively.

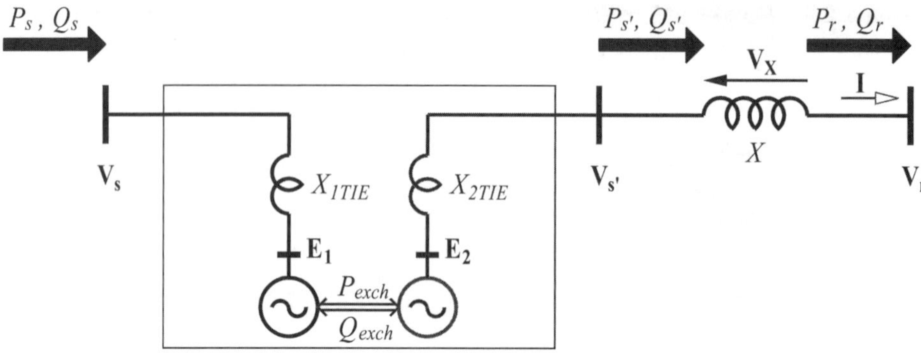

Figure 2-11. Scheme for implementing a shunt-connected compensating voltage.

to-point power transfer scheme for this purpose with shunt–shunt compensating units. The exchanged power by the shunt-connected compensating voltage, $\mathbf{E_2}$ (i.e., $E_2 \angle \delta_{s'} + \alpha_2$), with the line flows bidirectionally through another shunt-connected voltage, $\mathbf{E_1}$ (i.e., $E_1 \angle \delta_s + \alpha_1$). This is the basic principle of the ST with a shunt-connected compensating voltage.

2.1.3 Comparison Between Series-Connected and Shunt-Connected Compensating Voltages

A compensating voltage for modifying the transmission line voltage can be generated with one of three distinct methods by using (a) power electronics, (b) rotating machines, and (c) transformers and LTCs. A scheme with shunt–shunt compensating units is used in certain special cases for point-to-point transfer of power between two isolated networks with different voltages or phase angles. In an application in which power electronics or rotating machines are used, the frequencies of the two networks may be different and, therefore, it is referred to as an "asynchronous tie." A scheme with shunt–series compensating units is used for power flow control. The final choice of the PFC topology is dependent on the system requirements and the costs of both installation and operation.

When a series-connected compensating voltage is used, the modified sending-end voltage is $\mathbf{V_{s'}} = \mathbf{V_s} + \mathbf{V_{s's}}$. This is rewritten in the case of an autotransformer, shown in Figure 2-12a, as

$$\mathbf{V_{s'\text{-}se}} = \mathbf{V_s} + \mathbf{V_{comp}} \tag{2-110}$$

where $\mathbf{V_{s'\text{-}se}}$, $\mathbf{V_s}$, and $\mathbf{V_{comp}}$ are modified sending-end, sending-end, and compensating voltages. When a shunt-connected compensating voltage is used, the modified sending-end voltage ($\mathbf{V_{s'\text{-}sh}}$) is the compensating voltage. This is rewritten in the case of a two-winding transformer, shown in Figure 2-12b, as

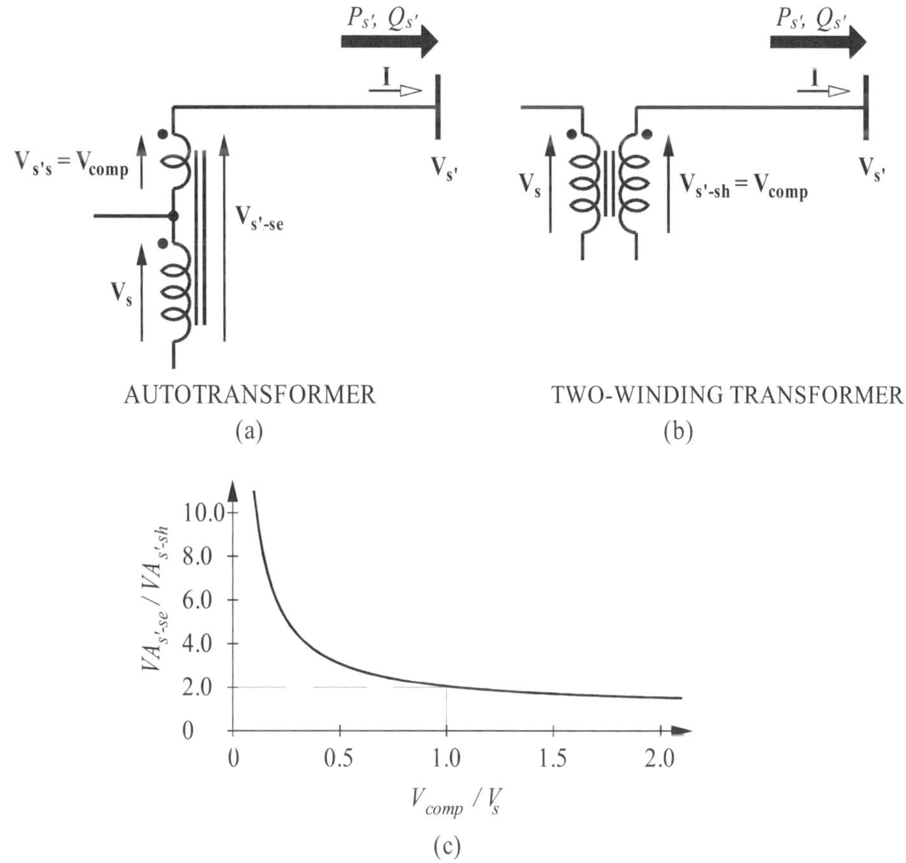

Figure 2-12. (a) Autotransformer. (b) Two-winding transformer. (c) Ratio of the transmitted VAs with series-compensated and shunt-compensated voltages as a function of normalized compensating voltage.

$$\mathbf{V_{s'\text{-}sh}} = \mathbf{V_{comp}} \tag{2-111}$$

When a line current (**I**) is desired using both types of compensation schemes, the ratio of the transmitted VAs ($VA_{s'\text{-}se}$) by using a series-connected compensating voltage and ($VA_{s'\text{-}sh}$) by using a shunt-connected compensating voltage, with substitutions from equations 2-110 and 2-111, is

$$\frac{VA_{s'-se}}{VA_{s'-sh}} = \frac{V_{s'\text{-}se}}{V_{s'\text{-}sh}} = \frac{V_S + V_{comp}}{V_{comp}} = 1 + \frac{V_S}{V_{comp}} = 1 + \frac{1}{V_{comp} / V_S} \tag{2-112}$$

The ratio ($VA_{s'\text{-}se}/VA_{s'\text{-}sh}$) is less than ∞ for V_{comp} greater than zero, but greater than one for V_{comp} less than ∞. Therefore, the PFC with the series compensating voltage always offers a large leverage between its own rating and the controlled transmission line VAs. It needs to be rated for only *a fraction of the transmitted VAs*. The ratio of the transmitted VAs with series-connected and shunt-connected compensating voltages as a function of normalized compensating voltage is plotted in Figure 2-12c. When the compensating voltage is one pu, the series-connected compensating voltage transmits twice the VAs that the shunt-connected compensating voltage transmits. When the series-connected compensating voltage is a fraction of the line voltage, the transmitted VAs ratio becomes greater than two.

2.2 IMPLEMENTATION OF POWER FLOW CONTROL CONCEPTS

The various methods of power flow control that have been established so far are summarized in this section. In order to control the flow of power in the transmission line, at least one parameter (voltage magnitude, its phase angle, or reactance) needs to be controlled.

2.2.1 Voltage Regulation

The various methods of voltage regulation can be categorized as (a) direct method and (b) indirect method.

2.2.1.1 Direct Method. The direct method of voltage regulation of a transmission line is to add a compensating voltage vectorially in or out of phase with the transmission line phase-to-neutral voltage at the point of compensation. The VRT generates a bipolar compensating voltage through autotransformer action and connects it in series with the transmission line as shown in Figure 2-13a.

The VRT as shown in Figure 2-13a is an autotransformer that consists of an exciter unit and a voltage regulating unit. The exciter unit consists of a three-phase (A, B, and C), Y-connected primary windings and is connected in shunt with the line. The three-phase primary windings are excited from the three-phase line voltages ($\mathbf{V}_{sA}$, $\mathbf{V}_{sB}$, and $\mathbf{V}_{sC}$). A three-phase, bipolar compensating voltages ($\mathbf{V}_{s'sA}$, $\mathbf{V}_{s'sB}$, and $\mathbf{V}_{s'sC}$) that are either in phase or out of phase with the corresponding phase-to-neutral voltage are generated from the induced secondary voltages. The voltage regulating unit consists of a total of six secondary windings (two windings in each phase). The magnitude of the compensating voltage ($\mathbf{V}_{s's}$) is varied with the use of LTCs. The line is regulated at a voltage ($\mathbf{V}_{s'}$) from the uncompensated voltage ($\mathbf{V}_{s}$). The corresponding phasor diagram is shown in Figure 2-13b.

Figure 2-14 shows the schematic diagram of a thyristor-controlled load tap changer (LTC). A transformer winding is tapped at various places. Each of the tapped points is connected to one side of a back-to-back thyristor (triac) switch. The other side of each triac switch is connected at point A. Depending on which thyristor is on, the voltage between points A and B can be varied between zero and the full-winding voltage with desired steps in between. In the mechanical contact version of this arrangement, the

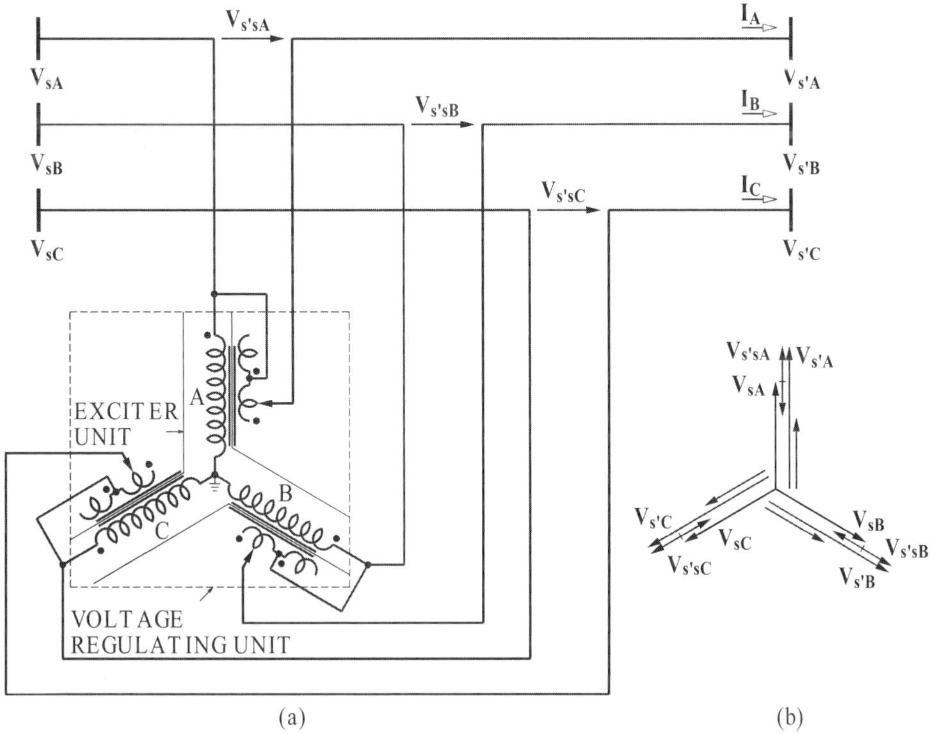

Figure 2-13. (a) Voltage regulating transformer. (b) Phasor diagram.

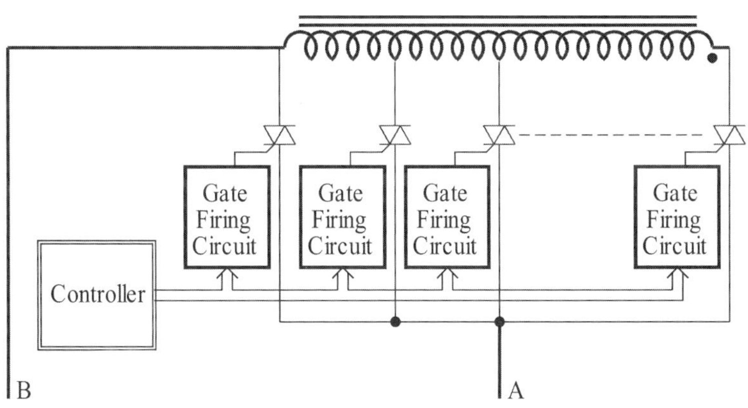

Figure 2-14. Thyristor-controlled load tap changer.

LTC connects with one of various taps to give a variable number of turns between the connected tap and one end of the winding.

There are other ways to regulate the line voltage, as described below.

2.2.1.2 Indirect Method.

The indirect way to regulate the line voltage is to connect a variable capacitor or a variable inductor in shunt with the transmission line. A shunt-connected capacitor raises the line voltage with its generated reactive power. A shunt-connected inductor absorbs reactive power from the line and lowers the line voltage, as explained below.

Consider a power system network, shown in Figure 2-15, that is represented at the point of compensation by a Thèvenin voltage ($\mathbf{V_{TH}}$) and a Thèvenin impedance ($\mathbf{Z_{TH}}$). When a compensating impedance is connected in shunt, the voltage across the compensating impedance is

$$V = \left| \frac{\mathbf{Z}}{\mathbf{Z_{TH}} + \mathbf{Z}} \right| V_{TH} \tag{2-113}$$

If $\mathbf{Z_{TH}}$ is represented by a reactance (jX_{Line}), equation 2-113 becomes

$$V = \frac{1}{\left| 1 + \dfrac{jX_{Line}}{\mathbf{Z}} \right|} V_{TH} \tag{2-114}$$

The compensating impedance ($\mathbf{Z}$) can be resistive (R), inductive ($jX_L = j\omega L$), or capacitive ($-jX_C = 1/j\omega C$).

For the case when $\mathbf{Z} = R$, the voltage across the compensating impedance becomes

$$V = \frac{1}{\left| 1 + \dfrac{jX_{Line}}{R} \right|} V_{TH} = \frac{1}{\sqrt{1 + \left(\dfrac{X_{Line}}{R} \right)^2}} V_{TH} \tag{2-115}$$

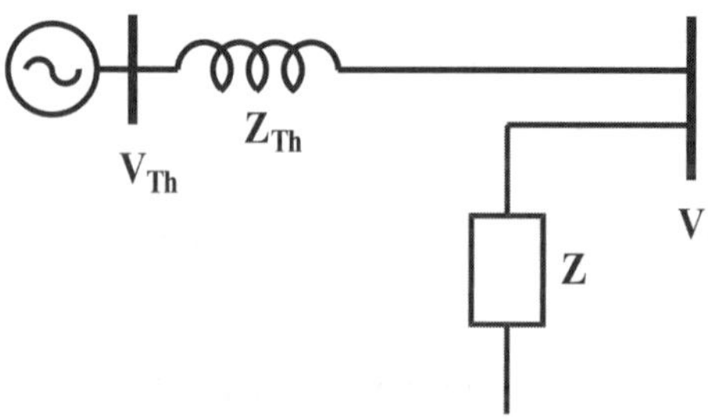

Figure 2-15. Equivalent transmission line with shunt compensation.

The denominator of this equation is always greater than 1, hence $V < V_{TH}$. Therefore, adding a shunt resistor results in lowering the line voltage.

For the case when $\mathbf{Z} = jX_L$, the voltage across the compensating impedance becomes

$$V = \frac{1}{\left|1 + \dfrac{jX_{Line}}{jX_L}\right|} V_{TH} = \frac{1}{1 + \dfrac{X_{Line}}{X_L}} V_{TH} \tag{2-116}$$

The denominator of this equation is always greater than 1, hence $V < V_{TH}$. Therefore, adding a shunt inductor also results in lowering the line voltage.

For the case when $\mathbf{Z} = -jX_C$, the voltage across the compensating impedance becomes

$$V = \frac{1}{\left|1 + \dfrac{jX_{Line}}{-jX_C}\right|} V_{TH} = \frac{1}{1 - \dfrac{X_{Line}}{X_C}} V_{TH} \tag{2-117}$$

The denominator of this equation is always less than 1, hence $V > V_{TH}$. Therefore, adding a shunt capacitor results in increasing the line voltage.

The shunt inductor or capacitor can be connected through mechanical or electronic switches. The various electronic topologies and their control methods are termed thyristor-switched capacitor (TSC), thyristor-switched reactor (TSR), and thyristor controlled reactor (TCR). The TSC connects fixed capacitors in a step-like manner in shunt with the line. The TSR connects fixed inductors in a step-like manner in shunt with the line. The TCR connects an inductor in shunt with the line through thyristor switches whose duty cycle can be varied, thereby making it appear like a variable inductor. A combination of TSC, TSR, and TCR is termed the static var compensator (SVC), as shown in Figure 2-16.

A capacitor with a parallel inductor offers a compensating reactance of

$$X_c = \frac{X_L X_C}{j(X_L - X_C)} = j\frac{X_L X_C}{(X_C - X_L)} \tag{2-118}$$

When $X_C > X_L$, X_c is inductive, and when $X_C < X_L$, X_c is capacitive.

A simple two-bus power system network model is shown in the single line diagram of Figure 2-17a. The network is driven by a source voltage ($\mathbf{V_{src}}$) behind a source reactance (X_{src}). The transmission line reactance is X and the sending- and receiving-end voltages are $\mathbf{V_s}$ and $\mathbf{V_r}$. The indirect way to implement a variable shunt capacitor or a variable shunt inductor is to generate a variable magnitude compensating voltage (E) in phase with the line voltage (V_s) at the point of compensation and to connect the compensating voltage in shunt with the line through a tie reactance (X_{TIE}). If the transmission line voltage (V_s), compensating voltage (E), and tie reactance (X_{TIE}) are thought of as the sending-end voltage, receiving-end voltage, and line reactance, re-

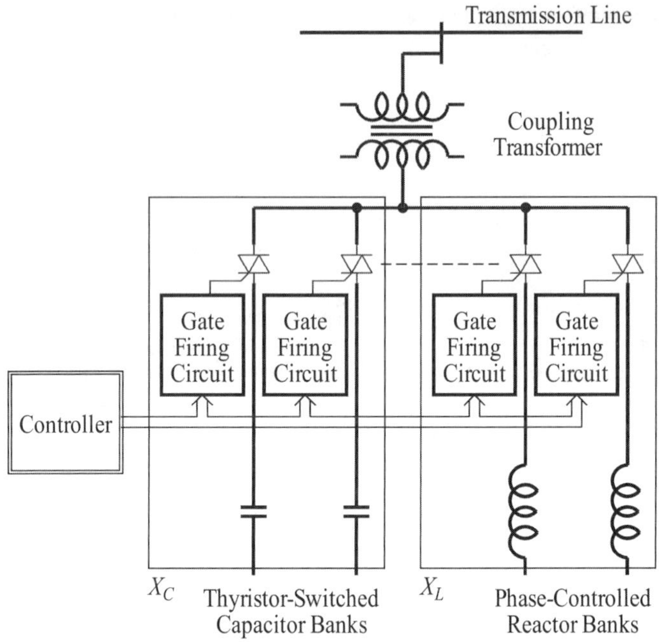

Figure 2-16. Static var compensator.

spectively, then the active power absorbed from the transmission line at the point of compensation is zero because the transmission line voltage and the compensating voltage are in phase ($\delta = 0°$). However, the compensating reactive power absorbed from the line can be expressed by rewriting equation 2-100b as

$$Q_{c-sh} = \frac{V_S E}{X_{TIE}}\left(\frac{V_S}{E} - 1\right) \qquad (2\text{-}119)$$

The concept of connecting a variable amplitude synchronous AC voltage in shunt with the AC line through inductive impedance has been practiced since the introduction of the synchronous condenser in which the internal voltage (E) of the synchronous machine is controlled by controlling the field current. The same concept has been implemented using power electronics based static synchronous compensator (STATCOM). The differences in power electronics based compensation system are there is no wear and tear of moving parts and the response time is a few ms whereas the exciter time constant of a 100-Mvar rated machine is measured in seconds.

A variable magnitude, shunt-connected compensating voltage is implemented with the use of the STATCOM as shown in Figure 2-17a. A net voltage ($\mathbf{V_s} - \mathbf{E}$) drives a current ($\mathbf{I_q}$) through the tie reactance (X_{TIE}) as shown in Figure 2-17b. *The current (I_q) always lags the driving voltage (V_s – E) by 90°.* Through control action, the magnitude of the compensating voltage (E) can be made higher or lower than the line

2.1.2.2 Power at the Receiving End. When the transmission line is compensated, the active and reactive power flows (P_r and Q_r) at the receiving end are

$$P_r = \frac{V_{s'} V_r}{X} \sin\left(\delta_{s'} - \delta_r\right) \tag{2-105a}$$

and

$$Q_r = \frac{V_{s'} V_r}{X} \left\{ \cos\left(\delta_{s'} - \delta_r\right) - \frac{V_r}{V_{s'}} \right\} \tag{2-105b}$$

Equations 2-105a and 2-105b can be rearranged as

$$P_r^2 + \left(Q_r + \frac{V_r^2}{X} \right)^2 = \left(\frac{V_{s'} V_r}{X} \right)^2 \tag{2-106}$$

or

$$\frac{P_r}{Q_r + \dfrac{V_r^2}{X}} = \tan\left(\delta_{s'} - \delta_r\right) \tag{2-107}$$

Then,

$$V_{s'} = \frac{X}{V_r} \sqrt{P_r^2 + \left(Q_r + \frac{V_r^2}{X} \right)^2} \tag{2-108}$$

$$\delta_{s'} = \tan^{-1} \frac{P_r}{Q_r + \dfrac{V_r^2}{X}} + \delta_r \tag{2-109}$$

For the active and reactive power flows (P_r and Q_r) at the receiving end, the required magnitude ($V_{s'}$) and phase angle ($\delta_{s'}$) of the modified sending-end voltage ($\mathbf{V}_{s'}$) are calculated from equations 2-108 and 2-109, respectively. Because of the prevailing line current being at any angle with the terminal voltage, the compensating voltage exchanges bidirectional active and reactive powers with the line. To deliver or absorb the active power to and from the line, an active source or a sink is required. To deliver or absorb the reactive power to and from the line, a capacitive or an inductive reactance is required. When only shunt reactive compensation is used, the exchanged active power is zero.

In order to implement a shunt-connected compensating voltage with a specific magnitude and phase angle, it is necessary to provide a bidirectional path for the exchanged power of the compensating voltage with the line. Figure 2-11 shows a point-

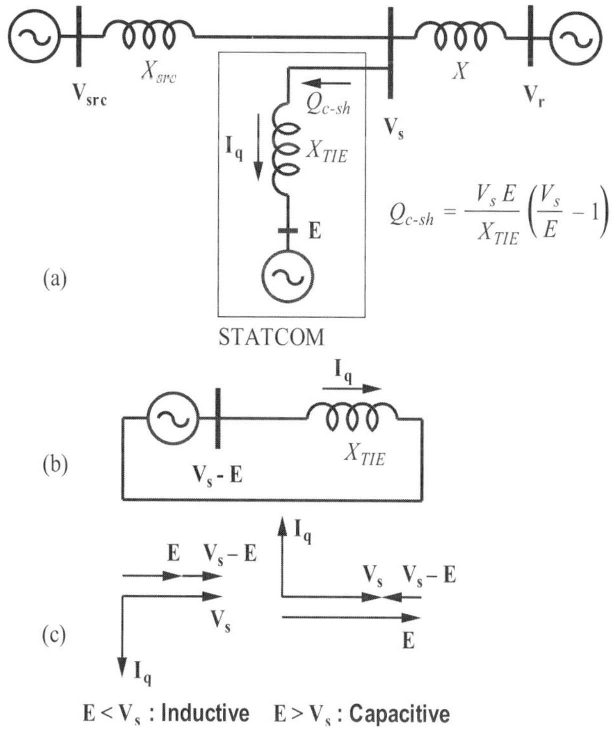

Figure 2-17. Concept of the STATCOM.

voltage (V_s). When $E < V_s$, as shown in Figure 2-17c, the current ($\mathbf{I_q}$) through the STATCOM lags the terminal voltage (V_s) and the line "sees" an inductive reactance connected at its terminal. The STATCOM is considered to be operating in an inductive mode and reactive power is absorbed at the point of compensation by the emulated inductor. Similarly, when $E > V_s$, as shown in Figure 2-17c, the current ($\mathbf{I_q}$) through the STATCOM leads the terminal voltage (V_s) and the line "sees" the STATCOM as a capacitive reactance. The STATCOM is considered to be operating in a capacitive mode and reactive power is delivered at the point of compensation by the emulated capacitor.

Figure 2-18 shows a schematic diagram of the STATCOM. The STATCOM consists of a HN-VSC (VSC1), a magnetic circuit (MC1), a shunt coupling transformer (SHNTR), a shunt breaker (SHNBRK), current and voltage sensors, and a control and protection unit. The operation of the STATCOM is such that the VSC produces a voltage that is almost in phase with the transmission line voltage. Through control action, the VSC output voltage is maintained to be smaller or larger than the line voltage. Therefore, the STATCOM essentially injects an almost sinusoidal reactive current of variable magnitude at the point of compensation. This reactive current, in turn, regulates the transmission line voltage.

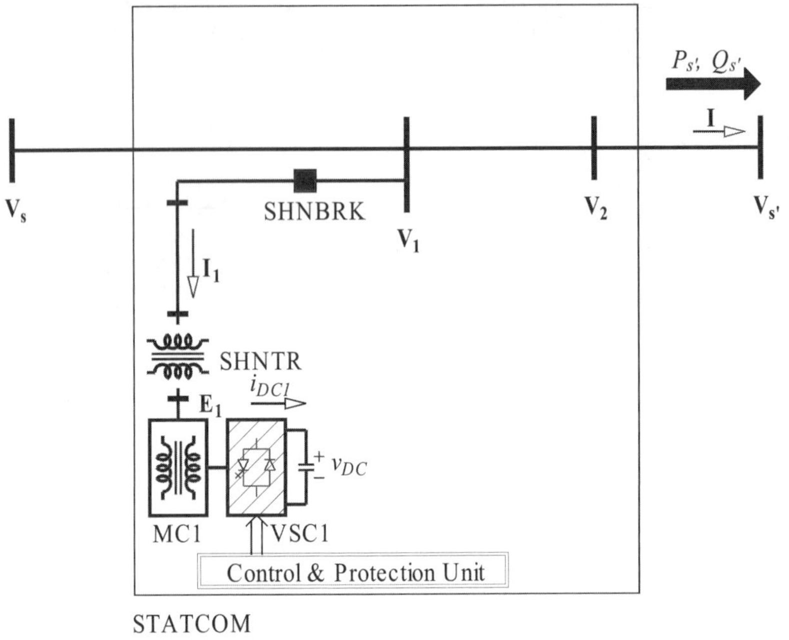

Figure 2-18. Static synchronous compensator (STATCOM).

In a HN-VSC, a number of square wave voltages are generated at fundamental frequency by operating the controllable semiconductor switches once per cycle of the fundamental frequency. These square wave voltages are combined using a magnetic circuit in order to produce a high-quality sinusoidal voltage with acceptable harmonic content. Note that there are many circuit topologies that can be used to create an electronic compensating voltage. However, only a few HN-VSC topologies are discussed in detail in this book. It should also be noted that a shunt-connected VSC1 always injects some harmonic current components into the power system at the point of compensation.

2.2.2 Phase Angle Regulation

The effective phase angle of the transmission line voltage is varied by using a phase angle regulator (PAR), which is also known as a phase shifting transformer (PST). The PAR connects a voltage in series with the transmission line and in quadrature with the phase-to-neutral voltage of the transmission line as shown in Figure 2-19a. The series-connected compensating voltage introduces a phase shift (ψ), shown in Figure 2-19b, whose magnitude (for a small change) in radians varies with the magnitude of the compensating voltage in pu where the phase-to-neutral voltage of the transmission line is the base voltage.

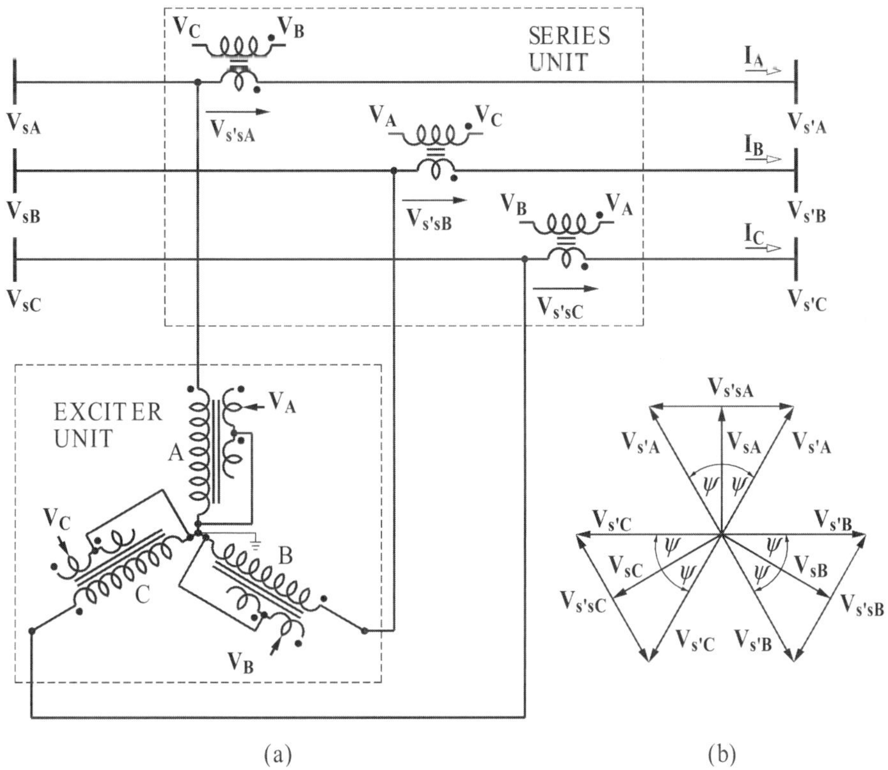

Figure 2-19. (a) Phase angle regulator. (b) Phasor diagram.

$$\psi \cong \tan \psi = \frac{V_{s's}}{V_s} \tag{2-120}$$

In a typical configuration, the PAR consists of two transformers. The first transformer (exciter unit) is called a regulating transformer. Its three-phase (A, B, and C), Y-connected primary windings are connected in shunt with the line. The three-phase primary windings are excited from the line voltages ($\mathbf{V_{sA}}$, $\mathbf{V_{sB}}$, and $\mathbf{V_{sC}}$) and three-phase voltages ($\mathbf{V_A}$, $\mathbf{V_B}$, and $\mathbf{V_C}$) are induced in the regulating transformer secondary windings. For series connection of these voltages, an electrical isolation is necessary. The second transformer (series unit) is called a series transformer and is excited from the phase-to-phase voltages ($\mathbf{V_B} - \mathbf{V_C}$, $\mathbf{V_C} - \mathbf{V_A}$, and $\mathbf{V_A} - \mathbf{V_B}$) of the regulating transformer. Three-phase, bipolar compensating voltages ($\mathbf{V_{s'sA}}$, $\mathbf{V_{s'sB}}$, and $\mathbf{V_{s'sC}}$) that are in quadrature with the corresponding phase-to-neutral voltage are induced in the secondary windings of the series transformer. The magnitudes of the compensating voltages are varied using the LTCs in the secondary windings of the regulating transformer. The induced voltages of the series transformer are connected in series with the transmission line. If the series transformer is a step-down transformer, the primary windings of the series transformer as well as the secondary windings of the regulating transformer are high-voltage and low-current rated, so that the LTCs on the secondary side of the regulating transformer can operate at a low current and can ride through a high fault current. Note that the PAR can be realized with one transformer as well. In this case, the LTCs are always required to carry a high line current as well as even a higher fault current.

2.2.3 Series Reactance Regulation

In a special case, the magnitude of the line voltage and its phase angle can also be varied together in such a way that the effective line reactance (X_{eff}) between the sending and receiving ends is controlled.

2.2.3.1 Direct Method. The direct way to regulate the effective line reactance is to connect a capacitor or an inductor in series with the transmission line. The power flow equations 2-105a and 2-105b can be rewritten as

$$P_r = \frac{V_s V_r}{X_{eff}} \sin(\delta_s - \delta_r) \tag{2-121a}$$

and

$$Q_r = \frac{V_s V_r}{X_{eff}} \left(\cos(\delta_s - \delta_r) - \frac{V_r}{V_s} \right) \tag{2-121b}$$

where X_{eff} (i.e., $X - X_c$) is the effective reactance of the transmission line between its two ends. The compensating reactance (X_c) is defined to be positive when it is a capacitor and negative when it is an inductor.

A series-connected capacitor increases the power flow in the line by decreasing the effective line reactance between its two ends. A series-connected inductor decreases the power flow in the line by increasing the effective line reactance between its two ends. Through the use of a thyristor-controlled series compensator (TCSC), a series-connected variable capacitor or a variable inductor can be implemented as shown in Figure 2-20. Each section of a TCSC is a parallel combination of a fixed capacitor and a thyristor-controlled reactor (TCR). Equation 2-118 defines the inductive or capacitive operation of each section of the TCSC.

2.2.3.2 Indirect Method. The indirect way to implement a variable series capacitor or a variable series inductor is to control the magnitude and phase angle of the transmission line voltage simultaneously with a series-connected compensating voltage in such a way that the effective line reactance is controlled. Through control action, the magnitude of the compensating voltage is varied and made lagging or leading the prevailing line current by 90° in order to emulate a variable capacitor or a variable inductor. A variable magnitude, series-connected, compensating voltage is implemented through the use of a static synchronous series compensator (SSSC).

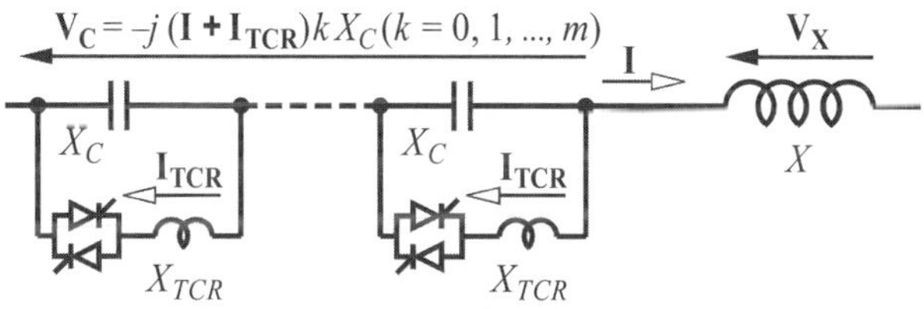

Figure 2-20. Thyristor-controlled series capacitor.

Figure 2-21a shows a simple power transmission system with the SSSC operated both in inductive and in capacitive modes. Figure 2-21b shows the related phasor diagram. Figure 2-21c shows that the line current (**I**) decreases from its uncompensated value when the series-connected compensating voltage ($\mathbf{V_q}$), in load convention, leads the line current by 90° to provide inductive reactance compensation. The line current (**I**) increases when the series-connected compensating voltage ($\mathbf{V_q}$) lags the line current by 90° to provide capacitive reactance compensation, as shown in Figure 2-21d.

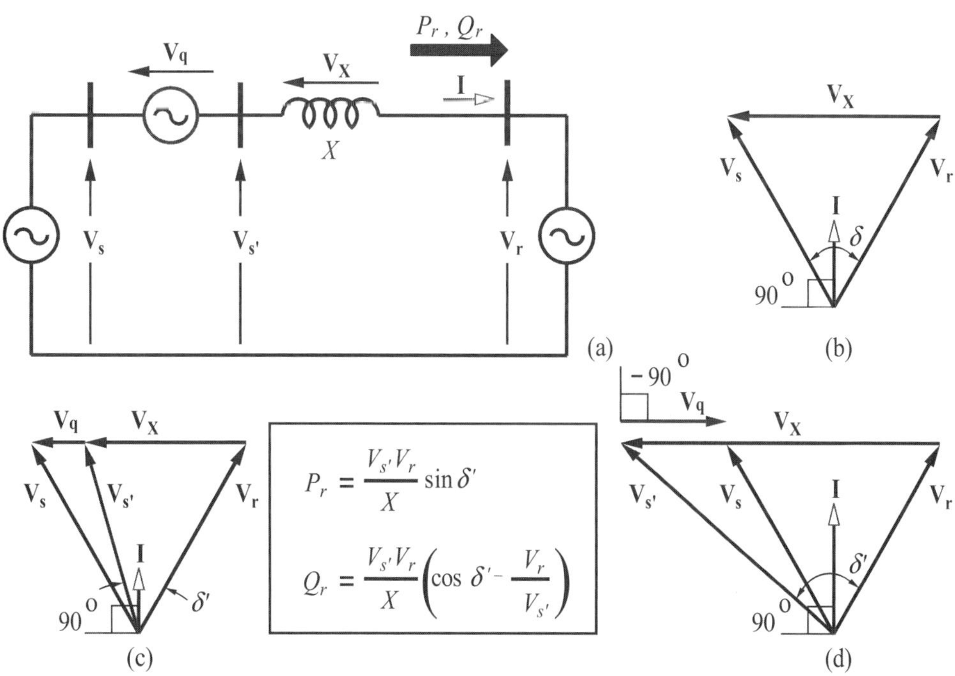

Figure 2-21. (a) Power transmission system and its series reactance emulator with a compensating voltage ($\mathbf{V_q}$) and phasor diagrams for (b) uncompensated line (c) inductively compensated line, and (d) capacitively compensated line.

The expressions for power flow at the receiving end of the line are shown, considering that the modified sending-end voltage is $\mathbf{V}_{s'}$.

Figure 2-22 shows a schematic diagram of the SSSC. The SSSC consists of a HN-VSC (VSC2), a magnetic circuit (MC2), a series coupling transformer (SRSTR), a series disconnect switch (SRSDS), an electronic bypass switch (ES), a series bypass breaker (SRSBRK), current and voltage sensors, and a control and protection unit. The operation of the SSSC is such that the VSC produces a voltage that is almost in quadrature with the prevailing line current. Through control action, the magnitude of the compensating voltage is varied to emulate a variable reactance in series with the line.

2.2.4 Independent Control of Active and Reactive Power Flows

In all the concepts presented thus far, only one parameter (voltage magnitude, its phase angle, or line reactance) is controlled and, therefore, both the active and reactive power flows are regulated simultaneously. For independent control of active and reactive power flows in the line, both the magnitude and the phase angle of the transmission line voltage must be controlled independently. This, in turn, regulates the voltage ($\mathbf{V}_X$) across the transmission line and hence the current and the power flow through the transmission line are regulated. A PFC controls the transmission line voltage by adding a series-connected compensating voltage to the original transmission line volt-

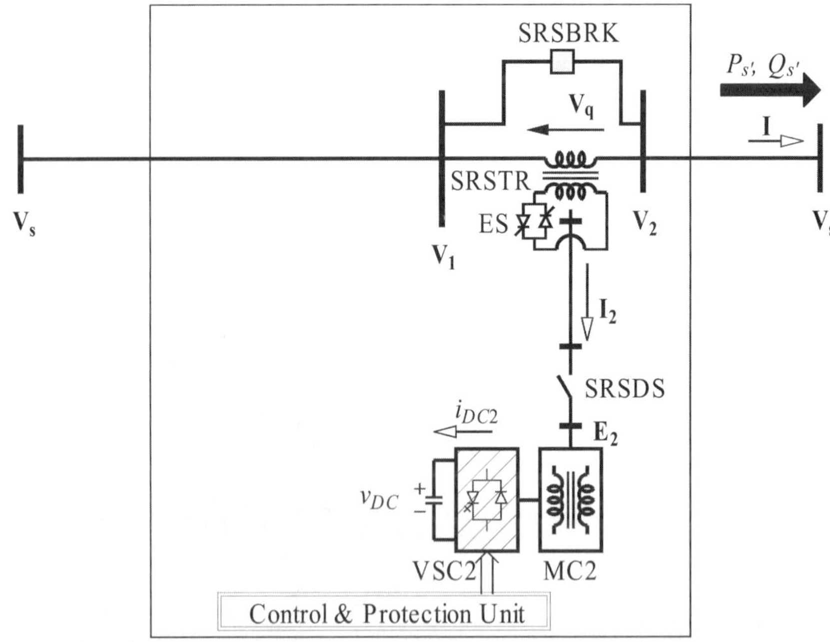

Figure 2-22. Static synchronous series compensator (SSSC).

age. There are several schemes to implement the concept of independent control of active and reactive power flows in a transmission line.

Figure 2-23 shows the basic building blocks of VSC-based PFCs namely, STATCOM and SSSC and their joint topology called a unified power flow controller (UPFC). When the STATCOM and the SSSC operate as stand-alone compensators, they only exchange reactive power (meaning the compensating voltage and the current flow through it are in quadrature) with the line and control only one power flow control parameter—line voltage by the STATCOM and line reactance by the SSSC, respectively. Of course, each of the STATCOM and SSSC also draws active power to provide its own losses that are a few percent of its rated power. When both the STAT-

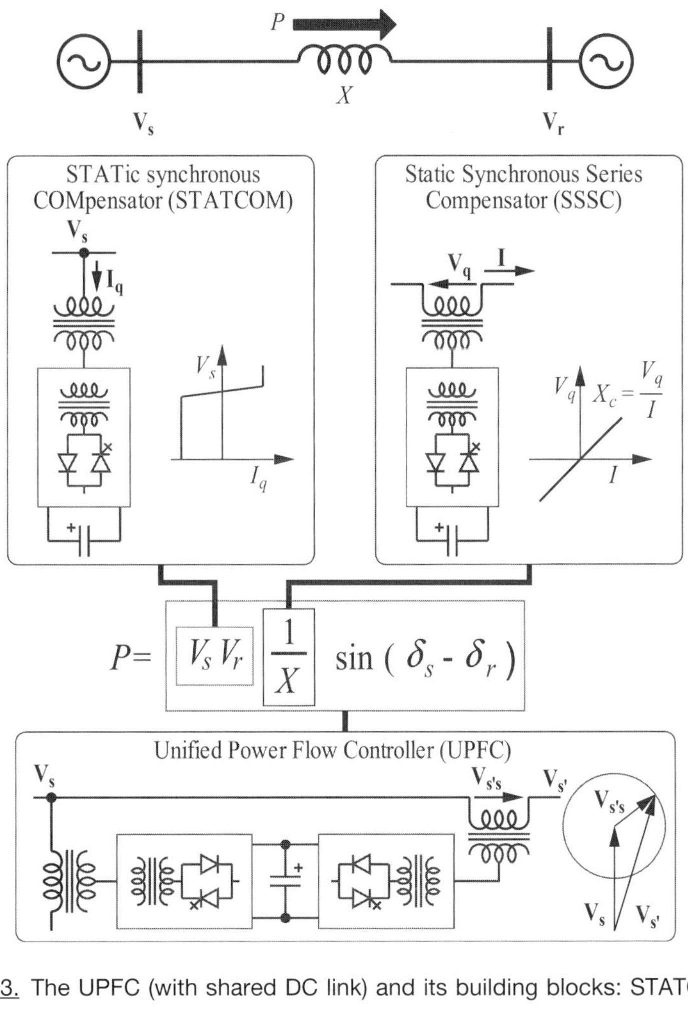

Figure 2-23. The UPFC (with shared DC link) and its building blocks: STATCOM and SSSC.

COM and the SSSC share the DC link, the series-connected compensating voltage of the SSSC is no longer restricted to be in quadrature with the prevailing line current. This unique DC link connection adds active power handling capability to both the STATCOM and the SSSC. The series-connected compensating voltage can be at any phase angle with the prevailing line current, thereby exchanging active power with the line bidirectionally. This exchanged active power flows freely back and forth through the DC link. The result is an emulation of an impedance in series with the line. The modified line voltage can be placed at any phase angle within the design limit to control the active and reactive power flows in the line independently. The hardware required for this recently developed topology is rated for only a fraction of the transmission line's power rating and the controlled power flow capability.

The concept of the UPFC is shown in Figure 2-24. The VSC1, **E** (i.e., $E \angle \delta_s + \alpha$), is connected to the transmission line voltage, $\mathbf{V_s}$ (i.e., $V_s \angle \delta_s$), with a tie reactance (X_{TIE}). The VSC2, $\mathbf{V_{s's}}$ (i.e., $V_{s's} \angle \delta_s + \beta$), is connected in series with the transmission line. The active power that flows through the shared link is P_{exch}. In addition, the two VSCs exchange reactive power with the respective transmission line independently.

2.2.4.1 *Unified Power Flow Controller.* Figure 2-25 shows the schematic diagram of the UPFC. The UPFC consists of two HN-VSCs (VSC1 and VSC2), two magnetic circuits (MC1 and MC2), a shunt coupling transformer (SHNTR), a series coupling transformer (SRSTR), a shunt breaker (SHNBRK), a series disconnect switch (SRSDS), an electronic bypass switch (ES), a DC link switch (DCLS), a series bypass breaker (SRSBRK), current and voltage sensors, and a control and protection unit. Each VSC is coupled with a transformer at its output. Both VSCs generate almost sinusoidal voltages.

When the DCLS is closed, the VSCs share the DC link. The series-connected, variable magnitude, compensating voltage that is at any phase angle with respect to the line current exchanges an active power as well as a reactive power with the line. This exchanged active power at the AC terminal of the series-connected VSC flows to the AC terminal of the shunt-connected VSC through the shared DC link. In addition, the shunt-connected VSC can deliver or absorb reactive power independently. When the

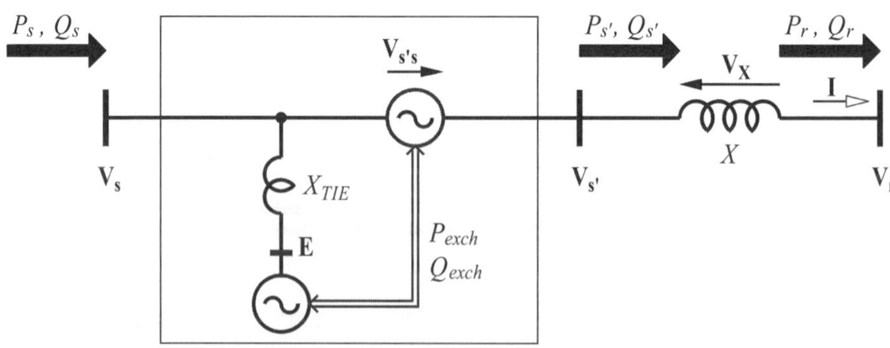

Figure 2-24. Concept of the unified power flow controller.

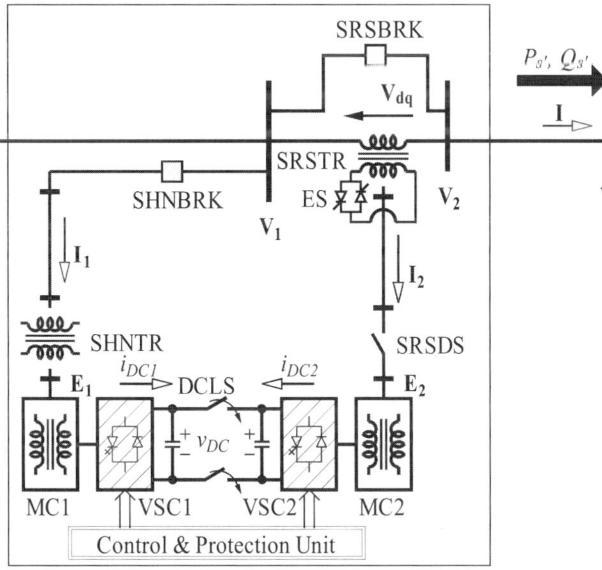

Figure 2-25. Unified power flow controller.

DCLS is open, the two VSCs can be operated as stand-alone, independent, reactive compensators, such as a shunt-connected compensator (STATCOM) and a series-connected compensator (SSSC) and, in turn, exchange almost exclusively reactive power with the line at their respective terminals.

Figure 2-26 shows the basic UPFC model consisting of the STATCOM and the SSSC with a shared DC link. The SSSC injects a compensating voltage ($\mathbf{V}_{s's}$) that is at

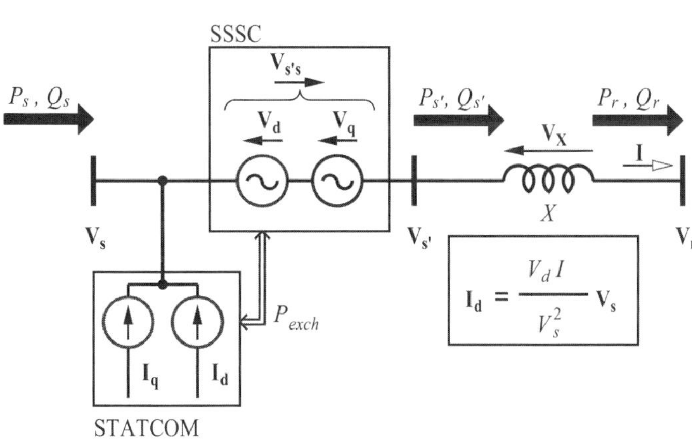

Figure 2-26. Basic UPFC model.

any phase angle with the prevailing line current (**I**). The series-connected compensating voltage ($\mathbf{V_{s's}}$) has active and reactive components and they are $\mathbf{V_d}$ and $\mathbf{V_q}$ with load convention. The component ($\mathbf{V_d}$) of the compensating voltage that is either in phase or out of phase with the line current emulates a positive or a negative resistor in series with the line. The remaining component ($\mathbf{V_q}$) that is in quadrature with the line current emulates either an inductor or a capacitor in series with the transmission line.

The compensating voltage ($\mathbf{V_{s's}}$) exchanges active and reactive powers (P_{exch} and Q_{exch}) that are defined as

$$P_{exch} = -\mathbf{V_{s's}} \bullet \mathbf{I} = \mathbf{V_{dq}} \bullet \mathbf{I} = V_d I = V_s I_d \tag{2-122a}$$

and

$$Q_{exch} = \left| -\mathbf{V_{s's}} \times \mathbf{I} \right| = \left| \mathbf{V_{dq}} \times \mathbf{I} \right| = V_q \tag{2-122b}$$

The active power (P_{exch}) exchanged with the line flows through the STATCOM and results in an active current flow ($\mathbf{I_d}$) through it. This current is either in phase or out of phase with the line voltage and emulates a positive or a negative resistor at the point of compensation in shunt with the line. However, the reactive power (Q_{exch}) exchanged with the line is independent of whether there is any reactive current ($\mathbf{I_q}$) flowing through the STATCOM. The loading effect (voltage change) of the power system network at the point of compensation due to (1) active current flow through the STATCOM and (2) load change in the line may be compensated by the independent control of the reactive current flow through the STATCOM. This reactive component ($\mathbf{I_q}$) that is in quadrature with the line voltage emulates an inductive or a capacitive reactance at the point of compensation in shunt with the line.

The fast switching capability of the VSCs enables the UPFC to provide fast dynamic response if properly designed. Even though the shunt–series type topology is ideal for a PFC, the VSC-based UPFC has two major drawbacks: its high installation and operating costs. Moreover, the fast dynamic response in the range of ms is not required in most utility applications. Traditionally, a mechanical LTC is used since its speed of operation provides adequate response time in most utility applications. If faster response is needed, the mechanical LTCs can be replaced with faster switches.

In the next section, a new FACTS controller is introduced that provides independent control of active and reactive power flows in a transmission line with the use of time-tested components, such as transformers and LTCs.

2.2.4.2 Sen Transformer. The schematic diagram of a Sen transformer (ST) and its operational phasor diagram are shown in Figure 2-27. The ST consists of an exciter unit and a compensating voltage unit that connects a compensating voltage in series with the line and, in turn, modifies the line voltage to a specific magnitude and phase angle within its design limit. While the series-connected compensating voltage ($\mathbf{V_{s's}}$) provides compensation in its entire range of operation of 360°, its exchanged active and reactive powers with the transmission line flow bidirectionally through the shunt-connected exciter unit, as shown in Figure 2-10.

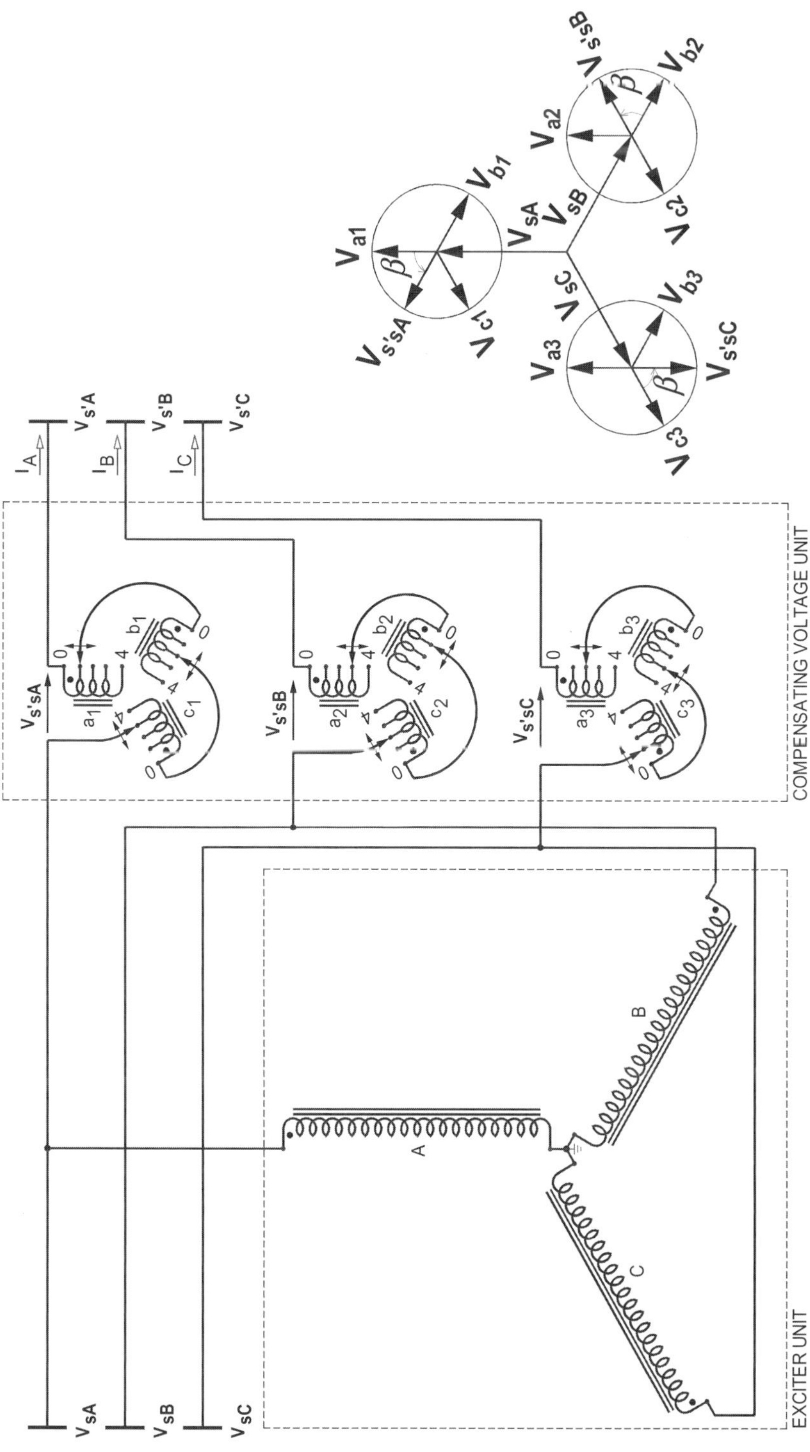

Figure 2-27. Sen transformer.

Figure 2-28 shows the basic ST model, which consists of a series-connected compensating voltage unit and a shunt-connected exciter unit. The compensating voltage unit produces a compensating voltage ($\mathbf{V_{s's}}$) that is at any phase angle with the prevailing line current (**I**). The series-connected compensating voltage ($\mathbf{V_{s's}}$) consists of active and reactive components, and they are $\mathbf{V_d}$ and $\mathbf{V_q}$ with load convention. The component ($\mathbf{V_d}$) of the compensating voltage that is either in phase or out of phase with the line current emulates either a positive or a negative resistor in series with the line. The remaining component ($\mathbf{V_q}$) that is in quadrature with the line current emulates either an inductor or a capacitor in series with the transmission line.

The compensating voltage ($\mathbf{V_{s's}}$) exchanges active and reactive powers (P_{exch} and Q_{exch}) that are defined as

$$P_{exch} = -\mathbf{V_{s's}} \bullet \mathbf{I} = \mathbf{V_{dq}} \bullet \mathbf{I} = V_d I = V_s I_d \qquad \text{(2-123a)}$$

and

$$Q_{exch} = \left|-\mathbf{V_{s's}} \times \mathbf{I}\right| = \left|\mathbf{V_{dq}} \times \mathbf{I}\right| = V_q I = V_s I_q \qquad \text{(2-123b)}$$

The active and reactive powers (P_{exch} and Q_{exch}), exchanged with the line, flow through the exciter unit, and result in active and reactive currents ($\mathbf{I_d}$ and $\mathbf{I_q}$) flowing through it. The active current ($\mathbf{I_d}$) that is either in phase or out of phase with the line voltage emulates a positive or a negative resistor at the point of compensation in shunt with the line. The reactive current ($\mathbf{I_q}$) that is in quadrature with the line voltage emulates an inductive or a capacitive reactance at the same point.

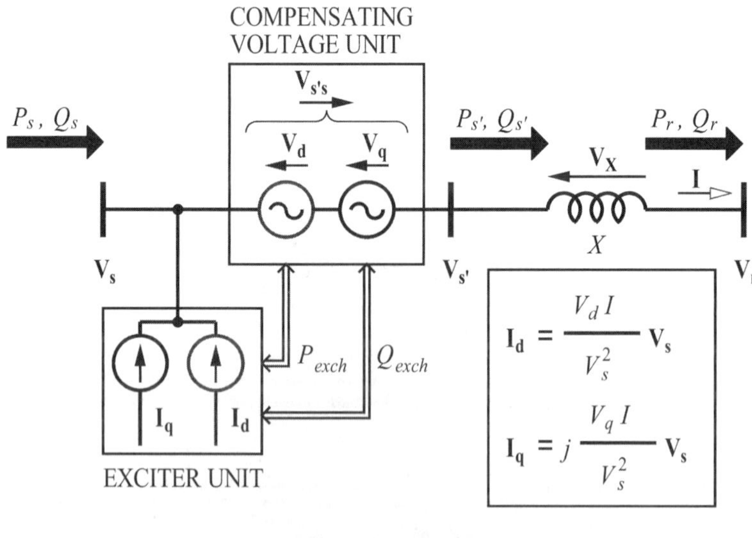

Figure 2-28. Basic ST model.

The active and reactive powers exchanged by the UPFC and the ST with the line are the same. In both the UPFC and ST, the exchanged active power (P_{exch}) contributes to the same amount of active current ($\mathbf{I_d}$) through the shunt connection. Therefore, when the series-connected compensating voltage unit emulates a $+R$, the shunt-connected exciter unit emulates a $-R$ and vice versa. While in case of the ST, the exchanged reactive power (Q_{exch}) contributes to an amount of reactive current ($\mathbf{I_q}$) through the shunt connection, the reactive current ($\mathbf{I_q}$) through the STATCOM is independent of the reactive power (Q_{exch}) exchanged by the series-connected compensating unit with the line. Therefore, in the ST, when the series-connected compensating unit emulates a capacitor, the shunt-connected exciter unit emulates an inductor and vice versa.

2.3 INTERLINE POWER FLOW CONCEPT

The ST or the UPFC is suitable for independent control of active and reactive power flows in a single line to which it is connected. However, several transmission lines are usually connected in close proximity to a shared voltage bus. Therefore, any change in the power flow in one line will affect the power flow in the other lines as well. The excessive power from one line cannot be directly transferred to a specific line. In a multiline transmission network, it would be advantageous to be able to transfer power from an overloaded to an underloaded line with minimum impact on the power flow in the other uncompensated lines.

A series-connected compensating voltage controls the magnitude and phase angle of the transmission line voltage independently and, therefore, regulates the active and reactive power flows in a transmission line independently. In the process, it exchanges active and reactive powers at the point of compensation with the line. The need for the exchanged active power demand can be met using a series–series power converter, which is composed of at least two DC-to-AC VSCs that are connected back-to-back at the shared DC link. Hence, it is called a back-to-back (BTB) SSSC. Each SSSC is connected in series with an AC transmission line. In a typical application, the BTB-SSSC can transfer active power from one or more transmission lines, referred to as "master" lines, to the others, referred to as "slave" lines. In addition, the BTB-SSSC can provide independent reactive compensation to each transmission line to which it is connected and thereby controls the effective reactance of each line between its two ends. The VSC-based PFC (BTB-SSSC), also known as interline power flow controller (IPFC), was demonstrated to be able to provide this operation. If the transmission lines that are involved in the compensation scheme are not connected to a shared voltage bus, they may be operated at different frequencies. This compensation scheme is then an "asynchronous tie."

Another FACTS controller, based on transformers and LTCs, is the multiline Sen transformer (MST) that provides independent control of active and reactive power flows in a transmission line and also transfers active and reactive powers selectively from one line to another in a multiline transmission system without affecting the power flows in the uncompensated lines.

2.3.1 Back-to-Back SSSC

A basic BTB-SSSC that provides compensation in a two-line transmission system is shown in Figure 2-29. Both lines are connected together at the sending end with voltage ($\mathbf{V_s}$). Line 1 has a series-connected compensating voltage ($\mathbf{V_{1s's}}$), line reactance (X_1), and a receiving-end voltage ($\mathbf{V_{1r}}$). Line 2 has a series-connected compensating voltage ($\mathbf{V_{2s's}}$), line reactance (X_2), and a receiving-end voltage ($\mathbf{V_{2r}}$).

The modified sending-end voltages for lines 1 and 2 are

$$\mathbf{V_{1s'}} = \mathbf{V_s} + \mathbf{V_{1s's}} \tag{2-124}$$

$$\mathbf{V_{2s'}} = \mathbf{V_s} + \mathbf{V_{2s's}} \tag{2-125}$$

Line 1 operates as the "master" line, for which the series-connected compensating voltage ($\mathbf{V_{1s's}}$) is of any magnitude within its allowable limit and at any phase angle with the sending-end voltage as well as the prevailing line current ($\mathbf{I_1}$). Therefore, the compensating voltage ($\mathbf{V_{1s's}}$) exchanges active and reactive powers (P_{1exch} and Q_{1exch}) with the line and emulates, in series with the transmission line, a capacitor that increases the power flow of the line or an inductor that decreases the power flow of the line and a positive resistor that absorbs active power from the line or a negative resistor that delivers active power to the line. The desired compensating voltage is actually an impedance emulator.

Line 2 operates as the "slave" line where the series-connected compensating voltage ($\mathbf{V_{2s's}}$) is of specific magnitude and phase angle with the prevailing line current ($\mathbf{I_2}$) and, therefore, exchanges, with the line, both active and reactive powers (P_{2exch} and Q_{2exch}), so as to satisfy the active and reactive powers demand (both delivered and absorbed) by the compensating voltage 1.

In an application, it may be desirable to provide series compensation to a number of transmission lines in such a way that the active power from certain master lines flows through certain slave lines while the reactive compensation in each line is indepen-

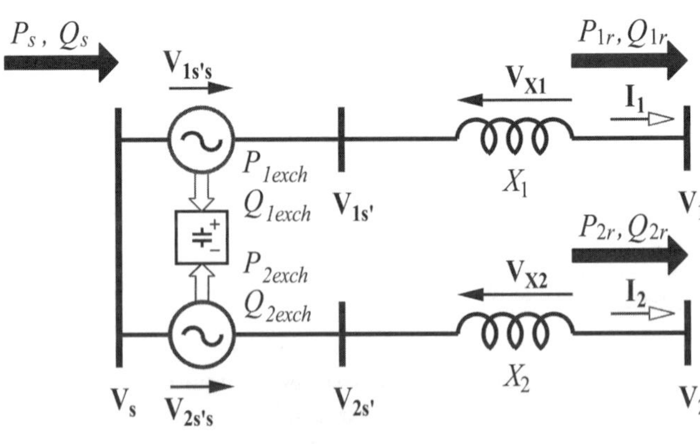

Figure 2-29. Basic BTB-SSSC.

dently provided. Such a scheme can be implemented with several lines compensated by the SSSCs that are connected to a shared DC link. Figure 2-30 shows a basic BTB-SSSC with two VSCs that are connected at their shared DC link. The compensating voltage ($\mathbf{V}_{1s's}$) from VSC1 is connected in series with transmission line 1. The compensating voltage ($\mathbf{V}_{2s's}$) from VSC2 is connected in series with transmission line 2. The active power that flows through the shared DC link is $P_{1exch} = -P_{2exch}$. In addition, the two VSCs exchange reactive power ($Q_{1exch} \neq Q_{2exch}$) with the respective transmission line independently.

The BTB-SSSC, shown in Figure 2-30, consists of two HN-VSCs (VSC1 and VSC2), two magnetic circuits (MC1 and MC2), two series coupling transformers (SRSTR1 and SRSTR2), two series disconnect switches (SRSDS1 and SRSDS2), two electronic bypass switches (ES1 and ES2), a DC link switch (DCLS), two series bypass breakers (SRSBK1 and SRSBK2), current and voltage sensors, and a control and protection unit.

When the DCLS is open, the two VSCs can be used as independent reactive compensators, such as a series-connected compensator, SSSC. Each compensator is coupled with a transformer at its output. The SSSC connects an almost sinusoidal voltage, of variable magnitude, in series with the transmission line. When the SSSCs operate as stand-alone controllers, they exchange almost exclusively reactive power with the line at their respective AC terminals.

When the DCLS is closed, the VSCs share the DC link. While operating both the VSCs together as a BTB-SSSC, the exchanged power at the terminal of each VSC can be reactive as well as active. The series-connected compensating voltage can be at any

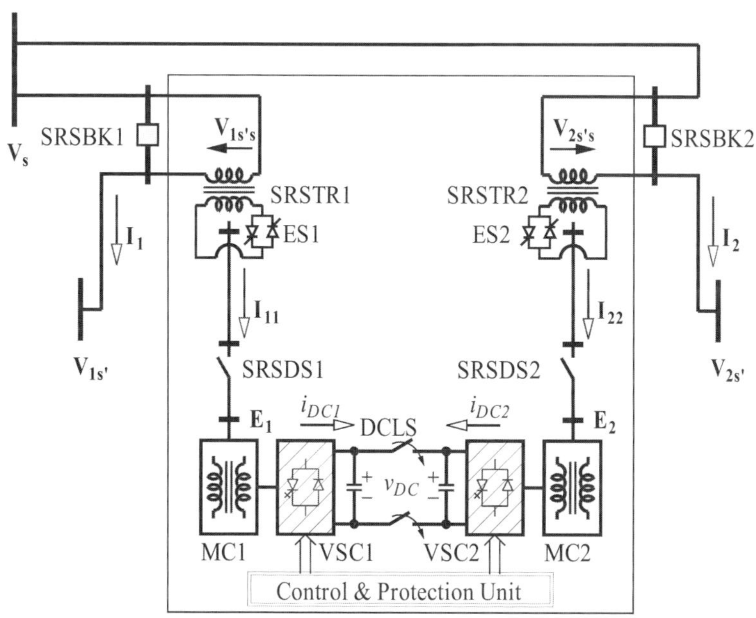

Figure 2-30. Implementation of a basic BTB-SSSC.

phase angle with respect to the prevailing line current. The exchanged active power at terminal of one VSC with the line flows to the terminal of the other VSC through the shared DC link. Note that the BTB-SSSC power circuit must be rated for the compensating (not transmitted) power in the AC lines to which it is connected.

If the transmission lines that are involved in the BTB-SSSC compensation scheme are not connected to a shared voltage bus, they are not required to be operated at the same frequency. This compensation scheme, shown in Figure 2-31, is then an “asynchronous tie.”

The modified sending-end voltages for lines 1 and 2 are

$$\mathbf{V}_{\mathbf{1s'}} = \mathbf{V}_{\mathbf{1s}} + \mathbf{V}_{\mathbf{1s'1s}} \tag{2-126}$$

$$\mathbf{V}_{\mathbf{2s'}} = \mathbf{V}_{\mathbf{2s}} + \mathbf{V}_{\mathbf{2s'2s}} \tag{2-127}$$

2.3.2 Multiline Sen Transformer

The alternative to the power flow control in a multiline transmission system with the use of VSC-based technology is the multiline Sen transformer (MST), shown in Figure 2-32, that is based on the established technology of transformer and LTCs and is capable of providing an effective power flow management for multiple transmission lines that share a common voltage bus.

The general power flow equations in an uncompensated transmission line are

$$P_{rn} = \frac{V_s V_r}{X} \sin(\delta_s - \delta_r) \tag{2-16a}$$

and

$$Q_{rn} = \frac{V_s V_r}{X} \left\{ \cos(\delta_s - \delta_r) - \frac{V_r}{V_s} \right\} \tag{2-16b}$$

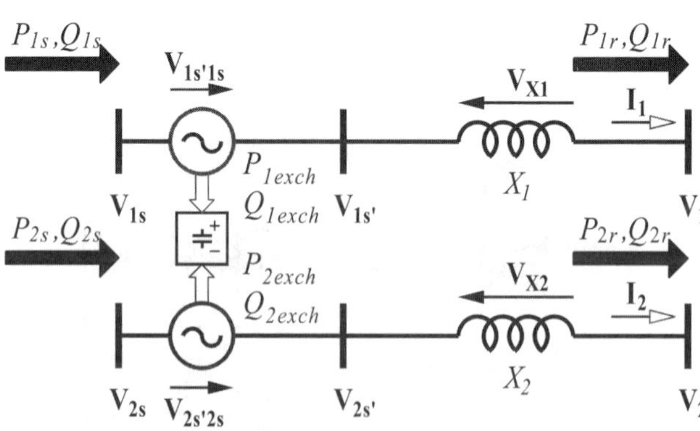

Figure 2-31. BTB-SSSC scheme as an asynchronous tie.

The power flow equations in the compensated transmission line 1 (“master” line) are

$$P_{1r} = \frac{V_{1s'}V_{1r}}{X_1}\sin(\delta_{1s'} - \delta_{1r}) \tag{2-128a}$$

and

$$Q_{1r} = \frac{V_{1s'}V_{1r}}{X_1}\left[\cos(\delta_{1s'} - \delta_{1r}) - \frac{V_{1r}}{V_{1s'}}\right] \tag{2-128b}$$

Equations 2-128a and 2-128b can be rearranged as

$$P_{1r}^2 + \left(Q_{1r} + \frac{V_{1r}^2}{X_1}\right)^2 = \left(\frac{V_{1s'}V_{1r}}{X_1}\right)^2 \tag{2-129}$$

and

$$\frac{P_{1r}}{Q_{1r} + \dfrac{V_{1r}^2}{X_1}} = \tan(\delta_{1s'} - \delta_{1r}) \tag{2-130}$$

The magnitude ($V_{1s'}$) of the modified sending-end voltage ($\mathbf{V}_{1s'}$) in line 1 is expressed by rearranging equation 2-129 as

$$V_{1s'} = \frac{X_1}{V_{1r}}\sqrt{P_{1r}^2 + \left(Q_{1r} + \frac{V_{1r}^2}{X_1}\right)^2} \tag{2-131}$$

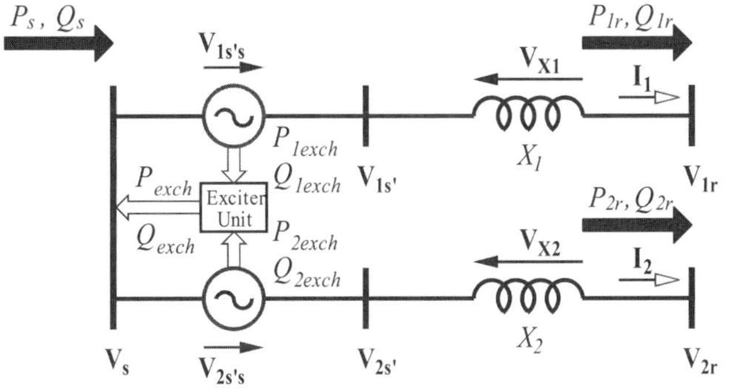

Figure 2-32. Implementation of a basic MST.

The phase angle ($\delta_{1s'}$) of the modified sending-end voltage ($\mathbf{V}_{1s'}$) in line 1 is expressed by rearranging equation 2-130 as

$$\delta_{1s'} = \tan^{-1} \frac{P_{1r}}{Q_{1r} + \dfrac{V_{1r}^2}{X_1}} + \delta_{1r} \tag{2-132}$$

Considering the sending-end voltage phasor as reference ($\delta_s = 0°$), ψ_1 (i.e., $\delta_{1s'} - \delta_s$) as the relative phase angle of the modified sending-end voltage in line 1, and δ_1 (i.e., $\delta_s - \delta_{1r}$) as the relative phase angle of the receiving-end voltage in line 1, equation 2-132 can be rewritten as

$$\psi_1 = \tan^{-1} \frac{P_{1r}}{Q_{1r} + \dfrac{V_{1r}^2}{X_1}} - \delta_1 \tag{2-133}$$

The modified sending-end voltage in line 1 is

$$V_{1s'} \angle \psi_1 = V_s \angle 0° + V_{1s's} \angle \beta_1 \tag{2-134}$$

When the real and imaginary parts of the both sides of equation 2-134 are separated, the result is

$$V_{1s'} \cos \psi_1 = V_s + V_{1s's} \cos \beta_1 \tag{2-135a}$$

and

$$V_{1s'} \sin \psi_1 = V_{1s's} \sin \beta_1 \tag{2-135b}$$

Therefore, the magnitude ($V_{1s's}$) and the relative phase angle (β_1) of the series-connected compensating voltage in line 1 are expressed by rearranging equations 2-135a and 2-135b as

$$V_{1s's} = \sqrt{V_{1s'}^2 + V_s^2 - 2V_{1s'} V_s \cos \psi_1} \tag{2-136a}$$

and

$$\beta_1 = \tan^{-1} \frac{V_{1s'} \sin \psi_1}{V_{1s'} \cos \psi_1 - V_s} \tag{2-136b}$$

The expressions for the active and reactive powers (P_{1exch} and Q_{1exch}) exchanged in line 1 can be written from equations 2-81 and 2-86 as

$$P_{1exch} = A_{1exch} \sin\left(\beta_1 - \tan^{-1} \frac{P_{1sn}}{Q_{1sn}}\right) \tag{2-137a}$$

and

$$Q_{1exch} = -\frac{V_{1s's}^2}{X_1} - A_{1exch} \cos\left(\beta_1 - \tan^{-1}\frac{P_{1sn}}{Q_{1sn}}\right) \tag{2-137b}$$

where

$$A_{1exch} = \frac{V_{1s's}V_{1Xn}}{X_1} \tag{2-138}$$

Note that the natural voltage (V_{1Xn}) across the transmission line 1 is

$$V_{1Xn} = \sqrt{V_s^2 + V_{1r}^2 - 2V_sV_{1r}\cos\delta_1} \tag{2-139}$$

If the multiline power flow concept is applied in two lines using compensating voltages generated by electrical machines or transformer and LTCs, it can be written from Figure 2-32 that

$$P_{1exch} + P_{2exch} = P_{exch} \tag{2-140a}$$

and

$$Q_{1exch} + Q_{2exch} = Q_{exch} \tag{2-140b}$$

If the multiline power flow concept is applied in two lines using compensating voltages generated by VSCs, it can be written from Figure 2-29 that

$$P_{1exch} + P_{2exch} = 0 \tag{2-141}$$

while the reactive powers exchanged with each line are independent of each other. The power exchanged with line 2 is the negative of the power exchanged with line 1. In other words, if line 1 is delivering power, line 2 is absorbing the same amount of power that line 1 is delivering and vice versa. Once the active and reactive powers (P_{2exch} and Q_{2exch}) exchanged with line 2 are known, the required compensating voltage, $\mathbf{V_{2s's}}$ (i.e., $V_{2s's}\angle\beta_2$), can be calculated as follows. The expressions for the active and reactive powers (P_{2exch} and Q_{2exch}) exchanged with line 2 can be written from equations 2-81 and 2-86 as

$$P_{2exch} = A_{2exch} \sin\left(\beta_2 - \tan^{-1}\frac{P_{2sn}}{Q_{2sn}}\right) \tag{2-142a}$$

and

$$Q_{2exch} = -\frac{V_{2s's}^2}{X_2} - A_{2exch} \cos\left(\beta_2 - \tan^{-1}\frac{P_{2sn}}{Q_{2sn}}\right) \tag{2-142b}$$

where

$$A_{2exch} = \frac{V_{2s's} V_{2Xn}}{X_2} \tag{2-143}$$

Note that the natural voltage (V_{2Xn}) across the transmission line 2 is

$$V_{2Xn} = \sqrt{V_s^2 + V_{2r}^2 - 2V_s V_{2r} \cos\delta_2} \tag{2-144}$$

Equations 2-142a and 2-142b can be rearranged as

$$P_{2exch}^2 + \left(Q_{2exch} + \frac{V_{2s's}^2}{X_2}\right)^2 = A_{2exch}^2 \tag{2-145}$$

and

$$\frac{P_{2exch}}{Q_{2exch} + \dfrac{V_{2s's}^2}{X_2}} = -\tan\left(\beta_2 - \tan^{-1}\frac{P_{2sn}}{Q_{2sn}}\right) \tag{2-146}$$

The magnitude ($V_{2s's}$) of the compensating voltage in line 2 is expressed by rearranging equations 2-143 and 2-145 as follows:

$$P_{2exch}^2 + Q_{2exch}^2 + 2Q_{2exch}\frac{V_{2s's}^2}{X_2} + \left(\frac{V_{2s's}^2}{X_2}\right)^2 - \left(\frac{V_{2s's}V_{2Xn}}{X_2}\right)^2 = 0$$

or

$$\left(V_{2s's}^2\right)^2 - \left(V_{2Xn}^2 - 2X_2Q_{2exch}\right)V_{2s's}^2 + X_2^2\left(P_{2exch}^2 + Q_{2exch}^2\right) = 0 \tag{2-147}$$

Solving the above quadratic equation results in

$$V_{2s's}^2 = \frac{\left(V_{2Xn}^2 - 2X_2Q_{2exch}\right) \pm \sqrt{\left(V_{2Xn}^2 - 2X_2Q_{2exch}\right)^2 - 4X_2^2\left(P_{2exch}^2 + Q_{2exch}^2\right)}}{2}$$

or

$$V_{2s's}^2 = \left(\frac{V_{2Xn}^2}{2} - X_2Q_{2exch}\right) \pm \sqrt{\left(\frac{V_{2Xn}^2}{2} - X_2Q_{2exch}\right)^2 - X_2^2\left(P_{2exch}^2 + Q_{2exch}^2\right)}$$

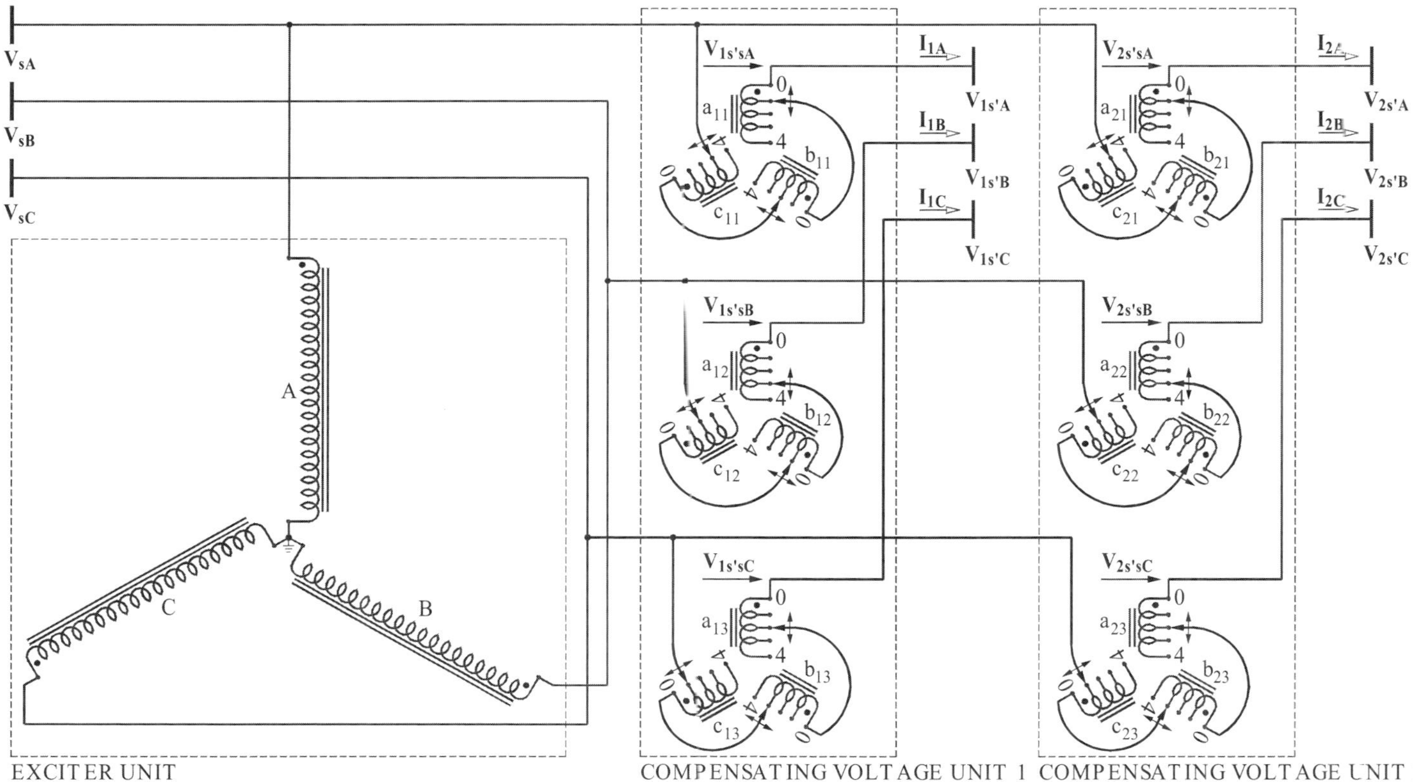

Figure 2-33. Multiline Sen transformer.

or

$$V_{2s's} = +\sqrt{\left(\frac{V_{2Xn}^2}{2} - X_2 Q_{2exch}\right) \mp \sqrt{\left(\frac{V_{2Xn}^2}{2} - X_2 Q_{2exch}\right)^2 - X_2^2\left(P_{2exch}^2 + Q_{2exch}^2\right)}} \tag{2-148}$$

The relative phase angle (β_2) of the compensating voltage ($\mathbf{V_{2s's}}$) is expressed by rearranging equation 2-146 as

$$\beta_2 = \tan^{-1}\frac{P_{2sn}}{Q_{2sn}} - \tan^{-1}\frac{P_{2exch}}{Q_{2exch} + \dfrac{V_{2s's}^2}{X_2}} \tag{2-149}$$

2.3.3 Back-to-Back STATCOM

For point-to-point transfer of power between two isolated networks or interconnection of two transmission lines with different voltages or phase angles (or frequencies), the magnitude and phase angle of the transmission line voltage can be controlled to specific values by a shunt-connected compensating voltage. The compensating voltage exchanges active and reactive powers at the point of compensation with the line. The need for the exchanged active power demand can be met using a shunt–shunt power converter consisting of at least two DC-to-AC VSCs that are connected back to back at the shared DC link. Hence, it is called a back-to-back (BTB) STATCOM. Each STATCOM is connected in shunt to an AC transmission line. In a typical application, the BTB-STATCOM transfers active power from one or more transmission lines, referred to as "master" lines, to the others, referred to as "slave" lines. In addition, the BTB-STATCOM provides independent reactive compensation to each transmission line to which it is connected and thereby regulates the voltage of the transmission line at the point of compensation. The VSC-based BTB-STATCOM was demonstrated to be able to provide this operation and operate as an "asynchronous tie."

Figure 2-34 shows a basic BTB-STATCOM with two VSCs connected at a shared DC link. The VSC1, $\mathbf{E_1}$ (i.e., $E_1 \angle\delta_s + \alpha_1$), is connected to the transmission line voltage, $\mathbf{V_s}$ (i.e., $V_s \angle\delta_s$), with a tie reactance (X_{1TIE}). The VSC2, $\mathbf{E_2}$ (i.e., $E_2 \angle\delta_{s'} + \alpha_2$), is connected to the transmission line voltage, $\mathbf{V_{s'}}$ (i.e., $V_{s'} \angle\delta_{s'}$), with a tie reactance (X_{2TIE}). Line 2 is operated as the "master" line in which the shunt-connected compensating voltage ($\mathbf{E_2}$) is of any magnitude within its allowable limit and at any phase angle with the modified sending-end voltage ($\mathbf{V_{s'}}$) to exchange active and reactive powers (P_{2exch} and Q_{2exch}) with the line. Line 1 is operated as the "slave" line in which the shunt-connected compensating voltage ($\mathbf{E_1}$) is of specific magnitude and phase angle with the sending-end voltage ($\mathbf{V_s}$) to exchange active and reactive powers (P_{1exch} and Q_{1exch}) at the point of compensation. The net active power that flows through the

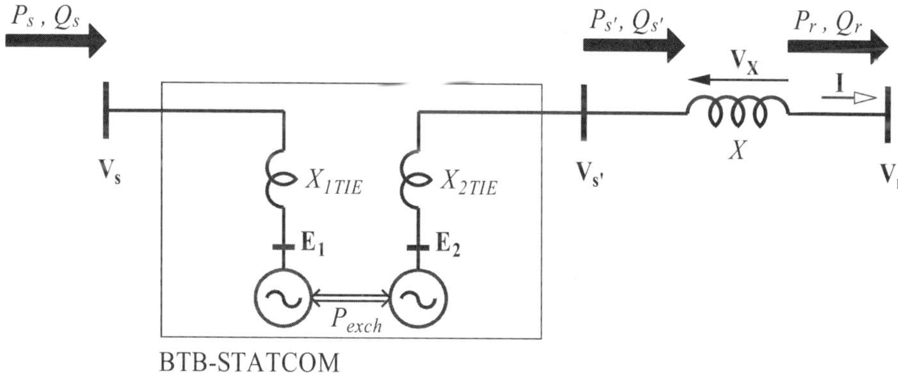

Figure 2-34. Concept of a back-to-back STATCOM.

shared DC link is P_{exch}. Each VSC exchanges reactive power ($Q_{1exch} \neq Q_{2exch}$) with its respective transmission line independently.

Figure 2-35 shows a basic BTB-STATCOM with two VSCs connected at their DC link. The BTB-STATCOM consists of two HN-VSCs (VSC1 and VSC2), two magnetic circuits (MC1 and MC2), two shunt coupling transformers (SHNTR1 and SHNTR2), two shunt breakers (SHNBK1 and SHNBK2), a DC link switch (DCLS), current and voltage sensors, and a control and protection unit.

When the DCLS is open, the two VSCs can be used as independent reactive compensators (STATCOMs). Each VSC is coupled with a transformer at its output. The

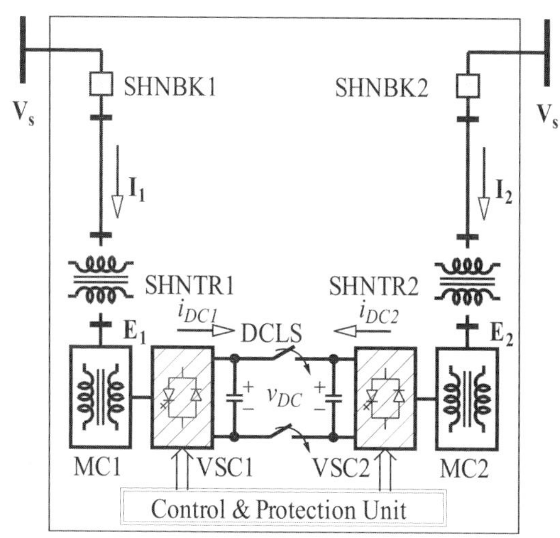

Figure 2-35. Back-to-back STATCOM.

STATCOM injects an almost sinusoidal current, of variable magnitude, in shunt with the transmission line. When the STATCOMs operate as stand-alone compensators, they exchange almost exclusively reactive power with the line.

When the DCLS is closed, the VSCs share the DC link. While operating both the VSCs together as a BTB-STATCOM, the exchanged power at the terminal of each VSC can be reactive as well as active. The shunt-connected compensating voltage can be at any phase angle with respect to the current flowing through it. The exchanged active power at the terminal of one VSC with the line flows to the terminal of the other VSC through the shared DC link. Note that the BTB-STATCOM power circuit must be rated for the full transmitted power between the sending and receiving transmission lines.

2.3.4 Generalized Power Flow Controller

The VSC-based PFCs that have been mentioned thus far (UPFC and BTB-SSSC) make use of at least two units and they are connected in shunt–series and series–series configurations. In a multiline transmission network, it would be advantageous to extend the functions of these PFCs to compensate more than two transmission lines in an area, so that the power flow management in all the compensated lines can be coordinated while interconnecting lines with different voltages, phase angles, or frequencies.

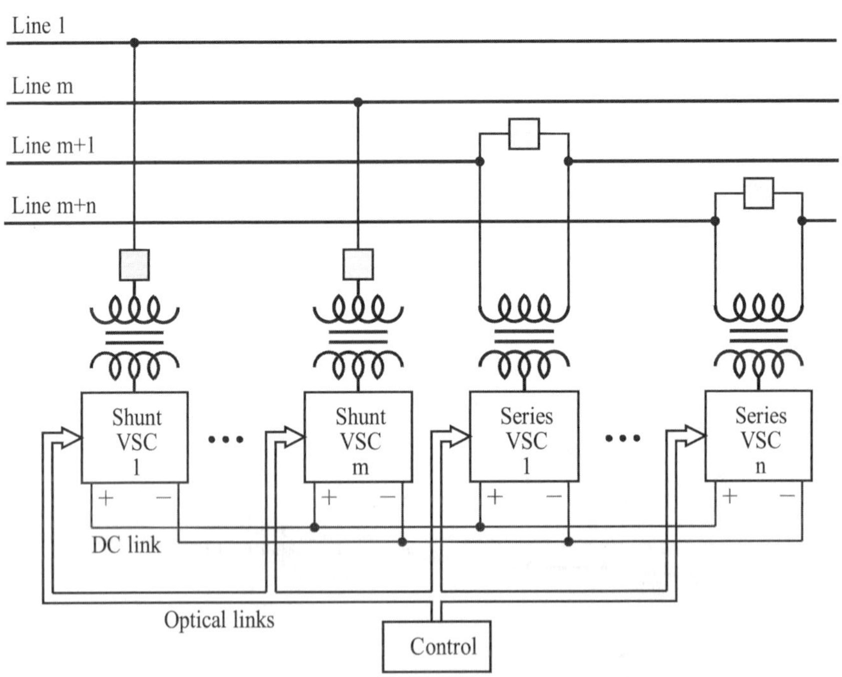

Figure 2-36. Generalized power flow controller.

The generalized power flow controller (GPFC), shown in Figure 2-36, consists of a number of VSCs, each of which is connected in shunt and/or series with a particular transmission line through a coupling transformer. All the VSCs in the GPFC are connected at their shared DC link. The compensating voltages are controlled in such a way that the active power can be transferred from one line to another through their shared DC link. In addition, each VSC provides independent reactive compensation at its AC terminal.

The overall control scheme of a GPFC is such that if the active power exchanged by the *m*th shunt VSC and *n*th series VSC with its AC line are $P_{mexch\text{-}sh}$ and $P_{nexch\text{-}se}$, respectively, the following active power balance is maintained, ignoring the losses in the VSCs:

$$P_{1exch\text{-}sh} + P_{2exch\text{-}sh} + \ldots + P_{mexch\text{-}sh} + P_{1exch\text{-}se} + P_{2exch\text{-}se} + \ldots + P_{nexch\text{-}se} = 0 \quad (2\text{-}150)$$

Each shunt-connected VSC may be connected to a transmission line through a tie inductor or coupling transformer. If a coupling transformer is used, its leakage reactance can replace the reactance of the tie inductor. In addition, a coupling transformer can keep the power electronics isolated from the high-voltage transmission line and also provide voltage matching, if needed. However, each series-connected VSC in a DC link scheme must be connected through a coupling transformer. The main purpose of the series coupling transformer is to convert a three-terminal (A-B-C) VSC into a six-terminal VSC with three isolated single-phase voltages that can be connected in series with a three-phase line. The coupling transformer may also be used for voltage matching, if desired.

3

MODELING PRINCIPLES

The best way to understand the subject matter of this book is to practice some mathematical modeling. There are many simulation software programs available for this purpose. However, the most widely used mathematical modeling software program in power system transient studies is based on the Electromagnetic Transients Program (EMTP). Various modeling techniques that are presented throughout this book can be easily translated into other simulation software. Only sample models are discussed in this book. However, a more comprehensive set of models is available at the following ftp address: ftp://ftp.wiley.com/public/sci_tech_med/facts_controllers. There are a maximum of eight characters in each filename, followed by the .DAT extension, such as `301NTWK1.DAT`. The first letter is reserved for Chapter number (3 in this case). The second and third letters are reserved for file number (01 in this case). The four letters from fourth through seventh are reserved for a description of the program file (network in this case), and the last letter is optional; in this case, it represents network #1. In this chapter, the modeling techniques are presented in detail.

3.1 THE MODELING IN EMTP

Figure 3-1 shows the block diagram of the modeling structure in EMTP. First, some general constants are defined. Next, the control or the transient analysis of control systems (TACS) section receives its input signals from the sensors or measuring switches. The TACS section generates the free-running phase-locked loop (PLL) angle. The control operations and calculations are performed in this section.

Introduction to FACTS Controllers. By Kalyan K. Sen and Mey Ling Sen

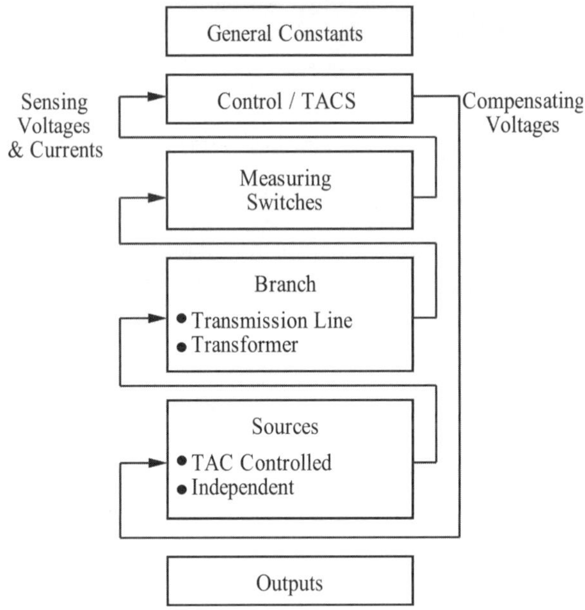

Figure 3-1. EMTP modeling structure.

The compensating voltages that are created in the TACS section are fed to a magnetic circuit, which is located in the branch section. The effects of a nonideal magnetic circuit, which includes leakage reactance, magnetic saturation, and so on, can be studied. The source section contains some independent voltage sources, which establish the power flow in a transmission line. The controlled and the independent sources are fed to the branch section, which contains the transmission line and the coupling transformer. The line voltage and the current through the compensating voltage are measured by the measuring switches. Finally, the output section is defined.

A simple two-bus power system network model, shown in the single line diagram of Figure 3-2, is used for verifying the operation of various FACTS controllers. The three-phase nominal power of 160 MVA and phase-to-phase nominal rms voltage of 138 kV are used. The base voltage (V_{base}) is chosen to be the phase-to-neutral peak voltage (V_p) and calculated as

$$V_{base} = V_p = \frac{138 \times 10^3}{\sqrt{3}} \sqrt{2} = 112{,}677 \text{ V} \tag{3-1}$$

The base current (I_{base}) is chosen to be the peak current (I_p) and calculated as

$$P_{base} = 3 \times V_{rms(L-N)} \times I_{rms} = 3 \frac{V_p}{\sqrt{2}} \frac{I_p}{\sqrt{2}} = 1.5 V_p I_p = 1.5 V_{base} I_{base} \tag{3-2}$$

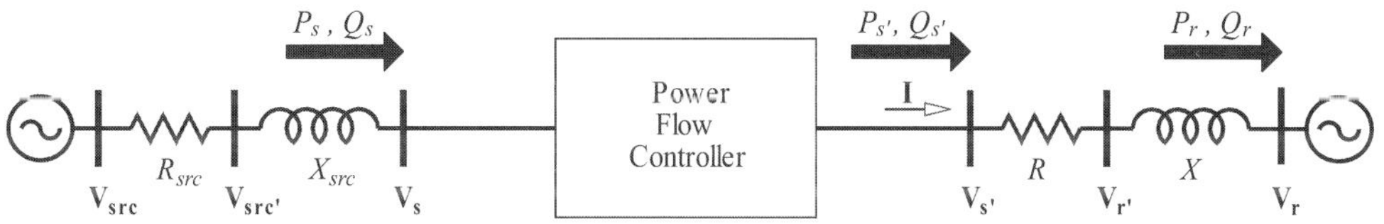

Figure 3-2. Two-bus network model.

or

$$I_{base} = \frac{P_{base}}{1.5\ V_{base}} = \frac{160 \times 10^6}{1.5 \times 112{,}677} = 946.66\ \text{A} \tag{3-3}$$

The base impedance (Z_{base}) is calculated as

$$Z_{base} = \frac{V_{base}}{I_{base}} = \frac{112{,}677}{946.66}\ \Omega = 119.02\ \Omega \tag{3-4}$$

The network is driven by a phase-to-phase source voltage ($\mathbf{V_{src}}$) of 138 kV-rms ($\sqrt{3}$ pu) and a source reactance (X_{src}) of 7.438 Ω (0.0625 pu) with a quality factor (Q = X_{src}/R_{src}) of 7.4. A series-connected compensating voltage ($\mathbf{V_{s's}}$) modifies the sending-end voltage ($\mathbf{V_s}$) to be the modified sending-end voltage ($\mathbf{V_{s'}}$). The transmission line consists of a reactance (X) of 22.316 Ω (0.1875 pu) with a quality factor (Q = X/R) of 7.4. The receiving-end voltage is to be 138 kV, lagging the source voltage by 20°. The series coupling transformer of the PFC consists of a leakage reactance of 17.853 Ω (0.15 pu) with a quality factor (Q) of 10. The uncompensated or natural active and reactive power flows (P_{rn} and Q_{rn}) at the receiving end of the line are 132 MW (0.82 pu) and –41 Mvar (–0.25 pu), respectively.

3.1.1 The Network Model

The network model shown in Figure 3-3 is used throughout this book. In the natural or uncompensated network, no PFC is connected. Later in the book, the network is studied with various PFCs connected to it. The simplified two-bus network reveals most of the controller's performance under both dynamic and steady-state conditions. The network file (`301NTWK1.DAT`) is given in Appendix C, Section 1.

The EMTP program is "column-specific," meaning that certain parameters should be placed within certain columns (e.g., the resistor values between columns 27 and 32) and there are 80 columns in each line. Special attention to right or left justification inside the specified columns is needed. In the two-line comment statements below, the column numbers are kept track of by reading the first digit in the top line and the second digit in the bottom line. To indicate a comment statement, a C in column 1 and a blank column 2 must appear. The input file structure and statements in the sample net-

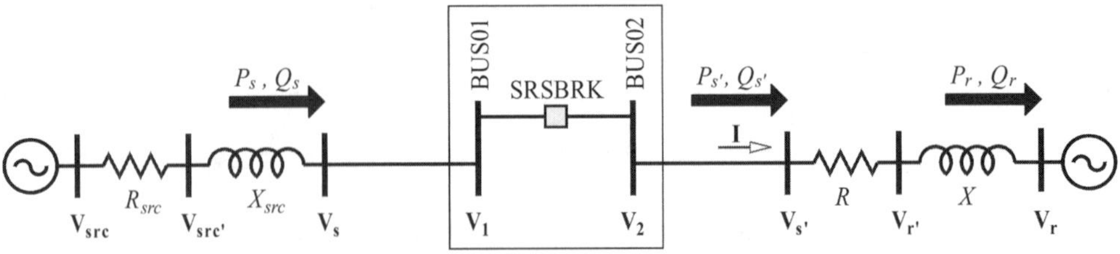

Figure 3-3. Two-bus network model in EMTP.

work with filename 301NTWK1.DAT are explained below. A reading of the EMTP program manual is recommended for further details.

The following statement indicates the start of the program:

```
BEGIN NEW DATA CASE
```

In the next statement, the time step, Δt = 16.666 μs, is entered in columns 1–8 and the simulation time, T_{max} = 150.00 ms is entered in columns 9–16:

```
C File Name: 301NTWK1.DAT
C A SIMPLE TWO-GENERATOR, SINGLE LINE POWER STSYEM NETWORK
C 00000001111111111222222222233333333334444444444555555555566666666667777777778
C 34567890123456789012345678901234567890123456789012345678901234567890123456789 0
C ----dt<---tmax<---Xopt<---Copt<--Toler<        ><        >
16.666-6150.00-3
```

The next statement indicates the miscellaneous data card:

```
C 00000001111111111222222222233333333334444444444555555555566666666667777777778
C 34567890123456789012345678901234567890123456789012345678901234567890123456789 0
C -Iprnt<--Iplot<-Idoubl<-KssOut<-MaxOut<        ><        ><---Icat<-Nenerg
      20      10                                                    1
```

Iprnt 20 in columns 7–8 denotes “printout every 20th point.” Iplot 10 in columns 15–16 denotes “plot every 10th point.” Icat 1 in column 64 denotes a plot file that is to be generated and saved.

In the next section are the data cards for TACS. An Itype of 99 is used in columns 1–2 preceding a variable name or constant in columns 3–8:

```
TACS HYBRID
99TWOPI  = 2.0 * PI
C BASE VOLTAGE, CURRENT, AND POWER
99Vbase    = 112676.528
99Ibase    = 946.662704
99Zbase    = Vbase / Ibase
99Pbase    = Vbase * Ibase
```

The next statement attaches the $INCLUDE$ file `302MEAS1.SWT` (see Appendix C, Section 2) that measures the voltages and currents:

```
$INCLUDE 302MEAS1.SWT
```

An Itype of 90 is used in columns 1–2 preceding a variable name in columns 3–8 for voltage measurement. An Itype of 91 is used in columns 1–2 preceding a variable name in columns 3–8 for current measurement.

The next statement attaches the $INCLUDE$ file `303PTCT.SCL` (see Appendix C, Section 3) that normalizes the measured voltages and currents:

```
$INCLUDE 303PTCT.SCL
```

The measuring points are at the BUS01 (point of compensation), BUS02 (modified sending end), and the receiving end. An Itype of 33 is used in columns 1–2 that is followed by a variable name in columns 3–8 for the values to be sent to the output. Other variable names, each of maximum 6 characters long can be listed in the designated columns, such as 9 through 14, 15 through 20, 21 through 26, and so on.

The line voltage (V_1) at BUS01 is expressed in terms of the three-phase instantaneous bus voltages as follows:

$$v_a = V_1 \cos\theta \tag{3-5a}$$

$$v_b = V_1 \cos(\theta - 2\pi/3) \tag{3-5b}$$

$$v_c = V_1 \cos(\theta + 2\pi/3) \tag{3-5c}$$

$$\begin{aligned}
& v_a^2 + v_b^2 + v_c^2 \\
&= V_1^2\left[\cos^2\theta + \cos^2(\theta - 2\pi/3) + \cos^2(\theta + 2\pi/3)\right] \\
&= \frac{V_1^2}{2}\left[2\cos^2\theta + 2\cos^2(\theta - 2\pi/3) + 2\cos^2(\theta + 2\pi/3)\right] \\
&= \frac{V_1^2}{2}\left[1 + \cos(2\theta) + 1 + \cos 2(\theta - 2\pi/3) + 1 + \cos 2(\theta + 2\pi/3)\right] \\
&= \frac{V_1^2}{2}\left[3 + \cos(2\theta) + \cos 2(\theta - 2\pi/3) + \cos 2(\theta + 2\pi/3)\right] \\
&= \frac{V_1^2}{2}\left[3 + \{\cos(2\theta) + \cos(2\theta + 2\pi/3) + \cos(2\theta - 2\pi/3)\}\right]
\end{aligned} \tag{3-6}$$

The three cosine waves that are of equal magnitude and $2\pi/3$ radian phase shifted from each other add up to be zero. Therefore,

$$v_a^2 + v_b^2 + v_c^2 = \frac{3}{2} V_1^2 \tag{3-7}$$

The expression for the line voltage (V_1) at BUS01 is

$$V_1 = \sqrt{\frac{2}{3}\left(v_a^2 + v_b^2 + v_c^2\right)} \tag{3-8}$$

The next statement attaches the $INCLUDE$ file `304OPLL.PLL` (see Appendix C, Section 4) that calculates the phase angle of the voltage at BUS01 and is referenced to the phase *a* voltage (v_{1a}):

```
$INCLUDE 304OPLL.PLL
```

The PLL angle is referenced to a cosine wave, meaning that during a complete cycle of a cosine wave, the phase angle (θ) linearly increases from 0 to 2π. From equation A-67a in Appendix A, it can be written that

$$v_{ds} = v_a = V_1 \cos\theta \tag{3-9}$$

$$\begin{aligned} v_{qs} &= -\frac{1}{\sqrt{3}}\left(v_a + 2v_c\right) = -\frac{V_1}{\sqrt{3}}\left(\cos\theta + 2\cos(\theta + 120^\circ)\right) \\ &= -\frac{V_1}{\sqrt{3}}\left(\cos\theta - \cos\theta - \sqrt{3}\sin\theta\right) = V_1 \sin\theta \end{aligned} \tag{3-10}$$

Therefore,

$$\theta = \tan^{-1}\left(\frac{\sin\theta}{\cos\theta}\right) = \tan^{-1}\left(\frac{v_{qs}}{v_{ds}}\right) \tag{3-11}$$

A small value of 10^{-8} is added to avoid a division by zero if $v_{ds} = 0$. If v_{ds} is negative, the phase angle is corrected by adding π to the calculated angle. If the calculated angle is negative, 2π is added to make the phase angle positive. If a sine-referenced phase angle is needed, the above mentioned cosine-referenced signal is phase shifted by adding $\pi/2$.

The next statement attaches the $INCLUDE$ file `305LINER.CMP` (see Appendix C, Section 5) that calculates the modified sending-end power, receiving-end power, and line resistance, which is presented later in the chapter:

```
$INCLUDE 305LINER.CMP
```

The next statement indicates the end of the TACS section:

```
BLANK RECORD ENDING TACS
```

In the following section are the branch data cards. In this section, columns 1–2 are left blank for the Itype. Columns 3–8 are for the first node name. Columns 9–14 are for the second node name. The resistor values in Ω are entered in columns 27–32. The inductor values in mH are entered in columns 33–38. The capacitor values in μF are entered in columns 39–44:

```
C 00000001111111111222222222233333333334444444444555555555566666666667777777778
C 34567890123456789012345678901234567890123456789012345678901234567890123456789
C <-Bus1<-Bus2<-Bus3<-Bus4<----R<----L<----C                                    V
  BUS01ABUS02A            0.001
  BUS01BBUS02B            0.001
  BUS01CBUS02C            0.001
```

BUS01A and BUS02A are connected for the time being with a short, which is represented by $R = 0.001\ \Omega$. These two nodes are reserved for connecting the series compensating voltage of the PFC. The next statement attaches the $INCLUDE$ file `306NTWK1.BRN` (see Appendix C, Section 6) that connects a two-bus transmission line and a source impedance:

```
$INCLUDE 306NTWK1.BRN
```

The source impedance for the *a* phase is entered in one line. The source impedances for the *b* and *c* phases are the same as that in the *a* phase. The line impedances with proper node names and resistor and inductor values are entered into the appropriate columns.

Figure 3-3 shows the source impedance in each phase consisting of a resistor (R_{src}) and a reactor (X_{src}) that are connected between VSRC and VS busses (VS and BUS01 busses are shorted). The transmission line impedance in each phase is represented by a resistor (R) and a reactor (X) and is connected between the VSP and the VR busses.

The impedance at the sending end (from the infinity bus to the point of compensation, i.e., the Thèvenin impedance) is represented by an inductive impedance of 6.25% of the base impedance that translates to an inductor, $L_{src} = 19.73$ mH $(0.0625 \times Z_{base}/\omega)$, where $\omega = 2 \times \pi \times f$ (system frequency, $f = 60$ Hz) and a resistor, $R_{src} = 1.0053\ \Omega$, where $R_{src} = \omega L_{src}/Q = 2 \times \pi \times 60 \times 0.01973/7.4$. The transmission line impedance is assumed to have a quality factor (Q) of 7.4.

The transmission line reactance at the receiving end is represented by an inductive impedance of 18.75% of the base impedance that translates to an inductor, $L = 59.19$ mH $(0.1875 \times Z_{base}/\omega)$ and a resistor, $R = 3.0159\ \Omega$, where $R = \omega L/Q = 2 \times \pi \times 60 \times 0.05919/7.4$.

The next statement indicates the end of the BRANCH section:

```
BLANK RECORD ENDING BRANCHES
```

The following section contains the switch data cards. There are different types of switches. The next statement attaches the $INCLUDE$ file `307MEAS2.SWT` (see Appendix C, Section 7) that measures branch currents:

```
$INCLUDE 307MEAS2.SWT
```

For a measuring switch, columns 1–2 are left blank. Nodename1 is typed in columns 3–8 and nodename2 in columns 9–14. The word MEASURING is typed in columns 55–63. If a measured current is to be sent to the plot file, Iout in column 80 should be 1.

The next statement indicates the end of the SWITCH section:

```
BLANK RECORD ENDING SWITCHES
```

The following section contains the source data cards. The next statement attaches the $INCLUDE$ file `308NTWK1.SRC` (see Appendix C, Section 8) that connects source voltages:

```
$INCLUDE 308NTWK1.SRC
```

An Itype of 14 in columns 1–2 is used for a sinusoidal voltage source. A node name is entered in columns 3–8. The peak value of the voltage source (112,676.528 V in the present case) is entered in columns 11–20. The value of the frequency (60 Hz in the present case) is entered in columns 21–30, and the phase angles of 0°, –120°, and 120° for the three phases of the sending-end voltage sources and –20°, –140°, and 100° for the three phases of the receiving-end voltage sources are entered in columns 31–40. A –1.0 in columns 67–70 is needed if an initial phasor solution is desired at the start of the simulation.

The next statement indicates the end of the SOURCE section:

```
BLANK RECORD ENDING SOURCES
```

The following four statements are required for a successful run of the simulation:

```
BLANK RECORD ENDING NODE VOLTAGE OUTPUT
BLANK RECORD ENDING PLOT
BEGIN NEW DATA CASE
BLANK RECORD ENDING ALL DATA CASES
```

The simulation results from the file `301NTWK1.DAT` (see Appendix C, Section 1) are shown in Figure 3-4. The lower-case variables are defined as instantaneous values. The PLL signal (reference phase angle, θ) is locked to the phase-*a* voltage at BUS01.

To study faults in power systems, a phase-*a*-to-ground fault was applied to the power system network and the corresponding main file is `309NTWK2.DAT` (see Appendix C, Section 9). The following statement in the SWITCH section starts a phase-to-ground fault at 50 ms by closing the switch and opening it after 100 ms to clear the fault when the current goes below 5 A:

```
C FAULT STARTS
C 00000001111111111222222222233333333334444444444555555555566666666667777777778
C 34567890123456789012345678901234567890123456789012345678901234567890123456789O
C <-Bus1<-Bus2<---Tclose<----Topen<-------Ie<----Flash<--Request<-----Target<--O
  BUS01A              .050      0.100         5.0
C FAULT ENDS
```

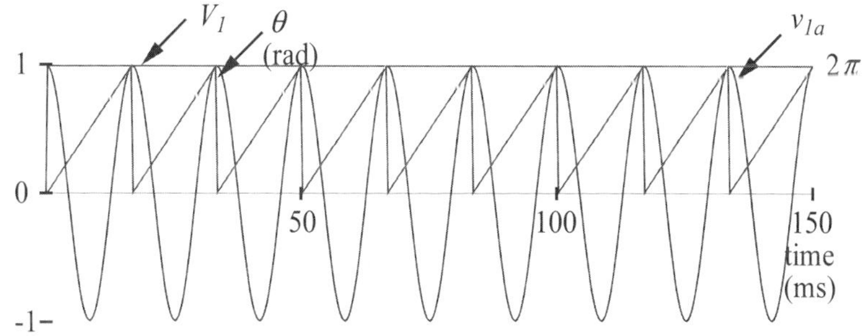

Figure 3-4. Phase *a* of line voltage (v_{1a}) at BUS01, its magnitude (V_1), and the PLL angle (θ).

The Ordinary PLL (OPLL) used in this section cannot preserve the PLL angle in the event there is a distortion at the point of measurement due to line faults. This problem can be overcome with the use of a Vector PLL (VPLL).

3.2 VECTOR PHASE-LOCKED LOOP (VPLL)

Figure 3-5 shows the block diagram of the Vector PLL (VPLL). The stationary frame components (v^s_{1d} and v^s_{1q}) of the line voltage (v_1) at BUS01 are calculated and converted to rotating frame components v_{1d} and v_{1q}. The quadrature component (v_{1q}) is regulated to be zero with the use of an error amplifier, the output of which is the synchronously rotating angular speed (ω) that produces the PLL angle (θ) after passing through an integrator. With this approach, the direct component (v_{1d}) is maintained to be the amplitude (V_1) of the line voltage (v_1) at BUS01 under steady-state condition. The next statement attaches the $INCLUDE$ file `310VPLL.PLL` (see Appendix

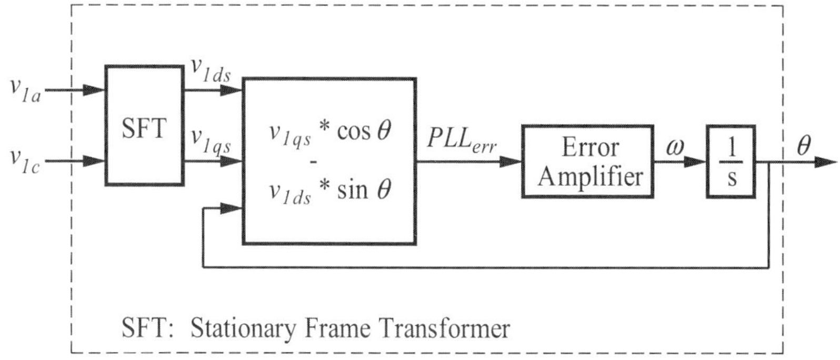

Figure 3-5. Block diagram of a vector PLL.

C, Section 10) that calculates the phase angle of the voltage at BUS01 and it is referenced to the phase-*a* voltage (v_{1a}):

```
$INCLUDE 310VPLL.PLL
```

Figure 3-6 shows that the VPLL is able to track the near original phase in the case of a phase-*a*-to-ground fault. Note that during a fault, the current through the line inductor changes rapidly, causing numerical instability in the simulation. To minimize this problem, a numerical snubber in the form of a resistor across each line inductor is used. The value of the snubber resistor should be high enough so that it does not cause spurious results in the simulation but facilitates the decaying of an oscillating current during a disturbance. The next statement attaches the $INCLUDE$ file `311NTWK2.BRN` (see Appendix C, Section 11) that connects a two-bus transmission line and a source impedance with snubber resistors across inductors:

```
$INCLUDE 311NTWK2.BRN
```

3.3 TRANSMISSION LINE STEADY-STATE RESISTANCE CALCULATOR

In some applications, it is necessary to know the value of the resistor (*R*) of the transmission line between two points. Figure 3-7 shows the block diagram of a scheme for the computation of transmission line resistance. The active power flows at the two points are measured, the difference of which can be assumed to be the loss across the transmission line resistor (*R*). The value of the resistor can be computed by dividing the power loss across the resistor by the square of the line current. This approach is good only for steady-state condition. Therefore, a low pass filter with narrow bandwidth is used to eliminate the undesired resistor values obtained from this measure-

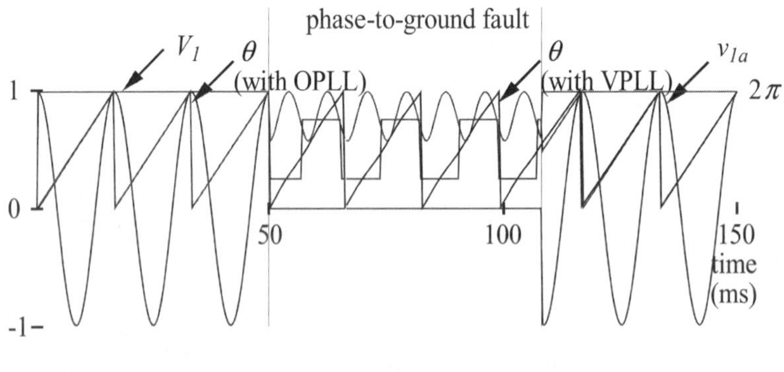

Figure 3-6. Phase a of line voltage (v_{1a}) at BUS01, its magnitude (V_1), and the PLL angle (θ) with and without a phase-to-ground fault.

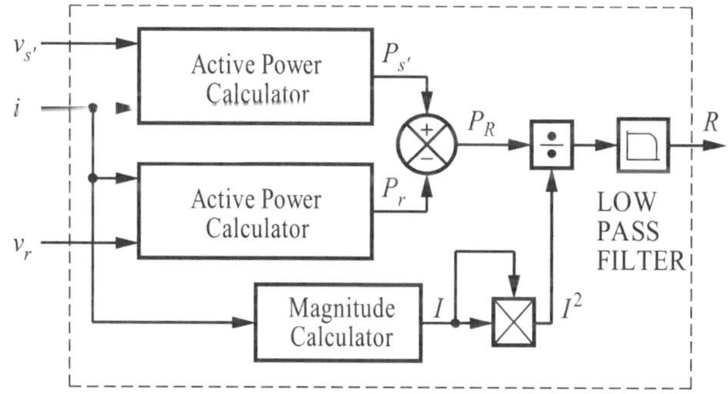

Figure 3-7. Block diagram for a steady-state resistance calculator.

ment. The calculated resistance is passed through a low-pass filter to ignore the change in resistance calculated during short electrical transients.

3.4 SIMULATION OF AN INDEPENDENT PFC IN A SINGLE LINE APPLICATION

An independent PFC regulates the active and reactive power flows in the transmission line independently by connecting a compensating voltage in series with the line as shown in Figure 2-10. The series compensating voltage exchanges active and reactive powers with the transmission line. In the case of the Sen transformer (ST), both the active and reactive powers exchanged by the series compensating voltage appear at the transmission line through the shunt-connected exciter unit. However, in the case of the unified power flow controller (UPFC), only the exchanged active power passes through the shunt unit. Both the shunt and series units exchange reactive power at their AC terminals independently. The models of the ST and the UPFC are shown in Figure 3-8. The corresponding main file is `315PQPFC.DAT` (see Appendix C, Section 15).

In this section, the effects of the ST and UPFC for power flow control in a transmission line are presented. The network model of Figure 3-3 is used with the short between BUS01 and BUS02 replaced, first with the ST and then with the UPFC. In each ST and UPFC, the compensating voltage is connected in series with the transmission line through a coupling transformer. The leakage reactance of the series coupling transformer of the PFC is represented by an inductive impedance of 15% of the base impedance that translates to an inductor, $L = 47.358$ mH ($0.15 \times Z_{base}/\omega$) and a resistor, $R = 1.7854\ \Omega$, where $R = \omega L/Q = 2 \times \pi \times 60 \times 0.047358/10$. The turns ratio of the series coupling transformer is considered to be 1:1 and the leakage reactance is distributed equally in both primary and secondary sides. The next statement attaches the $INCLUDE$ file `313TRAN2.TRN` (see Appendix C, Section 13) that connects a coupling transformer in series with the transmission line:

```
$INCLUDE 313TRAN2.TRN
```

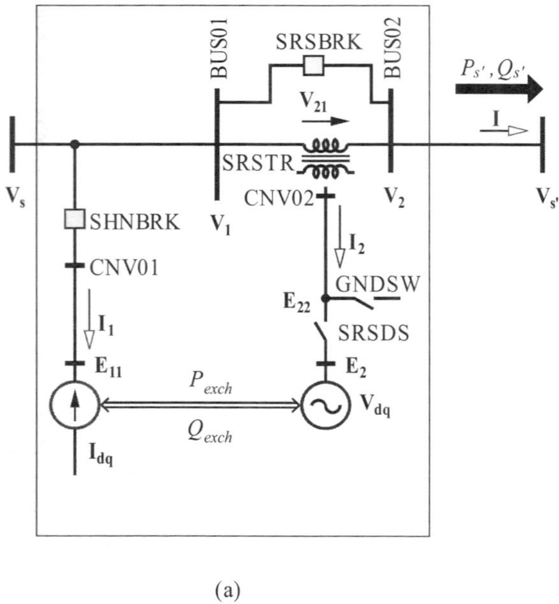

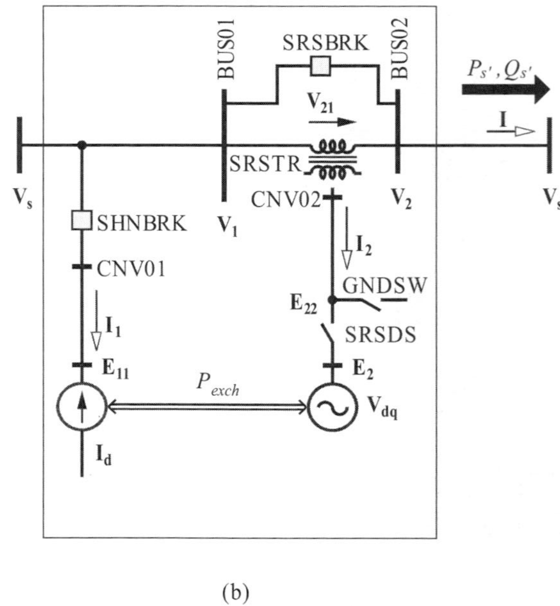

Figure 3-8. Simulation models of the (a) ST and (b) UPFC.

The term TRANSFORMER is a reserved word. SRSTRA in columns 39–44 is the name for the internal bus at the top of the magnetizing branch. This name uniquely identifies the transformer. The numbers 9999 in columns 13–16 terminates previous cards. The winding number is specified in columns 1–2. A 1 indicates the primary winding and a 2 indicates the secondary winding. BUS01A in columns 3–8 indicates nodename1. BUS02A in columns 9–14 indicates nodename2. A blank nodename indicates a ground. The winding resistance in ohms is specified in columns 27–32. The winding inductance in mH is specified in columns 33–38. Rated winding voltage in pu or number of turns is specified in columns 39–44.

A fixed value of 12 kV (0.15 pu) compensating voltage is series connected with its entire controllable range of $0° \leq \beta \leq 360°$. The next statement attaches the $INCLUDE$ file `316PQPFC.USR` (see Appendix C, Section 16) that defines the user inputs:

```
$INCLUDE 316PQPFC.USR
```

The next statement attaches the $INCLUDE$ file `317PQPFC.CON` (see Appendix C, Section 17) that defines the series-connected VSC2 open loop control:

```
$INCLUDE 317PQPFC.CON
```

Figure 3-9 shows the active, reactive, and apparent powers (P_{exch}, Q_{exch}, and S_{exch}) exchanged by the series compensating voltages of the ST and UPFC with the line. The maximum exchanged active, reactive, and apparent powers are 21 MW, 29 Mvar, and 30 MVA, respectively. In the case of the ST, both the active and reactive powers flow bidirectionally through the exciting shunt unit. In the case of the UPFC, only the active power flows bidirectionally through the STATCOM.

Consistent with Figure 2-9a, it is shown in Figure 3-9 that the reactive power exchanged by the series-connected compensating voltage is capacitive over the majority of the control range of 360°, except between 205° and 335°, when the exchanged reactive power is inductive. The shunt-connected exciter unit absorbs inductive reactive power when the series-connected compensating voltage unit delivers capacitive reactive power and vice versa. Also it should be noted that the compensating voltage unit absorbs from the line a maximum of 12 Mvar of inductive reactive power and delivers a maximum of 29 Mvar of capacitive reactive power. Therefore, the shunt unit absorbs from the line a maximum of 29 Mvar of inductive reactive power and delivers a maximum of 12 Mvar of capacitive reactive power. The shunt unit is inductive over the majority of the control range of 360°, except between 205° and 335°, when it is capacitive.

While the relative phase angle of the compensating voltage varies, the current and the power flow in the transmission line vary, as does the voltage across the source impedance. Therefore, the voltage phasor at the input of the PFC (i.e., at the bus, $\mathbf{V}_s$) varies its magnitude and phase angle from the uncompensated or natural line voltage ($\mathbf{V}_{sn}$). This variation from the natural voltage is highly dependent on the short-circuit impedance of the transmission line at the point of compensation.

Figure 3-10 shows the variations of the phase-to-phase voltage at the bus ($\mathbf{V}_s$) from its natural value of 136.7 kV ($\sqrt{3} \times 0.99$ pu) with the operation of the ST and UPFC. If

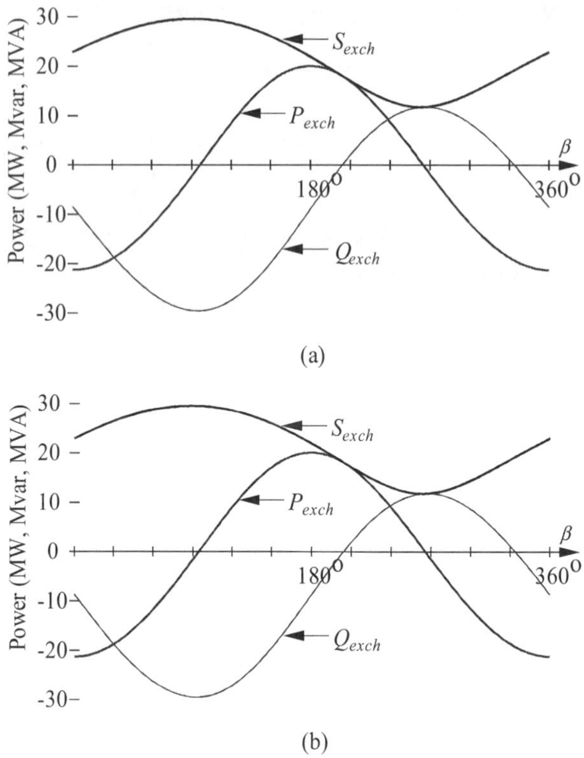

Figure 3-9. Variations of exchanged powers between the transmission line and the series unit of the (a) ST and (b) UPFC.

the point of compensation is near a "stiff" bus, the voltage variation at the point of compensation (i.e., at the bus, $\mathbf{V}_s$) is smaller and vice versa. In the case of the UPFC, the bidirectional flow of active power between the shunt and the series units emulates a $+R$ or a $-R$ and, in turn, causes a voltage sag or swell at the point of compensation. In the case of the ST, the additional reactive power flow between the shunt and the series units emulates an inductor or a capacitor, thus causing an additional voltage sag or

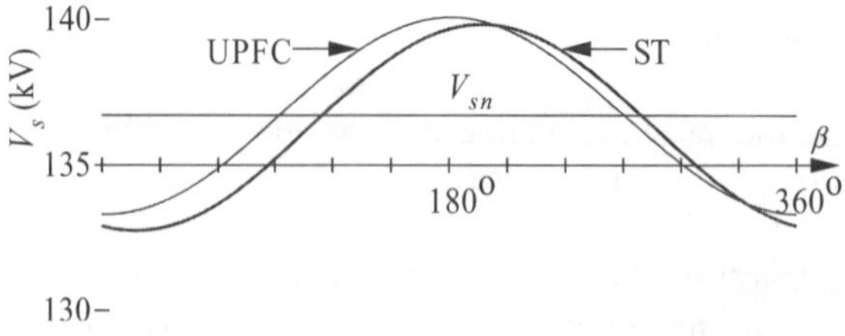

Figure 3-10. Variations of bus voltage with the operation of the ST and UPFC.

swell at the point of compensation. Note that during the control range of a series-connected compensating voltage between 205° and 335°, the voltage ($\mathbf{V}_s$) is higher in the case of the ST than in the case of the UPFC. In the rest of the region, the voltage ($\mathbf{V}_s$) is higher in the case of the UPFC than in the case of the ST.

Figure 3-11a shows that with the use of the ST, both the active and reactive power flows at the receiving end of the line under consideration vary by +57 and –58 MW, and by +60 and –63 Mvar, respectively, from the natural power flows (P_{rn} = 132 MW and Q_{rn} = –41 Mvar). Figure 3-11b shows that with the use of the UPFC, both the active and reactive power flows at the receiving end of the same line vary by ±58 MW and by ±60 Mvar, respectively, from the natural power flows.

The controllable range of active and reactive power flows with the operation of the ST and UPFC are presented in Figure 3-12. The ST and UPFC offer circular controllable areas that are almost identical. The radius of the controllable area increases with larger series compensating voltages. With proper series compensating voltage, the circular controllable area covers all four quadrants, resulting in a reversal of power flow from its natural direction. This phenomenon is presented in Chapter 8. Special opera-

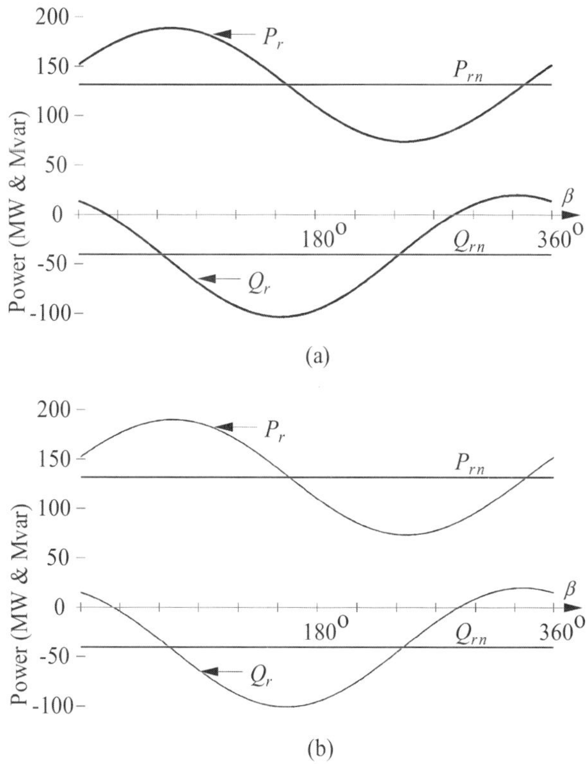

Figure 3-11. Variations of receiving-end power flow with the operation of the (a) ST and (b) UPFC.

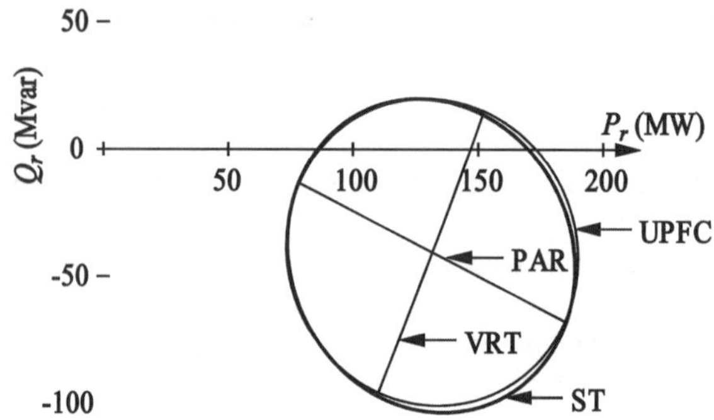

Figure 3-12. Comparison of active and reactive power flows at the receiving end of the transmission line with the operation of the VRT, PAR, UPFC, and ST.

tions of the ST with $\beta = 0°$ and $\beta = 90°$ show the functionalities of the VRT and PAR. Whereas the ST offers a circular controllable area, the PAR and the VRT offer linear control range.

The summary of the study results is given below.

- The variation of phase-to-phase voltage at the point of compensation is between 132.8 kV and 139.8 kV in the case of the ST, and between 133.3 kV and 140.1 kV in the case of the UPFC, whereas the uncompensated or natural voltage is 136.7 kV.
- Line active power flow changes from the natural flow of 132 MW by +57 and –58 MW in the case of the ST, and by ±58 MW in the case of the UPFC.
- Line reactive power flow changes from the natural flow of –41 Mvar by +60 and –63 Mvar in the case of the ST, and by ±60 Mvar in the case of the UPFC.
- Rating of the ST: 30 MVA.
- Rating of the UPFC: a series unit of 30 MVA and a shunt unit of 21 MVA.

Since the shunt and the series units of the UPFC are designed to be of the same rating, a 30 MVA-rated shunt unit is able to provide a maximum of 21 Mvar of reactive power while delivering the required 21 MW of active power. If this capability of supplying shunt reactive power is really needed, particularly when the short-circuit impedance at the point of compensation is high, a separate capacitor bank as a shunt reactive compensator may be used along with the ST.

4

TRANSFORMER-BASED FACTS CONTROLLERS

The transformer-based FACTS controllers are made from using two fundamental building blocks: voltage regulating transformers (VRTs) and phase angle regulators (PARs). The VRT primarily regulates the magnitude of the line voltage without significantly changing its original phase angle. The PAR primarily regulates the phase angle of the line voltage with little or no change in its original magnitude. The objective in this chapter is to study these traditional transformer-based FACTS controllers.

4.1 VOLTAGE REGULATING TRANSFORMER (VRT)

The VRT can be of two kinds: (a) autotransformer and (b) two-winding transformer, as shown in Figure 4-1. In the VRT, the primary winding with n_1 turns is excited from the input voltage and the secondary winding with n_2 turns is induced with a voltage. In an autotransformer, the secondary voltage is usually a fraction of the input voltage and the output voltage is the phasor sum of the primary and secondary voltages. The compensating secondary voltage is connected in series with the transmission line. There is no electric isolation between the input and output voltages. In a two-winding transformer, the input and output voltages are electrically isolated. The compensating voltage that is induced in the secondary winding is the output voltage and is connected in

Introduction to FACTS Controllers. By Kalyan K. Sen and Mey Ling Sen

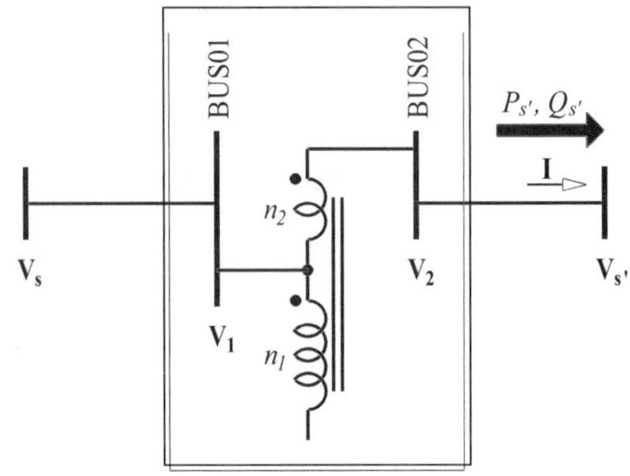

AUTOTRANSFORMER

(a)

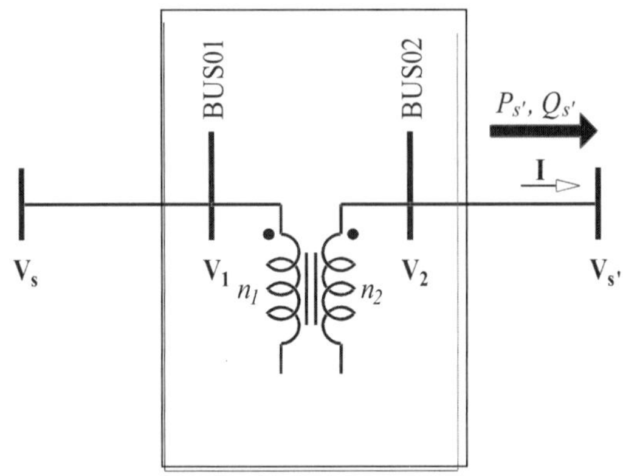

TWO-WINDING TRANSFORMER

(b)

Figure 4-1. (a) Autotransformer. (b) Two-winding transformer.

shunt with the transmission line. In both the autotransformer and the two-winding transformer, the active number of turns in the secondary windings are varied with the use of load tap changers (LTCs) to vary the compensating voltage.

The output voltage is

$$V_2 = V_1 \times \left(1 + \frac{n_2}{n_1}\right) \text{ in the case of an autotransformer} \tag{4-1a}$$

$$V_2 = V_1 \times \frac{n_2}{n_1} \text{ in the case of a two-winding transformer} \tag{4-1b}$$

4.1.1 Autotransformer

The autotransformer can be studied with the use of the mathematical model of the ST, as shown in Figure 3-8a. This main file `318PQPFC.DAT` (same as `315PQPFC.DAT` with a different series coupling transformer) is rerun with the relative phase angle of the series compensating voltage set to $\beta = 0°$ and its magnitude is varied from –0.15 pu to 0.15 pu in the user file `316PQPFC.USR` (see Appendix C, Section 16).

The leakage impedance of the series coupling transformer is deliberately set to nearly zero to preserve the ideal operation of the autotransformer. The next statement attaches the $INCLUDE$ file `314IDLT2.TRN` (see Appendix C, Section 14) that

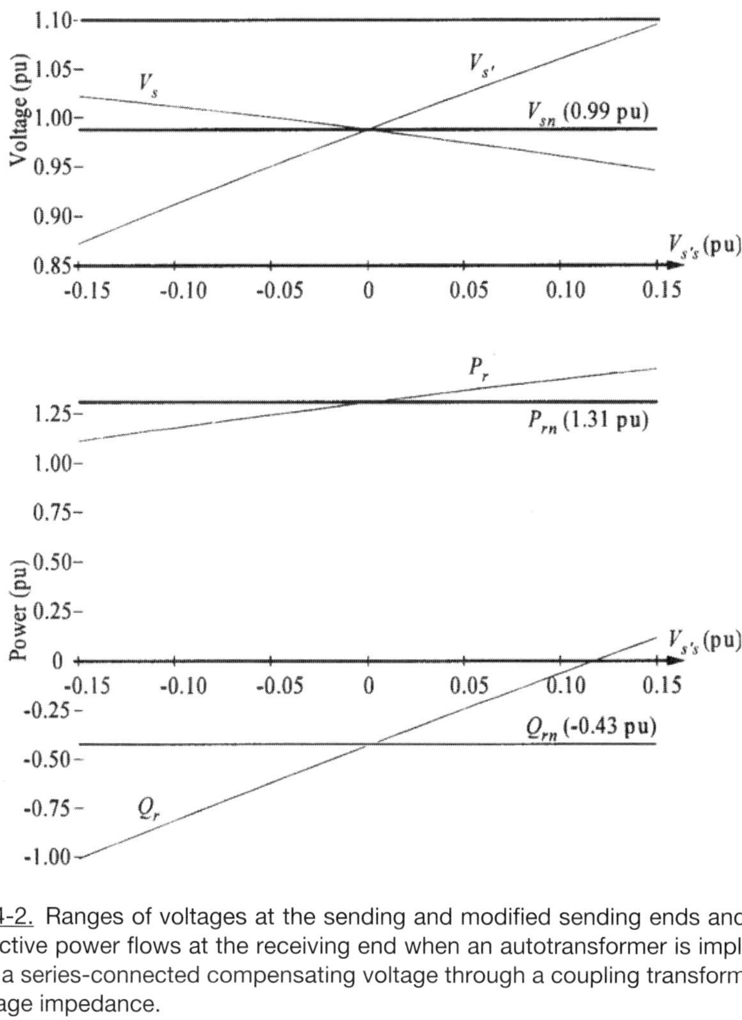

Figure 4-2. Ranges of voltages at the sending and modified sending ends and active and reactive power flows at the receiving end when an autotransformer is implemented with a series-connected compensating voltage through a coupling transformer with no leakage impedance.

connects a coupling transformer (with no leakage impedance) in series with the transmission line:

```
$INCLUDE 314IDLT2.TRN
```

The result is shown in Figure 4-2. Since the output voltage, $\mathbf{V}_{s'}$ (same as $\mathbf{V}_2$ at BUS02), and the input voltage, $\mathbf{V}_s$ (same as $\mathbf{V}_1$ at BUS01), are in phase, their scalar difference, $\mathbf{V}_{21}$ (same as $\mathbf{V}_{s's}$), is the series compensating voltage. A variable-magnitude, bipolar, series-connected compensating voltage ($V_{s's}$) increases or decreases $\mathbf{V}_{s'}$ from its uncompensated value, $\mathbf{V}_s$. It is discussed in Chapter 2 that an increase in the line voltage is caused by delivering capacitive reactive power to the line and a decrease in the line voltage is caused by absorbing inductive reactive power from the line at the point of compensation. Since a transformer neither generates nor absorbs reactive power, the primary winding must absorb inductive reactive power from the line, which will decrease the line voltage at BUS01 for the secondary winding to deliver capacitive reactive power to the line and vice versa. The figure shows that although the voltage ($\mathbf{V}_{s'}$) at the modified sending end is 0.15 pu more than the voltage ($\mathbf{V}_s$) at the sending end, the voltage ($\mathbf{V}_s$) is decreased to 0.95 pu from its natural value of 0.99 pu. The figure also shows that with a compensating voltage ($\mathbf{V}_{s's}$) of 0.15 pu and a negative polarity, the voltage ($\mathbf{V}_{s'}$) at the modified sending end is 0.15 pu less than the voltage ($\mathbf{V}_s$) at the sending end, but the voltage ($\mathbf{V}_s$) is increased to 1.02 pu from its natural value of 0.99 pu. Therefore, the range of $V_{s'}$ is 0.87 (i.e., 1.02 – 0.15) to 1.10 (i.e., 0.95 + 0.15) for a $V_{s's} = \mp 15\%$.

Increasing the voltage at the modified sending end increases the reactive power flow at the receiving end from its natural power flow and vice versa, whereas the receiving-end voltage is constant. The reactive power flow at the receiving end changes from –1.01 pu to 0.12 pu during the range of compensating voltage from –0.15 pu to 0.15 pu while the natural reactive power at the receiving end is –0.43 pu with no compensation. However, the active power flow at the receiving end changes as well in the range of 1.11 pu to 1.48 pu for the same range of variation in the compensating voltage, whereas the natural active power at the receiving end is 1.31 pu. Therefore, a $\mp$0.15 pu change in line voltage changes the reactive power flow by 1.13 pu and active power flow by 0.37 pu.

For an autotransformer with fixed turns ratio, the series-connected compensating voltage is a multiplier of the input voltage at BUS01, as simulated below. Note that the input voltage at BUS01 also varies with the line current. The main file `319PQPFC.DAT` (same as `318PQPFC.DAT` with different user file) is rerun with the relative phase angle of the series-connected compensating voltage set to $\beta = 0°$ and its magnitude is varied from –0.15 pu to 0.15 pu of the input voltage in the user file (`320PQPFC.USR`). The next statement attaches the $INCLUDE$ file `320PQPFC.USR` (see Appendix C, Section C-18) that defines the user inputs:

```
$INCLUDE 320PQPFC.USR
```

Figure 4-3 shows that a compensating voltage of $\mp$15% of the voltage ($\mathbf{V}_s$) at the sending end results in a voltage ($\mathbf{V}_{s'}$) at the modified sending end in the range of 0.87

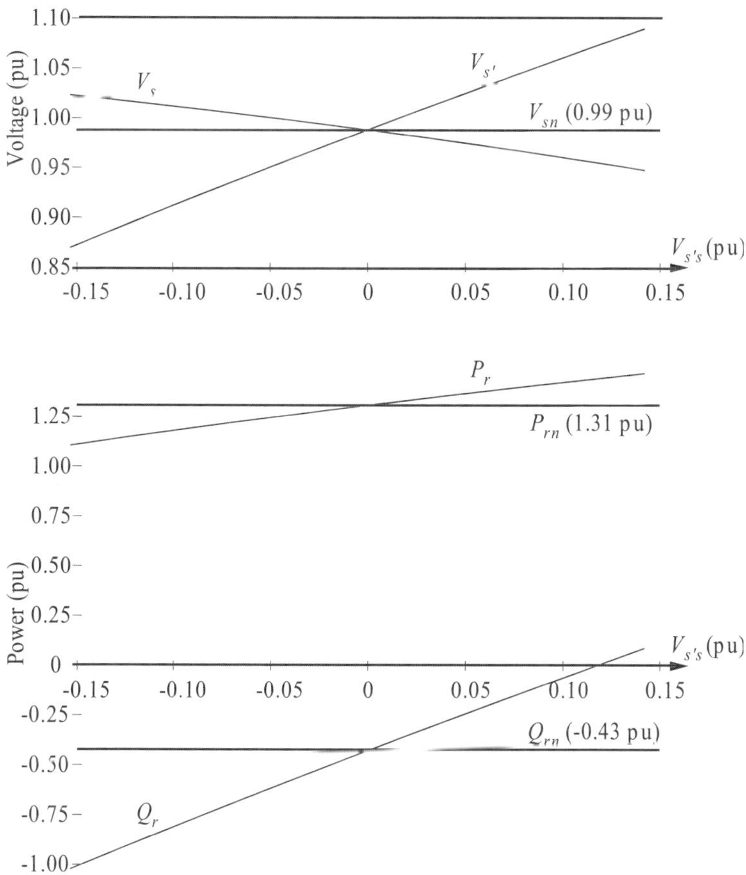

Figure 4-3. Ranges of voltages at the sending and modified sending ends and active and reactive power flows at the receiving end when an autotransformer is implemented with a series-connected compensating voltage through a coupling transformer with no leakage impedance and the series-connected compensating voltage is a multiplier of the input voltage.

to 1.09 pu. In the process, the voltage ($\mathbf{V_s}$) at the sending end varies in the range of 1.02 to 0.95 pu. The reactive power flow at the receiving end changes from –1.02 pu to 0.09 pu during the range of compensating voltage from –0.15 pu to 0.15 pu. However, the active power flow at the receiving end changes in the range of 1.10 pu to 1.47 pu for the same range of variation in the compensating voltage. Therefore, a $\mp$0.15 pu change in line voltage changes the reactive power flow by 1.11 pu and active power flow by 0.37 pu.

Next, an ideal autotransformer with no leakage reactance is run from the main file `401IDLAT.DAT` (see Appendix C, Section 24) with an actual autotransformer model given in file `402IDLAT.TRN`. The next statement attaches the $INCLUDE$ file

402IDLAT.TRN (see Appendix C, Section 25) that connects an autotransformer (with no leakage impedance) in series with the transmission line:

```
$INCLUDE 402IDLAT.TRN
```

Figure 4-4 shows the results in the variation of voltages ($\mathbf{V_s}$) at the sending end and ($\mathbf{V_{s'}}$) at the modified sending end, due to a compensating voltage that is $\mp 15\%$ of the voltage ($\mathbf{V_s}$) at the sending end using a TRANSFORMER model. The figure also shows the variation of active and reactive power flows at the receiving end. The results are identical to what was just shown using a series-connected compensating voltage through a coupling transformer with no leakage impedance.

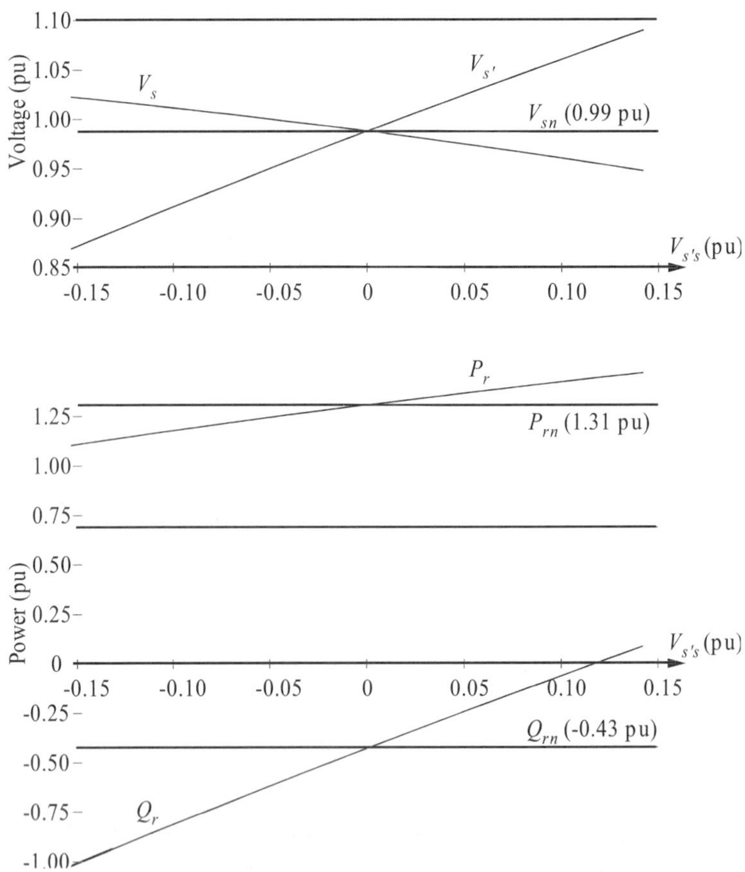

Figure 4-4. Ranges of voltages at the sending and modified sending ends, and active and reactive power flows at the receiving end with an actual autotransformer with no leakage impedance.

4.1.2 Two-Winding Transformer

An ideal two-winding transformer with no leakage reactance is run from the main file `403IDLTT.DAT` (same as `401IDLAT.DAT` with a change to a two-winding transformer). The next statement attaches the $INCLUDE$ file `404IDLTT.TRN` (see Appendix C, Section 26) that connects a two-winding transformer (with no leakage impedance) in shunt with the transmission line:

```
$INCLUDE 404IDLTT.TRN
```

Figure 4-5 shows the results from the variations of voltages ($\mathbf{V_s}$) at the sending end and ($\mathbf{V_{s'}}$) at the modified sending end due to changing the turns ratio (n_2/n_1) from 0.85

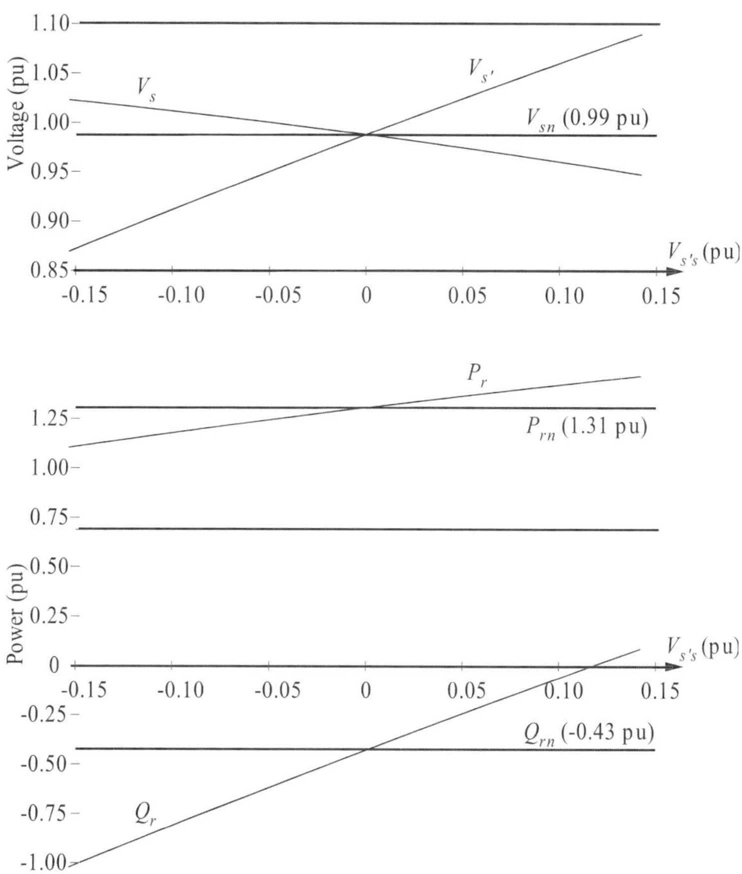

Figure 4-5. Ranges of voltages at the sending and modified sending ends and active and reactive power flows at the receiving end with an actual two-winding transformer with no leakage impedance.

to 1.15 while keeping n_1 constant at 1.00 using a TRANSFORMER model. The figure also shows the variation of active and reactive power flows at the receiving end. The results are identical to what was just shown using autotransformer models [both series-connected compensating voltage (Figure 4-3) and TRANSFORMER (Figure 4-4)]. The results show that the VRT changes the reactive power flow at the receiving end of the line much more than it changes the active power flow and that is consistent with the established theory in Chapter 2.

4.2 PHASE ANGLE REGULATOR (PAR)

The one-transformer Phase Angle Regulator (PAR) configuration and the corresponding phasor diagram are shown in Figure 4-6. The input voltage ($\mathbf{V_1}$) at BUS01 is applied to a Δ-connected, three-phase primary winding and a secondary voltage ($\mathbf{V_{21}}$)

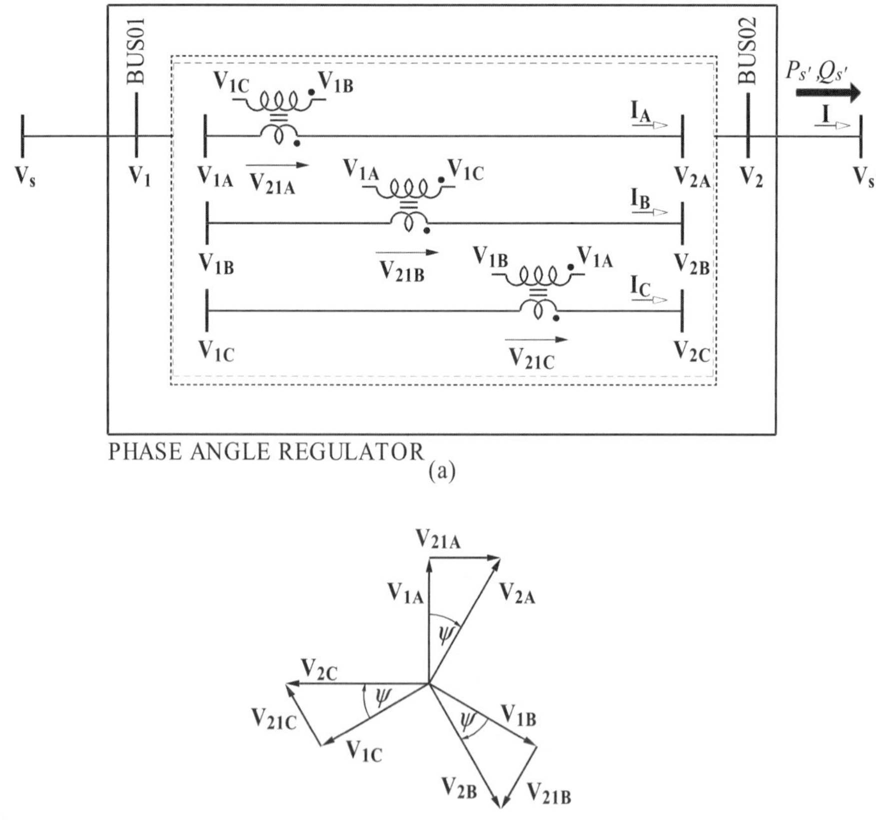

Figure 4-6. Phase angle regulator.

that is in quadrature with the primary voltage is connected in series with the transmission line. With proper polarities of the series-connected windings, the compensating secondary voltage is connected in series with the transmission line at 90° or –90° to the primary voltage. The resulting effect is that the phase angle of the line voltage is regulated.

An ideal PAR with no leakage reactance is run from the main file `405PAR1.DAT` (same as `401IDLAT.DAT` with a change to a phase angle regulating transformer). The next statement attaches the $INCLUDE$ file `406PAR1.TRN` (see Appendix C, Section 27) that connects a phase angle regulating transformer (with no leakage impedance) in shunt with the transmission line:

```
$INCLUDE 406PAR1.TRN
```

Figure 4-7 shows that a compensating voltage of ∓ 15% of the voltage ($\mathbf{V_s}$) at the sending end results in a voltage ($\mathbf{V_{s'}}$) at the modified sending end in the range of 1.005

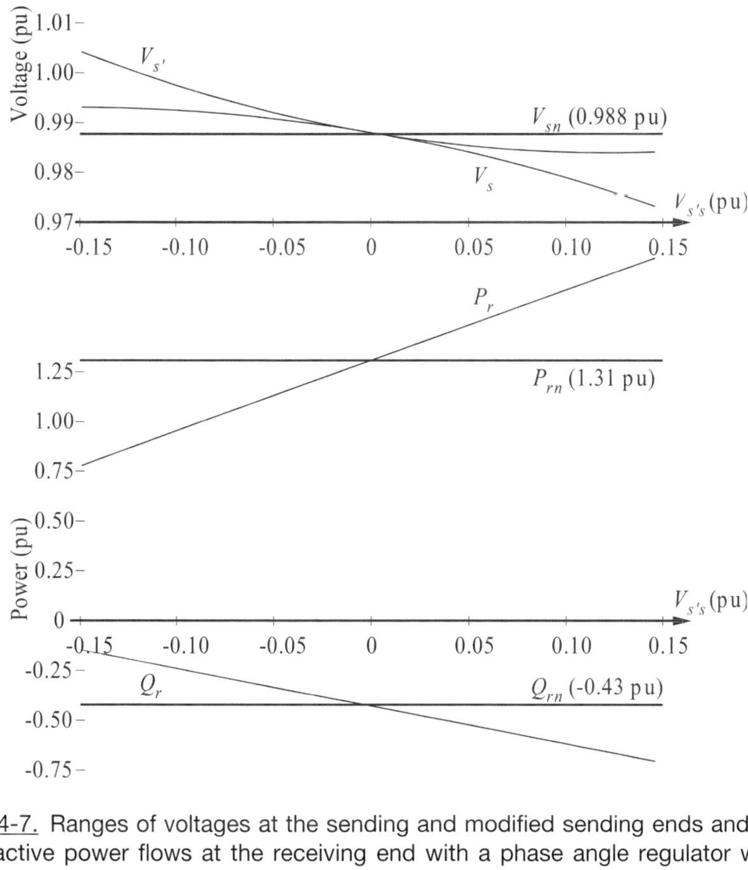

Figure 4-7. Ranges of voltages at the sending and modified sending ends and active and reactive power flows at the receiving end with a phase angle regulator with no leakage impedance.

to 0.986 pu. In the process, the voltage ($\mathbf{V}_s$) at the sending end varies in the range of 0.993 to 0.973 pu. The reactive power flow at the receiving end changes from –0.15 pu to –0.7 pu during the range of compensating voltage from –0.15 pu to 0.15 pu. However, the active power flow at the receiving end changes in the range of 0.78 pu to 1.82 pu for the same range of variation in the compensating voltage. Therefore, a $\mp$15% compensating voltage in quadrature with line voltage changes the reactive power flow by –0.55 pu and the active power flow by 1.04 pu.

The results show that the PAR changes the active power flow at the receiving end of the line much more than it changes the reactive power flow. In addition to changing the phase angle of the line voltage, the PAR also changes the magnitude of the line voltage. There are two contributing factors to this effect. First, the operation of the PAR changes the power flow of the line. This loading effect coupled with the Thèvenin impedance changes the voltage at the point of compensation. Second, when the compensating voltage is added or subtracted in quadrature with the line voltage, the resultant voltage is always larger than the original voltage. For a true implementa-

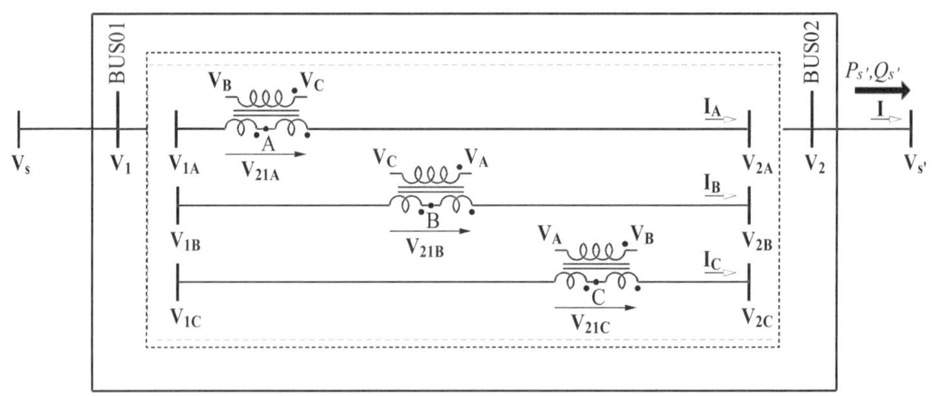

(a)

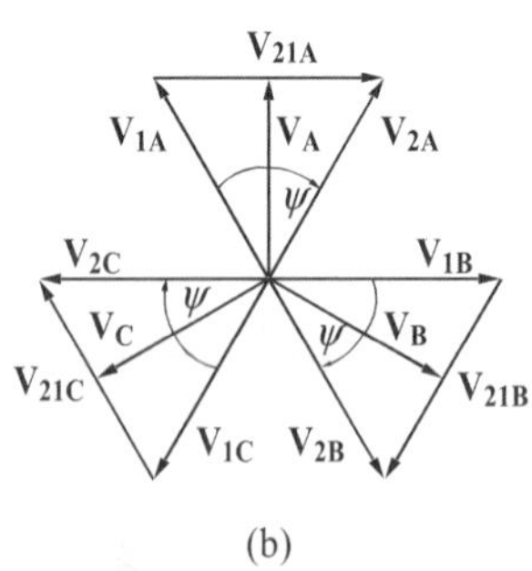

(b)

Figure 4-8. True phase angle regulator configuration.

tion of the PAR with the same voltage magnitude at the sending and the modified sending ends, the following configuration may be used.

Figure 4-8a shows the true PAR configuration. The three-phase compensating secondary windings are center-tapped at A, B, and C, respectively. The three-phase voltages ($\mathbf{V_A}$, $\mathbf{V_B}$, and $\mathbf{V_C}$) are applied to Δ-connected primary windings. The secondary voltage ($\mathbf{V_{21}}$) that is in phase with the primary line-to-line voltage, but in quadrature with the primary line-to-neutral voltage, is connected in series with the transmission line. With proper polarities of the series-connected windings, the compensating secondary voltage is connected in series with the transmission line at 90° or –90° with respect to the primary voltage. The resulting effect is that only the phase angle of the line voltage between BUS01 and BUS02 is changed, but the magnitude stays unchanged, as shown in the phasor diagram (Figure 4-8b).

An ideal PAR with no leakage reactance is run from the main file `407PAR2.DAT` (same as `401IDLAT.DAT` with a change to a phase angle regulating transformer). The

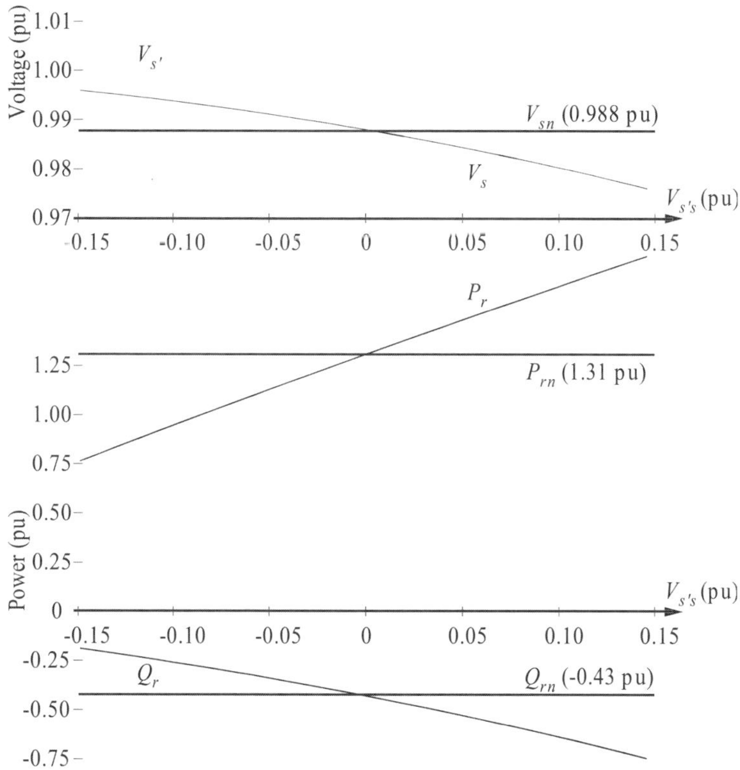

Figure 4-9. Ranges of voltages at the sending and modified sending ends and active and reactive power flows at the receiving end with a phase angle regulator with no leakage impedance and the Δ-connected primary windings are excited from the center-tapped series compensating windings.

next statement attaches the $INCLUDE$ file `408PAR2.TRN` (see Appendix C, Section 28) that connects a phase angle regulating transformer (with no leakage impedance) in shunt with the transmission line:

```
$INCLUDE 408PAR2.TRN
```

Figure 4-9 shows that a compensating voltage of $\mp$15% of the voltage ($\mathbf{V_{A\text{-}B\text{-}C}}$) at the midpoint of the compensating voltage windings results in a voltage ($\mathbf{V_{s'}}$) at the modified sending end in the range of 0.996 to 0.976 pu and an identical voltage ($\mathbf{V_s}$) at the sending end. The change in voltage at the point of compensation is solely due to the change in loading effect of the line. The reactive power flow at the receiving end changes from –0.19 pu to –0.75 pu during the range of compensating voltage from –0.15 pu to 0.15 pu. However, the active power flow at the receiving end changes in the range of 0.77 pu to 1.80 pu for the same range of variation in the compensating voltage. Therefore, a $\mp$15% compensating voltage in quadrature with line voltage changes the reactive power flow by –0.56 pu and active power flow by 1.03 pu.

5

MECHANICALLY SWITCHED FACTS CONTROLLERS

Various connections of mechanically switched capacitors and reactors in shunt and/or series with a transmission line allow the modification of the line voltage and, in the process, the power flow in the transmission line changes. The objective in this chapter is to show that the mechanically switched shunt- and series-connected capacitors and reactors can be used as FACTS controllers. The same operating points are studied using VSC-based FACTS controllers in Chapter 8.

5.1 SHUNT COMPENSATION

Both mechanically switched capacitors and reactors are studied in this section.

5.1.1 Mechanically Switched Capacitor (MSC)

It is discussed in Chapter 2 that a shunt-connected capacitor, shown in Figure 5-1, increases the line voltage at the point of compensation and also increases the delivery of reactive power to the receiving end of the transmission line significantly more than the delivery of active power. As an example, the shunt-connected reactance from the capacitor is adjusted until the reactive current drawn from the line at the point of com-

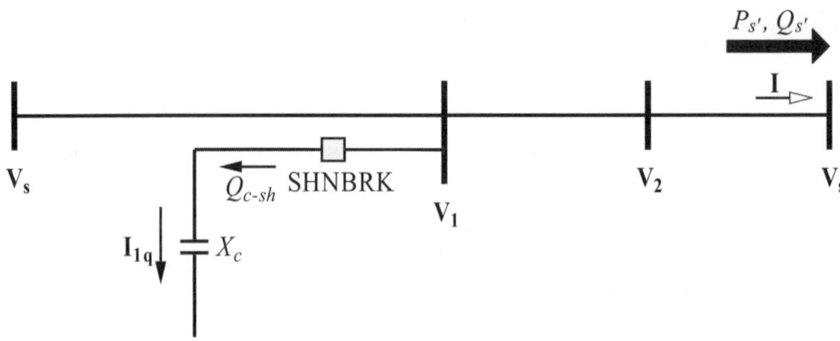

Figure 5-1. Shunt-connected mechanically switched capacitor.

pensation is one pu. The resulting line voltage is 1.035 pu. Therefore, the shunt-connected compensating reactance is

$$X_c = X_{pu} \times Z_{base} = 1.035 \times 119.02 = 123.19\ \Omega$$

where $X_{pu} = (V_{pu}/I_{pu}) = 1.035/1 = 1.035$ pu. Therefore, the shunt-connected capacitor $C = 1/(\omega X_c) = 1/(2 \times \pi \times 60 \times 123.19) = 21.532\ \mu$F.

A shunt-connected capacitive reactance is run from the main file `501SHREA.DAT` (see Appendix C, Section 29) with the user file `502SHREA.USR` (see Appendix C, Section 30), and the control file `503SHREA.CON` (see Appendix C, Section 31). The corresponding shunt capacitive reactance branch file is `321SHRE1.BRN` (see Appendix C, Section 19).

Figure 5-2 shows the digital simulation results from connecting a three-phase shunt capacitor bank to the transmission line. Between 0 and 50 ms, the shunt breaker (SNHBRK) stays open and the natural voltage at the point of compensation is 0.99 pu. At 50 ms, the SNHBRK closes and the shunt capacitor is connected to the line. The inrush current (i_{1a}) of phase *a* exceeds four pu. The instantaneous voltage of phase *a* exceeds 0.75 pu over the nominal value. Under steady-state conditions, the current (i_{1a}) of phase *a* through the shunt capacitor leads the line voltage (v_{1a}) of phase *a* by 90° with a magnitude of one pu. The line voltage becomes 1.03 pu at the point of compensation. Therefore, the compensating reactive power becomes –1.03 pu (capacitive). At 250 ms, the SNHBRK opens to disconnect the shunt capacitor from the line, the current through the shunt capacitor reduces to zero, and the line voltage at the point of compensation becomes the original 0.99 pu. When the capacitor disconnects, a spike in voltage occurs at the point of compensation due to the sudden change in flow of current through the line inductance.

Just prior to closing the switch, the voltage on one side of the contact of the shunt breaker (SHNBRK) is the capacitor voltage, which is zero. On the other side of the contact is the line voltage and it is different instantaneously on three different phases. The worst inrush current will occur in any phase if the contact closes on a peak voltage. The most desirable scenario is if the contact closes on zero voltage.

The inrush current through the capacitor can be limited by connecting the shunt capacitor through a series reactor, as shown in Figure 5-3. After some time, the series re-

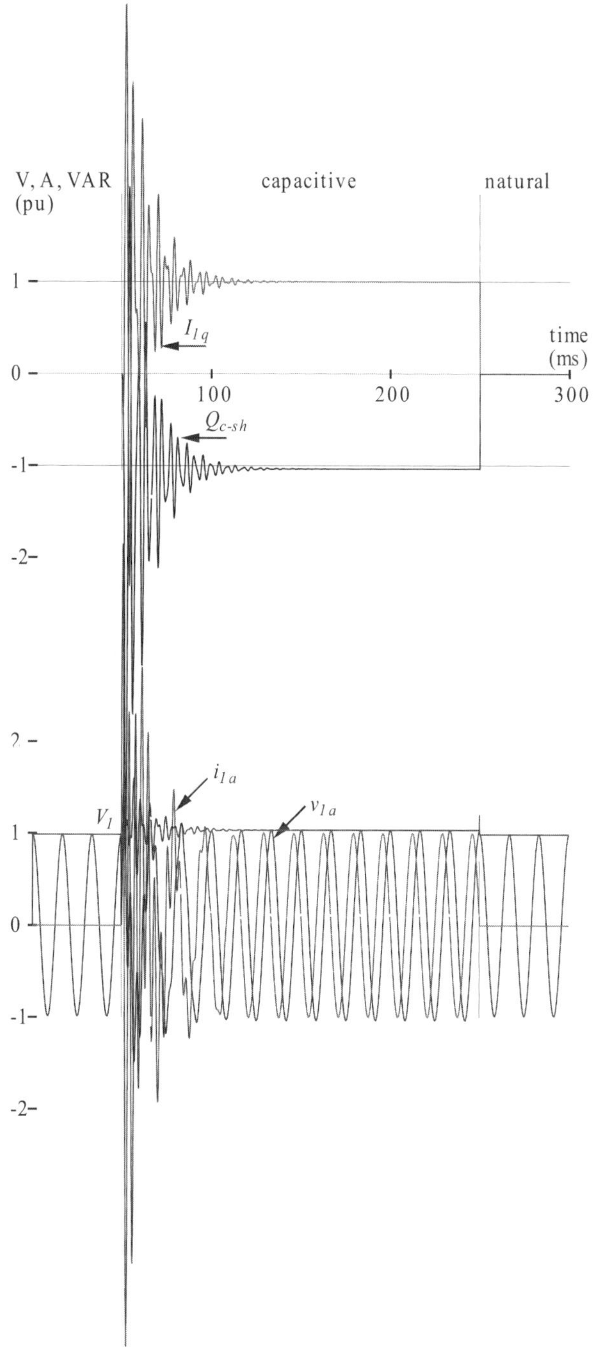

Figure 5-2. Exchanged reactive power and the resulting voltages and currents due to a shunt-connected mechanically switched capacitor to a transmission line.

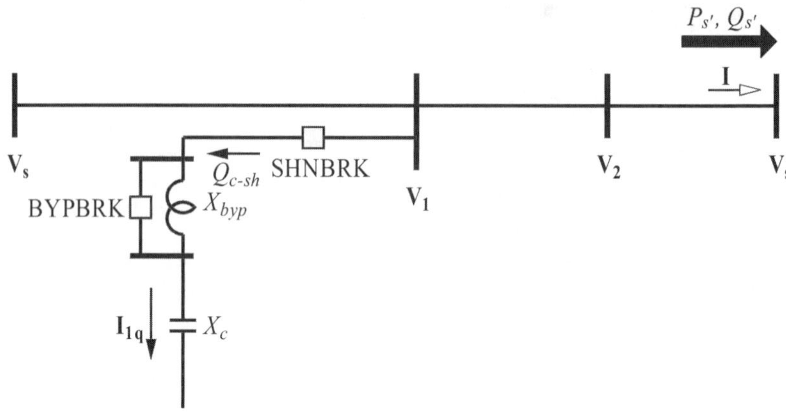

Figure 5-3. Shunt-connected mechanically switched capacitor with series reactor.

actor is shorted out by closing a bypass breaker (BYPBRK) and leaving only the shunt capacitor in the circuit. The bypass reactor (X_{byp}) is chosen to be 20% of the compensating reactance (X_c) with a quality factor (Q) of 10.

A shunt-connected capacitor with bypassed reactor is run from the main file `504SHREA.DAT` (same as `501SHREA.DAT` with a change of shunt reactance file) with the user file `505SHREA.USR` (see Appendix C, Section 32). The corresponding shunt capacitive reactance branch file is `322SHRE2.BRN` (see Appendix C, Section 20).

Figure 5-4 shows the digital simulation results when a three-phase shunt capacitor bank is connected to the transmission line through three-phase series reactors that are bypassed after a certain time. Between 0 and 50 ms, the shunt breaker (SNHBRK) stays open and the natural voltage at the point of compensation is 0.99 pu. At 50 ms, the SNHBRK closes and the shunt capacitor with the series reactor is connected to the line. The inrush current (i_{1a}) of phase a barely exceeds 3.2 pu. The instantaneous voltage (v_{1a}) of phase a reaches 0.18 pu over the nominal value. At 150 ms, the BYPBRK closes to short out the series reactor, causing some transients to occur in currents and voltages. After the transients die out, the current (i_{1a}) of phase a through the shunt capacitor leads the line voltage (v_{1a}) of phase a by 90° with a magnitude of one pu. From this point onward, the circuit operation is the same as before.

5.1.2 Mechanically Switched Reactor (MSR)

It is discussed in Chapter 2 that a shunt-connected reactor, shown in Figure 5-5, decreases the line voltage at the point of compensation and also decreases the delivery of reactive power to the receiving end of the transmission line significantly more than the delivery of active power.

The shunt-connected reactor is adjusted until the reactive current drawn from the line at the point of compensation is one pu. The resulting line voltage is 0.94 pu. The shunt-connected compensating reactance is

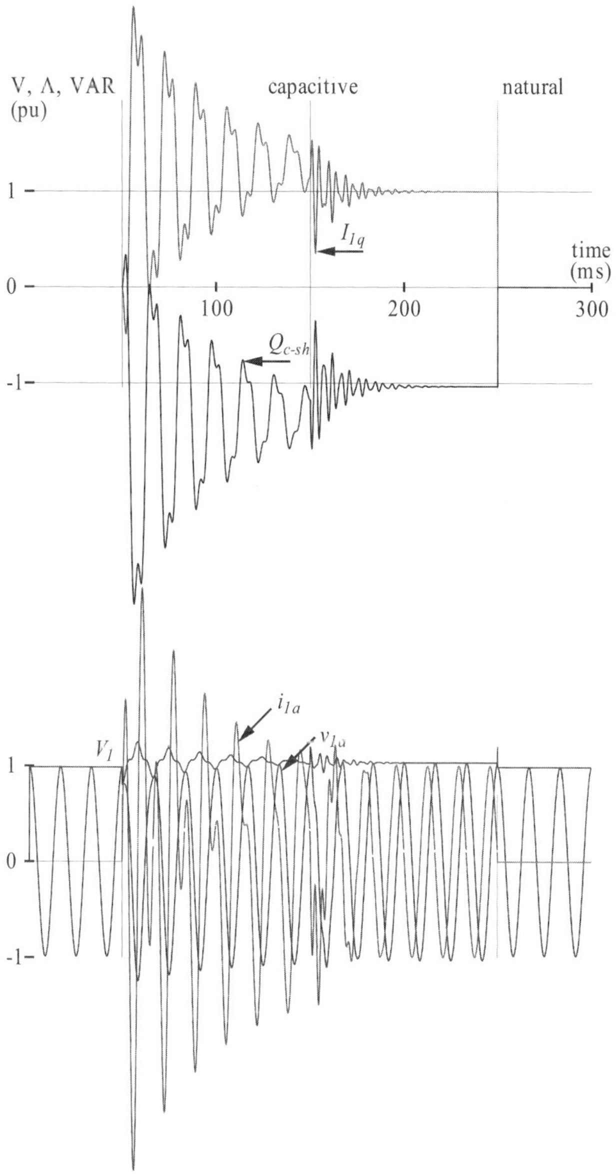

Figure 5-4. Exchanged reactive power and the resulting voltages and currents due to a shunt-connected mechanically switched capacitor (with a series reactor) to a transmission line.

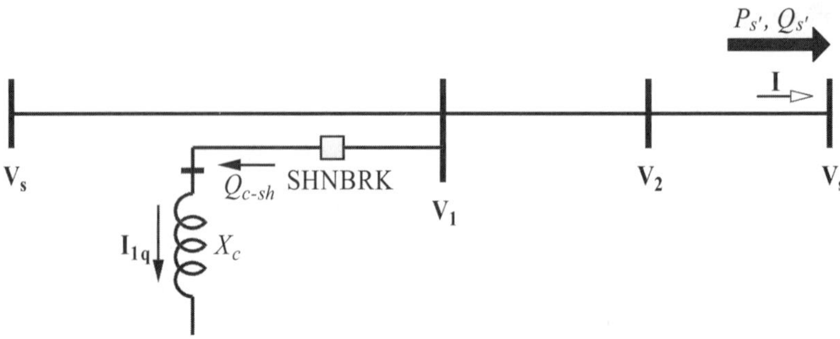

Figure 5-5. Shunt-connected mechanically switched reactor.

$$X_c = X_{pu} \times Z_{base} = 0.94 \times 119.02 = 111.89\ \Omega$$

where $X_{pu} = (V_{pu}/I_{pu}) = 0.94/1 = 0.94$ pu. Therefore, the shunt-connected reactor $L = X_c/\omega = 111.89/(2 \times \pi \times 60) = 296.78$ mH.

A resistor (R) connected in parallel to the reactor is $R = Q \times (\omega L) = 10 \times (2\pi \times 60 \times 296.78 \times 10^{-3}) = 1118.834\ \Omega$, where the quality factor, $Q = R/(\omega L)$, is considered to be 10. The equivalent series combination of resistor and reactor is

$$R_s = R\frac{1}{1+Q^2} = \frac{1118.834}{101} = 11.077\ \Omega$$

and

$$L_s = L\frac{Q^2}{1+Q^2} = 296.78\frac{100}{101} = 293.84 \text{ mH}$$

A shunt-connected reactor is run from the main file `501SHREA.DAT` with a different shunt reactor branch file `323SHRE3.BRN` (see Appendix C, Section 21).

Figure 5-6 shows the digital simulation results when a three-phase shunt reactor bank is connected to the transmission line. Between 0 and 50 ms, the shunt breaker (SNHBRK) stays open and the natural voltage at the point of compensation is 0.99 pu. At 50 ms, the SNHBRK closes and the shunt reactor is connected to the line. After the closing of the breaker, the current through the reactor builds up from its zero initial condition. Under steady-state conditions, the current (i_{1a}) of phase a through the shunt reactor lags the line voltage (v_{1a}) of phase a by 90° with a magnitude of one pu. The line voltage becomes 0.94 pu at the point of compensation. Therefore, the compensating reactive power becomes 0.94 pu. At 250 ms, the SNHBRK opens to disconnect the shunt reactor from the line. The current through the shunt reactor drops to zero, causing a spike voltage across the reactor. There is also a spike in voltage at the point of compensation due to the sudden change in flow of current through the line inductance. The line voltage at the point of compensation becomes the original 0.99 pu.

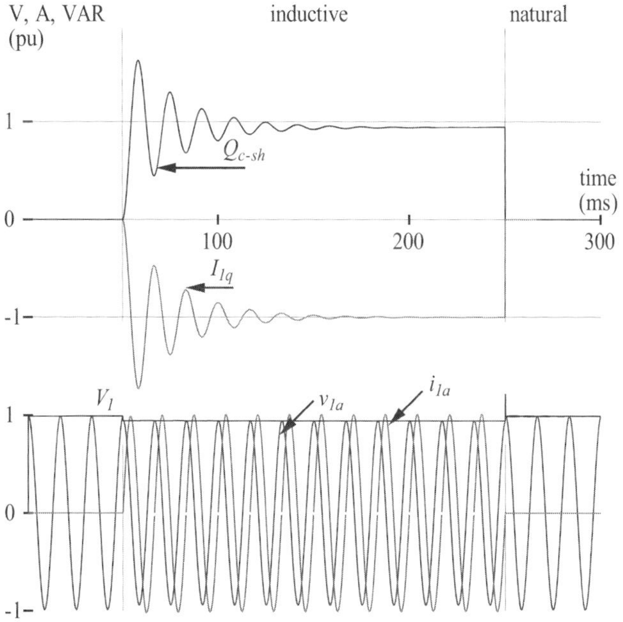

Figure 5-6. Exchanged reactive power and the resulting voltages and currents due to a shunt-connected reactor to a transmission line.

5.2 SERIES COMPENSATION

Both mechanically switched reactors and capacitors are studied in this section.

5.2.1 Mechanically Switched Reactor (MSR)

It is discussed in Chapter 2 that a series-connected reactor, shown in Figure 5-7, modifies the effective reactance of the transmission line between its two ends and regulates the active and reactive power flows in the line simultaneously.

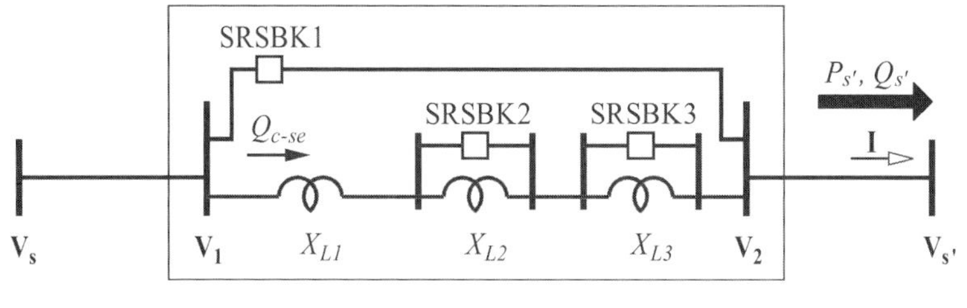

Figure 5-7. Series-connected mechanically switched reactor.

In this example, the series-connected reactors are inserted in three stages. All three series breakers (SRSBK1, SRSBK2, and SRSBK3) are closed initially, leaving the line with its natural active power flow of 1.31 pu and reactive power flow of –0.43 pu at the receiving end. Each of the three series-connected compensating reactances is $X_{L1} = X_{L2} = X_{L3} = 0.15 \times Z_{base} = 0.15 \times 119.025 = 17.854\ \Omega$. Therefore, the value of each compensating reactor is $L1 = X_{L1}/\omega = 17.854/(2 \times \pi \times 60) = 47.358\ \mu$H. The first compensating reactor (X_{L1}) has a quality factor (Q) of 10, whereas the other two reactors (X_{L2} and X_{L3}) have a quality factor of ∞. In Chapter 8, it is considered that the compensating reactance (X_{L1}) represents the leakage reactance of the series coupling transformer.

A series-connected reactor is run from the main file `506SEREA.DAT` (see Appendix C, Section 33) with the user file `507SEREA.USR` (see Appendix C, Section 34), and the control file `508SEREA.CON` (see Appendix C, Section 35). The corresponding series reactor branch file is `324SERE1.BRN` (see Appendix C, Section 22).

Figure 5-8 shows the digital simulation results when a three-phase series reactor bank is connected to the transmission line. The normally closed series breaker 1 (SRSBK1) opens at t = 25 ms and stays open for the next 275 ms to connect the reactor (X_{L1}) in series with the transmission line between BUS01 and BUS02. A voltage (v_{12a}) develops between BUS01 and BUS02. Both the active and reactive power flows (P_r and Q_r) in the line decrease. Further increases in the inductive compensation at 50 ms and 175 ms (by opening the series breakers SRSBK2 and SRSBK3) decrease the active and reactive power flows in the line simultaneously. These operating points are verified in series compensation with VSCs in Chapter 8. The compensating reactive power ($Q_{c\text{-}se}$) exchanged with the line is shown. At 300 ms, all the series breakers are closed, bypassing the series inductive compensation. The line current and power flow go back to the original uncompensated levels.

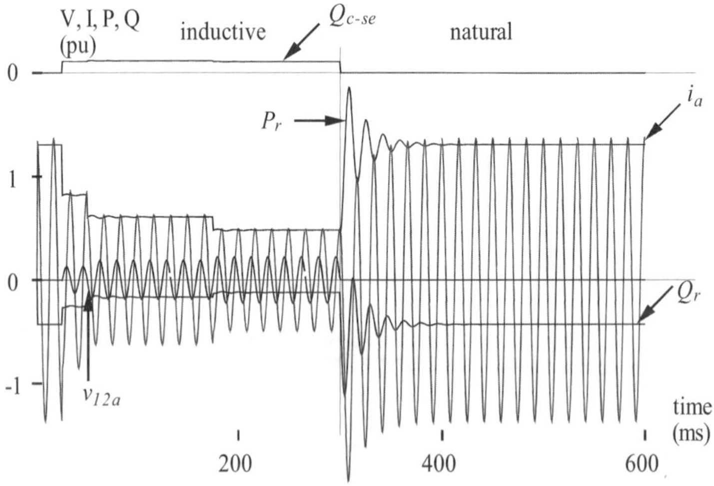

Figure 5-8. Effect of a series-connected reactor into the transmission line.

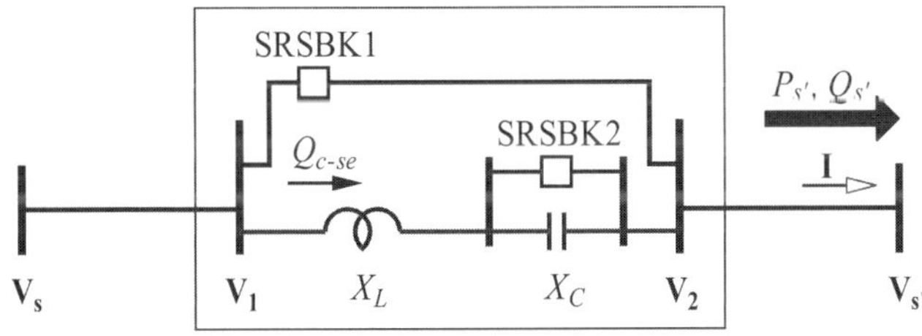

Figure 5-9. Series-connected mechanically switched capacitor with reactor.

5.2.2 Mechanically Switched Capacitor (MSC) with a Reactor

In this example, the series reactances from a three-phase reactor and capacitor banks, shown in Figure 5-9, are inserted in two stages. The two series breakers (SRSBK1 and SRSBK2) are closed initially, leaving the line with its natural active power flow of 1.31 pu and reactive power flow of –0.43 pu at the receiving end. The first series-connected compensating reactance is $X_L = 0.15 \times Z_{base} = 0.15 \times 119.025 = 17.854\ \Omega$ and its inductance is $L = X_L/\omega = 17.854/(2 \times \pi \times 60) = 47.358\ \mu\text{H}$. Its quality factor (Q) is 10. The second series-connected compensating reactance is $X_C = 0.15 \times Z_{base} = 0.15 \times 119.025 = 17.854\ \Omega$ and its capacitance is $C = 1/(\omega X_C) = 1/(17.854 \times 2 \times \pi \times 60) = 148.57\ \mu\text{F}$.

A series-connected inductive/capacitive reactance is run from the main file `509SEREA.DAT` that is same as `506SEREA.DAT` with a different series reactance file `325SERE2.BRN` (see Appendix C, Section 23).

Figure 5-8 shows the digital simulation results when a three-phase series reactor/capacitor banks are connected to the transmission line. The normally closed series break-

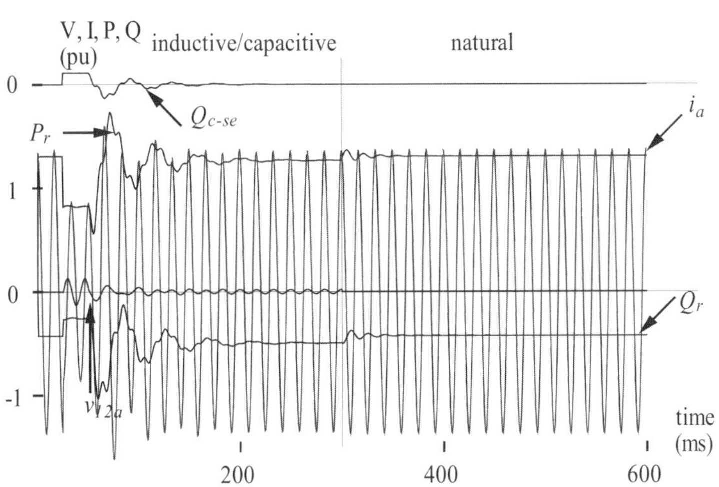

Figure 5-10. Effect of a series-connected reactor/capacitor in the transmission line.

er 1 (SRSBK1) opens at $t = 25$ ms and stays open for the next 275 ms to connect the reactor (X_L) in series with the transmission line between BUS01 and BUS02. A voltage (v_{12a}) develops between BUS01 and BUS02. Both active and reactive power flows (P_r and Q_r) in the line decrease. At 50 ms, series breaker 2 (SRSBK2) is opened, adding 15% capacitive compensation that nullifies the inductive compensation that is already in place. These operating points are verified in series compensation with VSCs in Chapter 8. The compensating reactive power ($Q_{c\text{-}se}$) exchanged with the line is shown. At 300 ms, all the series breakers are closed, bypassing the series inductive and capacitive compensations. The line current and power flow go back to the original uncompensated levels.

6

VOLTAGE-SOURCED CONVERTER (VSC)

In the 1980s, the first voltage-sourced converter (VSC)-based static converter was installed to regulate the voltage of a transmission line. In the 1990s, the application of VSCs was extended to independent control of active and reactive power flows in a transmission line. The VSCs were also installed in industrial applications, such as an arc furnace compensator, primarily for the control of voltage flicker at the utility interface bus. The VSC-based technology has also been used in an application such as DC transmission. A VSC produces an AC voltage of fundamental frequency along with an allowable level of harmonic components.

The semiconductor switches that are used in the implementation of a VSC are controllable, meaning that the switches can be turned on and off at a desired time. Some high-power-rated switches available for these applications are gate-turn-off (GTO) thyristors, integrated gate commutated thyristors (IGCT), and press-pack insulated gate bipolar transistors (IGBT). A GTO was generally used in slow switching applications, primarily due to its relatively long turn-on time and turn-off time. In a once-per-cycle switching application, a VSC employing GTO switches produces a number of time-shifted square wave voltages. The time period of these square wave voltages is the same as that of the fundamental frequency voltage. These square wave voltages are combined with a harmonic neutralized (HN) technique using a magnetic

Introduction to FACTS Controllers. By Kalyan K. Sen and Mey Ling Sen

circuit to produce a high-quality sinusoidal voltage with acceptable harmonic content. An IGBT's relatively short turn-on and turn-off times make it possible to be used in fast-switching applications. A VSC employing IGBT switches generally produces a pulse width modulated (PWM) square wave voltage, which contains a fundamental component and higher-order harmonic components, the frequencies of which are directly related to the switching frequency. A PWM voltage is usually filtered before it is used.

In this chapter, a number of DC-to-AC VSC topologies that are GTO-based and used in practical applications are discussed in detail; their simulation techniques are also presented. The number of switches, referred to as valves, in the VSC and the various configurations of the magnetic circuit depend on the desired quality of AC waveforms generated by the VSC. The VSC configurations that are described in this chapter are a few of many possible configurations that can be used to build a harmonic neutralized voltage-sourced converter (HN-VSC). Each switch in the HN-VSC may consist of several controllable semiconductor switch modules that are connected in a series or parallel combination to achieve the desired voltage and current ratings of the switch.

6.1 MODELING AN IDEAL VSC

The modeling of a VSC is presented using an EMTP simulation package. Figure 6-1a shows the block diagram of a DC-to-AC VSC that generates a three-phase AC voltage. Figure 6-1b shows a three-phase AC voltage set generated using an ideal VSC. A generally structured main file `601IDL21.DAT` (see Appendix C, Section 36) is run to generate an ideal sine wave from a DC voltage (v_{DC}) of $\pi/2$ pu using a VSC with a gain of $K_{VSC} = 2/\pi$. In the TACS/control section, a free-running PLL angle (θ) is generated from the power system frequency. The free-running phase angle (θ) is reset at every 2π radians to produce a saw-tooth-like angle (θ_1) in the range of 0 to 2π radians. The $INCLUDE file `602IDL21.VSC` (see Appendix C, Section 37) generates a three-phase sine wave set.

Although it is desirable to generate a fundamental frequency AC voltage using a VSC, the quality of the AC voltage, meaning the associated level of harmonic compo-

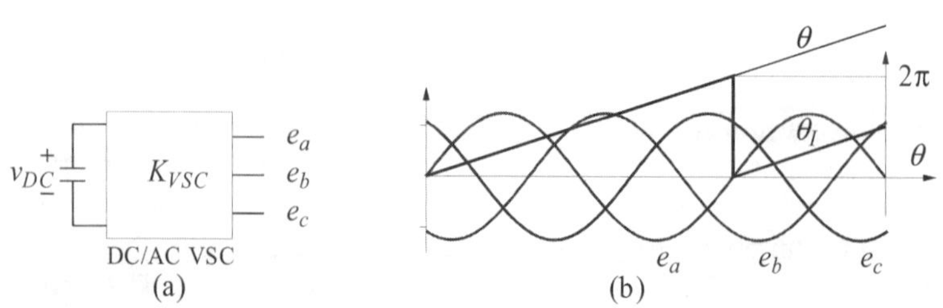

Figure 6-1. Generation of a three-phase ideal VSC voltage.

nents in it, depends on the topology of the VSC. The choice of the topology is very much influenced by the guideline set by the IEEE 519 standard that calls for limits on individual voltage and current distortions and total voltage and current harmonic distortions at the point of compensation for various line voltage levels. In the following section, a step-by-step approach is presented showing how to generate an AC voltage that is acceptable in transmission line and industrial applications.

6.2 DC-TO-AC VSC

From the early days of FACTS controllers, used in transmission and industrial applications, DC-to-AC VSCs using the HN technique were the first choices because of their lowest possible losses among all available topologies. A HN-VSC consists of a set of poles (comprising of GTOs, diodes, and snubbers), a DC capacitor, a magnetic circuit, and a control and protection unit. In order to produce a sine-wave-like voltage using a HN-VSC, there are three required stages. First, the control unit must produce a set of gate pulses; second, the gate pulses operate a set of poles, which produce a set of square wave voltages; and third, the square wave voltages are combined with a magnetic circuit to produce a sinusoidal voltage. A HN-VSC with a finite pulse number produces a high-quality sinusoidal voltage with acceptable harmonic content. If a VSC produces a pure sine voltage of fundamental frequency, it can be thought of as a VSC of infinite pulse number.

6.2.1 Generation of a Square Wave Voltage with a Two-Level Pole

The fundamental building block of a HN-VSC pole is a pair of DC capacitors and a set of switches, which can be used to generate a square wave voltage.

Figure 6-2 shows a single-phase VSC circuit, referred to as a pole, which consists of a positive switch (A+) and a negative switch (A–). The pole is connected across a pair of capacitors that are charged with a total DC voltage of v_{DC}. When the switch A+ is on, the midpoint of the pole is tied to the positive DC terminal; when the switch A–

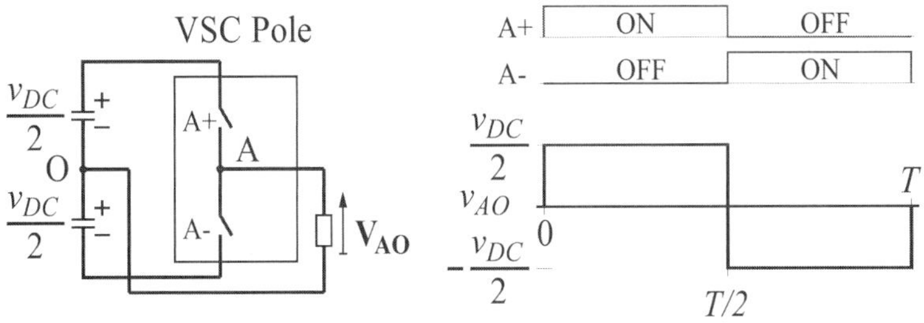

Figure 6-2. VSC pole and its output voltage.

is on, the midpoint of the pole is tied to the negative DC terminal. When the switches are closed and opened alternately, the pole output voltage (v_{AO}) at the midpoint of pole A with respect to the midpoint (O) of the DC link capacitors is a square wave with two levels ($+v_{DC}/2$ and $-v_{DC}/2$). Therefore, this pole circuit is referred to as a two-level pole. The time period, T (i.e., $1/f$), is equally divided between the positive and the negative half-cycles, where f is the fundamental frequency of the square wave voltage.

The pole output voltage that contains harmonic components (n) is expressed in a general form as

$$v_{AO} = a_0 + \sum_{n=1}^{\infty} V_{AO,n} \sin(n\theta + \alpha_{An}) \tag{6-1}$$

where θ is the free-running phase angle. The amplitude ($V_{AO,n}$) and the relative phase angle (α_{An}) with respect to the PLL angle (θ) are given by

$$V_{AO,n} = \sqrt{a_n^2 + b_n^2} \tag{6-2a}$$

and

$$\alpha_{An} = \tan^{-1}\frac{a_n}{b_n} \tag{6-2b}$$

The DC component (a_0) in v_{AO} is calculated from equation A-90 (see Appendix A) as

$$a_0 = \frac{1}{2\pi}\int_0^{2\pi} v_{AO}\, d\theta = \frac{1}{2\pi}\frac{v_{DC}}{2}\left[\int_0^{\pi} d\theta - \int_{\pi}^{2\pi} d\theta\right]$$

$$= \frac{1}{2\pi}\frac{v_{DC}}{2}\left[\theta\big|_0^{\pi} - \theta\big|_{\pi}^{2\pi}\right] = \frac{1}{2\pi}\frac{v_{DC}}{2}[\pi - 2\pi + \pi] = 0$$

$$a_0 = 0 \tag{6-3}$$

The amplitudes (a_n) for $n = 1, 2, 3, \ldots$ of the cosine components in v_{AO} are calculated from equation A-91 as

$$a_n = \frac{1}{\pi}\int_0^{2\pi} v_{AO}\cos(n\theta)\, d\theta = \frac{1}{\pi}\frac{v_{DC}}{2}\left[\int_0^{\pi}\cos(n\theta)\, d\theta - \int_{\pi}^{2\pi}\cos(n\theta)\, d\theta\right]$$

$$= \frac{1}{n\pi}\frac{v_{DC}}{2}\left[\sin(n\theta)\big|_0^{\pi} - \sin(n\theta)\big|_{\pi}^{2\pi}\right] = \frac{1}{n\pi}\frac{v_{DC}}{2}[0] = 0$$

$$a_n = 0 \tag{6-4}$$

The amplitudes (b_n) for $n = 1, 2, 3, \ldots$ of the sine components in v_{AO} are calculated from equation A-92 as

$$b_n = \frac{1}{\pi}\int_0^{2\pi} v_{AO}\sin(n\theta)\,d\theta = \frac{1}{\pi}\frac{v_{DC}}{2}\left[\int_0^{\pi}\sin(n\theta)\,d\theta - \int_{\pi}^{2\pi}\sin(n\theta)\,d\theta\right]$$

$$= \frac{1}{n\pi}\frac{v_{DC}}{2}\left[-\cos(n\theta)\big|_0^{\pi} + \cos(n\theta)\big|_{\pi}^{2\pi}\right]$$

$$= \frac{1}{n\pi}\frac{v_{DC}}{2}\left[-\cos(n\pi) + 1 + \cos(2n\pi) - \cos(n\pi)\right] = \frac{1}{n\pi}\frac{v_{DC}}{2}\left[2 - 2\cos(n\pi)\right]$$

$$= \frac{v_{DC}}{n\pi}\left[1 - \cos(n\pi)\right] = \frac{v_{DC}}{n\pi}\left[1 - (-1)^n\right] = \frac{2v_{DC}}{n\pi}$$

when n is an odd integer. Otherwise, it is zero.

$$b_n = \frac{2v_{DC}}{n\pi} \tag{6-5}$$

The amplitude ($V_{AO,n}$) and the relative phase angle (α_{An}) with respect to the PLL angle (θ) are

$$V_{AO,n} = \sqrt{a_n^2 + b_n^2} = b_n = \frac{2v_{DC}}{n\pi} \tag{6-6a}$$

and

$$\alpha_{An} = \tan^{-1}\frac{a_n}{b_n} = 0 \tag{6-6b}$$

for any odd n.

The expression for v_{AO} shown in equation 6-1 can be rewritten as

$$v_{AO} = \frac{2v_{DC}}{\pi}\sum_{n=1}^{\infty}\frac{1}{n}\sin(n\theta) \tag{6-7}$$

for any odd n.

The amplitude of the fundamental component (for $n = 1$) is

$$V_{AO,1} = \frac{2v_{DC}}{\pi} \tag{6-8}$$

From equations 6-6a and 6-8, the amplitudes of harmonic components (n) normalized to the fundamental component are

$$\frac{V_{AO,n}}{V_{AO,1}} = \frac{1}{n} \tag{6-9}$$

for any odd n.

6.2.1.1 Modeling a Single-Phase VSC and Simulation Results. The modeling of a single-phase VSC can be done by using either ideal switches or TAC-controlled switches. The VSC pole circuit with a two-level pole is shown in Figure 6-3. The two switches in a pole are discussed in detail in Chapter 7.

A main file `6072POLE.DAT` (see Appendix C, Section 38) is run to generate a square wave voltage using a two-level pole. The $INCLUDE$ file `6082POLE.GAT` (see Appendix C, Section 39) generates the gating pulses for the two switches. The $INCLUDE$ file `6092POLE.POL` (see Appendix C, Section 40) generates the ideal pole voltage using ideal switches. The functions of an ideal pole circuit can be implemented mathematically by combining the DC link voltage with appropriate switching functions. The $INCLUDE$ file `6102POLE.MAG` (see Appendix C, Section 41) emulates an ideal 1:1 transformer. The functions of an ideal magnetic circuit can be implemented mathematically by combining various time-shifted square wave voltages. During the transition from on to off and vice versa, the ideal switches and the magnetic circuit exhibit voltage and current with no transients.

The gating signals generated in the $INCLUDE$ file `6082POLE.GAT` can be used to operate two TAC-controlled switches as shown in the main file `6112POLE.DAT` (see Appendix C, Section 42). When "ON," an ideal switch carries current with no voltage across it; when "OFF," it carries no current with the open circuit voltage appearing across it.

Figure 6-4 shows the pole voltages and currents generated by the above two models

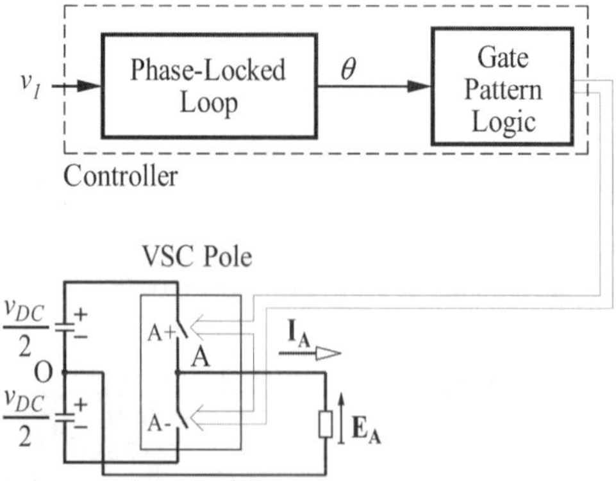

Figure 6-3. Model of a VSC pole in EMTP.

with poles constructed with ideal switches and TAC-controlled switches. They are virtually identical.

In the rest of the chapter, the poles and the magnetic circuits are modeled as ideal. However, if desired, the generated gating signals can be used to trigger the switches of the poles whose output voltages can then be fed to magnetic circuits to produce HN-VSC output voltages.

6.2.2 Six-Pulse VSC with Two-Level Poles

A six-pulse VSC requires the generation of three square wave voltages using the three poles A, B, and C. Figure 6-5a shows the three square wave voltages (v_{AO}, v_{BO}, and v_{CO}) and the corresponding fundamental components ($v_{AO,1}$, $v_{BO,1}$, and $v_{CO,1}$). The fundamental pole voltage phasors ($\mathbf{V_{AO,1}}$, $\mathbf{V_{BO,1}}$, and $\mathbf{V_{CO,1}}$), shown in Figure 6-5b, are at 0° (reference), –120°, and –240°, respectively. The A–B–C set of square wave voltages are said to be placed at 0°.

Figure 6-6 shows three poles A, B, and C that are connected across a shared DC link with a pair of capacitors and the related voltages. Note that the pole outputs are connected to a three-phase load whose neutral point (N) is not connected to the midpoint of the DC link capacitor.

The poles A, B, and C are operated in such a way that the pole voltages (v_{AO}, v_{BO}, and v_{CO}) are time-shifted from one another by one-third of the time period (T) of the pole voltage. The B and C poles' output voltages are generally expressed as

$$v_{BO} = a_0 + \sum_{n=1}^{\infty} V_{BO,n} \sin(n\theta + \alpha_{Bn}) \tag{6-10}$$

and

$$v_{CO} = a_0 + \sum_{n=1}^{\infty} V_{CO,n} \sin(n\theta + \alpha_{Cn}) \tag{6-11}$$

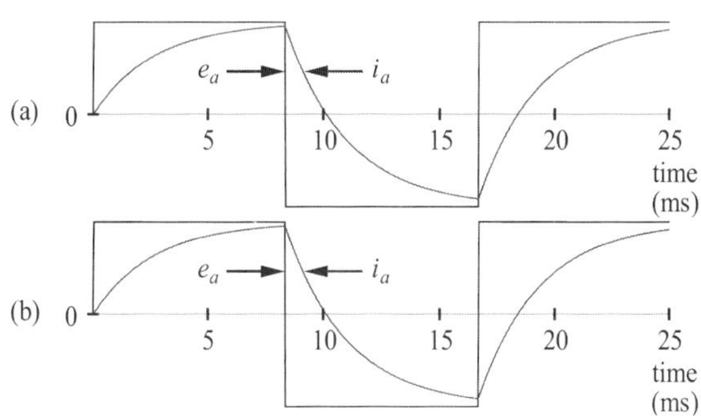

Figure 6-4. Voltages and currents from the simulation of a VSC pole with (a) ideal switches and (b) TAC-controlled switches.

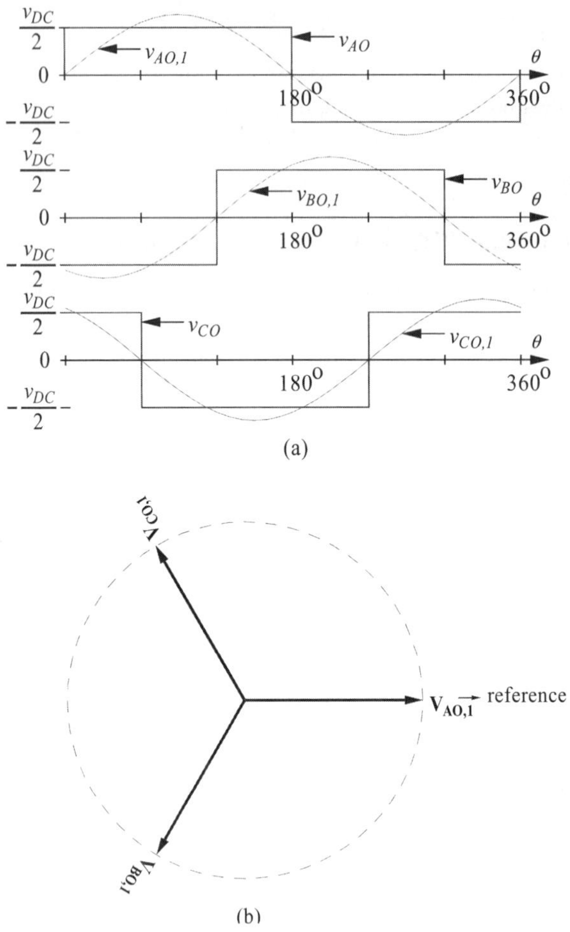

Figure 6-5. Pole voltages and fundamental voltage phasors in a six-pulse VSC configuration. (a) Three 120° phase-shifted, fundamental pole voltages and the corresponding square wave voltages. (b) Phasor diagram of fundamental pole voltage phasors in a six-pulse VSC configuration.

where

$$V_{BO,n} = V_{CO,n} = \sqrt{a_n^2 + b_n^2} \tag{6-12a}$$

and

$$\alpha_{Bn} = \alpha_{Cn} = \tan^{-1}\frac{a_n}{b_n} \tag{6-12b}$$

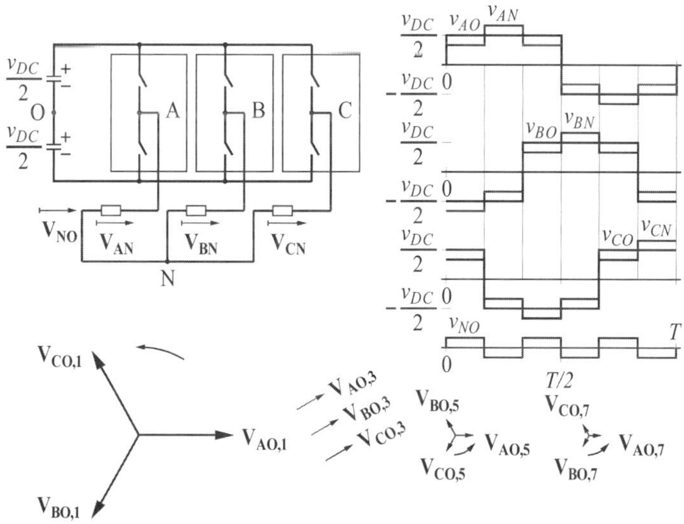

Figure 6-6. Six-pulse VSC and its output voltages.

The DC component (a_0) in v_{BO} is calculated from equation A-90 (see Appendix A) as

$$a_0 = \frac{1}{2\pi}\int_0^{2\pi} v_{BO}\,d\theta = \frac{1}{2\pi}\frac{v_{DC}}{2}\left[-\int_0^{2\pi/3} d\theta + \int_{2\pi/3}^{5\pi/3} d\theta - \int_{5\pi/3}^{2\pi} d\theta\right]$$

$$= \frac{1}{2\pi}\frac{v_{DC}}{2}\left[-\theta\Big|_0^{2\pi/3} + \theta\Big|_{2\pi/3}^{5\pi/3} - \theta\Big|_{5\pi/3}^{2\pi}\right] = \frac{1}{2\pi}\frac{v_{DC}}{2}\left[-\frac{2\pi}{3} + \frac{5\pi}{3} - \frac{2\pi}{3} - 2\pi + \frac{5\pi}{3}\right] = 0$$

$$a_0 = 0 \tag{6-13}$$

The amplitudes (a_n) for $n = 1, 2, 3, \ldots$ of the cosine components in v_{BO} are calculated from equation A-91 as

$$a_n = \frac{1}{\pi}\int_0^{2\pi} v_{BO}\cos(n\theta)\,d\theta$$

$$= \frac{1}{\pi}\frac{v_{DC}}{2}\left[-\int_0^{2\pi/3}\cos(n\theta)d\theta + \int_{2\pi/3}^{5\pi/3}\cos(n\theta)d\theta - \int_{5\pi/3}^{2\pi}\cos(n\theta)d\theta\right]$$

$$= \frac{1}{n\pi}\frac{v_{DC}}{2}\left[-\sin(n\theta)\Big|_0^{2\pi/3} + \sin(n\theta)\Big|_{2\pi/3}^{5\pi/3} - \sin(n\theta)\Big|_{5\pi/3}^{2\pi}\right]$$

$$= \frac{1}{n\pi}\frac{v_{DC}}{2}\left[-\sin\left(n\frac{2\pi}{3}\right)+\sin\left(n\frac{5\pi}{3}\right)-\sin\left(n\frac{2\pi}{3}\right)+\sin\left(n\frac{5\pi}{3}\right)\right]$$

$$= \frac{1}{n\pi}\frac{v_{DC}}{2}\left[-2\sin\left(n\frac{2\pi}{3}\right)+2\sin\left(n\frac{5\pi}{3}\right)\right]$$

$$= \frac{v_{DC}}{n\pi}\left[\sin\left(n\frac{5\pi}{3}\right)-\sin\left(n\frac{2\pi}{3}\right)\right]$$

$$= \frac{2v_{DC}}{n\pi}\cos\left(n\frac{7\pi}{6}\right)\sin\left(n\frac{\pi}{2}\right)$$

$$a_n = \frac{2v_{DC}}{n\pi}\cos\left(n\frac{7\pi}{6}\right)\sin\left(n\frac{\pi}{2}\right) \tag{6-14}$$

The amplitudes (b_n) for $n = 1, 2, 3, \ldots$ of the sine components in v_{BO} are calculated from equation A-92 as

$$b_n = \frac{1}{\pi}\int_0^{2\pi} v_{BO}\sin(n\theta)\,d\theta$$

$$= \frac{1}{\pi}\frac{v_{DC}}{2}\left[-\int_0^{2\pi/3}\sin(n\theta)d\theta + \int_{2\pi/3}^{5\pi/3}\sin(n\theta)d\theta - \int_{5\pi/3}^{2\pi}\sin(n\theta)d\theta\right.$$

$$= \frac{1}{n\pi}\frac{v_{DC}}{2}\left[\cos(n\theta)\Big|_0^{2\pi/3} - \cos(n\theta)\Big|_{2\pi/3}^{5\pi/3} + \cos(n\theta)\Big|_{5\pi/3}^{2\pi}\right]$$

$$= \frac{1}{n\pi}\frac{v_{DC}}{2}\left[\cos\left(n\frac{2\pi}{3}\right)-1-\cos\left(n\frac{5\pi}{3}\right)+\cos\left(n\frac{2\pi}{3}\right)+1-\cos\left(n\frac{5\pi}{3}\right)\right]$$

$$= \frac{v_{DC}}{n\pi}\left[\cos\left(n\frac{2\pi}{3}\right)-\cos\left(n\frac{5\pi}{3}\right)\right] = \frac{2v_{DC}}{n\pi}\sin\left(n\frac{7\pi}{6}\right)\sin\left(n\frac{\pi}{2}\right)$$

$$b_n = \frac{2v_{DC}}{n\pi}\sin\left(n\frac{7\pi}{6}\right)\sin\left(n\frac{\pi}{2}\right) \tag{6-15}$$

The amplitude ($V_{BO,n}$) and the relative phase angle (α_{Bn}) with respect to the angle (θ) are

$$V_{BO,n} = \sqrt{a_n^2 + b_n^2} = \sqrt{\left[\frac{2}{n\pi}v_{DC}\sin\left(n\frac{\pi}{2}\right)\right]^2\left[\cos^2\left(n\frac{7\pi}{6}\right)+\sin^2\left(n\frac{7\pi}{6}\right)\right]} = \frac{2v_{DC}}{n\pi} \tag{6-16a}$$

and

$$\alpha_{Bn} = \tan^{-1}\frac{a_n}{b_n} = \tan^{-1}\frac{1}{\tan\left(n\frac{7\pi}{6}\right)}$$

for any odd n. But

$$\tan\alpha_{Bn} = \frac{1}{\tan\left(n\frac{7\pi}{6}\right)}$$

or

$$\frac{\sin\alpha_{Bn}}{\cos\alpha_{Bn}} = \frac{\cos\left(n\frac{7\pi}{6}\right)}{\sin\left(n\frac{7\pi}{6}\right)}$$

or

$$\sin\alpha_{Bn}\sin\left(n\frac{7\pi}{6}\right) = \cos\alpha_{Bn}\cos\left(n\frac{7\pi}{6}\right)$$

or

$$\cos\alpha_{Bn}\cos\left(n\frac{7\pi}{6}\right) - \sin\alpha_{Bn}\sin\left(n\frac{7\pi}{6}\right) = 0$$

or

$$\cos\left(\alpha_{Bn} + n\frac{7\pi}{6}\right) = 0 = \cos\left(n\frac{\pi}{2}\right)$$

for any odd n. Then,

$$\alpha_{Bn} + n\frac{7\pi}{6} = n\frac{\pi}{2}$$

or

$$\alpha_{Bn} = n\left(-\frac{2\pi}{3}\right) \tag{6-16b}$$

The expression for v_{BO} shown in equation 6-10 can be rewritten as

$$v_{BO} = \frac{2v_{DC}}{\pi}\sum_{n=1}^{\infty}\frac{1}{n}\sin n\left(\theta - \frac{2\pi}{3}\right) \tag{6-17}$$

for any odd n. The amplitude of the fundamental component (for $n = 1$) is

$$V_{BO,1} = \frac{2v_{DC}}{\pi} \tag{6-18}$$

From equations 6-16a and 6-18, the amplitudes of harmonic components (n) normalized to the fundamental component are

$$\frac{V_{BO,n}}{V_{BO,1}} = \frac{1}{n} \tag{6-19}$$

for any odd n. The DC component (a_0) in v_{CO} is calculated from equation A-90 (see Appendix A) as

$$a_0 = \frac{1}{2\pi}\int_0^{2\pi} v_{CO}\,d\theta = \frac{1}{2\pi}\frac{v_{DC}}{2}\left[\int_0^{\pi/3} d\theta - \int_{\pi/3}^{4\pi/3} d\theta + \int_{4\pi/3}^{2\pi} d\theta\right]$$

$$= \frac{1}{2\pi}\frac{v_{DC}}{2}\left[\theta\Big|_0^{\pi/3} - \theta\Big|_{\pi/3}^{4\pi/3} + \theta\Big|_{4\pi/3}^{2\pi}\right] = \frac{1}{2\pi}\frac{v_{DC}}{2}\left[\frac{\pi}{3} - \frac{4\pi}{3} + \frac{\pi}{3} + 2\pi - \frac{4\pi}{3}\right] = 0$$

$$a_0 = 0 \tag{6-20}$$

The amplitudes (a_n) for $n = 1, 2, 3, \ldots$ of the cosine components in v_{CO} are calculated from equation A-91 as

$$a_n = \frac{1}{\pi}\int_0^{2\pi} v_{CO}\cos(n\theta)\,d\theta$$

$$= \frac{1}{\pi}\frac{v_{DC}}{2}\left[\int_0^{\pi/3}\cos(n\theta)d\theta - \int_{\pi/3}^{4\pi/3}\cos(n\theta)d\theta + \int_{4\pi/3}^{2\pi}\cos(n\theta)\,d\theta\right]$$

$$= \frac{1}{n\pi}\frac{v_{DC}}{2}\left[\sin(n\theta)\Big|_0^{\pi/3} - \sin(n\theta)\Big|_{\pi/3}^{4\pi/3} + \sin(n\theta)\Big|_{4\pi/3}^{2\pi}\right]$$

$$= \frac{1}{n\pi}\frac{v_{DC}}{2}\left[\sin\left(n\frac{\pi}{3}\right) - \sin\left(n\frac{4\pi}{3}\right) + \sin\left(n\frac{\pi}{3}\right) - \sin\left(n\frac{4\pi}{3}\right)\right]$$

$$= \frac{1}{n\pi}\frac{v_{DC}}{2}\left[2\sin\left(n\frac{\pi}{3}\right) - 2\sin\left(n\frac{4\pi}{3}\right)\right]$$

$$= \frac{v_{DC}}{n\pi}\left[\sin\left(n\frac{\pi}{3}\right) - \sin\left(n\frac{4\pi}{3}\right)\right] = \frac{2v_{DC}}{n\pi}\cos\left(n\frac{5\pi}{6}\right)\sin\left(n\frac{-\pi}{2}\right)$$

$$a_n = \frac{2v_{DC}}{n\pi}\cos\left(n\frac{5\pi}{6}\right)\sin\left(n\frac{-\pi}{2}\right) \tag{6-21}$$

The amplitudes (b_n) for $n = 1, 2, 3, \ldots$ of the sine components in v_{CO} are calculated from equation A-92 as

$$b_n = \frac{1}{\pi}\int_0^{2\pi} v_{CO} \sin(n\theta)\, d\theta$$

$$= \frac{1}{\pi}\frac{v_{DC}}{2}\left[\int_0^{\pi/3} \sin(n\theta)d\theta - \int_{\pi/3}^{4\pi/3} \sin(n\theta)d\theta + \int_{4\pi/3}^{2\pi} \sin(n\theta)\, d\theta\right]$$

$$= \frac{1}{n\pi}\frac{v_{DC}}{2}\left[-\cos(n\theta)\Big|_0^{\pi/3} + \cos(n\theta)\Big|_{\pi/3}^{4\pi/3} - \cos(n\theta)\Big|_{4\pi/3}^{2\pi}\right]$$

$$= \frac{1}{n\pi}\frac{v_{DC}}{2}\left[-\cos\left(n\frac{\pi}{3}\right) + 1 + \cos\left(n\frac{4\pi}{3}\right) - \cos\left(n\frac{\pi}{3}\right) - 1 + \cos\left(n\frac{4\pi}{3}\right)\right]$$

$$= \frac{v_{DC}}{n\pi}\left[\cos\left(n\frac{4\pi}{3}\right) - \cos\left(n\frac{\pi}{3}\right)\right] = \frac{2v_{DC}}{n\pi}\sin\left(n\frac{5\pi}{6}\right)\sin\left(-n\frac{\pi}{2}\right)$$

$$b_n = \frac{2v_{DC}}{n\pi}\sin\left(n\frac{5\pi}{6}\right)\sin\left(n\frac{-\pi}{2}\right) \tag{6-22}$$

The amplitude ($V_{CO,n}$) and the relative phase angle (α_{Cn}) with respect to the PLL angle (θ) are

$$V_{CO,n} = \sqrt{a_n^2 + b_n^2} = \sqrt{\left[\frac{2}{n\pi}v_{DC}\sin\left(-n\frac{\pi}{2}\right)\right]^2\left[\cos^2\left(n\frac{5\pi}{6}\right) + \sin^2\left(n\frac{5\pi}{6}\right)\right]} = \frac{2v_{DC}}{n\pi} \tag{6-23a}$$

and

$$\alpha_{Cn} = \tan^{-1}\frac{a_n}{b_n} = \tan^{-1}\frac{1}{\tan\left(n\frac{5\pi}{6}\right)}$$

for any odd n. But

$$\tan\alpha_{Cn} = \frac{1}{\tan\left(n\frac{5\pi}{6}\right)}$$

or

$$\frac{\sin\alpha_{Cn}}{\cos\alpha_{Cn}} = \frac{\cos\left(n\frac{5\pi}{6}\right)}{\sin\left(n\frac{5\pi}{6}\right)}$$

or

$$\sin\alpha_{Cn}\sin\left(n\frac{5\pi}{6}\right)=\cos\alpha_{Cn}\cos\left(n\frac{5\pi}{6}\right)$$

$$\cos\alpha_{Cn}\cos\left(n\frac{5\pi}{6}\right)-\sin\alpha_{Cn}\sin\left(n\frac{5\pi}{6}\right)=0$$

or

$$\cos\left(\alpha_{Cn}+n\frac{5\pi}{6}\right)=0=\cos\left(-n\frac{\pi}{2}\right)$$

for any odd n. Then,

$$\alpha_{Cn}+n\frac{5\pi}{6}=-n\frac{\pi}{2}$$

or

$$\alpha_{Cn}=n\left(-\frac{4\pi}{3}\right) \qquad (6\text{-}23b)$$

The expression for v_{CO} shown in equation 6-11 can be rewritten as

$$v_{CO}=\frac{2v_{DC}}{\pi}\sum_{n=1}^{\infty}\frac{1}{n}\sin n\left(\theta-\frac{4\pi}{3}\right)=\frac{2v_{DC}}{\pi}\sum_{n=1}^{\infty}\frac{1}{n}\sin n\left(\theta+\frac{2\pi}{3}\right) \qquad (6\text{-}24)$$

for any odd n.

The amplitude of the fundamental component (for $n = 1$) is

$$V_{CO,1}=\frac{2v_{DC}}{\pi} \qquad (6\text{-}25)$$

From equations 6-23a and 6-25, the amplitudes of harmonic components (n) normalized to the fundamental component are

$$\frac{V_{CO,n}}{V_{CO,1}}=\frac{1}{n} \qquad (6\text{-}26)$$

for any odd n.

The three 120° phase-shifted square wave voltages are rewritten as

$$v_{AO}=\frac{2v_{DC}}{\pi}\sum_{n=1}^{\infty}\frac{1}{n}\sin n\theta \qquad (6\text{-}27a)$$

$$v_{BO} = \frac{2v_{DC}}{\pi}\sum_{n=1}^{\infty}\frac{1}{n}\sin n(\theta - 120°) \tag{6-27b}$$

$$v_{CO} = \frac{2v_{DC}}{\pi}\sum_{n=1}^{\infty}\frac{1}{n}\sin n(\theta + 120°) \tag{6-27c}$$

for any odd harmonic n. Note that all three square wave voltages have the same harmonic amplitude of $(2/n\pi)v_{DC}$.

When $n = 1$, the fundamental components in the three square wave voltages are

$$v_{AO,1} = \frac{2}{\pi}v_{DC}\sin\theta \tag{6-28a}$$

$$v_{BO,1} = \frac{2}{\pi}v_{DC}\sin(\theta - 120°) \tag{6-28b}$$

$$v_{CO,1} = \frac{2}{\pi}v_{DC}\sin(\theta + 120°) \tag{6-28c}$$

These three voltages are in A–B–C sequence and are referred to as positive sequence voltages.

When $n = 3$, the third harmonic components in the three square wave voltages are

$$v_{AO,3} = \frac{2v_{DC}}{3\pi}\sin 3\theta \tag{6-29a}$$

$$v_{BO,3} = \frac{2v_{DC}}{3\pi}\sin 3(\theta - 120°) = \frac{2v_{DC}}{3\pi}\sin 3\theta \tag{6-29b}$$

$$v_{CO,3} = \frac{2v_{DC}}{3\pi}\sin 3(\theta + 120°) = \frac{2v_{DC}}{3\pi}\sin 3\theta \tag{6-29c}$$

These three voltages are the same and referred to as zero sequence voltages.

When $n = 5$, the fifth harmonic components in the three square wave voltages are

$$v_{AO,5} = \frac{2v_{DC}}{5\pi}\sin 5\theta \tag{6-30a}$$

$$v_{BO,5} = \frac{2v_{DC}}{5\pi}\sin 5(\theta - 120°) = \frac{2v_{DC}}{5\pi}\sin(5\theta + 120°) \tag{6-30b}$$

$$v_{CO,5} = \frac{2v_{DC}}{5\pi}\sin 5(\theta + 120°) = \frac{2v_{DC}}{5\pi}\sin(5\theta - 120°) \tag{6-30c}$$

These three voltages are in A-C-B sequence and are referred to as negative sequence voltages.

When $n = 7$, the seventh harmonic components in the three square wave voltages are

$$v_{AO,7} = \frac{2v_{DC}}{7\pi}\sin 7\theta \tag{6-31a}$$

$$v_{BO,7} = \frac{2v_{DC}}{7\pi}\sin 7(\theta - 120°) = \frac{2v_{DC}}{7\pi}\sin(7\theta - 120°) \tag{6-31b}$$

$$v_{CO,7} = \frac{2v_{DC}}{7\pi}\sin 7(\theta + 120°) = \frac{2v_{DC}}{7\pi}\sin(7\theta + 120°) \tag{6-31c}$$

These three voltages are in A–B–C sequence and are referred to as positive sequence voltages.

The three square wave voltages, time-shifted by one-third of the time period, are a combination of a fundamental and odd harmonic components. The fundamental component is a set of three-phase positive sequence voltages. Among the harmonic components, the $n = 6k + 1$ components are in positive sequence; the $n = 6k - 1$ components are in negative sequence, where $k = 1, 2, 3$, and so on; and the $n = 3k$ (referred to as triplens) are in zero sequence, where $k = 1, 3, 5$, and so on.

The sum of the three voltages in equation 6-27 results in

$$v_{AO} + v_{BO} + v_{CO} = v_{DC}\sum_{n=1}^{\infty}\frac{2}{n\pi}\left[\sin(n\theta) + \sin n(\theta - 120°) + \sin n(\theta + 120°)\right]$$

for any odd n

$$= v_{DC}\sum_{k=0}^{\infty}\frac{2}{3(2k+1)\pi}\left[3\sin 3(2k+1)\theta\right]$$

Therefore, the zero sequence component voltages are

$$v_{NO} = \frac{v_{AO} + v_{BO} + v_{CO}}{3} = v_{DC}\sum_{k=0}^{\infty}\frac{2}{3(2k+1)\pi}\left[\sin 3(2k+1)\theta\right] \tag{6-32}$$

Since the neutral point is not connected to the midpoint of the DC link capacitor, the zero sequence components of each pole current have no return path and, therefore, the zero sequence components (v_{NO}) of each pole voltage appear between the neutral point and the midpoint of the DC link capacitor. Figure 6-6 shows that the fundamental phasors ($\mathbf{V}_{\mathbf{AO,1}}$, $\mathbf{V}_{\mathbf{BO,1}}$, and $\mathbf{V}_{\mathbf{CO,1}}$) are 120° apart. The three-phase load voltages between phases to neutral are

$$v_{AN} = v_{AO} - v_{NO} \tag{6-33a}$$

$$v_{BN} = v_{BO} - v_{NO} \tag{6-33b}$$

$$v_{CN} = v_{CO} - v_{NO} \tag{6-33c}$$

The load voltages consist of only a fundamental component and harmonic components (n) with a normalized amplitude of ($1/n$), where $n = 6k \pm 1$ for $k = 1, 2, 3$, and so on. Note that the fundamental components in any pole voltage (with respect to the midpoint of the DC link capacitor) and in the corresponding phase voltage are the same ($\mathbf{V}_{\mathbf{AO,1}} = \mathbf{V}_{\mathbf{AN,1}}$, $\mathbf{V}_{\mathbf{BO,1}} = \mathbf{V}_{\mathbf{BN,1}}$, and $\mathbf{V}_{\mathbf{CO,1}} = \mathbf{V}_{\mathbf{CN,1}}$). These fundamental voltages will be referred to as $\mathbf{V}_{\mathbf{A,1}}$, $\mathbf{V}_{\mathbf{B,1}}$, and $\mathbf{V}_{\mathbf{C,1}}$, respectively. Since the phase voltages (v_{AN}, v_{BN}, and v_{CN}) exhibit a six-step waveform, they are referred to as six-pulse voltages. Therefore, the VSC is referred to as a six-pulse VSC.

The three-phase load voltages are

$$v_{AN} = \frac{2v_{DC}}{\pi} \sum_{n=1}^{\infty} \frac{1}{n} \sin(n\theta) \tag{6-34a}$$

$$v_{BN} = \frac{2v_{DC}}{\pi} \sum_{n=1}^{\infty} \frac{1}{n} \sin n(\theta - 120°) \tag{6-34b}$$

$$v_{CN} = \frac{2v_{DC}}{\pi} \sum_{n=1}^{\infty} \frac{1}{n} \sin n(\theta + 120°) \tag{6-34c}$$

which contain fundamental components and harmonic components $n = 6k \pm 1$ for $k = 1, 2, 3$, and so on.

The output voltages exhibit fundamental components and harmonic components (n) with amplitudes of

$$V_{AN,n} = V_{BN,n} = V_{CN,n} = \frac{2v_{DC}}{n\pi} \tag{6-35}$$

where $n = 1$ and $6k \pm 1$ for $k = 1, 2, 3$, and so on.

The amplitudes of harmonic components (n) normalized to the fundamental component are

$$\frac{V_{AN,n}}{V_{AN,1}} = \frac{V_{BN,n}}{V_{AN,1}} = \frac{V_{CN,n}}{V_{AN,1}} = \frac{1}{n} \tag{6-36}$$

where $n = 6k \pm 1$ for $k = 1, 2, 3$, and so on.

The amplitudes of the fundamental components normalized to the DC voltage (gain of the VSC) are

$$\frac{V_{AN,1}}{v_{DC}} = \frac{V_{BN,1}}{v_{DC}} = \frac{V_{CN,1}}{v_{DC}} = \frac{2}{\pi} \tag{6-37}$$

6.2.2.1 Modeling a Six-Pulse VSC with Two-Level Poles. Figure 6-7 shows the circuit modeled in EMTP to generate a three-phase, six-pulse voltage using two-level poles. The points A, B, and C in Figure 6-6 are renamed in Figure 6-7 as A1, B1, and C1, respectively, referring to VSC1. A PLL angle (θ_1) for VSC1 is established. From the VSC1 PLL angle (θ_1), the pole A1 phase angle (`ang1A1`) is established. The gating signal (`pos1A1`) for the positive switch (A1+) is positive during 0° ≤ `pos1A1` < 180°; otherwise, `pos1A1` is zero. The gating signal (`neg1A1`) for the negative switch (A1–) is positive during 180° ≤ `neg1A1` < 360°; otherwise, `neg1A1` is zero. A switch stays "ON" during a positive gating signal; otherwise, it stays "OFF." The pole A1 voltage is the positive half of the DC link voltage during the time when the gating signal (`pos1A1`) is positive. The pole A1 voltage is the negative half of the DC link voltage during the time when the gating signal (`neg1A1`) is positive. Therefore, the pole A1 voltage is established. Note that the pole gating signals (`pos1A1`, `neg1A1`, etc.) can be applied to operate actual switches in the SWITCH section of the program to reflect a six-pulse VSC. The pole B1 phase angle (`ang1B1`) and the pole C1 phase angle (`ang1C1`) lag the VSC1 PLL angle (θ_1) by 120° and 240°, respectively. The square wave pole voltages (`v1A1`, `v1B1`, and `v1C1`) are applied to the primary side of an optional Y–Y transformer, as shown in Figure 6-7. The applied three-phase voltages (`v1A1N`, `v1B1N`, and `v1C1N`) on the primary side are the square wave voltages (`v1A1`, `v1B1`, and `v1C1`) minus the zero sequence voltage (`v1N1`). The per-turn voltage is reflected on the secondary side. Considering the turns ratio of 1:1, three-phase, six-pulse voltages (`e1a1`, `e1b1`, and `e1c1`) are obtained. Note that the signals (v_{AN}, v_{BN}, and v_{CN}) in Figure 6-6 are renamed as `e1a`, `e1b`, and `e1c`, respectively. The

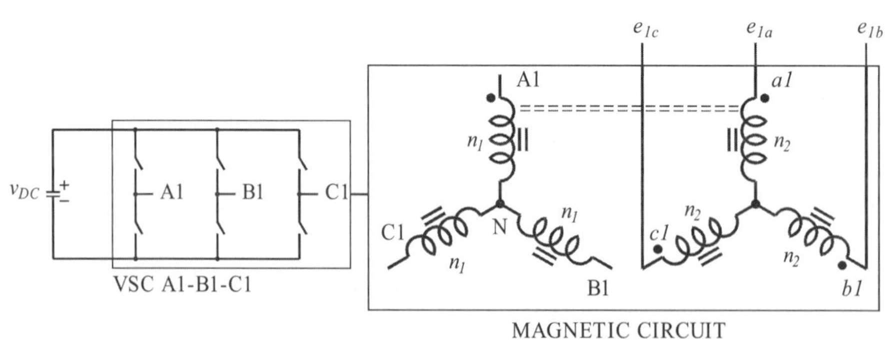

Figure 6-7. Six-pulse VSC model in EMTP.

modeling of a single-phase VSC with ideal switches and a magnetic circuit is presented below.

A main file `61206P21.DAT` (see Appendix C, Section 43) is run to generate a six-pulse voltage using two-level poles. The $INCLUDE$ file `61306P21.GAT` (see Appendix C, Section 44) generates the gating pulses for the switches. The $INCLUDE$ file `61406P21.POL` (see Appendix C, Section 45) generates the ideal pole voltage using ideal switches. The $INCLUDE$ file `61506P21.MAG` (see Appendix C, Section 46) emulates an ideal Y–Y transformer.

In order to generate a higher quality AC voltage, multiple six-pulse VSCs need to be used. The two key points in the design of a multipulse VSC are the following.

1. The displacement angle between two consecutive six-pulse VSCs in an m six-pulse VSC configuration is $2\pi/6m$.
2. The configuration of the magnetic circuit is such that if a VSC pole voltage is time-shifted by an equivalent fundamental angle of $-\theta$, the fundamental component of the pole voltage gets a phase shift by an angle of $+\theta$ and vice versa.

6.2.3 12-Pulse HN-VSC with Two-Level Poles

A 12-pulse HN-VSC requires the generation of two three-phase sets (A–B–C and D–E–F) of square wave voltages that are defined by

$$v_{AO} = \frac{2v_{DC}}{\pi} \sum_{n=1}^{\infty} \frac{1}{n} \sin(n\theta) \tag{6-27a}$$

$$v_{BO} = \frac{2v_{DC}}{\pi} \sum_{n=1}^{\infty} \frac{1}{n} \sin n(\theta - 120^\circ) \tag{6-27b}$$

$$v_{CO} = \frac{2v_{DC}}{\pi} \sum_{n=1}^{\infty} \frac{1}{n} \sin n(\theta + 120^\circ) \tag{6-27c}$$

and

$$v_{DO} = \frac{2v_{DC}}{\pi} \sum_{n=1}^{\infty} \frac{1}{n} \sin n\left(\theta - \frac{\pi}{6}\right) \tag{6-38a}$$

$$v_{EO} = \frac{2v_{DC}}{\pi} \sum_{n=1}^{\infty} \frac{1}{n} \sin n\left(\theta - \frac{\pi}{6} - 120^\circ\right) \tag{6-38b}$$

$$v_{FO} = \frac{2v_{DC}}{\pi} \sum_{n=1}^{\infty} \frac{1}{n} \sin n\left(\theta - \frac{\pi}{6} + 120^\circ\right) \tag{6-38c}$$

for any odd harmonic n. The displacement angle between two consecutive six-pulse VSCs in a multipulse VSC configuration is $2\pi/6m$ (i.e., 30°), where m (i.e., 2) is the total number of six-pulse VSCs that are used.

Figure 6-8, parts a and c, show the two sets of three square wave voltages (v_{AO}, v_{BO}, and v_{CO}; and v_{DO}, v_{EO}, and v_{FO}) and the corresponding fundamental components ($v_{A,1}$, $v_{B,1}$, and $v_{C,1}$; and $v_{D,1}$, $v_{E,1}$, and $v_{F,1}$). The fundamental pole voltage phasors ($\mathbf{V_{A,1}}$, $\mathbf{V_{B,1}}$, and $\mathbf{V_{C,1}}$), shown in Figure 6-8b, are at 0° (reference), –120°, and –240°, respectively. The A–B–C set of square wave voltages are said to be placed at 0°. The fundamental pole voltage phasors ($\mathbf{V_{D,1}}$, $\mathbf{V_{E,1}}$, and $\mathbf{V_{F,1}}$), shown in Figure 6-8d, are at –30°, –150°, and 90°, respectively. The D–E–F set of square wave voltages are said to be placed at –30°.

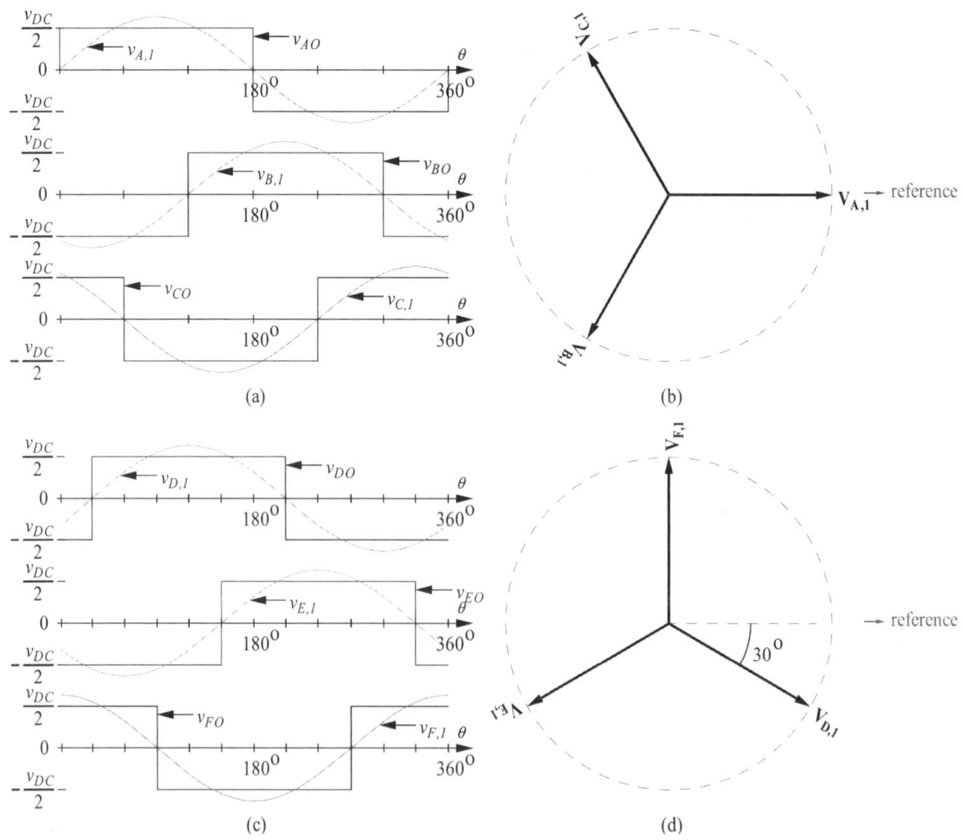

Figure 6-8. Pole voltages and fundamental voltage phasors in a 12-pulse HN-VSC configuration. (a) Three 120° phase-shifted fundamental pole voltages and the corresponding square wave voltages at 0°. (b) Phasor diagram of fundamental pole voltage phasors in A–B–C six-pulse VSC. (c) Three 120° phase-shifted fundamental pole voltages and the corresponding square wave voltages at –30°. (d) Phasor diagram of fundamental pole voltage phasors in D–E–F six-pulse VSC.

Figure 6-9 shows two six-pulse VSCs (A–B–C and D–E–F) that are connected across a shared DC link with a pair of capacitors and the related fundamental voltage phasors. The pole voltages are combined by a magnetic circuit to form a three-phase (*a–b–c*) voltage. The poles A, B, C, D, E, and F are operated in such a way that the pole voltages (v_{AO}, v_{BO}, and v_{CO}) of the A–B–C VSC are time-shifted from one another by one-third of the time period (*T*) of the pole voltage. The pole voltages (v_{DO}, v_{EO}, and v_{FO}) of the D–E–F VSC are time-shifted with respect to the corresponding A, B, and C pole voltages by one-twelfth of the time period, equivalent to a fundamental angle of –30°. Therefore, the fundamental phasors in a group ($\mathbf{V_{A,1}}$, $\mathbf{V_{B,1}}$, and $\mathbf{V_{C,1}}$ of the A–B–C VSC; and $\mathbf{V_{D,1}}$, $\mathbf{V_{E,1}}$, and $\mathbf{V_{F,1}}$ of the D–E–F VSC) are 120° apart and the fundamental voltage phasor set of the D–E–F VSC lags the fundamental voltage phasor set of the A–B–C VSC by 30°. Note that the fundamental components in the square wave voltages ($\mathbf{V_{DO,1}}$, $\mathbf{V_{EO,1}}$, and $\mathbf{V_{FO,1}}$) of the D–E–F VSC with respect to the midpoint of the DC link capacitor are referred to as $\mathbf{V_{D,1}}$, $\mathbf{V_{E,1}}$, and $\mathbf{V_{F,1}}$.

The configuration of the magnetic circuit is such that if a VSC pole voltage is time-shifted by an equivalent fundamental angle of –30°, the fundamental component of the pole voltage gets phase-shifted by an angle of +30°. The simplest way to give the fundamental components in the D–E–F pole voltages a +30° phase shift is to apply these voltages to the terminals of a Δ winding. The square wave voltages from the A–B–C VSC are fed to the primary windings of a Y–Y transformer and those from the D–E–F VSC are fed to the primary windings of a Δ–Y transformer. In order to keep the same volts per turn in both Y and Δ windings, the Δ-connected primary windings have $\sqrt{3}$ times the turns (n_1) of the Y-connected primary windings. Both sets of secondary Y windings are connected in series in the respective phases and the final output voltage is connected to a three-phase load (*a–b–c*).

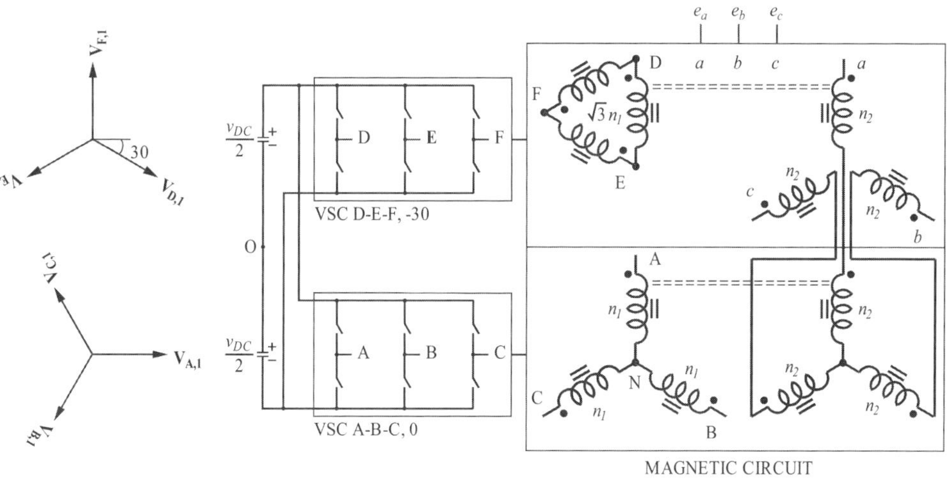

Figure 6-9. 12-pulse HN-VSC configuration (all angles are in degrees).

When $n = 1$, the positive sequence fundamental components in the A–B–C and D–E–F pole voltages are

$$v_{A,1} = \frac{2v_{DC}}{\pi}\sin(n\theta) \tag{6-39a}$$

$$v_{B,1} = \frac{2v_{DC}}{\pi}\sin(\theta - 120°) \tag{6-39b}$$

$$v_{C,1} = \frac{2v_{DC}}{\pi}\sin(\theta + 120°) \tag{6-39c}$$

$$v_{D,1} = \frac{2v_{DC}}{\pi}\sin(\theta - 30°) \tag{6-40a}$$

$$v_{E,1} = \frac{2v_{DC}}{\pi}\sin(\theta - 120° - 30°) = \frac{2v_{DC}}{\pi}\sin(\theta - 150°) \tag{6-40b}$$

$$v_{F,1} = \frac{2v_{DC}}{\pi}\sin(\theta + 120° - 30°) = \frac{2v_{DC}}{\pi}\sin(\theta + 90°) \tag{6-40c}$$

All positive sequence components in the pole-to-pole voltage, $\mathbf{V_{DE}}$ (i.e., $V_D \angle 0° - V_E \angle{-120°} = V_{DE} \angle 30°$), lead the pole voltage ($\mathbf{V_D}$) by 30°. However, all negative sequence components in the pole-to-pole voltage, $\mathbf{V_{DE}}$ (i.e., $V_D \angle 0° - V_E \angle{+120°} = V_{DE} \angle{-30°}$), lags the pole voltage ($\mathbf{V_D}$) by 30°. Therefore, the modified D–E–F pole voltages induced per turn in the Y-connected secondary windings are

$$v^*_{D,1} = \frac{v_{D,1} - v_{E,1}}{\sqrt{3}} = \frac{2v_{DC}}{\pi}\sin(\theta - 30° + 30°) = \frac{2v_{DC}}{\pi}\sin\theta \tag{6-41a}$$

$$v^*_{E,1} = \frac{v_{E,1} - v_{F,1}}{\sqrt{3}} = \frac{2v_{DC}}{\pi}\sin(\theta - 150° + 30°) = \frac{2v_{DC}}{\pi}\sin(\theta - 120°) \tag{6-41b}$$

$$v^*_{F,1} = \frac{v_{F,1} - v_{D,1}}{\sqrt{3}} = \frac{2v_{DC}}{\pi}\sin(\theta + 90° + 30°) = \frac{2v_{DC}}{\pi}\sin(\theta + 120°) \tag{6-41c}$$

Among the harmonic components, each of the $6k + 1$ components (for $k = 1, 2, 3$, etc.) of the pole-to-pole voltage ($v_{DE} = v_D - v_E$) is of a positive sequence and leads the respective component of the pole voltage ($v_{D,6k+1}$) by 30°. Similarly, each of the $6k - 1$ components (for $k = 1, 2, 3$, etc.) of the pole-to-pole voltage ($v_{DE} = v_D - v_E$) is of a negative sequence and lags the respective component of the pole voltage ($v_{D,6k-1}$) by 30°.

When n = 5, the fifth harmonic components in the A–B–C, D–E–F, and modified D–E–F (D*–E*–F*) voltages are

$$v_{A,5} = \frac{2v_{DC}}{5\pi}\sin(5\theta) \tag{6-42a}$$

$$v_{B,5} = \frac{2v_{DC}}{5\pi}\sin 5(\theta - 120°) = \frac{2v_{DC}}{5\pi}\sin(5\theta + 120°) \tag{6-42b}$$

$$v_{C,5} = \frac{2v_{DC}}{5\pi}\sin 5(\theta + 120°) = \frac{2v_{DC}}{5\pi}\sin(5\theta - 120°) \tag{6-42c}$$

$$v_{D,5} = \frac{2v_{DC}}{5\pi}\sin 5(\theta - 30°) = \frac{2v_{DC}}{5\pi}\sin(5\theta - 150°) \tag{6-43a}$$

$$v_{E,5} = \frac{2v_{DC}}{5\pi}\sin 5(\theta - 120° - 30°) = \frac{2v_{DC}}{5\pi}\sin(5\theta - 30°) \tag{6-43b}$$

$$v_{F,5} = \frac{2v_{DC}}{5\pi}\sin 5(\theta + 120° - 30°) = \frac{2v_{DC}}{5\pi}\sin(5\theta + 90°) \tag{6-43c}$$

$$v^*_{D,5} = \frac{v_{D,5} - v_{E,5}}{\sqrt{3}} = \frac{2v_{DC}}{5\pi}\sin(5\theta - 150° - 30°) = \frac{2v_{DC}}{5\pi}\sin(5\theta + 180°) \tag{6-44a}$$

$$v^*_{E,5} = \frac{v_{E,5} - v_{F,5}}{\sqrt{3}} = \frac{2v_{DC}}{5\pi}\sin(5\theta - 30° - 30°) = \frac{2v_{DC}}{5\pi}\sin(5\theta - 60°) \tag{6-44b}$$

$$v^*_{F,5} = \frac{v_{F,5} - v_{D,5}}{\sqrt{3}} = \frac{2v_{DC}}{5\pi}\sin(5\theta + 90° - 30°) = \frac{2v_{DC}}{5\pi}\sin(5\theta + 60°) \tag{6-44c}$$

Note that the negative sequence, fifth harmonic component pole-to-pole voltage ($v_{DE,5} = v_{D,5} - v_{E,5}$) lags the pole voltage ($v_{D,5}$) by 30°.

When n = 7, the seventh harmonic components in the A–B–C, D–E–F, and modified D–E–F (D*–E*–F*) voltages are

$$v_{A,7} = \frac{2v_{DC}}{7\pi}\sin(7\theta) \tag{6-45a}$$

$$v_{B,7} = \frac{2v_{DC}}{7\pi}\sin 7(\theta - 120°) = \frac{2v_{DC}}{7\pi}\sin(7\theta - 120°) \tag{6-45b}$$

$$v_{C,7} = \frac{2v_{DC}}{7\pi}\sin 7(\theta + 120°) = \frac{2v_{DC}}{7\pi}\sin(7\theta + 120°) \qquad (6\text{-}45c)$$

$$v_{D,7} = \frac{2v_{DC}}{7\pi}\sin 7(\theta - 30°) = \frac{2v_{DC}}{7\pi}\sin(7\theta + 150°) \qquad (6\text{-}46a)$$

$$v_{E,7} = \frac{2v_{DC}}{7\pi}\sin 7(\theta - 120° - 30°) = \frac{2v_{DC}}{7\pi}\sin(7\theta + 30°) \qquad (6\text{-}46b)$$

$$v_{F,7} = \frac{2v_{DC}}{7\pi}\sin 7(\theta + 120° - 30°) = \frac{2v_{DC}}{7\pi}\sin(7\theta - 90°) \qquad (6\text{-}46c)$$

$$v^*_{D,7} = \frac{v_{D,7} - v_{E,7}}{\sqrt{3}} = \frac{2v_{DC}}{7\pi}\sin(7\theta + 150° + 30°) = \frac{2v_{DC}}{7\pi}\sin(7\theta + 180°) \qquad (6\text{-}47a)$$

$$v^*_{E,7} = \frac{v_{E,7} - v_{F,7}}{\sqrt{3}} = \frac{2v_{DC}}{7\pi}\sin(7\theta + 30° + 30°) = \frac{2v_{DC}}{7\pi}\sin(7\theta + 60°) \qquad (6\text{-}47b)$$

$$v^*_{F,7} = \frac{v_{F,7} - v_{D,7}}{\sqrt{3}} = \frac{2v_{DC}}{7\pi}\sin(7\theta - 90° + 30°) = \frac{2v_{DC}}{7\pi}\sin(7\theta - 60°) \qquad (6\text{-}47c)$$

Note that the positive sequence, seventh harmonic component pole-to-pole voltage ($v_{DE,7} = v_{D,7} - v_{E,7}$) leads the pole voltage ($v_{D,7}$) by 30°.

When $n = 11$, the eleventh harmonic components in the A–B–C, D–E–F, and modified D–E–F (D*–E*–F*) voltages are

$$v_{A,11} = \frac{2v_{DC}}{11\pi}\sin(11\theta) \qquad (6\text{-}48a)$$

$$v_{B,11} = \frac{2v_{DC}}{11\pi}\sin 11(\theta - 120°) = \frac{2v_{DC}}{11\pi}\sin(11\theta + 120°) \qquad (6\text{-}48b)$$

$$v_{C,11} = \frac{2v_{DC}}{11\pi}\sin 11(\theta + 120°) = \frac{2v_{DC}}{11\pi}\sin(11\theta - 120°) \qquad (6\text{-}48c)$$

$$v_{D,11} = \frac{2v_{DC}}{11\pi}\sin 11(\theta - 30°) = \frac{2v_{DC}}{11\pi}\sin(11\theta + 30°) \qquad (6\text{-}49a)$$

$$v_{E,11} = \frac{2v_{DC}}{11\pi}\sin 11(\theta - 120° - 30°) = \frac{2v_{DC}}{11\pi}\sin(11\theta + 150°) \qquad (6\text{-}49b)$$

$$v_{F,11} = \frac{2v_{DC}}{11\pi}\sin 11(\theta + 120° - 30°) = \frac{2v_{DC}}{11\pi}\sin(11\theta - 90°) \quad (6\text{-}49c)$$

$$v^*_{D,11} = \frac{v_{D,11} - v_{E,11}}{\sqrt{3}} = \frac{2v_{DC}}{11\pi}\sin(11\theta + 30° - 30°) = \frac{2v_{DC}}{11\pi}\sin 11\theta \quad (6\text{-}50a)$$

$$v^*_{E,11} = \frac{v_{E,11} - v_{F,11}}{\sqrt{3}} = \frac{2v_{DC}}{11\pi}\sin(11\theta + 150° - 30°) = \frac{2v_{DC}}{11\pi}\sin(11\theta + 120°) \quad (6\text{-}50b)$$

$$v^*_{F,11} = \frac{v_{F,11} - v_{D,11}}{\sqrt{3}} = \frac{2v_{DC}}{11\pi}\sin(11\theta - 90° - 30°) = \frac{2v_{DC}}{11\pi}\sin(11\theta - 120°) \quad (6\text{-}50c)$$

Note that the negative sequence, eleventh harmonic component pole-to-pole voltage ($v_{DE,11} = v_{D,11} - v_{E,11}$) lags the pole voltage ($v_{D,11}$) by 30°.

When $n = 13$, the thirteenth harmonic components in the A–B–C, D–E–F, and modified D–E–F (D*–E*–F*) voltages are

$$v_{A,13} = \frac{2v_{DC}}{13\pi}\sin(13\theta) \quad (6\text{-}51a)$$

$$v_{B,13} = \frac{2v_{DC}}{13\pi}\sin 13(\theta - 120°) = \frac{2v_{DC}}{13\pi}\sin(13\theta - 120°) \quad (6\text{-}51b)$$

$$v_{C,13} = \frac{2v_{DC}}{13\pi}\sin 13(\theta + 120°) = \frac{2v_{DC}}{13\pi}\sin(13\theta + 120°) \quad (6\text{-}51c)$$

$$v_{D,13} = \frac{2v_{DC}}{13\pi}\sin 13(\theta - 30°) = \frac{2v_{DC}}{13\pi}\sin(13\theta - 30°) \quad (6\text{-}52a)$$

$$v_{E,13} = \frac{2v_{DC}}{13\pi}\sin 13(\theta - 120° - 30°) = \frac{2v_{DC}}{13\pi}\sin(13\theta - 150°) \quad (6\text{-}52b)$$

$$v_{F,13} = \frac{2v_{DC}}{13\pi}\sin 13(\theta + 120° - 30°) = \frac{2v_{DC}}{13\pi}\sin(13\theta + 90°) \quad (6\text{-}52c)$$

$$v^*_{D,13} = \frac{v_{D,13} - v_{E,13}}{\sqrt{3}} = \frac{2v_{DC}}{13\pi}\sin(13\theta - 30° + 30°) = \frac{2v_{DC}}{13\pi}\sin 13\theta \quad (6\text{-}53a)$$

$$v^*_{E,13} = \frac{v_{E,13} - v_{F,13}}{\sqrt{3}} = \frac{2v_{DC}}{13\pi}\sin(13\theta - 150° + 30°) = \frac{2v_{DC}}{13\pi}\sin(13\theta - 120°) \quad (6\text{-}53b)$$

$$v^*_{F,13} = \frac{v_{F,13} - v_{D,13}}{\sqrt{3}} = \frac{2v_{DC}}{13\pi}\sin(13\theta + 90° + 30°) = \frac{2v_{DC}}{13\pi}\sin(13\theta + 120°) \quad (6\text{-}53c)$$

Note that the positive sequence, thirteenth harmonic component pole-to-pole voltage ($v_{DE13} = v_{D13} - v_{E13}$) leads the pole voltage (v_{D13}) by 30°.

When n = 17, the seventeenth harmonic components in the A–B–C, D–E–F, and modified D–E–F (D*–E*–F*) voltages are

$$v_{A,17} = \frac{2v_{DC}}{17\pi}\sin(17\theta) \quad (6\text{-}54a)$$

$$v_{B,17} = \frac{2v_{DC}}{17\pi}\sin 17(\theta - 120°) = \frac{2v_{DC}}{17\pi}\sin(17\theta + 120°) \quad (6\text{-}54b)$$

$$v_{C,17} = \frac{2v_{DC}}{17\pi}\sin 17(\theta + 120°) = \frac{2v_{DC}}{17\pi}\sin(17\theta - 120°) \quad (6\text{-}54c)$$

$$v_{D,17} = \frac{2v_{DC}}{17\pi}\sin 17(\theta - 30°) = \frac{2v_{DC}}{17\pi}\sin(17\theta - 150°) \quad (6\text{-}55a)$$

$$v_{E,17} = \frac{2v_{DC}}{17\pi}\sin 17(\theta - 120° - 30°) = \frac{2v_{DC}}{17\pi}\sin(17\theta - 30°) \quad (6\text{-}55b)$$

$$v_{F,17} = \frac{2v_{DC}}{17\pi}\sin 17(\theta + 120° - 30°) = \frac{2v_{DC}}{17\pi}\sin(17\theta + 90°) \quad (6\text{-}55c)$$

$$v^*_{D,17} = \frac{v_{D,17} - v_{E,17}}{\sqrt{3}} = \frac{2v_{DC}}{17\pi}\sin(17\theta - 150° - 30°) = \frac{2v_{DC}}{17\pi}\sin(17\theta + 180°) \quad (6\text{-}56a)$$

$$v^*_{E,17} = \frac{v_{E,17} - v_{F,17}}{\sqrt{3}} = \frac{2v_{DC}}{17\pi}\sin(17\theta - 30° - 30°) = \frac{2v_{DC}}{17\pi}\sin(17\theta - 60°) \quad (6\text{-}56b)$$

$$v^*_{F,17} = \frac{v_{F,17} - v_{D,17}}{\sqrt{3}} = \frac{2v_{DC}}{17\pi}\sin(17\theta + 90° - 30°) = \frac{2v_{DC}}{17\pi}\sin(17\theta + 60°) \quad (6\text{-}56c)$$

Note that the negative sequence, seventeenth harmonic component pole-to-pole voltage ($v_{DE17} = v_{D17} - v_{E17}$) lags the pole voltage (v_{D17}) by 30°.

When n = 19, the nineteenth harmonic components in the A–B–C, D–E–F, and modified D–E–F (D*–E*–F*) voltages are

$$v_{A,19} = \frac{2v_{DC}}{19\pi}\sin(19\theta) \tag{6-57a}$$

$$v_{B,19} = \frac{2v_{DC}}{19\pi}\sin 19(\theta - 120^\circ) = \frac{2v_{DC}}{19\pi}\sin(19\theta - 120^\circ) \tag{6-57b}$$

$$v_{C,19} = \frac{2v_{DC}}{19\pi}\sin 19(\theta + 120^\circ) = \frac{2v_{DC}}{19\pi}\sin(19\theta + 120^\circ) \tag{6-57c}$$

$$v_{D,19} = \frac{2v_{DC}}{19\pi}\sin 19(\theta - 30^\circ) = \frac{2v_{DC}}{19\pi}\sin(19\theta + 150^\circ) \tag{6-58a}$$

$$v_{E,19} = \frac{2v_{DC}}{19\pi}\sin 19(\theta - 120^\circ - 30^\circ) = \frac{2v_{DC}}{19\pi}\sin(19\theta + 30^\circ) \tag{6-58b}$$

$$v_{F,19} = \frac{2v_{DC}}{19\pi}\sin 19(\theta + 120^\circ - 30^\circ) = \frac{2v_{DC}}{19\pi}\sin(19\theta - 90^\circ) \tag{6-58c}$$

$$v^*_{D,19} = \frac{v_{D,19} - v_{E,19}}{\sqrt{3}} = \frac{2v_{DC}}{19\pi}\sin(19\theta + 150^\circ + 30^\circ) = \frac{2v_{DC}}{19\pi}\sin(19\theta + 180^\circ) \tag{6-59a}$$

$$v^*_{E,19} = \frac{v_{E,19} - v_{F,19}}{\sqrt{3}} = \frac{2v_{DC}}{19\pi}\sin(19\theta + 30^\circ + 30^\circ) = \frac{2v_{DC}}{19\pi}\sin(19\theta + 60^\circ) \tag{6-59b}$$

$$v^*_{F,19} = \frac{v_{F,19} - v_{D,19}}{\sqrt{3}} = \frac{2v_{DC}}{19\pi}\sin(19\theta - 90^\circ + 30^\circ) = \frac{2v_{DC}}{19\pi}\sin(19\theta - 60^\circ) \tag{6-59c}$$

Note that the positive sequence, nineteenth harmonic component pole-to-pole voltage ($v_{DE19} = v_{D19} - v_{E19}$) leads the pole voltage (v_{D19}) by 30°.

When $n = 23$, the twenty-third harmonic components in the A–B–C, D–E–F, and modified D–E–F (D*–E*–F*) voltages are

$$v_{A,23} = \frac{2v_{DC}}{23\pi}\sin(23\theta) \tag{6-60a}$$

$$v_{B,23} = \frac{2v_{DC}}{23\pi}\sin 23(\theta - 120^\circ) = \frac{2v_{DC}}{23\pi}\sin(23\theta + 120^\circ) \tag{6-60b}$$

$$v_{C,23} = \frac{2v_{DC}}{23\pi}\sin 23(\theta + 120^\circ) = \frac{2v_{DC}}{23\pi}\sin(23\theta - 120^\circ) \tag{6-60c}$$

$$v_{D,23} = \frac{2v_{DC}}{23\pi}\sin 23(\theta - 30^\circ) = \frac{2v_{DC}}{23\pi}\sin(23\theta + 30^\circ) \tag{6-61a}$$

$$v_{E,23} = \frac{2v_{DC}}{23\pi}\sin 23(\theta - 120^\circ - 30^\circ) = \frac{2v_{DC}}{23\pi}\sin(23\theta + 150^\circ) \tag{6-61b}$$

$$v_{F,23} = \frac{2v_{DC}}{23\pi}\sin 23(\theta + 120^\circ - 30^\circ) = \frac{2v_{DC}}{23\pi}\sin(23\theta - 90^\circ) \tag{6-61c}$$

$$v^*_{D,23} = \frac{v_{D,23} - v_{E,23}}{\sqrt{3}} = \frac{2v_{DC}}{23\pi}\sin(23\theta + 30^\circ - 30^\circ) = \frac{2v_{DC}}{23\pi}\sin 23\theta \tag{6-62a}$$

$$v^*_{E,23} = \frac{v_{E,23} - v_{F,23}}{\sqrt{3}} = \frac{2v_{DC}}{23\pi}\sin(23\theta + 150^\circ - 30^\circ) = \frac{2v_{DC}}{23\pi}\sin(23\theta + 120^\circ) \tag{6-62b}$$

$$v^*_{F,23} = \frac{v_{F,23} - v_{D,23}}{\sqrt{3}} = \frac{2v_{DC}}{23\pi}\sin(23\theta - 90^\circ - 30^\circ) = \frac{2v_{DC}}{23\pi}\sin(23\theta - 120^\circ) \tag{6-62c}$$

Note that the negative sequence, twenty-third harmonic component pole-to-pole voltage ($v_{DE23} = v_{D23} - v_{E23}$) lags the pole voltage (v_{D23}) by 30°.

When $n = 25$, the twenty-fifth harmonic components in the A–B–C, D–E–F, and modified D–E–F (D*–E*–F*) voltages are

$$v_{A,25} = \frac{2v_{DC}}{25\pi}\sin(25\theta) \tag{6-63a}$$

$$v_{B,25} = \frac{2v_{DC}}{25\pi}\sin 25(\theta - 120^\circ) = \frac{2v_{DC}}{25\pi}\sin(25\theta - 120^\circ) \tag{6-63b}$$

$$v_{C,25} = \frac{2v_{DC}}{25\pi}\sin 25(\theta + 120^\circ) = \frac{2v_{DC}}{25\pi}\sin(25\theta + 120^\circ) \tag{6-63c}$$

$$v_{D,25} = \frac{2v_{DC}}{25\pi}\sin 25(\theta - 30^\circ) = \frac{2v_{DC}}{25\pi}\sin(25\theta - 30^\circ) \tag{6-64a}$$

$$v_{E,25} = \frac{2v_{DC}}{25\pi}\sin 25(\theta - 120^\circ - 30^\circ) = \frac{2v_{DC}}{25\pi}\sin(25\theta - 150^\circ) \tag{6-64b}$$

$$v_{F,25} = \frac{2v_{DC}}{25\pi}\sin 25(\theta + 120^\circ - 30^\circ) = \frac{2v_{DC}}{25\pi}\sin(25\theta + 90^\circ) \tag{6-64c}$$

$$v^*_{D,25} = \frac{v_{D,25} - v_{E,25}}{\sqrt{3}} = \frac{2v_{DC}}{25\pi}\sin(25\theta - 30° + 30°) = \frac{2v_{DC}}{25\pi}\sin 25\theta \quad (6\text{-}65a)$$

$$v^*_{E,25} = \frac{v_{E,25} - v_{F,25}}{\sqrt{3}} = \frac{2v_{DC}}{25\pi}\sin(25\theta - 150° + 30°) = \frac{2v_{DC}}{25\pi}\sin(25\theta - 120°) \quad (6\text{-}65b)$$

$$v^*_{F,25} = \frac{v_{F,25} - v_{D,25}}{\sqrt{3}} = \frac{2v_{DC}}{25\pi}\sin(25\theta + 90° + 30°) = \frac{2v_{DC}}{25\pi}\sin(25\theta + 120°) \quad (6\text{-}65c)$$

Note that the positive sequence, twenty-fifth harmonic component pole-to-pole voltage ($v_{DE25} = v_{D25} - v_{E25}$) leads the pole voltage (v_{D25}) by 30°.

The results are summarized in Table 6-1. It shows the final phase angles (after appropriate phase shifts) of the first 25 harmonic components of the A and D pole voltages. In the secondary Y-windings, the pole voltages from the A–B–C VSC exhibit a six-pulse waveform with harmonic components $n = 6k \pm 1$ for $k = 1, 2, 3$, and so on. Similarly, in the other secondary Y-windings, the pole voltages from the D–E–F VSC exhibit a six-pulse waveform with harmonic components $n = 6k \pm 1$ for $k = 1, 2, 3$, and so on. Each secondary Y-winding induces the same magnitudes of fundamental and harmonic components. The fundamental and harmonic components ($n = 6k \pm 1$, $k = 2, 4, 6$, etc.) in each of the two series-connected secondary windings are in phase while the harmonic components ($n = 6k \pm 1$, $k = 1, 3, 5$, etc.) are in opposite phase, thereby averaging out in the final output voltage. The resulting output voltage exhibits only a fundamental component and harmonic components (n) with normalized amplitudes ($1/n$), where $n = 12k \pm 1$ for $k = 1, 2, 3$, and so on. The number of turns (n_2) of the secondary Y-windings is chosen to match the final output voltage of the VSC to the voltage of the circuit, where it will be connected. The following waveforms are analyzed with turns ratio (n_2/n_1) of 0.5. Therefore, the three-phase output voltages of a 12-pulse VSC are

$$e_a = 2\frac{n_2}{n_1}\frac{2v_{DC}}{\pi}\sum_{n=1}^{\infty}\frac{1}{n}\sin(n\theta) \quad (6\text{-}66a)$$

Table 6-1. Phase angles of phasors out of A and D poles of a 12-pulse VSC

n	Time shift pole A	Phase shift	Final phase angle	Time shift pole D	Phase shift	Final phase angle
5	5 × (0)	0	0	5 × (–30°)	–30°	180°
7	7 × (0)	0	0	7 × (–30°)	+30°	180°
11	11 × (0)	0	0	11 × (–30°)	–30°	0
13	13 × (0)	0	0	13 × (–30°)	+30°	0
17	17 × (0)	0	0	17 × (–30°)	–30°	180°
19	19 × (0)	0	0	19 × (–30°)	+30°	180°
23	23 × (0)	0	0	23 × (–30°)	–30°	0
25	25 × (0)	0	0	25 × (–30°)	+30°	0

$$e_b = 2\frac{n_2}{n_1}\frac{2v_{DC}}{\pi}\sum_{n=1}^{\infty}\frac{1}{n}\sin n(\theta - 120°) \tag{6-66b}$$

$$e_c = 2\frac{n_2}{n_1}\frac{2v_{DC}}{\pi}\sum_{n=1}^{\infty}\frac{1}{n}\sin n(\theta + 120°) \tag{6-66c}$$

which contain fundamental components and harmonic components $n = 12k \pm 1$ for $k =$ 1, 2, 3, and so on. For $n_2/n_1 = 0.5$, the fundamental component of the 12-pulse output voltage and that of each pole output voltage are of the same nominal value. Normally, a VSC of a given power rating is designed in such a way that the current and voltage utilization of the semiconductor switches are maximized. It is most likely that the VSC output voltage is lower than the transmission line voltage. Therefore, the turns ratio of the windings in the magnetic circuit is adjusted to match the line voltage at the point of compensation.

Figure 6-10a shows the three-phase, 12-pulse voltage (e_a, e_b, and e_c) and the corresponding fundamental components ($e_{a,1}$, $e_{b,1}$, and $e_{c,1}$). The fundamental pole voltage phasors ($\mathbf{E_{a,1}}$, $\mathbf{E_{b,1}}$, and $\mathbf{E_{c,1}}$), shown in Figure 6-10b, are at 0° (reference), –120°, –240°, respectively.

The output voltages exhibit fundamental components and harmonic components (n) with amplitudes of

$$E_{a,n} = E_{b,n} = E_{c,n} = 2\frac{n_2}{n_1}\frac{2v_{DC}}{n\pi} \tag{6-67}$$

where $n = 1$ and $12k \pm 1$ for $k = 1, 2, 3$, and so on. The amplitudes of harmonic components (n) normalized to the fundamental component are

$$\frac{E_{a,n}}{E_{a,1}} = \frac{E_{b,n}}{E_{b,1}} = \frac{E_{c,n}}{E_{c,1}} = \frac{1}{n} \tag{6-68}$$

where $n = 12k \pm 1$ for $k = 1, 2, 3$, and so on.

The amplitudes of the fundamental components normalized to the DC voltage (gain of the VSC) are

$$\frac{E_{a,1}}{v_{DC}} = \frac{E_{b,1}}{v_{DC}} = \frac{E_{c,1}}{v_{DC}} = \frac{n_2}{n_1}\frac{4}{\pi} \tag{6-69}$$

6.2.3.1 Graphical Presentation of the Cancellation Technique of the Fifth and the Seventh Harmonic Components

The cancellation technique of the fifth and the seventh harmonic components from the load voltage (e_a) is presented graphically in Figure 6-11. The phasors of the positive sequence fundamental component ($v_{A,1}$), the negative sequence fifth harmonic component ($v_{A,5}$), and the positive sequence seventh harmonic component ($v_{A,7}$) of the phase A voltage (v_A) of the A–B–C VSC are all in phase, as shown in Figure 6-11a. Since the

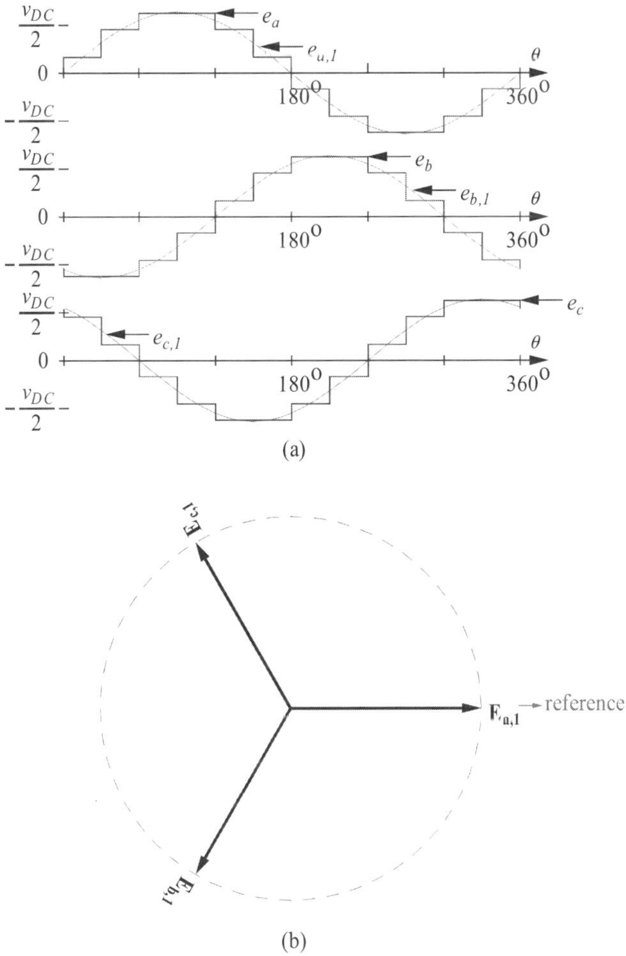

Figure 6-10. (a) Three 120° phase-shifted fundamental output voltages and the corresponding 12-pulse voltages. (b) Phasor diagram of fundamental output voltage phasors in a 12-pulse HN-VSC configuration.

phase D positive sequence fundamental voltage ($v_{D,1}$) lags the phase A positive sequence fundamental voltage ($v_{A,1}$) by 30°, the negative sequence fifth harmonic component ($v_{D,5}$) lags the negative sequence fifth harmonic component ($v_{A,5}$) by 150° and the positive sequence seventh harmonic component ($v_{D,7}$) lags the positive sequence seventh harmonic component ($v_{A,7}$) by 210° or leads the positive sequence seventh harmonic component ($v_{A,7}$) by 150°. The phasors of the positive sequence fundamental component ($v_{D,1}$), the negative sequence fifth harmonic component ($v_{D,5}$), and the positive sequence seventh harmonic component ($v_{D,7}$) of the phase D voltage (v_D) of the D–E–F VSC are shown in Figure 6-11b.

In the DE winding, the pole-to-pole voltage ($v_{DE} = v_D - v_E$) exhibits the positive sequence fundamental and harmonic components ($6k + 1$ for $k = 1, 2, 3$, etc.) and the negative sequence harmonic components ($6k - 1$ for $k = 1, 2, 3$, etc.). The positive sequence components of the pole-to-pole voltage (v_{DE}) lead the respective positive sequence components of the pole voltage (v_D) by 30° and the negative sequence components lag the respective negative sequence components by 30°, as shown in Figure

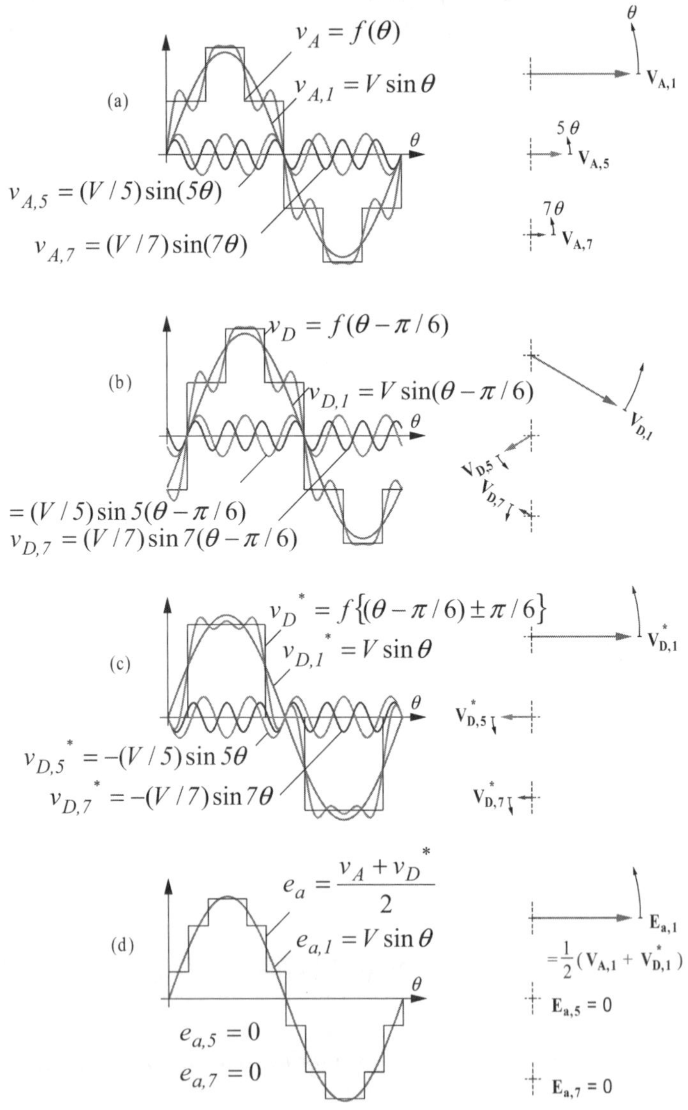

Figure 6-11. Graphical presentation of the cancellation technique of the fifth and the seventh harmonic components (all angles are in radians).

6-11c. These phase-shifted per-turn voltages are referred to as modified voltages. The fundamental component ($v^*_{D,1}$) of the modified phase D voltage (v^*_D) is in phase with the fundamental component ($v_{A,1}$) of the phase A voltage (v_A). But, the fifth harmonic component ($v^*_{D,5}$) and the seventh harmonic component ($v^*_{D,7}$) of the modified phase D voltage (v^*_D) are in opposite phase with the fifth harmonic component ($v_{A,5}$) and the seventh harmonic component ($v_{A,7}$) of the phase A voltage (v_A), respectively. Finally, when the modified phase D voltage (v^*_D) is added to the phase A voltage (v_A) by the series connection of the secondary windings, the fifth and the seventh harmonic components cancel out and do not appear at the output phase-*a* 12-pulse voltage, as shown in Figure 6-11d. In the same way, it can be shown that the seventeenth, nineteenth, twenty-ninth, thirty-first, and so on harmonic components also cancel out in the output voltage. The phase-*a* voltage exhibits only a fundamental component and any odd harmonic component (n) with a normalized amplitude of ($1/n$), where $n = 12k \pm 1$ for $k = 1, 2, 3$, and so on.

6.2.3.2 Modeling a 12-Pulse HN-VSC with Two-Level Poles. The points A, B, C, D, E, and F in Figure 6-9 are referred to as A1, B1, C1, D1, E1, and F1, respectively, referring to VSC1. From the VSC1 PLL angle (θ_1), the pole A1 phase angle (`ang1A1`) is established. The gating signal (`pos1A1`) for the positive switch (A1+) is positive during 0° ≤ `pos1A1` < 180°; otherwise, `pos1A1` is zero. The gating signal (`neg1A1`) for the negative switch (A1–) is positive during 180° ≤ `neg1A1` < 360°; otherwise, `neg1A1` is zero. A switch stays "ON" during a positive gating signal; otherwise, it stays "OFF." The pole A1 voltage is the positive half of the DC link voltage during the time when the gating signal (`pos1A1`) is positive. The pole A1 voltage is the negative half of the DC link voltage during the time when the gating signal (`neg1A1`) is positive. Therefore, the pole A1 voltage is established. The pole B1 phase angle (`ang1B1`) and the pole C1 phase angle (`ang1C1`) lag the VSC1 PLL angle (θ_1) by 120° and 240°, respectively. The pole D1 phase angle (`ang1D1`), pole E1 phase angle (`ang1E1`), and the pole F1 phase angle (`ang1F1`) lag the VSC1 PLL angle (θ_1) by 30°, 150°, and 270°, respectively. Once each pole's instantaneous phase angle in the range of 0° and 360° is defined, the gating signals for its positive and negative switches as well as the pole output voltage are defined. The square wave voltages from the six poles are passed through the magnetic circuits to produce the final three-phase, 12-pulse output voltage. Note that the signals (e_a, e_b, and e_c) in Figure 6-9 are renamed as `e1a`, `e1b`, and `e1c`, respectively. The square wave pole voltages (`v1A1`, `v1B1`, and `v1C1`) are applied to a Y-connected primary side of a Y–Y transformer. The applied three-phase voltages (`v1A1N`, `v1B1N`, and `v1C1N`) on the primary side are the square wave voltages (`v1A1`, `v1B1`, and `v1C1`) minus the zero sequence voltage (`v1N`). The square wave pole voltages (`v1D1`, `v1E1`, and `v1F1`) are applied to a Δ-connected primary side of a Δ–Y transformer. The applied three-phase voltages (`v1D1E1`, `v1E1F1`, and `v1F1D1`) on the primary side are the pole-to-pole voltage. The per-turn voltage is reflected on the secondary side of the transformer. Considering the proper turns ratio (0.5 in the case of Figure 6-9) and the series connection of the secondary voltages, the three-phase, 12-pulse voltages (`e1a1`, `e1b1`, and `e1c1`) are obtained.

The 12-pulse, two-level HN-VSC is modeled in the main file `62412P21.DAT` that has the same setup as the six-pulse, two-level VSC in the file `61206P21.DAT`, except that the three $INCLUDE$ files are replaced with `62512P21.GAT` (see Appendix C, Section 47) for generating gating pulses for the switches, `62612P21.POL` (see Appendix C, Section 48) for generating ideal pole voltages, and `62712P21.MAG` (see Appendix C, Section 49) for implementing the ideal magnetic circuit.

The presence of 12-pulse harmonic components in the VSC output voltage may not be acceptable in many applications. As a result, a VSC with a higher pulse output voltage is considered.

6.2.4 24-Pulse HN-VSC with Two-Level Poles

A 24-pulse HN-VSC requires the generation of four three-phase sets (A1–B1–C1, A2–B2–C2, D1–E1–F1, and D2–E2–F2) of square wave voltages that are defined by

$$v_{A1O} = \frac{2v_{DC}}{\pi}\sum_{n=1}^{\infty}\frac{1}{n}\sin n\left(\theta + \frac{\pi}{8}\right) \tag{6-70a}$$

$$v_{B1O} = \frac{2v_{DC}}{\pi}\sum_{n=1}^{\infty}\frac{1}{n}\sin n\left(\theta + \frac{\pi}{8} - 120^\circ\right) \tag{6-70b}$$

$$v_{C1O} = \frac{2v_{DC}}{\pi}\sum_{n=1}^{\infty}\frac{1}{n}\sin n\left(\theta + \frac{\pi}{8} + 120^\circ\right) \tag{6-70c}$$

$$v_{A2O} = \frac{2v_{DC}}{\pi}\sum_{n=1}^{\infty}\frac{1}{n}\sin n\left(\theta + \frac{\pi}{24}\right) \tag{6-71a}$$

$$v_{B2O} = \frac{2v_{DC}}{\pi}\sum_{n=1}^{\infty}\frac{1}{n}\sin n\left(\theta + \frac{\pi}{24} - 120^\circ\right) \tag{6-71b}$$

$$v_{C2O} = \frac{2v_{DC}}{\pi}\sum_{n=1}^{\infty}\frac{1}{n}\sin n\left(\theta + \frac{\pi}{24} + 120^\circ\right) \tag{6-71c}$$

$$v_{D1O} = \frac{2v_{DC}}{\pi}\sum_{n=1}^{\infty}\frac{1}{n}\sin n\left(\theta - \frac{\pi}{24}\right) \tag{6-72a}$$

$$v_{E1O} = \frac{2v_{DC}}{\pi}\sum_{n=1}^{\infty}\frac{1}{n}\sin n\left(\theta - \frac{\pi}{24} - 120^\circ\right) \tag{6-72b}$$

$$v_{F1O} = \frac{2v_{DC}}{\pi}\sum_{n=1}^{\infty}\frac{1}{n}\sin n\left(\theta-\frac{\pi}{24}+120^\circ\right) \tag{6-72c}$$

$$v_{D2O} = \frac{2v_{DC}}{\pi}\sum_{n=1}^{\infty}\frac{1}{n}\sin n\left(\theta-\frac{\pi}{8}\right) \tag{6-73a}$$

$$v_{E2O} = \frac{2v_{DC}}{\pi}\sum_{n=1}^{\infty}\frac{1}{n}\sin n\left(\theta-\frac{\pi}{8}-120^\circ\right) \tag{6-73b}$$

$$v_{F2O} = \frac{2v_{DC}}{\pi}\sum_{n=1}^{\infty}\frac{1}{n}\sin n\left(\theta-\frac{\pi}{8}+120^\circ\right) \tag{6-73c}$$

for any odd harmonic n. The displacement angle between two consecutive six-pulse VSCs in a multipulse VSC configuration is $2\pi/6m$ (i.e., 15°), where m (i.e., 4) is the total number of six-pulse VSCs that are used. The four phasors in each of the three-phase group (A1–A2–D1–D2, B1–B2–E1–E2, and C1–C2–F1–F2) are centered on 0°, –120°, and –240°, respectively. Figure 6-12a shows the two sets of three square wave voltages (v_{A1O}, v_{B1O}, and v_{C1O}; and v_{A2O}, v_{B2O}, and v_{C2O}) at 22.5° and 7.5°, respectively. The fundamental pole voltage phasors ($\mathbf{V}_{A1,1}$, $\mathbf{V}_{B1,1}$, and $\mathbf{V}_{C1,1}$; and $\mathbf{V}_{A2,1}$, $\mathbf{V}_{B2,1}$, and $\mathbf{V}_{C2,1}$), shown in Figure 6-12b, are at 22.5°, 22.5° – 120° (i.e., –97.5°), and 22.5° – 240° (i.e., –217.5°); and 7.5°, 7.5° – 120° (i.e., –112.5°), and 7.5° – 240° (i.e., –232.5°), respectively. Figure 6-12c shows the two sets of three square wave voltages (v_{D1O}, v_{E1O}, and v_{F1O}; and v_{D2O}, v_{E2O}, and v_{F2O}) at –7.5° and –22.5°, respectively. The fundamental pole voltage phasors ($\mathbf{V}_{D1,1}$, $\mathbf{V}_{E1,1}$, and $\mathbf{V}_{F1,1}$; and $\mathbf{V}_{D2,1}$, $\mathbf{V}_{E2,1}$, and $\mathbf{V}_{F2,1}$), shown in Figure 6-12d, are at –7.5°, –7.5° – 120° (i.e., –127.5°), and –7.5° – 240° (i.e., –247.5°); and –22.5°, –22.5° – 120° (i.e., –142.5°), and –22.5° – 240° (i.e., –262.5°), respectively.

Figure 6-13 shows four six-pulse VSCs (A1–B1–C1, A2–B2–C2, D1–E1–F1, and D2–E2–F2) that are connected across a shared DC link with a pair of capacitors and the related fundamental voltage phasors. The pole voltages are combined by a magnetic circuit to form a three-phase (a–b–c) voltage. The poles (A1, B1, C1, A2, B2, C2, D1, E1, F1, D2, E2, and F2) are operated in such a way that the pole voltages in a six-pulse VSC group (v_{A1O}, v_{B1O}, and v_{C1O} of the A1–B1–C1 VSC; v_{A2O}, v_{B2O}, and v_{C2O} of the A2–B2–C2 VSC; v_{D1O}, v_{E1O}, and v_{F1O} of the D1–E1–F1 VSC; and v_{D2O}, v_{E2O}, and v_{F2O} of the D2–E2–F2 VSC) are time shifted from one another by one-third of the time period (T) of the pole voltage. Furthermore, the four sets of three square wave voltages (v_{A1O}, v_{B1O}, and v_{C1O}; v_{A2O}, v_{B2O}, and v_{C2O}; v_{D1O}, v_{E1O}, and v_{F1O}; and v_{D2O}, v_{E2O}, and v_{F2O}) are placed at 22.5°, 7.5°, –7.5°, and –22.5°, respectively.

The configuration of the magnetic circuit is such that if a VSC pole voltage is time shifted by an equivalent fundamental angle of $-\theta$, the fundamental component of the pole voltage gets a phase shift by an angle of $+\theta$. In the process, all positive sequence

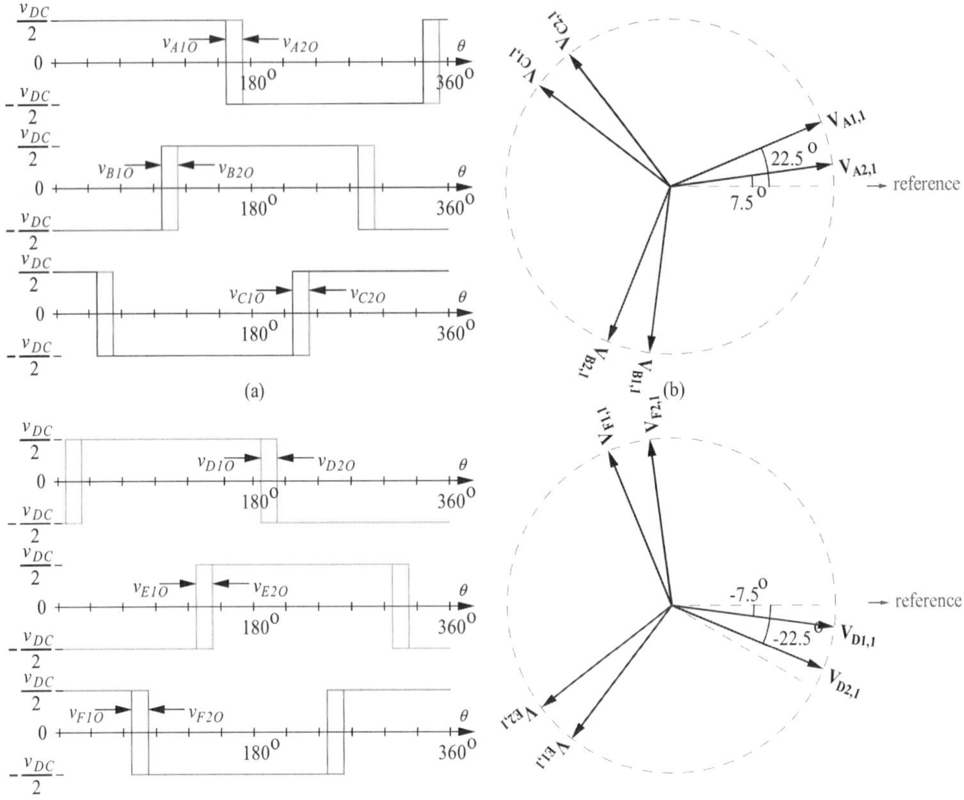

Figure 6-12. Pole voltages and fundamental voltage phasors in a 24-pulse HN-VSC configuration. (a) Two sets of square wave voltages at 22.5° and 7.5°. (b) Phasor diagram of fundamental pole voltage phasors in A1–B1–C1 six-pulse VSC and A2–B2–C2 six-pulse VSC. (c) Two sets of square wave voltages at –7.5° and –22.5°. (d) Phasor diagram of fundamental pole voltage phasors in D1–E1–F1 six-pulse VSC and D2–E2–F2 six-pulse VSC.

$6k + 1$ harmonic components (for $k = 1, 2, 3$, etc.) get phase-shifted by an angle of $+\theta$ as well. However, all negative sequence $6k - 1$ harmonic components (for $k = 1, 2, 3$, etc.) get a phase shift by an angle of $-\theta$.

All the time-shifted pole voltages of the A1–B1–C1 VSC get phase-shifted by an angle of –22.5° for the positive sequence voltages and by an angle of 22.5° for the negative sequence voltages. Similarly, all the time-shifted pole voltages of the A2–B2–C2, D1–E1–F1, and D2–E2–F2 VSCs get phase-shifted by –7.5°, +7.5°, and +22.5°, respectively, for the positive sequence voltages and by +7.5°, –7.5°, and –22.5° for the negative sequence voltages. The simplest way to give the fundamental components in the A1–B1–C1 pole voltages a –22.5° phase shift is to add a B1-phase voltage to A1, C1-phase voltage to B1, and A1-phase voltage to C1, or subtract a C1-

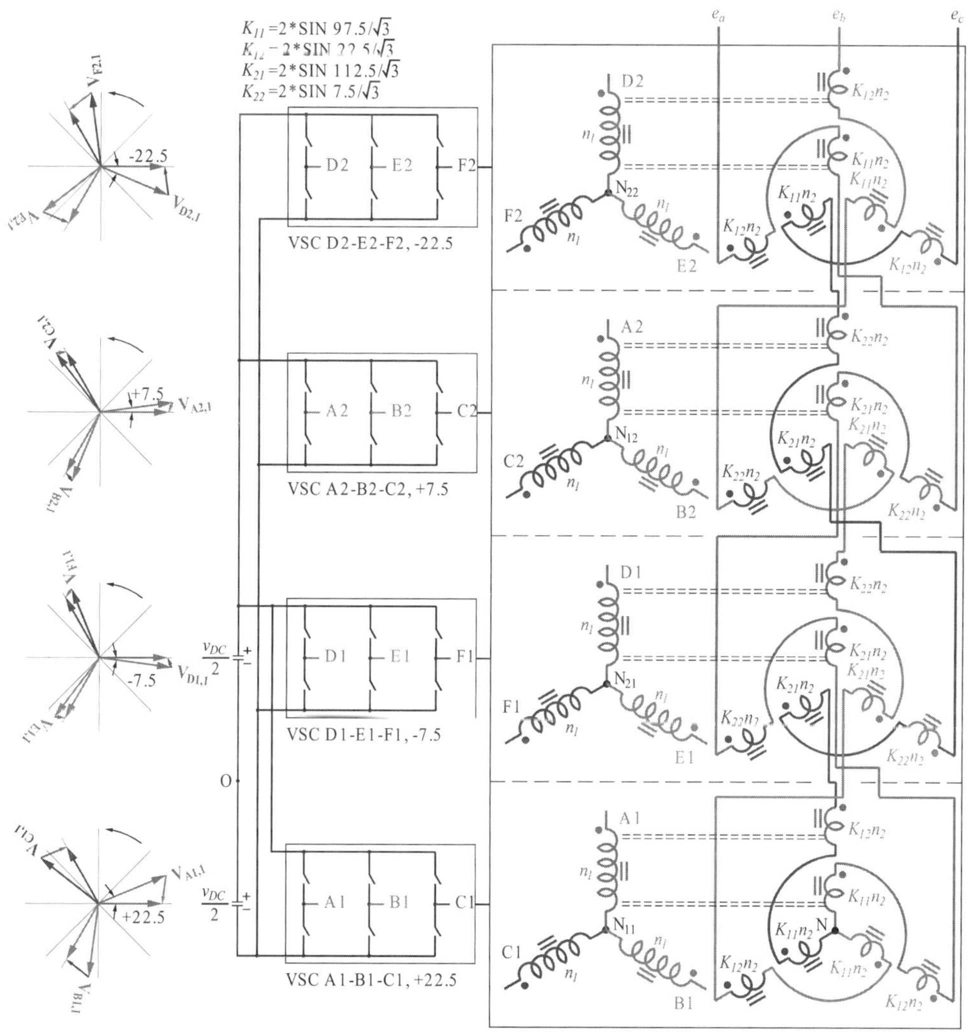

Figure 6-13. 24-pulse HN-VSC configuration (all angles are in degrees).

phase voltage from A1, A1-phase voltage from B1, and B1-phase voltage from C1 using zig-zag windings. Similarly, the fundamental components in the A2–B2–C2 pole voltages get a –7.5° phase shift when a B1-phase voltage, C1-phase voltage, and A1-phase voltage are added with A1-, B1-, and C1-phase voltages, respectively, using zig-zag windings. The fundamental components in the D1–E1–F1 pole voltages get a +7.5° phase shift and those in the D2–E2–F2 pole voltages get a +22.5° phase shift when a F1-phase voltage, D1-phase voltage, and E1-phase voltage are added with D1, E1-, and F1-phase voltages, respectively, using zig-zag windings. The secondary sides of the zig-zag transformers are connected in series in the respective phases and the final output voltage is connected to a three-phase load (*a–b–c*).

The modified pole voltage (v^*_{A1O}) is

$$v^*_{A1O} = \frac{n_2}{n_1}\frac{2v_{DC}}{\pi}\sum_{n=1}^{\infty}\frac{1}{n}\left\{K_{11}\sin n\left(\theta+\frac{\pi}{8}\right)+K_{12}\sin n\left(\theta+\frac{\pi}{8}-120^\circ\right)\right\}$$

$$= \frac{n_2}{n_1}\frac{2v_{DC}}{\pi}\sum_{n=1}^{\infty}\frac{1}{n}\left\{K_{11}\left[\sin(n\theta)\cos\left(n\frac{\pi}{8}\right)+\cos(n\theta)\sin\left(n\frac{\pi}{8}\right)\right]\right.$$

$$\left.+K_{12}\left[\sin(n\theta)\cos\left(-n\frac{13\pi}{24}\right)+\cos(n\theta)\sin\left(-n\frac{13\pi}{24}\right)\right]\right\}$$

$$= \frac{n_2}{n_1}\frac{2v_{DC}}{\pi}\sum_{n=1}^{\infty}\frac{1}{n}\left\{\sin(n\theta)\left[K_{11}\cos\left(n\frac{\pi}{8}\right)+K_{12}\cos\left(-n\frac{13\pi}{24}\right)\right]\right.$$

$$\left.+\cos(n\theta)\left[K_{11}\sin\left(n\frac{\pi}{8}\right)+K_{12}\sin\left(-n\frac{13\pi}{24}\right)\right]\right\}$$

$$= \frac{n_2}{n_1}\frac{2v_{DC}}{\pi}\sum_{n=1}^{\infty}\frac{1}{n}\sin(n\theta+\alpha_{A1n}) \tag{6-74}$$

where

$$\cos\alpha_{A1n} = K_{11}\cos\left(n\frac{\pi}{8}\right)+K_{12}\cos\left(-n\frac{13\pi}{24}\right) \tag{6-75a}$$

and

$$\sin\alpha_{A1n} = K_{11}\sin\left(n\frac{\pi}{8}\right)+K_{12}\sin\left(-n\frac{13\pi}{24}\right) \tag{6-75b}$$

It is desired that $\alpha_{A1n} = 0°$. Therefore,

$$1 = K_{11}\cos\left(n\frac{\pi}{8}\right)+K_{12}\cos\left(-n\frac{13\pi}{24}\right) \tag{6-76a}$$

and

$$0 = K_{11}\sin\left(n\frac{\pi}{8}\right)+K_{12}\sin\left(-n\frac{13\pi}{24}\right) \tag{6-76b}$$

In matrix form,

$$\begin{bmatrix} \cos\left(n\frac{\pi}{8}\right) & \cos\left(-n\frac{13\pi}{24}\right) \\ \sin\left(n\frac{\pi}{8}\right) & \sin\left(-n\frac{13\pi}{24}\right) \end{bmatrix} \begin{bmatrix} K_{11} \\ K_{12} \end{bmatrix} = \begin{bmatrix} 1 \\ 0 \end{bmatrix} \tag{6-77}$$

or

$$\begin{bmatrix} K_{11} \\ K_{12} \end{bmatrix} = \begin{bmatrix} \cos\left(n\frac{\pi}{8}\right) & \cos\left(-n\frac{13\pi}{24}\right) \\ \sin\left(n\frac{\pi}{8}\right) & \sin\left(-n\frac{13\pi}{24}\right) \end{bmatrix}^{-1} \begin{bmatrix} 1 \\ 0 \end{bmatrix}$$

$$= \frac{-1}{\sin\frac{2\pi}{3}} \begin{bmatrix} \sin\left(-n\frac{13\pi}{24}\right) & -\cos\left(-n\frac{13\pi}{24}\right) \\ -\sin\left(n\frac{\pi}{8}\right) & \cos\left(n\frac{\pi}{8}\right) \end{bmatrix} \begin{bmatrix} 1 \\ 0 \end{bmatrix} = \frac{2}{\sqrt{3}} \begin{bmatrix} \sin\left(n\frac{13\pi}{24}\right) \\ \sin\left(n\frac{\pi}{8}\right) \end{bmatrix} \tag{6-78}$$

or

$$K_{11} = \frac{2}{\sqrt{3}} \sin\left(n\frac{13\pi}{24}\right) \tag{6-79a}$$

and

$$K_{12} = \frac{2}{\sqrt{3}} \sin\left(n\frac{\pi}{8}\right) \tag{6-79b}$$

Therefore,

$$v_{A1O}^{*} = \frac{n_2}{n_1} \frac{2v_{DC}}{\pi} \sum_{n=1}^{\infty} \frac{1}{n} \sin(n\theta) \tag{6-80}$$

Similarly, the modified pole voltage (v_{A2O}^{*}) is

$$v_{A2O}^{*} = \frac{n_2}{n_1} \frac{2v_{DC}}{\pi} \sum_{n=1}^{\infty} \frac{1}{n} \left\{ K_{21} \sin n\left(\theta + \frac{\pi}{24}\right) + K_{22} \sin n\left(\theta + \frac{\pi}{24} - 120^{\circ}\right) \right\}$$

$$= \frac{n_2}{n_1} \frac{2v_{DC}}{\pi} \sum_{n=1}^{\infty} \frac{1}{n} \left\{ K_{21} \left[\sin(n\theta)\cos\left(n\frac{\pi}{24}\right) + \cos(n\theta)\sin\left(n\frac{\pi}{24}\right) \right] \right.$$

$$\left. + K_{22} \left[\sin(n\theta)\cos\left(-n\frac{15\pi}{24}\right) + \cos(n\theta)\sin\left(-n\frac{15\pi}{24}\right) \right] \right\}$$

$$= \frac{n_2}{n_1} \frac{2v_{DC}}{\pi} \sum_{n=1}^{\infty} \frac{1}{n} \left\{ \sin(n\theta) \left[K_{21} \cos\left(n\frac{\pi}{24} \right) + K_{22} \cos\left(-n\frac{15\pi}{24} \right) \right] \right.$$

$$\left. + \cos(n\theta) \left[K_{21} \sin\left(n\frac{\pi}{24} \right) + K_{22} \sin\left(-n\frac{15\pi}{24} \right) \right] \right\}$$

$$= \frac{n_2}{n_1} \frac{2v_{DC}}{\pi} \sum_{n=1}^{\infty} \frac{1}{n} \sin(n\theta + \alpha_{A2n}) \tag{6-81}$$

where

$$\cos \alpha_{A2n} = K_{21} \cos\left(n\frac{\pi}{24} \right) + K_{22} \cos\left(n\frac{\pi}{24} - n\frac{2\pi}{3} \right) \tag{6-82a}$$

$$\sin \alpha_{A2n} = K_{21} \sin\left(n\frac{\pi}{24} \right) + K_{22} \sin\left(n\frac{\pi}{24} - n\frac{2\pi}{3} \right) \tag{6-82b}$$

It is desired that $\alpha_{A2n} = 0°$. Therefore,

$$1 = K_{21} \cos\left(n\frac{\pi}{24} \right) + K_{22} \cos\left(n\frac{\pi}{24} - n\frac{2\pi}{3} \right) \tag{6-83a}$$

and

$$0 = K_{21} \sin n\frac{\pi}{24} + K_{22} \sin\left(n\frac{\pi}{24} - n\frac{2\pi}{3} \right) \tag{6-83b}$$

In matrix form,

$$\begin{bmatrix} \cos\left(n\frac{\pi}{24} \right) & \cos\left(-n\frac{15\pi}{24} \right) \\ \sin\left(n\frac{\pi}{24} \right) & \sin\left(-n\frac{15\pi}{24} \right) \end{bmatrix} \begin{bmatrix} K_{21} \\ K_{22} \end{bmatrix} = \begin{bmatrix} 1 \\ 0 \end{bmatrix} \tag{6-84}$$

or

$$\begin{bmatrix} K_{21} \\ K_{22} \end{bmatrix} = \begin{bmatrix} \cos\left(n\frac{\pi}{24} \right) & \cos\left(-n\frac{15\pi}{24} \right) \\ \sin\left(n\frac{\pi}{24} \right) & \sin\left(-n\frac{15\pi}{24} \right) \end{bmatrix}^{-1} \begin{bmatrix} 1 \\ 0 \end{bmatrix}$$

$$= \frac{-1}{\sin\frac{2\pi}{3}} \begin{bmatrix} \sin\left(-n\frac{15\pi}{24} \right) & -\cos\left(-n\frac{15\pi}{24} \right) \\ -\sin\left(n\frac{\pi}{24} \right) & \cos\left(n\frac{\pi}{24} \right) \end{bmatrix} \begin{bmatrix} 1 \\ 0 \end{bmatrix} = \frac{2}{\sqrt{3}} \begin{bmatrix} \sin\left(n\frac{15\pi}{24} \right) \\ \sin\left(n\frac{\pi}{24} \right) \end{bmatrix} \tag{6-85}$$

or

$$K_{21} = \frac{2}{\sqrt{3}} \sin\left(n \frac{15\pi}{24} \right) \tag{6-86a}$$

and

$$K_{22} = \frac{2}{\sqrt{3}} \sin\left(n \frac{\pi}{24} \right) \tag{6-86b}$$

Therefore,

$$\overset{*}{v}_{A2O} = \frac{n_2}{n_1} \frac{2v_{DC}}{\pi} \sum_{n=1}^{\infty} \frac{1}{n} \sin(n\theta) \tag{6-87}$$

$$\overset{*}{v}_{DIO} = \frac{n_2}{n_1} \frac{2v_{DC}}{\pi} \sum_{n=1}^{\infty} \frac{1}{n} \left\{ K_{21} \sin n\left(\theta + \frac{\pi}{24} \right) + K_{22} \sin n\left(\theta + \frac{\pi}{24} + 120^\circ \right) \right\}$$

$$= \frac{n_2}{n_1} \frac{2 v_{DC}}{\pi} \sum_{n=1}^{\infty} \frac{1}{n} \sin(n\theta) \tag{6-88}$$

$$\overset{*}{v}_{D2O} = \frac{n_2}{n_1} \frac{2v_{DC}}{\pi} \sum_{n=1}^{\infty} \frac{1}{n} \left\{ K_{11} \sin n\left(\theta + \frac{\pi}{8} \right) + K_{12} \sin n\left(0 + \frac{\pi}{8} + 120^\circ \right) \right\}$$

$$= \frac{n_2}{n_1} \frac{2 v_{DC}}{\pi} \sum_{n=1}^{\infty} \frac{1}{n} \sin(n\theta) \tag{6-89}$$

When the time-shifted pole voltage sets from two six-pulse VSCs (A1–B1–C1 and D1–E1–F1) with a phase displacement of 30° between them are fed to two Y-windings with n_1 turns and the output voltages (after properly phase-shifted with the use of zig-zag windings) are combined by connecting the corresponding phases in series, a 12-pulse HN waveform with harmonic components $n = 12k \pm 1$ for $k = 1, 2, 3$, and so on is obtained. In the process, the six-pulse harmonic components ($n = 6k \pm 1$ for $k = 1, 3, 5$, etc.) from the A1–B1–C1 and D1–E1–F1 VSCs average out and do not appear in the 12-pulse *a1–b1–c1* voltage. The three-phase, 12-pulse voltages are

$$e_{a1} = \frac{n_2}{n_1} \left(\overset{*}{v}_{A1O} + \overset{*}{v}_{D1O} \right) = 2 \frac{n_2}{n_1} \frac{2v_{DC}}{\pi} \sum_{n=1}^{\infty} \frac{1}{n} \sin(n\theta) \tag{6-90a}$$

$$e_{b1} = \frac{n_2}{n_1} \left(\overset{*}{v}_{B1O} + \overset{*}{v}_{E1O} \right) = 2 \frac{n_2}{n_1} \frac{2v_{DC}}{\pi} \sum_{n=1}^{\infty} \frac{1}{n} \sin n(\theta - 120^\circ) \tag{6-90b}$$

$$e_{c1} = \frac{n_2}{n_1} \left(\overset{*}{v}_{C1O} + \overset{*}{v}_{F1O} \right) = 2 \frac{n_2}{n_1} \frac{2v_{DC}}{\pi} \sum_{n=1}^{\infty} \frac{1}{n} \sin n(\theta + 120^\circ) \tag{6-90c}$$

which contain fundamental components and harmonic components $n = 12k \pm 1$ for $k =$ 1, 2, 3, and so on.

Similarly, the time-shifted pole voltage sets from the remaining two six-pulse VSCs (A2–B2–C2 and D2–E2–F2) with a phase displacement of 30° between them are fed to two Y-windings and the output voltages (after properly phase-shifted with the use of zig-zag windings) are combined by connecting the corresponding phases in series, a second 12-pulse HN waveform with harmonic components $n = 12k \pm 1$ for $k = 1, 2, 3,$ and so on is obtained. In the process, the six-pulse harmonic components ($n = 6k \pm 1$ for $k = 1, 3, 5$, etc.) from the A2–B2–C2 and D2–E2–F2 VSCs average out and do not appear in the 12-pulse *a2–b2–c2* voltage. The three-phase, 12-pulse voltages are

$$e_{a2} = \frac{n_2}{n_1}\left(\overset{*}{v}_{A2O} + \overset{*}{v}_{D2O} \right) = 2\frac{n_2}{n_1}\frac{2v_{DC}}{\pi}\sum_{n=1}^{\infty}\frac{1}{n}\sin(n\theta) \tag{6-91a}$$

$$e_{b2} = \frac{n_2}{n_1}\left(\overset{*}{v}_{B2O} + \overset{*}{v}_{E2O} \right) = 2\frac{n_2}{n_1}\frac{2v_{DC}}{\pi}\sum_{n=1}^{\infty}\frac{1}{n}\sin n(\theta - 120°) \tag{6-91b}$$

$$e_{c2} = \frac{n_2}{n_1}\left(\overset{*}{v}_{C2O} + \overset{*}{v}_{F2O} \right) = 2\frac{n_2}{n_1}\frac{2v_{DC}}{\pi}\sum_{n=1}^{\infty}\frac{1}{n}\sin n(\theta + 120°) \tag{6-91c}$$

which contain fundamental components and harmonic components $n = 12k \pm 1$ for $k =$ 1, 2, 3, and so on. Note that both 12-pulse HN waveforms are placed at 0°.

Each 12-pulse waveform contains the same magnitudes of fundamental and harmonic components. However, the harmonic components ($n = 12k \pm 1$ for $k = 1, 3, 5$, etc.) of the *a1–b1–c1* 12-pulse HN waveform are in opposite phase, whereas the harmonic components ($n = 12k \pm 1$ for $k = 2, 4, 6$, etc.) are in phase with the corresponding harmonic components of the *a2–b2–c2* 12-pulse HN waveform. Therefore, when all the output voltages from each six-pulse VSC are combined by connecting the corresponding phases in series, a 24-pulse HN waveform is obtained, as shown in Figure 6-13. The resulting output voltage exhibits a fundamental component of amplitude $(2/\pi)v_{DC}$, where v_{DC} is the DC link capacitor voltage, and harmonic components (n) with normalized amplitudes ($1/n$), where $n = 24k \pm 1$ for $k = 1, 2, 3$, and so on.

Therefore, the three-phase output voltages of a 24-pulse VSC are

$$e_a = e_{a1} + e_{a2} = 4\frac{n_2}{n_1}\frac{2v_{DC}}{\pi}\sum_{n=1}^{\infty}\frac{1}{n}\sin(n\theta) \tag{6-92a}$$

$$e_b = e_{b1} + e_{b2} = 4\frac{n_2}{n_1}\frac{2v_{DC}}{\pi}\sum_{n=1}^{\infty}\frac{1}{n}\sin n(\theta - 120°) \tag{6-92b}$$

$$e_c = e_{c1} + e_{c2} = 4\frac{n_2}{n_1}\frac{2v_{DC}}{\pi}\sum_{n=1}^{\infty}\frac{1}{n}\sin n(\theta + 120°) \tag{6-92c}$$

which contain fundamental components and harmonic components $n = 24k \pm 1$ for $k =$ 1, 2, 3, and so on.

The output voltages exhibit fundamental components and harmonic components (n) with amplitudes of

$$E_{a,n} = E_{b,n} = E_{c,n} = 4\frac{n_2}{n_1}\frac{2v_{DC}}{n\pi} \tag{6-93}$$

where $n = 1$ and $24k \pm 1$ for $k = 1, 2, 3$, and so on.

The amplitudes of harmonic components (n) normalized to the fundamental component are

$$\frac{E_{a,n}}{E_{a,1}} = \frac{E_{b,n}}{E_{b,1}} = \frac{E_{c,n}}{E_{c,1}} = \frac{1}{n} \tag{6-94}$$

where $n = 24k \pm 1$ for $k = 1, 2, 3$, and so on.

The amplitudes of the fundamental components normalized to the DC voltage (gain of the VSC) are

$$\frac{E_{a,1}}{v_{DC}} = \frac{E_{b,1}}{v_{DC}} = \frac{E_{c,1}}{v_{DC}} = \frac{n_2}{n_1}\frac{8}{\pi} \tag{6-95}$$

The results are summarized in Tables 6-2a and 6-2b, which show the final phase angles (after appropriate phase shifts) of the first 25 harmonic components of the A1, D1, A2, and D2 pole voltages. In the secondary zig-zag windings, the pole voltages from the A1–B1–C1 VSC and D1–E1–F1 VSC exhibit two six-pulse waveforms with harmonic components $n = 6k \pm 1$ for $k = 1, 2, 3$, and so on. Similarly, the pole voltages from the A2–B2–C2 VSC and D2–E2–F2 VSC exhibit two six-pulse waveforms with harmonic components $n = 6k \pm 1$ for $k = 1, 2, 3$, and so on. Each six-pulse waveform contains the same magnitudes of the fundamental and harmonic components. In each 12-pulse HN waveforms, the fifth, seventh, seventeenth, nineteenth and so on components from two six-pulse waveforms are in opposite phase. Therefore, when the secondary zig-zag wind-

Table 6-2a. Phase angles of *a1–b1–c1* 12-pulse VSC phasors

n	Time shift pole A1	Phase shift	Final phase angle	Time shift pole D1	Phase shift	Final phase angle
5	5(+22.5°)	+22.5°	+135°	5(–7.5°)	–7.5°	–45°
7	7(+22.5°)	–22.5°	+135°	7(–7.5°)	+7.5°	–45°
11	11(+22.5°)	+22.5°	–90°	11(–7.5°)	–7.5°	–90°
13	13(+22.5°)	–22.5°	–90°	13(–7.5°)	+7.5°	–90°
17	17(+22.5°)	+22.5°	+45°	17(–7.5°)	–7.5°	–135°
19	19(+22.5°)	–22.5°	+45°	19(–7.5°)	+7.5°	–135°
23	23(+22.5°)	+22.5°	180°	23(–7.5°)	–7.5°	180°
25	25(+22.5°)	–22.5°	180°	25(–7.5°)	+7.5°	180°

Table 6-2a. Phase angles of *a2–b2–c2* 12-pulse VSC phasors

n	Time shift pole A2	Phase shift	Final phase angle	Time shift pole D2	Phase shift	Final phase angle
5	5(+7.5°)	+7.5°	+45°	5(–22.5°)	–22.5°	–135°
7	7(+7.5°)	–7.5°	+45°	7(–22.5°)	+22.5°	–135°
11	11(+7.5°)	+7.5°	+90°	11(–22.5°)	–22.5°	+90°
13	13(+7.5°)	–7.5°	+90°	13(–22.5°)	+22.5°	+90°
17	17(+7.5°)	+7.5°	+135°	17(–22.5°)	–22.5°	–45°
19	19(+7.5°)	–7.5°	+135°	19(–22.5°)	+22.5°	–45°
23	23(+7.5°)	+7.5°	180°	23(–22.5°)	–22.5°	180°
25	25(+7.5°)	–7.5°	180°	25(–22.5°)	+22.5°	180°

ings on the load side are connected in series in their respective phases, the harmonic components ($n = 6k \pm 1$, $k = 1, 3, 5$, etc.) cancel out in the final phase-a voltage that exhibits a 12-pulse HN waveform. Moreover, the eleventh, thirteenth, thirty-fifth, thirty-seventh, and so on components from two 12-pulse HN waveforms are in opposite phase. Therefore, when all the secondary zig-zag windings are connected in series in the respective phases, the harmonic components ($n = 12k \pm 1$, $k = 1, 3, 5$, etc.) cancel out in the final phase-a voltage that exhibits a 24-pulse HN waveform. The resulting output voltage exhibits only a fundamental component and harmonic components (n) with normalized amplitudes ($1/n$), where $n = 24k \pm 1$ for $k = 1, 2, 3$, and so on.

The above waveforms are analyzed with turns ratio (n_2/n_1) of 0.25. The values of K_{11}, K_{12}, K_{21}, and K_{22} (Figure 6-13) are calculated using equations 6-79a and 6-79b and 6-86a and 6-86b with a substitution of $n = 1$.

In each 12-pulse VSC, the averaging process keeps the fundamental voltage at the same nominal value as that of a pole voltage and also cancels the harmonic components ($n = 6k \pm 1$, $k = 1, 3, 5$, etc.) to produce a 12-pulse HN waveform. In the 24-pulse VSC, another averaging process keeps the fundamental voltage at the same nominal value as that of a pole voltage and also cancels the harmonic components ($n = 12k \pm 1$, $k = 1, 3, 5$, etc.) to produce a 24-pulse HN waveform. These two-time averaging accounts for a factor of 0.25, which is n_2/n_1.

Figure 6-14, parts a and c, show the two sets of three-phase, 12-pulse voltages (e_{a1}, e_{b1}, and e_{c1}; and e_{a2}, e_{b2}, and e_{c2}) and the corresponding fundamental components ($e_{a1,1}$, $e_{b1,1}$, and $e_{c1,1}$; and $e_{a2,1}$, $e_{b2,1}$, and $e_{c2,1}$). The fundamental voltage phasors ($\mathbf{E_{a1,1}}$, $\mathbf{E_{b1,1}}$, and $\mathbf{E_{c1,1}}$; and $\mathbf{E_{a2,1}}$, $\mathbf{E_{b2,1}}$, and $\mathbf{E_{c2,1}}$), shown in Figure 6-14, parts b and d, are at 0° (reference), –120°, and –240°, respectively.

Figure 6-15a shows the three-phase, 24-pulse voltages (e_a, e_b, and e_c) and the corresponding fundamental components ($e_{a,1}$, $e_{b,1}$, and $e_{c,1}$). The fundamental pole voltage phasors ($\mathbf{E_{a,1}}$, $\mathbf{E_{b,1}}$, and $\mathbf{E_{c,1}}$), shown in Figure 6-15b, are at 0° (reference), –120°, and –240°, respectively.

6.2.4.1 Modeling a 24-Pulse HN-VSC with Two-Level Poles. The 24-pulse, two-level HN-VSC is modeled in the main file `63624P21.DAT` that has the

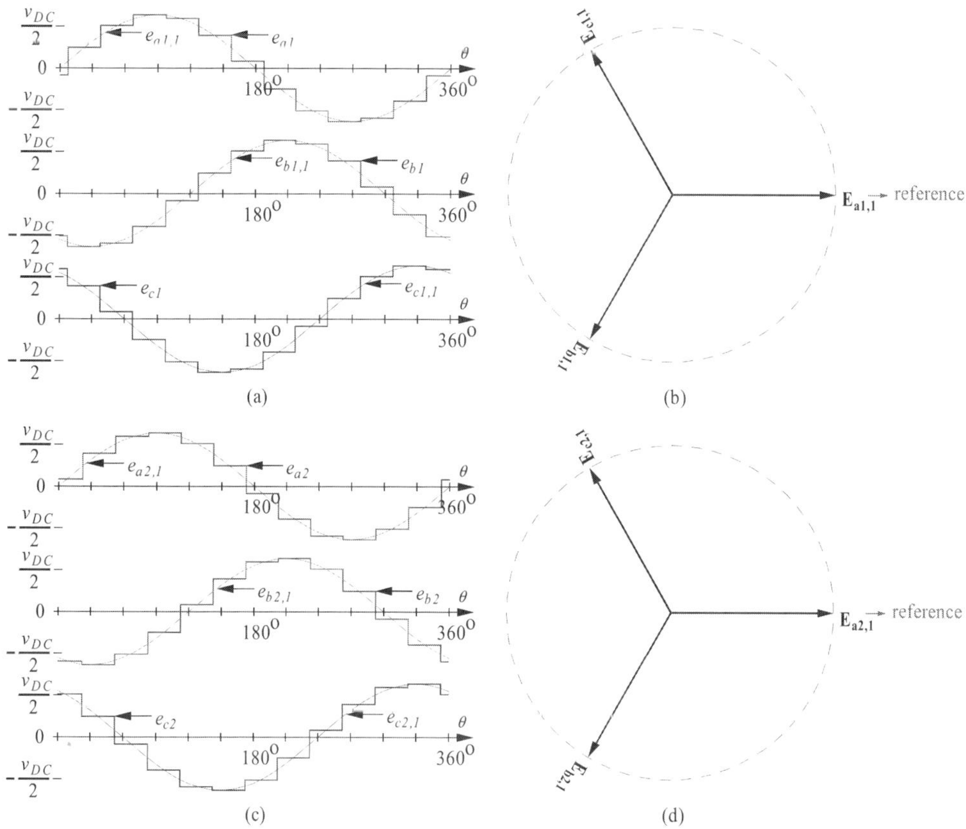

Figure 6-14. Two 12-pulse voltages and fundamental voltage phasors in a 24-pulse HN-VSC configuration. (a) Three 120° phase-shifted fundamental voltages and the corresponding 12-pulse voltages at 0°. (b) Phasor diagram of fundamental voltage phasors in *a1–b1–c1* 12-pulse HN-VSC. (c) Three 120° phase-shifted fundamental voltages and the corresponding 12-pulse voltages at 0°. (d) Phasor diagram of fundamental voltage phasors in *a2–b2–c2* 12-pulse HN-VSC.

same setup as the six-pulse, two-level VSC in the file `61206P21.DAT`, except that the three $INCLUDE$ files are replaced with `63724P21.GAT` (see Appendix C, Section 50) for generating gating pulses for the switches, `63824P21.POL` (see Appendix C, Section 51) for generating ideal pole voltages, and `63924P21.MAG` (see Appendix C, Section 52) for implementing the ideal magnetic circuit.

Figure 6-16 shows the output voltages from the simulations of a 12- and a 24-pulse HN-VSC. A 24-pulse or higher order VSC is more than adequate in most applications to meet the power quality standard.

Figure 6-17 shows the normalized harmonic components of the output voltages from a 12-pulse and a 24-pulse HN-VSC as a function of harmonic order (n).

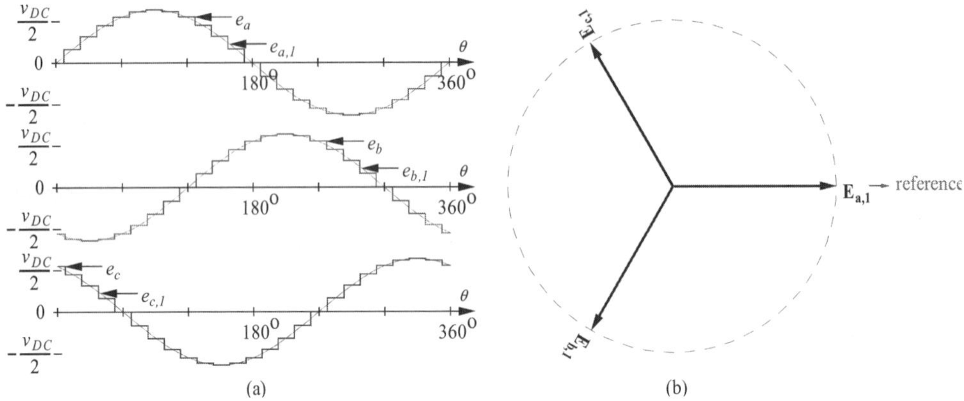

Figure 6-15. (a) Three 120° phase-shifted fundamental output voltages and the corresponding 24-pulse voltages. (b) Phasor diagram of fundamental output voltage phasors in a 24-pulse HN-VSC configuration.

6.2.5 24-Pulse QHN-VSC with Two-Level Poles

The square wave voltages generated by the six-pulse VSCs can be combined using HN techniques to generate a high-quality sinusoidal voltage with acceptable harmonic content. A true p-pulse HN-VSC generates the first dominant harmonic in the spectrum as $p - 1$. The magnetic circuit for a VSC with 24 pulse and higher is quite cumbersome.

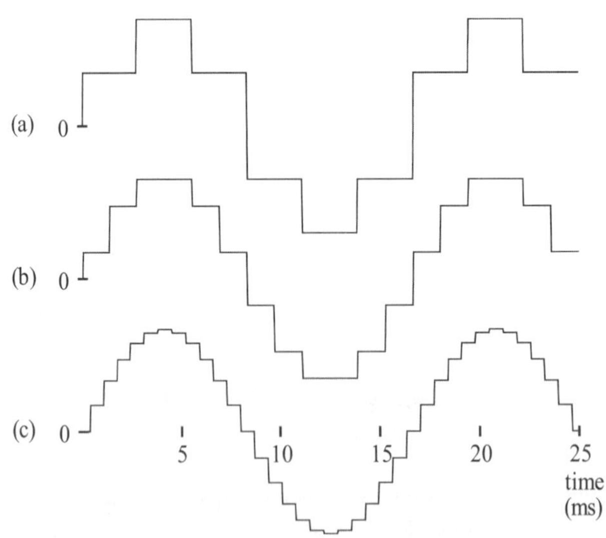

Figure 6-16. Output voltages from the simulation of (a) a 6-pulse HN-VSC, (b) a 12-pulse HN-VSC, and (c) a 24-pulse HN-VSC.

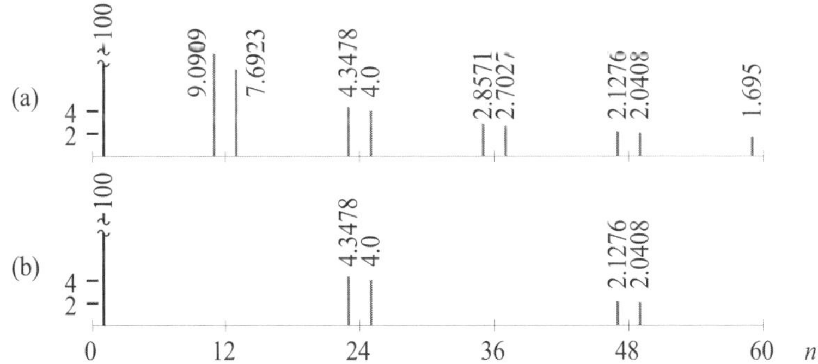

Figure 6-17. Normalized harmonic components of output voltages from (a) a 12-pulse HN-VSC and (b) a 24-pulse HN-VSC as a function of harmonic order (n).

Its nonstandard features makes it unattractive because of its higher cost. Therefore, an alternate solution is to use a magnetic circuit for a quasiharmonic neutralized (QHN) VSC.

A 24-pulse QHN-VSC requires the generation of four three-phase sets (A1–B1–C1, A2–B2–C2, D1–E1–F1, and D2–E2–F2) of square wave voltages that are defined by

$$v_{A1O} = \frac{2v_{DC}}{\pi} \sum_{n=1}^{\infty} \frac{1}{n} \sin n\left(\theta + \frac{\pi}{24}\right) \tag{6-96a}$$

$$v_{B1O} = \frac{2v_{DC}}{\pi} \sum_{n=1}^{\infty} \frac{1}{n} \sin n\left(\theta + \frac{\pi}{24} - 120^\circ\right) \tag{6-96b}$$

$$v_{C1O} = \frac{2v_{DC}}{\pi} \sum_{n=1}^{\infty} \frac{1}{n} \sin n\left(\theta + \frac{\pi}{24} + 120^\circ\right) \tag{6-96c}$$

$$v_{A2O} = \frac{2v_{DC}}{\pi} \sum_{n=1}^{\infty} \frac{1}{n} \sin n\left(\theta - \frac{\pi}{24}\right) \tag{6-97a}$$

$$v_{B2O} = \frac{2v_{DC}}{\pi} \sum_{n=1}^{\infty} \frac{1}{n} \sin n\left(\theta - \frac{\pi}{24} - 120^\circ\right) \tag{6-97b}$$

$$v_{C2O} = \frac{2v_{DC}}{\pi} \sum_{n=1}^{\infty} \frac{1}{n} \sin n\left(\theta - \frac{\pi}{24} + 120^\circ\right) \tag{6-97c}$$

$$v_{D1O} = \frac{2v_{DC}}{\pi} \sum_{n=1}^{\infty} \frac{1}{n} \sin n\left(\theta - \frac{3\pi}{24}\right) \tag{6-98a}$$

$$v_{E1O}=\frac{2v_{DC}}{\pi}\sum_{n=1}^{\infty}\frac{1}{n}\sin n\left(\theta-\frac{3\pi}{24}-120^\circ\right) \tag{6-98b}$$

$$v_{F1O}=\frac{2v_{DC}}{\pi}\sum_{n=1}^{\infty}\frac{1}{n}\sin n\left(\theta-\frac{3\pi}{24}+120^\circ\right) \tag{6-98c}$$

$$v_{D2O}=\frac{2v_{DC}}{\pi}\sum_{n=1}^{\infty}\frac{1}{n}\sin n\left(\theta-\frac{5\pi}{24}\right) \tag{6-99a}$$

$$v_{E2O}=\frac{2v_{DC}}{\pi}\sum_{n=1}^{\infty}\frac{1}{n}\sin n\left(\theta-\frac{5\pi}{24}-120^\circ\right) \tag{6-99b}$$

$$v_{F2O}=\frac{2v_{DC}}{\pi}\sum_{n=1}^{\infty}\frac{1}{n}\sin n\left(\theta-\frac{5\pi}{24}+120^\circ\right) \tag{6-99c}$$

for any odd harmonic n. The displacement angle between two consecutive six-pulse VSCs in a multipulse VSC configuration is $2\pi/6m$ (i.e., 15°), where m (i.e., 4) is the total number of six-pulse VSCs that are used. The two phasors in the A–B–C three-phase groups (A1–A2, B1–B2, and C1–C2) are centered on 0°, –120°, and –240°, respectively, with a displacement angle ($\theta_{displacement}$) of ±7.5°. The two phasors in the D–E–F three-phase groups (D1–D2, E1–E2, and F1–F2) are centered on –30°, –150°, and +90°, respectively, with a displacement angle ($\theta_{displacement}$) of ±7.5°. Figure 6-18a shows the two sets of three square wave voltages (v_{A1O}, v_{B1O}, and v_{C1O}; and v_{A2O}, v_{B2O}, and v_{C2O}) at 7.5° and –7.5°, respectively. The fundamental pole voltage phasors ($\mathbf{V}_{\mathbf{A1,1}}$, $\mathbf{V}_{\mathbf{B1,1}}$, and $\mathbf{V}_{\mathbf{C1,1}}$; and $\mathbf{V}_{\mathbf{A2,1}}$, $\mathbf{V}_{\mathbf{B2,1}}$, and $\mathbf{V}_{\mathbf{C2,1}}$), shown in Figure 6-18b, are at 7.5°, 7.5° – 120° (i.e., –112.5°), and 7.5° – 240° (i.e., –232.5°); and –7.5°, –7.5° – 120° (i.e., –127.5°), and –7.5° – 240° (i.e., –247.5°), respectively. Figure 6-18c shows the two sets of three square wave voltages (v_{D1O}, v_{E1O}, and v_{F1O}; and v_{D2O}, v_{E2O}, and v_{F2O}) at –22.5° and –37.5°, respectively. The fundamental pole voltage phasors ($\mathbf{V}_{\mathbf{D1,1}}$, $\mathbf{V}_{\mathbf{E1,1}}$, and $\mathbf{V}_{\mathbf{F1,1}}$; and $\mathbf{V}_{\mathbf{D2,1}}$, $\mathbf{V}_{\mathbf{E2,1}}$, and $\mathbf{V}_{\mathbf{F2,1}}$), shown in Figure 6-18d, are at 7.5° – 30° (i.e., –22.5°), 7.5° – 30° – 120° (i.e., –142.5°), and 7.5° – 30° – 240° (i.e., –262.5°); and –7.5° – 30° (i.e., –37.5°), –7.5° – 30° – 120° (i.e., –157.5°), and –7.5° – 30° – 240° (i.e., –277.5°), respectively. The following waveforms are analyzed with a turns ratio (n_2/n_1) of 0.25.

Figure 6-19 shows four six-pulse VSCs (A1–B1–C1, A2–B2–C2, D1–E1–F1, and D2–E2–F2) that are connected across a shared DC link with a pair of capacitors and the related fundamental voltage phasors. The pole voltages are combined by a magnetic circuit to form a three-phase (*a*–*b*–*c*) voltage. The poles A1, B1, C1, A2, B2, C2, D1, E1, F1, D2, E2, and F2 are operated in such a way that the pole voltages in a six-pulse VSC group, for example, v_{A1O}, v_{B1O}, and v_{C1O} of the A1–B1–C1 VSC; v_{A2O}, v_{B2O}, and v_{C2O} of the A2–B2–C2 VSC; v_{D1O}, v_{E1O}, and v_{F1O} of the D1–E1–F1 VSC; and v_{D2O}, v_{E2O}, and v_{F2O} of the D2–E2–F2 VSC, are time-shifted from one another by one-

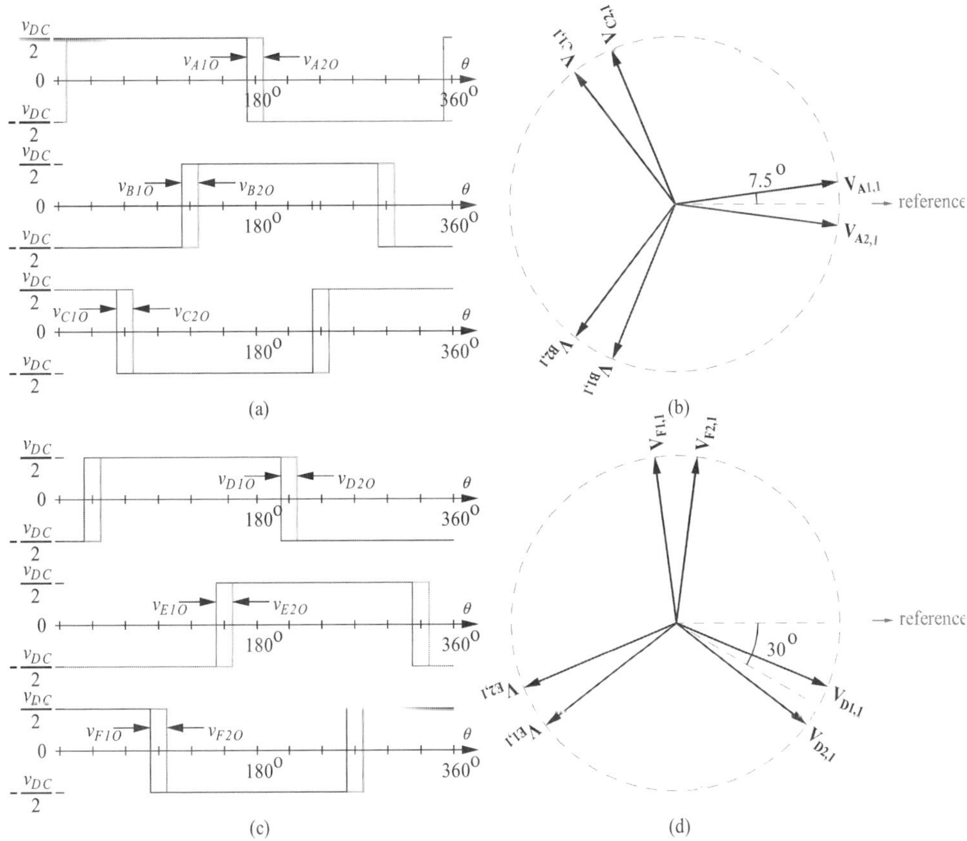

Figure 6-18. Pole voltages and fundamental voltage phasors in a 24-pulse QHN-VSC configuration. (a) Two sets of square wave voltages at 7.5° and –7.5°. (b) Phasor diagram of fundamental pole voltage phasors in A1–B1–C1 six-pulse VSC and A2–B2–C2 six-pulse VSC. (c) Two sets of square wave voltages at –22.5° and –37.5°. (d) Phasor diagram of fundamental pole voltage phasors in D1–E1–F1 six-pulse VSC and D2–E2–F2 six-pulse VSC.

third of the time period (T) of the pole voltage. Furthermore, the four sets of three square wave voltages (v_{A1O}, v_{B1O}, and v_{C1O}; v_{A2O}, v_{B2O}, and v_{C2O}; v_{D1O}, v_{E1O}, and v_{F1O}; and v_{D2O}, v_{E2O}, and v_{F2O}) are placed at 7.5°, –7.5°, –22.5°, and –37.5°, respectively.

When the pole voltages from the VSCs A1–B1–C1 and D1–E1–F1 are magnetically combined, as described in Figure 6-9, the output voltage exhibits a 12-pulse HN waveform with harmonic components $n = 12k \pm 1$ for $k = 1, 2, 3$, and so on. In the 12-pulse HN-VSC of Figure 6-9, there are two sets of square wave voltages: one set from the A–B–C VSC is at 0° and the other set from the D–E–F VSC is at –30°; the *a–b–c* 12-pulse voltage set is at 0°. In the 24-pulse QHN-VSC of Figure 6-19, there are two sets of square wave voltages: one set from the A1–B1–C1 VSC is at 7.5° and the other

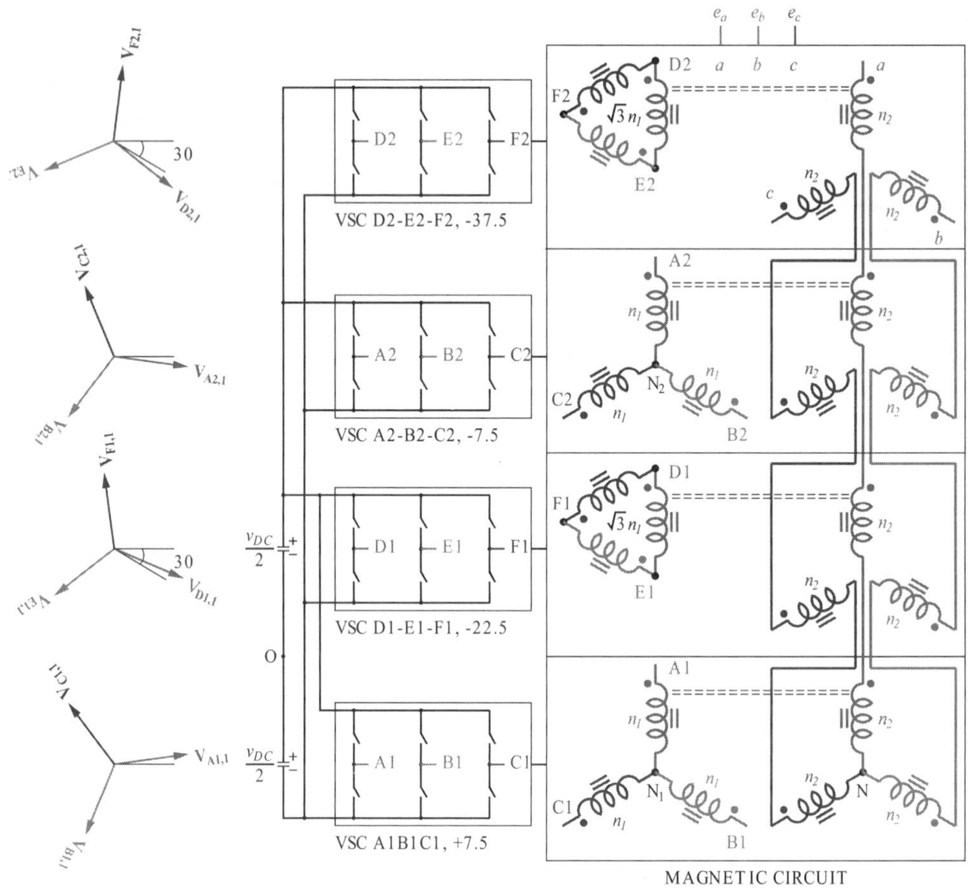

Figure 6-19. 24-pulse QHN-VSC configuration (all angles are in degrees).

set from the D1–E1–F1 VSC is at –22.5°; therefore, the *a1–b1–c1* 12-pulse voltage set is at 7.5°. There are additional two sets of square wave voltages: one set from the A2–B2–C2 VSC is at –7.5° and the other set from the D1–E1–F1 VSC is at –37.5°; therefore, the *a2–b2–c2* 12-pulse voltage set is at –7.5°. Each 12-pulse waveform contains the same magnitudes of fundamental and harmonic components ($n = 12k \pm 1$ for $k = 1, 2, 3$, etc.). The two sets of three-phase, 12-pulse voltages are

$$e_{a1} = 2\frac{n_2}{n_1}\frac{2v_{DC}}{\pi}\sum_{n=1}^{\infty}\frac{1}{n}\sin n\left(\theta + \frac{\pi}{24}\right) \tag{6-100a}$$

$$e_{b1} = 2\frac{n_2}{n_1}\frac{2v_{DC}}{\pi}\sum_{n=1}^{\infty}\frac{1}{n}\sin n\left(\theta + \frac{\pi}{24} - 120°\right) \tag{6-100b}$$

$$e_{c1} = 2\frac{n_2}{n_1}\frac{2v_{DC}}{\pi}\sum_{n=1}^{\infty}\frac{1}{n}\sin n\left(\theta + \frac{\pi}{24} + 120°\right) \tag{6-100c}$$

and

$$e_{a2} = 2\frac{n_2}{n_1}\frac{2v_{DC}}{\pi}\sum_{n=1}^{\infty}\frac{1}{n}\sin n\left(\theta - \frac{\pi}{24}\right) \tag{6-101a}$$

$$e_{b2} = 2\frac{n_2}{n_1}\frac{2v_{DC}}{\pi}\sum_{n=1}^{\infty}\frac{1}{n}\sin n\left(\theta - \frac{\pi}{24} - 120°\right) \tag{6-101b}$$

$$e_{c2} = 2\frac{n_2}{n_1}\frac{2v_{DC}}{\pi}\sum_{n=1}^{\infty}\frac{1}{n}\sin n\left(\theta - \frac{\pi}{24} + 120°\right) \tag{6-101c}$$

which contain fundamental components and harmonic components $n = 12k \pm 1$ for $k =$ 1, 2, 3, and so on.

Figure 6-20, parts a and c, show the two sets of three-phase, 12-pulse voltages (e_{a1}, e_{b1}, and e_{c1}; and e_{a2}, e_{b2}, and e_{c2}) and the corresponding fundamental components ($e_{a1,1}$, $e_{b1,1}$, and $e_{c1,1}$; and $e_{a2,1}$, $e_{b2,1}$, and $e_{c2,1}$). The fundamental voltage phasors ($\mathbf{E_{a1,1}}$, $\mathbf{E_{b1,1}}$, and $\mathbf{E_{c1,1}}$; and $\mathbf{E_{a2,1}}$, $\mathbf{E_{b2,1}}$, and $\mathbf{E_{c2,1}}$), shown in Figure 6-20, parts b and d, are at 7.5°, –112.5°, and –232.5°; and at –7.5°, –127.5°, and –247.5°, respectively.

When all the output voltages from each six-pulse VSC are combined by connecting the corresponding phases in series as shown in Figure 6-19, the resulting output voltages are

$$e_a = e_{a1} + e_{a2} = 2\frac{n_2}{n_1}\frac{2v_{DC}}{\pi}\sum_{n=1}^{\infty}\frac{1}{n}\left\{\sin n\left(\theta + \frac{\pi}{24}\right) + \sin n\left(\theta - \frac{\pi}{24}\right)\right\}$$

$$= 4\frac{n_2}{n_1}\frac{2v_{DC}}{\pi}\sum_{n=1}^{\infty}\frac{1}{n}\cos\left(\frac{n\pi}{24}\right)\sin n\theta$$

$$e_a = 4\frac{n_2}{n_1}\frac{2v_{DC}}{\pi}\sum_{n=1}^{\infty}\frac{1}{n}\cos\left(\frac{n\pi}{24}\right)\sin(n\theta) \tag{6-102a}$$

$$e_b = e_{b1} + e_{b2} = 2\frac{n_2}{n_1}\frac{2v_{DC}}{\pi}\sum_{n=1}^{\infty}\frac{1}{n}\left\{\sin n\left(\theta + \frac{\pi}{24} - 120°\right) + \sin n\left(\theta - \frac{\pi}{24} - 120°\right)\right\}$$

$$= 4\frac{n_2}{n_1}\frac{2v_{DC}}{\pi}\sum_{n=1}^{\infty}\frac{1}{n}\cos\left(\frac{n\pi}{24}\right)\sin n(\theta - 120°)$$

$$e_b = 4\frac{n_2}{n_1}\frac{2v_{DC}}{\pi}\sum_{n=1}^{\infty}\frac{1}{n}\cos\left(\frac{n\pi}{24}\right)\sin n(\theta - 120°) \tag{6-102b}$$

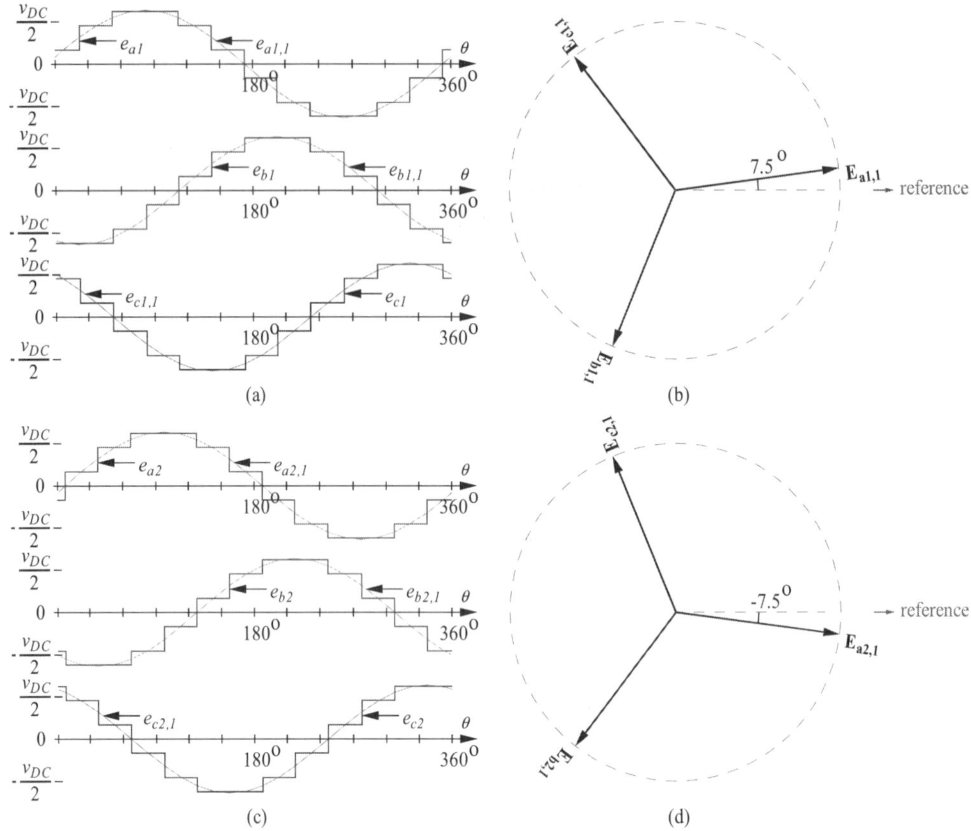

Figure 6-20. Two 12-pulse voltages and fundamental voltage phasors in a 24-pulse QHN-VSC configuration. (a) Three 120° phase-shifted fundamental voltages and the corresponding 12-pulse voltages at 7.5°. (b) Phasor diagram of fundamental voltage phasors in *a1–b1–c1* 12-pulse HN-VSC. (c) Three 120° phase-shifted fundamental voltages and the corresponding 12-pulse voltages at –7.5°. (d) Phasor diagram of fundamental voltage phasors in *a2–b2–c2* 12-pulse HN-VSC.

$$e_c = e_{c1} + e_{c2} = 2\frac{n_2}{n_1}\frac{2v_{DC}}{\pi}\sum_{n=1}^{\infty}\frac{1}{n}\left\{\sin n\left(\theta + \frac{\pi}{24} + 120°\right) + \sin n\left(\theta - \frac{\pi}{24} + 120°\right)\right\}$$

$$= 4\frac{n_2}{n_1}\frac{2v_{DC}}{\pi}\sum_{n=1}^{\infty}\frac{1}{n}\cos\left(\frac{n\pi}{24}\right)\sin n(\theta + 120°)$$

$$e_c = 4\frac{n_2}{n_1}\frac{2v_{DC}}{\pi}\sum_{n=1}^{\infty}\frac{1}{n}\cos\left(\frac{n\pi}{24}\right)\sin n(\theta + 120°) \tag{6-102c}$$

which contain fundamental components and harmonic components $n = 12k \pm 1$ for k = 1, 2, 3, and so on, and exhibit a 24-pulse QHN waveform.

Figure 6-21a shows the three-phase, 24-pulse voltage (e_a, e_b, and e_c) and the corresponding fundamental components ($e_{a,1}$, $e_{b,1}$, and $e_{c,1}$). The fundamental pole voltage phasors ($\mathbf{E_{a,1}}$, $\mathbf{E_{b,1}}$, and $\mathbf{E_{c,1}}$), shown in Figure 6-21b, are at 0° (reference), –120°, and –240°, respectively.

The output voltages exhibit fundamental components and harmonic components (n) with amplitudes of

$$E_{a,n} = E_{b,n} = E_{c,n} = 4\frac{n_2}{n_1}\frac{2v_{DC}}{n\pi}\cos\left(\frac{n\pi}{24}\right) \tag{6-103}$$

where $n = 1$ and $12k \pm 1$ for $k = 1, 2, 3$, and so on.

The amplitudes of harmonic components (n) normalized to the fundamental component are

$$\frac{E_{a,n}}{E_{a,1}} = \frac{E_{b,n}}{E_{b,1}} = \frac{E_{c,n}}{E_{c,1}} = \frac{\cos(n\pi/24)}{n\cos(\pi/24)} \tag{6-104}$$

where $n = 12k \pm 1$ for $k = 1, 2, 3$, and so on.

The amplitudes of the fundamental components normalized to the DC voltage (gain of the VSC) are

$$\frac{E_{a,1}}{v_{DC}} = \frac{E_{b,1}}{v_{DC}} = \frac{E_{c,1}}{v_{DC}} = \frac{n_2}{n_1}\frac{8}{\pi}\cos\left(\frac{\pi}{24}\right) \tag{6-105}$$

6.2.5.1 *Modeling a 24-Pulse QHN-VSC with Two-Level Poles.* The 24-pulse, two-level QHN-VSC is run from the main file `64824Q21.DAT` that has the

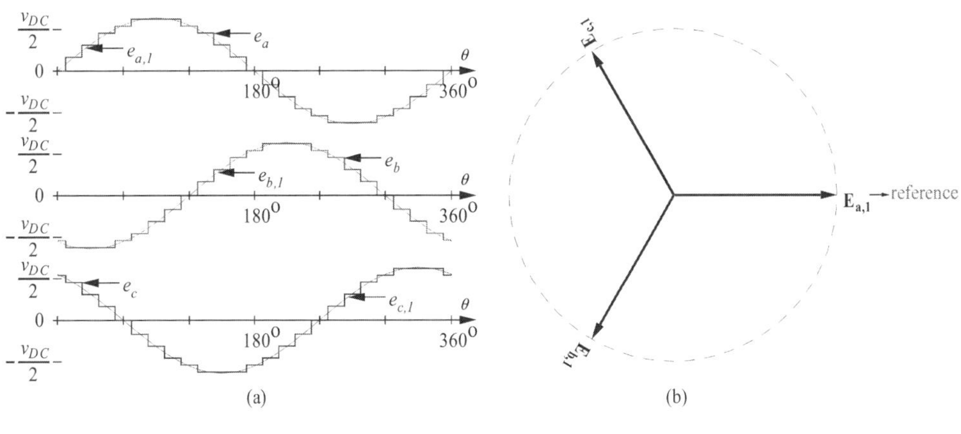

Figure 6-21. (a) Three 120° phase-shifted fundamental output voltages and the corresponding 24-pulse voltages. (b) Phasor diagram of fundamental output voltage phasors in a 24-pulse QHN-VSC configuration.

same setup as the 24-pulse, two-level HN-VSC in file `63624P21.DAT`, except that the two $INCLUDE$ files are replaced with `64924Q21.GAT` (see Appendix C, Section 53) for generating gating pulses for the switches and `65024Q21.MAG` (see Appendix C, Section 54) for implementing the ideal magnetic circuit.

6.2.6 48-Pulse QHN-VSC with Two-Level Poles

The idea of the QHN-VSC can be extended to higher pulse numbers. A 48-pulse QHN-VSC requires the generation of eight three-phase sets (A1–B1–C1, A2–B2–C2, A3–B3–C3, A4–B4–C4, D1–E1–F1, D2–E2–F2, D3–E3–F3, and D4–E4–F4) of square wave voltages that are defined by

$$v_{A1O} = \frac{2v_{DC}}{\pi}\sum_{n=1}^{\infty}\frac{1}{n}\sin n\left(\theta + \frac{3\pi}{48}\right) \tag{6-106a}$$

$$v_{B1O} = \frac{2v_{DC}}{\pi}\sum_{n=1}^{\infty}\frac{1}{n}\sin n\left(\theta + \frac{3\pi}{48} - 120^\circ\right) \tag{6-106b}$$

$$v_{C1O} = \frac{2v_{DC}}{\pi}\sum_{n=1}^{\infty}\frac{1}{n}\sin n\left(\theta + \frac{3\pi}{48} + 120^\circ\right) \tag{6-106c}$$

$$v_{A2O} = \frac{2v_{DC}}{\pi}\sum_{n=1}^{\infty}\frac{1}{n}\sin n\left(\theta + \frac{\pi}{48}\right) \tag{6-107a}$$

$$v_{B2O} = \frac{2v_{DC}}{\pi}\sum_{n=1}^{\infty}\frac{1}{n}\sin n\left(\theta + \frac{\pi}{48} - 120^\circ\right) \tag{6-107b}$$

$$v_{C2O} = \frac{2v_{DC}}{\pi}\sum_{n=1}^{\infty}\frac{1}{n}\sin n\left(\theta + \frac{\pi}{48} + 120^\circ\right) \tag{6-107c}$$

$$v_{A3O} = \frac{2v_{DC}}{\pi}\sum_{n=1}^{\infty}\frac{1}{n}\sin n\left(\theta - \frac{\pi}{48}\right) \tag{6-108a}$$

$$v_{B3O} = \frac{2v_{DC}}{\pi}\sum_{n=1}^{\infty}\frac{1}{n}\sin n\left(\theta - \frac{\pi}{48} - 120^\circ\right) \tag{6-108b}$$

$$v_{C3O} = \frac{2v_{DC}}{\pi}\sum_{n=1}^{\infty}\frac{1}{n}\sin n\left(\theta - \frac{\pi}{48} + 120^\circ\right) \tag{6-108c}$$

$$v_{A4O} = \frac{2v_{DC}}{\pi}\sum_{n=1}^{\infty}\frac{1}{n}\sin n\left(\theta - \frac{3\pi}{48}\right) \tag{6-109a}$$

$$v_{B4O} = \frac{2v_{DC}}{\pi} \sum_{n=1}^{\infty} \frac{1}{n} \sin n\left(\theta - \frac{3\pi}{48} - 120°\right) \tag{6-109b}$$

$$v_{C4O} = \frac{2v_{DC}}{\pi} \sum_{n=1}^{\infty} \frac{1}{n} \sin n\left(\theta - \frac{3\pi}{48} + 120°\right) \tag{6-109c}$$

$$v_{D1O} = \frac{2v_{DC}}{\pi} \sum_{n=1}^{\infty} \frac{1}{n} \sin n\left(\theta - \frac{5\pi}{48}\right) \tag{6-110a}$$

$$v_{E1O} = \frac{2v_{DC}}{\pi} \sum_{n=1}^{\infty} \frac{1}{n} \sin n\left(\theta - \frac{5\pi}{48} - 120°\right) \tag{6-110b}$$

$$v_{F1O} = \frac{2v_{DC}}{\pi} \sum_{n=1}^{\infty} \frac{1}{n} \sin n\left(\theta - \frac{5\pi}{48} + 120°\right) \tag{6-110c}$$

$$v_{D2O} = \frac{2v_{DC}}{\pi} \sum_{n=1}^{\infty} \frac{1}{n} \sin n\left(\theta - \frac{7\pi}{48}\right) \tag{6-111a}$$

$$v_{E2O} = \frac{2v_{DC}}{\pi} \sum_{n=1}^{\infty} \frac{1}{n} \sin n\left(\theta - \frac{7\pi}{48} - 120°\right) \tag{6 111b}$$

$$v_{F2O} = \frac{2v_{DC}}{\pi} \sum_{n=1}^{\infty} \frac{1}{n} \sin n\left(\theta - \frac{7\pi}{48} + 120°\right) \tag{6-111c}$$

$$v_{D3O} = \frac{2v_{DC}}{\pi} \sum_{n=1}^{\infty} \frac{1}{n} \sin n\left(\theta - \frac{9\pi}{48}\right) \tag{6-112a}$$

$$v_{E3O} = \frac{2v_{DC}}{\pi} \sum_{n=1}^{\infty} \frac{1}{n} \sin n\left(\theta - \frac{9\pi}{48} - 120°\right) \tag{6-112b}$$

$$v_{F3O} = \frac{2v_{DC}}{\pi} \sum_{n=1}^{\infty} \frac{1}{n} \sin n\left(\theta - \frac{9\pi}{48} + 120°\right) \tag{6-112c}$$

$$v_{D4O} = \frac{2v_{DC}}{\pi} \sum_{n=1}^{\infty} \frac{1}{n} \sin n\left(\theta - \frac{11\pi}{48}\right) \tag{6-113a}$$

$$v_{E4O} = \frac{2v_{DC}}{\pi} \sum_{n=1}^{\infty} \frac{1}{n} \sin n\left(\theta - \frac{11\pi}{48} - 120°\right) \tag{6-113b}$$

$$v_{F4O} = \frac{2v_{DC}}{\pi}\sum_{n=1}^{\infty}\frac{1}{n}\sin n\left(\theta - \frac{11\pi}{48} + 120°\right) \tag{6-113c}$$

for any odd harmonic n. The displacement angle between two consecutive six-pulse VSCs in a multipulse VSC configuration is $2\pi/6m$ (i.e., 7.5°), where m (i.e., 8) is the total number of six-pulse VSCs that are used. The four phasors in the A–B–C three-phase groups (A1–A2–A3–A4, B1–B2–B3–B4, and C1–C2–C3–C4) are centered on 0°, –120°, and –240°, respectively, with displacement angles, $\theta_{displacement}$, of ±3.75° and ±11.25°. The four phasors in the D–E–F three-phase group (D1–D2–D3–D4, E1–E2–E3–E4, and F1–F2–F3–F4) are centered on –30°, –150°, and +90°, respectively, with displacement angles, $\theta_{displacement}$, of ±3.75° and ±11.25°.

Figure 6-22a shows the four sets of three square wave voltages (v_{A1O}, v_{B1O}, and v_{C1O}; v_{A2O}, v_{B2O}, and v_{C2O}; v_{A3O}, v_{B3O}, and v_{C3O}; and v_{A4O}, v_{B4O}, and v_{C4O}) at 11.25°, 3.75°, –3.75°, and –11.25°, respectively. The fundamental pole voltage phasors ($\mathbf{V_{A1,1}}$, $\mathbf{V_{B1,1}}$, and $\mathbf{V_{C1,1}}$; $\mathbf{V_{A2,1}}$, $\mathbf{V_{B2,1}}$, and $\mathbf{V_{C2,1}}$; $\mathbf{V_{A3,1}}$, $\mathbf{V_{B3,1}}$, and $\mathbf{V_{C3,1}}$; and $\mathbf{V_{A4,1}}$, $\mathbf{V_{B4,1}}$, and $\mathbf{V_{C4,1}}$), shown in Figure 6-22b, are at 11.25°, 11.25° – 120° (i.e., –108.75°), and 11.25° – 240° (i.e., –228.75°); 3.75°, 3.75° – 120° (i.e., –116.25°), and 3.75° – 240° (i.e., –236.25°); –3.75°, –3.75° – 120° (i.e., –123.75°), and –3.75° – 240° (i.e., –243.75°); and –11.25°, –11.25° – 120° (i.e., –131.25°), and –11.25° – 240° (i.e., –251.25°), respectively. Figure 6-22c shows the four sets of three square wave voltages (v_{D1O}, v_{E1O}, and v_{F1O}; v_{D2O}, v_{E2O}, and v_{F2O}; v_{D3O}, v_{E3O}, and v_{F3O}; and v_{D4O}, v_{E4O}, and v_{F4O}) at –18.75°, –26.25°, –33.75°, and –41.25°, respectively. The fundamental pole voltage phasors ($\mathbf{V_{D1,1}}$, $\mathbf{V_{E1,1}}$, and $\mathbf{V_{F1,1}}$; $\mathbf{V_{D2,1}}$, $\mathbf{V_{E2,1}}$, and $\mathbf{V_{F2,1}}$; $\mathbf{V_{D3,1}}$, $\mathbf{V_{E3,1}}$, and $\mathbf{V_{F3,1}}$; and $\mathbf{V_{D4,1}}$, $\mathbf{V_{E4,1}}$, and $\mathbf{V_{F4,1}}$), shown in Figure 6-22d, are at 11.25° – 30° (i.e., –18.75°), 11.25° – 30° – 120° (i.e., –138.75°), and 11.25° – 30° – 240° (i.e., –258.75°); 3.75° – 30° (i.e., –26.25°), 3.75° – 30° – 120° (i.e., –146.25°), and 3.75° – 30° – 240° (i.e., –266.25°); –3.75° – 30° (i.e., –33.75°), –3.75° – 30° – 120° (i.e., –153.75°), and –3.75° – 30° – 240° (i.e., –273.75°); and –11.25° – 30° (i.e., –41.25°), –11.25° – 30° – 120° (i.e., –161.25°), and –11.25° – 30° – 240° (i.e., –281.25°), respectively. The following waveforms are analyzed with turns ratio (n_2/n_1) of 0.125.

Figure 6-23 shows eight six-pulse VSCs (A1–B1–C1, A2–B2–C2, A3–B3–C3, A4–B4–C4, D1–E1–F1, D2–E2–F2, D3–E3–F3, and D4–D4–D4) that are connected across a shared DC link with a pair of capacitors and the related fundamental voltage phasors. The pole voltages are combined by a magnetic circuit to form a three-phase (*a–b–c*) voltage. The poles A1, B1, C1, A2, B2, C2, A3, B3, C3, A4, B4, C4, D1, E1, F1, D2, E2, F2, D3, E3, F3, D4, E4, and F4 are operated in such a way that the pole voltages in a six-pulse VSC group, for example v_{A1O}, v_{B1O}, and v_{C1O} of the A1–B1–C1 VSC; v_{A2O}, v_{B2O}, and v_{C2O} of the A2–B2–C2 VSC; v_{A3O}, v_{B3O}, and v_{C3O} of the A3–B3–C3 VSC; v_{A4O}, v_{B4O}, and v_{C4O} of the A4–B4–C4 VSC; v_{D1O}, v_{E1O}, and v_{F1O} of the D1–E1–F1 VSC; v_{D2O}, v_{E2O}, and v_{F2O} of the D2–E2–F2 VSC; v_{D3O}, v_{E3O}, and v_{F3O} of the D3–E3–F3 VSC; v_{D4O}, v_{E4O}, and v_{F4O} of the D4–E4–F4 VSC are time-shifted from one another by one-third of the time period (T) of the pole voltage. Furthermore, the eight sets of three square wave voltages (v_{A1O}, v_{B1O}, and v_{C1O}; v_{A2O}, v_{B2O}, and v_{C2O}; v_{A3O}, v_{B3O}, and v_{C3O}; v_{A4O}, v_{B4O}, and v_{C4O}; v_{D1O}, v_{E1O}, and v_{F1O}; v_{D2O}, v_{E2O}, and v_{F2O};

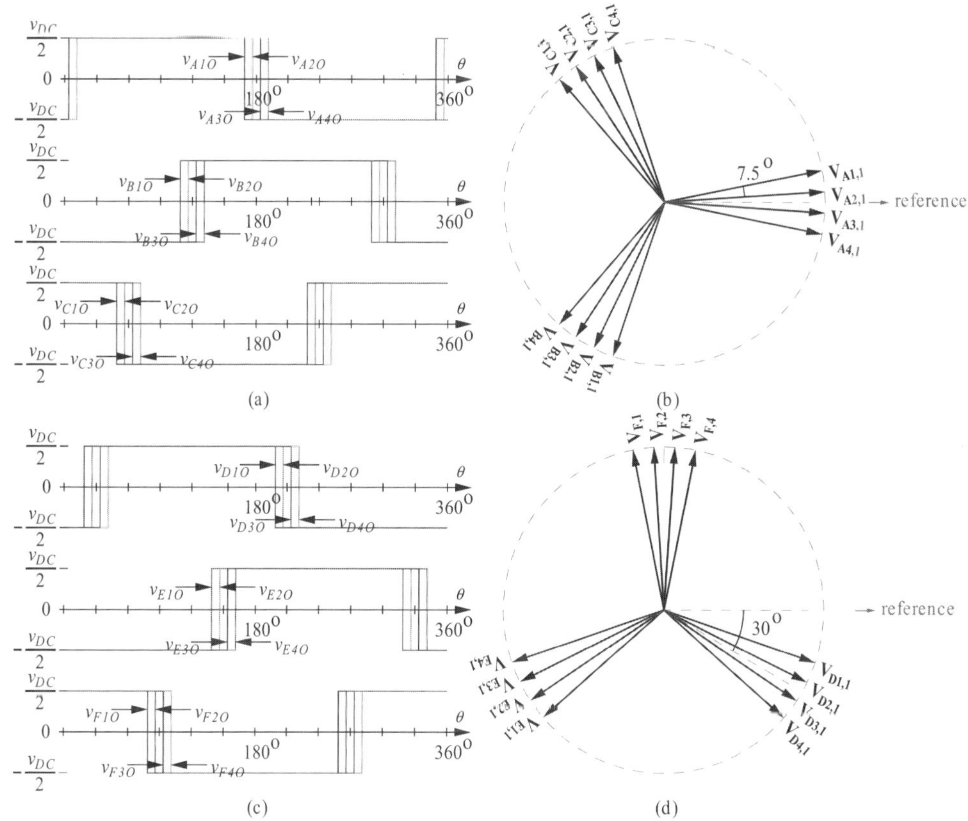

Figure 6-22. Pole voltages and fundamental voltage phasors in a 48-pulse QHN-VSC configuration. (a) Four sets of square wave voltages at 11.25°, 3.75°, –3.75°, and –11.25°. (b) Phasor diagram of fundamental pole voltage phasors in A1–B1–C1 six-pulse VSC, A2–B2–C2 six-pulse VSC, A3–B3–C3 six-pulse VSC, and A4–B4–C4 six-pulse VSC. (c) Four sets of square wave voltages at –18.75°, –26.25°, –33.75°, and –41.25°. (d) Phasor diagram of fundamental pole voltage phasors in D1–E1–F1 six-pulse VSC, D2–E2–F2 six-pulse VSC, D3–E3–F3 six-pulse VSC, and D4–E4–F4 six-pulse VSC.

v_{D3O}, v_{E3O}, and v_{F3O}; and v_{D4O}, v_{E4O}, and v_{F4O}) are placed at 11.25°, 3.75°, –3.75°, –11.25°, –18.25°, –26.25°, –33.25°, and –41.25°, respectively.

When the pole voltages from the VSCs A1–B1–C1 and D1–E1–F1 are magnetically combined, as described in Figure 6-9, the output voltage exhibits a 12-pulse HN waveform with harmonic components $n = 12k \pm 1$ for $k = 1, 2, 3$, and so on. In the 12-pulse HN-VSC of Figure 6-9, there are two sets of square wave voltages: one set from the A–B–C VSC is at 0° and the other set from the D–E–F VSC is at –30°; the *a–b–c* 12-pulse voltage set is at 0°. In the 48-pulse QHN-VSC of Figure 6-23, there are four sets of 12-pulse waveforms. The first 12-pulse waveform is composed of two sets of square wave voltages: one set from the A1–B1–C1 VSC is at 11.25° and the other set

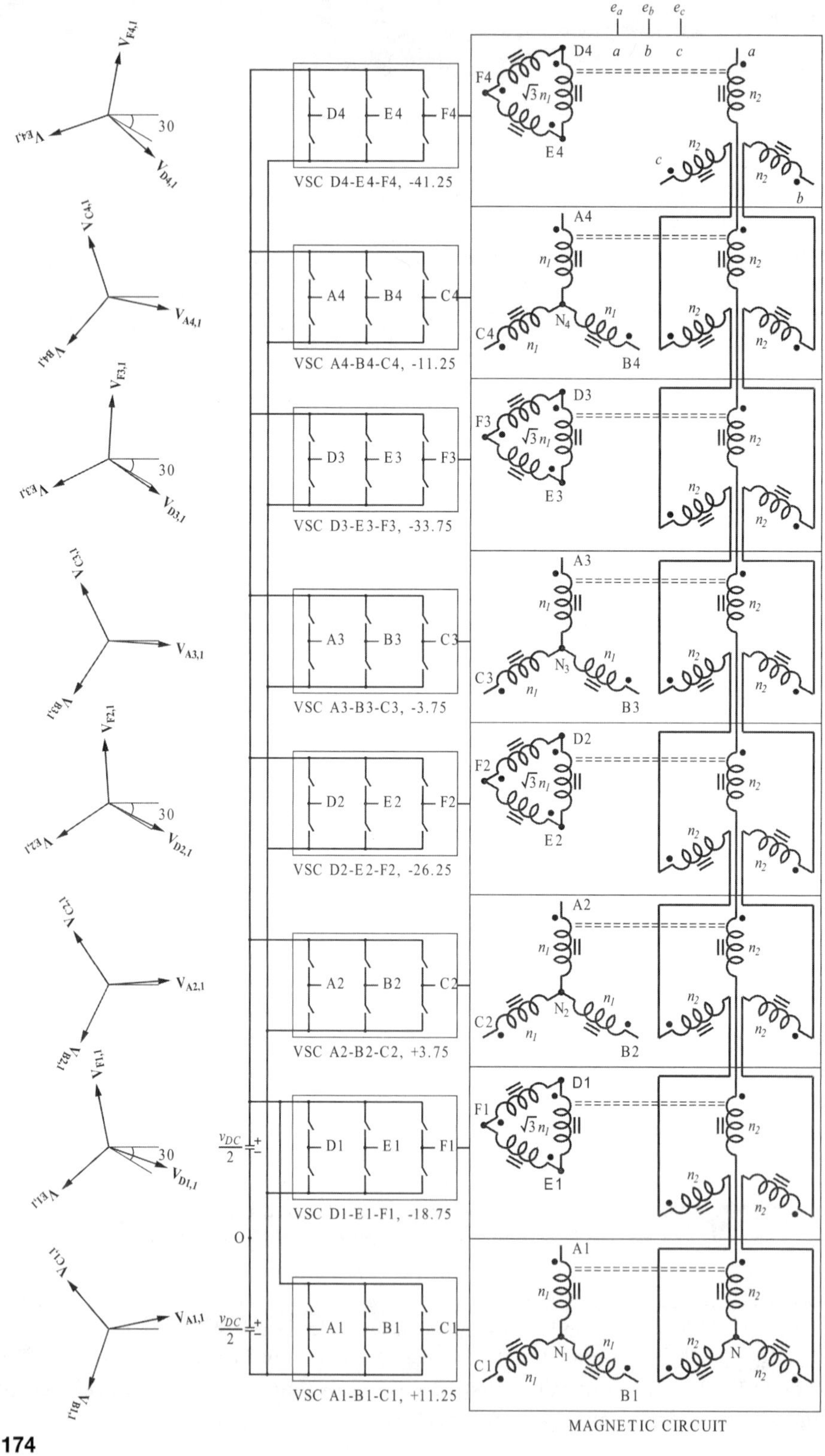

e_a e_b e_c
VSC D4-E4-F4, -41.25
VSC A4-B4-C4, -11.25
VSC D3-E3-F3, -33.75
VSC A3-B3-C3, -3.75
VSC D2-E2-F2, -26.25
VSC A2-B2-C2, +3.75
VSC D1-E1-F1, -18.75
VSC A1-B1-C1, +11.25
$\frac{v_{DC}}{2}$
O
$\sqrt{3}n_1$
n_1
n_2
MAGNETIC CIRCUIT

from the D1–E1–F1 VSC is at –18.75°; therefore, the *a1–b1–c1* 12-pulse voltage set is at 11.25°. The second 12-pulse waveform is composed of two sets of square wave voltages: one set from the A2–B2–C2 VSC is at 3.75° and the other set from the D2–E2–F2 VSC is at –26.25°; therefore, the *a2–b2–c2* 12-pulse voltage set is at 3.75°. The third 12-pulse waveform is composed of two sets of square wave voltages: one set from the A3–B3–C3 VSC is at –3.75° and the other set from the D3–E3–F3 VSC is at –33.75°; therefore, the *a3–b3–c3* 12-pulse voltage set is at –3.75°. The fourth 12-pulse waveform is composed of two sets of square wave voltages: one set from the A4–B4–C4 VSC is at –11.25° and the other set from the D4–E4–F4 VSC is at –41.25°; therefore, the *a4–b4–c4* 12-pulse voltage set is at –11.25°. Each 12-pulse waveform contains the same magnitudes of fundamental and harmonic components ($n = 12k \pm 1$ for $k = 1, 2, 3$, etc.). The four sets of three-phase, 12-pulse voltages are

$$e_{a1} = 2\frac{n_2}{n_1}\frac{2v_{DC}}{\pi}\sum_{n=1}^{\infty}\frac{1}{n}\sin n\left(\theta+\frac{3\pi}{48}\right) \tag{6-114a}$$

$$e_{b1} = 2\frac{n_2}{n_1}\frac{2v_{DC}}{\pi}\sum_{n=1}^{\infty}\frac{1}{n}\sin n\left(\theta+\frac{3\pi}{48}-120^\circ\right) \tag{6-114b}$$

$$e_{c1} = 2\frac{n_2}{n_1}\frac{2v_{DC}}{\pi}\sum_{n=1}^{\infty}\frac{1}{n}\sin n\left(\theta+\frac{3\pi}{48}+120^\circ\right) \tag{6-114c}$$

$$e_{a2} = 2\frac{n_2}{n_1}\frac{2v_{DC}}{\pi}\sum_{n=1}^{\infty}\frac{1}{n}\sin n\left(\theta+\frac{\pi}{48}\right) \tag{6-115a}$$

$$e_{b2} = 2\frac{n_2}{n_1}\frac{2v_{DC}}{\pi}\sum_{n=1}^{\infty}\frac{1}{n}\sin n\left(\theta+\frac{\pi}{48}-120^\circ\right) \tag{6-115b}$$

$$e_{c2} = 2\frac{n_2}{n_1}\frac{2v_{DC}}{\pi}\sum_{n=1}^{\infty}\frac{1}{n}\sin n\left(\theta+\frac{\pi}{48}+120^\circ\right) \tag{6-115c}$$

$$e_{a3} = 2\frac{n_2}{n_1}\frac{2v_{DC}}{\pi}\sum_{n=1}^{\infty}\frac{1}{n}\sin n\left(\theta-\frac{\pi}{48}\right) \tag{6-116a}$$

$$e_{b3} = 2\frac{n_2}{n_1}\frac{2v_{DC}}{\pi}\sum_{n=1}^{\infty}\frac{1}{n}\sin n\left(\theta-\frac{\pi}{48}-120^\circ\right) \tag{6-116b}$$

$$e_{c3} = 2\frac{n_2}{n_1}\frac{2v_{DC}}{\pi}\sum_{n=1}^{\infty}\frac{1}{n}\sin n\left(\theta-\frac{\pi}{48}+120^\circ\right) \tag{6-116c}$$

Figure 6-23. 48-pulse QHN-VSC configuration (all angles are in degrees).

and

$$e_{a4} = 2\frac{n_2}{n_1}\frac{2v_{DC}}{\pi}\sum_{n=1}^{\infty}\frac{1}{n}\sin n\left(\theta - \frac{3\pi}{48}\right) \tag{6-117a}$$

$$e_{b4} = 2\frac{n_2}{n_1}\frac{2v_{DC}}{\pi}\sum_{n=1}^{\infty}\frac{1}{n}\sin n\left(\theta - \frac{3\pi}{48} - 120°\right) \tag{6-117b}$$

$$e_{c4} = 2\frac{n_2}{n_1}\frac{2v_{DC}}{\pi}\sum_{n=1}^{\infty}\frac{1}{n}\sin n\left(\theta - \frac{3\pi}{48} + 120°\right) \tag{6-117c}$$

which contain fundamental components and harmonic components $n = 12k \pm 1$ for $k =$ 1, 2, 3, and so on.

Figure 6-24a shows the two sets of three-phase, 12-pulse voltages (e_{a1}, e_{b1}, and e_{c1}; and e_{a2}, e_{b2}, and e_{c2}). The fundamental voltage phasors ($\mathbf{E_{a1,1}}$, $\mathbf{E_{b1,1}}$, and $\mathbf{E_{c1,1}}$; and $\mathbf{E_{a2,1}}$, $\mathbf{E_{b2,1}}$, and $\mathbf{E_{c2,1}}$), shown in Figure 6-24b, are at 11.25°, –108.75°, and –228.75°; and at 3.75°, –116.25°, and –236.25°, respectively. Figure 6-24c shows the two sets of three-phase, 12-pulse voltages (e_{a3}, e_{b3}, and e_{c3}; and e_{a4}, e_{b4}, and e_{c4}). The fundamental voltage phasors ($\mathbf{E_{a3,1}}$, $\mathbf{E_{b3,1}}$, and $\mathbf{E_{c3,1}}$; and $\mathbf{E_{a4,1}}$, $\mathbf{E_{b4,1}}$, and $\mathbf{E_{c4,1}}$), shown in Figure 6-24d, are at –3.75°, –123.75°, and –243.75°; and at –11.25°, –131.25°, and –251.25°, respectively.

When the output voltages from two 12-pulse VSCs are combined by connecting the corresponding phases in series, as shown in Figure 6-23, the resulting output voltages are

$$v_{a1} = e_{a1} + e_{a3} = 2\frac{n_2}{n_1}\frac{2v_{DC}}{\pi}\sum_{n=1}^{\infty}\frac{1}{n}\left\{\sin n\left(\theta + \frac{3\pi}{48}\right) + \sin n\left(\theta - \frac{\pi}{48}\right)\right\}$$

$$v_{a1} = 4\frac{n_2}{n_1}\frac{2v_{DC}}{\pi}\sum_{n=1}^{\infty}\frac{1}{n}\cos\left(\frac{n\pi}{24}\right)\sin n\left(\theta + \frac{\pi}{48}\right) \tag{6-118a}$$

$$v_{b1} = e_{b1} + e_{b3} = 2\frac{n_2}{n_1}\frac{2v_{DC}}{\pi}\sum_{n=1}^{\infty}\frac{1}{n}\left\{\sin n\left(\theta + \frac{3\pi}{48} - 120°\right) + \sin n\left(\theta - \frac{\pi}{48} - 120°\right)\right\}$$

$$v_{b1} = 4\frac{n_2}{n_1}\frac{2v_{DC}}{\pi}\sum_{n=1}^{\infty}\frac{1}{n}\cos\left(\frac{n\pi}{24}\right)\sin n\left(\theta + \frac{\pi}{48} - 120°\right) \tag{6-118b}$$

$$v_{c1} = e_{c1} + e_{c3} = 2\frac{n_2}{n_1}\frac{2v_{DC}}{\pi}\sum_{n=1}^{\infty}\frac{1}{n}\left\{\sin n\left(\theta + \frac{3\pi}{48} + 120°\right) + \sin n\left(\theta - \frac{\pi}{48} + 120°\right)\right\}$$

$$v_{c1} = 4\frac{n_2}{n_1}\frac{2v_{DC}}{\pi}\sum_{n=1}^{\infty}\frac{1}{n}\cos\left(\frac{n\pi}{24}\right)\sin n\left(\theta + \frac{\pi}{48} + 120°\right) \tag{6-118c}$$

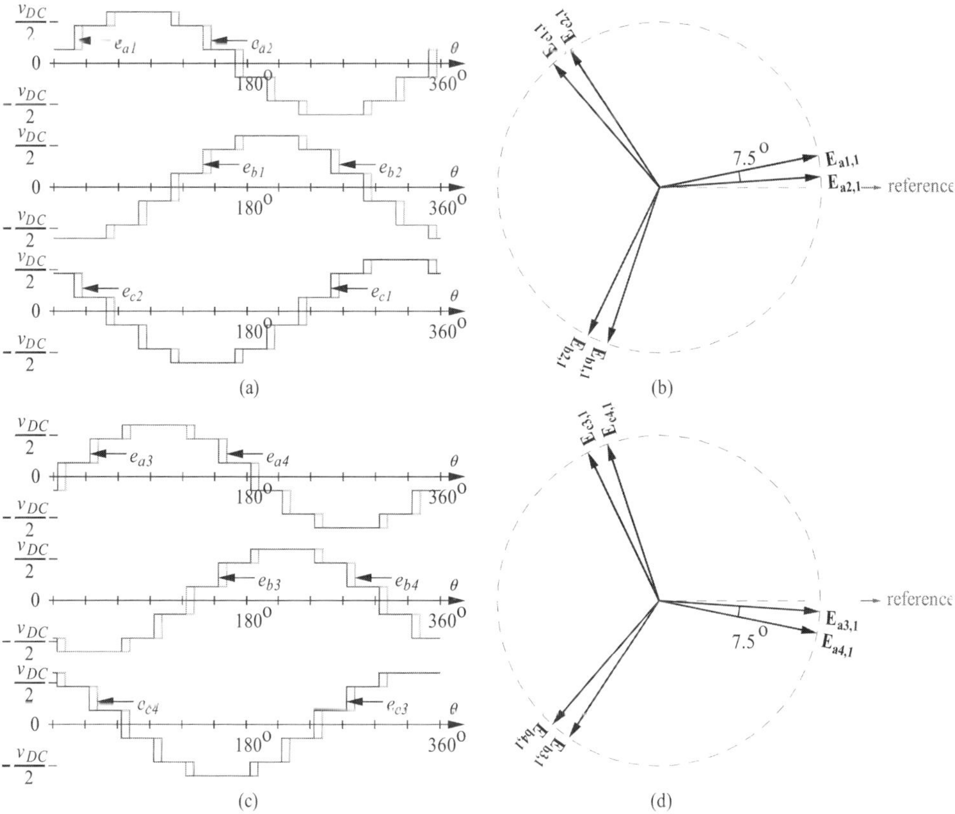

Figure 6-24. Four 12-pulse voltages and fundamental voltage phasors in a 48-pulse QHN-VSC configuration. (a) Two 12-pulse voltages at 11.25° and 3.75°, respectively. (b) Phasor diagram of fundamental voltage phasors in *a1–b1–c1* and *a2–b2–c2* 12-pulse HN-VSCs. (c) Two 12-pulse voltages at –3.75° and –11.25°, respectively. (d) Phasor diagram of fundamental voltage phasors in *a3–b3–c3* and *a4–b4–c4* 12-pulse HN-VSCs.

$$v_{a2} = e_{a2} + e_{a4} = 2\frac{n_2}{n_1}\frac{2v_{DC}}{\pi}\sum_{n=1}^{\infty}\frac{1}{n}\left\{\sin n\left(\theta + \frac{\pi}{48}\right) + \sin n\left(\theta - \frac{3\pi}{48}\right)\right\}$$

$$v_{a2} = 4\frac{n_2}{n_1}\frac{2v_{DC}}{\pi}\sum_{n=1}^{\infty}\frac{1}{n}\cos\left(\frac{n\pi}{24}\right)\sin n\left(\theta - \frac{\pi}{48}\right) \tag{6-119a}$$

$$v_{b2} = e_{b2} + e_{b4} = 2\frac{n_2}{n_1}\frac{2v_{DC}}{\pi}\sum_{n=1}^{\infty}\frac{1}{n}\left\{\sin n\left(\theta + \frac{\pi}{48} - 120^\circ\right) + \sin n\left(\theta - \frac{3\pi}{48} - 120^\circ\right)\right\}$$

$$v_{b2} = 4\frac{n_2}{n_1}\frac{2v_{DC}}{\pi}\sum_{n=1}^{\infty}\frac{1}{n}\cos\left(\frac{n\pi}{24}\right)\sin n\left(\theta - \frac{\pi}{48} - 120^\circ\right) \tag{6-119b}$$

$$v_{c2} = e_{c2} + e_{c4} = 2\frac{n_2}{n_1}\frac{2v_{DC}}{\pi}\sum_{n=1}^{\infty}\frac{1}{n}\left\{\sin n\left(\theta + \frac{\pi}{48} + 120°\right) + \sin n\left(\theta - \frac{3\pi}{48} + 120°\right)\right\}$$

$$v_{c2} = 4\frac{n_2}{n_1}\frac{2v_{DC}}{\pi}\sum_{n=1}^{\infty}\frac{1}{n}\cos\left(\frac{n\pi}{24}\right)\sin n\left(\theta - \frac{\pi}{48} + 120°\right) \tag{6-119c}$$

which contain fundamental components and harmonic components $n = 12k \pm 1$ for $k =$ 1, 2, 3, and so on, and exhibit a 24-pulse QHN waveform. Note that equations 6-102, 6-118, and 6-119 represent 24-pulse QHN waveforms. In the case where the three-phase voltage set in equation 6-102 is placed at 0°, the three-phase voltage sets in equations 6-118 and 6-119 are placed at 3.75° and –3.75°, respectively.

Figure 6-25, parts a and c, show the two sets of three-phase, 24-pulse voltages (v_{a1}, v_{b1}, and v_{c1}; and v_{a2}, v_{b2}, and v_{c2}) and the corresponding fundamental components ($v_{a1,1}$, $v_{b1,1}$, and $v_{c1,1}$; and $v_{a2,1}$, $v_{b2,1}$, and $v_{c2,1}$). The fundamental voltage phasors ($\mathbf{V_{a1,1}}$, $\mathbf{V_{b1,1}}$, and $\mathbf{V_{c1,1}}$; and $\mathbf{V_{a2,1}}$, $\mathbf{V_{b2,1}}$, and $\mathbf{V_{c2,1}}$), shown in Figure 6-25, parts b and d, are at 3.75°, –116.25°, and –236.25°; and at –3.75°, –123.75°, and –243.75°, respectively.

When all the output voltages from each six-pulse VSC are combined by connecting the corresponding phases in series, as shown in Figure 6-23, the resulting output voltages are

$$e_a = v_{a1} + v_{a2} = 4\frac{n_2}{n_1}\frac{2v_{DC}}{\pi}\sum_{n=1}^{\infty}\frac{1}{n}\cos\left(\frac{n\pi}{24}\right)\left\{\sin n\left(\theta + \frac{\pi}{48}\right) + \sin n\left(\theta - \frac{\pi}{48}\right)\right\}$$

$$e_a = 8\frac{n_2}{n_1}\frac{2v_{DC}}{\pi}\sum_{n=1}^{\infty}\frac{1}{n}\cos\left(\frac{n\pi}{24}\right)\cos\left(\frac{n\pi}{48}\right)\sin n\theta \tag{6-120a}$$

$$e_b = v_{b1} + v_{b2} = 4\frac{n_2}{n_1}\frac{2v_{DC}}{\pi}\sum_{n=1}^{\infty}\frac{1}{n}\cos\left(\frac{n\pi}{24}\right)\left\{\sin n\left(\theta + \frac{\pi}{48} - 120°\right) + \sin n\left(\theta - \frac{\pi}{48} - 120°\right)\right\}$$

$$e_b = 8\frac{n_2}{n_1}\frac{2v_{DC}}{\pi}\sum_{n=1}^{\infty}\frac{1}{n}\cos\left(\frac{n\pi}{24}\right)\cos\left(\frac{n\pi}{48}\right)\sin n(\theta - 120°) \tag{6-120b}$$

$$e_c = v_{c1} + v_{c2} = 4\frac{n_2}{n_1}\frac{2v_{DC}}{\pi}\sum_{n=1}^{\infty}\frac{1}{n}\cos\left(\frac{n\pi}{24}\right)\left\{\sin n\left(\theta + \frac{\pi}{48} + 120°\right) + \sin n\left(\theta - \frac{\pi}{48} + 120°\right)\right\}$$

$$e_c = 8\frac{n_2}{n_1}\frac{2v_{DC}}{\pi}\sum_{n=1}^{\infty}\frac{1}{n}\cos\left(\frac{n\pi}{24}\right)\cos\left(\frac{n\pi}{48}\right)\sin n(\theta + 120°) \tag{6-120c}$$

which contain fundamental components and harmonic components $n = 12k \pm 1$ for $k =$ 1, 2, 3, and so on, and exhibit a 48-pulse QHN waveform. The three-time averaging accounts for a factor of 0.125, which is n_2/n_1.

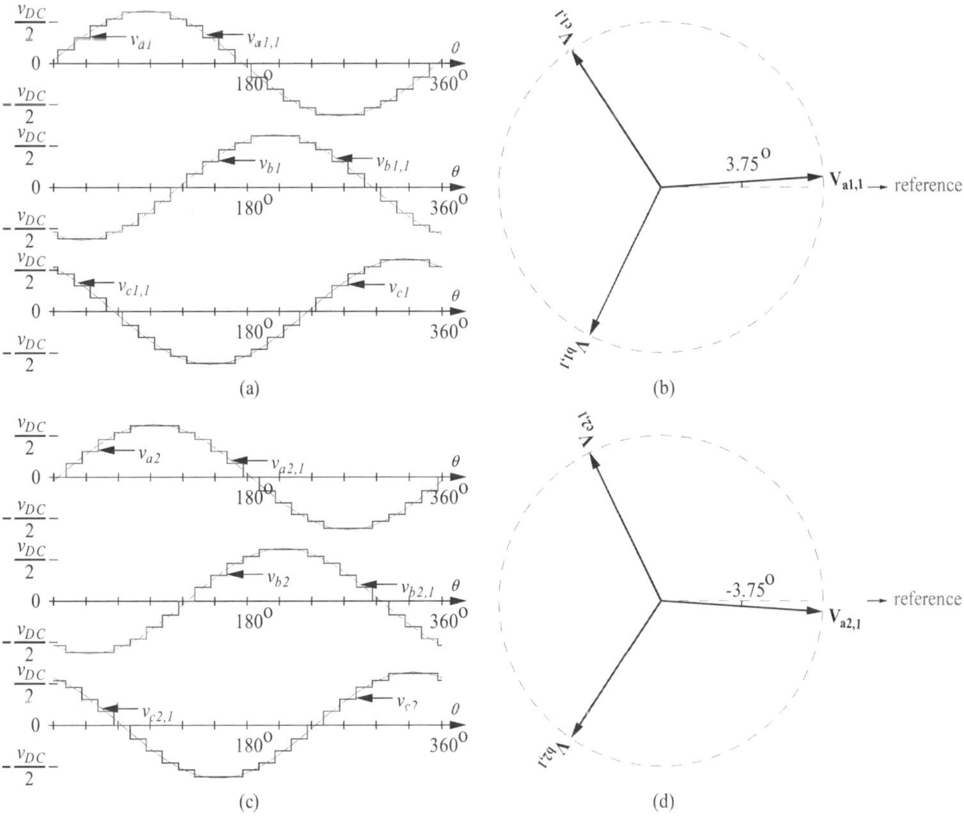

Figure 6-25. Two 24-pulse voltages and fundamental voltage phasors in a 48-pulse QHN-VSC configuration. (a) Three 120° phase-shifted fundamental voltages and the corresponding 24-pulse voltages at 3.75°. (b) Phasor diagram of fundamental voltage phasors in *a1–b1–c1* 24-pulse QHN-VSC. (c) Three 120° phase-shifted fundamental voltages and the corresponding 12-pulse voltages at –3.75°. (d) Phasor diagram of fundamental voltage phasors in *a2–b2–c2* 24-pulse QHN-VSC.

Figure 6-26a shows the three-phase, 48-pulse voltage (e_a, e_b, and e_c) and the corresponding fundamental components ($e_{a,1}$, $e_{b,1}$, and $e_{c,1}$). The fundamental pole voltage phasors ($\mathbf{E_{a,1}}$, $\mathbf{E_{b,1}}$, and $\mathbf{E_{c,1}}$), shown in Figure 6-26b, are at 0° (reference), –120°, and –240°, respectively.

The output voltages exhibit fundamental components and harmonic components with amplitudes of

$$E_{a,n} = E_{b,n} = E_{c,n} = 8\frac{n_2}{n_1}\frac{2v_{DC}}{n\pi}\cos\left(\frac{n\pi}{24}\right)\cos\left(\frac{n\pi}{48}\right) \qquad (6\text{-}121)$$

where $n = 1$ and $12k \pm 1$ for $k = 1, 2, 3$, and so on.

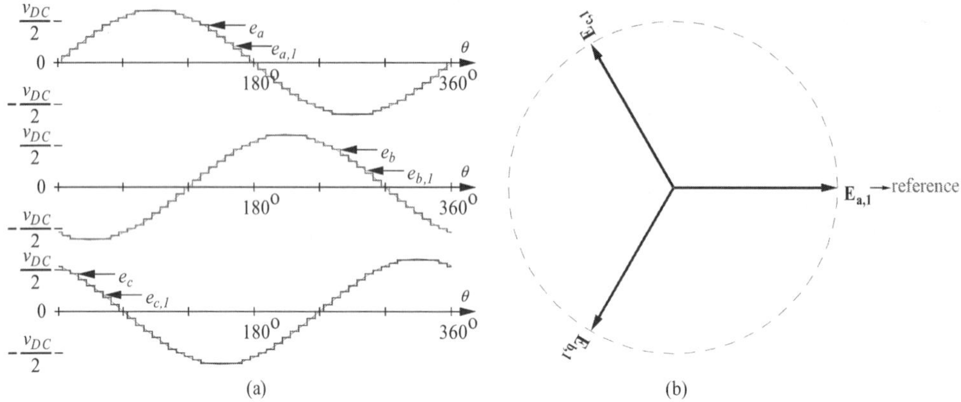

Figure 6-26. (a) Three 120° phase-shifted fundamental output voltages and the corresponding 48-pulse voltages. (b) Phasor diagram of fundamental output voltage phasors in a 48-pulse QHN-VSC configuration.

The amplitudes of harmonic components (n) normalized to the fundamental component are

$$\frac{E_{a,n}}{E_{a,1}} = \frac{E_{b,n}}{E_{b,1}} = \frac{E_{c,n}}{E_{c,1}} = \frac{\cos(n\pi/24)\cos(n\pi/48)}{n\cos(\pi/24)\cos(\pi/48)} \tag{6-122}$$

where $n = 12k \pm 1$ for $k = 1, 2, 3$, and so on.

The amplitudes of the fundamental components normalized to the DC voltage (gain of the VSC) are

$$\frac{E_{a,1}}{v_{DC}} = \frac{E_{b,1}}{v_{DC}} = \frac{E_{c,1}}{v_{DC}} = \frac{n_2}{n_1}\frac{16}{\pi}\cos\left(\frac{\pi}{24}\right)\cos\left(\frac{\pi}{48}\right) \tag{6-123}$$

6.2.6.1 Modeling a 48-Pulse QHN-VSC with Two-Level Poles. The 48-pulse, two-level QHN-VSC is run from the main file `65748Q21.DAT` that has the same setup as the six-pulse, two-level VSC in file `61206P21.DAT`, except that the three $INCLUDE$ files are replaced with `65848Q21.GAT` (see Appendix C, Section 55) for generating gating pulses for the switches, `65948P21.POL` (see Appendix C, Section 56) for generating ideal pole voltages, and `66048Q21.MAG` (see Appendix C, Section 57) for implementing the ideal magnetic circuit.

Figure 6-27 shows the output voltages from a 12-, 24-, and 48-pulse HN/QHN-VSC. A 24-pulse or higher order VSC is more than adequate in most applications to meet the power quality standard.

Figure 6-28 shows the normalized harmonic components of the output voltages from a 12-pulse HN-VSC, a 24-pulse QHN-VSC, and a 48-pulse QHN-VSC as a function of harmonic order (n).

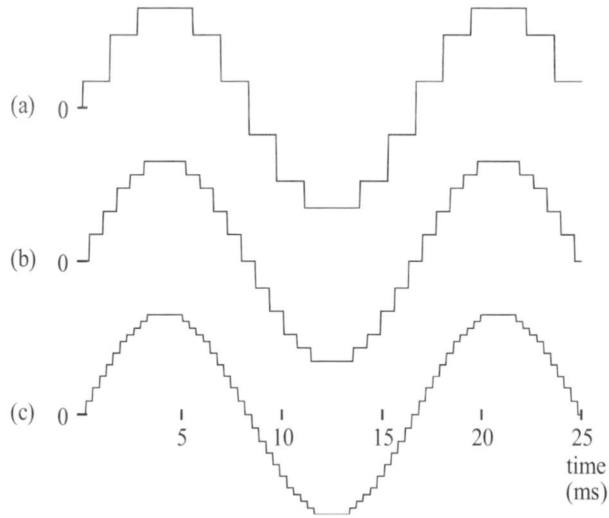

Figure 6-27. Output voltages from the simulation of (a) a 12-pulse HN-VSC, (b) a 24-pulse QHN-VSC, and (c) a 48-pulse QHN-VSC.

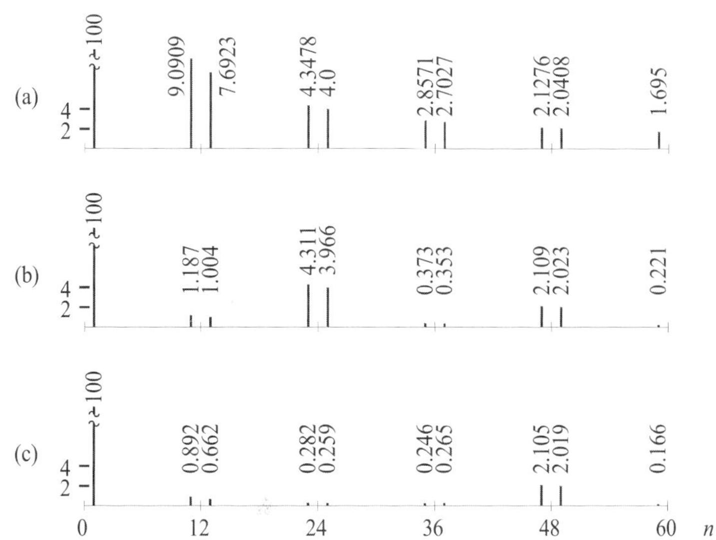

Figure 6-28. Normalized harmonic components of output voltages from (a) a 12-pulse HN-VSC, (b) a 24-pulse QHN-VSC, and (c) a 48-pulse QHN-VSC as a function of harmonic order (n).

The QHN-VSC circuits have been discussed so far are made of two-level poles, which produce a square wave voltage with two levels: a positive level and a negative level. The magnitude of the AC voltage is related to the magnitude of the DC voltage by the gain of the VSC. In certain applications, more than one VSC may operate from a shared DC link capacitor and one VSC is required to produce a variable magnitude AC voltage from zero to the maximum value within the design limit. One way to achieve this goal would be to use a DC-to-DC converter that converts a DC voltage from one level to another and use a multipulse HN-VSC with two-level poles as discussed in this chapter. The alternate way to generate a variable AC voltage from a fixed DC voltage is to use a three-level pole. A three-level pole produces a variable width square wave voltage. As a special case, a three-level pole produces a square wave voltage when used as a two-level pole. Next, a QHN-VSC with three-level poles is described.

6.2.7 Generation of a Quasisquare Wave Voltage with a Three-Level Pole

Figure 6-29 shows the single phase VSC circuit, referred to as a three-level "pole," which consists of a positive switch, A+, a negative switch, A–, and an AC switch, A_{AC}. Each switch may be made out of several controllable semiconductor switches, such as GTOs that are connected in a series or parallel combination to achieve the desired voltage and current ratings of the switch. The pole is connected across a pair of capacitors that are charged with a total DC voltage of v_{DC}. When the switch A+ is on, the midpoint of the pole is tied to the positive DC terminal; when the switch A_{AC} is on, the midpoint of the pole is tied to the midpoint of the DC link capacitor; and when the switch A– is on, the midpoint of the pole is tied to the negative DC terminal. When the switches are closed and opened alternately, the pole output voltage (v_{AO}) at the midpoint of pole A with respect to the midpoint (O) of the DC link capacitor is a quasisquare wave, which has three levels ($+v_{DC}/2$, 0, and $-v_{DC}/2$). Therefore, this pole circuit is referred to as a three-level pole. The time period, T (i.e., $1/f$), is equally divided between the positive and the negative half-cycles, where f is the fundamental frequency of the quasisquare wave.

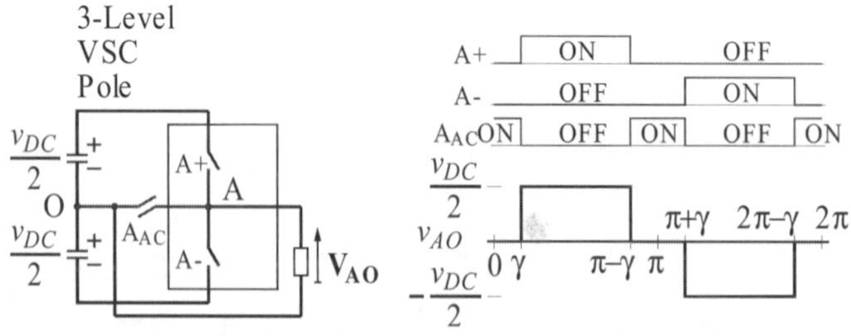

Figure 6-29. Three-level VSC "pole" and its output voltage (all angles are in radians).

The pole output voltage that contains harmonic components (n) is expressed in a general form as

$$v_{AO} = a_0 + \sum_{n=1}^{\infty} V_{AO,n} \sin(n\theta + \alpha_{An}) \tag{6-1}$$

where θ is the free-running phase angle. The amplitude ($V_{AO,n}$) and the relative phase angle (α_{An}) with respect to the PLL angle (θ) are given by

$$V_{AO,n} = \sqrt{a_n^2 + b_n^2} \tag{6-2a}$$

and

$$\alpha_{An} = \tan^{-1} \frac{a_n}{b_n} \tag{6-2b}$$

The DC component (a_0) in v_{AO} is calculated from equation A-90 (see Appendix A) as

$$a_0 = \frac{1}{2\pi} \int_0^{2\pi} v_{AO} d\theta = \frac{1}{2\pi} \frac{v_{DC}}{2} \left[\int_{\gamma}^{\pi-\gamma} d\theta - \int_{\pi+\gamma}^{2\pi-\gamma} d\theta \right]$$

$$= \frac{1}{2\pi} \frac{v_{DC}}{2} \left[\theta\Big|_{\gamma}^{\pi-\gamma} - \theta\Big|_{\pi+\gamma}^{2\pi-\gamma} \right] = \frac{1}{2\pi} \frac{v_{DC}}{2} [\pi - \gamma - \gamma - (2\pi - \gamma) + \pi + \gamma] = 0$$

$$a_0 = 0 \tag{6-124}$$

The amplitudes (a_n) for $n = 1, 2, 3, \ldots$ of the cosine components in v_{AO} are calculated from equation A-91 as

$$a_n = \frac{1}{\pi} \int_0^{2\pi} v_{AO} \cos(n\theta)\, d\theta = \frac{1}{\pi} \frac{v_{DC}}{2} \left[\int_{\gamma}^{\pi-\gamma} \cos(n\theta)\, d\theta - \int_{\pi+\gamma}^{2\pi-\gamma} \cos(n\theta)\, d\theta \right]$$

$$= \frac{1}{n\pi} \frac{v_{DC}}{2} \left[\sin(n\theta)\Big|_{\gamma}^{\pi-\gamma} - \sin(n\theta)\Big|_{\pi+\gamma}^{2\pi-\gamma} \right]$$

$$= \frac{1}{n\pi} \frac{v_{DC}}{2} [\sin n(\pi - \gamma) + \sin n(\pi + \gamma) - \sin n\gamma - \sin n(2\pi - \gamma)]$$

$$= \frac{1}{n\pi} \frac{v_{DC}}{2} [2\sin(n\pi)\cos(n\gamma) - 2\sin(n\pi)\cos n(\pi - \gamma)] = 0$$

$$a_n = 0 \tag{6-125}$$

The amplitudes (b_n) for $n = 1, 2, 3, \ldots$ of the sine components in v_{AO} are calculated from equation A-92 as

$$b_n = \frac{1}{\pi}\int_0^{2\pi} v_{AO}\sin(n\theta)\,d\theta = \frac{1}{\pi}\frac{v_{DC}}{2}\left[\int_{\gamma}^{\pi-\gamma}\sin(n\theta)\,d\theta - \int_{\pi+\gamma}^{2\pi-\gamma}\sin(n\theta)d\theta\right]$$

$$= \frac{1}{n\pi}\frac{v_{DC}}{2}\left[-\cos(n\theta)\Big|_{\gamma}^{\pi-\gamma} + \cos(n\theta)\Big|_{\pi+\gamma}^{2\pi-\gamma}\right]$$

$$= \frac{1}{n\pi}\frac{v_{DC}}{2}\left[-\cos n(\pi-\gamma) - \cos n(\pi+\gamma) + \cos n\gamma + \cos n(2\pi-\gamma)\right]$$

$$= \frac{1}{n\pi}\frac{v_{DC}}{2}\left[-2\cos(n\pi)\cos(n\gamma) + 2\cos(n\pi)\cos n(\pi-\gamma)\right]$$

$$= \frac{1}{n\pi}v_{DC}\left[\cos n(\pi-\gamma) - \cos(n\gamma)\right]\cos(n\pi)$$

$$= \frac{1}{n\pi}v_{DC}(-1)^n\left[2\sin\left(n\frac{\pi}{2}\right)\sin n\left(-\frac{\pi}{2}+\gamma\right)\right]$$

$$= \frac{2}{n\pi}v_{DC}(-1)^n\sin\left(n\frac{\pi}{2}\right)\left[\sin n\left(-\frac{\pi}{2}\right)\cos(n\gamma) + \cos n\left(-\frac{\pi}{2}\right)\sin(n\gamma)\right]$$

for any odd n,

$$= \frac{2}{n\pi}v_{DC}(-1)^{n+1}\sin\left(n\frac{\pi}{2}\right)\sin\left(n\frac{\pi}{2}\right)\left[\cos n\gamma\right] = \frac{2v_{DC}}{n\pi}\cos(n\gamma) \qquad (6\text{-}126)$$

for any odd n

$$b_n = \frac{2v_{DC}}{n\pi}\cos(n\gamma) \qquad (6\text{-}127)$$

The amplitude ($V_{AO,n}$) and the relative phase angle (α_{An}) with respect to the PLL angle (θ) are

$$V_{AO,n} = \sqrt{a_n^2 + b_n^2} = b_n = \frac{2v_{DC}}{n\pi}\cos(n\gamma) \qquad (6\text{-}128a)$$

and

$$\alpha_{An} = \tan^{-1}\frac{a_n}{b_n} = 0 \qquad (6\text{-}128b)$$

for any odd n.

The expression for v_{AO} shown in equation 6-1 can be rewritten as

$$v_{AO} = \frac{2v_{DC}}{\pi} \sum_{n=1}^{\infty} \frac{1}{n} \cos(n\gamma)\sin(n\theta) \tag{6-129}$$

for any odd n.

The quasisquare wave voltage contains a positive sequence fundamental component and all the odd harmonic components, such as the zero-sequence third, the negative-sequence fifth, and the positive-sequence seventh, and so on.

The amplitude of the fundamental component (for $n = 1$) is

$$V_{AO,1} = \frac{2v_{DC}}{\pi} \cos\gamma \tag{6-130}$$

From equations 6-128a and 6-130, the amplitudes of harmonic components (n) normalized to the fundamental component are

$$\frac{V_{AO,n}}{V_{AO,1}} = \frac{\cos(n\gamma)}{n\cos\gamma} \tag{6-131}$$

for any odd n.

6.2.8 Six-pulse VSC with Three-Level Poles

Figure 6-30 shows three poles A, B, and C that are connected across a shared DC link with a pair of capacitors, and the pole outputs are connected to a three-phase load whose neutral point (N) is not connected to the midpoint of the DC link capacitor. A

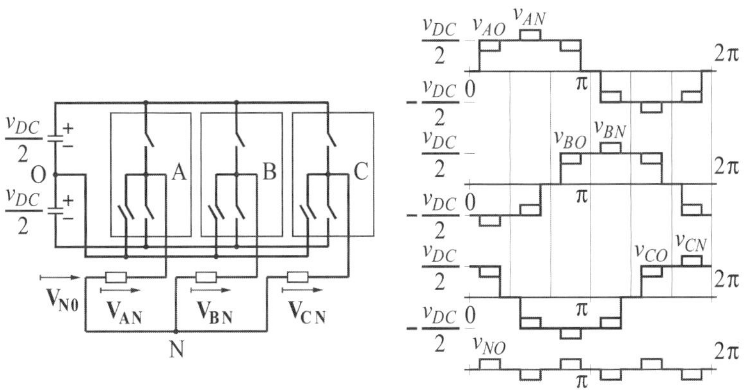

Figure 6-30. Six-pulse VSC with three-level poles and its output voltages (all angles are in radians).

six-pulse VSC requires the generation of three square wave voltages (v_{AO}, v_{BO}, and v_{CO}) as shown in Figure 6-30.

The poles A, B, and C are operated in such a way that the pole voltages (v_{AO}, v_{BO}, and v_{CO}) are time-shifted from one another by one-third of the time period (T) of the pole voltage. Therefore, the fundamental phasors ($\mathbf{V}_{A,1}$, $\mathbf{V}_{B,1}$, and $\mathbf{V}_{C,1}$) are 120° apart. The B and C poles' output voltages are generally expressed as

$$v_{BO} = a_0 + \sum_{n=1}^{\infty} V_{BO,n} \sin(n\theta + \alpha_{Bn}) \tag{6-10}$$

and

$$v_{CO} = a_0 + \sum_{n=1}^{\infty} V_{CO,n} \sin(n\theta + \alpha_{Cn}) \tag{6-11}$$

where

$$V_{BO,n} = V_{CO,n} = \sqrt{a_n^2 + b_n^2} \tag{6-12a}$$

and

$$\alpha_{Bn} = \alpha_{Cn} = \tan^{-1} \frac{a_n}{b_n} \tag{6-12b}$$

The DC component (a_0) in v_{BO} is calculated from equation A-90 (see Appendix A) as

$$a_0 = \frac{1}{2\pi} \int_0^{2\pi} v_{BO}\, d\theta = \frac{1}{2\pi} \frac{v_{DC}}{2} \left[- \int_0^{2\pi/3-\gamma} d\theta + \int_{2\pi/3+\gamma}^{5\pi/3-\gamma} d\theta - \int_{5\pi/3+\gamma}^{2\pi} d\theta \right]$$

$$= \frac{1}{2\pi} \frac{v_{DC}}{2} \left[-\theta \Big|_0^{2\pi/3-\gamma} + \theta \Big|_{2\pi/3+\gamma}^{5\pi/3-\gamma} - \theta \Big|_{5\pi/3+\gamma}^{2\pi} \right]$$

$$= \frac{1}{2\pi} \frac{v_{DC}}{2} \left[-\frac{2\pi}{3} + \gamma + \frac{5\pi}{3} - \gamma - \frac{2\pi}{3} - \gamma - 2\pi + \frac{5\pi}{3} + \gamma \right] = 0$$

$$a_0 = 0 \tag{6-132}$$

The amplitudes (a_n) for $n = 1, 2, 3, \ldots$ of the cosine components in v_{BO} are calculated from equation A-91 as

$$a_n = \frac{1}{\pi} \int_0^{2\pi} v_{BO} \cos(n\theta)\, d\theta$$

$$= \frac{1}{\pi} \frac{v_{DC}}{2} \left[- \int_0^{2\pi/3-\gamma} \cos(n\theta) d\theta + \int_{2\pi/3+\gamma}^{5\pi/3-\gamma} \cos(n\theta) d\theta - \int_{5\pi/3+\gamma}^{2\pi} \cos(n\theta)\, d\theta \right]$$

$$= \frac{1}{n\pi}\frac{v_{DC}}{2}\left[-\sin(n\theta)\Big|_0^{2\pi/3-\gamma} + \sin(n\theta)\Big|_{2\pi/3+\gamma}^{5\pi/3-\gamma} - \sin(n\theta)\Big|_{5\pi/3+\gamma}^{2\pi}\right]$$

$$= \frac{1}{n\pi}\frac{v_{DC}}{2}\left[-\sin n\left(\frac{2\pi}{3}-\gamma\right) - \sin n\left(\frac{2\pi}{3}+\gamma\right) + \sin n\left(\frac{5\pi}{3}-\gamma\right) + \sin n\left(\frac{5\pi}{3}+\gamma\right)\right]$$

$$= \frac{1}{n\pi}\frac{v_{DC}}{2}\left[-2\sin\left(n\frac{2\pi}{3}\right)\cos(n\gamma) + 2\sin\left(n\frac{5\pi}{3}\right)\cos(n\gamma)\right]$$

$$= \frac{1}{n\pi}v_{DC}\left[\sin\left(n\frac{5\pi}{3}\right) - \sin\left(n\frac{2\pi}{3}\right)\right]\cos(n\gamma)$$

$$= \frac{2v_{DC}}{n\pi}\cos\left(n\frac{7\pi}{6}\right)\sin\left(n\frac{\pi}{2}\right)\cos(n\gamma)$$

$$a_n = \frac{2v_{DC}}{n\pi}\cos\left(n\frac{7\pi}{6}\right)\sin\left(n\frac{\pi}{2}\right)\cos(n\gamma) \tag{6-133}$$

The amplitudes (b_n) for $n = 1, 2, 3, \ldots$ of the sine components in v_{BO} are calculated from equation A-92 as

$$b_n = \frac{1}{\pi}\int_0^{2\pi} v_{BO}\sin(n\theta)\, d\theta$$

$$= \frac{1}{\pi}\frac{v_{DC}}{2}\left[-\int_0^{2\pi/3-\gamma}\sin(n\theta)d\theta + \int_{2\pi/3+\gamma}^{5\pi/3-\gamma}\sin(n\theta)d\theta - \int_{5\pi/3+\gamma}^{2\pi}\sin(n\theta)\,d\theta\right]$$

$$= \frac{1}{n\pi}\frac{v_{DC}}{2}\left[\cos(n\theta)\Big|_0^{2\pi/3-\gamma} - \cos(n\theta)\Big|_{2\pi/3+\gamma}^{5\pi/3-\gamma} + \cos(n\theta)\Big|_{5\pi/3+\gamma}^{2\pi}\right]$$

$$= \frac{1}{n\pi}\frac{v_{DC}}{2}\left[\cos n\left(\frac{2\pi}{3}-\gamma\right) - 1 - \cos n\left(\frac{5\pi}{3}-\gamma\right) + \cos n\left(\frac{2\pi}{3}+\gamma\right) + 1 - \cos n\left(\frac{5\pi}{3}+\gamma\right)\right]$$

$$= \frac{1}{n\pi}\frac{v_{DC}}{2}\left[\cos n\left(\frac{2\pi}{3}-\gamma\right) + \cos n\left(\frac{2\pi}{3}+\gamma\right) - \cos n\left(\frac{5\pi}{3}-\gamma\right) - \cos n\left(\frac{5\pi}{3}+\gamma\right)\right]$$

$$= \frac{1}{n\pi}\frac{v_{DC}}{2}\left[2\cos\left(n\frac{2\pi}{3}\right)\cos(n\gamma) - 2\cos\left(n\frac{5\pi}{3}\right)\cos(n\gamma)\right]$$

$$= \frac{1}{n\pi}v_{DC}\left[\cos\left(n\frac{2\pi}{3}\right) - \cos\left(n\frac{5\pi}{3}\right)\right]\cos(n\gamma)$$

$$= \frac{2v_{DC}}{n\pi} \sin\left(n\frac{7\pi}{6}\right) \sin\left(n\frac{\pi}{2}\right) \cos(n\gamma)$$

$$b_n = \frac{2v_{DC}}{n\pi} \sin\left(n\frac{7\pi}{6}\right) \sin\left(n\frac{\pi}{2}\right) \cos(n\gamma) \tag{6-134}$$

The amplitude ($V_{BO,n}$) and the relative phase angle (α_{Bn}) with respect to the PLL angle (θ) are

$$V_{BO,n} = \sqrt{a_n^2 + b_n^2} = \sqrt{\left[\frac{2}{n\pi} v_{DC} \sin\left(n\frac{\pi}{2}\right) \cos(n\gamma)\right]^2 \left[\cos^2\left(n\frac{7\pi}{6}\right) + \sin^2\left(n\frac{7\pi}{6}\right)\right]}$$

$$= \frac{2v_{DC}}{n\pi} \cos(n\gamma) \tag{6-135a}$$

and

$$\alpha_{Bn} = \tan^{-1}\frac{a_n}{b_n} = \tan^{-1}\frac{1}{\tan\left(n\frac{7\pi}{6}\right)} \quad \text{or} \quad \alpha_{Bn} = n\left(-\frac{2\pi}{3}\right) \tag{6-135b}$$

for any odd n.

The expression for v_{BO} shown in equation 6-10 can be rewritten as

$$v_{BO} = \frac{2v_{DC}}{\pi} \sum_{n=1}^{\infty} \frac{1}{n} \cos(n\gamma) \sin n\left(\theta - \frac{2\pi}{3}\right) \tag{6-136}$$

for any odd n.

The quasisquare wave voltage contains a positive sequence fundamental component and all the odd harmonic components, such as the zero-sequence third, the negative-sequence fifth, and the positive-sequence seventh, and so on.

The amplitude of the fundamental component (for $n = 1$) is

$$V_{BO,1} = \frac{2v_{DC}}{\pi} \cos\gamma \tag{6-137}$$

From equations 6-135a and 6-137, the amplitudes of harmonic components (n) normalized to the fundamental component are

$$\frac{V_{BO,n}}{V_{BO,1}} = \frac{\cos(n\gamma)}{n\cos\gamma} \tag{6-138}$$

for any odd n.

The DC component (a_0) in v_{CO} is calculated from equation A-90 (see Appendix A) as

$$a_0 = \frac{1}{2\pi}\int_0^{2\pi} v_{CO}d\theta = \frac{1}{2\pi}\frac{v_{DC}}{2}\left[\int_0^{\pi/3-\gamma} d\theta - \int_{\pi/3+\gamma}^{4\pi/3-\gamma} d\theta + \int_{4\pi/3+\gamma}^{2\pi} d\theta\right]$$

$$= \frac{1}{2\pi}\frac{v_{DC}}{2}\left[\theta\Big|_0^{\pi/3-\gamma} - \theta\Big|_{\pi/3+\gamma}^{4\pi/3-\gamma} + \theta\Big|_{4\pi/3+\gamma}^{2\pi}\right]$$

$$= \frac{1}{2\pi}\frac{v_{DC}}{2}\left[\frac{\pi}{3} - \gamma - \frac{4\pi}{3} + \gamma + \frac{\pi}{3} + \gamma + 2\pi - \frac{4\pi}{3} - \gamma)\right] = 0$$

$$a_0 = 0 \tag{6-139}$$

The amplitudes (a_n) for $n = 1, 2, 3, \ldots$ of the cosine components in v_{CO} are calculated from equation A-91 as

$$a_n = \frac{1}{\pi}\int_0^{2\pi} v_{CO}\cos(n\theta)\,d\theta$$

$$= \frac{1}{\pi}\frac{v_{DC}}{2}\left[\int_0^{\pi/3-\gamma} \cos(n\theta)d\theta - \int_{\pi/3+\gamma}^{4\pi/3-\gamma} \cos(n\theta)d\theta + \int_{4\pi/3+\gamma}^{2\pi} \cos(n\theta)\,d\theta\right]$$

$$= \frac{1}{n\pi}\frac{v_{DC}}{2}\left[\sin(n\theta)\Big|_0^{\pi/3-\gamma} - \sin(n\theta)\Big|_{\pi/3+\gamma}^{4\pi/3-\gamma} + \sin(n\theta)\Big|_{4\pi/3+\gamma}^{2\pi}\right]$$

$$= \frac{1}{n\pi}\frac{v_{DC}}{2}\left[\sin n\left(\frac{\pi}{3} - \gamma\right) + \sin n\left(\frac{\pi}{3} + \gamma\right) - \sin n\left(\frac{4\pi}{3} - \gamma\right) - \sin n\left(\frac{4\pi}{3} + \gamma\right)\right]$$

$$= \frac{1}{n\pi}\frac{v_{DC}}{2}\left[2\sin\left(n\frac{\pi}{3}\right)\cos(n\gamma) - 2\sin\left(n\frac{4\pi}{3}\right)\cos(n\gamma)\right]$$

$$= \frac{1}{n\pi}v_{DC}\left[\sin\left(n\frac{\pi}{3}\right) - \sin\left(n\frac{4\pi}{3}\right)\right]\cos(n\gamma)$$

$$= \frac{2v_{DC}}{n\pi}\cos\left(n\frac{5\pi}{6}\right)\sin\left(n\frac{-\pi}{2}\right)\cos(n\gamma)$$

$$a_n = \frac{2v_{DC}}{n\pi}\cos\left(n\frac{5\pi}{6}\right)\sin\left(n\frac{-\pi}{2}\right)\cos(n\gamma) \tag{6-140}$$

The amplitudes (b_n) for $n = 1, 2, 3, \ldots$ of the sine components in v_{CO} are calculated from equation A-92 as

$$b_n = \frac{1}{\pi}\int_0^{2\pi} v_{CO}\sin(n\theta)\,d\theta$$

$$= \frac{1}{\pi}\frac{v_{DC}}{2}\left[\int_0^{\pi/3-\gamma}\sin(n\theta)d\theta - \int_{\pi/3+\gamma}^{4\pi/3-\gamma}\sin(n\theta)d\theta + \int_{4\pi/3+\gamma}^{2\pi}\sin(n\theta)\,d\theta\right]$$

$$= \frac{1}{n\pi}\frac{v_{DC}}{2}\left[-\cos(n\theta)\Big|_0^{\pi/3-\gamma} + \cos(n\theta)\Big|_{\pi/3+\gamma}^{4\pi/3-\gamma} - \cos(n\theta)\Big|_{4\pi/3+\gamma}^{2\pi}\right]$$

$$= \frac{1}{n\pi}\frac{v_{DC}}{2}\left[-\cos n\left(\frac{\pi}{3}-\gamma\right)+1+\cos n\left(\frac{4\pi}{3}-\gamma\right)-\cos n\left(\frac{\pi}{3}+\gamma\right)-1+\cos n\left(\frac{4\pi}{3}+\gamma\right)\right]$$

$$= \frac{1}{n\pi}\frac{v_{DC}}{2}\left[-\cos n\left(\frac{\pi}{3}-\gamma\right)-\cos n\left(\frac{\pi}{3}+\gamma\right)+\cos n\left(\frac{4\pi}{3}-\gamma\right)+\cos n\left(\frac{4\pi}{3}+\gamma\right)\right]$$

$$= \frac{1}{n\pi}\frac{v_{DC}}{2}\left[-2\cos\left(n\frac{\pi}{3}\right)\cos(n\gamma) + 2\cos\left(n\frac{4\pi}{3}\right)\cos(n\gamma)\right]$$

$$= \frac{1}{n\pi}v_{DC}\left[-\cos\left(n\frac{\pi}{3}\right)+\cos\left(n\frac{4\pi}{3}\right)\right]\cos(n\gamma)$$

$$= \frac{2v_{DC}}{n\pi}\sin\left(n\frac{5\pi}{6}\right)\sin\left(-n\frac{\pi}{2}\right)\cos(n\gamma)$$

$$b_n = \frac{2v_{DC}}{n\pi}\sin\left(n\frac{5\pi}{6}\right)\sin\left(n\frac{-\pi}{2}\right)\cos(n\gamma) \tag{6-141}$$

The amplitude ($V_{CO,n}$) and the relative phase angle (α_{Cn}) with respect to the PLL angle (θ) are

$$V_{CO,n} = \sqrt{a_n^2 + b_n^2} = \sqrt{\left[\frac{2}{n\pi}v_{DC}\sin\left(-n\frac{\pi}{2}\right)\cos(n\gamma)\right]^2\left[\cos^2\left(n\frac{5\pi}{6}\right)+\sin^2\left(n\frac{5\pi}{6}\right)\right]}$$

$$= \frac{2v_{DC}}{n\pi}\cos(n\gamma) \tag{6-142a}$$

and

$$\alpha_{Cn} = \tan^{-1}\frac{a_n}{b_n} = \tan^{-1}\frac{1}{\tan\left(n\frac{5\pi}{6}\right)}$$

or

$$\alpha_{Cn} = n\left(-\frac{4\pi}{3}\right) \tag{6-142b}$$

for any odd n.

The expression for v_{CO} shown in equation 6-11 can be rewritten as

$$v_{CO} = \frac{2v_{DC}}{\pi}\sum_{n=1}^{\infty}\frac{1}{n}\cos(n\gamma)\sin n\left(\theta - \frac{4\pi}{3}\right) = \frac{2v_{DC}}{\pi}\sum_{n=1}^{\infty}\frac{1}{n}\cos(n\gamma)\sin n\left(\theta + \frac{2\pi}{3}\right) \tag{6-143}$$

for any odd n.

The quasisquare wave voltage contains a positive-sequence fundamental component and all the odd harmonic components, such as the zero-sequence third, the negative-sequence fifth, and the positive-sequence seventh, and so on.

The amplitude of the fundamental component (for $n = 1$) is

$$V_{CO,1} = \frac{2v_{DC}}{\pi}\cos\gamma \tag{6-144}$$

From equations 6-142a and 6-144, the amplitudes of harmonic components (n) normalized to the fundamental component are

$$\frac{V_{CO,n}}{V_{CO,1}} = \frac{\cos(n\gamma)}{n\cos\gamma} \tag{6-145}$$

for any odd n.

The three 120° phase-shifted quasisquare wave voltages are rewritten as

$$v_{AO} = \frac{2v_{DC}}{\pi}\sum_{n=1}^{\infty}\frac{1}{n}\cos(n\gamma)\sin n\theta \tag{6-146a}$$

$$v_{BO} = \frac{2v_{DC}}{\pi}\sum_{n=1}^{\infty}\frac{1}{n}\cos(n\gamma)\sin n(\theta - 120°) \tag{6-146b}$$

$$v_{CO} = \frac{2v_{DC}}{\pi}\sum_{n=1}^{\infty}\frac{1}{n}\cos(n\gamma)\sin n(\theta + 120°) \tag{6-146c}$$

Since the zero-sequence components of each pole current have no return path to the midpoint of the DC link capacitor, the zero-sequence components of each pole voltage, $v_{NO} = (v_{AO} + v_{BO} + v_{CO})/3$, appear between the neutral point and the midpoint of the DC link capacitor. Therefore, each phase of the load voltages ($v_{AN} = v_{AO} - v_{NO}$, $v_{BN} = v_{BO} - v_{NO}$, and $v_{CN} = v_{CO} - v_{NO}$) consists of only a fundamental component and harmonic components (n) with normalized amplitudes, $[\cos(n\gamma)]/(n \cos \gamma)$, where $n = 6k \pm 1$ for $k = 1, 2, 3$, and so on.

The three-phase load voltages are

$$v_{AN} = \frac{2v_{DC}}{\pi} \sum_{n=1}^{\infty} \frac{1}{n} \cos(n\gamma) \sin n\theta \tag{6-147a}$$

$$v_{BN} = \frac{2v_{DC}}{\pi} \sum_{n=1}^{\infty} \frac{1}{n} \cos(n\gamma) \sin n(\theta - 120°) \tag{6-147b}$$

$$v_{CN} = \frac{2v_{DC}}{\pi} \sum_{n=1}^{\infty} \frac{1}{n} \cos(n\gamma) \sin n(\theta + 120°) \tag{6-147c}$$

which contain fundamental components and harmonic components $n = 6k \pm 1$ for $k = 1, 2, 3$, and so on.

The output voltages exhibit fundamental components and harmonic components (n) with amplitudes of

$$V_{AN,n} = V_{BN,n} = V_{CN,n} = \frac{2v_{DC}}{n\pi} \cos(n\gamma) \tag{6-148}$$

where $n = 1$ and $6k \pm 1$ for $k = 1, 2, 3$, and so on.

The amplitudes of harmonic components (n) normalized to the fundamental component are

$$\frac{V_{AN,n}}{V_{AN,1}} = \frac{V_{BN,n}}{V_{AN,1}} = \frac{V_{CN,n}}{V_{AN,1}} = \frac{\cos(n\gamma)}{n \cos \gamma} \tag{6-149}$$

where $n = 6k \pm 1$ for $k = 1, 2, 3$, and so on.

The amplitudes of the fundamental components normalized to the DC voltage (gain of the VSC) are

$$\frac{V_{AN,1}}{v_{DC}} = \frac{V_{BN,1}}{v_{DC}} = \frac{V_{CN,1}}{v_{DC}} = \frac{2}{\pi} \cos \gamma \tag{6-150}$$

For a $v_{DC} = \pi/2$ pu, the amplitudes of fundamental and harmonic components ($v_{AN,n}$) are shown in Figure 6-31 as a function of dead angle (γ) for possible values of γ between 0° and 90°. For $\gamma = 0°$, the fundamental as well as all the harmonic components have the highest possible amplitudes. At $\gamma = 0°$, a three-level pole acts as a two-level pole, which switches from $+v_{DC}/2$ to $-v_{DC}/2$ and offers the maximum possible DC-to-AC utilization. However, at $\gamma > 0°$, a three-level pole, which switches from

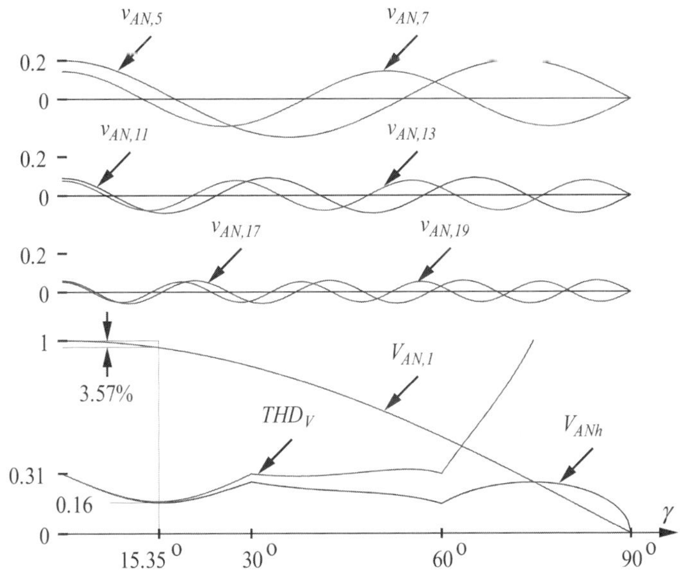

Figure 6-31. Variations of amplitudes of fundamental and harmonic components, harmonic content, and total harmonic distortion factor as a function of (γ), which is the dead angle of a three-level pole.

$+v_{DC}/2$ to 0 to $-v_{DC}/2$, produces an output waveform with improved quality, but at a reduced fundamental voltage. This is just what is needed if a VSC needs to produce a variable AC voltage from the shared DC link capacitor. For $\gamma > 0°$, the fundamental amplitude decreases monotonically to zero when γ reaches 90°. However, the amplitudes of the harmonic components vary sinusoidally. When the amplitude of a particular harmonic component is negative, it means that the harmonic component is in opposite phase with the fundamental component. A measure of the harmonic content defined as

$$V_{ANh} = \sqrt{\Sigma V_{AN,\,n}^2} \tag{6-151}$$

where $n = 6k \pm 1$ for $k = 1, 2, 3$, and so on, is also shown in Figure 6-31. A measure of the total harmonic distortion, THD_V, of the voltage, defined as

$$THD_V = V_{ANh}/V_{AN,1} \tag{6-152}$$

is also shown in Figure 6-31. At $\gamma_{optimum} = 15.35°$, the fundamental amplitude decreases by 3.57% compared to what is obtained from a two-level operation of the poles, but the harmonic content is reduced from 31% that is obtained from a two-level operation of the poles in a six-pulse VSC to 16%, resulting in the smallest THD_V.

6.2.9 12-Pulse HN-VSC with Three-Level Poles

A 12-pulse HN-VSC requires the generation of two three-phase sets (A–B–C and D–E–F) of square wave voltages, the placements of which are shown in Figure 6-8. The zero crossings (with positive slopes) of the voltages out of the A–B–C poles are at 0°, –120°, and –240°, respectively. The zero crossings (with positive slopes) of the voltages out of the D–E–F poles are at –30°, –150°, and –270°, respectively.

Figure 6-32 shows two six-pulse VSCs (A–B–C and D–E–F), which are operated from a shared DC link with a pair of capacitors. The six-pulse voltage from the A–B–C VSC is fed to a Y–Y transformer and that from the D–E–F VSC is fed to a Δ–Y transformer. The secondary sides of the Y-connected transformers are connected in series in the respective phases and the final output voltage is connected to a three-phase load (*a*–*b*–*c*). The displacement angle between two consecutive six-pulse VSCs in a multi-pulse VSC configuration is $2\pi/6m$ (i.e., 30°), where m (i.e., 2) is the total number of six-pulse VSCs that are used. The square wave voltages of the second six-pulse VSC with D, E, and F poles are time-shifted with respect to the corresponding A, B, and C pole voltages of the first six-pulse VSC by one-twelfth of the time period, equivalent of a fundamental angle of –30°. The pole voltage fundamental phasors ($\mathbf{V_{DO,1}}$, $\mathbf{V_{EO,1}}$, and $\mathbf{V_{FO,1}}$) are 120° apart and the fundamental voltage phasor set of the D–E–F VSC lags the fundamental voltage phasor set of the A–B–C VSC by 30°. Note that the fundamental components in the square wave voltages (with respect to the midpoint of the DC link capacitor) and the phase voltage are the same ($\mathbf{V_{DO,1}} = \mathbf{V_{DN,1}} = \mathbf{V_{D,1}}$, $\mathbf{V_{EO,1}} = \mathbf{V_{EN,1}} = \mathbf{V_{E,1}}$, and $\mathbf{V_{FO,1}} = \mathbf{V_{FN,1}} = \mathbf{V_{F,1}}$). The A winding and DE winding on the VSC side are excited with per-turn fundamental component voltages, which are of the same magnitude and in phase, and the fifth and the seventh harmonic components, each of which are of same magnitude but in opposite phase. Therefore, when the transformer

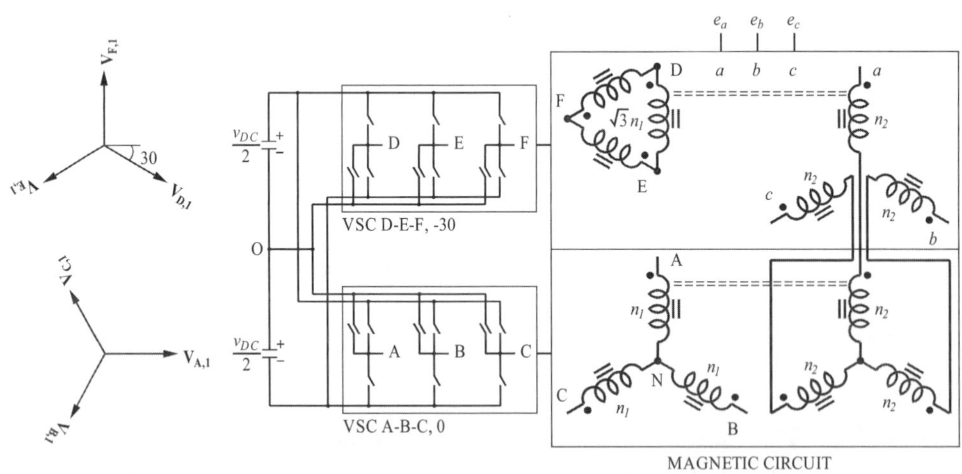

Figure 6-32. 12-pulse HN-VSC configuration with three-level poles (all angles are in degrees).

windings on the line side are connected in series, the phase-a voltage exhibits only a fundamental component and 12-pulse harmonic components. The secondary windings can have any turns ratio other than 0.5 to increase or decrease the output voltage.

The pole voltages from the A–B–C VSC exhibit a six-pulse harmonic neutralized waveform with harmonic components $n = 6k \pm 1$ for $k = 1, 2, 3$, and so on. Similarly, the pole voltages from the D–E–F VSC exhibit a six-pulse harmonic neutralized waveform whose harmonic components ($n = 6k \pm 1$ for $k = 1, 2, 3$, etc.) have the same magnitudes as the corresponding harmonic components of the A–B–C VSC's six-pulse harmonic neutralized waveform. However, the harmonic components ($n = 6k \pm 1$ for $k = 1, 3, 5$, etc.) are in opposite phases, whereas the harmonic components ($n = 6k \pm 1$ for $k = 2, 4, 6$, etc.) are in phase with the corresponding harmonic components of the A–B–C VSC's six-pulse harmonic neutralized waveform. Therefore, when all the output voltages from each six-pulse VSC are combined by connecting the corresponding phases in series, a 12-pulse harmonic neutralized waveform is obtained. The resulting output voltage exhibits a fundamental component and harmonic components (n), where $n = 12k \pm 1$ for $k = 1, 2, 3$, and so on. Note that the output voltage of a 12-pulse VSC with three-level poles is referred to as a 12-pulse waveform when the poles are operated with dead angle $\gamma = 0°$.

The three-phase load voltages are

$$e_a = 2\frac{n_2}{n_1}\frac{2v_{DC}}{\pi}\sum_{n=1}^{\infty}\frac{1}{n}\cos(n\gamma)\sin(n\theta) \tag{6-153a}$$

$$e_b = 2\frac{n_2}{n_1}\frac{2v_{DC}}{\pi}\sum_{n=1}^{\infty}\frac{1}{n}\cos(n\gamma)\sin n(\theta - 120°) \tag{6-153b}$$

$$e_c = 2\frac{n_2}{n_1}\frac{2v_{DC}}{\pi}\sum_{n=1}^{\infty}\frac{1}{n}\cos(n\gamma)\sin n(\theta + 120°) \tag{6-153c}$$

which contain fundamental components and harmonic components $n = 12k \pm 1$ for $k = 1, 2, 3$, and so on.

The output voltages exhibit fundamental components and harmonic components (n) with amplitudes of

$$E_{a,n} = E_{b,n} = E_{c,n} = 2\frac{n_2}{n_1}\frac{2v_{DC}}{n\pi}\cos(n\gamma) \tag{6-154}$$

where $n = 1$ and $12k \pm 1$ for $k = 1, 2, 3$, and so on.

The amplitudes of harmonic components (n) normalized to the fundamental component are

$$\frac{E_{a,n}}{E_{a,1}} = \frac{E_{b,n}}{E_{b,1}} = \frac{E_{c,n}}{E_{c,1}} = \frac{\cos(n\gamma)}{n\cos\gamma} \tag{6-155}$$

where $n = 6k \pm 1$ for $k = 1, 2, 3$, and so on.

The amplitudes of the fundamental components normalized to the DC voltage (gain of the VSC) are

$$\frac{E_{a,1}}{v_{DC}} = \frac{E_{b,1}}{v_{DC}} = \frac{E_{c,1}}{v_{DC}} = \frac{n_2}{n_1}\frac{4}{\pi}\cos\gamma \tag{6-156}$$

6.2.9.1 Modeling a 12-Pulse HN-VSC with Three-Level Poles. A main file `67512P31.DAT` (see Appendix C, Section 58) is run to generate a 12-pulse voltage using three-level poles. The $INCLUDE$ file `67612P31.GAT` (see Appendix C, Section 59) generates the gating pulses for the switches. The two additional $INCLUDE$ files—`62612P21.POL` and `62712P21.MAG`—for generating the ideal pole voltages and implementing the ideal magnetic circuit are the same as those used in a 12-pulse HN-VSC with two-level poles.

6.2.10 24-Pulse QHN-VSC with Three-Level Poles

The presence of 12-pulse harmonic components in the VSC output voltage may not be acceptable in many applications. Therefore, a VSC with a higher pulse output voltage should be considered. A 24-pulse QHN-VSC requires the generation of four three-phase sets of square wave voltages using A1–B1–C1, A2–B2–C2, D1–E1–F1, and D2–E2–F2 VSCs as discussed earlier. The displacement angle between two consecutive six-pulse VSCs in a multipulse VSC configuration is $2\pi/6m$ (i.e., 15°), where m (i.e., 4) is the total number of six-pulse VSCs that are used. The two phasors in the A–B–C three-phase groups (A1–A2, B1–B2, and C1–C2) are centered on 0°, –120°, and –240°, respectively. The two phasors in the D–E–F three-phase groups (D1–D2, E1–E2, and F1–F2) are centered on –30°, –150°, and +90°, respectively. The fundamental voltage phasor set of the A1–B1–C1 VSC is time-shifted by an equivalent fundamental angle of +7.5°. The zero-crossing positive voltages out of the A1–B1–C1 poles start at 7.5°, 7.5° – 120° (i.e., –112.5°), and 7.5° – 240° (i.e., –232.5°), respectively. The fundamental voltage phasor set of the A2–B2–C2 VSC is time-shifted by an equivalent fundamental angle of –7.5°. The zero-crossing positive voltages out of the A2–B2–C2 poles start at –7.5°, –7.5° – 120° (i.e., –127.5°), and –7.5° – 240° (i.e., –247.5°), respectively. The fundamental voltage phasor set of the D1–E1–F1 VSC is time-shifted by an equivalent fundamental angle of –22.5°. The zero-crossing positive voltages out of the D1–E1–F1 poles start at –22.5°, –22.5° – 120° (i.e., –142.5°), and –22.5° – 240° (i.e., –262.5°), respectively. The fundamental voltage phasor set of the D2–E2–F2 VSC is time-shifted by an equivalent fundamental angle of –37.5°. The zero-crossing positive voltages out of the D2–E2–F2 poles start at –37.5°, –37.5° – 120° (i.e., –157.5°), and –37.5° – 240° (i.e., –277.5°), respectively.

A 24-pulse QHN-VSC configuration, shown in Figure 6-33, consists of four six-pulse VSCs (A1–B1–C1, A2–B2–C2, D1–E1–F1, and D2–E2–F2) that are operated from a shared DC link with a pair of capacitors that are charged with a total DC voltage of v_{DC}. On the AC side, the output voltages are combined by a magnetic circuit to form a three-phase (a–b–c) voltage. The poles A1, B1, C1, A2, B2, C2, D1, E1, F1,

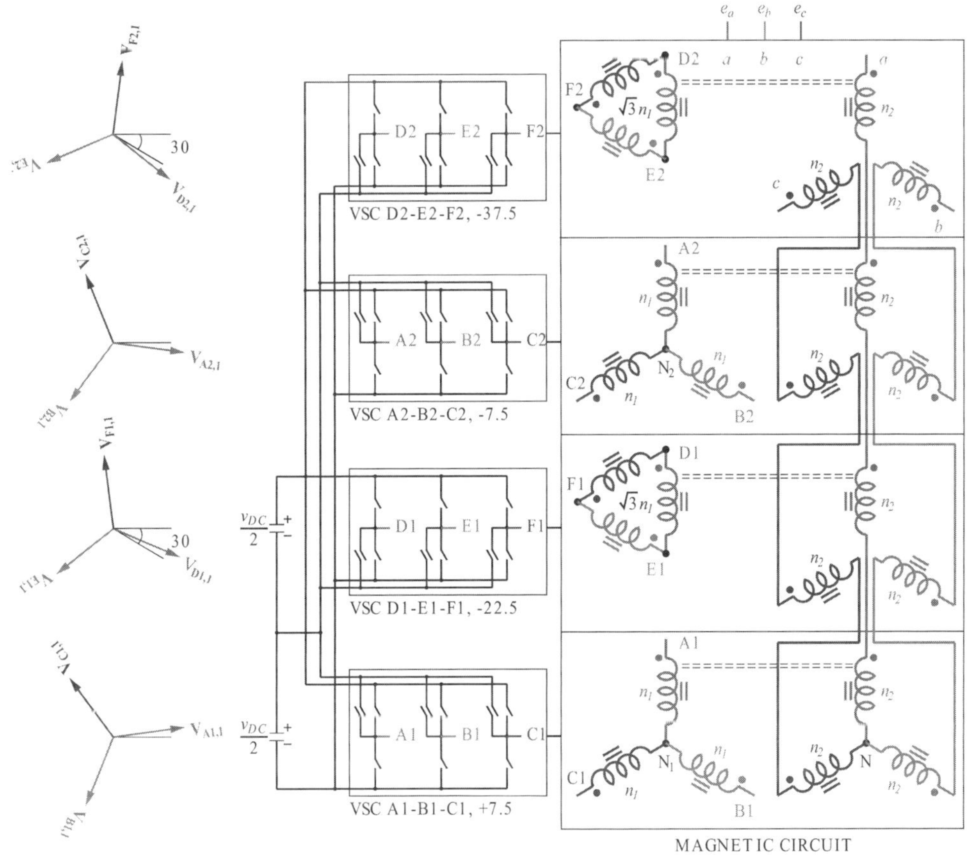

Figure 6-33. 24-pulse QHN-VSC configuration with three-level poles (all angles are in degrees).

D2, E2, and F2 are operated in such a way that the pole voltage fundamental phasors in a group, for example, $\mathbf{V}_{A1,1}$, $\mathbf{V}_{B1,1}$, and $\mathbf{V}_{C1,1}$ of the A1–B1–C1 VSC; $\mathbf{V}_{A2,1}$, $\mathbf{V}_{B2,1}$, and $\mathbf{V}_{C2,1}$ of the A2–B2–C2 VSC; $\mathbf{V}_{D1,1}$, $\mathbf{V}_{E1,1}$, and $\mathbf{V}_{F1,1}$ of the D1–E1–F1 VSC; and $\mathbf{V}_{D2,1}$, $\mathbf{V}_{E2,1}$, and $\mathbf{V}_{F2,1}$ of the D2–E2–F2 VSC, are 120° apart.

When the pole voltages from the VSCs A1–B1–C1 and D1–E1–F1 are magnetically combined, as described in Figure 6-9, the output voltage exhibits a 12-pulse HN waveform with harmonic components $n = 12k \pm 1$ for $k = 1, 2, 3$, and so on. In the 12-pulse HN-VSC of Figure 6-9, there are two sets of square wave voltages: one set from the A–B–C VSC is at 0° and the other set from the D–E–F VSC is at –30°; the final voltage set is at 0°. In the 24-pulse QHN-VSC of Figure 6-30, there are two sets of square wave voltages: one set from the A1–B1–C1 VSC is at 7.5° and the other set from the D1–E1–F1 VSC is at –22.5°; the corresponding 12-pulse voltage set is at 7.5°. When the pole voltages from the VSCs A2–B2–C2 and D2–E2–F2 are magnetically combined, the output voltage exhibits a 12-pulse HN waveform whose harmonic components ($n = 12k \pm 1$ for $k = 1, 2, 3$, etc.) have the same magnitudes as the corresponding harmonic components of the first 12-pulse HN waveform. However, in the second 12-pulse VSC, there are two sets of square wave voltages: one set from the A2–B2–C2 VSC is at –7.5° and the other set from the D1–E1–F1 VSC is at –37.5°; the corresponding 12-pulse voltage set is at –7.5°.

Two sets of three-phase, 12-pulse voltages are

$$e_{a1} = 2\frac{n_2}{n_1}\frac{2v_{DC}}{\pi}\sum_{n=1}^{\infty}\frac{1}{n}\cos(n\gamma)\sin n\left(\theta + \frac{\pi}{24}\right) \tag{6-157a}$$

$$e_{b1} = 2\frac{n_2}{n_1}\frac{2v_{DC}}{\pi}\sum_{n=1}^{\infty}\frac{1}{n}\cos(n\gamma)\sin n\left(\theta + \frac{\pi}{24} - 120^\circ\right) \tag{6-157b}$$

$$e_{c1} = 2\frac{n_2}{n_1}\frac{2v_{DC}}{\pi}\sum_{n=1}^{\infty}\frac{1}{n}\cos(n\gamma)\sin n\left(\theta + \frac{\pi}{24} + 120^\circ\right) \tag{6-157c}$$

and

$$e_{a2} = 2\frac{n_2}{n_1}\frac{2v_{DC}}{\pi}\sum_{n=1}^{\infty}\frac{1}{n}\cos(n\gamma)\sin n\left(\theta - \frac{\pi}{24}\right) \tag{6-158a}$$

$$e_{b2} = 2\frac{n_2}{n_1}\frac{2v_{DC}}{\pi}\sum_{n=1}^{\infty}\frac{1}{n}\cos(n\gamma)\sin n\left(\theta - \frac{\pi}{24} - 120^\circ\right) \tag{6-158b}$$

$$e_{c2} = 2\frac{n_2}{n_1}\frac{2v_{DC}}{\pi}\sum_{n=1}^{\infty}\frac{1}{n}\cos(n\gamma)\sin n\left(\theta - \frac{\pi}{24} + 120^\circ\right) \tag{6-158c}$$

which contain fundamental components and harmonic components $n = 12k \pm 1$ for $k = 1, 2, 3$, and so on.

When all the output voltages from each six-pulse VSC are combined by connecting the corresponding phases in series, as shown in Figure 6-16, the resulting output voltages are

$$e_a = 4\frac{n_2}{n_1}\frac{2v_{DC}}{\pi}\sum_{n=1}^{\infty}\frac{1}{n}\cos(n\gamma)\cos\left(\frac{n\pi}{24}\right)\sin n\theta \tag{6-159a}$$

$$e_b = 4\frac{n_2}{n_1}\frac{2v_{DC}}{\pi}\sum_{n=1}^{\infty}\frac{1}{n}\cos(n\gamma)\cos\left(\frac{n\pi}{24}\right)\sin n(\theta - 120^\circ) \tag{6-159b}$$

$$e_c = 4\frac{n_2}{n_1}\frac{2v_{DC}}{\pi}\sum_{n=1}^{\infty}\frac{1}{n}\cos(n\gamma)\cos\left(\frac{n\pi}{24}\right)\sin n(\theta + 120^\circ) \tag{6-159c}$$

which contain fundamental components and harmonic components $n = 12k \pm 1$ for $k =$

1, 2, 3, and so on, and exhibit a 24-pulse QHN waveform.

The output voltages exhibit fundamental components and harmonic components (n) with amplitudes of

$$E_{a,n} = E_{b,n} = E_{c,n} = 4\frac{n_2}{n_1}\frac{2v_{DC}}{n\pi}\cos(n\gamma)\cos\left(\frac{n\pi}{24}\right) \tag{6-160}$$

where $n = 1$ and $12k \pm 1$ for $k = 1, 2, 3$, and so on, and γ is the period during which each AC switch operates in each quarter cycle.

The amplitudes of harmonic components (n) normalized to the fundamental component are

$$\frac{E_{a,n}}{E_{a,1}} = \frac{E_{b,n}}{E_{b,1}} = \frac{E_{c,n}}{E_{c,1}} = \frac{\cos(n\pi/24)\cos(n\gamma)}{n\cos(\pi/24)\cos\gamma} \tag{6-161}$$

where $n = 12k \pm 1$ for $k = 1, 2, 3$, and so on.

The amplitudes of the fundamental components normalized to the DC voltage (gain of the VSC) are

$$\frac{E_{a,1}}{v_{DC}} = \frac{E_{b,1}}{v_{DC}} = \frac{E_{c,1}}{v_{DC}} = \frac{n_2}{n_1}\frac{8}{\pi}\cos\gamma\cos\left(\frac{\pi}{24}\right) = K_{VSC}\cos\gamma \tag{6-162}$$

where $K_{VSC} = (4 \times n_2/n_1) \times (2/\pi)\cos(\pi/24)$ is the fundamental AC amplitude-to-DC gain factor for a 24-pulse QHN-VSC with two-level poles.

Note that the output voltage of a 24-pulse QHN-VSC with three-level poles is referred to as a 24-pulse QHN waveform when the poles are operated with dead angle $\gamma = 0°$. If a 24-pulse QHN-VSC with three-level poles is operated with $\gamma = \pi/48$, the resulting output voltage is the same as that is obtained from a 48-pulse QHN-VSC. In conclusion, a QHN-VSC produces, in addition to the first dominant harmonic in the spectrum the same as the pulse number of the VSC, a residual harmonic content of $12k \pm 1$ components for $k = 1, 2, 3$ and so on, and, of course, a fundamental voltage.

Figure 6-34, parts a and b, show the normalized harmonic components of the output voltages from a 24-pulse QHN-VSC and a 48-pulse QHN-VSC as a function of harmonic order (n). The variations of fundamental component, harmonic content, and total harmonic distortion factor of a 24-pulse QHN-VSC with three-level poles as a function of γ is shown in Figure 6-34c. At $\gamma_{optimum} = 3.82°$, the fundamental amplitude decreases by 1.08% compared to what is obtained from a two-level operation of the poles, but the harmonic content is reduced to 3.91% from 7.7%, obtained from a two-level operation of the poles in a 24-pulse VSC, resulting in the smallest THD_V.

6.2.10.1 Modeling a 24-Pulse QHN-VSC with Three-Level Poles. A main file `68124Q31.DAT` (see Appendix C, Section 60) is run to generate a 24-pulse voltage using three-level poles. The $INCLUDE$ file `68224Q31.GAT` (see Appendix

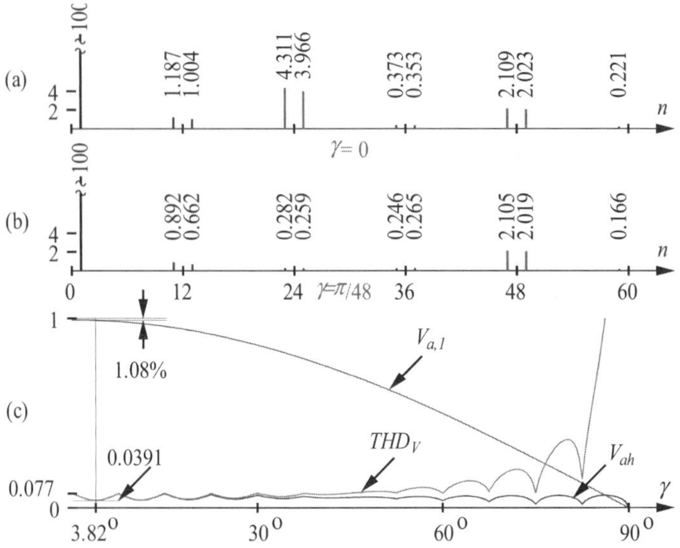

Figure 6-34. Normalized harmonic components of output voltages from (a) a 24-pulse QHN-VSC and (b) a 48-pulse QHN-VSC as a function of harmonic order. (c) Variations of amplitude of fundamental component, harmonic content, and THD as a function of γ, which is the dead angle of a three-level pole.

C, Section 61) generates the gating pulses for the switches. The two additional $INCLUDE$ files—`63824P21.POL` and `65024Q21.MAG`—for generating the ideal pole voltages and implementing the ideal magnetic circuit are the same as those used in a 24-pulse QHN-VSC with two-level poles.

Although a QHN-VSC scheme is implemented with standard transformers, the number of transformers needed makes the scheme quite heavy. The QHN-VSC schemes that are in use today are a mixture of standard and nonstandard transformers. This technique has been demonstrated in transmission line applications with VSC ratings in the range of 40 to 160 MVA.

6.2.11 Alternate Configuration for a QHN-VSC

A HN/QHN VSC can be built with multiple six-pulse VSCs with two-level poles whose output voltages are combined through multiple transformers that form a high-quality output voltage. If the two-level poles are replaced with three-level poles, the requirements of the magnetic circuit becomes less. For example, a 24-pulse QHN VSC with three-level poles that are operated with a fixed dead angle ($\gamma = 3.75°$) and a 48-pulse QHN VSC with two-level poles produce identical waveforms. In theory, by increasing the use of power electronics, the use of magnetics can be reduced to a minimum of just a coupling transformer at the output of a VSC. The choice of topology depends on the feasibility of the implementation of a particular compensation function, overall losses, and complexity.

A p-pulse QHN-VSC generates, in addition to the first dominant harmonic, $p - 1$, in the harmonic spectrum, a residual harmonic content of $12k + 1$ components for $k = 1, 2, 3$ and so on, and, of course, a fundamental voltage. A 12-pulse HN-VSC uses two six-pulse VSCs that are placed at 0° and –30°, respectively. In 24-pulse or higher order QHN-VSCs, half of the number of six-pulse VSCs are placed on 0° and the other half of the number of six-pulse VSCs are placed on –30°. The configurations as shown in the previous section use Y–Y and Δ–Y transformers for every 12-pulse VSC. The primary Y-windings of the Y–Y transformer are connected to a six-pulse VSC that generates the three-phase voltage at $\theta_{displacement}$, $\theta_{displacement} - 120°$, and $\theta_{displacement} - 240°$. The primary Δ-windings of the Δ–Y transformer are connected to a six-pulse VSC that generates the three-phase voltage at $\theta_{displacement} - 30°$, $\theta_{displacement} - 150°$, and $\theta_{displacement} + 90°$. In each of the three-phase, secondary Y-windings, there exists a fundamental component voltage and $6k \pm 1$ harmonic components for $k = 1, 2, 3$, and so on. Each six-pulse waveform consists of the same magnitudes of fundamental and harmonic components. However, the harmonic components for $k = 1, 3, 5$, and so on are in opposite phase, whereas harmonic components for $k = 2, 4, 6$, and so on, and the fundamental components are in phase in the two series-connected Y-windings. The voltages out of two series-connected windings go through an averaging process, thereby eliminating harmonic components for $k = 1, 3, 5$, and so on, and exhibiting fundamental and $12k \pm 1$ harmonic components for $k = 1, 2, 3$, and so on.

Using 12-pulse HN-VSCs as building blocks, a higher-order QHN-VSC can be made up by connecting a number of 12-pulse HN-VSCs in series as shown in the last section for 24-pulse and 48-pulse QHN-VSCs. Each 12-pulse HN-VSC provides an averaging of two six-pulse voltages. In a 24-pulse QHN-VSC, an additional averaging of the two 12-pulse voltages takes place. In a 48-pulse QHN-VSC, another averaging of the two 24-pulse voltages takes place. The same function can be implemented by averaging first the 0°-group pole voltages and then the –30°-group pole voltages with the use of multiple interphase transformers (IPTs). The only one-time use of a Y–Y transformer and a Δ–Y transformer (for implementing the necessary +30° shift of the D–E–F pole voltages) results in an additional averaging.

6.2.11.1 *Interphase Transformer (IPT).* An interphase transformer (IPT), shown in Figure 6-35, consists of two windings that are connected in series and wound on a single core. The two pole voltages (v_1 and v_2) are applied at the two free ends of the IPT windings and the output voltage (v_O) is at the common connection point of the two windings. If each of the windings consists of n_1 turns, the per-turn voltage in the two windings are related by

$$\frac{v_1 - v_O}{n_1} = \frac{v_O - v_2}{n_1} \tag{6-163}$$

The above equation can be rearranged as

$$v_O = \frac{v_1 + v_2}{2} \tag{6-164}$$

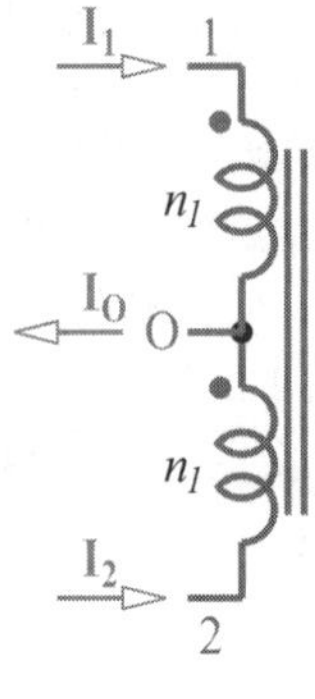

Figure 6-35. Interphase transformer.

Note that the output current ($\mathbf{I_O}$) of the IPT is the sum of the two input currents ($\mathbf{I_1}$ and $\mathbf{I_2}$). Therefore, an IPT averages the input voltages and sums the input currents.

6.2.11.2 24-Pulse QHN-VSC with IPTs. A 24-pulse QHN-VSC requires the generation of four three-phase sets (A1–B1–C1, A2–B2–C2, D1–E1–F1, and D2–E2–F2) of square wave voltages that are given in equations 6-96, 6-97, 6-98, and 6-99, respectively. The displacement angle between two consecutive six-pulse VSCs in a multipulse VSC configuration is $2\pi/6m$ (i.e., 15°), where m (i.e., 4) is the total number of six-pulse VSCs that are used. The two phasors in the A–B–C three-phase groups (A1–A2, B1–B2, and C1–C2) are centered on 0°, –120°, and –240°, respectively, with a displacement angle ($\theta_{displacement}$) of ±7.5°. The two phasors in the D–E–F three-phase groups (D1–D2, E1–E2, and F1–F2) are centered on –30°, –150°, and +90°, respectively, with a displacement angle ($\theta_{displacement}$) of ±7.5°. Figure 6-18a shows the two sets of three square wave voltages (v_{A1O}, v_{B1O}, and v_{C1O}; and v_{A2O}, v_{B2O}, and v_{C2O}) at 7.5° and –7.5°, respectively. The fundamental pole voltage phasors ($\mathbf{V_{A1,1}}$, $\mathbf{V_{B1,1}}$, and $\mathbf{V_{C1,1}}$; and $\mathbf{V_{A2,1}}$, $\mathbf{V_{B2,1}}$, and $\mathbf{V_{C2,1}}$), shown in Figure 6-18b, are at 7.5°, 7.5° – 120° (i.e., –112.5°), and 7.5° – 240° (i.e., –232.5°); and –7.5°, –7.5° – 120° (i.e., –127.5°), and –7.5° – 240° (i.e., –247.5°), respectively. Figure 6-18c shows the two sets of three square wave voltages (v_{D1O}, v_{E1O}, and v_{F1O}; and v_{D2O}, v_{E2O}, and v_{F2O}) at –22.5° and –37.5°, respectively. The fundamental pole voltage phasors ($\mathbf{V_{D1,1}}$, $\mathbf{V_{E1,1}}$, and $\mathbf{V_{F1,1}}$; and $\mathbf{V_{D2,1}}$, $\mathbf{V_{E2,1}}$, and $\mathbf{V_{F2,1}}$), shown in Figure 6-18d, are at 7.5° –30° (i.e., –22.5°), 7.5° –30° – 120° (i.e., –142.5°), and 7.5° –30° – 240° (i.e., –262.5°); and –7.5° –30° (i.e., –37.5°), –7.5° –30° – 120° (i.e., –157.5°), and –7.5° –30° – 240° (i.e., –277.5°), respectively.

Figure 6-36 shows four six-pulse VSCs (A1–B1–C1, A2–B2–C2, D1–E1–F1, and D2–E2–F2) that are connected across a shared DC link with a pair of capacitors and the related fundamental voltage phasors. The pole voltages are combined by a magnetic circuit to form a three-phase (a–b–c) voltage. The poles A1, B1, C1, A2, B2, C2, D1, E1, F1, D2, E2, and F2 are operated in such a way that the pole voltages in

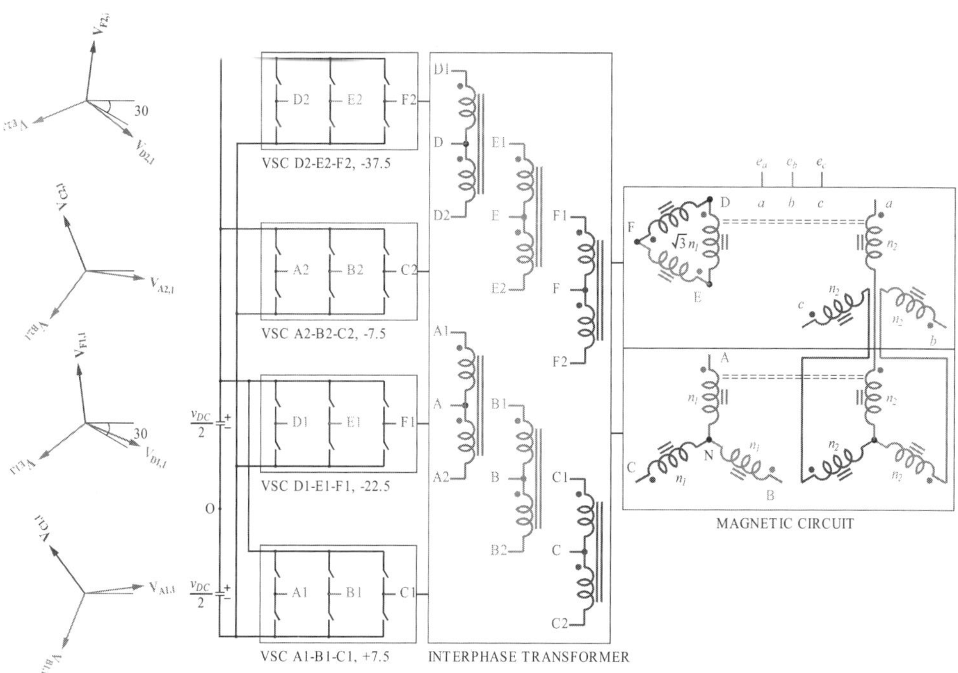

Figure 6-36. A 24-pulse QHN-VSC configuration with interphase transformers (all angles are in degrees).

a six-pulse VSC group, for example, v_{A1O}, v_{B1O}, and v_{C1O} of the A1–B1–C1 VSC; v_{A2O}, v_{B2O}, and v_{C2O} of the A2–B2–C2 VSC; v_{D1O}, v_{E1O}, and v_{F1O} of the D1–E1–F1 VSC; and v_{D2O}, v_{E2O}, and v_{F2O} of the D2–E2–F2 VSC, are time-shifted from one another by one-third of the time period (T) of the pole voltage. Furthermore, the four sets of three square wave voltages (v_{A1O}, v_{B1O}, and v_{C1O}; v_{A2O}, v_{B2O}, and v_{C2O}; v_{D1O}, v_{E1O}, and v_{F1O}; and v_{D2O}, v_{E2O}, and v_{F2O}) are placed at 7.5°, –7.5°, –22.5°, and –37.5°, respectively.

When the pole voltages from the VSCs A1–B1–C1 and A2–B2–C2 are averaged with the use of IPTs as described in the previous section, the output voltages are three quasisquare wave voltages (v_{AO}, v_{BO}, and v_{CO}) as shown in Figure 6-37a and the corresponding fundamental components are $v_{A,1}$, $v_{B,1}$, and $v_{C,1}$. The fundamental pole voltage phasors ($\mathbf{V_{A,1}}$, $\mathbf{V_{B,1}}$, and $\mathbf{V_{C,1}}$), shown in Figure 6-37b, are at 0°, –120°, and –240°, respectively. Similarly, when the pole voltages from the VSCs D1–E1–F1 and D2–E2–F2 are averaged with the use of IPTs, the output voltages are three quasisquare wave voltages (v_D, v_E, and v_F) at –30°, as shown in Figure 6-37c, and the corresponding fundamental components are $v_{D,1}$, $v_{E,1}$, and $v_{F,1}$. The fundamental pole voltage phasors ($\mathbf{V_{D,1}}$, $\mathbf{V_{E,1}}$, and $\mathbf{V_{F,1}}$), shown in Figure 6-37d, are at –30°, –150°, and 90°, respectively.

The two sets of three quasisquare wave voltages are

$$v_{AO} = \frac{v_{A1O} + v_{A2O}}{2} = \frac{v_{DC}}{\pi} \sum_{n=1}^{\infty} \frac{1}{n} \left\{ \sin n\left(\theta + \frac{\pi}{24}\right) + \sin n\left(\theta - \frac{\pi}{24}\right) \right\}$$

$$v_{AO} = \frac{2v_{DC}}{\pi} \sum_{n=1}^{\infty} \frac{1}{n} \cos\left(\frac{n\pi}{24}\right) \sin(n\theta) \tag{6-165a}$$

$$v_{BO} = \frac{v_{B1O} + v_{B2O}}{2} = \frac{v_{DC}}{\pi} \sum_{n=1}^{\infty} \frac{1}{n} \left\{ \sin n\left(\theta + \frac{\pi}{24} - 120^\circ\right) + \sin n\left(\theta - \frac{\pi}{24} - 120^\circ\right) \right\}$$

$$v_{BO} = \frac{2v_{DC}}{\pi} \sum_{n=1}^{\infty} \frac{1}{n} \cos\left(\frac{n\pi}{24}\right) \sin n(\theta - 120^\circ) \tag{6-165b}$$

$$v_{CO} = \frac{v_{C1O} + v_{C2O}}{2} = \frac{v_{DC}}{\pi} \sum_{n=1}^{\infty} \frac{1}{n} \left\{ \sin n\left(\theta + \frac{\pi}{24} + 120^\circ\right) + \sin n\left(\theta - \frac{\pi}{24} + 120^\circ\right) \right\}$$

$$v_{CO} = \frac{2v_{DC}}{\pi} \sum_{n=1}^{\infty} \frac{1}{n} \cos\left(\frac{n\pi}{24}\right) \sin n(\theta + 120^\circ) \tag{6-165c}$$

and

$$v_{DO} = \frac{v_{D1O} + v_{D2O}}{2} = \frac{v_{DC}}{\pi} \sum_{n=1}^{\infty} \frac{1}{n} \left\{ \sin n\left(\theta - \frac{3\pi}{24}\right) + \sin n\left(\theta - \frac{5\pi}{24}\right) \right\}$$

$$v_{DO} = \frac{2v_{DC}}{\pi} \sum_{n=1}^{\infty} \frac{1}{n} \cos\left(\frac{n\pi}{24}\right) \sin n\left(\theta - \frac{\pi}{6}\right) \tag{6-166a}$$

$$v_{EO} = \frac{v_{E1O} + v_{E2O}}{2} = \frac{v_{DC}}{\pi} \sum_{n=1}^{\infty} \frac{1}{n} \left\{ \sin n\left(\theta - \frac{3\pi}{24} - 120^\circ\right) + \sin n\left(\theta - \frac{5\pi}{24} - 120^\circ\right) \right\}$$

$$v_{EO} = \frac{2v_{DC}}{\pi} \sum_{n=1}^{\infty} \frac{1}{n} \cos\left(\frac{n\pi}{24}\right) \sin n\left(\theta - \frac{\pi}{6} - 120^\circ\right) \tag{6-166b}$$

$$v_{FO} = \frac{v_{F1O} + v_{F2O}}{2} = \frac{v_{DC}}{\pi} \sum_{n=1}^{\infty} \frac{1}{n} \left\{ \sin n\left(\theta - \frac{3\pi}{24} + 120^\circ\right) + \sin n\left(\theta - \frac{5\pi}{24} + 120^\circ\right) \right\}$$

$$v_{FO} = \frac{2v_{DC}}{\pi} \sum_{n=1}^{\infty} \frac{1}{n} \cos\left(\frac{n\pi}{24}\right) \sin n\left(\theta - \frac{\pi}{6} + 120^\circ\right) \tag{6-166c}$$

In the 12-pulse HN-VSC of Figure 6-9, there are two sets of square wave voltages: one set from the A–B–C VSC is at 0° and the other set from the D–E–F VSC is at –30°; the *a–b–c* 12-pulse voltage set is at 0°. In the 24-pulse QHN-VSC with IPTs of Figure 6-36, there are two sets of quasisquare wave voltages: one set from the A–B–C IPT is at 0° and the other set from the D–E–F IPT is at –30°; therefore, the *a–b–c* 24-pulse voltage set is at 0°. When the two sets of three quasisquare wave voltages—one at 0° and one at –30°—are applied to the primary windings of Y–Y and Δ–Y transformers and the secondary Y-windings are connected in series, the output voltages are

$$e_a = 2\frac{n_2}{n_1}\frac{2v_{DC}}{\pi}\sum_{n=1}^{\infty}\frac{1}{n}\cos\left(\frac{n\pi}{24}\right)\sin(n\theta) \quad \text{(6-167a)}$$

$$e_b = 2\frac{n_2}{n_1}\frac{2v_{DC}}{\pi}\sum_{n=1}^{\infty}\frac{1}{n}\cos\left(\frac{n\pi}{24}\right)\sin n(\theta - 120°) \quad \text{(6-167b)}$$

$$e_c = 2\frac{n_2}{n_1}\frac{2v_{DC}}{\pi}\sum_{n=1}^{\infty}\frac{1}{n}\cos\left(\frac{n\pi}{24}\right)\sin n(\theta + 120°) \quad \text{(6-167c)}$$

which contain fundamental components and harmonic components $n = 12k \pm 1$ for $k =$ 1, 2, 3, and so on, and exhibit a 24-pulse QHN waveform. The following waveforms are analyzed with turns ratio (n_2/n_1) of 0.5.

Figures 6-37, parts a and c, show the two sets of three quasisquare wave voltages (v_{AO}, v_{BO}, and v_{CO}; and v_{DO}, v_{EO}, and v_{FO}) and the corresponding fundamental components ($v_{A,1}$, $v_{B,1}$, and $v_{C,1}$; and $v_{D,1}$, $v_{E,1}$, and $v_{F,1}$). The fundamental voltage phasors ($\mathbf{V}_{A,1}$, $\mathbf{V}_{B,1}$, and $\mathbf{V}_{C,1}$; and $\mathbf{V}_{D,1}$, $\mathbf{V}_{E,1}$, and $\mathbf{V}_{F,1}$), shown in Figure 6-37, parts b and d, are at 0° (reference), –120°, and –240°; and at –30°, –150°, and +90°, respectively.

Figure 6-21a shows the three-phase, 24-pulse voltage (e_a, e_b, and e_c) and the corresponding fundamental components ($e_{a,1}$, $e_{b,1}$, and $e_{c,1}$). The fundamental pole voltage phasors ($\mathbf{E}_{a,1}$, $\mathbf{E}_{b,1}$, and $\mathbf{E}_{c,1}$), shown in Figure 6-21b, are at 0° (reference), –120°, and –240°, respectively.

The voltages across one of the two windings of the IPTs are shown in Figure 6-38.

6.2.11.3 Modeling a 24-Pulse QHN-VSC with Two-Level Poles and IPTs. The 24-pulse, two-level QHN-VSC is run from the main file `69324Q21.DAT` that has the same setup as the 24-pulse, two-level QHN-VSC in the file `64824Q21.DAT`, except that one $INCLUDE$ file is replaced with `69424Q21.MAG` (see Appendix C, Section 62) for implementing the ideal magnetic circuit that includes IPTs.

6.2.12 Realizable Pole Circuits

A pole circuit may be realized in many ways. The schematic diagrams of the single-phase, two-level pole, shown in Figure 6-2, and a three-level pole, shown in Figure 6-

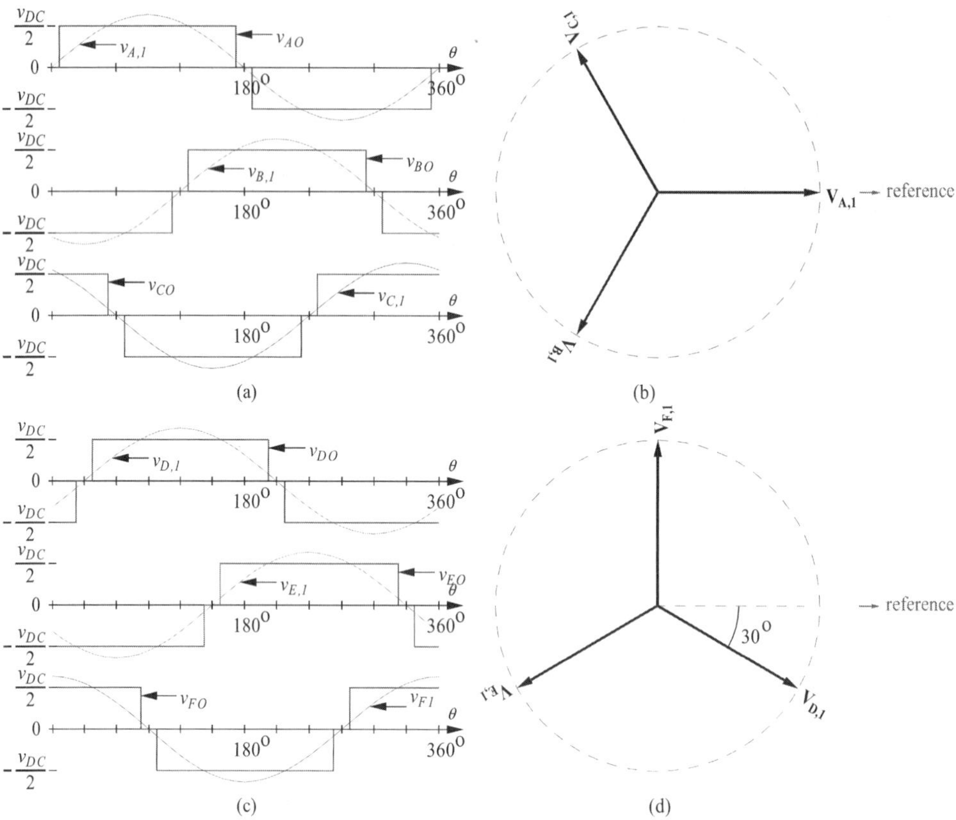

Figure 6-37. Two sets of IPT output voltages and fundamental voltage phasors in a 24-pulse QHN-VSC configuration. (a) Three 120° phase-shifted fundamental voltages and the corresponding quasisquare wave voltages at 0°. (b) Phasor diagram of fundamental voltages in A–B–C IPT output voltage. (c) Three 120° phase-shifted fundamental voltages and the corresponding quasisquare wave voltages at –30°. (d) Phasor diagram of fundamental voltages in D–E–F IPT output voltage.

29, may be realized with semiconductor devices, such as GTO thyristors and diodes, as shown in Figure 6-39. The basic switch is made by using a GTO with an antiparallel diode. Two switches (G1 with D1 and G2 with D2) constitute a two-level pole, as shown in Figure 6-39a. A three-level pole consists of a two-level pole and an additional AC switch that is made by using two switches (G3 with D3 and G4 with D4), as shown in Figure 6-39b. The AC switch conducts a bidirectional current either through G3 and D4 or G4 and D3. Each switch may be made by using a number of semiconductor devices, connected in series and/or parallel, depending on the voltage and current ratings of the switch.

A commonly used three-level pole is shown in Figure 6-40. For a positive output voltage, the two upper switches (G1 with D1 and G3 with D3) are turned on. For a

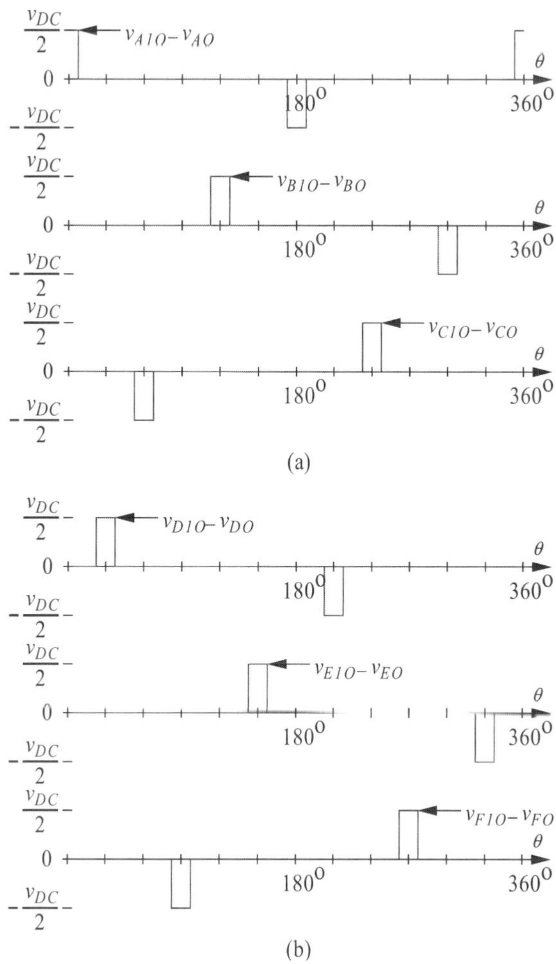

Figure 6-38. Interphase transformer voltages.

negative output voltage, the two lower switches (G2 with D2 and G4 with D4) are turned on. For zero output voltage, the two inner switches (G1 with D1 and G2 with D2) are turned on along with two neutral clamp diodes (d1 and d2).

The detailed analysis of a two-level pole, as shown in Figure 6-39a, is presented in Chapter 7.

6.2.13 Considerations for a HN-VSC

The following considerations must be taken into account when contemplating the use of a VSC with HN technique:

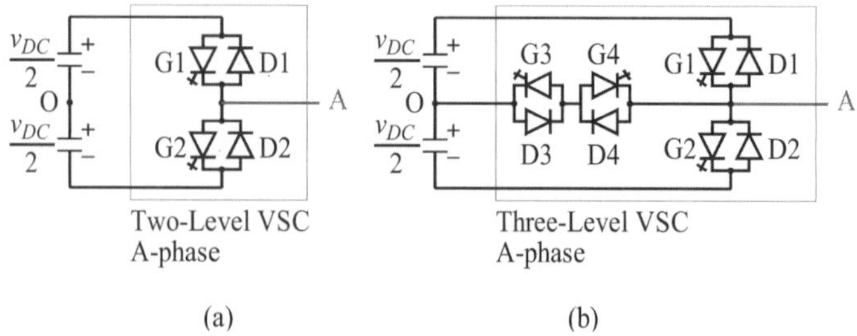

Figure 6-39. Single phase DC-to-AC VSC with (a) two levels and (b) three levels.

- Since there are a certain number of stages required to build the HN-VSC that provides a low harmonic output voltage and current, a HN-VSC is economical only for larger ratings, for example, 40 MVA and higher.
- It has been demonstrated that a HN-VSC is suitable for utility applications in which a 10–20 cycle response time is adequate. In an application in which a disturbed load, particularly an AC arc furnace, needs to be compensated, the HN technique mandates that the magnetic circuit be sufficiently overrated that the VSC is functional during all operating moments. Without generous core design, the magnetic circuit often saturates and makes the VSC unavailable when it is needed most, making this technique unreliable.
- A HN-VSC produces a positive-sequence fundamental voltage. The negative-sequence components, resulting from an imbalance in the AC system, can freely pass through the VSC. During a fault, the operation of the VSC may be inhibited due to high instantaneous negative-sequence current flow through the semiconductor switches.

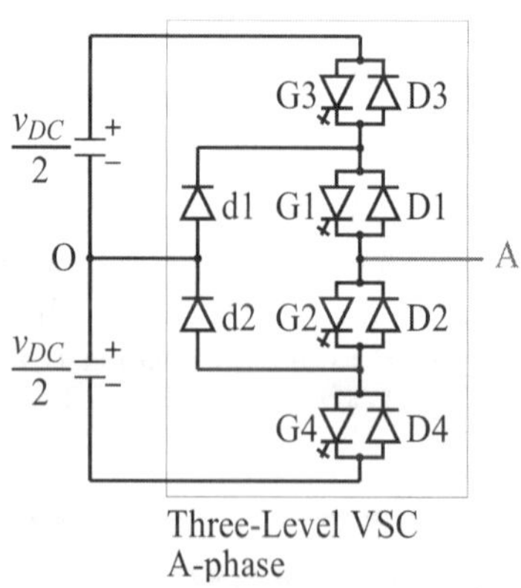

Figure 6-40. Single phase DC-to-AC VSC with three levels.

- All the poles in a HN-VSC operate from the same DC voltage. With once-per-cycle switching, the poles produce a set of time-shifted square wave voltages with equal amplitudes but different phases. If a VSC is required to provide a compensating voltage that is different in each phase, an alternate topology must be considered.

6.2.14 DC-to-AC VSC Operated with PWM Technique

A VSC that produces voltages that are independent in each phase is operated with the pulse width modulation (PWM) technique. The fundamental building block of a VSC is shown in Figure 6-41a. A DC capacitor is connected to one side of the VSC and generates a three-phase AC voltage at its output. The VSC is made up of a set of semiconductor switches, such as IGBTs as shown in Figure 6-41b. The semiconductor switches in a VSC are grouped into three poles: Pole A, Pole B, and Pole C. Each pole consists of an upper switch and a lower switch. Each switch has an antiparallel diode across it. The operation of the PWM-VSC is described below.

As shown in Figure 6-42, a fundamental frequency reference signal is compared with a high-frequency triangular carrier signal (v_{tri}). When the reference signal is high-

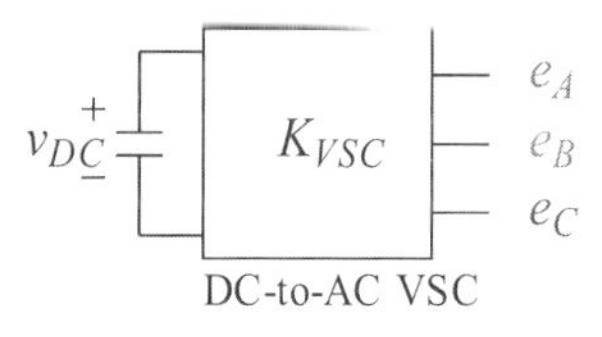

(a)

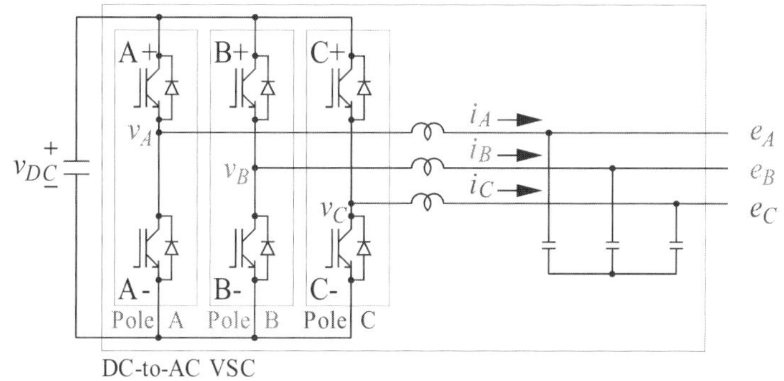

(b)

Figure 6-41. DC-to-AC VSC operated with the PWM technique.

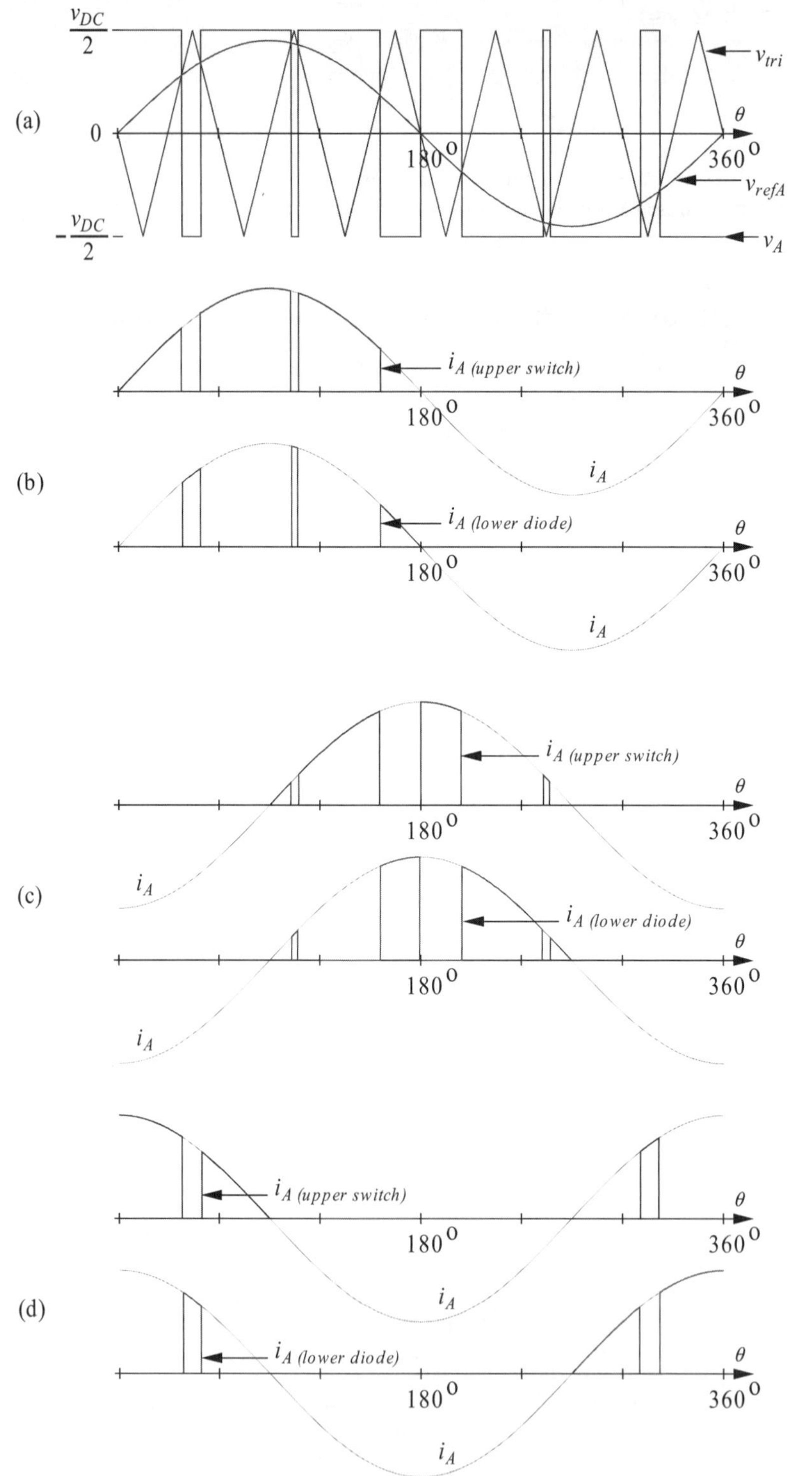

$\frac{v_{DC}}{2}$
(a)
0
$-\frac{v_{DC}}{2}$
v_{tri}
θ
180^0
360^0
v_{refA}
v_A
(b)
i_A (upper switch)
i_A
i_A (lower diode)
(c)
i_A (upper switch)
i_A (lower diode)
(d)
i_A (upper switch)
i_A (lower diode)

er than the carrier signal, the upper switch, A+, is turned on; otherwise, the lower switch, A–, is turned on. The generated PWM square wave voltage contains a fundamental component, the amplitude of which is the same as that of the fundamental frequency reference signal and harmonic components with the largest amplitude at the carrier frequency. Therefore, the amplitude of the fundamental component in the generated voltage can be varied by varying the amplitude of the fundamental frequency reference signal. The frequency of the dominant harmonic component in the generated voltage can be varied by varying the carrier frequency and, thereby, the switching frequency of the semiconductor devices. It is desirable to increase the frequency of the dominant harmonic component in the PWM voltage. By increasing the switching frequency, the dominant harmonic component can be moved outside the range of concern. For example, if each switch is operated at 25 times the reference signal frequency, the output voltage contains a reference signal frequency and a first dominant harmonic component of 25th order. Increasing the switching frequency is not an attractive option because it increases the switching loss. The compromise is to use moderately high switching frequency that generates components of about the 20th harmonic order and a harmonic trapping filter, as shown in Figure 6-41b.

6.3 DISCUSSION

For independent control of active and reactive power flows in a transmission line, the magnitude and phase angle of the line voltage must be controlled independently by using a series-connected compensating voltage, which is of variable magnitude and phase angle. The compensating voltage is connected in series with the line through a transformer that provides electrical isolation and voltage matching. The compensating voltage is an AC voltage and the current flowing through it can be at any phase angle with respect to the phase angle of its terminal voltage, thereby emulating a capacitor or an inductor and a positive or a negative resistor. The compensating voltage essentially emulates an impedance at its terminal. A VSC, by its nature, is capable of absorbing or delivering reactive power at its AC terminal. However, for the exchange of active power at the AC terminals, the VSC must be an AC-to-AC VSC that is composed of either two DC-to-AC VSCs with a shared DC link or two AC-to-AC VSCs with a shared AC link.

When both types of DC and AC shared-linked VSC topologies are considered, the DC shared-link topology is found to be comparatively less costly. The ST is even more cost-effective than a DC-linked AC-to-AC VSC.

Figure 6-42. (a) Carrier voltage, pole A reference voltage and pole A output voltage. (b) Currents through upper switch, lower diode, and pole A for unity power factor. (c) Currents through upper switch, lower diode, and pole A for zero power factor lagging. (d) Currents through upper switch, lower diode, and pole A for zero power factor leading.

7

TWO-LEVEL POLE DESIGN

Voltage-sourced converters (VSCs) consist of high-power, solid-state switches that are made by using semiconductor devices such as GTOs and the newer IGCTs and IGBTs for utility level equipment. Designers of these types of switches must understand the detailed operation of the individual switch element in order to make appropriate design trade-offs. Device manufacturers create mathematical models of semiconductor devices in great detail. However, to meet the switch designers' first-order simulation needs, an approximate terminal characteristic model of high-power semiconductor device is quite adequate. In trade-off analysis, the switch designers are really interested in the detailed characteristics of the commutation process to assess items such as device stress caused by overvoltage and overcurrent, power loss in the semiconductor devices and other circuit components, the effectiveness of a particular snubber design, and the layout of a circuit to keep the stray inductances within an acceptable level. A final design can only be accepted after numerous iterations have taken place. Thus, the designers can greatly benefit from a simulation tool that can provide answers to these questions in an efficient and accurate manner.

In this chapter, a detailed mathematical model of a high-power, solid-state, switch composed of GTO thyristors is used in an analysis of a high-power VSC two-level pole circuit. The input variables to the mathematical model are the main circuit components, snubber circuit stray inductances, line stray inductances, the reverse recovery charge of the diodes, and the switching characteristics of the GTOs and diodes. The model is used to examine critical operating characteristics, such as peak voltage across and peak current through the GTOs, as part of a trade-off analysis. Comprehensive data for the turn-on and turn-off operation of the switch are tabulated. The same analysis can be carried out with switches that are made by using other types of semiconductor devices.

7.1 A THREE-PHASE, SIX-PULSE VSC WITH TWO-LEVEL POLES

The circuit diagram of a three-phase, six-pulse VSC with two-level poles is shown in Figure 7-1. Each single-phase VSC is commonly referred to as a pole. Each pole consists of two switches, such as A+ and A– in pole A. When switch A+ is closed, the pole A output voltage is $+V_{DC}/2$ and when switch A– is closed, the pole A output voltage is $-V_{DC}/2$. Since each pole output voltage in this configuration switches between $+V_{DC}/2$ and $-V_{DC}/2$, it is referred to as a two-level pole.

Figure 7-2a shows the three pole voltages that are 120° apart and their fundamental components. The three-phase pole currents for absorbing and generating VARs are shown in Figure 7-2, parts b and c. Note that a component with the current entering into its positive terminal and leading the terminal voltage by 90° generates VARs as in the case of a capacitor. Also, a component with the current entering into its positive terminal and lagging the terminal voltage by 90° absorbs VARs as in the case of an inductor. Since the direction of current is leaving the poles, the leading currents in Figure 7-2b are for absorbing VARs and the lagging currents in Figure 7-2c are for generating VARs.

Each of the two switches in a pole (e.g., A+ and A– in pole A) can be composed of a GTO and an antiparallel diode, as shown in Figure 7-3. The two-level voltage of pole A and its fundamental component are shown in Figure 7-4a. The pole currents for absorbing and generating VARs are shown in Figure 7-4, parts b and c. During the positive output voltage, the positive current flows through the GTO (G1) and the negative current flows through the diode (D1). During the negative output voltage,

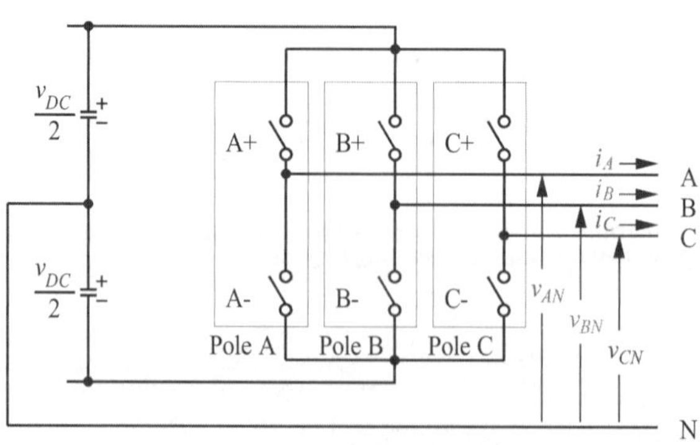

Figure 7-1. Three-phase, six-pulse VSC with two-level poles.

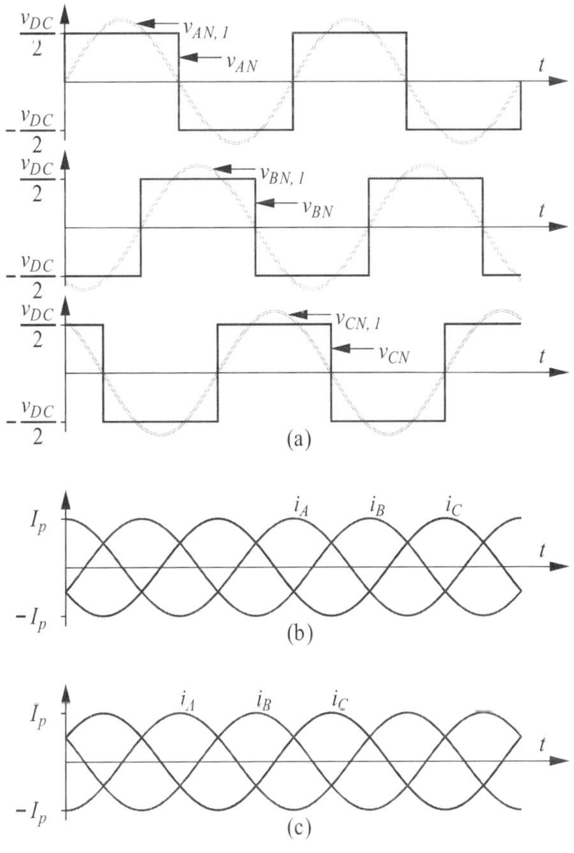

Figure 7-2. (a) Pole output to neutral square wave voltages with their fundamental components. (b) Pole output currents during absorbing VARs. (c) Pole output currents during generating VARs.

the positive current flows through the diode (D2) and the negative current flows through the GTO (G2).

During absorbing VARs when the pole output voltage transitions from negative to positive, the positive current transitions from diode (D2) to GTO (G1) at the peak value and when the pole output voltage transitions from positive to negative, the negative current transitions from diode (D1) to GTO (G2) at the peak value. During generating VARs when the pole output voltage transitions from negative to positive, the negative current transitions from GTO (G2) to diode (D1) at the peak value and when the pole output voltage transitions from positive to negative, the positive current transitions from GTO (G1) to diode (D2) at the peak value. While generating VARs, the GTO experiences a forced turn-off while conducting peak current, making it necessary to use semiconductor devices with forced turn-off capability.

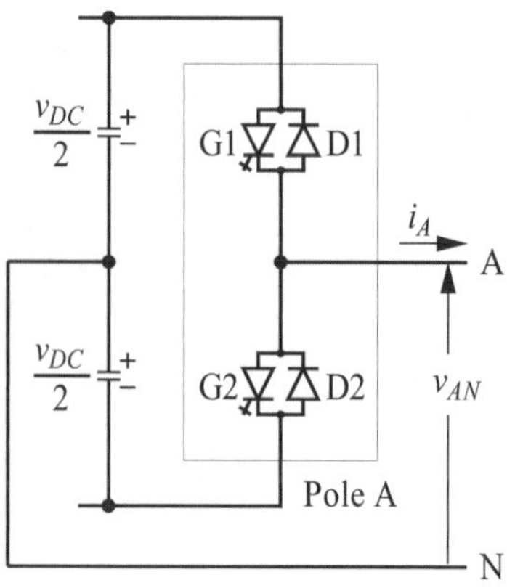

Figure 7-3. Realization of a two-level pole A with GTOs and antiparallel diodes.

While absorbing VARs when the pole output voltage transitions, the current transitions from a diode to a GTO that is turned on with a gate signal. While generating VARs when the pole output voltage transitions, the current transitions from a GTO to a diode and then to its antiparallel GTO, such as GTO (G1) to diode (D2) to GTO (G2). The question is "when should the GTO (G2) be turned on?" If it is turned on at the time the GTO (G1) is turned off, there may be a period of a time when the GTO (G1) is not quite turned off and the GTO (G2) on its way to be turned on. During this time period, both GTOs may be in between on and off states that may cause a short circuit at the DC link capacitor, resulting in a destructive shoot-through current in the capacitor and the GTOs. Therefore, it is necessary to delay the turn-on of GTO (G2) following

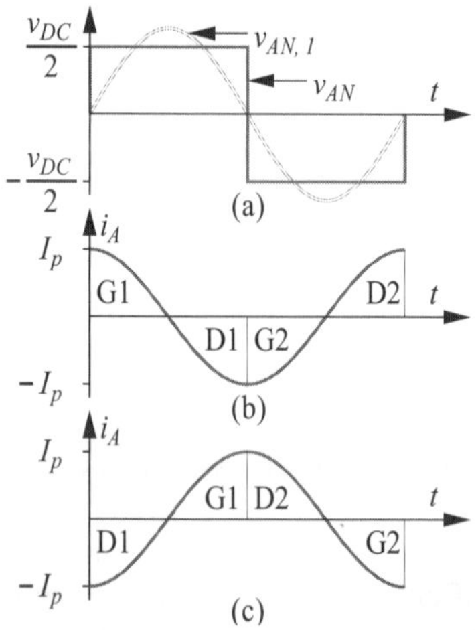

Figure 7-4. (a) Pole A to neutral voltage and its fundamental component. (b) Pole A current during absorbing VARs. (c) Pole A current during generating VARs.

the turn-off of GTO (G1) by a delay time (t_{delay}). The effect of this delay time is investigated.

It is sufficient to study the transition processes from a GTO to a diode and a diode to a GTO at peak current. It is true that a pole operating at other than zero power factor transitions at other than peak current. However, a designer should study the worst operating point at zero power factor. For the following analysis, two operating conditions—commutating peak current from diode (D2) to GTO (G1) and from GTO (G1) to diode (D2)—are considered.

In a six-pulse VSC with two-level poles, it is not necessary to access the midpoint of the DC link capacitor. The configuration of a typical six-pulse VSC with two-level poles is shown in Figure 7-5. The detailed analysis of one of the three poles is described in the next section.

7.2 ANALYSIS OF A POLE

The details of a pole circuit are shown in Figure 7-6. The reference directions of currents and polarities of voltage for computation purposes are also shown. The two halves (upper and lower) of the pole are symmetric. Each of the two switches (upper and lower) consists of a single GTO (G1 or G2) with an antiparallel diode (D1 or D2) across the GTO. The line stray inductances are *LS1* and *LS4*. The turn-on snubber circuits include *di*/*dt* limiting inductors *L1* and *L2*. These inductors have shunt loops, each consisting of a free-wheeling diode (DS1 or DS3), a stray inductance (*LS2* or *LS5*), and a series resistor (*R1* or *R4*). The turn-off snubber circuits include *dv*/*dt* limiting capacitors (*C1* and *C3*), diodes (DS2 and DS4), and the snubber capacitor discharge resistors (*R2* and *R5*). The snubber capacitor leakage inductors are *LS3* and *LS6*. The snubber circuits of the snubber diodes (DS2 and DS4) are *R3*-*C2* and *R6*-*C4*, respectively. The pole circuit is analyzed by freezing the instantaneous DC link voltage (v_{DC}) to V_{DC} and output current (i_O) to I_O that is equal to the peak current (I_p) at the output of the pole.

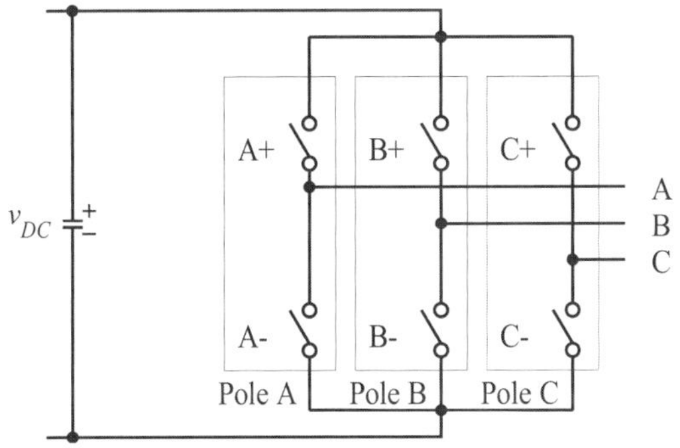

Figure 7-5. Three-phase, six-pulse VSC and its poles.

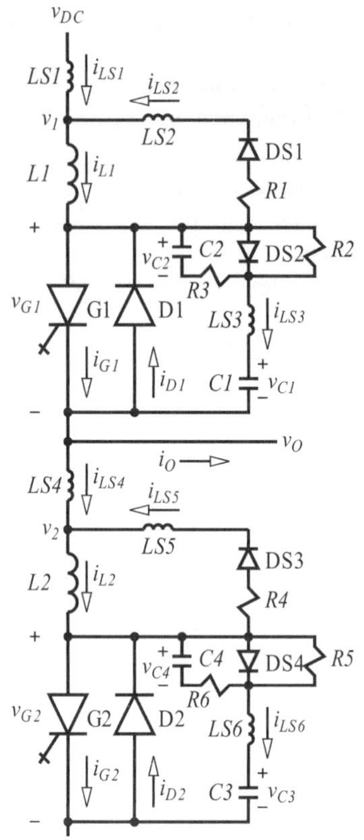

Figure 7-6. Details of a two-level pole circuit.

7.2.1 Device Characteristics

The switching characteristics of the diodes (D1 and D2) in this model are shown in Figure 7-7a. When the diode conducts a current (i_D) in the forward direction, the voltage (v_D) across the diode is zero. When the diode current reverses at $t = t_x$, the voltage continues to remain at zero value until the reverse recovery charge (Q_a) is swept away during $t_x \le t \le t_y$. The reverse current through the diode is determined by the dynamics of the circuit. For $t > t_y$, the reverse current (i_D) of the diode rises to zero in a nonlinear fashion. The slope of i_D is hypothesized by the expression

$$
\begin{aligned}
Di_D &= a(t-t_y)^2 + b(t-t_y) \qquad && t_y \le t \le t_z \\
&= 0 && t_z \le t \le t_y
\end{aligned}
\tag{7-1}
$$

where

$D \equiv d/dt$
$a = -y_m/T^2$

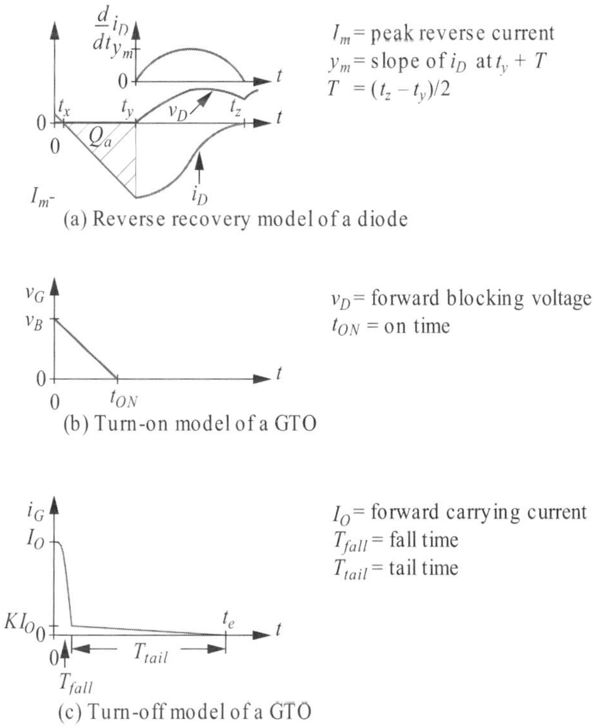

Figure 7-7. Switching characteristics of a diode and a GTO.

y_m = slope of i_D at $t_y + T$
$T = (t_z - t_y)/2 = -0.75\ I_m/y_m$
I_m = peak reverse current
$b = -2aT$

Integrating equation 7-1, the diode current (i_D) can be obtained as

$$i_D = \frac{1}{3}a(t-t_y)^3 + \frac{1}{2}b(t-t_y)^2 + I_m \qquad t_y \le t \le t_z$$
$$= 0 \qquad t_z \le t \qquad (7\text{-}2)$$

The maximum slope (y_m) of i_D at $t_y + T$ is considered to be four times the average slope of i_D between t_x and t_y. The diode starts to develop the reverse voltage (v_D) at time t_y.

The GTO switching characteristics (turn-on and turn-off) are shown in Figure 7-7, parts b and c. During turn-on, the GTO acts as a variable voltage source (v_G) that falls linearly from the forward blocking voltage (v_B) to zero. The switching characteristic during turn-on is defined by

$$v_G = v_B\left(1-\frac{t}{t_{ON}}\right) \qquad 0 \le t \le t_{ON}$$
$$= 0 \qquad t_{ON} \le t \tag{7-3}$$

where v_B = forward blocking voltage and t_{ON} = turn-on time of a GTO.

During turn-off, the GTO acts as a variable current source (i_G) that falls nonlinearly from the forward conducting current (I_O) to a low value at the end of the fall time (T_{fall}); then the current tapers off at a much slower rate until the end of the tail time (T_{tail}). The switching characteristic during turn-off is defined by

$$i_G = \frac{1}{3}a_1 t^3 + \frac{1}{2}b_1 t^2 + I_O \qquad 0 \le t \le T_{fall}$$
$$= \frac{1}{3}a_2(t-T_{fall})^3 + \frac{1}{2}b_2(t-T_{fall})^2 + KI_O \qquad T_{fall} \le t \le t_e$$
$$= 0 \qquad t_e \le t \tag{7-4}$$

where

I_O = forward conducting current
T_{fall} = fall time = $-1.5(1-K)I_O/Di_{G(Tfall)}$
T_{tail} = tail time = $-1.5KI_O/Di_{G(Ttail)}$
$Di_{G(Tfall)}$ = slope of i_G at the midpoint of T_{fall}
$Di_{G(Ttail)}$ = slope of i_G at the midpoint of T_{tail}
$a_1 = -4Di_{G(Tfall)}/T_{fall}^2$
$b_1 = -a_1 T_{fall}$
$a_2 = -4Di_{G(Ttail)}/T_{tail}^2$
$b_2 = -a_2 T_{tail}$
K = ratio of the GTO currents at the end of T_{fall} and that at the beginning of T_{fall}.

7.2.2 Mathematical Model

The circuit behavior is governed by the following equations:

$$i_{LS1} = i_{LS4} + I_O \Rightarrow Di_{LS1} = Di_{LS4} \tag{7-5}$$

$$Di_{LS1} = \frac{V_{DC} - v_1}{LS1} \tag{7-6}$$

$$Di_{LS4} = \frac{v_O - v_2}{LS4} \tag{7-7}$$

$$Di_{L1} = \frac{v_1 - v_O - v_{G1}}{L1} \tag{7-8}$$

$$Di_{L2} = \frac{v_2 - v_{G2}}{L2} \tag{7-9}$$

where i_{LS1} and i_{LS4} are the currents through the line stray inductances (*LS1* and *LS4*), i_{L1} and i_{L2} are the currents through the snubber inductors (*L1* and *L2*), v_O is the output voltage at the midpoint of the pole circuit, and v_{G1} and v_{G2} are the voltages across the GTOs (G1 and G2), respectively,

$$i_{LS1} + i_{LS2} = i_{L1} \Rightarrow Di_{LS1} + Di_{LS2} = Di_{L1} \tag{7-10}$$

$$i_{LS4} + i_{LS5} = i_{L2} \Rightarrow Di_{LS4} + Di_{LS5} = Di_{L2} \tag{7-11}$$

$$Di_{LS2} = \frac{v_O + v_{G1} - v_1 - R_1 i_{LS2}}{LS2} \tag{7-12}$$

$$Di_{LS5} = \frac{v_{G2} - v_2 - R_4 i_{LS5}}{LS5} \tag{7-13}$$

where i_{LS2} and i_{LS5} are the currents through the snubber inductors loop stray inductances *LS2* and *LS5*,

$$Dv_{C1} = \frac{i_{LS3}}{C1} \tag{7-14}$$

$$Dv_{C2} = \frac{R2(i_{LS3} - i_{DS2}) - v_{C2}}{(R2 + R3)C2} \tag{7-15}$$

$$Di_{LS3} = \frac{v_{G1} - \{v_{C2} + R3(i_{LS3} - i_{DS2})\}R2/(R2 + R3) - v_{C1}}{LS3} \tag{7-16}$$

$$Dv_{C3} = \frac{i_{LS6}}{C3} \tag{7-17}$$

$$Dv_{C4} = \frac{R5(i_{LS6} - i_{DS4}) - v_{C4}}{(R5 + R6)C4} \tag{7-18}$$

$$Di_{LS6} = \frac{v_{G2} - \{v_{C4} + R6(i_{LS6} - i_{DS4})\}R5/(R5 + R6) - v_{C3}}{LS6} \tag{7-19}$$

where v_{C1} and v_{C3} are the voltages across the snubber capacitors (*C1* and *C3*), v_{C2} and v_{C4} are the voltages across the snubber capacitors (*C2* and *C4*), i_{LS3} and i_{LS6} are the currents through the snubber capacitors stray inductances (*LS3* and *LS6*), and i_{DS2} and i_{DS4} are the forward currents through the snubber diodes (DS2 and DS4), respectively.

Equations 7-5 through 7-7 can be rearranged as

$$v_O = v_2 + (V_{DC} - v_1)\frac{LS4}{LS1} \tag{7-20}$$

Before attempting to solve the state equations, the voltages (v_{G1} and v_{G2}) across the GTOs (G1 and G2) and three node voltages (v_O, v_1, and v_2) need to be calculated. The state variables are calculated using Adams-Bashforth's fourth-order numerical integration formula (see Appendix A):

$$f_{n+1} = f_n + \left(\frac{55}{24}Df_n - \frac{59}{24}Df_{n-1} + \frac{37}{24}Df_{n-2} - \frac{9}{24}Df_{n-3}\right)h \tag{A-97}$$

where f_{n+1} and f_n are the values of state variable f at the beginning of time steps $n + 1$ and n, respectively. The initial value of the state variable f at the beginning of the $(n + 1)$th step is the final value of the same state variable f at the end of the nth step where h is the step size. Df_n, Df_{n-1}, Df_{n-2}, and Df_{n-3} are differential coefficients of state variable f at the beginning of time steps n, $n - 1$, $n - 2$, and $n - 3$, respectively.

7.2.3 Analysis of the Model

To analyze the pole circuit model, two modes of operation need to be studied. These modes are commutating output current (I_O) from the antiparallel diode (D2) to the GTO (G1) and commutating output current (I_O) from the GTO (G1) to the GTO (G2) to the antiparallel diode (D2). In the latter mode of operation, three different conditions are studied when the GTO (G2) is turned on after the GTO (G1) is turned off with various delay times (t_{delay}), such as 2 μs, 0, and 20 μs, respectively. With a large delay time, such as $t_{delay} = 20$ μs, the output current (I_O) at full load commutates naturally from the GTO (G1) to the antiparallel diode (D2). Each mode is studied with the two extreme values of the output current (I_O), such as full load when $I_O = I_p$, and no load when $I_O = 0$.

The VSC nameplate rating is given in Table 7-1. Therefore, the DC link voltage (V_{DC}) is 1900 V. The output current (I_O) at full load is 4000 A. The results are indicated in pu, with one pu current as the nominal output current ($I_O = 4000$ A) and the one pu voltage as the nominal DC link voltage ($V_{DC} = 1900$ V).

The circuit parameters are given in Table 7-2. The turn-on time (t_{ON}) of a GTO is typically 1 μs. The reverse recovery charge (Q_a) of an antiparallel diode is assumed to be linearly proportional to the peak forward current that flows through the antiparallel diode before the diode commutates. A snubber diode is assumed to have the

Table 7-1. Namplate rating of a VSC

VSC continuous rating	50	MVA
Overcurrent rating (115%)	57.5	MVA
VSC type	48-pulse	
Pole type	two-level	
VSC gain	$\frac{2}{\pi}\cos\left(\frac{\pi}{24}\right)\cos\left(\frac{\pi}{48}\right)=0.63$	
Steady-state device current	4000	A peak
Six-pulse VSC rating	$\frac{57500000}{8}=7187500$	VA
Steady-state device voltage	$\frac{7187500}{1.5\times 4000}=1198$	V peak
Steady-state DC link voltage	$\frac{1198}{0.63}=1900$	V

similar type of characteristics as an antiparallel diode. The reverse recovery charge (Q_a) of an antiparallel diode is a linear function of the peak forward current through the antiparallel diode with a slope of 1 μC/A up to 1000 A of peak forward current, and beyond that value of peak forward current, the slope changes to 0.2 μC/A. The reverse recovery charge (Q_a) of a snubber diode is a linear function of the peak forward current through the snubber diode with a slope of 0.6 μC/A up to 1000 A of peak forward current, and beyond that value of peak forward current, the slope changes to 0.2 μC/A. The slope ($Di_{G1(Tfall)}$) of the GTO current (i_{G1}) at the midpoint of T_{fall} is –4000 A/μs and that ($Di_{G1(Ttail)}$) at the midpoint of T_{tail} is –75 A/μs. The ratio (K) of the GTO currents at the end of T_{fall} to that at the beginning of T_{fall} is 0.1.

7.2.3.1 Mode 1 of Operation. In this mode of operation, the output current (I_O) commutates from the antiparallel diode (D2) to the GTO (G1), as shown in Figure

Table 7-2. Circuit parameters in a two-level VSC pole

Snubber inductances	*L1, L2*	1.73	μH
Snubber capacitances	*C1, C3*	6	μF
Snubber inductor loop resistances	*R1, R4*	0.15	Ω
Snubber capacitor discharge resistances	*R2, R5*	5	Ω
Snubber diode snubber resistances	*R3, R6*	1.5	Ω
Snubber diode snubber capacitances	*C2, C4*	0.2	μF
Line stray inductances	*LS1, LS4*	0.17	μH
Snubber inductor loop stray inductances	*LS2, LS5*	0.1	μH
Snubber capacitor stray inductances	*LS3, LS6*	0.085	μH

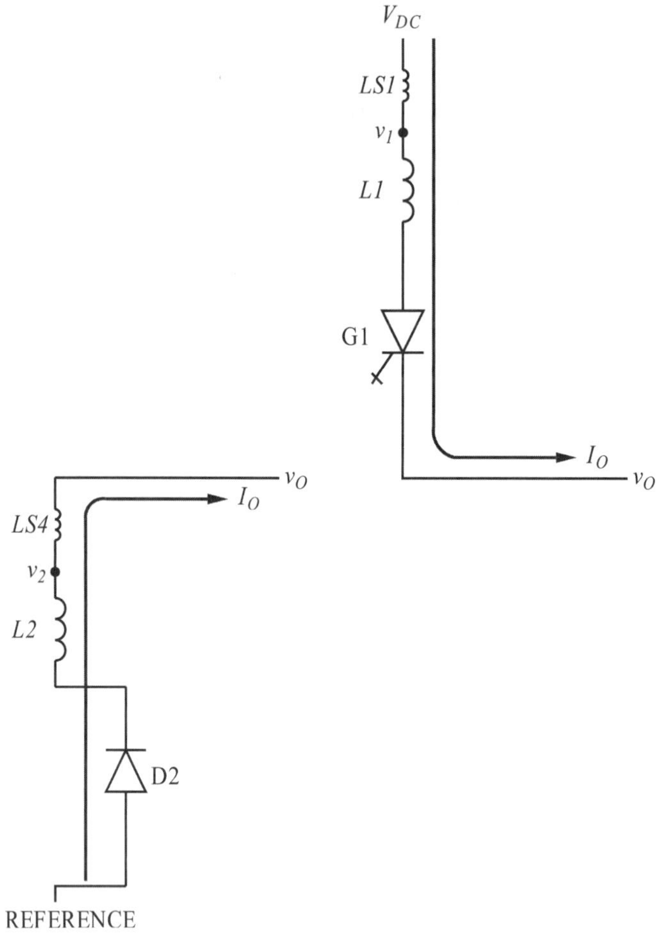

Figure 7-8. Commutation from D2 to G1.

7-8. The waveforms in this mode of operation are shown in Figure 7-9a for full load conduction ($I_O = I_p$) and in Figure 7-9b for no load conduction ($I_O = 0$). Under steady-state conditions (Figure 7-10a), the output current (I_O) flows through the diode (D2), the snubber inductor (*L2*), and the line stray inductance (*LS4*). The capacitor (*C1*) is charged to the GTO (G1) forward blocking voltage ($v_B = V_{DC}$), the capacitor (*C3*) is discharged completely, and the output voltage (v_O) is zero.

At time t_0 (Figure 7-10b), under full load condition ($I_O = I_p$), the GTO (G1) is turned on. The snubber capacitor (*C1*) starts to discharge through the snubber capacitor stray inductance (*LS3*), the snubber resistor (*R2*), and the GTO (G1). The snubber diode snubber capacitor (*C2*) charges very quickly with a negative polarity and then the voltage (v_{C2}) across the capacitor (*C2*) discharges through the resistors (*R3* and

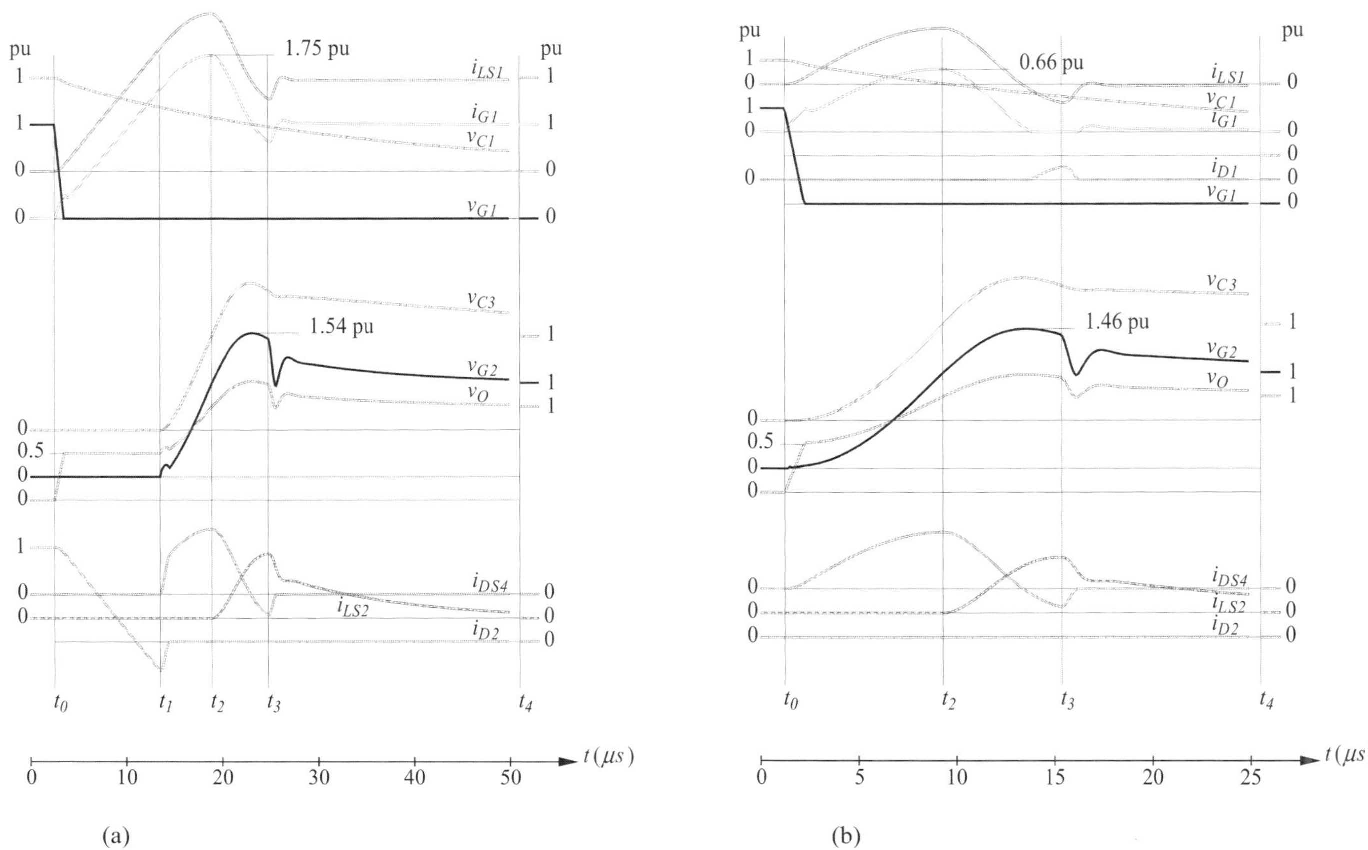

Figure 7-9. Transients from commutation from D2 tc G1 under (a) full load and (b) no load conditions.

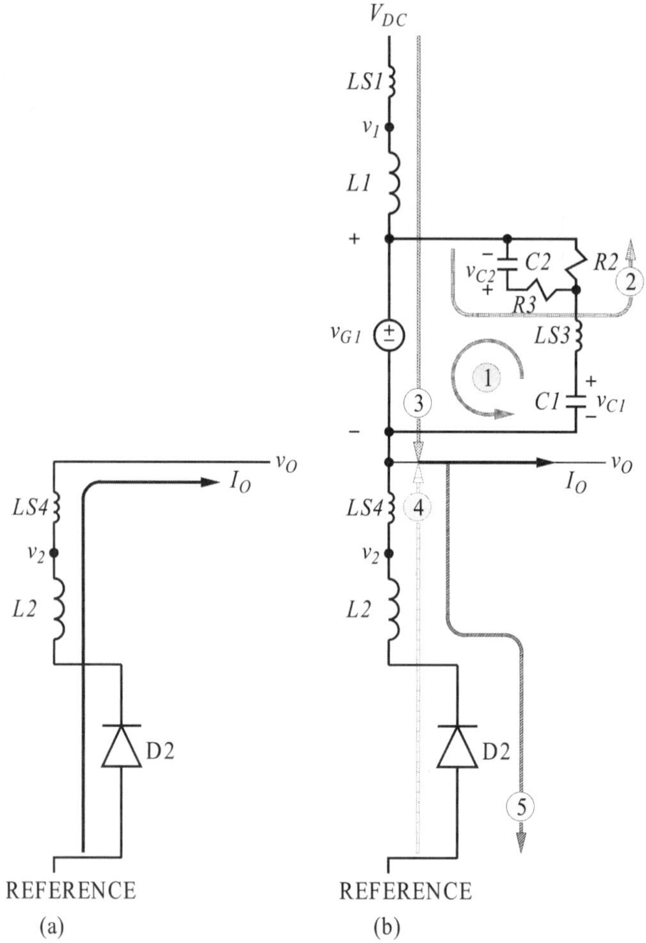

Figure 7-10(a, b). Transients from commutation from D2 to G1 during (a) $t < t_0$ and (b) $t_0 \leqq t < t_1$.

$R2$). The governing equations that define the state voltages (v_{C1} and v_{C2}) and the state current (i_{LS3}) are given in equations 7-14 through 7-16 with $i_{DS2} = 0$ and the initial conditions $v_{C1}(t_0) = V_{DC}$, $v_{C2}(t_0) = 0$, and $i_{LS3}(t_0) = 0$.

During the time between t_0 and t_1, the voltage (v_{G1}) across the GTO (G1), falls linearly to zero according to equation (7-3) and stays at zero. Since $i_{LS1} = i_{L1}$ and therefore, $Di_{LS1} = Di_{L1}$, equations 7-6 and 7-8 can be rearranged as

$$v_1 = \frac{V_{DC} L1 + (v_O + v_{G1}) LS1}{L1 + LS1} \tag{7-21}$$

Since $i_{LS4} = i_{L2}$ and therefore $Di_{LS4} = Di_{L2}$, equations 7-7 and 7-9 can be rearranged as

$$v_2 = \frac{v_O L2 + v_{G2} LS4}{L2 + LS4} \tag{7-22}$$

Substituting equations 7-21 and 7-22 into equation 7-20, it can be written that

$$v_O = \frac{(V_{DC} - v_{G1})(L2 + LS4) + v_{G2}(L1 + LS1)}{L1 + L2 + LS1 + LS4} \tag{7-23}$$

where $v_{D2} = 0 = v_{G2}$, with v_{D2} being the forward voltage drop across the diode (D2).

The state equations that characterize the behavior of the circuit are 7-6 and 7-7 and the initial conditions are $i_{LS1}(t_0) = 0$ and $i_{LS4}(t_0) = -I_O$.

The output voltage (v_O) rises to half of the value of the DC link voltage (V_{DC}) and stays constant. The current (i_{LS1}) through the line stray inductance *LS1* rises. The current (i_{LS4}) through the line stray inductance *LS4* reduces. The GTO (G1) carries the sum of the line stray inductance (*LS1*) current (i_{LS1}) and the snubber capacitor (*C1*) discharge current (i_{LS3}). The current (i_{LS4}), which is the current (i_{D2}) through the diode (D2) reverses its direction.

At time t_1 (Figure 7-10c), the current (i_{D2}) through the diode (D2) reaches its maximum negative value and all the reverse recovery charge (Q_a) of the diode (D2) is swept away. During the time between t_1 and t_2, the current (i_{D2}) through the diode (D2), increases from its maximum negative value to zero according to equation 7 2 and stays at zero. The current (i_{LS6}) through the snubber capacitor stray inductance (*LS6*) is defined by

$$i_{LS6} = i_{LS4} + i_{D2} \Rightarrow Di_{LS6} = Di_{LS4} + Di_{D2} \tag{7-24}$$

The state equations that characterize the behavior of the circuit are 7-6, 7-7, 7-17 and 7-19, where $i_{LS6} = i_{DS4}$ and $v_{C4}(t_1) = 0$, and the initial conditions are $i_{LS1}(t_1)$, $i_{LS4}(t_1)$, $v_{C3}(t_1) = 0$, and $i_{LS6}(t_1) = 0$.

From equations 7-7, 7-19, 7-22 through 7-24, it can be written that

$$v_{G2} = \frac{(V_{DC} - v_{G1})LS6 + (v_{C3} + LS6\, Di_{D2})(L1 + L2 + LS1 + LS4)}{L1 + L2 + LS1 + LS4 + LS6} \tag{7-25}$$

where Di_{D2} can be obtained from equation 7-1. The expressions for v_1, v_2, and v_O are given in equations 7-21 through 7-23.

The snubber capacitor (*C3*) starts to charge up. The voltage (v_{G2}) across the GTO (G2) rises and so does the output voltage (v_O). After the reverse current (i_{D2}) becomes zero, all of current (i_{LS4}) charges the snubber capacitor (*C3*).

At time t_2 (Figure 7-10d), the output voltage (v_O) exceeds the DC voltage (v_{DC}). The polarities of the voltages across the snubber inductors (*L1* and *L2*) reverse. These reverse polarity voltages turn the diodes (*DS1* and *DS3*) on in the snubber shunt loops. Part of the energy in the snubber inductors (*L1* and *L2*) gets trapped and this energy is dissipated in the shunt loop series resistors (*R1* and *R4*).

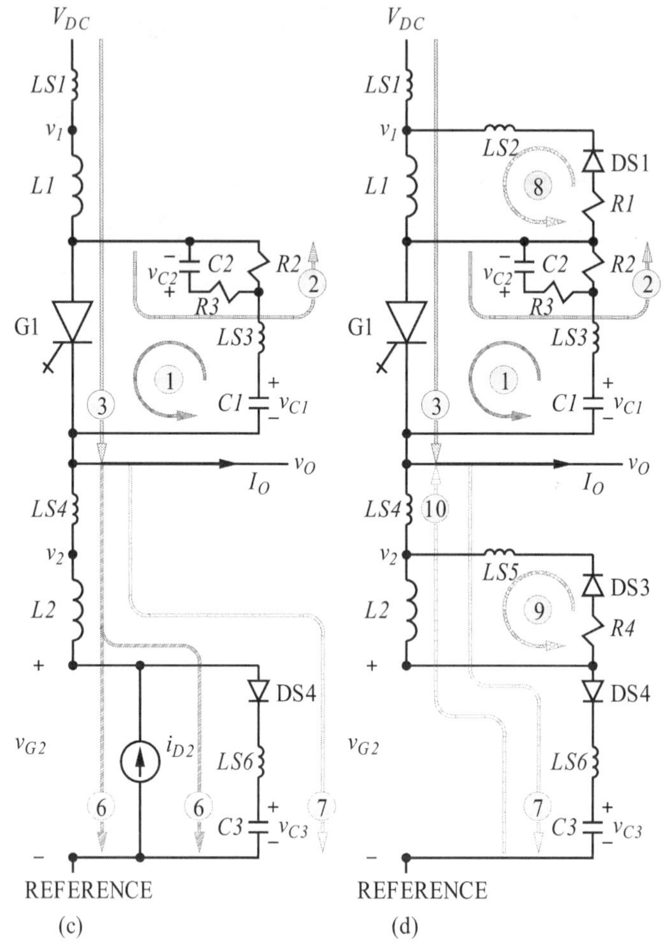

Figure 7-10(c, d). Transients from commutation from D2 to G1 during (c) $t_1 \leqq t < t_2$ and (d) $t_2 \leqq t < t_3$.

The state equations that characterize the behavior of the circuit are 7-6 through 7-9, 7-12, 7-13, and 7-17, where $i_{LS6} = i_{DS4}$ and $v_{C4}(t_2) = 0$ and the initial conditions are $i_{LS1}(t_2)$, $i_{LS4}(t_2)$, $i_{L1}(t_2) = i_{LS1}(t_2)$, $i_{L2}(t_2) = i_{LS4}(t_2)$, $i_{LS2}(t_2) = 0$, $i_{LS5}(t_2) = 0$, and $v_{C3}(t_2)$.

From equations 7-6, 7-8, 7-10, and 7-12, it can be written that

$$v_1 = K_1\left[V_{DC} L1\, LS2 + \{(v_O + v_{G1})(L1 + LS2) - R1\, L1\, i_{LS2}\} LS1\right] \quad (7\text{-}26)$$

where $K_1 = \dfrac{1}{L1\, LS2 + LS1\,(L1 + LS2)}$.

From equations 7-7, 7-9, 7-11, and 7-13, it can be written that

$$v_2 = K_2\left[v_O L2\,LS5 + \left\{v_{G2}\,(L2 + LS5) - R4\,L2\,i_{LS5}\right\}LS4\right] \tag{7-27}$$

where $K_2 = \dfrac{1}{L2\,LS5 + LS4\,(L2 + LS5)}$.

Substituting v_1 and v_2 from equations 7-26 and 7-27 into equation 7-20, it can be written that

$$v_O = K_3\left[K_1\left\{(V_{DC} - v_{G1})(L1 + LS2) + R1\,L1\,i_{LS2}\right\} + K_2\left\{v_{G2}\,(L2 + LS5) - R4\,L2\,i_{LS5}\right\}\right] \tag{7-28}$$

where $K_3 = \dfrac{1}{K_1\,(L1 + LS2) + K_2\,(L2 + LS5)}$.

Substituting v_2 from equation 7-27 and v_O from equation 7-28, equations 7-7, 7-19, and 7-24 can be rearranged as

$$v_{G2} =$$

$$K_4\left[v_{C3} + LS6\,Di_{D2} + K_2 LS6\left\{\begin{array}{l} K_1K_3\,(L2 + LS5)\left[(V_{DC} - v_{G1})(L1 + LS2) + R1\,L1\,i_{LS2}\right] \\ + \left[1 - K_2K_3\,(L2 + LS5)\right]R4\,L2\,i_{LS5} \end{array}\right\}\right] \tag{7-29}$$

where

$$K_4 = \frac{1}{1 + K_2\,(L2 + LS5)LS6\left\{1 - K_2\,(L2 + LS5)\,K_3\right\}}.$$

The rates of change in currents (i_{LS1} and i_{LS4}) in the line stray inductances (*LS1* and *LS4*) are faster than those in the snubber inductors (*L1* and *L2*). During the time between t_2 and t_3, the current (i_{LS4}), which is also the current (i_{LS6}) through *LS6* reverses its direction.

At time t_3 (Figure 7-10e), the current (i_{DS4}) through the diode (*DS4*) reaches its maximum negative value and all the reverse recovery charge (Q_a) of the diode (*DS4*) is swept away. During the time between t_3 and t_4, the current (i_{DS4}) through the diode (*DS4*) increases from its maximum negative value to zero according to equation 7-2 and stays at zero.

The state equations that characterize the behavior of the circuit are 7-6 through 7-9, 7-12, 7-13, 7-17, and 7-18 and the initial conditions are $i_{LS1}(t_3)$, $i_{LS4}(t_3)$, $i_{L1}(t_3)$, $i_{L2}(t_3)$, $i_{LS2}(t_3)$, $i_{LS5}(t_3)$, $v_{C3}(t_3)$, and $v_{C4}(t_3) = 0$.

Since $i_{LS4} = i_{LS6}$ and therefore $Di_{LS4} = Di_{LS6}$, substituting Di_{LS4} from equation 7-7, v_2 from equation 7-27, and v_O from equation 7-28 into equation 7-19, it can be written that

$$v_{G2} = K_4\left[v_{C3} + \left\{v_{C4} + R6\left(i_{LS6} - i_{DS4}\right)\right\}\frac{R5}{R5+R6}\right.$$

$$\left. + K_2 LS6\left\{\begin{matrix} K_1K_3\left(L2+LS5\right)\left[\left(V_{DC}-v_{G1}\right)\left(L1+LS2\right)+R1\,L1\,i_{LS2}\right] \\ +\left[1-K_2K_3\left(L2+LS5\right)\right]R4\,L2\,i_{LS5}\end{matrix}\right\}\right] \quad (7\text{-}30)$$

The expressions for v_1, v_2, and v_O are given in equations 7-26 through 7-28.

Since the resistor (*R6*) is much smaller than the resistor (*R5*), the snubber diode snubber capacitor (*C4*) charges very quickly with a negative polarity that, in turn, creates a sharp dip in the voltage (v_{G2}) across the GTO (G2). The recovery of the voltage (v_{G2}) from this dip can have a high *dv*/*dt* to turn the GTO (G2) on. The voltage (v_{C4}) of the snubber diode snubber capacitor (*C4*) discharges through the resistors (*R6* and *R5*). The voltage (v_{C3}) of the snubber capacitor (*C3*) in excess of the DC link voltage (V_{DC}) discharges through the DC voltage source (V_{DC}). Eventually, the snubber shunt loop currents (i_{LS2} and i_{LS5}) decay to zero. The snubber capacitor (*C1*) discharges completely, the snubber capacitor (*C3*) charges to the GTO (G2) forward blocking voltage (V_{DC}), and the output voltage (v_O) reaches V_{DC}.

At time t_4 (Figure 7-10f), the pole circuit reaches the steady-state condition and the output current (I_O) flows through the line stray inductance (*LS1*), the snubber inductor (L1), and the GTO (G1).

Under no load or light load conditions, during the time between t_2 and t_4, the current through the GTO (G1) can swing between positive and negative values for a few times until the transients die out. This negative current can only flow through the antiparallel diode (D1).

7.2.3.2 Mode 2 of Operation. In this mode of operation, the output current (I_O) commutates from the GTO (G1) to the GTO (G2) to the antiparallel diode (D2) as shown in Figure 7-11. This mode of operation has been analyzed with three different conditions when the GTO (G2) is turned on after the GTO (G1) is turned off with various delay times (t_{delay}), such as 2 μs, 0, and 20 μs, respectively.

OPERATION WITH $t_{delay} = 2$ μS. The waveforms in this mode of operation are shown in Figure 7-12 for full load condition ($I_O = I_p$). Under steady-state condition (Figure 7-13a), the output current (I_O) flows through the line stray inductance (*LS1*), the snubber inductor (*L1*), and the GTO (G1). The capacitor (*C3*) is charged to the GTO (G2) forward blocking voltage ($v_B = V_{DC}$), the capacitor (*C1*) is discharged completely, and the output voltage (v_O) equals to the DC link voltage (V_{DC}).

At time t_0 (Figure 7-13b), the GTO (G1) is turned off. The current (i_{G1}) through the GTO (G1) starts to fall as shown in Figure 7-7c. The current (i_{LS3}) through the snubber capacitor stray inductance (*LS3*) is defined by

$$i_{LS3} = i_{LS1} - i_{G1} \Rightarrow Di_{LS3} = Di_{LS1} - Di_{G1} \quad (7\text{-}31)$$

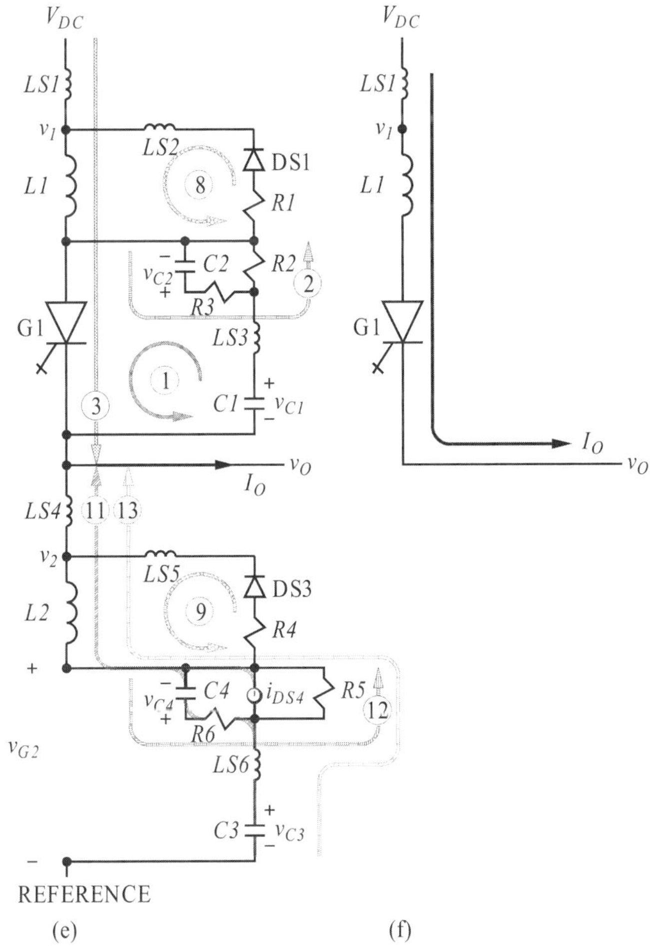

Figure 7-10(e, f). Transients from commutation from D2 to G1 during (e) $t_3 \leqq t < t_4$, and (f) $t_4 \leqq t$.

The state equations that characterize the behavior of the circuit are 7-6 through 7-9, 7-12 through 7-14, and 7-16 through 7-19, where $i_{LS3}(t_0) = i_{DS2}(t_0)$ and $v_{C2}(t_0) = 0$ and the initial conditions are $i_{LS1}(t_0) = I_O$, $i_{LS4}(t_0) = 0$, $i_{L1}(t_0) = I_O$, $i_{L2}(t_0) = 0$, $i_{LS2}(t_0) = 0$, $i_{LS5}(t_0) = 0$, $v_{C1}(t_0) = 0$, $i_{LS3}(t_0) = 0$, $v_{C3}(t_0) = V_{DC}$, $v_{C4}(t_0) = 0$, and $i_{LS6}(t_0) = 0$.

From equations 7-6, 7-26, 7-28, 7-16, and 7-31, it can be written that

$$v_{G1} = \frac{K_5}{1-K_6}\left[v_{C1} - LS3\, Di_{G1} + K_1 LS3 \left\{ \begin{matrix} \left[1-K_1K_3(L1+LS2)\right]\left[V_{DC}(L1+LS2)+R1\,L1\,i_{LS2}\right] \\ -K_2K_3(L1+LS2)\left[V_{G2}(L2+LS5)-R4\,L2\,i_{LS5}\right] \end{matrix} \right\} \right] \tag{7-32}$$

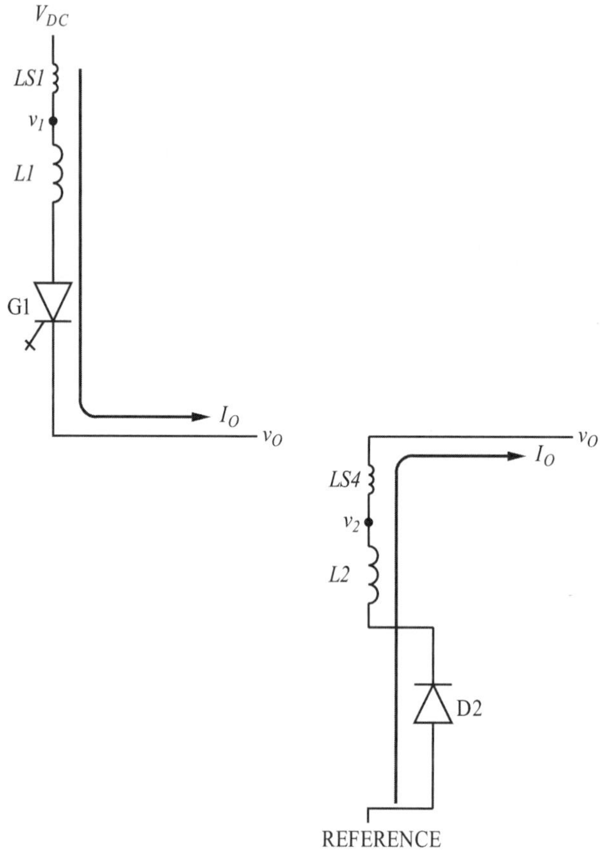

Figure 7-11. Commutation from G1 to G2 to D2.

where $K_5 = \dfrac{1}{1 + K_1 (L1 + LS2) LS3}$, $K_6 = K_1^2 K_3 K_5 LS3 (L1 + LS2)^2$, and Di_{G1} can be obtained from equation 7-1.

Setting $i_{DS4} = 0$ in equation 7-30 and substituting for v_{G1} from equation 7-32, it can be written that

$$v_{G2} = \frac{K_4}{1 - K_6 - K_6 K_7} \left[\begin{array}{l} \left\{ v_{C3} + \left(v_{C4} + R6\, i_{LS6} \right) \dfrac{R5}{R5 + R6} \right\} (1 - K_6) \\ + K_2 LS6 \left\{ \begin{array}{l} K_1 K_3 K_5 (L2 + LS5) \left[\left(V_{DC} + LS3\, Di_{G1} - v_{C1} \right)(L1 + LS2) + R1\, L1\, i_{LS2} \right] \\ - \left[K_2 K_3 (L2 + LS5) + K_6 - 1 \right] R4\, L2\, i_{LS5} \end{array} \right\} \end{array} \right] \tag{7-33}$$

where $K_7 = K_2^2 K_3 K_4 LS6 (L2 + LS5)^2$.

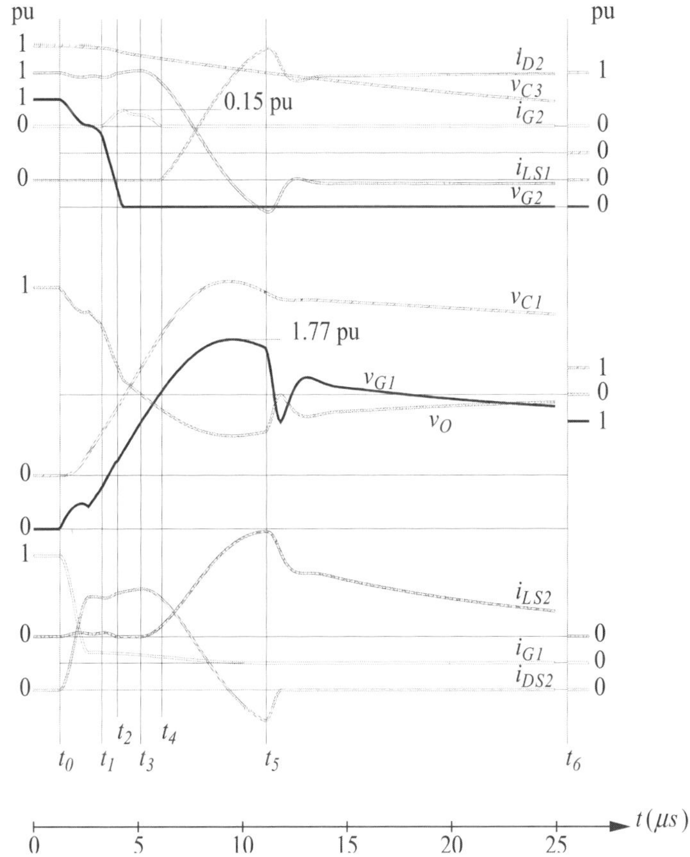

Figure 7-12. Commutation from G1 to G2 to D2 under full load with t_{delay} = 2 μs.

The expressions for v_1, v_2, and v_O are given in equations 7-26 through 7-28.

The current (i_{LS1}) through the line stray inductance (*LS1*) starts to drop and so does the output voltage (v_O). The difference between the line stray inductance (*LS1*) current (i_{LS1}) and the GTO (G1) current (i_{G1}) charges the snubber capacitor (*C1*) through the snubber diode (*DS2*) and the snubber capacitor stray inductance (*LS3*). The voltage (v_{C3}) of the snubber capacitor (*C3*) discharges through the output voltage (v_O). The snubber diode snubber capacitor (*C4*) charges very quickly with a negative polarity and then the voltage (v_{C4}) across the capacitor (*C4*) discharges through the resistors (*R6* and *R5*). The current (i_{LS4}) through the line stray inductance (*LS4*) is the discharge current (i_{LS6}) from the snubber capacitor (*C3*). The polarities of the voltages across the snubber inductors (*L1* and *L2*) reverse; these reverse polarity voltages turn the diodes (*DS1* and *DS3*) on in the snubber shunt loops. Part of the energy in the snubber inductors (*L1* and *L2*) gets trapped and this energy is dissipated in the shunt loop series resistors (*R1* and *R4*).

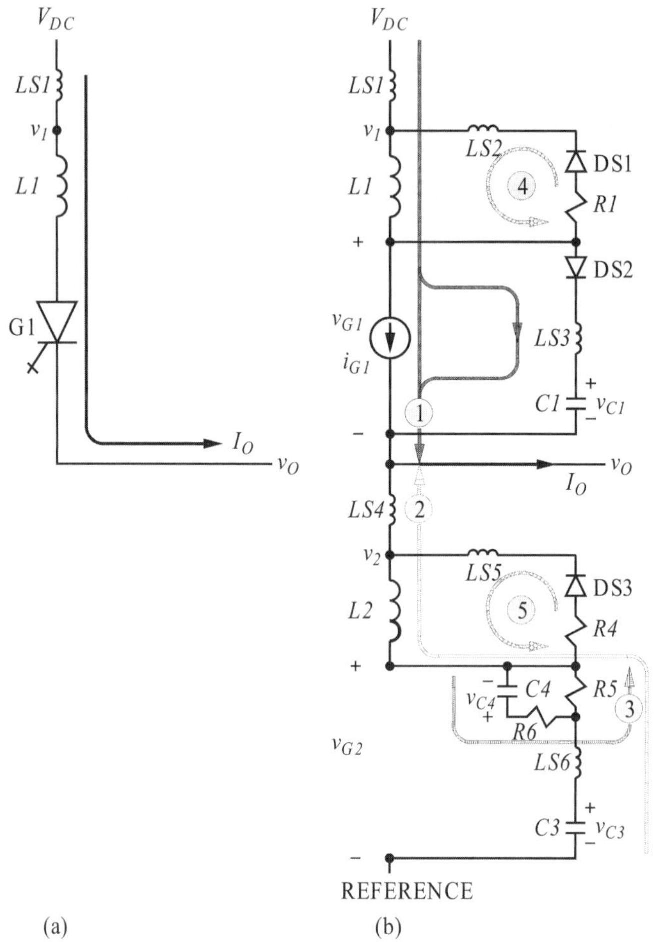

Figure 7-13(a, b). Commutation from G1 to G2 to D2 under full load with t_{delay} = 2 μs during (a) $t < t_0$, and (b) $t_0 \leqq t < t_1$.

At time t_1 (Figure 7-13c), the GTO (G2) is turned on. The voltage (v_{G2}) across the GTO (G2) falls linearly from its forward blocking voltage (v_B) to zero according to equation 7-3 and stays at zero. The state equations that characterize the behavior of the circuit are 7-6 through 7-9, 7-12 through 7-14, and 7-17 through 7-19, where $i_{LS3}(t_1) = i_{DS2}(t_1)$ and $v_{C2}(t_1) = 0$ and the initial conditions are $i_{LS1}(t_1)$, $i_{LS4}(t_1)$, $i_{L1}(t_1)$, $i_{L2}(t_1)$, $i_{LS2}(t_1)$, $i_{LS5}(t_1)$, $v_{C1}(t_1)$, $v_{C3}(t_1)$, $v_{C4}(t_1)$, and $i_{LS6}(t_1)$. The expressions for v_1, v_2, v_O, and v_{G1}, are given in equations 7-26 through 7-28 and 7-32. The rates of change in currents (i_{LS1} and i_{LS4}) in the line stray inductances (*LS1* and *LS4*) are faster than those in the snubber inductors (*L1* and *L2*).

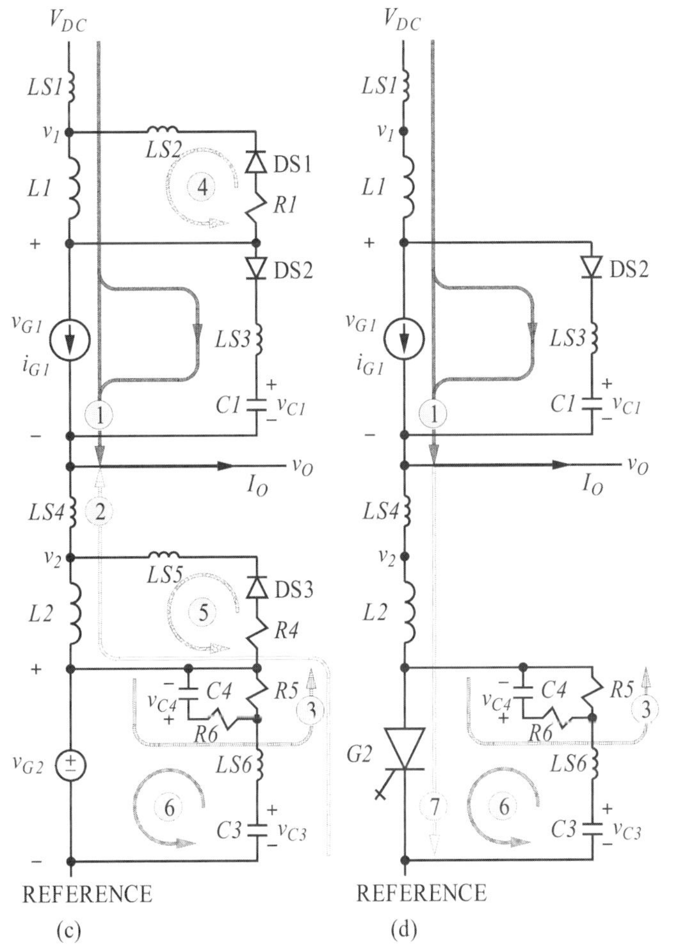

Figure 7-13(c, d). Commutation from G1 to G2 to D2 under full load with t_{delay} = 2 μs during (c) $t_1 \leqq t < t_2$, and (d) $t_2 \leqq t < t_3$.

The snubber capacitor (*C3*) discharges through the snubber capacitor stray inductance (*LS6*), the snubber resistor (*R5*), and the GTO (G2) and also through the line stray inductance (*LS4*). During the time between t_1 and t_2, the polarities of the voltages across the snubber inductors (*L1* and *L2*) reverse. These reverse polarity voltages bring the snubber shunt loop currents (i_{LS2} and i_{LS5}) down to zero and then turn the diodes (*DS1* and *DS3*) off in the snubber shunt loops.

At time t_2 (Figure 7-13d), the currents (i_{LS1} and i_{L1}) become equal. The state equations that characterize the behavior of the circuit are 7-6, 7-7, 7-14, and 7-17 through 7-19, where $i_{LS3}(t_2) = i_{DS2}(t_2)$ and $v_{C2}(t_2) = 0$ and the initial conditions are $i_{LS1}(t_2)$,

$i_{LS4}(t_2)$, $v_{C1}(t_2)$, $v_{C3}(t_2)$, $v_{C4}(t_2)$, and $i_{LS6}(t_2)$. The expressions for v_1, v_2, and v_O are given in equations 7-21 through 7-23.

From equations 7-6, 7-21, 7-23, 7-16, and 7-31, it can be written that

$$v_{G1} = \frac{(V_{DC} - v_{G2})LS3 + (v_{C1} - LS3\ Di_{G1})(L1 + L2 + LS1 + LS4)}{L1 + L2 + LS1 + LS4 + LS3} \tag{7-34}$$

During the time between t_2 and t_3, the output voltage (v_O) falls to zero. The currents (i_{LS1} and i_{LS4}) rise because of the positive voltages across the line stray inductances (*LS1* and *LS4*). The GTO (G2) carries the sum of the line stray inductance (*LS4*) current (i_{LS4}) and the snubber capacitor (*C3*) discharge current (i_{LS6}).

At time t_3 (Figure 7-13e), the output voltage (v_O) becomes negative. The polarities of the voltages across the snubber inductors (*L1* and *L2*) reverse; these reverse polarity voltages turn the diodes (*DS1* and *DS3*) on in the snubber shunt loops. The direction of line stray inductance (*LS4*) current (i_{LS4}) reverses. The state equations that characterize the behavior of the circuit are 7-6 through 7-9, 7-12 through 7-14, and 7-17 through 7-19, where $i_{LS3}(t_3) = i_{DS2}(t_3)$ and $v_{C2}(t_3) = 0$, and the initial conditions are $i_{LS1}(t_3)$, $i_{LS4}(t_3)$, $i_{L1}(t_3) = i_{LS1}(t_3)$, $i_{L2}(t_3) = i_{LS4}(t_3)$, $i_{LS2}(t_3) = 0$, $i_{LS5}(t_3) = 0$, $v_{C1}(t_3)$, $v_{C3}(t_3)$, $v_{C4}(t_3)$, and $i_{LS6}(t_3)$. The expressions for v_1, v_2, v_O, and v_{G1} are given in equations 7-26 through 7-28 and 7-32.

At time t_4, (Figure 7-13f), the current (i_{G2}) through the GTO (G2) becomes negative and that turns the GTO (G2) off and turns the diode (D2) on. The state equations that characterize the behavior of the circuit are 7-6 through 7-9, 7-12 through 7-14, and 7-17 through 7-19, where $i_{LS3}(t_4) = i_{DS2}(t_4)$, $v_{C2}(t_4) = 0$, and $v_{D2}(t_4) = 0 = v_{G2}(t_4)$, and the initial conditions are $i_{LS1}(t_4)$, $i_{LS4}(t_4)$, $i_{L1}(t_4)$, $i_{L2}(t_4)$, $i_{LS2}(t_4)$, $i_{LS5}(t_4)$, $v_{C1}(t_4)$, $v_{C3}(t_4)$, $v_{C4}(t_4)$, and $i_{LS6}(t_4)$. The expressions for v_1, v_2, v_O, and v_{G1} are given in equations 7-26 through 7-28 and 7-32.

During the time between t_4 and t_5, the current (i_{G1}) goes to zero according to equation 7-4 and stays at zero. After the current (i_{G1}) becomes zero, all of current (i_{LS1}) charges the snubber circuit (*C1*). The current (i_{LS3}) reverses its direction.

At time t_5 (Figure 7-13g), the current (i_{DS2}) through the diode (*DS2*) reaches its maximum negative value and all the reverse recovery charge (Q_a) of the diode (*DS2*) is swept away. During the time between t_5 and t_6, the current (i_{DS2}) through the diode (*DS2*) increases from its maximum negative value to zero according to equation 7-2 and stays at zero. The state equations that characterize the behavior of the circuit are 7-6 through 7-9, 7-12 through 7-15, and 7-17 through 7-19, where $v_{D2}(t_5) = 0 = v_{G2}(t_5)$ and the initial conditions are $i_{LS1}(t_5)$, $i_{LS4}(t_5)$, $i_{L1}(t_5)$, $i_{L2}(t_5)$, $i_{LS2}(t_5)$, $i_{LS5}(t_5)$, $v_{C1}(t_5)$, $v_{C2}(t_5) = 0$, $v_{C3}(t_5)$, $v_{C4}(t_5)$, and $i_{LS6}(t_5)$. The expressions for v_1, v_2, and v_O are given in equations 7-26 through 7-28.

From equations 7-6, 7-26, 7-28, 7-16, and 7-31, it can be written that

$$v_{G1} = \frac{K_5}{1-K_6}\left[\begin{array}{l} v_{C1} - LS3\ Di_{G1} + \left\{ v_{C2} + R3\,(i_{LS3} - i_{DS2})\dfrac{R2}{R2+R3} \right\} \\ +K_1 LS3 \left\{ \begin{array}{l} [1 - K_1 K_3 (L1 + LS2)][V_{DC}(L1 + LS2) + R1\, L1\, i_{LS2}] \\ -K_2 K_3 (L1 + LS2)[V_{G2}(L2 + LS5) - R4\, L2\, i_{LS5}] \end{array} \right\} \end{array}\right] \tag{7-35}$$

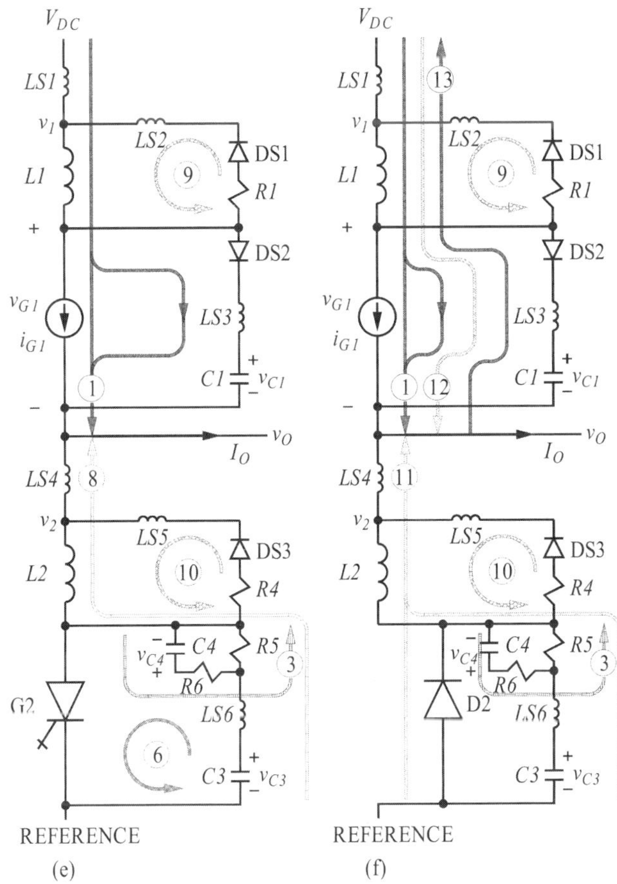

Figure 7-13(e, f). Commutation from G1 to G2 to D2 under full load with $t_{delay} = 2\ \mu s$ during (e) $t_3 \leqq t < t_4$, and (f) $t_4 \leqq t < t_5$.

As the resistor (*R3*) is much smaller than the resistor (*R2*), the snubber diode snubber capacitor (*C2*) charges very quickly with a negative polarity that, in turn, creates a sharp dip in the voltage (v_{G1}) across the GTO (G1). The recovery of the voltage (v_{G1}) from this dip can have a high *dv*/*dt* to turn the GTO (G1) on. The voltage (v_{C2}) of the snubber diode snubber capacitor (*C2*) discharges through the resistors (*R3* and *R2*). The voltage (v_{C1}) of the snubber capacitor (*C1*) in excess of the DC link voltage (V_{DC}) discharges through the DC voltage source (V_{DC}). Eventually, the snubber shunt loop currents (i_{LS2} and i_{LS5}) decay to zero. The snubber capacitor (*C3*) discharges completely, the snubber capacitor (*C1*) charges to the GTO (G1) forward blocking voltage (V_{DC}), and the output voltage (v_O) reaches to zero.

At time t_6 (Figure 7-13h), the pole circuit reaches the steady-state condition and the output current (I_O) flows through the diode (D2), the snubber inductor (*L2*), and the line stray inductance (*LS4*).

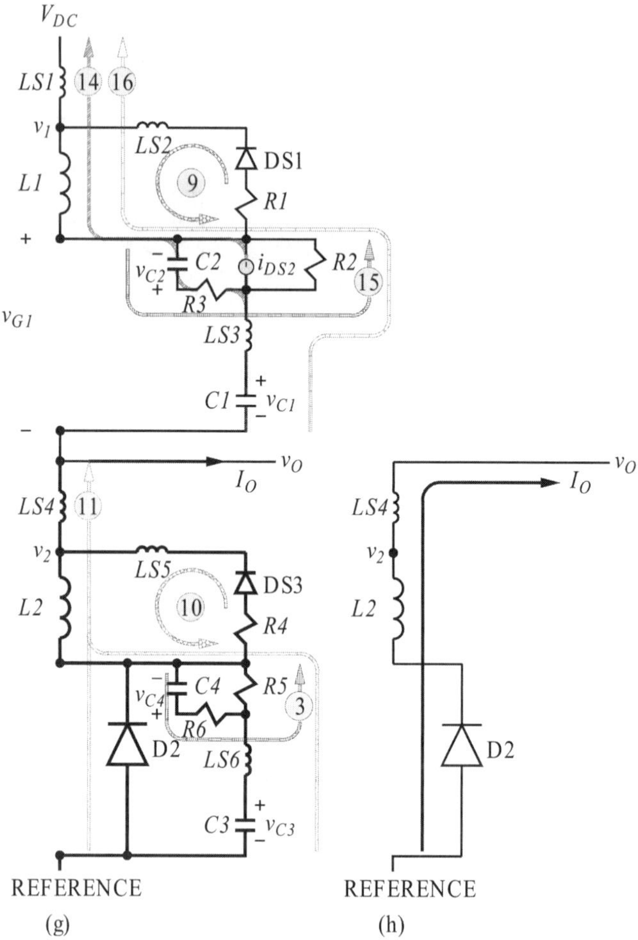

Figure 7-13(g, h). Commutation from G1 to G2 to D2 under full load with t_{delay} = 2 μs during (g) $t_5 \leqq t < t_6$, and (h) $t_6 \leqq t$.

Operation with $t_{delay} = 0$. The waveforms in this mode of operation are shown in Figure 7-14a for full load conduction ($I_O = I_p$) and in Figure 7-14b for no load condition ($I_O = 0$). Under steady-state condition (Figure 7-15a), the output current (I_O) flows through the line stray inductance (*LS1*), the snubber inductor (*L1*), and the GTO (G1). The capacitor (*C3*) is charged to the GTO (G2) forward blocking voltage ($v_B = V_{DC}$), the capacitor (*C1*) is discharged completely, and the output voltage (v_O) equals the DC link voltage (V_{DC}).

At time t_0 (Figure 7-15b), under full load condition ($I_O = I_p$), the GTO (G1) is turned off and the GTO (G2) is turned on. The current (i_{G1}) through the GTO (G1) starts to fall, as shown in Figure 7-7c. The snubber capacitor (*C3*) starts to discharge

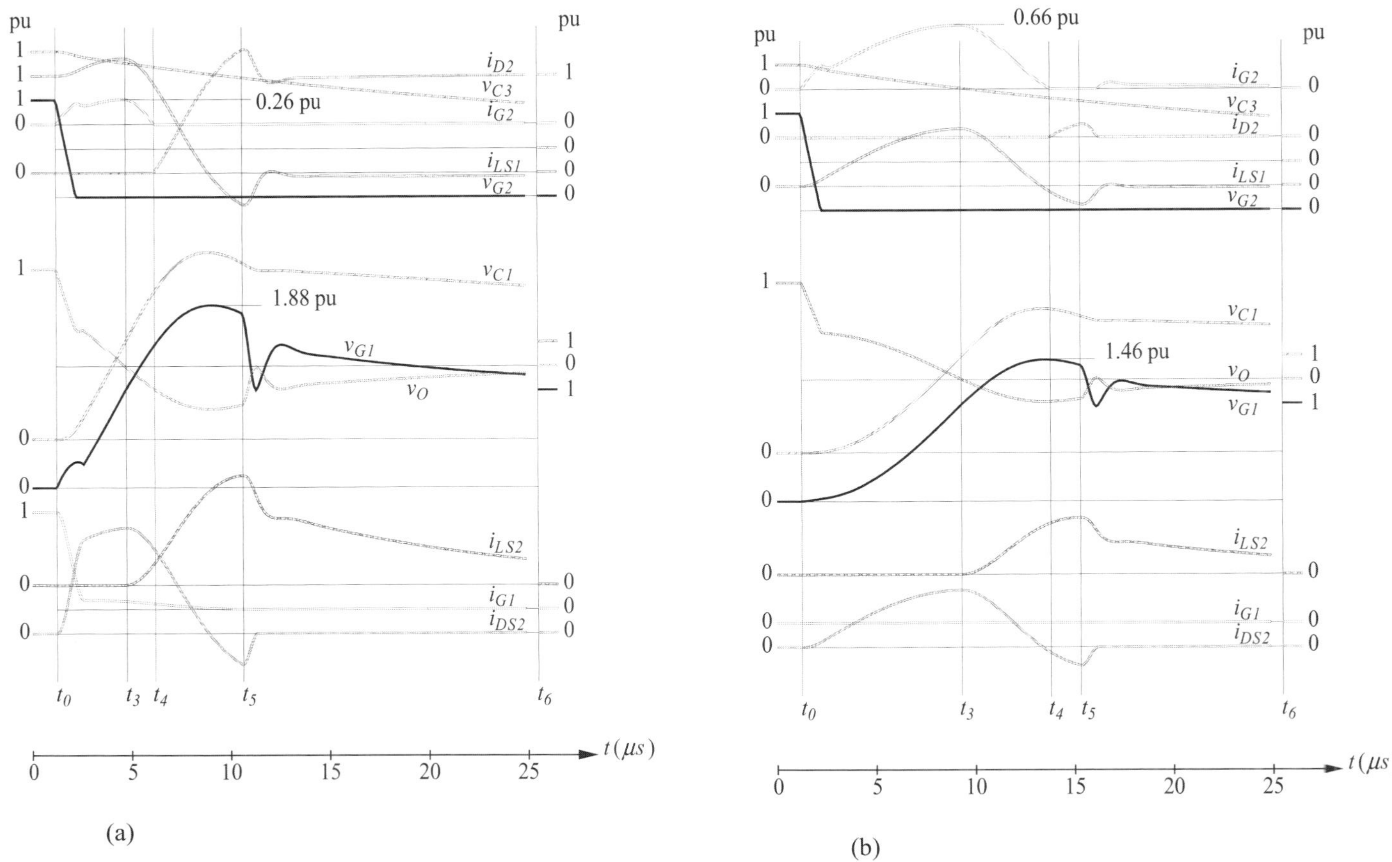

Figure 7-14. Transients from commutation from G1 to G2 to D2 under (a) full load and (b) no load conditions $t_{delay} = 0$.

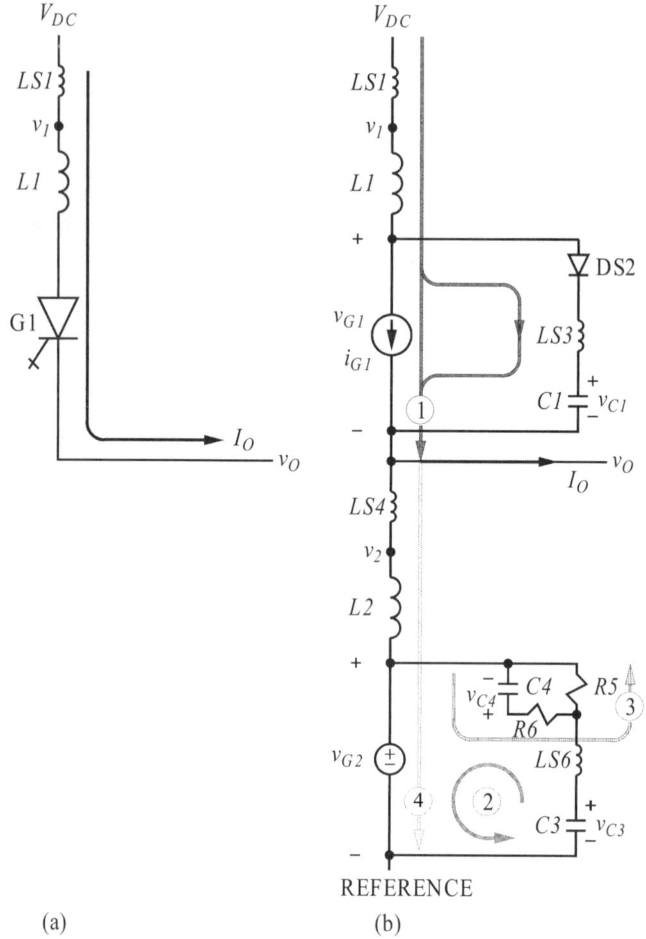

Figure 7-15. Commutation from G1 to G2 to D2 with $t_{delay} = 0$ μs during (a) $t < t_0$, and (b) $t_0 \leqq t < t_3$.

through the snubber capacitor stray inductance (*LS6*), the snubber resistor (*R5*), and the GTO (G2). The snubber diode snubber capacitor (*C4*) charges very quickly with negative polarity and then the voltage (v_{C4}) across the capacitor (*C4*) discharges through the resistors (*R6* and *R5*). During the time between t_0 and t_3, the voltage (v_{G2}) across the GTO (G2) falls linearly from its forward blocking voltage (v_B) to zero according to equation 7-3 and stays at zero. The state equations that characterize the behavior of the circuit are 7-6, 7-7, 7-14, and 7-16 through 7-19, where $i_{LS3} = i_{DS2}$ and $v_{C2} = 0$, and the initial conditions are $i_{LS1}(t_0) = I_O$, $i_{LS4}(t_0) = 0$, $v_{C1}(t_0) = 0$, $i_{LS3}(t_0) = 0$, $v_{C3}(t_0) = V_{DC}$, $v_{C4}(t_0) = 0$, and $i_{LS6}(t_0) = 0$.

The expressions for v_1, v_2, v_O, and v_{G1} are given by equations 7-21 through 7-23 and 7-32. The output voltage (v_O) falls to zero. The currents (i_{LS1} and i_{LS4}) rise because of the positive voltages across the line stray inductances (*LS1* and *LS4*). The GTO (G2) carries the sum of the line stray inductance (*LS4*) current (i_{LS4}) and the snubber capacitor (*C3*) discharge current (i_{LS6}). The difference between the line stray inductance (*LS1*) current (i_{LS1}) and the GTO (G1) current (i_{G1}) charges the snubber capacitor (*C1*) through the snubber diode (DS2) and the snubber capacitor stray inductance (*LS3*). The description of the circuit behavior during the time between t_3 and t_6 is as given before.

Under no load or light load condition, during the time between t_4 and t_6, the current through the GTO (G2) can swing between positive and negative values for a few times until the transients die out. This negative current can only flow through the antiparallel diode (D2).

Operation with $t_{delay} = 20$ μs. The waveforms in this mode of operation are shown in Figure 7-16 for full load conduction ($I_O = I_p$). Under steady-state condition (Figure 7-15a), the output current (I_O) flows through the line stray inductance (*LS1*), the snub-

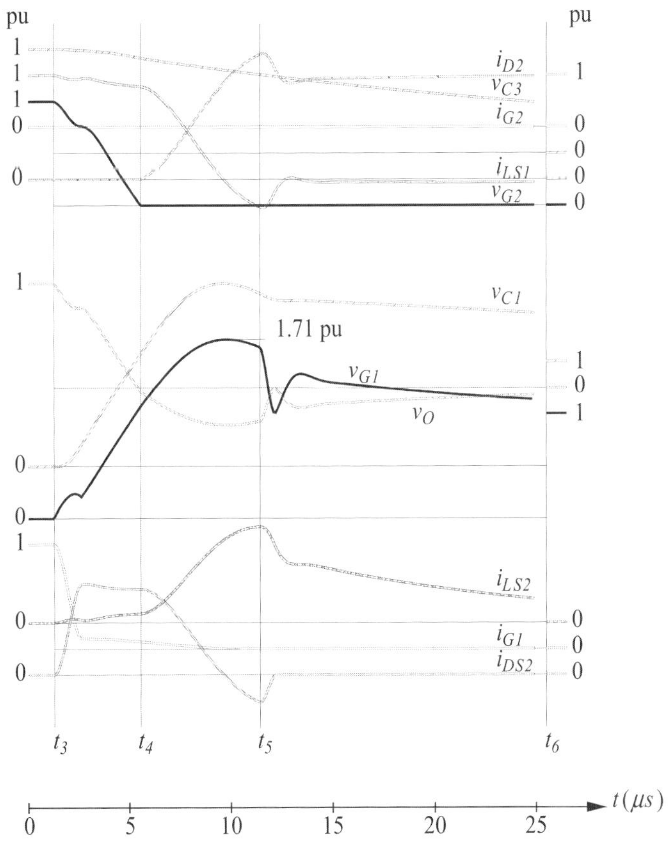

Figure 7-16. Commutation from G1 to G2 to D2 under full load with $t_{delay} = 20$ μs.

ber inductor (*L1*), and the GTO (G1). The capacitor (*C3*) is discharged to the GTO (G2) forward blocking voltage ($v_B = V_{DC}$), the capacitor (*C1*) is discharged completely, and the output voltage (v_O) equals to the DC link voltage (V_{DC}).

At time t_3, the GTO (G1) is turned off. The description of the circuit behavior during the time between t_3 and t_6 is as given before. Because of the large delay time (t_{delay} = 20 μs), the output current (I_O) at full load commutes naturally from GTO (G1) to diode (D2). The turn-on of the GTO (G2) has no impact on the circuit behavior.

7.2.4 Results

To examine the situation when the worst voltage stress across and current stress through a GTO occur, both types of commutation processes, from the diode (D2) to the GTO (G1) and from the GTO (G1) to the GTO (G2) to the diode (D2), are studied with various delay times (t_{delay}). The results are given in Table 7-3.

Under no load operation, the voltage stress across the GTO (G2) with the GTO (G1) turn-on is the same as that across the GTO (G1) with the GTO (G1) turn-off and the GTO (G2) turn-on with no delay time and the current stress through the GTO (G1) with the GTO (G1) turn-on is the same as that through the GTO (G2) with the GTO (G1) turn-off and the GTO (G2) turn-on with no delay time.

Table 7-3. Voltage and current stresses when a GTO turns on and off

		Peak voltage across the GTO, G1 (pu)	Peak voltage across the GTO, G2 (pu)	Peak current through the GTO, G1 (pu)	Peak current through the GTO, G2 (pu)
GTO (G1) turn-on	Full load Figure 7-09a		1.54	1.75	
	No load Figure 7-09b		1.46	0.66	
GTO (G1) turn-off and GTO (G2) turn-on with t_{delay}= 2 μs	Full load Figure 7-12	1.77			0.15
GTO (G1) turn-off and GTO (G2) turn-on with t_{delay}= 0	Full load Figure 7-14a	1.88			0.26
	No load Figure 7-14b	1.46			0.66
GTO (G1) turn-off and GTO (G2) turn-on with t_{delay}= 20 μs	Full load Figure 7-16	1.71			0

The results are indicated in pu with one pu current as the nominal output current (I_O = 4000 A) and the one pu voltage as the nominal DC link voltage (V_{DC} = 1900 V).

The worst overvoltage of 1.88 pu occurs across the GTO (G1) when the output current (I_O) commutates from the GTO (G1) to the GTO (G2) to the diode (D2) under full load condition with $t_{delay} = 0$. The worst overcurrent of 1.75 pu occurs through the GTO (G1) when the output current (I_O) commutates from the diode (D2) to the GTO (G1) under full load condition. Therefore, in the pole circuit under consideration, the GTO must be derated at least by a factor of two in terms of its voltage and current ratings. In real-world applications, each switch consists of a number of GTOs that are connected in series. Considering the need for continuous operation of the VSC with an allowable number of failed GTOs, each GTO is further derated in terms of its voltage rating.

Since the voltage across and the current through each element of the circuit is calculated in the simulation, the total loss in the pole circuit can be calculated for trade-off analysis of components selection, heat sink sizing, and determining the overall efficiency of the pole circuit. The single GTO switch model can serve as the basis to extend the simulation of the circuit, shown in Figure 7-6, to represent a higher power switch built with a stack of GTO modules that are connected in series although there may be only one *di*/*dt* limiting inductor serving all the GTO modules. Each GTO module has its own antiparallel diode and turn-off snubber circuit, as shown in Figure 7-6. Failure of any one GTO in the stack will make the upper and lower halves of the VSC pole asymmetric. Circuit behavior under this asymmetric condition can be examined as well as the variations caused by differences in the dynamic switching characteristics of the individual GTOs in the stack. The acceptable dynamic sharing of each GTO may be obtained by proper selection of GTOs with similar characteristics. Since the leakage resistance of each GTO is different, static sharing is ensured by connecting a high resistance (on the order of 100 KΩ) across each GTO.

8

VSC-BASED FACTS CONTROLLERS

The voltage-sourced converter (VSC)-based flexible alternating current transmission systems (FACTS) controllers are made by using two fundamental building blocks: static synchronous compensator (STATCOM) and static synchronous series compensator (SSSC). The main part of each controller is a VSC that generates a compensating voltage. In this chapter, only harmonic neutralized voltage-sourced converters (HN-VSC) are used. In an HN-VSC, a number of square wave voltages are generated at fundamental frequency by operating the semiconductor switches once per cycle. These square wave voltages are combined in a magnetic circuit in order to produce a high-quality sinusoidal voltage with acceptable harmonic content. The number of switches in the VSC and the various configurations of the magnetic circuit depend on the desired quality of the AC waveforms generated by the VSC.

In the STATCOM, a compensating voltage is connected in shunt with the transmission line through a tie impedance. The voltages at both sides of the tie impedance, the line voltage and the compensating voltage, are never of the same value on an instantaneous basis due to the harmonic presence in the compensating voltage. The fundamental components of these two voltages are in phase under steady-state conditions. The STATCOM behaves as a current source, the value of which is the ratio of the difference between the two voltages (line voltage and compensating voltage) and the tie impedance. Often, a shunt coupling transformer is used between the VSC and the line to keep the STATCOM isolated from the high-voltage transmission line. The coupling transformer is also used to match the compensating voltage with the line voltage. The leakage impedance of the coupling transformer acts as the tie impedance. In the SSSC, a compensating voltage is connected in series with the transmission line through a coupling transformer.

The STATCOM is shown in Figure 8-1. The STATCOM consists of a HN-VSC (VSC1), a magnetic circuit (MC1), a shunt coupling transformer (SHNTR), a shunt circuit breaker (SHNBRK), current and voltage sensors, and a control and protection unit. The STATCOM injects a line frequency, variable magnitude, fundamental current and some unavoidable harmonic currents at the point of compensation. The current through the STATCOM is almost in quadrature with the line voltage, thereby emulating an inductive or a capacitive reactance at the point of compensation. Thus, a stand-alone STATCOM exchanges almost exclusively reactive power at the point of compensation. The primary function of the STATCOM is to control the reactive current that flows through it. The emulated variable inductive and capacitive reactances at the point of compensation regulate the line voltage.

The SSSC is shown in Figure 8-2. The SSSC consists of a HN-VSC (VSC2), a magnetic circuit (MC2), a series coupling transformer (SRSTR), a series circuit breaker (SRSBRK), current and voltage sensors, and a control and protection unit. The SSSC injects a line frequency, variable magnitude fundamental voltage, and some unavoidable harmonic voltages in series with the line. The compensating voltage is almost in quadrature with the prevailing line current, thereby emulating an inductive or a capacitive reactance in series with the line. Thus, a stand-alone SSSC exchanges almost exclusively reactive power with the line. The series-connected compensating voltage modifies the magnitude and phase angle of the line voltage.

When the STATCOM and SSSC operate as stand-alone compensators, they provide the control of two parameters: line voltage and line reactance. The control of ei-

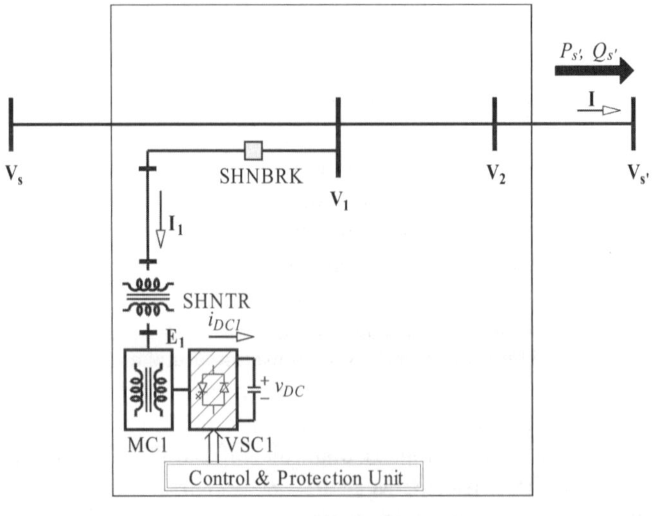

Figure 8-1. Static synchronous compensator (STATCOM).

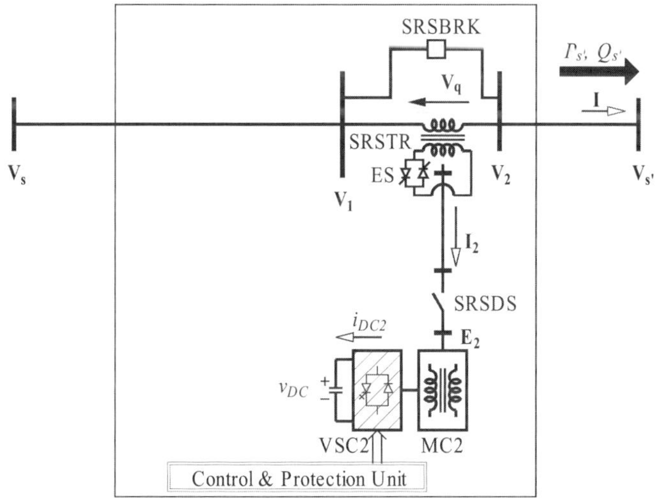

Figure 8-2. Static synchronous series compensator (SSSC).

ther parameter regulates both the active and reactive power flows in the line simultaneously. In order to regulate both the active and reactive power flows in the line independently, the line voltage needs to be controlled to a specific magnitude and phase angle.

A series-connected compensating voltage can modify the transmission line voltage to a specific magnitude and phase angle within the design limit. This requires the compensating voltage to be at any phase angle with the prevailing line current. Therefore, the compensating voltage emulates, in series with the transmission line, a capacitor that increases the power flow of the line or an inductor that decreases the power flow of the line and a positive resistor that absorbs active power from the line or a negative resistor that delivers active power to the line. The desired compensating voltage is actually an impedance emulator that is able to exchange both active and reactive powers with the line.

When the STATCOM and the SSSC are connected back to back at a shared DC link, the series-connected compensating voltage of the SSSC is no longer restricted to be always in quadrature with the prevailing line current. The compensating voltage in series with the transmission line can be at any phase angle with respect to the line current; therefore, the exchanged power at the terminal of the SSSC can be active as well as reactive. The compensating voltage component that is in phase with the prevailing line current exchanges with the line active power that flows back and forth through the STATCOM to the point of compensation. In addition, the STATCOM carries a reactive current to regulate the bus voltage independently. This configuration is referred to as the unified power flow controller (UPFC). The concept of a shared DC link between a shunt-connected VSC and a series-connected VSC was first introduced in the active power line conditioner (APLC) for distribution power level applications. The same concept was later implemented in the UPFC for transmission power level applications.

The UPFC is shown in Figure 8-3. The UPFC consists of two HN-VSCs (VSC1 and VSC2), two magnetic circuits (MC1 and MC2), a shunt coupling transformer (SHNTR), a series coupling transformer (SRSTR), a shunt circuit breaker (SHNBRK), a series circuit breaker (SRSBRK), current and voltage sensors, and a control and protection unit. The VSCs are connected to a shared DC link through a DC link switch (DCLS). The coupling transformers include a lumped leakage reactance that is an equivalent of all the leakage reactances of the intermediate magnetic circuits and the main transformers. The lumped leakage reactance is 15% of the VSC rating and divided equally on both primary and secondary sides.

In this chapter, the STATCOM, SSSC, and UPFC are described in great detail. Each of these VSC-based FACTS controllers is modeled using the Electromagnetic Transients Program (EMTP) simulation package, as discussed in Chapter 3. The control or the transient analysis of control systems (TACS) section coordinates the generation of the VSC output voltage that requires the combined effect of three items: (a) gating signals, (b) pole switches, and (c) a magnetic circuit. The TACS section generates the free-running gating signals for the pole switches. At this point, one may assume that the pole switches are ideal in the sense that the gating signals turn the pole switches on and off instantly. Assuming also that the magnetic circuit is ideal in the sense that there is no loss of any kind in the magnetic circuit and its coupling is perfect, one may combine the ideal pole voltages mathematically to produce a three-phase HN-VSC output voltage set that can be fed to the source section. The voltage (v_{DC}) across the shared DC link capacitor is maintained by the instantaneous power balance equations at both AC and DC sides of the two VSCs. This modeling technique gives sufficient insight to the operation of the power circuit that produces a three-phase voltage set.

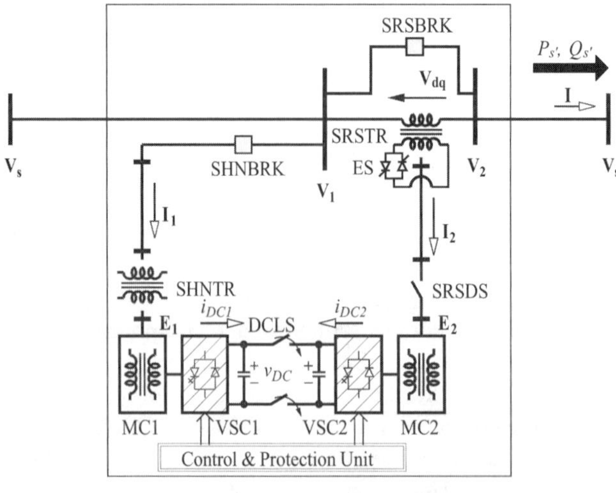

Figure 8-3. Unified power flow controller (UPFC).

In an alternate modeling technique, the switches can be located in the switch section and be modeled with a number of force-commutated semiconductor switches that are connected in series/parallel, each having an antiparallel diode and appropriate snubber circuits. The pole output voltages are then fed to a magnetic circuit, which produces a three-phase voltage set. The magnetic circuit can also serve as the coupling transformer for matching the voltages between the line and VSC sides. Thus, there is no need for an additional coupling transformer.

The modeling may be done at various levels. For example, to study the functionality of a FACTS controller on an elaborate power system network, a FACTS controller with a simplified model consisting of sinusoidal compensating voltages and detailed control and protection schemes may be adequate. For magnetic circuit and switch designers, the primary focus should be on the modeling of the detailed power circuit. In that case, it is not so important to model an elaborate power system network. A simple power system network with a few transmission lines and generators will serve the purpose, at least initially, in the design stage of the FACTS controllers.

The models, described in this chapter, characterize the behavior of the FACTS controllers as viewed from the network terminal, especially the dynamic behavior during disturbances and faults in the power system. The models provide the implementation of the power flow control using the series-connected VSC2 and line voltage control using the shunt-connected VSC1. The control is implemented in the synchronously rotating reference frame. The model represents the behavior of the FACTS controller in the presence of negative sequence and harmonic line voltages and currents. A comprehensive model must include protective limits on series voltage injection, and overcurrent and overvoltage limits for both VSCs. The modeling techniques, described in this chapter, are useful tools to the designers of FACTS controllers.

Figure 8-4 shows the modeling structure of a FACTS controller. The inputs from the voltage and current measuring units, the breaker status, and the compensation lev-

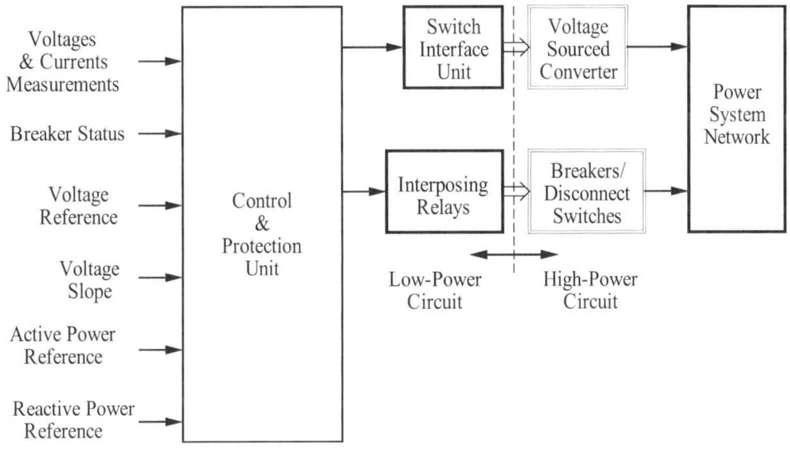

Figure 8-4. Modeling structure of a FACTS controller.

els—voltage reference, voltage slope (also known as droop), and the active and reactive power references—are fed to the control and protection unit. The output of the control and protection unit is a set of low-voltage optical signals. These optical signals are fed to the switch interface unit, which converts them into electrical gating signals that turn ON and OFF the semiconductor switches of the high-power VSC, which is connected to the power system network through its coupling transformer. The output of the control and protection unit also has another set of low-voltage electrical signals that operate the interposing relays, which operate the high-power breakers and disconnect switches that connect and disconnect the FACTS controllers to the power system network.

Each model is a virtual representation of an actual installation at its terminal. The control and protection unit implements the basic control and protection schemes. The gating signals for the pole switches are generated as the simulation progresses. The ideal pole output voltages are mathematically combined to produce a three-phase HN-VSC voltage. The DC link capacitor is modeled by a single lumped capacitor sized in accordance with other model parameters, including the transformer leakage reactance, to preserve the same dynamics in the model as in the real system. The voltage (v_{DC}) across the shared DC link capacitor is maintained by the instantaneous power balance equations at both AC and DC sides of the VSCs.

In system transients studies with idealized VSCs, special care should be taken in modeling the rectifier circuit that charges the DC link capacitor to at least the peak-to-peak of the AC bus voltage at the point of compensation when the VSC1 gating is inhibited. At the moment when the gating is inhibited, the ideal VSC may be replaced by a rectifier circuit with appropriate snubber circuits. The stored energy in the tie reactor continues to overcharge the DC link capacitor. When the VSC1 gating begins, the DC link capacitor discharges to the appropriate level. The voltage variation across the DC link capacitor should be correctly represented with the operation of the voltage clamp that discharges the DC link capacitor partially when a certain overvoltage threshold is exceeded.

The EMTP models and accompanying descriptions are powerful tools that provide the sequence of how a VSC is commissioned step-by-step and used as a FACTS controller. The models include all the necessary components: a VSC with a DC link capacitor, a magnetic circuit, and a realizable controller. The control system in the model has been implemented on a per-unit (pu) basis. Therefore, it is easy to integrate the model in a power system network of any voltage and current level.

The objective of this chapter is to demonstrate that the simulation results of a VSC-based FACTS controller can be used to accurately predict the behavior of the controller at an installation site. This demonstration is substantiated with the comparison of results from the computer simulation and actual field measurements. The operation of the FACTS controllers is verified with the models connected to a simple two-bus network, as discussed in Chapter 3 in Figure 3-2. This model is used to verify the response of the controller following a step change in the reference of the control inputs. The FACTS controller models can be quite beneficial for the utility system planners and they can easily be incorporated in a larger network to obtain validation of the simulation results and the field measurements.

8.1 SHUNT COMPENSATION

In Chapter 5, it is discussed that a shunt-connected capacitor or inductor to the line increases or decreases the voltage at the point of compensation to the line with the generated or absorbed reactive power. The same objective of regulating the line voltage can be met with the use of a shunt-connected compensating voltage through an inductive impedance. A synchronous machine is used to implement this method when the machine runs as a motor and its back emf (**E**) that is in phase with the line voltage ($\mathbf{V}_s$) is controlled by the amount of its field current.

8.1.1 Shunt Reactive Current Injection

Figure 8-5 shows a shunt reactive current injection at the point of compensation. The main file `801CSC1.DAT` (see Appendix C, Section 63) contains a modified program from Chapter 5, where a mechanically switched capacitor and inductor are used. In this program, a one pu current from a current source is injected in each of the three phases in both capacitive and inductive modes. The current sources are defined in the TACS section with an Itype of 99 and in the source section with an Itype of 60. The –1 in columns 7 and 8 denotes a current source. In the user file `802CSC1.USR` (see Appendix C, Section 64), the magnitude and the polarity of the compensating current source are varied to emulate a capacitor or an inductor in shunt with the transmission line. In the control file `803CSC1.CON` (see Appendix C, Section 65), the current sources in the time domain are defined. The polarity of the shunt-connected current sources (i_{1qa}, i_{1qb}, and i_{1qc}) are reversed because of the EMTP's convention of a current source that is opposite of what is drawn in Figure 8-5.

Figure 8-6 shows the digital simulation results from the injection of a shunt reactive current into the transmission line. Between 0 and 50 ms, the shunt breaker (SNHBRK) is open. The natural voltage at the point of compensation is 0.99 pu. At 50 ms, the SNHBRK closes while the reactive current source is initialized to zero. At 125 ms, the reactive current source is set to one pu capacitive, which means that the reactive current (i_{1a}) of phase a leads the line voltage (v_{1a}) of phase a by 90°. The line voltage becomes 1.03 pu at the point of compensation. At 175 ms, the reactive current source is

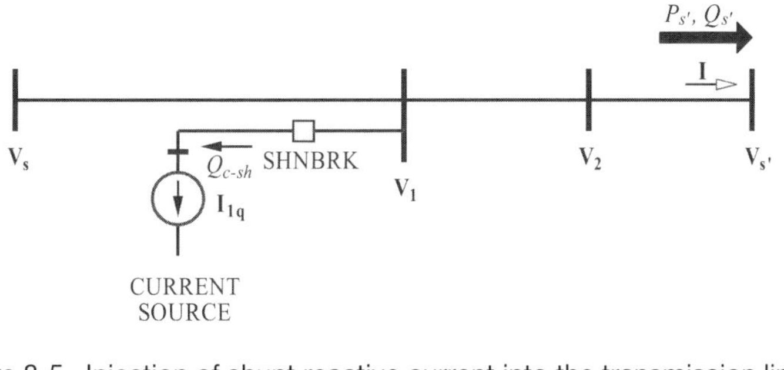

Figure 8-5. Injection of shunt reactive current into the transmission line.

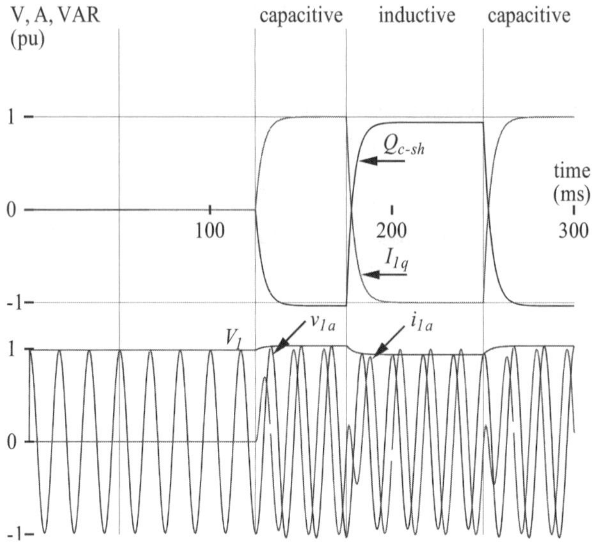

Figure 8-6. Exchanged reactive power and the resulting voltages and currents due to an injection of shunt current into the transmission line.

set to one pu inductive, which means that the reactive current (i_{1a}) of phase *a* lags the line voltage (v_{1a}) of phase *a* by 90°. The line voltage then becomes 0.94 pu. At 250 ms, the reactive current source is returned to one pu capacitive. While emulating a capacitive reactance at 1.03 pu line voltage and drawing one pu current, the current source delivers 1.03 pu reactive power to the transmission line. While emulating an inductive reactance at 0.94 pu line voltage and drawing one pu current, the current source absorbs 0.94 pu reactive power from the transmission line.

8.1.2 Shunt-Connected Compensating Voltage Source Behind an Impedance

Figure 8-7 shows that an in-phase compensating voltage ($\mathbf{E_1}$) is connected to the line voltage ($\mathbf{V_1}$) at BUS01 through a tie reactance (X_{TIE}). The difference between the line voltage ($\mathbf{V_1}$) and the compensating voltage ($\mathbf{E_1}$) causes a reactive current ($\mathbf{I_{1q}}$) to flow through the tie reactance. Through control action, the magnitude of the compensating voltage (E_1) can be made higher or lower than the line voltage (V_1). When $E_1 < V_1$, the compensating voltage acts as an inductive reactance connected to the line and absorbs reactive power at the point of compensation. When $E_1 > V_1$, the compensating voltage acts as a capacitive reactance connected to the line and delivers reactive power at the point of compensation.

In the main file `804VSC1.DAT` (see Appendix C, Section 66), a sinusoidal compensating voltage that is in phase with the line voltage is connected in shunt with the transmission line at BUS01 through a tie reactance. In the user file `805VSC1.USR` (see

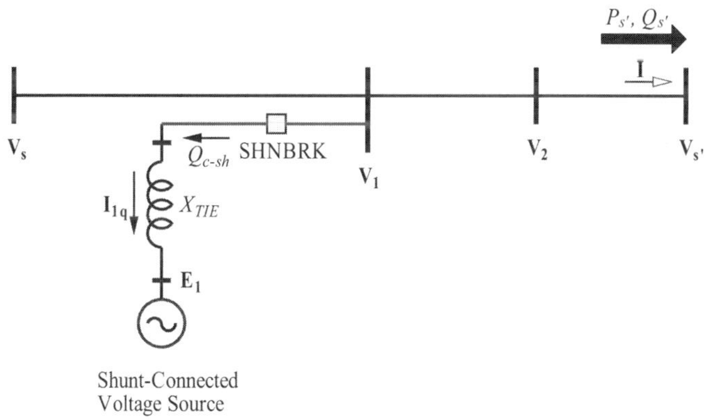

Figure 8-7. Transmission line voltage control using a shunt-connected compensating voltage.

Appendix C, Section 67), the magnitude of the compensating voltage is varied to emulate a capacitor or an inductor in shunt with the transmission line. In the control file `806VSC1.CON` (see Appendix C, Section 68), the three-phase compensating voltages in time domain are defined.

Figure 8-8 shows the digital simulation results from the shunt-connected compensating voltage (E_1) that is in phase with the line voltage. Between 0 and 50 ms, the shunt breaker (SNHBRK) is open. The natural voltage at the point of compensation is 0.99 pu. At 50 ms, the SNHBRK closes and the compensating voltage magnitude is

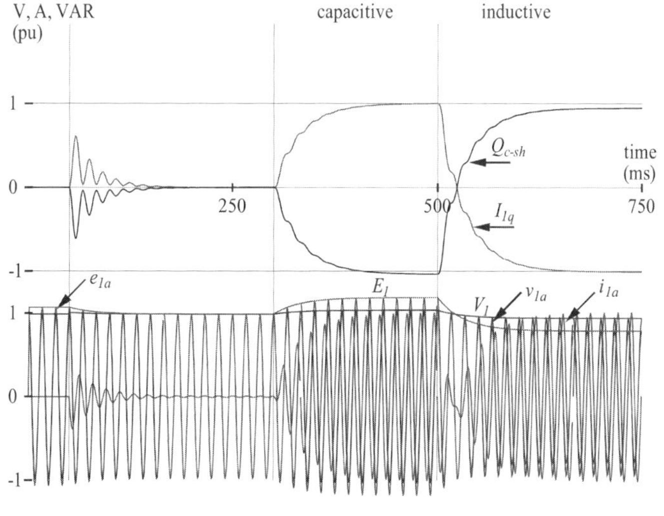

Figure 8-8. Exchanged reactive power and the resulting voltages and currents due to a shunt-connected voltage into the transmission line through a tie reactance.

lowered from the initialized value of 1.068 pu to set the reactive current flow to zero. The compensating voltage acts transiently in capacitive mode, which is explained in later simulation results. At 300 ms, the magnitude of the compensating voltage is adjusted to 1.187 pu to produce one pu capacitive current, which means that the reactive current (i_{1a}) of phase *a* leads the line voltage (v_{1a}) of phase *a* by 90°. The line voltage becomes 1.03 pu at the point of compensation. At 500 ms, the magnitude of the compensating voltage is adjusted to 0.787 pu to produce one pu inductive current, which means the reactive current (i_{1a}) of phase *a* lags the line voltage (v_{1a}) of phase *a* by 90°. The line voltage then becomes 0.94 pu. While emulating a capacitive reactance at 1.03 pu line voltage and drawing one pu current, the current source delivers 1.03 pu reactive power to the transmission line. While emulating an inductive reactance at 0.94 pu line voltage and drawing one pu current, the current source absorbs 0.94 pu reactive power from the transmission line.

8.1.3 Shunt-Connected Compensating Voltage Behind a Coupling Transformer

In power system applications, a coupling transformer, SHNTR, shown in Figure 8-9, is used for electric isolation of the shunt-connected compensating voltage to the line and also for voltage matching with the line. The above-mentioned tie impedance is replaced with a shunt-connected transformer whose total leakage impedance on both primary and secondary sides is of the same value as the tie impedance. The transformer model is given in an $INCLUDE$ file, `312TRAN1.TRN` (see Appendix C, Section 12). The leakage impedance (*R*, *X*) of the transformer is 15% of the base impedance. Since the transformer winding ratio (n_1:n_2) is considered to be 1:1, the leakage impedance (*R*, *X*) is divided equally across both windings.

The simulation results from the main file `807VSC2.DAT` (same as `804VSC1.DAT`, except that the tie impedance is replaced with a shunt coupling transformer) are identi-

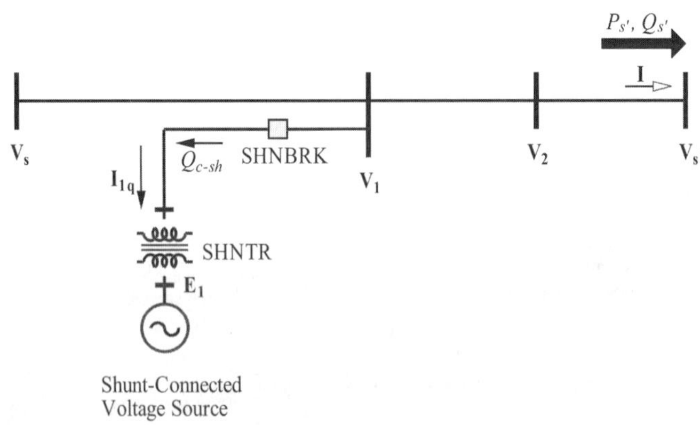

Figure 8-9. Transmission line voltage control using a shunt-connected compensating voltage through a coupling transformer.

cal to that from the file 804VSC1.DAT. For the same compensating voltage values, the line voltages are regulated at the same values as in the previous case, when a tie impedance is used.

Next, we consider that the compensating voltage is generated through a VSC, as shown in Figure 8-10. The VSC has a gain of $2/\pi$ and it generates a sinusoidal (i.e., infinite pulse number) voltage. The STATCOM model is given in the three files, namely (a) main file 808VSCDC.DAT, (b) user input file 808VSCDC.USR, and (c) control file 808VSCDC.CON (see Appendix C, Sections 69 through 71). The DC capacitor voltage is specified in the user file. The VSC is considered to be of infinite pulse.

Figure 8-11 shows the simulation results. Between 0 and 50 ms, the shunt breaker (SNHBRK) is open. The DC link capacitor is precharged. The VSC output compensating voltage (e_{1a}) of phase *a* and the line voltage (v_{1a}) of phase *a* are in phase. At 50 ms, the SNHBRK closes and the DC link capacitor voltage (v_{DC}) is reduced to a level, so that the VSC does not supply any reactive current (I_{1q}) to the line. The excess voltage from the DC link capacitor discharges as an emulated capacitor. Since the current through the VSC is zero, the VSC output voltage (e_{1a}) of phase *a* and the line voltage (v_{1a}) of phase *a* have equal amplitudes. Next, the voltage of the DC capacitor is maintained at 1.86 pu and 1.23 pu. The corresponding AC voltages are 1.18 pu and 0.78 pu, respectively. The VSC currents are one pu capacitive and inductive, respectively. The resulting line voltage is 1.03 pu and 0.94 pu, respectively. These simulation results are identical to that shown in Figure 8-8 with the one exception of the DC link voltage.

Next, the closed loop control that regulates the DC capacitor voltage will be implemented for the same operating points.

8.1.4 Static Synchronous Compensator (STATCOM)

The STATCOM model in EMTP is shown in Figure 8-12. A controller that controls the reactive current through the VSC is implemented. The controller of the STAT-

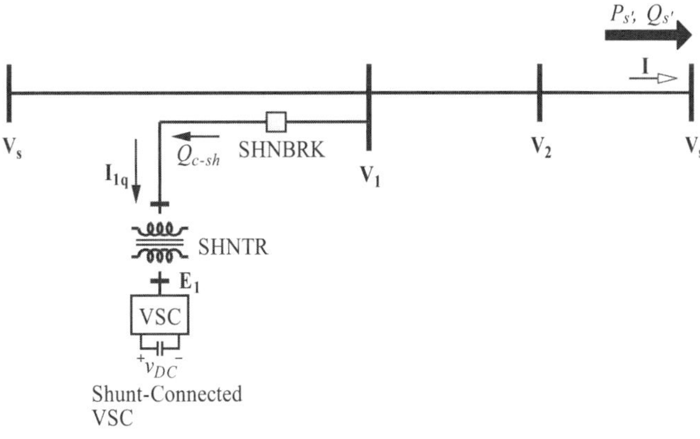

Figure 8-10. Transmission line voltage control using a shunt-connected VSC.

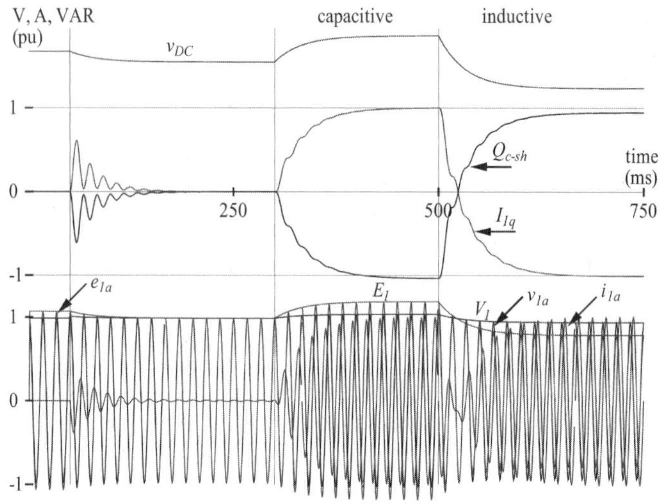

Figure 8-11. Exchanged reactive power and the resulting voltages and currents due to a shunt-connected VSC into the transmission line through a tie reactance.

COM is used to operate the VSC in such a way that the phase angle (α) between the line voltage ($\mathbf{V}_1$) and the compensating voltage ($\mathbf{E}_1$) of the VSC is dynamically adjusted. When the compensating voltage lags the line voltage, the VSC draws active power from the line. When the compensating voltage leads the line voltage, the VSC delivers active power to the line. This active power regulates the DC capacitor voltage (v_{DC}) to the right level that produces the right amount of compensating voltage at the output of

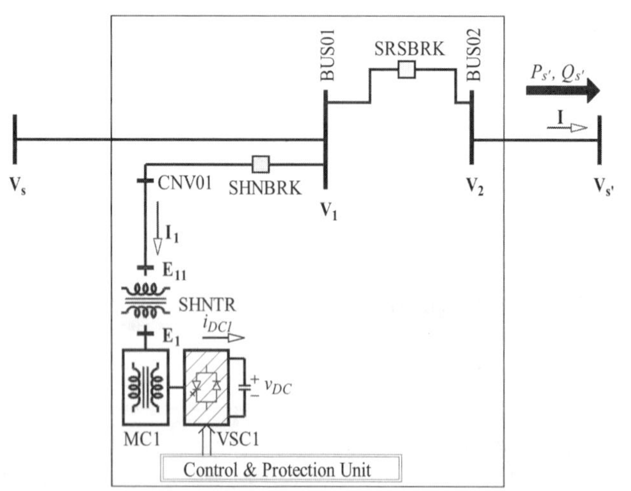

Figure 8-12. STATCOM model in EMTP.

the VSC. A compensating voltage that is larger than the line voltage generates VARs, thereby emulating a capacitor at the point of compensation. A compensating voltage that is smaller than the line voltage absorbs VARs, thereby emulating an inductor at the point of compensation. Thus, the primary controller of the VSC controls the reactive current flow through the STATCOM and regulates the line voltage at the point of compensation.

8.1.4.1 Control of STATCOM. Figure 8-13 shows the reactive current control block diagram of the STATCOM. An instantaneous three-phase set of line voltages (v_1) at BUS01 is used to calculate the PLL angle (θ), which is phase locked to the phase *a* of the line voltage (v_{1a}). An instantaneous three-phase set of measured currents (i_1) through the VSC is decomposed into its active or direct component (I_{1d}) and reactive or quadrature component (I_{1q}). The quadrature component is compared with the desired reference value (I_{1q}^*) and the error is passed through an error amplifier, which produces a relative phase angle (α) of the compensating voltage ($\mathbf{E_1}$) with respect to the line voltage ($\mathbf{V_1}$). The absolute phase angle (θ_1) of the compensating voltage is calculated by adding the relative phase angle of the compensating voltage with respect to the PLL angle (θ).

The reference quadrature component (I_{1q}^*) of the current through the VSC is defined to be either positive if the STATCOM is emulating a capacitive reactance or negative if it is emulating an inductive reactance. The DC link capacitor voltage (v_{DC}) is dynamically adjusted in relationship with the compensating voltage. The control scheme described above shows the implementation of the inner current control loop, which regulates the reactive current flow through the VSC regardless of the line voltage. However, if it is desired to regulate the line voltage, an outer voltage control loop must be implemented. The outer voltage control loop automatically determines the reference reactive current for the inner current control loop, which, in turn, regulates the line voltage.

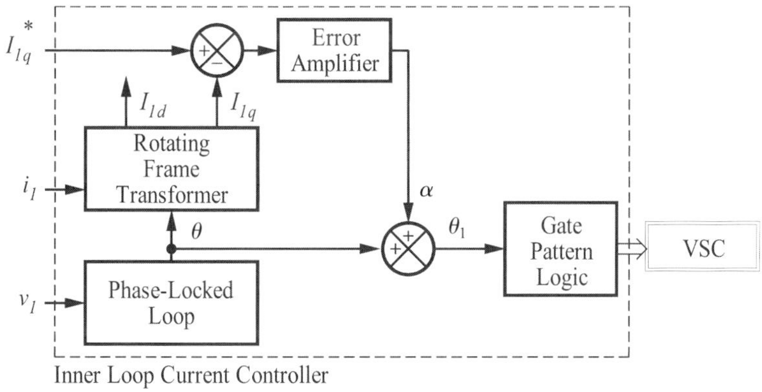

Figure 8-13. Reactive current control block diagram of the STATCOM.

Figure 8-14 shows the voltage control block diagram of the STATCOM. Using the reference PLL angle (θ), the instantaneous three-phase set of measured line voltage (v_1) at BUS01 is decomposed into its active or direct component (V_{1d}) and reactive or quadrature component (V_{1q}). The magnitude of the BUS01 voltage (V_{1dq}) is calculated, which is compared with the desired reference value, V_1^* (adjusted by the droop factor, K_{droop}), and the error is passed through an error amplifier that produces the reference current (I_{1q}^*) for the inner current control loop. The droop factor (K_{droop}) is defined as the allowable voltage error at the rated reactive current flow through the STATCOM.

8.1.4.2 Modeling of STATCOM in EMTP and Simulation Results. The STATCOM model with an inner current control loop is run from the main file `811STAT1.DAT` that is same as `808VSCDC.DAT`, except that the $INCLUDE$ user input file is `812STATI.USR` (see Appendix C, Section 72) and the control file is `813STATI.CON` (see Appendix C, Section 73). The voltage control model is run from the main file `814STAT2.DAT` that is same as `811STAT1.DAT`, except that the $INCLUDE$ user input file is `815STATV.USR` (see Appendix C, Section 74) and the control file is `816STATV.CON` (see Appendix C, Section 75). The VSC is considered to be of infinite pulse.

Figure 8-15 shows the digital simulation results from the reactive current control operation of the STATCOM. Between 0 and 50 ms, the shunt breaker (SNHBRK) is open. The DC link capacitor is precharged. The VSC output sinusoidal (infinite-pulse) compensating voltage (e_{1a}) of phase *a* and the line voltage (v_{1a}) of phase *a* are in phase. At 50 ms, the SNHBRK closes and the reference quadrature current (I_{1q}^*) of the VSC is set to zero. Since the current through the VSC is zero, the VSC output voltage (e_{1a}) of phase *a* and the line voltage (v_{1a}) of phase *a* have equal amplitudes. At 125 ms, the ref-

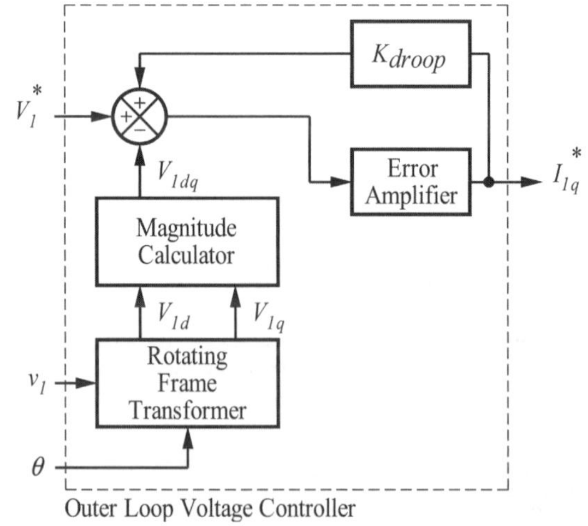

Figure 8-14. Voltage control block diagram of the STATCOM.

erence quadrature current (I_{1q}^*) of the VSC is set to one pu capacitive, and the VSC current (i_{1a}) of phase *a* leads the line voltage (v_{1a}) of phase *a* by almost 90°. The VSC output voltage (e_1) is greater than the line voltage set (v_1). At 175 ms, the reference quadrature current (I_{1q}^*) of the VSC is set to one pu inductive and the VSC current (i_{1a}) of phase *a* lags the line voltage (v_{1a}) of phase *a* by almost 90°. The VSC output voltage (e_1) is less than the line voltage set (v_1). At 250 ms, the reference quadrature current (I_{1q}^*) of the VSC is again set to one pu capacitive and the transition takes place in a subcycle time. Note that the VSC output voltage and the line voltage are in phase. In reality, the STATCOM would encounter power losses in the switches and in the magnetic circuit. Therefore, there will always be a small part of active current component (I_{1d}) flowing into the VSC and the VSC output voltage will lag the line voltage by a small phase angle under steady-state conditions. The relative phase angle (α) of the VSC output voltage with respect to the line voltage is dynamically adjusted, so that the VSC maintains a proper DC link capacitor voltage. Figure 8-16 shows the expanded view of two sections of Figure 8-15. During the capacitive mode of operation, the VSC output voltage is larger than the line voltage and during the inductive mode of operation, the VSC output voltage is smaller than the line voltage.

Figure 8-17 shows the digital simulation results from the voltage control operation of the STATCOM. At the beginning of the simulation, when the shunt breaker (SHNBRK) is open, the nominal line voltage at BUS01 is 0.988 pu. If the controller's set reference voltage (V_1^*) is higher than the line voltage (V_{1dq}) at BUS01, the controller operates the VSC in a capacitive mode to deliver up to a maximum of one pu current,

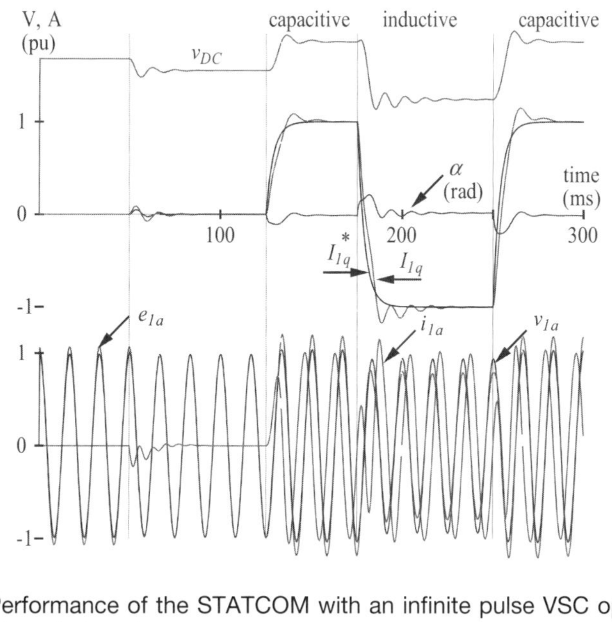

Figure 8-15. Performance of the STATCOM with an infinite pulse VSC operating in a reactive current control mode.

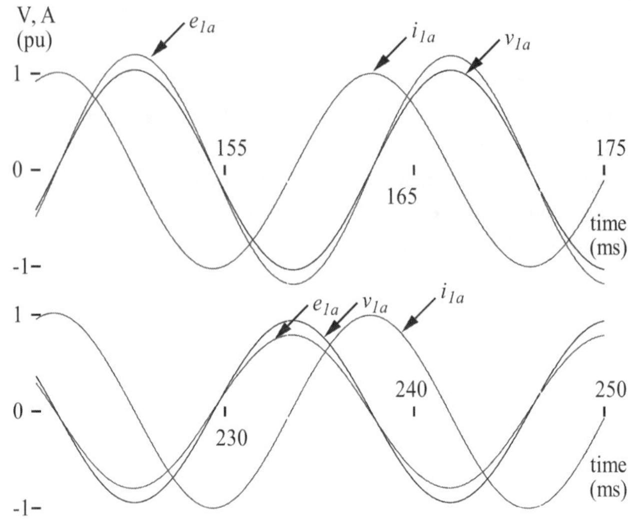

Figure 8-16. Waveforms of the STATCOM with an infinite pulse VSC operating in a reactive current control mode.

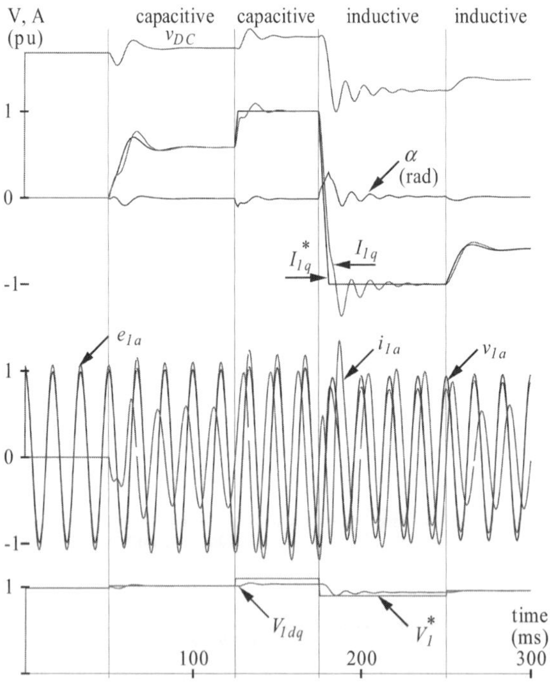

Figure 8-17. Performance of the STATCOM with an infinite pulse VSC operating in a voltage control mode.

if necessary. Similarly, if the controller's set reference voltage (V_1^*) is lower than the line voltage (V_{1dq}) at BUS01, the controller operates the VSC in an inductive mode to deliver up to a maximum of one pu current, if necessary. At 50 ms, the SHNBRK closes and the reference line voltage (V_1^*) at BUS01 is set to 1.015 pu, which results in a capacitive mode of operation of the STATCOM and the VSC current (i_{Ia}) of phase *a* leads the line voltage (v_{Ia}) of phase *a* by almost 90°. At 125 ms, the reference line voltage (V_1^*) is set to 1.10 pu. The controller operates the VSC in such a way that the current through the STATCOM is regulated at a capacitive current limit of one pu that results in a line voltage of 1.03 pu. At 175 ms, the reference line voltage (V_1^*) is set to 0.90 pu. The controller operates the VSC in such a way that the current through the STATCOM is regulated at an inductive current limit of one pu, which results in a line voltage of 0.94 pu. The VSC current (i_{Ia}) of phase *a* lags the line voltage (v_{Ia}) of phase *a* by almost 90°. At 250 ms, the reference line voltage (V_1^*) is set to 0.96 pu, which results in an inductive mode of operation of the STATCOM. Note that when a outer voltage control loop is used to produce the reference current (I_{Iq}^*) for the inner current control loop, the STATCOM takes somewhat longer time to respond than it would take if only the inner current control loop were used with user specified reference current (I_{Iq}^*).

The previous test case for reactive current control and voltage control are rerun for the STATCOM with 12- and 24-pulse HN-VSCs. The STATCOM model with the inner current control loop is run from the main file `817STAT3.DAT` and the outer voltage control loop is run from the main file `818STAT4.DAT` for a 12-pulse HN-VSC. For a 24-pulse HN-VSC, the STATCOM model with the inner current control loop is run from the main file `819STAT5.DAT` and the outer voltage control loop is run from the main file `820STAT6.DAT`. The results are shown in Figures 8-18, 8-19, and 8-20 and Figures 8-21, 8-22, and 8-23, respectively. The VSC voltage and current show the presence of harmonic components.

For a higher pulse HN-VSC, the magnetic circuit consists of specially designed transformers and, therefore, it becomes expensive. One alternate solution is to use standard Y–Δ transformer to build a quasiharmonic neutralized voltage-sourced converter (QHN-VSC). The previous test case for reactive current control was rerun for the STATCOM with 24- and 48-pulse QHN-VSCs. The STATCOM model with the inner current control loop is run from the main file `821STAT7.DAT` for a 24-pulse QHN-VSC. For a 48-pulse QHN-VSC, the STATCOM model with the inner current control loop is run from the main file `822STAT8.DAT`. The results are shown in Figures 8-24 and 8-25 and Figures 8-26 and 8-27, respectively. For VSCs with higher pulse configuration, the generated harmonic components are lower.

8.2 SERIES COMPENSATION

Electric power flow through an alternating current transmission line is a function of the line impedance, the magnitudes of the sending- and receiving-end voltages, and the phase angle between these voltages. A single-line diagram of a simple transmission line with an inductive reactance (X), connecting two voltage sources ($\mathbf{V}_s$) at the send-

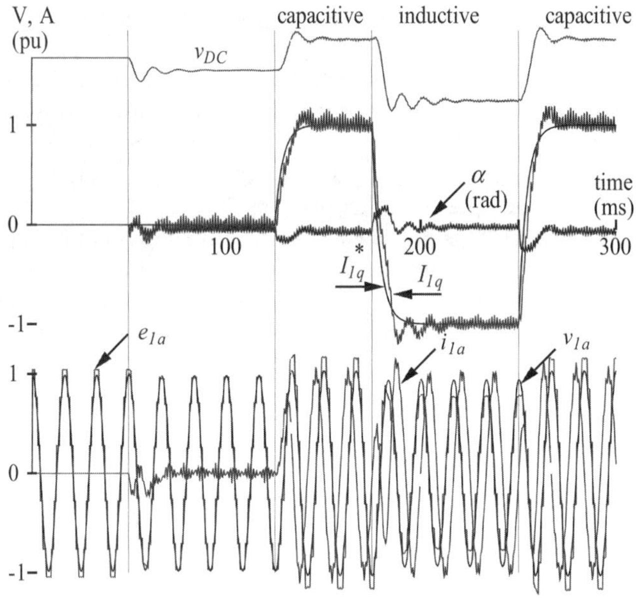

Figure 8-18. Performance of the STATCOM with a 12-pulse HN-VSC operating in a reactive current control mode.

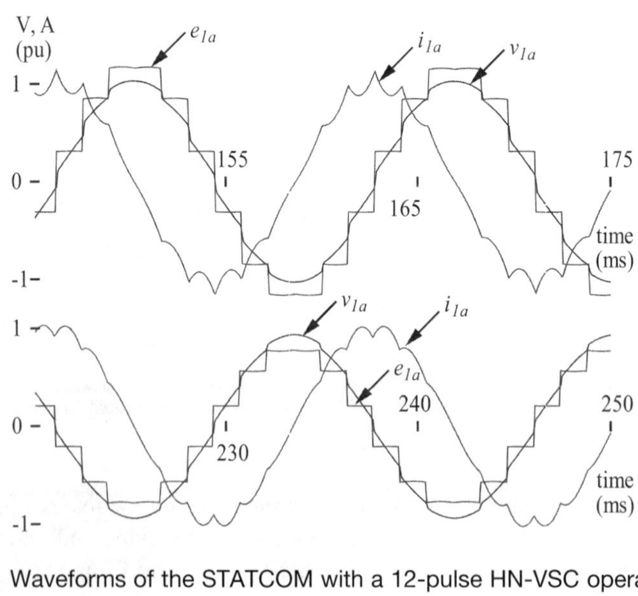

Figure 8-19. Waveforms of the STATCOM with a 12-pulse HN-VSC operating in a reactive current control mode.

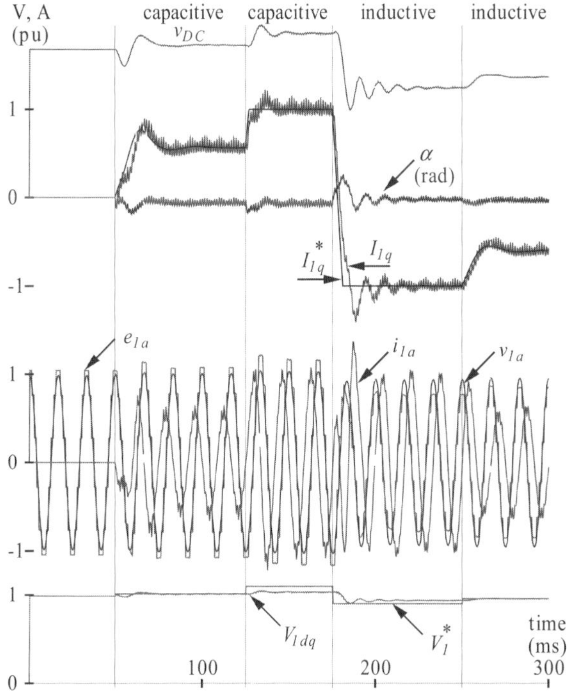

Figure 8-20. Performance of the STATCOM with a 12-pulse HN-VSC operating in a voltage control mode.

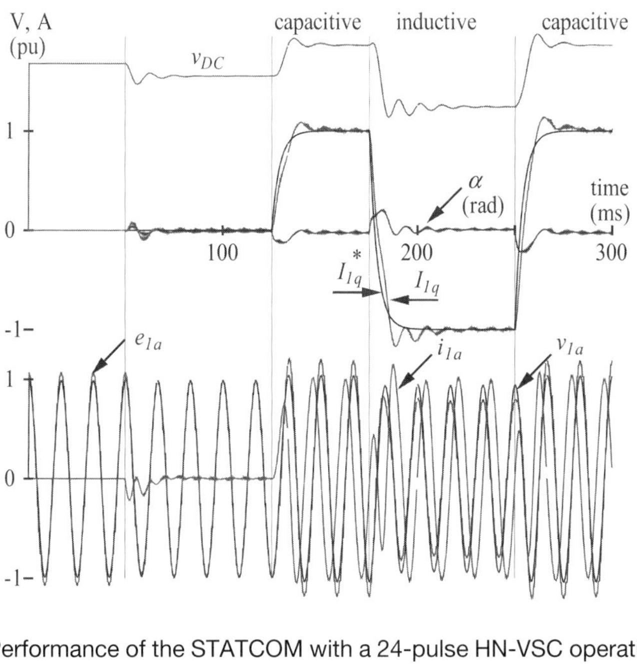

Figure 8-21. Performance of the STATCOM with a 24-pulse HN-VSC operating in a reactive current control mode.

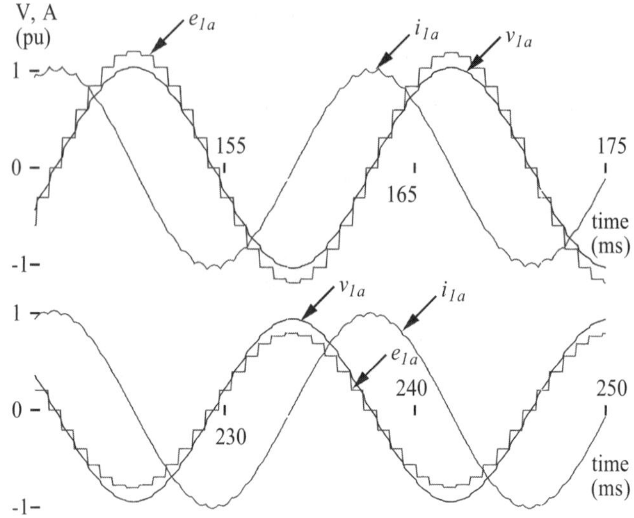

Figure 8-22. Waveforms of the STATCOM with a 24-pulse HN-VSC operating in a reactive current control mode.

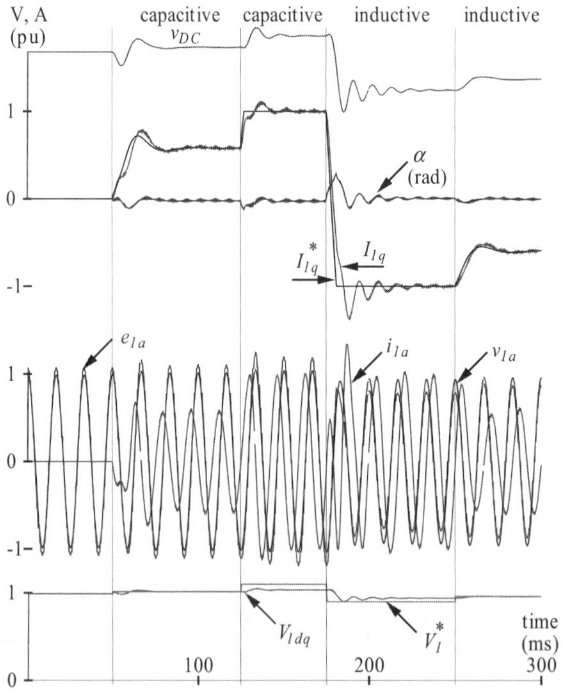

Figure 8-23. Performance of the STATCOM with a 24-pulse HN-VSC operating in a voltage control mode.

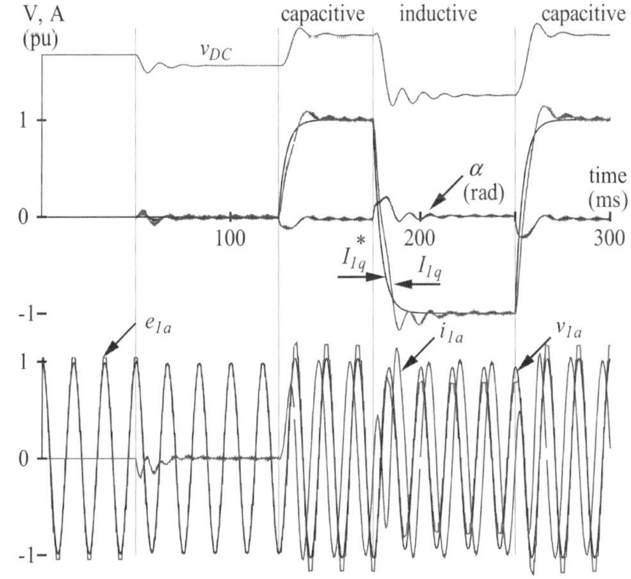

Figure 8-24. Performance of the STATCOM with a 24-pulse QHN-VSC operating in a reactive current control mode.

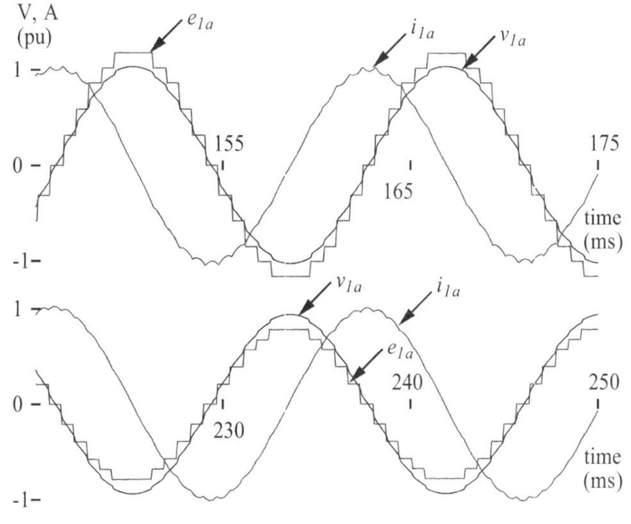

Figure 8-25. Waveforms of the STATCOM with a 24-pulse QHN-VSC operating in a reactive current control mode.

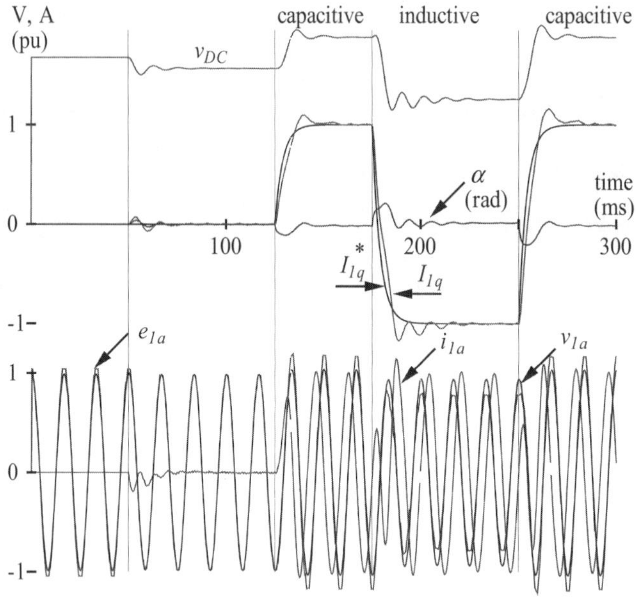

Figure 8-26. Performance of the STATCOM with a 48-pulse QHN-VSC operating in a reactive current control mode.

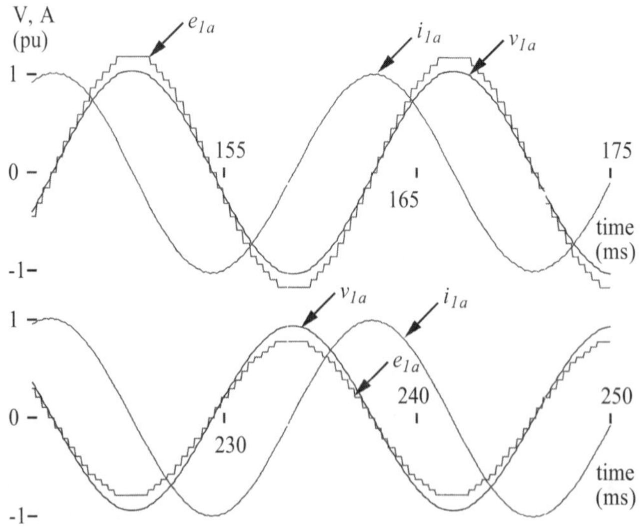

Figure 8-27. Waveforms of the STATCOM with a 48-pulse QHN-VSC operating in a reactive current control mode.

ing end and ($\mathbf{V_r}$) at the receiving end, respectively, are shown in Figure 8-28a and the corresponding phasor diagram in Figure 8-28b. For simplicity, a line with a reactance (X) is considered in the analysis. The natural active and reactive power flows (P_{rn} and Q_{rn}) at the receiving end of the line are

$$P_{rn} = \frac{V_s V_r}{X}\sin(\delta_s - \delta_r) = \frac{V^2}{X}\sin\delta \tag{8-1a}$$

and

$$Q_{rn} = \frac{V_s V_r}{X}\cos(\delta_s - \delta_r) - \frac{V_r^2}{X} = \frac{V^2}{X}(\cos\delta - 1) \tag{8-1b}$$

where V_s and V_r are the magnitudes and δ_s and δ_r are the phase angles of the voltage sources $\mathbf{V_s}$ and $\mathbf{V_r}$, respectively. The voltage magnitudes are chosen such that $V_s = V_r = V$ and the difference between the phase angles is δ (i.e., $\delta_s - \delta_r$).

The impedance of the transmission line is inductive. The power flow can be decreased by inserting an additional inductive reactance in series with the transmission line, thereby increasing the effective reactance of the transmission line between its two ends. The power flow can also be increased by inserting an additional capacitive reactance in series with the transmission line, thereby decreasing the effective reactance of the transmission line between its two ends.

Traditionally, in order to control the power flow of a transmission line, the effective line reactance is controlled by using fixed or thyristor-controlled series capacitors or inductors. With the use of the reactance compensation controller, a static synchronous series compensator (SSSC) connects an almost sinusoidal compensating voltage, of variable magnitude, in series with a transmission line. The compensating voltage is almost in quadrature with the prevailing line current and emulates an inductive or a capacitive reactance in series with the transmission line. A small part of the compensating voltage that is in phase with the line current accounts for the losses in the VSC.

The compensating voltage, leading the line current, emulates an inductive reactance in series with the transmission line, causing the power flow as well as the line current to decrease as the level of compensation increases, and the SSSC is considered to be

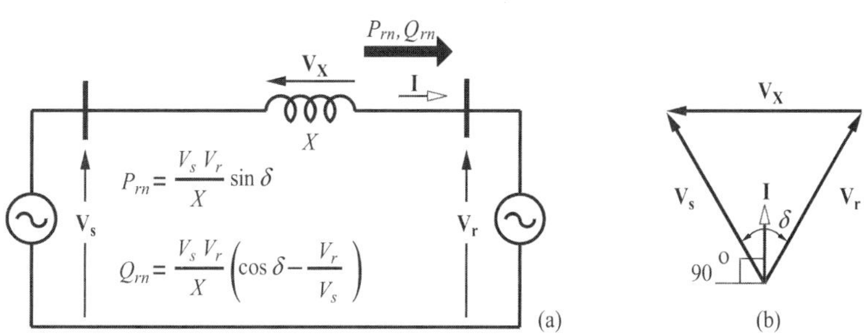

Figure 8-28. Elementary power transmission system.

operating in an inductive mode. The compensating voltage, lagging the line current, emulates a capacitive reactance in series with the transmission line, causing the power flow as well as the line current to increase as the level of compensation increases, and the SSSC is considered to be operating in a capacitive mode.

The SSSC, limited by its voltage and current ratings, is capable of emulating a compensating reactance (X_c) both as inductive and capacitive in series with the transmission line inductive reactance (X). Therefore, the expressions for power flow given in equations 8-1a and 8-1b become

$$P_r = \frac{V^2}{X_{eff}} \sin\delta \tag{8-2a}$$

and

$$Q_r = \frac{V^2}{X_{eff}} (\cos\delta - 1) \tag{8-2b}$$

where X_{eff} is the effective reactance of the transmission line between its two ends, including the compensating reactance emulated by the compensating voltage. The compensating reactance (X_c) is defined to be negative when the SSSC is operated in an inductive mode and positive when the SSSC is operated in a capacitive mode. The effective line reactance (X_{eff}) has a positive value when being inductive and a negative value when being capacitive.

Figure 8-29 shows an example of a simple power transmission system with the SSSC operated both in inductive and in capacitive modes along with the related pha-

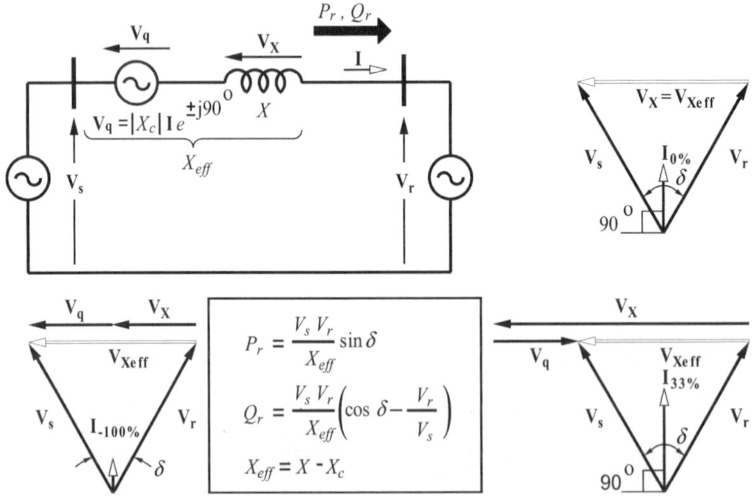

Figure 8-29. SSSC operated in inductive and capacitive modes and the related phasor diagrams.

sor diagrams. The line current decreases from $\mathbf{I}_{0\%}$ to $\mathbf{I}_{-100\%}$, when the inductive reactance compensation ($-X_c/X$) increases from 0% to 100%. The line current increases from $\mathbf{I}_{0\%}$ to $\mathbf{I}_{33\%}$, when the capacitive reactance compensation (X_c/X) increases from 0% to 33%.

The normalized effective reactance of the transmission line is

$$X_{eff}/X = (X - X_c)/X = 1 - X_c/X \tag{8-3}$$

From equations 8-1a and 8-1b, 8-2a and 8-2b, and 8-3, the expressions for the normalized power flow in the transmission line can be written as

$$\frac{P_r}{P_{rn}} = \frac{Q_r}{Q_{rn}} = \frac{X}{X_{eff}} = \frac{1}{1 - X_c/X} \tag{8-4}$$

The effects of the compensating reactance (X_c) on the normalized effective reactance of the transmission line and the normalized power flow in the transmission line are shown in Figure 8-30. When the compensating reactance is inductive, the effective reactance (X_{eff}) increases and the active and reactive power flows (P_r and Q_r) decrease as the reactance compensation ($-X_c/X$) increases. When the compensating reactance is capacitive, the effective reactance (X_{eff}) decreases and the active and reactive power flows (P_r and Q_r) increase as the reactance compensation (X_c/X) increases.

Figure 8-31 shows a single-line diagram of a power transmission system with the SSSC emulating a reactance by a compensating voltage ($\mathbf{V}_q$) that is connected in series with the transmission line and the related phasor diagrams. Through control action, the

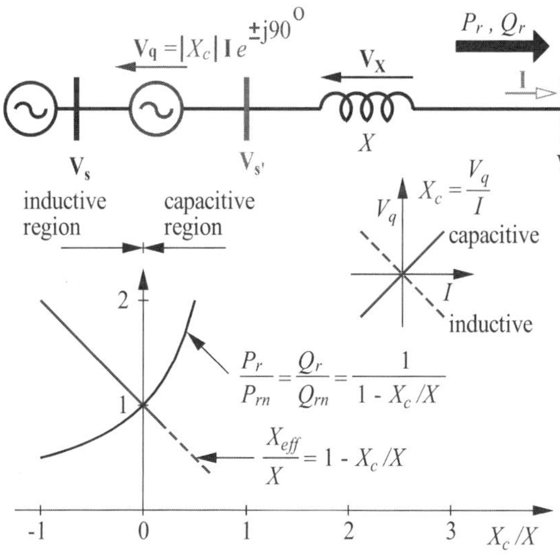

Figure 8-30. Effect of compensating reactance on power flow and effective reactance.

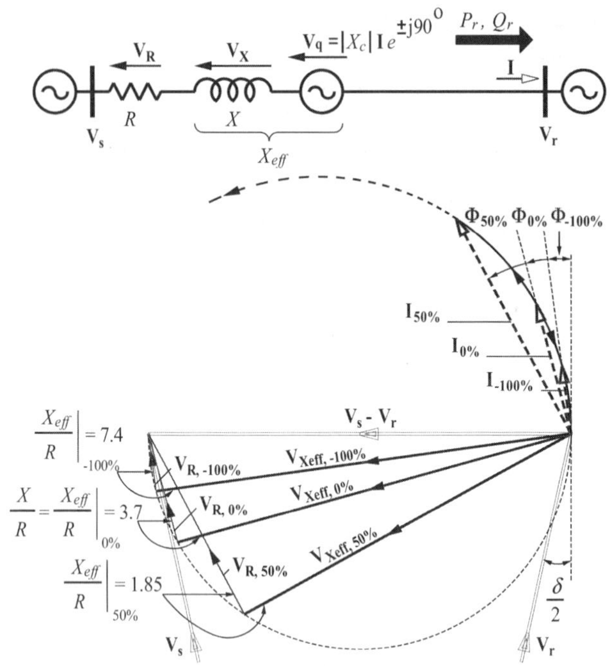

Figure 8-31. Reactance compensation of a lossy transmission line.

magnitude of the compensating voltage (V_q) and its phase angle with relationship to the prevailing line current can be regulated. When the compensating voltage emulates an inductive reactance connected to the line, it absorbs reactive power from the line. When the compensating voltage emulates a capacitive reactance connected to the line, it delivers reactive power to the line.

Consider an uncompensated ($X_c = 0$) transmission line with a X/R ratio of 3.7, delivering power from voltage source $\mathbf{V_s}$ to voltage source $\mathbf{V_r}$. In order to increase the active and reactive power flows (P_r and Q_r) through the transmission line, the SSSC must emulate a capacitive reactance in series with the transmission line.

The trajectory of line current from $\mathbf{I_{0\%}}$ with no reactance compensation to $\mathbf{I_{50}}$ with 50% capacitive reactance compensation, ($X_c/X = 0.5$) is shown in Figure 8-31. As the capacitive reactance compensation further increases, the line current and the power flow increase as indicated by the dashed line. The trajectory of the line current from $\mathbf{I_{0\%}}$ with no reactance compensation to $\mathbf{I_{-100\%}}$ with 100% inductive reactance compensation ($X_c/X = -1.0$) is also shown in Figure 8-31. As the inductive reactance compensation further increases, the line current and the power flow decrease toward a minimum value. The implementation of the reactance emulation technique using the SSSC is described in the next section.

The main file `823SSSCO.DAT` contains a modified program from Chapter 5, where a mechanically switched capacitor and inductor are used. In this program, a sinusoidal compensating voltage that is in quadrature with the prevailing line current is connected in series with the transmission line between BUS01 and BUS02 through a coupling transformer. The three-phase compensating voltages are defined in the TACS section with an Itype of 99 and in the source section with an Itype of 60. In the user input file `824SSSCO.USR` (see Appendix C, Section 76), the magnitude of the compensating voltage is varied to emulate an inductor or a capacitor in series with the transmission line. In the control file `825SSSCO.CON` (see Appendix C, Section 77), the three-phase compensating voltages in time domain are defined in the open loop manner. The compensating voltage ($\mathbf{V_q}$) with polarities negative and positive is implemented to emulate inductive reactance and capacitive reactance, respectively. Figure 8-32 shows the simulation results.

8.2.1 Static Synchronous Series Compensator (SSSC)

The SSSC model in EMTP is shown Figure 8-33. The compensating voltage is almost in quadrature with the prevailing line current and is proportional to the line current with the emulated reactance being the constant of proportionality. The controller of the SSSC is used to operate the VSC in such a way that the phase angle of the compensating voltage of the VSC is dynamically adjusted to regulate the DC capacitor voltage to the right level that produces the right amount of compensating voltage at the output of the VSC. The compensating voltage influences the power flow in the line.

8.2.2 Control of SSSC

A controller that emulates a compensating reactance is considered. Figure 8-34 shows the control block diagram of the SSSC. An instantaneous three-phase set of

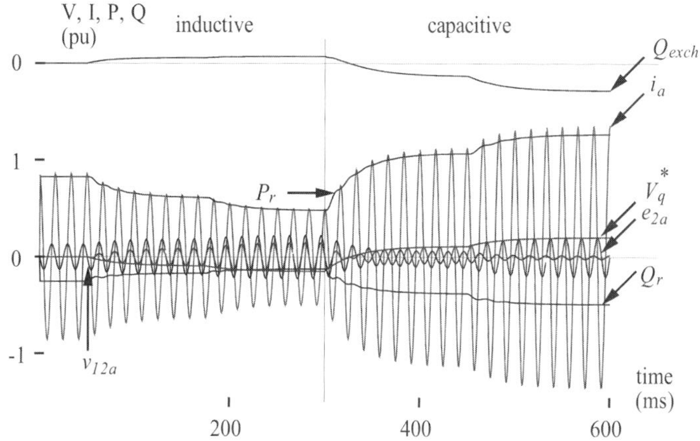

Figure 8-32. Exchanged reactive power and the resulting voltages and currents due to a series-connected compensating voltage into the transmission line through a coupling transformer.

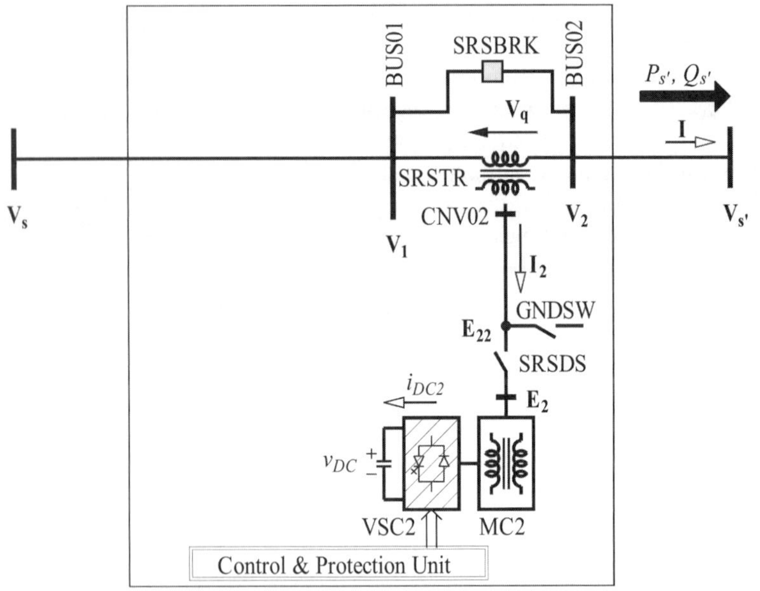

Figure 8-33. Static Synchronous Series Compensator.

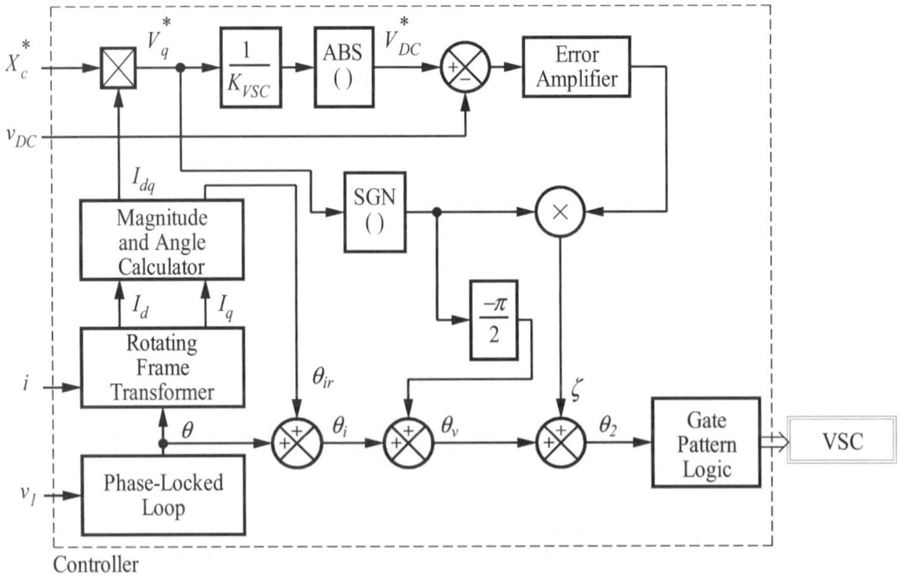

Figure 8-34. Reactance control block diagram of the static synchronous series compensator.

line voltages (v_1) at BUS01 is used to calculate the PLL angle (θ), which is phase locked to the phase a of the line voltage (v_{1a}). Using the reference PLL angle (θ), the instantaneous three-phase set of measured line current (i) is first decomposed into its active or direct component (I_d) and reactive or quadrature component (I_q), and then the amplitude (I_{dq}) and the relative phase angle (θ_{ir}) of the line current with respect to the PLL angle (θ) are calculated. The phase angle (θ_i) of the line current is calculated by adding the relative phase angle (θ_{ir}) of the line current and the PLL angle (θ). The calculated amplitude (I) of the line current multiplied by the compensating reactance demand (X_c^*) is the compensating voltage amplitude demand (V_q^*). The phase angle (θ_v) of this compensating voltage demand is either $\theta_i + 90°$ if the demanding compensating reactance is inductive or $\theta_i - 90°$ if the demanding compensating reactance is capacitive. The DC link capacitor voltage is dynamically regulated in relationship with the compensating voltage amplitude demand. The compensating voltage amplitude demand (V_q^*) and the DC link capacitor voltage demand (V_{DC}^*) are related by the VSC's fundamental AC amplitude-to-DC gain factor, ($K_{VSC} = 2/\pi$ in case of a HN-VSC). The DC link capacitor voltage demand (V_{DC}^*) and the measured DC voltage (v_{DC}) are compared and the error is passed through an error amplifier which produces a relative phase angle (ζ). The absolute phase angle (θ_2) of the VSC2 output voltage is calculated by adding the relative phase angle (ζ) of the DC voltage regulator and the phase angle (θ_v) of the compensating voltage demand. The compensating reactance demand (X_c^*) is either negative if the SSSC is emulating an inductive reactance or positive if it is emulating a capacitive reactance. In another application, the compensating voltage amplitude demand (V_q^*) may directly be specified and the SSSC will apply the desired compensating voltage almost in quadrature with the line current, as was shown in the previous section.

8.2.3 Modeling of SSSC in EMTP and Simulation Results

The SSSC model with DC link voltage control loop is given in the following three files, namely (a) main file `826SSSC1.DAT`, (b) user file `827SSSCC.USR`, and (c) control file `828SSSCC.CON` (see Appendix C, Sections 78 through 80). The VSC is considered to be of infinite pulse.

Figure 8-35 shows the digital simulation results when the SSSC emulates a reactance in series with the transmission line. At the beginning of the operation, the series bypass breaker (SRSBRK) and the ground switch (GNDSW) are open and the series disconnect switch (SRSDS) is closed. The VSC (VSC2) injects no voltage. The DC link capacitor voltage (v_{DC}) is zero. At 50 ms, an inductive reactance compensation of 0.15 pu is requested. The VSC2 output sinusoidal (infinite pulse) voltage (e_{2a}) of phase a leads the line current (i_a) by almost 90°. At 175 ms, the inductive reactance demand is increased to 0.3 pu. As the inductive reactance demand increases, the line current (i_a) decreases. As a result, the active and reactive power flows (P_r and Q_r) at the receiving end of the line decrease. At 300 ms, a capacitive reactance compensation of 0.1 pu is requested. The VSC2 output voltage (e_{2a}) lags the line current (i_a) by almost 90°. At 450 ms, the capacitive reactance demand is increased to 0.15 pu. As the capacitive reactance demand increases, the line current (i_a)

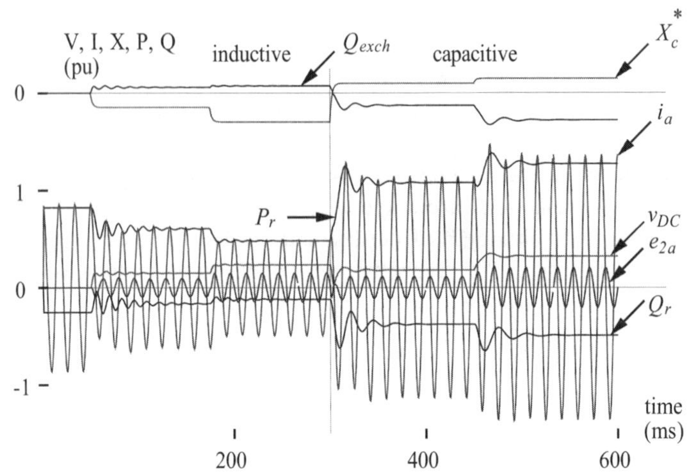

Figure 8-35. Performance of the SSSC with an infinite pulse VSC operating in inductive and capacitive modes.

and the active and reactive power flows (P_r and Q_r) in the line increase. The reactive power (Q_{exch}) exchanged at the terminal of the VSC2 is shown in the figure. In reality, the SSSC would encounter power losses in the switches and in the magnetic circuit. Therefore, there will always be a small part of active current component (I_d) flowing into the VSC2 and the VSC2 output voltage will be almost 90° out of phase with the line current. The instantaneous DC link capacitor voltage is proportional to the amplitude of the VSC2 output voltage.

Therefore, when the SSSC emulates a reactance in series with the transmission line, the power flow in the transmission line always decreases if the emulated reactance is inductive and the power flow always increases if the emulated reactance is capacitive.

Figure 8-36 shows the expanded view of the two sections of Figure 8-35. The VSC output voltage shows the presence of only fundamental components.

The previous test case for reactance control was rerun for the SSSC with a 24-pulse HN-VSC. The corresponding main file is `829SSSC2.DAT`. The results are shown in Figures 8-37 and 8-38, respectively. The VSC voltage and current show the presence of harmonic components. For VSCs with higher pulse configuration, the generated harmonic components are lower.

Figure 8-38 shows the expanded view of the two sections of Figure 8-37. The VSC output voltage shows the presence of 24-pulse harmonic components.

The SSSC injects an almost sinusoidal voltage, of variable magnitude, in series with a transmission line. The compensating voltage is almost in quadrature with the prevailing line current and thereby emulates a reactance in series with the transmission line. A compensating voltage, lagging the line current, emulates a capacitor that increases the power flow of the line. A compensating voltage, leading the line current, emulates an inductor that decreases the power flow of the line. If the emulated capaci-

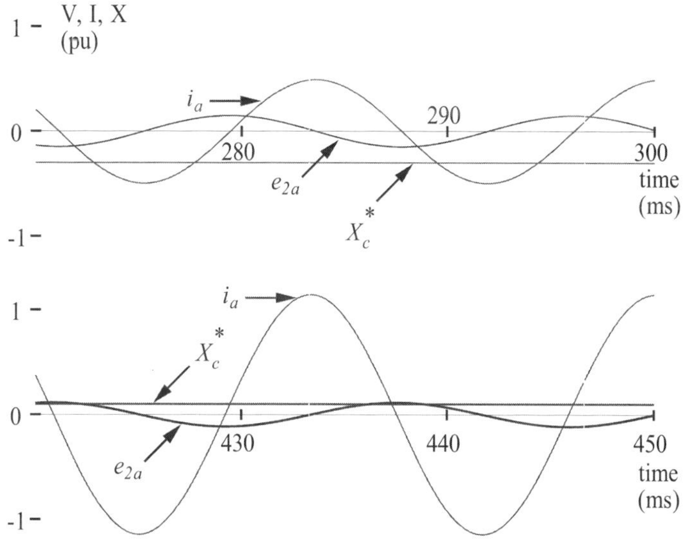

Figure 8-36. Waveforms from the SSSC with an infinite pulse VSC operating in inductive and capacitive modes.

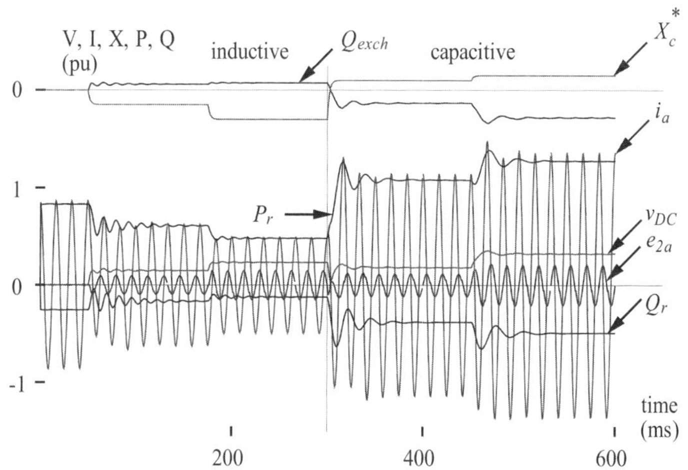

Figure 8-37. Performance of the SSSC with a 24-pulse HN-VSC operating in inductive and capacitive modes.

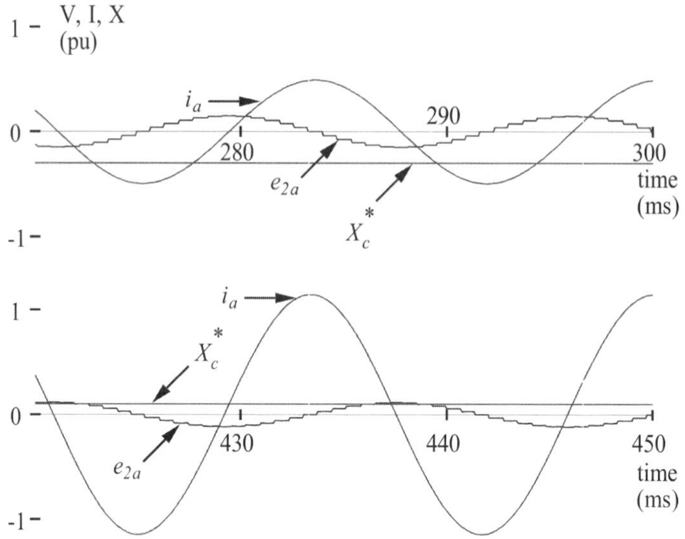

Figure 8-38. Waveforms from the SSSC with a 24-pulse HN-VSC operating in inductive and capacitive modes.

tive reactance is more than the transmission line inductive reactance, the effective reactance of the transmission line between its two ends becomes capacitive and the direction of power flow in the transmission line reverses.

8.2.4 Stable Reversal of Power Flow

The reactance control method, just implemented, is based on emulating a reactance in series with the transmission line. This control method is used to operate the VSC, so that the series-connected compensating voltage is proportional to the prevailing line current with the emulated reactance being the constant of proportionality. For this control algorithm to work successfully, a line current must exist. Also, in the reactance control method, the polarity of the reactance is defined before the desired control action is to take place. If it is defined to be inductive, the most that can be achieved is the reduction of the line current to somewhat lower than the normal uncompensated value. The line current can never be brought to nearly zero by the controller since the successful operation of the control algorithm depends on the existence of the line current. If the emulated reactance is to be defined as capacitive, the line current at first increases. If the emulated capacitive reactance demand is higher than the line inductive reactance, the effective line reactance becomes capacitive and the power flow in the line reverses. But reversing the power flow while the line current is high causes a number of problems. During the transition, when the power flow reverses, the transmission line may operate beyond its maximum thermal capability. Also, during the emulation of a higher capacitive reactance, there exists a point at which the inductive reactance of the transmission line and the capacitive reactance emulated by the SSSC become equal, which can cause instability in the power flow of the transmission line. In a practical implementation, if the current through the SSSC exceeds its rating, the SSSC will be bypassed. The reactance control method provides the basic features of the SSSC in terms of controlling the power flow in the transmission line. However, the undesirable feature of this control method appears during a reversal of power flow in the transmission line, when the line current should be brought to nearly zero and be increased gradually in the opposite direction; instead, a high-current reverses, causing even higher current transients, as shown in Figure 8-44.

The desirable way to reverse the power flow in a transmission line is to implement a control method, which reduces the line current to nearly zero and reverses with absolute stability. That means, in the process of reversing the power flow in the transmission line, this control method should first emulate an inductive reactance of higher and higher value, then change to emulate a capacitive reactance of very high value, and gradually reduce the capacitive reactance value to increase the power flow in the reverse direction. The reactance control method does not allow a near-zero line current operating condition because this method relies primarily on the existence of the line current for its operation. The voltage control method is used to connect a compensating voltage in series with the transmission line with any amount of bidirectional line current, including near zero. This compensating voltage is either in phase or in opposite phase with the voltage across the transmission line effective reactance. This control method offers all the desirable features offered by the reactance control method in terms of controlling the power flow in the transmission line. In addition, the voltage control method offers an absolute stability in the power flow, causing the power to go through near zero while changing its direction of flow.

Two control methods—reactance control and voltage control—are compared in the following section. The latter is shown to have a superior characteristic when a reversal of power flow in the transmission line occurs. Both control methods have successfully been implemented in digital simulation. The simulation also includes a three-phase compensating voltage of variable magnitude in series with a three-phase transmission line with a resistance (R) and an inductive reactance (X) in each phase connecting a three-phase voltage source at each end. The simulation results show the basic functions of the SSSC in terms of controlling the power flow in the transmission line and the superior characteristics of the voltage control method.

8.2.4.1 Reactance Control Method. Figure 8-39 shows a single-line diagram of a simple power transmission system with the SSSC operating with a capacitive reactance control method along with the related phasor diagrams. The line current increases from an uncompensated value of $\mathbf{I}_{0\%}$ to a compensated value of $\mathbf{I}_{33\%}$ when the reactance compensation (X_c/X) increases from 0% to 33%. For a larger reactance compensation ($X_c/X = 300\%$), the line current ($\mathbf{I}_{300\%}$) flows in the reverse direction and decreases further as the compensation increases. Figure 8-40 shows the SSSC operating with an inductive reactance control method along with the related phasor diagram. The line current decreases from $\mathbf{I}_{0\%}$ to $\mathbf{I}_{-100\%}$ when the reactance compensation ($-X_c/X$) increases from 0% to 100%.

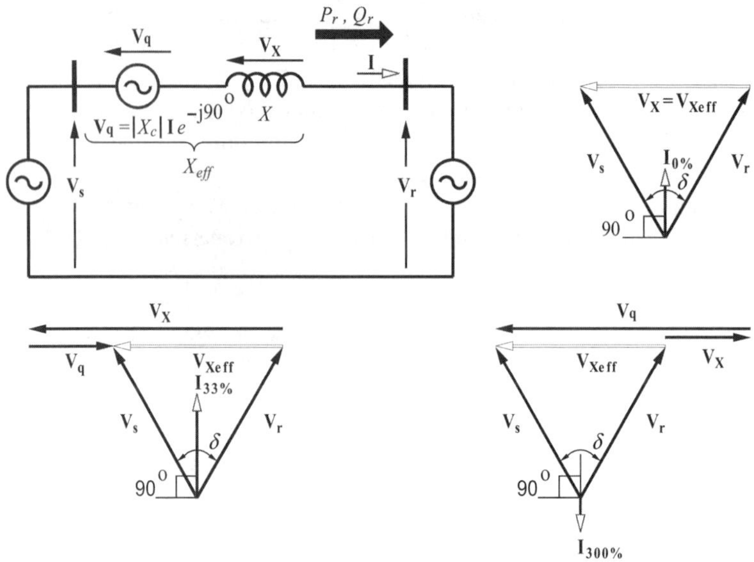

Figure 8-39. SSSC operating with a capacitive reactance control method and the related phasor diagrams.

The effects of the compensating reactance (X_c) on the normalized power flow in the transmission line and the normalized effective reactance of the transmission line are shown in Figure 8-41. In an operation of the SSSC with a capacitive reactance control method, the power flow (P_r) increases monotonically as the reactance compensation (X_c/X) increases during the period when $0 < X_c/X < 1$. When $X_c/X = 1$, the power flow (P_r) can possibly reach an extremely high value limited by the line resistance only. Finally, the power flow (P_r) decreases monotonically in the reverse direction as the reactance compensation (X_c/X) increases during the period when $X_c/X > 1$. In an operation of the SSSC with an inductive reactance control method, the power flow (P_r) decreas-

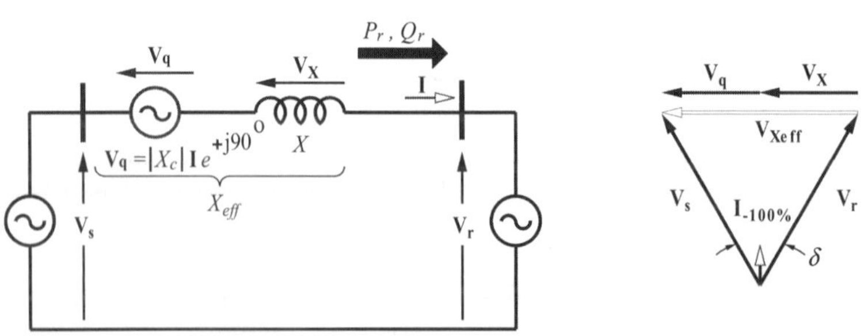

Figure 8-40. SSSC operating with an inductive reactance control method and the related phasor diagram.

es monotonically as the reactance compensation ($-X_c/X$) increases when $X_c/X < 0$. The points to be made here are

- The SSSC is capable of controlling the power flow in a transmission line when operating with a reactance control method.
- In an operation with a capacitive reactance control method, this power flow can be reversed.

However, while changing direction, the power flow (P_r) goes through two simultaneous extremes: one is a very high positive value when the transmission line effective reactance (X_{eff}) is nearly zero inductive and the next one is a very high negative value when the effective reactance (X_{eff}) is nearly zero capacitive. The fact that these two extremes are practically limited by the transmission line resistance (R) is described below.

Figure 8-42 shows a single-line diagram of a power transmission system with the SSSC operating with the reactance control method along with the related phasor diagrams. Consider an uncompensated ($X_c = 0$) transmission line with a X/R ratio of 3.7, delivering power from voltage source ($\mathbf{V_s}$) to voltage source ($\mathbf{V_r}$). For a change in operating point with a compensation level of $X_c/X = 0.5$, the trajectory of line current from $\mathbf{I}_{0\%}$

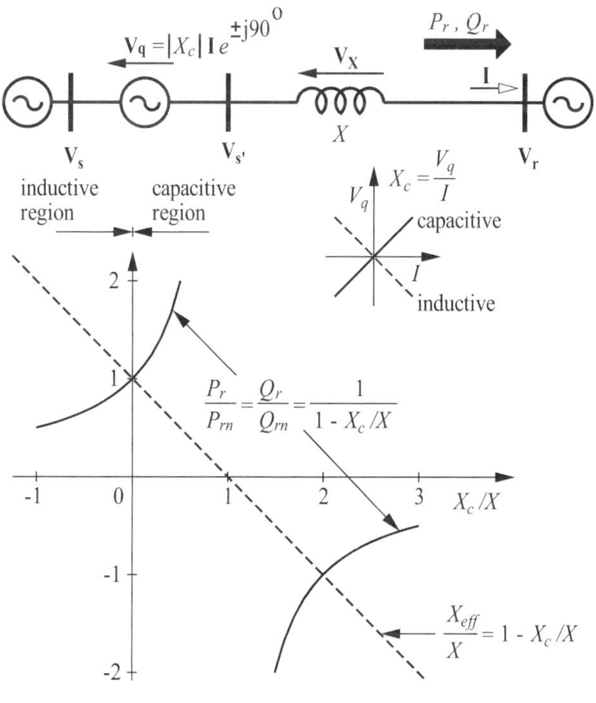

Figure 8-41. Effect of compensating reactance on effective reactance and power flow.

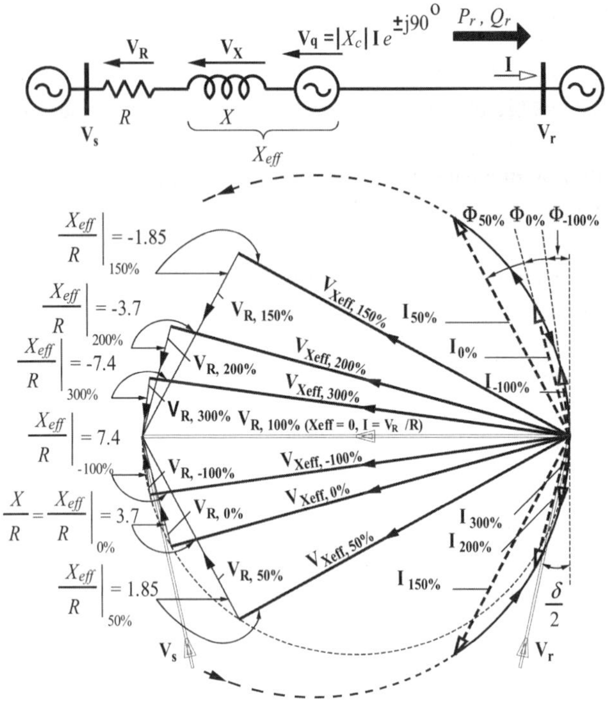

Figure 8-42. SSSC operating with reactance control method.

with no reactance compensation to $\mathbf{I}_{50\%}$ with 50% capacitive reactance compensation (X_c/X) is shown in the figure, where Φ is the phase angle between the voltage ($\mathbf{V}_{\mathbf{Xeff}}$) across the transmission line effective reactance and the voltage, $\mathbf{V}_t$ (i.e., $\mathbf{V}_s - \mathbf{V}_r$), across the transmission line's two ends. For most transmission lines operating today, 50% capacitive reactance compensation would make the transmission line operate near its maximum thermal capability. In a hypothetical case, further increase in the capacitive reactance compensation (X_c/X) causes the line current as well as the power flow to increase greatly. When $X_c/X = 1$, the voltage ($\mathbf{V_R}$) across the transmission line resistance (R) becomes equal to the difference ($\mathbf{V}_s - \mathbf{V}_r$) between the voltage sources at both ends of the transmission line. Only this voltage difference and the line resistance limit the line current. In order to reverse the power flow (P_r), the capacitive reactance compensation (X_c/X) has to be increased further. During the transition when power flow (P_r) reverses, the transmission line is operated beyond its maximum thermal limit and the line current trajectory is shown by the outer dashed line. Further increase in the capacitive reactance compensation (X_c/X) decreases the power flow (P_r) in the reverse direction, as shown by the trajectory of the reverse line current. The trajectory of line current from $\mathbf{I}_{0\%}$ with no reactance compensation to $\mathbf{I}_{-100\%}$ with 100% inductive reactance compensation ($-X_c/X$) is also shown in the figure. As the inductive reactance compensation ($-X_c/X$) further increases, the line current and the power flow decrease toward zero.

Figure 8-43 shows the reactance control scheme of the SSSC. The PLL angle (θ) is phase locked to the phase a of the sending-end voltage (v_{sa}). Using the reference PLL angle (θ), the instantaneous three-phase set of measured line current (i) is first decomposed into its active or direct component (I_d) and reactive or quadrature component (I_q), and then the amplitude (I_{dq}) and the relative phase angle (θ_{ir}) of the line current with respect to the PLL angle (θ) are calculated. The phase angle of the line current (i) is calculated by adding the relative phase angle (θ_{ir}) of the line current and the PLL angle (θ). The calculated amplitude (I) of the line current multiplied by the absolute value of the compensating reactance demand ($|X_c^*|$) is the compensating voltage amplitude demand ($|V_q^*|$). The phase of this compensating voltage is in quadrature with the line current, either leading if the demanding compensating reactance is inductive or lagging if the demanding compensating reactance is capacitive. A positive compensating reactance emulates a capacitor and a negative compensating reactance emulates an inductor. Note that an indirect way to implement the reactance control method would be to specify the compensating voltage demand directly to the controller. The ratio of the compensating voltage and the prevailing line current is the emulated reactance. A positive compensating voltage emulates a capacitor and a negative compensating voltage emulates an inductor.

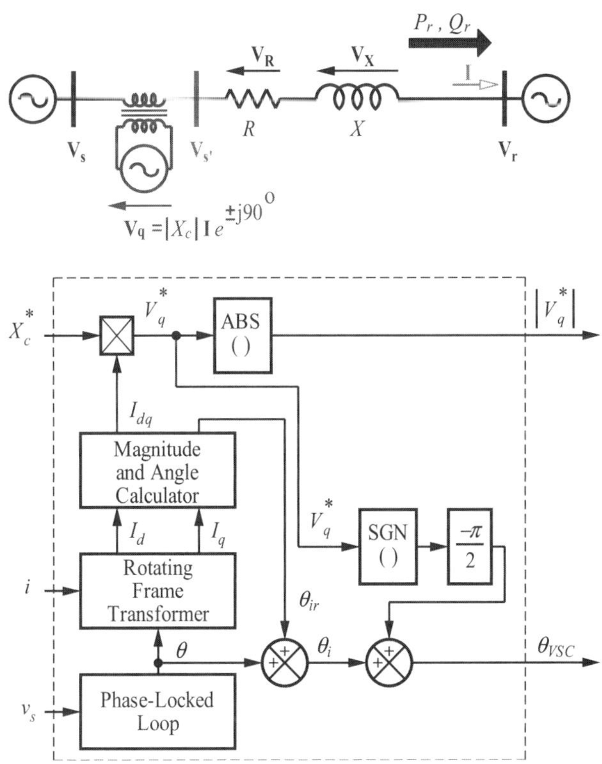

Figure 8-43. Reactance control scheme of the SSSC.

The corresponding main file is `830SSSC3.DAT` that calls the two files, namely (a) user file `831SSSCI.USR` and (b) control file `832SSSCI.CON` (see Appendix C, Sections 81 through 82). The VSC is considered to be of infinite pulse.

Figure 8-44 shows the results from the digital simulation of the SSSC operating with a reactance control method. In an operation with an inductive reactance control method, the compensating voltage (v_{qa}) leads the line current (i_a). As the inductive reactance demand increases, the line current (i_a) and the power flow (P_r) in the transmission line decrease toward zero. In an operation with a capacitive reactance control method, the compensating voltage (v_{qa}) lags the line current (i_a). As the capacitive reactance demand increases, the line current (i_a) and the power flow (P_r) increase. At an operating point, when the power flow in the line is near nominal value, the compensating capacitive reactance demand is set to a value that results in nominal power flow in the transmission line in the reverse direction. The operating points in the simulation are selected to keep the steady-state active power flow within the nominal value.

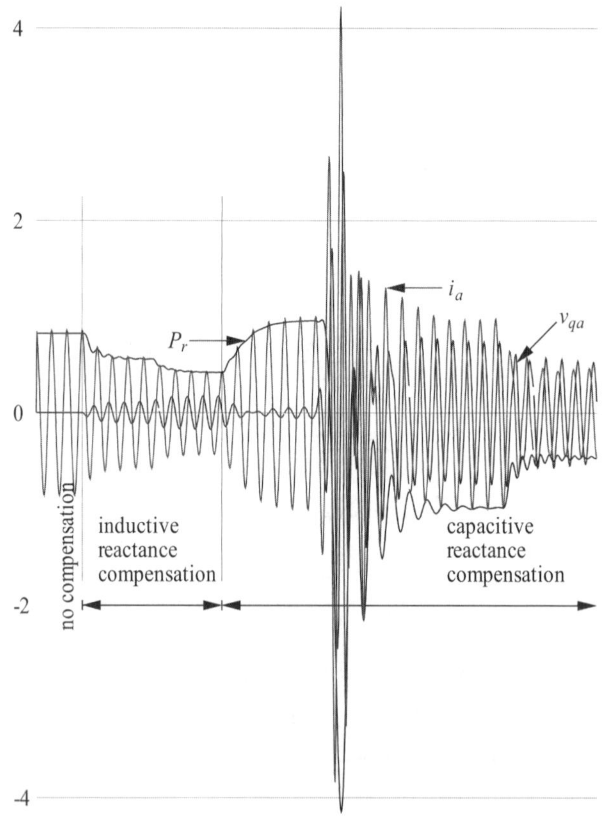

Figure 8-44. Performance of the SSSC with reactance control method.

Therefore, in an operation of the SSSC with a reactance control method, the power flow (P_r) in the transmission line always decreases when the reactance compensation is inductive. Also, the power flow (P_r) increases, at first, when the reactance compensation is capacitive. Further increase in the capacitive reactance compensation increases the power flow (P_r) to the maximum value before decreasing in the reverse direction, thereby operating the transmission line beyond its maximum thermal limit and through a possible instability during a power flow reversal.

8.2.4.2 Voltage Control Method. In operation of the SSSC with a voltage control method, a compensating voltage ($\mathbf{V_q}$) is connected in series with the transmission line, either in phase or in opposite phase with the voltage ($\mathbf{V_{Xeff}}$) across the transmission line effective reactance (X_{eff}).

Figure 8-45 shows a single-line diagram of a simple power transmission system with the SSSC operating with a voltage control method when the compensating voltage is in phase with the voltage ($\mathbf{V_{Xeff}}$) across the transmission line effective reactance (X_{eff}) along with the related phasor diagrams. The line current decreases from $\mathbf{I_{0\%}}$ to $\mathbf{I_{-100\%}}$ when the voltage compensation ($-V_q/V_{Xeff}$) increases from 0% to 50%, which is analogous to an increase in the inductive reactance compensation ($-X_c/X$) from 0% to 100%. The relationship between the voltage compensation (V_q/V_{Xeff}) and the reactance compensation (X_c/X) is explained in the following section. When fully compensated (i.e., at $-V_q/V_{Xeff} = 100\%$), the line current is zero. With further increase in the voltage

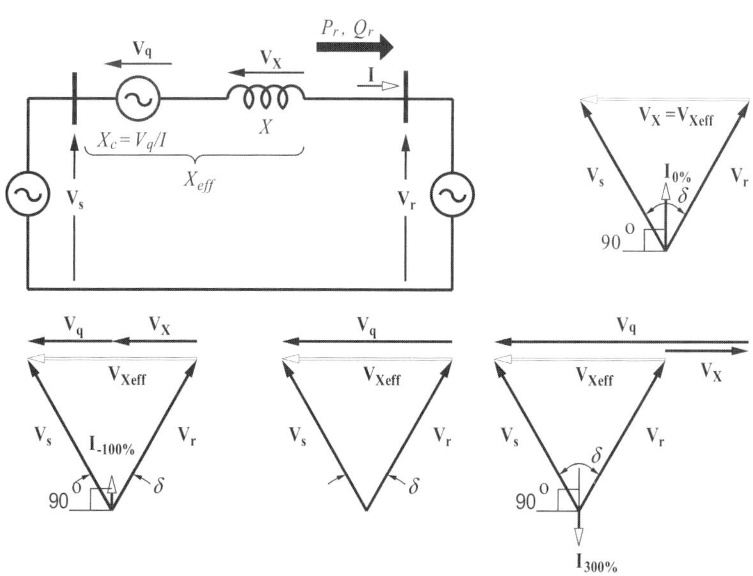

Figure 8-45. SSSC operating with a voltage control method with a compensating voltage in phase with voltage across transmission line effective reactance and the related phasor diagrams.

compensation (i.e., at $-V_q/V_{Xeff}$=150%), the line current ($\mathbf{I}_{300\%}$) flows in the reverse direction and any further increase in the voltage compensation ($-V_q/V_{Xeff}$) only increases this reverse line current. Once again, a voltage compensation ($-V_q/V_{Xeff}$) of 150% is analogous to a capacitive reactance compensation (X_c/X) of 300%, which will be explained later in the chapter.

Figure 8-46 shows the SSSC operating with a voltage control method when the compensating voltage is in opposite phase with the voltage ($\mathbf{V}_{\mathbf{Xeff}}$) across the transmission line effective reactance (X_{eff}) and the related phasor diagram. The line current increases from $\mathbf{I}_{0\%}$ to $\mathbf{I}_{33\%}$ when the voltage compensation (V_q/V_{Xeff}) increases from 0% to 50%, which is analogous to an increase in the capacitive reactance compensation (X_c/X) from 0% to 33%. It can be noted from the phasor diagram that the voltage ($\mathbf{V}_{\mathbf{Xeff}}$) across the transmission line effective reactance (X_{eff}) is finite and is the difference between the voltage sources ($\mathbf{V}_s$ and $\mathbf{V}_r$). Also, this voltage phasor remains constant, whether or not there is any voltage compensation, in a system in which the sending- and receiving-end voltages are fixed. Moreover, this finite difference prevents the two opposing voltage phasors—one is the compensating voltage ($\mathbf{V}_q$) and the other one is the voltage ($\mathbf{V}_X$) across the transmission line inductive reactance (X)—from being equal, thereby making the transmission line immune from resonance.

From the phasor diagram in Figure 8-46,

$$I_{33\%}X_{eff} = V_{Xeff} \tag{8-5}$$

$$I_{33\%}X = V_X = V_{Xeff} + V_q = V_{Xeff}(1 + V_q / V_{Xeff}) \tag{8-6}$$

Equations 8-5 and 8-6 can be rearranged as

$$\frac{X_{eff}}{X} = \frac{1}{1 + V_q/V_{Xeff}} \tag{8-7}$$

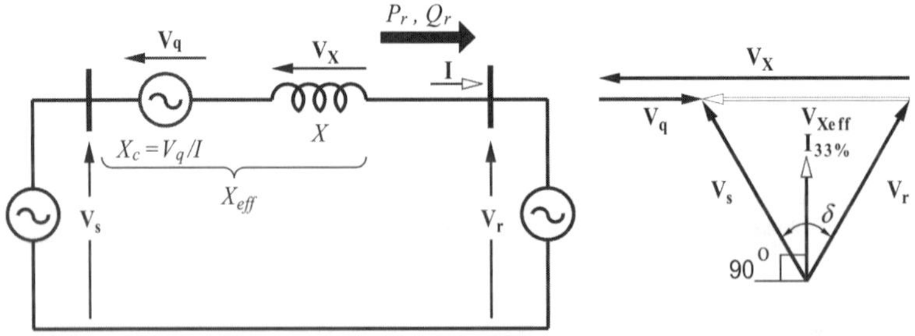

Figure 8-46. SSSC operating with a voltage control method with a compensating voltage in opposite phase with voltage across transmission line effective reactance and the related phasor diagram.

The compensating voltage (V_q) has a positive value when it is in opposite phase with the voltage across the transmission line effective reactance and has a negative value when it is in phase with this voltage.

Combining equations 8-3 and 8-7, the relationship between the compensating reactance and the compensating voltage can be found:

$$\frac{X_c}{X} = \frac{V_q / V_{Xeff}}{1 + V_q / V_{Xeff}} \tag{8-8}$$

The expression for power flow in the transmission line can be written from equations 8-2a and 8-2b as

$$P_r = \frac{V^2}{V_{Xeff}} \sin\delta = \frac{X}{X_{eff}} \frac{V^2}{X} \sin\delta \tag{8-9a}$$

and

$$Q_r = \frac{V^2}{X_{eff}} (\cos\delta - 1) = \frac{X}{X_{eff}} \frac{V^2}{X} (\cos\delta - 1) \tag{8-9b}$$

Equations 8-9a and 8-9b can be rewritten with a substitution from equation 8-7 as

$$P_r = \frac{V^2}{X} (1 + V_q / V_{Xeff}) \sin\delta \tag{8-10a}$$

and

$$Q_r = \frac{V^2}{X} (1 + V_q / V_{Xeff})(\cos\delta - 1) \tag{8-10b}$$

From equations 8-1, 8-10a, and 8-10b, the expression for the normalized power flow in the transmission line is

$$\frac{P_r}{P_{rn}} = \frac{Q_r}{Q_{rn}} = 1 + V_q / V_{Xeff} \tag{8-11}$$

The expression for the normalized power flow in the transmission line given by equations 8-4 and 8-11 are basically the same. In an operation of the SSSC with a reactance control method, the control parameter is X_c/X, which is essentially V_q/V_X. The power expression given by equation 8-4 has a pole at $X_c/X = 1$, causing the power flow (P_r) to be unstable at resonance, whereas, in an operation of the SSSC with a voltage control method, the control parameter is V_q/V_{Xeff}. The power expression given by equation 8-11 has a zero at $V_q/V_{Xeff} = -1$, causing an absolute stability at the instant when power flow (P_r) reverses.

The effects of the compensating voltage ($\mathbf{V_q}$) on the normalized power flow in the transmission line, the normalized effective reactance, and the normalized compensating reactance of the transmission line are shown in Figure 8-47. When the compensating voltage is in phase with the voltage across the transmission line effective reactance, the effective line reactance increases monotonically as the voltage compensation ($-V_q/V_{Xeff}$) increases during $0 > V_q/V_{Xeff} > -1$. When fully compensated (i.e., at $V_q/V_{Xeff} = -1$), the effective line reactance reaches infinity. Finally, the effective line reactance decreases monotonically in the reverse direction as the voltage compensation ($-V_q/V_{Xeff}$) increases during $V_q/V_{Xeff} < -1$. When the compensating voltage is in opposite phase with the voltage across the transmission line effective reactance, the effective line reactance monotonically decreases as the voltage compensation (V_q/V_{Xeff}) increases during $V_q/V_{Xeff} > 0$. The points to be made here are

- The SSSC is capable of controlling the power flow in a transmission line when operating with voltage control method.
- When the compensating voltage is in phase with the voltage across the transmission line effective reactance, this power flow can be reversed.

However, while changing direction, the power flow (P_r) goes through zero, resulting in an absolute stability in the power flow (P_r) of the transmission line. During the transi-

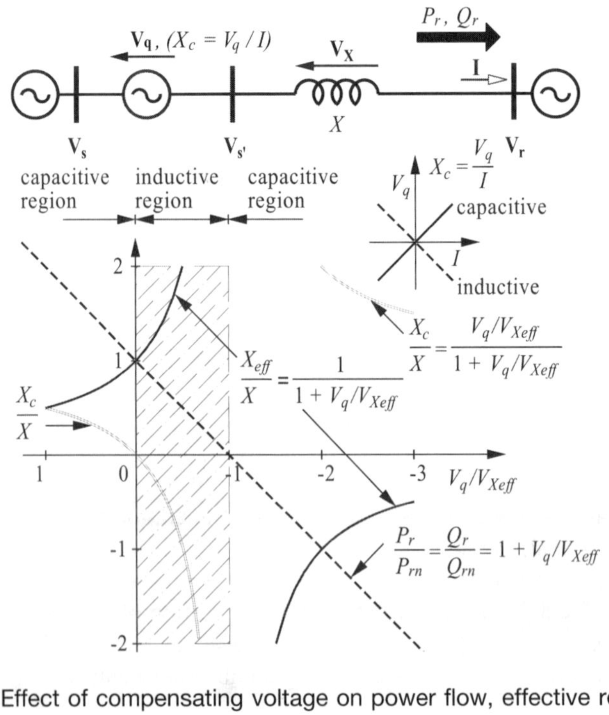

Figure 8-47. Effect of compensating voltage on power flow, effective reactance, and compensating reactance.

tion period when the power flow (P_r) reverses, the effective reactance (X_{eff}) goes through two simultaneous extremes: one is a very high inductive value when the power flow (P_r) in the transmission line is nearly zero positive and the next one is a very high capacitive value when the power flow (P_r) is nearly zero negative. The fact is that in the region where $0 > V_q/V_{Xeff} > -1$, the SSSC acts like a reactance compensator operating with an inductive reactance control method. In the rest of the region, the SSSC acts like a reactance compensator operating with a capacitive reactance control method.

Figure 8-48 shows a single-line diagram of a power transmission system with the SSSC operating with a voltage control method and the related phasor diagrams. Consider the previous example of a transmission line with a X/R ratio of 3.7 and without any voltage compensation (i.e., $V_q = 0$), delivering power from voltage source $\mathbf{V_s}$ to voltage source $\mathbf{V_r}$. For a change in operating point with a compensation level of $V_q/V_{Xeff} = 1$, which is equivalent to $X_c/X = 0.5$, the trajectory of line current $\mathbf{I_{0\%}}$ with no compensation to $\mathbf{I_{50\%}}$ with an equivalent of 50% capacitive reactance compensation (X_c/X) is shown in the figure. Further increase in the voltage compensation (V_q/V_{Xeff}) in the same direction only increases the power flow (P_r) in the transmission line. To decrease the power flow (P_r), the SSSC operating with a voltage control method injects a compensating voltage ($\mathbf{V_q}$) in phase with the voltage ($\mathbf{V_{Xeff}}$) across the transmission

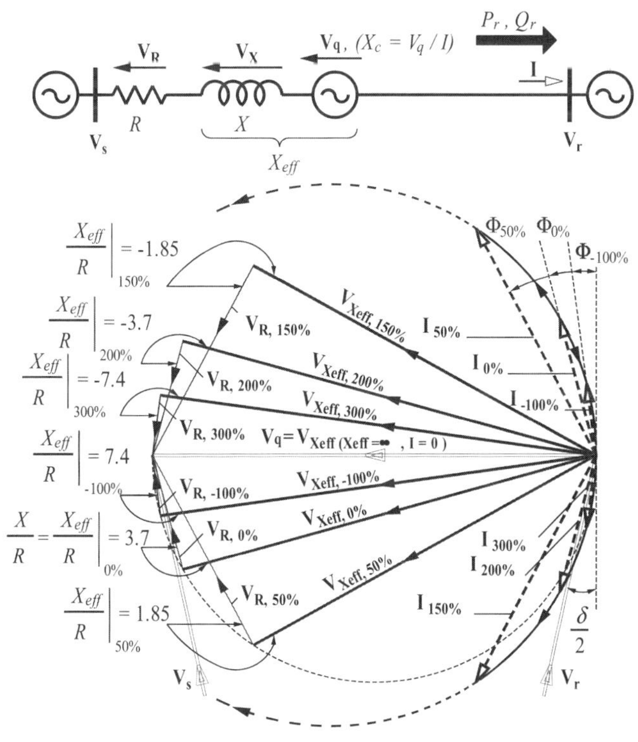

Figure 8-48. SSSC operating with voltage control method.

line effective reactance (X_{eff}). The resulting line current ($\mathbf{I}_{-100\%}$) is in quadrature, lagging the compensating voltage. The trajectory of line current from $\mathbf{I}_{0\%}$ with no compensation to $\mathbf{I}_{-100\%}$ with an equivalent of 100% inductive reactance compensation ($-X_c/X$) is also shown in the figure. When the compensating voltage ($\mathbf{V}_q$) equals the voltage ($\mathbf{V}_{Xeff}$) across the transmission line effective reactance (X_{eff}), the line current, as well as the power flow, ceases. At this point, the compensating voltage emulates an infinite inductive reactance in series with the transmission line. Further increase in the compensating voltage in the same direction causes the line current and the power flow to increase in the reverse direction. At this point, the compensating voltage changes its role from being a very large inductive reactance to a very large capacitive reactance. The trajectory of reverse line current during the reverse power flow is also shown in the figure. Therefore, the SSSC operated with a voltage control method, as described, always maintains the line current within safe limits, as shown by the solid trajectory, regardless of the direction of power flow (P_r).

Figure 8-49 shows the voltage control scheme of the SSSC. The PLL angle (θ) is phase locked to the phase a of the sending-end voltage (v_{sa}). In an operation of the

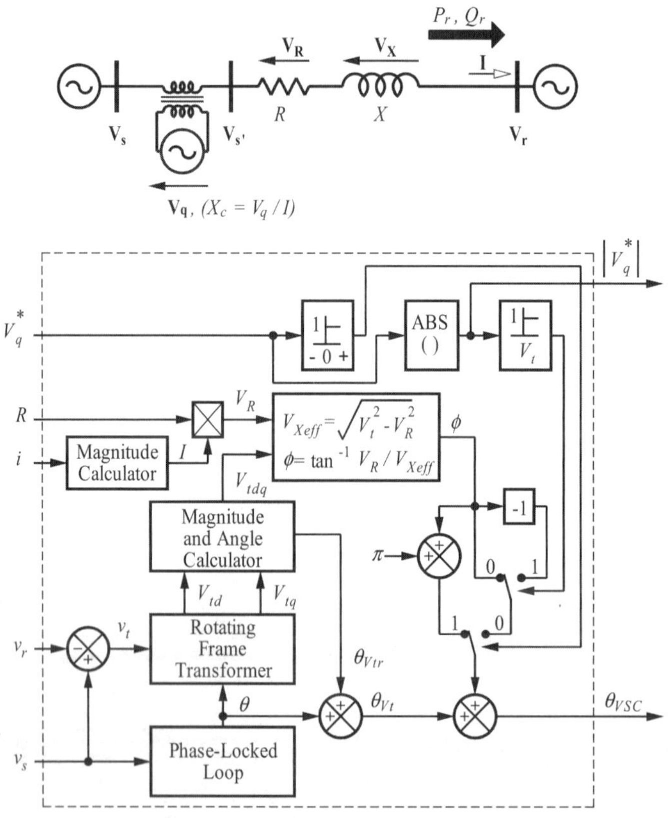

Figure 8-49. Voltage control scheme of the SSSC.

SSSC with a voltage control method, the compensating voltage demand, which must be within the limit of the SSSC rating, is specified. As was shown earlier that in the operation of the SSSC with a reactance control method, the compensating voltage demand is proportional to the prevailing line current. Using the reference PLL angle (θ), the instantaneous three-phase set of measured transmission line voltage (i.e., $v_t = v_s - v_r$) is first decomposed into its active or direct component (v_{td}) and reactive or quadrature component (v_{tq}), and then the amplitude (V_{tdq}) and the relative phase angle (θ_{Vtr}) of the transmission line voltage with respect to the PLL angle (θ) are calculated. The phase angle (θ_{Vt}) of the transmission line voltage, $\mathbf{V_t}$ (i.e., $\mathbf{V_s} - \mathbf{V_r}$), which is the difference of voltages between both ends of the transmission line, is calculated by adding the relative phase angle (θ_{Vtr}) of the transmission line voltage with respect to the sending-end voltage and the PLL angle (θ). The calculated amplitude (I) of the line current multiplied by the transmission line resistance (R) is the voltage (V_R). The phase angle of the voltage ($\mathbf{V_{Xeff}}$) across the transmission line effective reactance (X_{eff}) is determined by the voltage ($\mathbf{V_R}$) and the transmission line voltage ($\mathbf{V_t}$). The compensating voltage demand is V_q^*. The negative or positive polarity of the compensating voltage demand determines whether the compensating voltage ($\mathbf{V_q}$) is in phase or in opposite phase with the voltage ($\mathbf{V_{Xeff}}$) across the transmission line effective reactance.

The corresponding main file is `833SSSC4.DAT` that calls the two files, namely (a) user file `834SSSCV.USR` and (b) control file `835SSSCV.CON` (see Appendix C, Sections 83 through 84). The VSC is considered to be of infinite pulse.

Figure 8-50 shows the results from the digital simulation of the SSSC operating with the voltage control method. When the compensating voltage is in opposite phase with the voltage across the transmission line effective reactance, the line current (i_a) leads the compensating voltage (v_{qa}), causing the SSSC to operate as an equivalent capacitive reactance. As the compensating voltage demand increases in the same direc-

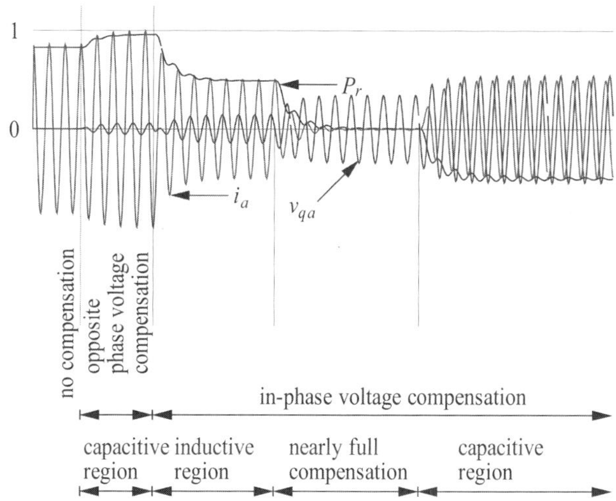

Figure 8-50. Performance of the SSSC with the voltage control method.

tion, the line current (i_a) and the power flow (P_r) in the transmission line increase monotonically. When the compensating voltage is in phase with the voltage across the transmission line effective reactance, the line current (i_a) lags the compensating voltage (v_{qa}) at first, causing the SSSC to operate as an equivalent inductive reactance. As the compensating voltage demand continues to increase in the same direction, the line current (i_a) and the power flow (P_r) in the transmission line decrease toward zero. The compensating voltage demand can further be increased in the same direction, resulting in a reversal of power flow (P_r) in the transmission line. The line current (i_a) now reverses and leads the compensating voltage (v_{qa}), causing the SSSC to appear as an equivalent capacitive reactance.

Therefore, in an operation of the SSSC with the voltage control method, the power flow (P_r) in the transmission line always increases when the compensating voltage is in opposite phase with the voltage across the transmission line effective reactance. Also, the power flow (P_r) decreases, at first, from the uncompensated value when the compensating voltage is in phase with the voltage across the transmission line effective reactance. Further increase in the compensating voltage in the same direction decreases the power flow (P_r) to nearly zero before increasing it in the reverse direction. Therefore, a stable bidirectional power flow in the transmission line is achieved.

In the voltage control scheme of Figure 8-49, the resistance (R) of the transmission line is one of the input parameters. The computation of line resistance under steady-state conditions is explained in Chapter 3. Under transient conditions, when a series-connected SSSC changes its compensation level, the energy balance at the both ends of the transmission line also includes the differential energy storage in the transmission line reactance, the leakage reactance of the series coupling transformer, and the SSSC itself. The transmission line resistance remains practically unchanged during the relatively short electrical transients. Therefore, a low-pass filter with narrow bandwidth is used to eliminate the undesired resistance values obtained from this measurement. The resistance values obtained due to seasonal changes will not be affected by the low-pass filter due to the high thermal time constant of the transmission line.

8.3 SHUNT–SERIES COMPENSATION USING A UNIFIED POWER FLOW CONTROLLER (UPFC)

Electric power flow through an alternating current transmission line is a function of the line impedance, the magnitudes of the sending- and receiving-end voltages, and the phase angle between these voltages. Essentially, the power flow is dependent on the voltage across the line impedance. Figure 8-51(a) shows a single-line diagram of a simple transmission line with an inductive reactance (X) and a compensating voltage ($\mathbf{V}_{s's}$) connecting a sending-end voltage source ($\mathbf{V}_s$) and a receiving-end voltage source ($\mathbf{V}_r$). The voltage across the transmission line reactance (X) is

$$\mathbf{V}_X = \mathbf{V}_s + \mathbf{V}_{s's} - \mathbf{V}_r = j\mathbf{I}X \tag{8-12}$$

where $\mathbf{I}$ is the current in the transmission line.

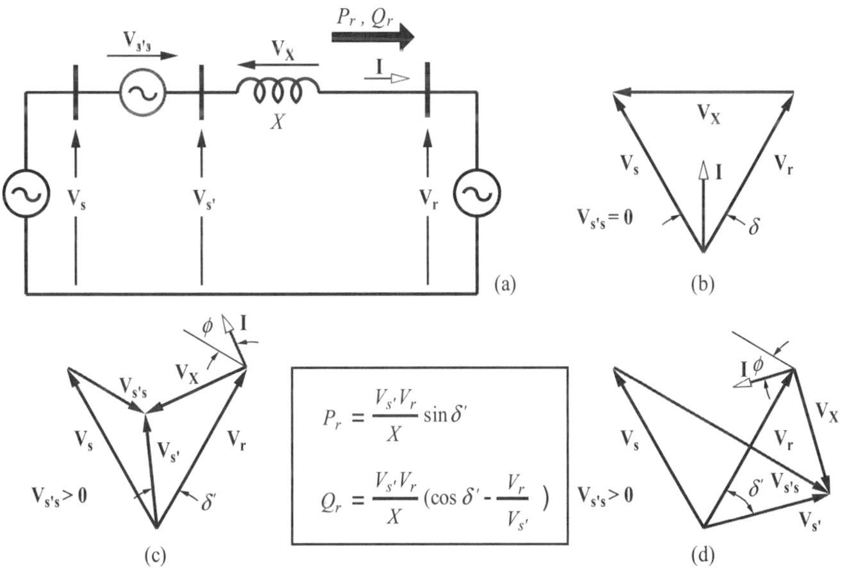

Figure 8-51. UPFC operating with an open loop compensating voltage and the related phasor diagrams.

The voltage ($\mathbf{V_X}$) across the transmission line can be controlled by controlling the compensating voltage ($\mathbf{V_{s's}}$) in series with the transmission line; consequently, the line current and the power flow in the line are regulated. Consider the case where $\mathbf{V_{s's}} = 0$ as shown in Figure 8-51b. The transmission line's sending-end voltage ($\mathbf{V_s}$) leads the receiving-end voltage ($\mathbf{V_r}$) by a phase angle (δ). The resulting current in the line is **I** and the active and reactive power flows at the receiving end are P_r and Q_r, respectively. With a compensating voltage ($\mathbf{V_{s's}}$) in series with the transmission line, the transmission line's modified sending-end voltage ($\mathbf{V_{s'}}$) still leads the receiving-end voltage ($\mathbf{V_r}$), but by a different phase angle (δ') as shown in Figure 8-51c. The resulting line current and the amount of power flow change. With a larger amount of compensating voltage ($\mathbf{V_{s's}}$) in series with the transmission line, the sending-end voltage ($\mathbf{V_{s'}}$) now lags the receiving-end voltage ($\mathbf{V_r}$) by a different phase angle (δ'), as shown in Figure 8-51d. The resulting line current and the power flow now reverse.

The phase angle (ϕ) between the negative compensating voltage (i.e., $-\mathbf{V_{s's}} = +\mathbf{V_{dq}}$) and the line current (**I**), as shown in Figure 8-51, parts c and d, can be between 0 and 2π. The component of the negative compensating voltage that is in or out of phase with the line current emulates a positive or a negative resistance in series with the transmission line. The remaining component that is in quadrature with the line current emulates an inductive or a capacitive reactance in series with the transmission line. The active and reactive powers (P_{exch} and Q_{exch}) exchanged by the SSSC with the line are

$$P_{exch} = -\mathbf{V_{s's}} \bullet \mathbf{I} = \mathbf{V_{dq}} \bullet \mathbf{I} = V_{dq} I \cos\varphi \tag{8-13a}$$

and

$$Q_{exch} = \left|-\mathbf{V_{s's}} \times \mathbf{I}\right| = \left|\mathbf{V_{dq}} \times \mathbf{I}\right| = V_{dq} I \sin\varphi \tag{8-13b}$$

The active power is absorbed from or delivered to the line through the STATCOM, which injects a current at the point of compensation. The current injected by the STATCOM has an active or direct component (I_d), which is in phase or in opposite phase with the line voltage. The current injected by the STATCOM also has a reactive or quadrature component (I_q), which is in quadrature with the line voltage, thereby emulating an inductive or a capacitive reactance at the point of compensation. This reactive current can be independently controlled, which, in turn, regulates the line voltage at the point of compensation.

The UPFC model in EMTP is shown Figure 8-52. The UPFC consists of two VSCs that are connected through a shared DC link. The configurations of the two VSCs used are identical for practical reasons, such as storing spare parts, maintenance, and so on. Each VSC is coupled with a transformer at its output. The first VSC, known as the STATCOM, injects an almost sinusoidal current, of variable magnitude, at the point of compensation in shunt with the line. The second VSC, known as the SSSC, injects an almost sinusoidal voltage, of variable magnitude, at the point of compensation in series with the line.

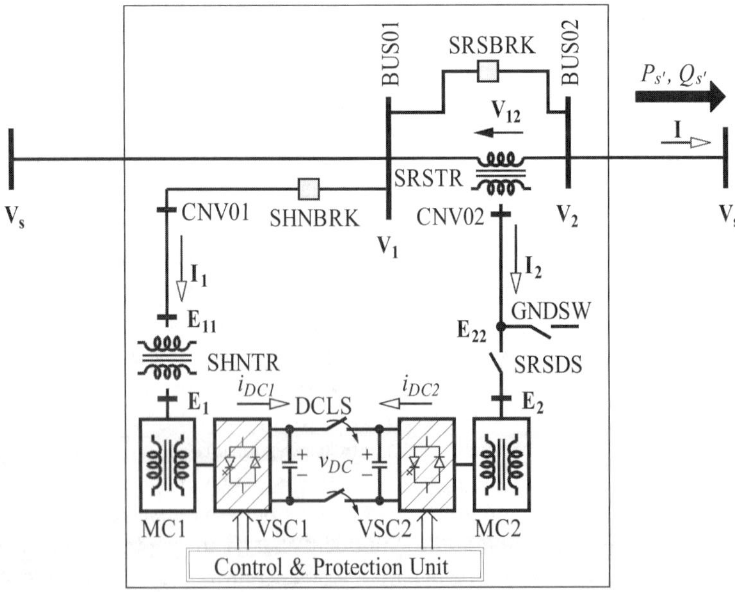

Figure 8-52. UPFC model in an EMTP.

8.3.1 Control of UPFC

The control of the UPFC can be divided into two parts: the control of the STATCOM and the control of the SSSC. When the STATCOM and the SSSC operate as stand-alone FACTS controllers, they exchange almost exclusively reactive power at their terminals. During the stand-alone operations, the SSSC injects a voltage in quadrature with the line current, thereby emulating an inductive or a capacitive reactance at the point of compensation in series with the line, and the STATCOM injects a reactive current, thereby also emulating a reactance at the point of compensation in shunt with the line.

While operating both the VSCs together as the UPFC, the series-connected compensating voltage can be at any phase angle with respect to the prevailing line current. Therefore, the exchanged power at the terminal of each VSC can be reactive as well as active. The exchanged active power at the terminal of one VSC with the line flows to the terminal of the other VSC through the shared DC link. The exchanged reactive powers at both the AC terminals of the UPFC are independent of each other.

The controller of the STATCOM is used to operate the VSC in such a way that the phase angle between the VSC output voltage and the line voltage is dynamically adjusted, so that the STATCOM generates or absorbs desired VARs at the point of compensation. Figure 8-13 shows the reactive current control block diagram of the STATCOM. Figure 8-15 shows the digital simulation results from the reactive current control operation of the STATCOM while the SSSC is bypassed by the series breaker (SRSBRK).

The SSSC can be operated in many different modes such as (a) voltage injection, (b) phase angle regulation, (c) line impedance emulation, (d) automatic power flow control, and so on. In each mode of operation, the final outcome is such that the SSSC injects a voltage in series with the transmission line. The control block diagram of the SSSC in an open loop voltage injection mode is shown in Figure 8-53.

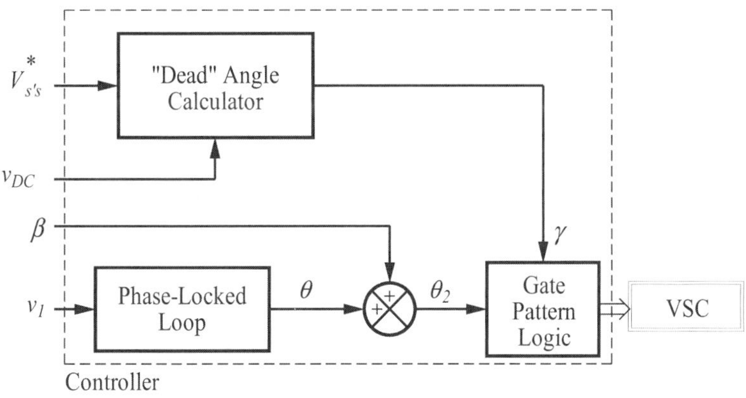

Figure 8-53. Control block diagram of the SSSC in an open loop voltage injection mode of operation of the UPFC.

The desired peak fundamental voltage ($V^*_{s's}$) at the output of the VSC and its relative phase angle (β) with respect to the reference PLL angle (θ) are specified. The absolute phase angle (θ_2) of the VSC output voltage is calculated by adding the relative phase angle (β) of the VSC output voltage and the PLL angle (θ). The dead angle (γ) of each pole is calculated from equation 6-162 as

$$\gamma = \cos^{-1}\frac{V_{AO,1}}{K_{VSC}v_{DC}}$$

or

$$\gamma = \cos^{-1}\frac{V^*_{s's}}{K_{VSC}v_{DC}} \tag{8-14}$$

where K_{VSC} is the fundamental AC amplitude-to-DC gain factor for a VSC with two-level poles.

8.3.2 Modeling of UPFC in EMTP and Simulation Results

The UPFC model with open loop voltage injection mode of operation is given in the three files, namely (a) main file `836UPFC1.DAT`, (b) user file `837UPFC.USR`, and (c) control file `838UPFC.CON` (see Appendix C, Sections 85 through 87). The VSCs are considered to be of infinite pulse.

Figure 8-54 shows the digital simulation results from the open loop voltage injection mode of operation of the SSSC while the STATCOM is operated to deliver no reactive current. At the beginning of the operation, the series bypass breaker (SRSBRK) and the series disconnect switch (SRSDS) are open and the ground switch (GNDSW) is closed. The HN-VSC (VSC2) generates no compensating voltage. The voltage (v_{12a}) at the terminal of the series coupling transformer (SRSTR) is the voltage across its leakage reactance. The active and reactive compensating powers ($P_{c\text{-}se}$ and $Q_{c\text{-}se}$) exchanged at the terminal of the series coupling transformer (SRSTR) are mostly reactive due to the high quality factor of the leakage reactance. The shunt breaker (SHNBRK) is open. The DC link capacitor is precharged.

At 50 ms, the SHNBRK closes and the quadrature current demand (I^*_{1q}) of the VSC1 is set to zero. At 100 ms, a series compensating voltage at the VSC2 side is set at 0.2 pu with a relative phase angle $\beta = 300°$ leading the reference PLL angle (θ). The series VSC2 output voltage (e_{2a}) of phase a leads the line current (i_a) by a phase angle (ϕ). Since $\phi > 90°$, the SSSC emulates a negative resistance in addition to an inductive reactance in series with the transmission line. The active power that is delivered to the line by the series VSC2 flows from BUS01 through the STATCOM. The shunt VSC1 output voltage (e_{1a}) of phase a is almost in phase with the current (i_{1a}) flowing through it. The voltage (v_{2a}) of phase a at BUS02 leads the voltage (v_{ra}) of phase a at the receiving end. The active power (P_r) delivered at the receiving end decreases. The reactive power (Q_r) delivered at the receiving end becomes inductive. At 200 ms, the series-connected compensating voltage is maintained at 0.2 pu while the relative phase

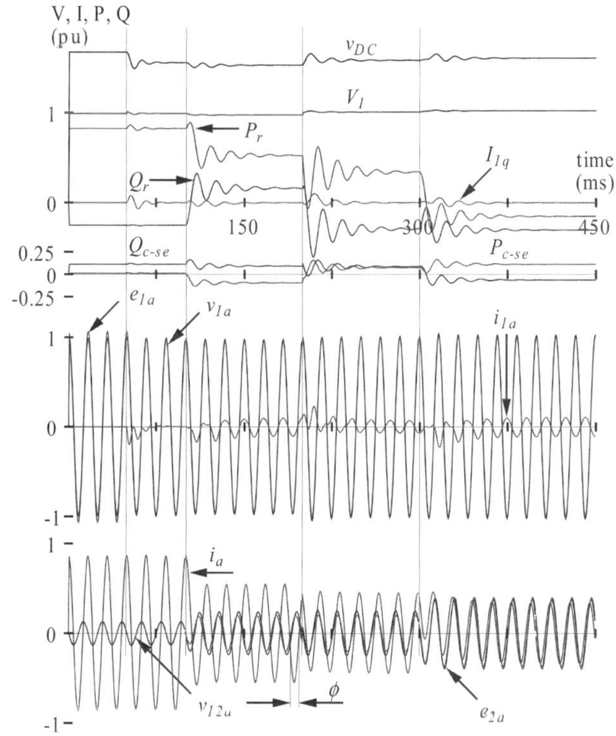

Figure 8-54. Performance of the UPFC with an infinite pulse VSC2 operating in an open loop voltage injection mode while regulating a zero reactive current through an infinite pulse VSC1.

angle (β) is changed to 240°. The active power that is absorbed from the line by the series VSC2 flows to BUS01 through the STATCOM. The shunt VSC1 output voltage (e_{1a}) of phase *a* is almost 180° out of phase with the current (i_{1a}) flowing through it. The reactive power (Q_r) delivered at the receiving end becomes capacitive. At 300 ms, the series-connected compensating voltage is increased to 0.4 pu while the relative phase angle (β) stays at 240°. The voltage (v_{2a}) of phase *a* at BUS02 lags the voltage (v_{ra}) of phase *a* and the active power flow (P_r) at the receiving end reverses. The series VSC2 output voltage (e_{2a}) of phase *a* lags the line current (i_a) by a phase angle (ϕ). The reactive compensating power ($Q_{c\text{-}se}$) exchanged at the terminal of the series coupling transformer (SRSTR) becomes capacitive.

Figure 8-55 shows the expanded view of two sections of Figure 8-54. The VSC2 voltage and current show the presence of only fundamental components.

The previous test case was rerun for the UPFC with two 24-pulse QHN-VSCs with three-level poles: one operating as the STATCOM with a fixed dead angle (γ) to produce a 48-pulse waveform and the other operating as the SSSC to produce a variable output voltage. The corresponding main file is `839UPFC2.DAT`. The results are shown

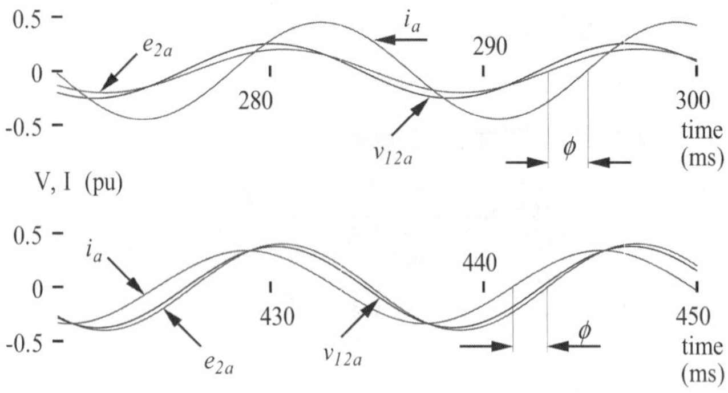

Figure 8-55. Waveforms from the UPFC with an infinite-pulse VSC2 operating in an open loop voltage injection mode.

in Figures 8-56 and 8-57, respectively. The VSC2 voltage and current show the presence of harmonic components.

In the above test cases, the STATCOM was operated in a reactive current control mode with the reactive current through the STATCOM set to zero. The only current that passed through the STATCOM was either in phase or out of phase with the BUS01 voltage, depending on whether the SSSC was delivering or absorbing active power to or from the line. The BUS01 voltage was not regulated. However, the remaining current carrying capability of the STATCOM can be utilized to inject reactive current to the line at the point of compensation and, in the process, the line voltage can be regulated. Figure 8-58 shows the previous test case with the voltage control mode of operation of the STATCOM. The STATCOM carries additional reactive current and regulates the line voltage at the point of compensation (BUS01).

8.3.3 Test Results

Figure 8-59 shows the closed loop automatic power flow control block diagram of the UPFC. The control can be divided into two parts: the control of the VSC1 as a shunt compensator and the control of the VSC2 as a series compensator. An instantaneous three-phase set of transmission line voltages (v_1) at BUS01 is used to calculate the reference PLL angle (θ), which is phase locked to the phase a of the bus voltage (v_{1a}). Using the reference PLL angle (θ), the instantaneous three-phase set of measured line voltage (v_1) at BUS01 is decomposed into its active or direct component (V_{1d}) and reactive or quadrature component (V_{1q}). The magnitude of the BUS01 voltage (V_{1dq}) is calculated. Using the reference PLL angle (θ), an instantaneous three-phase set of measured VSC1 current (i_1) is decomposed into its direct component (I_{1d}) and quadrature component (I_{1q}), and an instantaneous three-phase set of measured line current (i) is decomposed into its direct component (I_d) and quadrature component (I_q).

The control for the VSC1 works in such a way that the desired bus voltage magnitude reference, V_1^* (adjusted by the droop factor, K_{droop}), is compared with the BUS01

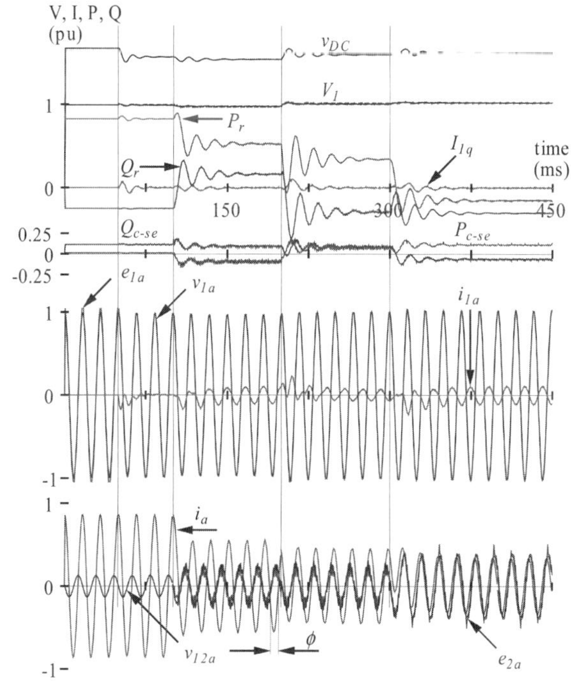

Figure 8-56. Performance of the UPFC with a 24-pulse QHN-VSC2 with three-level poles operating in an open loop voltage injection mode while regulating a zero reactive current through a 24-pulse QHN-VSC1 with three-level poles operating at a fixed dead angle to act as a 48-pulse QHN.

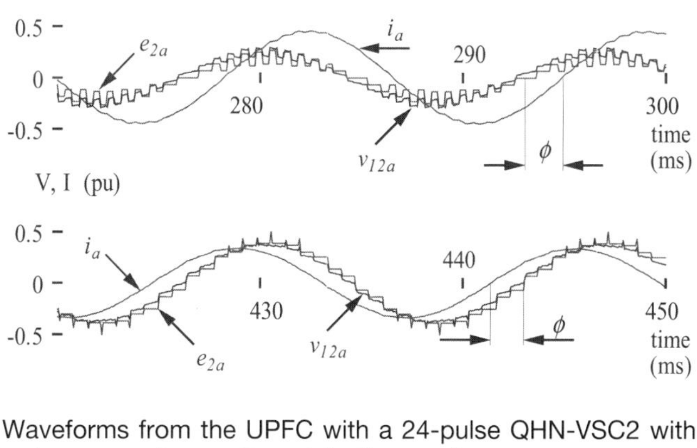

Figure 8-57. Waveforms from the UPFC with a 24-pulse QHN-VSC2 with three-level poles operating in an open loop voltage injection mode.

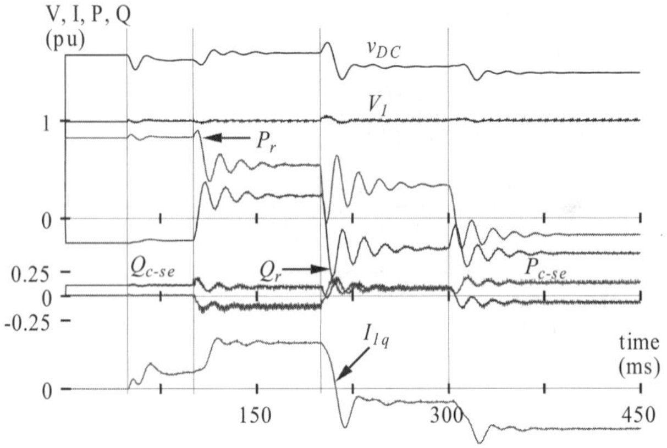

Figure 8-58. Performance of the UPFC with a 24-pulse QHN-VSC2 with three-level poles operating in an open loop voltage injection mode while regulating the BUS01 voltage with a 24-pulse QHN-VSC1 with three-level poles operating at a fixed dead angle to act as a 48-pulse QHN.

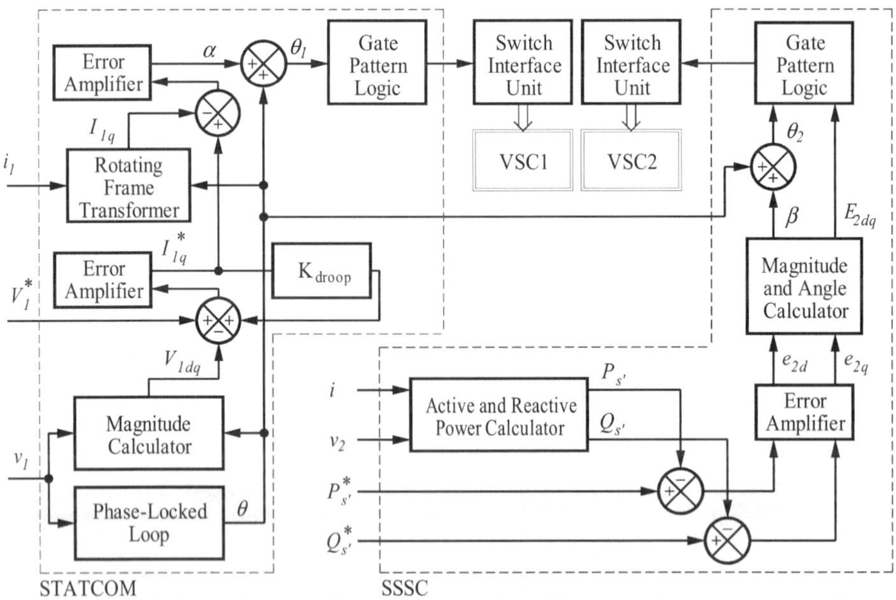

Figure 8-59. Closed loop automatic power flow control block diagram of the unified power flow controller.

voltage magnitude (V_{1dq}) using an outer voltage control loop and the error is passed through an error amplifier, which produces the reference quadrature component (I_{1q}^*) of the VSC1 current. The droop factor (K_{droop}) is defined as the allowable voltage error at the rated reactive current flow through the VSC1. The reference quadrature component (I_{1q}^*) is compared with the measured quadrature component (I_{1q}) of the VSC1 current using an inner reactive current control loop and the error is passed through an error amplifier, which produces a relative phase angle (α) of the VSC1 voltage with respect to the BUS01 voltage. The absolute phase angle (θ_1) of the VSC1 voltage is calculated by adding the relative phase angle (α) of the VSC1 voltage with respect to the PLL angle (θ) that is the reference phase angle of the BUS01 voltage. This is called voltage control mode of operation. The DC link capacitor voltage (v_{DC}) is dynamically adjusted in relation to the VSC1's AC terminal voltage. Note that for a particular bus voltage magnitude demand (V_1^*), if the reference quadrature component (I_{1q}^*) exceeds the rated current of the VSC, then $|I_{1q}^*|$ is limited to $\sqrt{(1 - I_{1d}^2)}$ and the bus voltage is regulated to an intermediate value. The controllable range of the bus voltage can easily be determined by operating the VSC1 with the inner reactive current control loop and varying I_{1q}^* from $-\sqrt{(1 - I_{1d}^2)}$ to $\sqrt{(1 - I_{1d}^2)}$. The reference quadrature component (I_{1q}^*) of the VSC1 current is defined to be either positive if the VSC1 is emulating a capacitive reactance or negative if it is emulating an inductive reactance. This is called reactive current control mode of operation. In this operation, the outer voltage control loop is deactivated.

The automatic power flow control algorithm works in such a way that the desired active and reactive power flows ($P_s^{*\prime}$ and $Q_s^{*\prime}$) are compared with the measured active and reactive power flows (P_s' and Q_s') and the errors are passed through an error amplifier, which produces, with respect to the reference PLL angle (θ), the direct and quadrature components of the series-connected compensating voltage (e_{2d} and e_{2q}). Next, the magnitude of the voltage (E_{2dq}) at the output of the VSC2 and its relative phase angle (β) with respect to the reference PLL angle (θ) are calculated. The absolute phase angle (θ_2) of the VSC2 voltage is calculated by adding the relative phase angle (β) of the VSC2 voltage and the PLL angle (θ). Note that for a particular active and reactive power flows (P_s^* and Q_s^*) demand, if the magnitude of the voltage (E_{2dq}) exceeds the rated voltage of the VSC, then E_{2dq} is limited to its design rating and the active and reactive power flows are regulated to an intermediate value. The controllable range of the active and reactive power flows can easily be determined with an open loop control of the rated compensating voltage (E_{2dq}) within its entire 360° range. This is called open loop voltage injection mode of operation. In this operation, the closed loop automatic power flow control algorithm is deactivated.

In the following test results, step changes in the reference of the control inputs and the subsequent response have been used. Figure 8-60 shows the changes in shunt-compensated reactive power due to the operation of the ±160 MVA, 138 kV rated STATCOM to regulate the bus voltage (V_1) at various levels. Note that the SSSC is not producing any compensating voltage in series with the line. The STATCOM is operated in voltage control mode to regulate the BUS01 voltage at 0.985 pu, 1.02 pu, 0.95 pu,

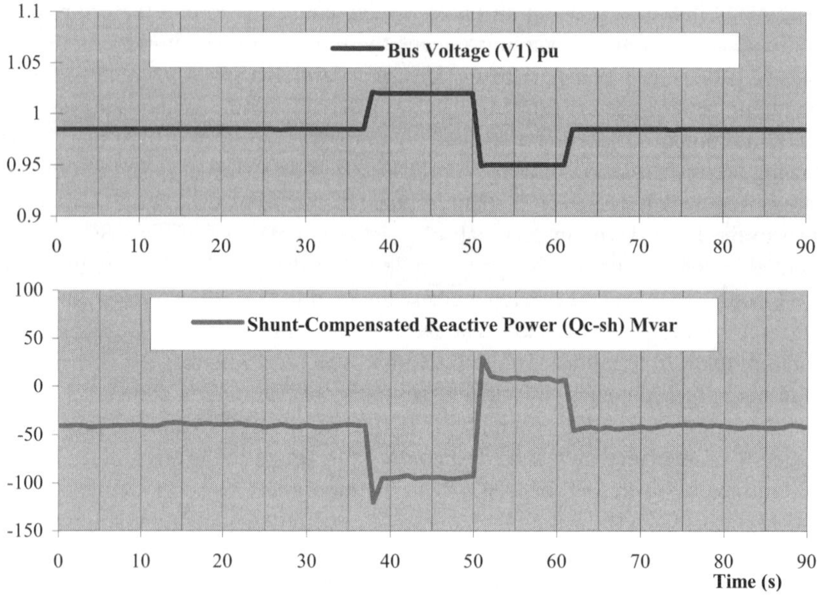

Figure 8-60. STATCOM test results for voltage control mode of operation (Sen and Keri-2002).

and 0.985 pu levels, respectively. The exchanged reactive power by the STATCOM changes from 42 Mvar capacitive to 93 Mvar capacitive to 7 Mvar inductive to, finally, back to 42 Mvar capacitive.

Figure 8-61 shows the regulation of active power at the modified sending end at unity power factor while maintaining one pu voltage at BUS01 using a ±160 MVA, 138 kV rated STATCOM and a ±160 MVA, 13.33 kV rated SSSC together as the UPFC. The active power flow at the modified sending end is regulated at –145 MW, –65 MW, –240 MW, and –145 MW levels, respectively, while holding the reactive power flow at the same point at zero. The STATCOM exchanges appropriate amount of reactive power with the line to hold the BUS01 voltage at one pu level.

Figure 8-62 shows the regulation of active power only at the modified sending end while holding the reactive power flow at the same point unchanged and maintaining one pu voltage at BUS01 using a ±160 MVA, 138 kV rated STATCOM and a ±160 MVA, 13.33 kV rated SSSC together as the UPFC. The active power flow at the modified sending end is regulated at –150 MW, –240 MW, –70 MW, and –150 MW levels, respectively, while holding the reactive power flow at the same point at 35 Mvar. The STATCOM exchanges appropriate amount of reactive power with the line to hold the BUS01 voltage at one pu level.

Figure 8-63 shows the regulation of reactive power only at the modified sending end while holding the active power flow at the same point unchanged and maintaining one pu voltage at BUS01 using a ±160 MVA, 138 kV rated STATCOM and a ±160

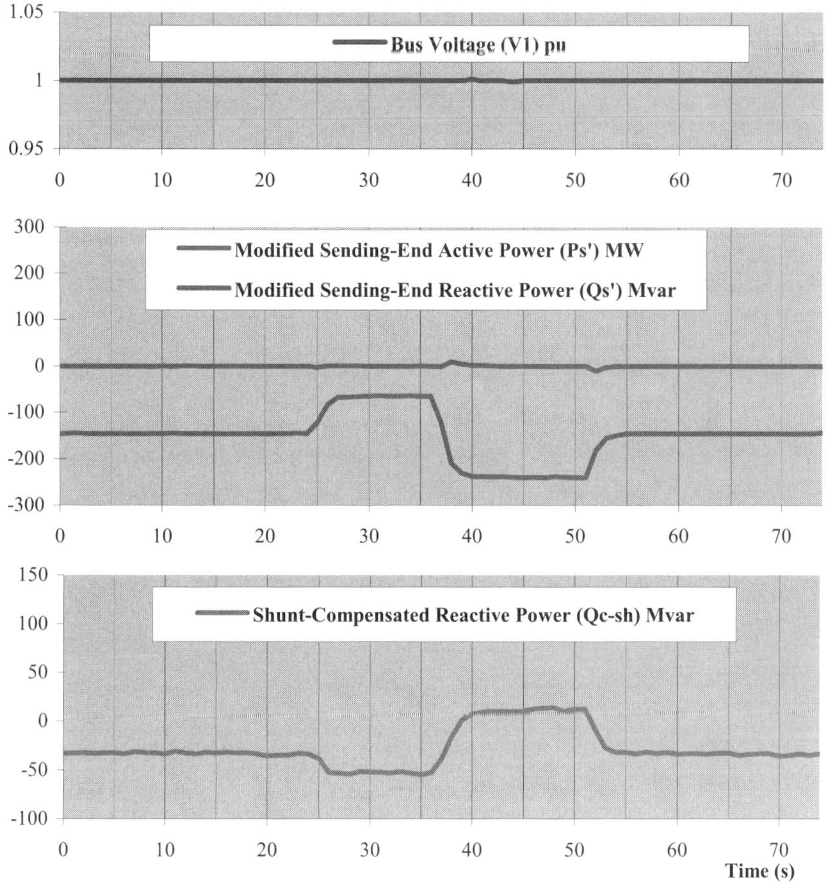

Figure 8-61. UPFC test results for automatic power flow control mode of operation with maintaining unity power factor and one pu bus voltage while regulating the line active power (Sen and Keri-2002).

MVA, 13.33 kV rated SSSC together as the UPFC. The reactive power flow at the modified sending end is regulated at 35 Mvar, –30 Mvar, 100 Mvar, and 0 Mvar levels, respectively, while holding the active power flow at the same point at –150 MW. The STATCOM exchanges appropriate amount of reactive power with the line to hold the BUS01 voltage at one pu level.

Figure 8-64 shows the changes in active and reactive power flows at the modified sending end due to the operation of the ±160 MVA, 13.33 kV rated SSSC to emulate compensating reactance at various levels. Note that the STATCOM is not operating. Therefore, the BUS01 voltage at the point of compensation is not regulated. The SSSC is operated in reactance control mode to regulate the active and reactive power flows at the modified sending end to be –105 MW, –7 Mvar; –170 MW, 10 Mvar; –235 MW,

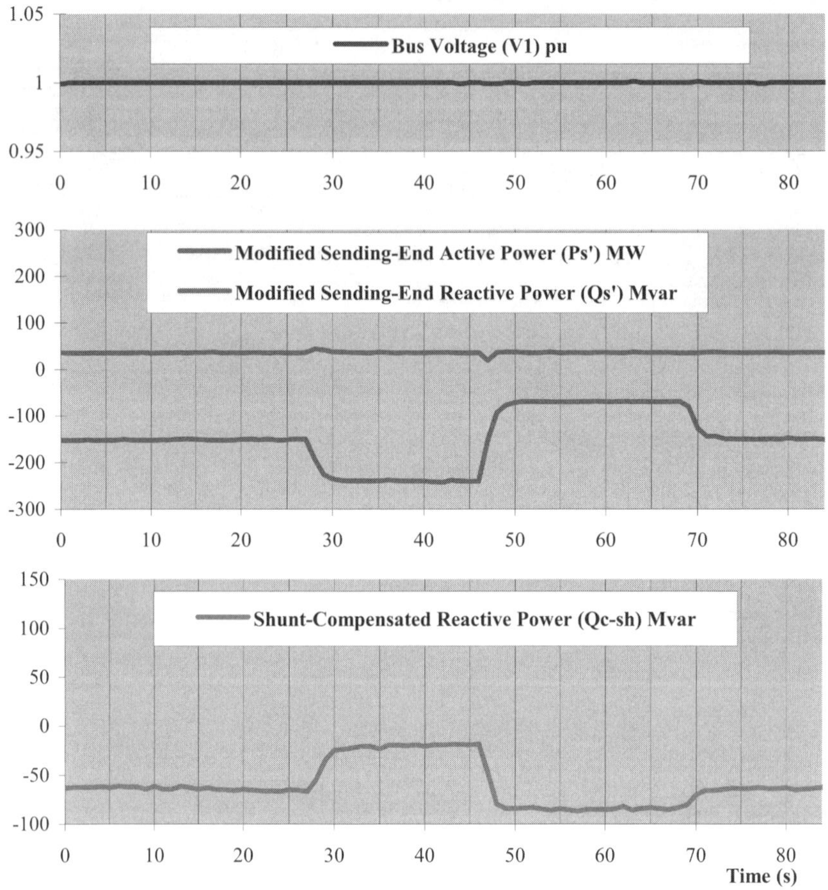

Figure 8-62. UPFC test results for automatic power flow control mode of operation while regulating only active power of the line and maintaining one pu bus voltage (Sen and Keri-2002).

50 Mvar; –190 MW, 20 Mvar, respectively. Note that the active and reactive power flows at the modified sending end change simultaneously.

8.3.4 Protection of UPFC

The general protection scheme of the UPFC is shown in Figure 8-65. When the VSCs are operated as stand-alone compensators, the basic protection scheme is as follows: The VSC1 stops gating when the instantaneous overcurrent through it exceeds a set value or the bus voltage drops below a set value. The VSC2 goes into an electronic bypass mode when the instantaneous overcurrent through it exceeds a set

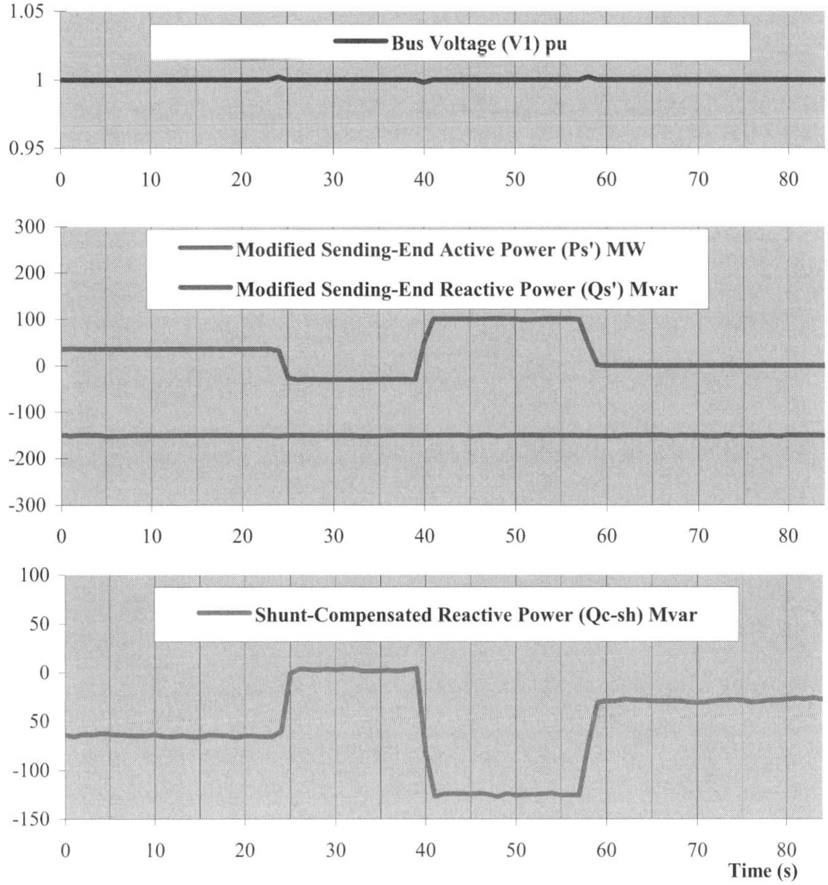

Figure 8-63. UPFC test results for automatic power flow control mode of operation while regulating only reactive power of the line and maintaining one pu bus voltage (Sen and Keri-2002).

value. When both VSCs are operated together as the UPFC, the VSC2 is bypassed when the VSC1 stops gating. Since the active power exchanged by the VSC2 with the line flows bidirectionally through the DC link capacitor and the VSC1, the VSC2 must be bypassed when the VSC1 has stopped operating. The DC link capacitor is prevented from being excessively charged by switching on a resistive clamp circuit when the DC voltage rises above a set value and by switching off when the DC voltage falls below a set value.

The bypass of VSC2 with the use of an electronic bypass on the secondary side (VSC side) of the series coupling transformer may be represented using the VSC2 switches. The behavior of the UPFC during disturbances is in large part determined

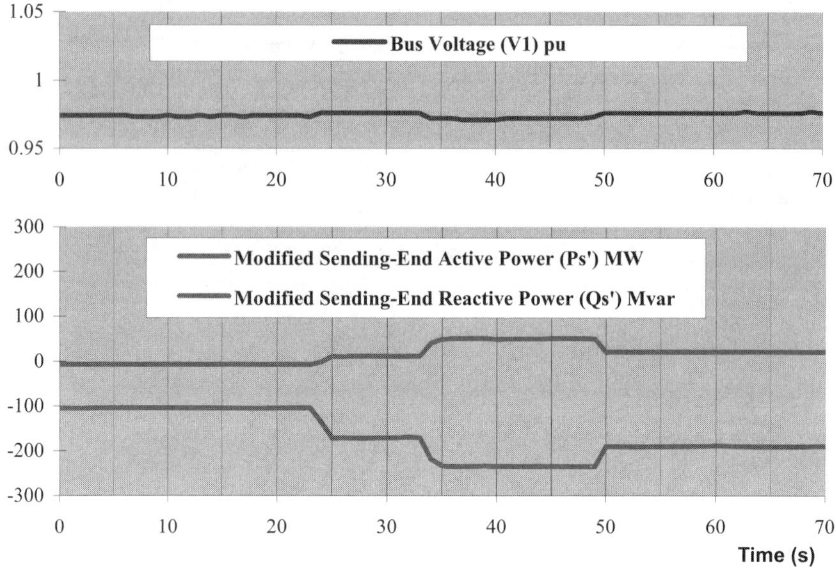

Figure 8-64. SSSC test results for reactance control mode of operation (Sen and Keri-2002).

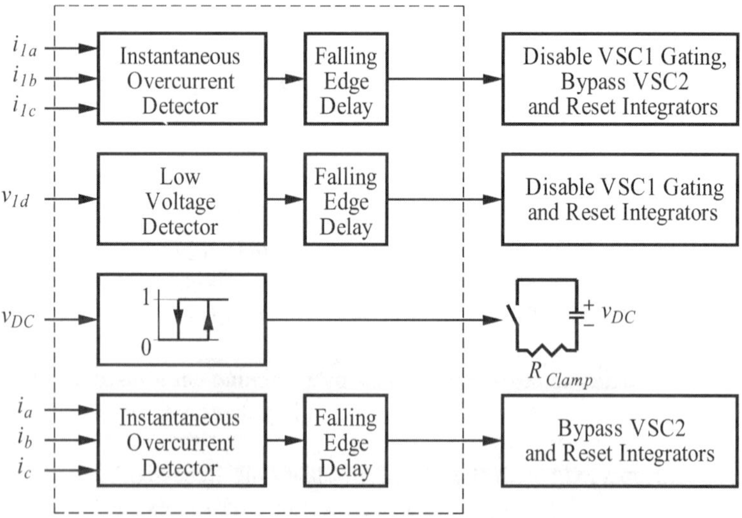

Figure 8-65. Protection scheme of the UPFC.

by a number of threshold levels and delay times. These numbers determine the current and voltage thresholds and timing delays associated with protective action and they are set in the first instance to values that seem reasonable for the actual equipment.

The integrators in the error amplifiers are reset at the time when a VSC stops operating normally. The integrators start from zero initial conditions when the VSC returns to normal operating condition.

9

SEN TRANSFORMER

Electric power flow through an AC transmission line is a function of the line impedance, the magnitudes of the sending-end voltage ($\mathbf{V_s}$) and the receiving-end voltage ($\mathbf{V_r}$), and the phase angle (δ) between these voltages. Figure 9-1 shows a simple transmission line represented with only a reactance (X) and the related expressions for active and reactive power flows at the receiving end of the line.

Power flow control parameters (voltage magnitude, its phase angle, and line reactance) can be regulated with the use of the following, now considered conventional, equipment:

- Voltage regulating transformer (VRT), shunt or parallel-connected switched inductor/capacitor, static var compensator (SVC), or static synchronous compensator (STATCOM) for voltage regulation
- Phase angle regulator (PAR) or phase shifting transformer (PST) for phase angle regulation
- Thyristor-controlled series capacitor (TCSC) for series reactance regulation

By regulating any one of the power flow control parameters using one of the above conventional solutions, both the active and reactive power flows in the transmission line are affected simultaneously.

Introduction to FACTS Controllers. By Kalyan K. Sen and Mey Ling Sen

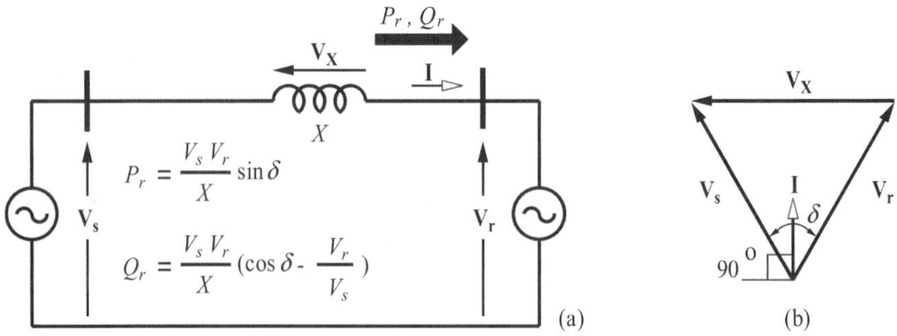

Figure 9-1. (a) Power transmission system. (b) Phasor diagram.

Consider that the point of compensation in the transmission line is at its sending end. Assuming that there are no changes in the transmission line's impedance and the voltage at the receiving end, a power flow controller (PFC) can control the active and reactive power flows (P_r and Q_r) to be a particular pair of values by modifying the sending-end voltage to a specific magnitude and phase angle. This can be accomplished by connecting a compensating voltage in series with the line using a shunt–series power converter. The shunt–series-type converters have been installed in several locations to be used as unified power flow controllers (UPFCs). A ±160 MVA-rated UPFC demonstrated for the first time that the active and reactive power flows in a transmission line could be regulated independently while maintaining a fixed line voltage at the point of compensation.

In certain special cases for point-to-point transfer of power between two isolated networks or interconnection of two power lines with different voltages or phase angles (or frequencies), the transmission line voltage can be controlled to a specific magnitude and phase angle by connecting a compensating voltage in shunt with the line using a shunt–shunt power converter. A shunt–shunt power converter has been installed to be used as a back-to-back static synchronous compensator (BTB-STATCOM). A ±36 MVA-rated BTB-STATCOM asynchronous tie demonstrated for the first time that the active power could be transferred from one transmission line to another through the BTB-STATCOM while at the same time independent shunt reactive compensation could be provided at the points of connection with the transmission lines.

Regardless of whether the compensating voltage is series or shunt connected, it is at any phase angle with the current flowing through it and, therefore, it exchanges active and reactive powers at the point of compensation with the transmission line. In a VSC-based compensation scheme, a shunt–series or shunt–shunt power converter facilitates the bidirectional power flow through its shared DC link. It is important to note that in the shunt–series power converter, the series-connected compensating voltage carries the full line current and its terminal voltage rating is a fractional amount of the line voltage. Therefore, the rating of the shunt–series power converter is only a fraction of the transmitted power of the line. In the shunt–shunt power converter, the shunt-connected compensating voltage also carries the full line current, but its terminal voltage rating is the same as the full line voltage. Therefore, the rating of the shunt–shunt power converter is the same as the transmitted power of the line. For the same amount of controllable transmitted power, the shunt–shunt power converter requires higher-power-rated power electronics hardware than what is required in a shunt–series power converter. For the same amount of transmitted power, the shunt–shunt power converter costs more than the shunt–series power converter. Therefore, the preferred topology of a PFC uses a shunt–series configuration. A shunt–shunt configuration is, however, the preferred choice for the interconnection of two nearby transmission lines with different voltages or phase angles (or frequencies).

The VSC-based technology has the capability of providing a fast dynamic response, but this capability is not required in most utility applications, where the need is to regulate the line voltage and the power flow in the line(s) in a relatively slow manner. Although the UPFC has a more versatile power flow controllability than any previously developed PFC, it has two major drawbacks: its high installation and operating costs. In order to address these two issues, a new concept in FACTS controllers, the Sen transformer (ST), is proposed. This new concept is more reliable and more cost-effective and it is based on the proven technology of transformers and LTCs. The ST is a new family of power flow controlling transformers that provides the same independent control of active and reactive power flows as the UPFC, albeit at a reduced dynamic rate, but with an estimated 5:1 reduction in equipment cost and a 10:1 improvement in operational cost. The objective in this chapter is to describe the evolution of the ST and introduce it as the most cost-effective, independent active and reactive PFC in a utility application.

9.1 EXISTING SOLUTIONS

It is well established that for optimum operation of a transmission line, independent control of active and reactive power flows in the line is necessary. These require the transmission line voltage to be modified to a specific magnitude and phase angle. For many decades, the regulations of voltage magnitude and its phase angle have been accomplished using the VRT and PAR.

9.1.1 Voltage Regulation

The VRT is an autotransformer, shown in Figure 9-2a, that provides three-phase, bipolar compensating voltages ($\mathbf{V}_{s'sA}$, $\mathbf{V}_{s'sB}$, and $\mathbf{V}_{s'sC}$). The uncompensated transmission line's three-phase sending-end voltages ($\mathbf{V}_{sA}$, $\mathbf{V}_{sB}$, and $\mathbf{V}_{sC}$) are regulated at $\mathbf{V}_{s'A}$, $\mathbf{V}_{s'B}$, and $\mathbf{V}_{s'C}$ when the compensating voltage is connected in series with the line at the point of compensation. Since the compensating voltage is a fraction of the line voltage, the rating of the autotransformer is only a fraction of the transmitted power.

As shown in Figure 9-2a, the VRT operates as an autotransformer. The exciter unit consists of a three-phase (A, B, and C), Y-connected primary windings that are connected in shunt with the line. The three-phase primary windings are excited from the three-phase line voltages ($\mathbf{V}_{sA}$, $\mathbf{V}_{sB}$, and $\mathbf{V}_{sC}$). A three-phase, bipolar compensating

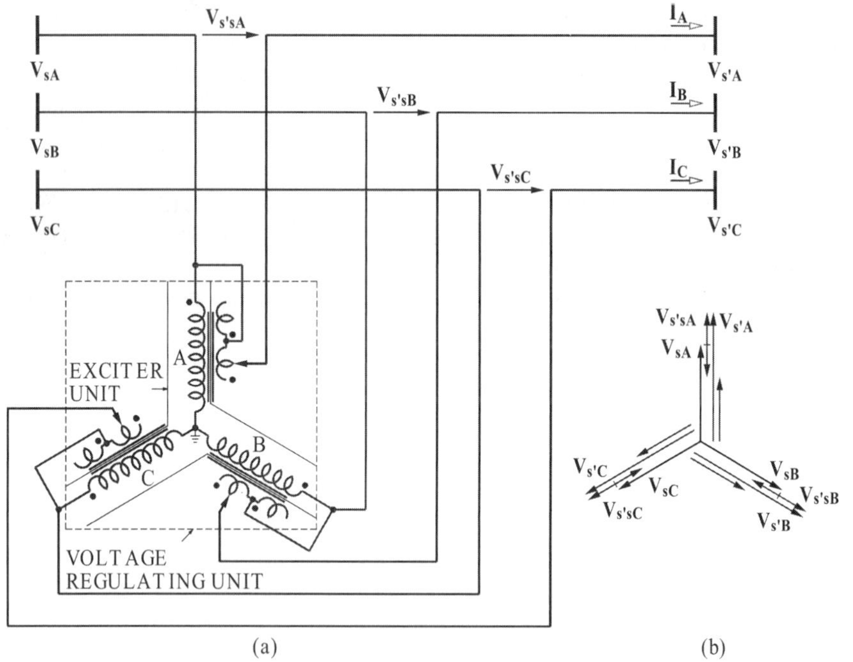

Figure 9-2. (a) Voltage regulating transformer (autotransformer). (b) Phasor diagram.

voltages ($\mathbf{V}_{s'sA}$, $\mathbf{V}_{s'sB}$, and $\mathbf{V}_{s'sC}$) that are either in phase or out of phase with the corresponding phase-to-neutral voltage are generated from the induced secondary voltages. The voltage regulating unit consists of a total of six secondary windings (two windings in each phase). With the use of LTCs, the magnitude of the compensating voltage ($\mathbf{V}_{s's}$) is varied. The line is regulated at a voltage ($\mathbf{V}_{s'}$) from the uncompensated voltage ($\mathbf{V}_s$). Figure 9-2b shows the related phasor diagram.

The uncompensated transmission line's three-phase sending-end voltages ($\mathbf{V}_{sA}$, $\mathbf{V}_{sB}$, and $\mathbf{V}_{sC}$) can also be modified to $\mathbf{V}_{s'A}$, $\mathbf{V}_{s'B}$, and $\mathbf{V}_{s'C}$ with the use of a shunt-connected compensating voltages from a two-winding transformer, as shown in Figure 9-3a, where the three-phase compensating voltages are $\mathbf{V}_{s'A}$, $\mathbf{V}_{s'B}$, and $\mathbf{V}_{s'C}$. Since the compensating voltage is the line voltage, the rating of the two-winding transformer is the same as the transmitted power.

As shown in Figure 9-3a, the VRT operates as a two-winding transformer. The exciter unit consists of three-phase (A, B, and C), Y-connected primary windings that are connected in shunt with the line. The three-phase primary windings are excited from the three-phase line voltages ($\mathbf{V}_{sA}$, $\mathbf{V}_{sB}$, and $\mathbf{V}_{sC}$). Three-phase compensating voltages ($\mathbf{V}_{s'A}$, $\mathbf{V}_{s'B}$, and $\mathbf{V}_{s'C}$) that are in phase with the corresponding phase-to-neutral voltage are generated from the induced secondary voltages. The voltage regulating unit consists of a total of three secondary windings (one winding in each phase). With the use of LTCs, the magnitude of the compensating voltage ($\mathbf{V}_{s'}$) is varied. The line is regu-

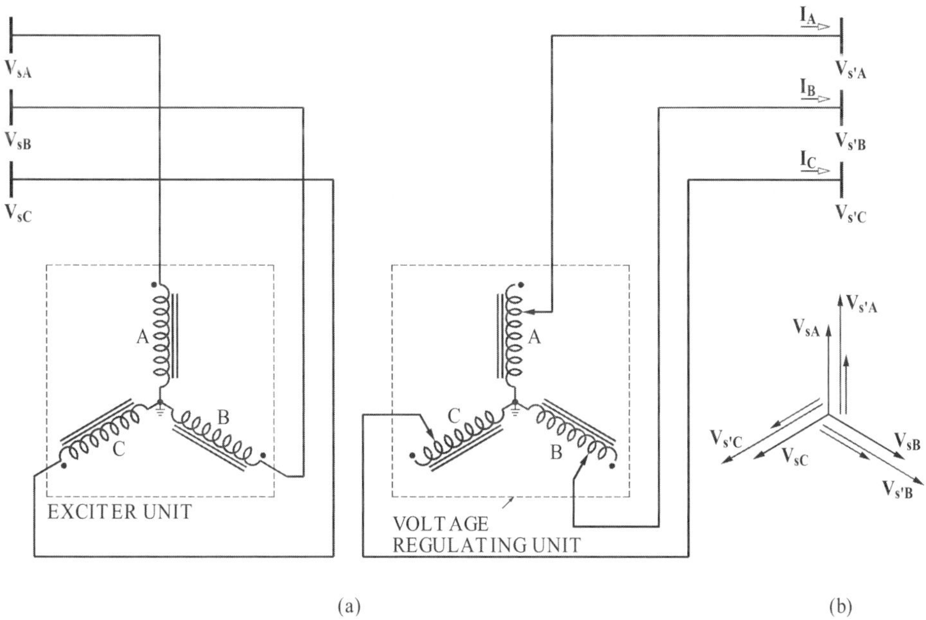

Figure 9-3. (a) Voltage regulating transformer (two-winding transformer). (b) Phasor diagram.

lated at a voltage ($\mathbf{V_{s'}}$) from the uncompensated voltage ($\mathbf{V_s}$). Figure 9-3b shows the related phasor diagram.

9.1.2 Phase Angle Regulation

The effective phase angle of the transmission line voltage is varied by using the PAR. The PAR connects a voltage in series with the transmission line and in quadrature with the phase-to-neutral voltage, as shown in Figure 9-4a. In a typical configuration, the PAR consists of two transformers. The first transformer (exciter unit) is called a regulating transformer. Its three-phase (A, B, and C), Y-connected primary windings are connected in shunt with the line. The three-phase primary windings are excited from the line voltages ($\mathbf{V_{sA}}$, $\mathbf{V_{sB}}$, and $\mathbf{V_{sC}}$) and three-phase voltages ($\mathbf{V_A}$, $\mathbf{V_B}$, and $\mathbf{V_C}$) are induced in the regulating transformer secondary windings. For series connection of this voltage, an electrical isolation is necessary. The second transformer (series unit) is called a series transformer that is excited from the phase-to-phase voltages ($\mathbf{V_B} - \mathbf{V_C}$, $\mathbf{V_C} - \mathbf{V_A}$, and $\mathbf{V_A} - \mathbf{V_B}$) of the regulating transformer. Three-phase, bipolar compensating voltages ($\mathbf{V_{s'sA}}$, $\mathbf{V_{s'sB}}$, and $\mathbf{V_{s'sC}}$) that are in quadrature with the phase-to-neutral voltages are induced in the secondary windings of the series transformer. The magnitude of the compensating voltage is varied using the LTCs in the secondary windings of the regulating transformer. The induced voltage

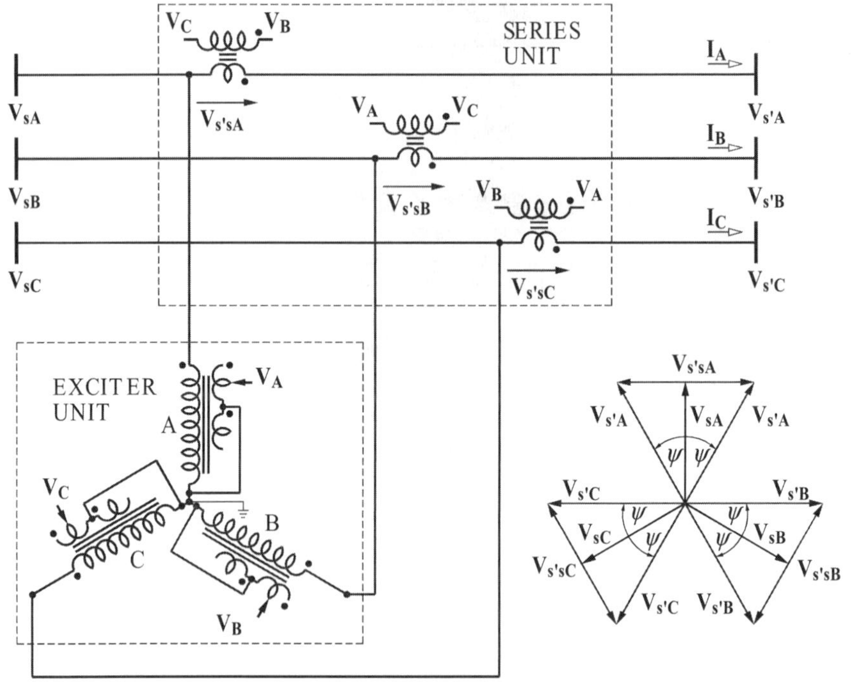

Figure 9-4. (a) Phase angle regulator. (b) Phasor diagram.

of the series transformer is connected in series with the transmission line. Figure 9-4b shows the related phasor diagram.

The VRT (autotransformer or two-winding transformer) can only modify the magnitude of the transmission line voltage, but not the phase angle of the voltage. The phase angle remains almost the same as that of the original voltage before compensation. The PAR can only modify the phase angle of the transmission line voltage, but not the magnitude of the voltage. The magnitude remains almost the same as that of the original voltage before compensation. For a PFC to regulate the active and reactive power flows in the line independently, it needs to modify the line voltage, $\mathbf{V_s}$ (i.e., $V_s \angle \delta_s$), to a specific magnitude ($V_{s'}$) and phase angle ($\delta_{s'}$).

9.2 DESIRED SOLUTION

A series-connected compensating voltage ($\mathbf{V_{s's}}$), shown in Figure 9-5a, modifies the magnitude and phase angle of the transmission line voltage independently. For the desired amount of active and reactive power flows in the line, the compensating voltage has to be a specific magnitude ($V_{s's}$) and phase angle (β) with respect to the line voltage.

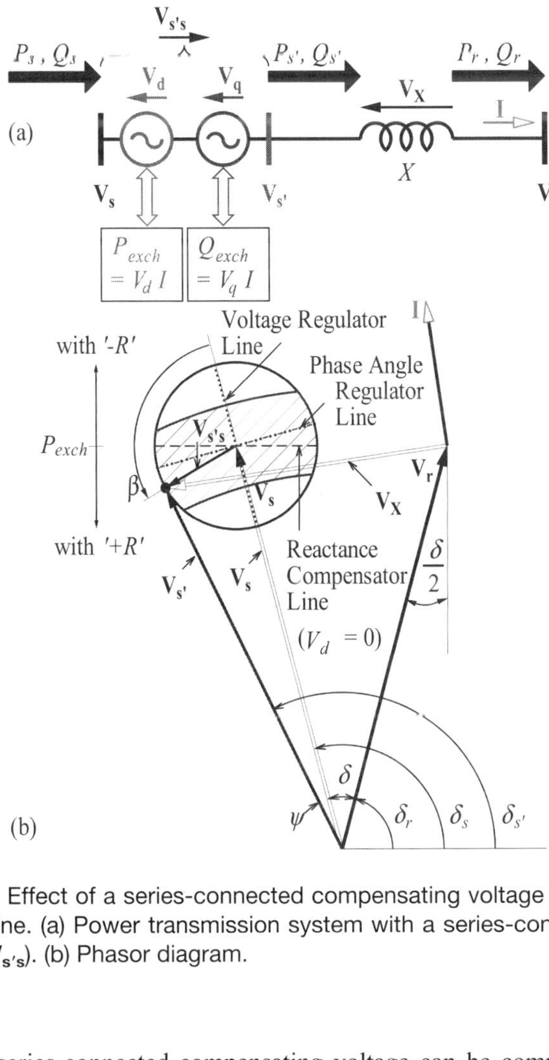

Figure 9-5(a, b). Effect of a series-connected compensating voltage on power flow in a transmission line. (a) Power transmission system with a series-connected compensating voltage ($\mathbf{V}_{s's}$). (b) Phasor diagram.

The desired series-connected compensating voltage can be composed of two orthogonal voltages: one regulates the magnitude of the line voltage and other regulates the phase angle of the line voltage with the use of the VRT and PAR. The ST combines the functions of the VRT and PAR in a single unit that results in a reduced amount of hardware from what is required for separate VRT and PAR.

Consider the simple power transmission system with a sending-end voltage, $\mathbf{V}_s$ (magnitude, V_s, and phase angle, δ_s), a receiving-end voltage, $\mathbf{V}_r$ (magnitude, V_r, and phase angle, δ_r), the voltage ($\mathbf{V}_X$) across the line reactance (X), and a compensating voltage ($\mathbf{V}_{s's}$), as shown in Figure 9-5a. When the compensating voltage, $\mathbf{V}_{s's}$ (magnitude, $V_{s's}$ and phase angle, $\delta_s + \beta$), is added in series with the transmission line, the modified sending-end voltage becomes $\mathbf{V}_{s'}$ (i.e., $\mathbf{V}_s + \mathbf{V}_{s's}$). Figure 9-5b shows the phasor diagram related to the series-connected compensating voltage with a fixed magnitude of 0.2 pu and its entire controllable range of $0° \leq \beta \leq 360°$. The difference voltage ($\mathbf{V_{s'}} - \mathbf{V_r}$) provides the voltage ($\mathbf{V_X}$) across the line reactance (X). As the relative phase angle (β) is varied over its full 360° range, the end of phasor ($\mathbf{V_{s's}}$) moves along a circle with its center located at the end of phasor ($\mathbf{V_s}$). The rotation of phasor ($\mathbf{V_{s's}}$) modulates the magnitude and phase angle of phasor ($\mathbf{V_X}$).

In the transmission system considered above, the magnitude ($V_{s'}$) and phase angle ($\delta_{s'}$) of the modified sending-end voltage ($\mathbf{V_{s'}}$) are modulated between $V_s + V_{s's}$ and $V_s - V_{s's}$ and between $\delta_s + \psi_{max}$ and $\delta_s - \psi_{max}$, respectively, where $V_{s's} = 0.2$ pu and $\psi_{max} = 11.54°$, as shown in Figure 9-5c. Note that the line voltage is allowed to be controlled within a narrow band, for example, 10%, of the nominal value.

The compensating voltage ($\mathbf{V_{s's}}$) is at any phase angle with the prevailing line current ($\mathbf{I}$) and, therefore, exchanges with the line both active power ($P_{exch} = V_d I$) and reactive power ($Q_{exch} = V_q I$), where $\mathbf{V_d}$ and $\mathbf{V_q}$ are the respective active or direct and reactive or quadrature components of the compensating voltage with load convention. These exchanged active and reactive powers (P_{exch} and Q_{exch}) are also sinusoidal functions of relative phase angle (β), as shown in Figure 9-5d. For a given magnitude of a compensating voltage, the exchanged capacitive power (Q_{exch}) is larger than its inductive counterpart due to the fact that the capacitive compensation produces a larger line current. The compensating voltage, being at any phase angle with the prevailing line current, emulates a series capacitor (C) or an inductor (L) and a positive resistor ($+R$) or a negative resistor ($-R$).

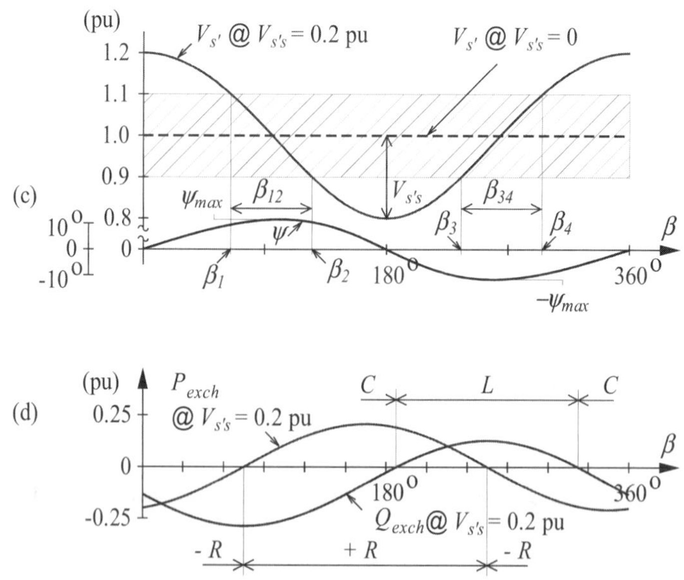

Figure 9-5 (c, d). Effect of a series-connected compensating voltage on power flow in a transmission line. (c) Variations of the line voltage magnitude and relative phase angle (ψ). (d) Exchanged active and reactive powers as a function of relative phase angle (β).

The active and reactive power flows (P_r and Q_r) at the receiving end vary with the relative phase angle (β) in a sinusoidal manner, as shown in Figure 9-5e. The relationship between Q_r and P_r is a circle centered at (P_{rn}, Q_{rn}) with a radius of A_r, where P_{rn} and Q_{rn} are the natural active and reactive power flows at the receiving end of the transmission line without any compensation and A_r is the maximum power flow enhancement. The active and reactive power flows at the receiving end of the line can be controlled within the range defined by the P–Q plot in Figure 9-5f by choosing the magnitude ($V_{s's}$) of the compensating voltage ($\mathbf{V}_{s's}$) within its rated value and the relative phase angle (β) between 0° and 360°. If the magnitude ($V_{s's}$) of the compensating voltage is increased, the controllable range of the power flows in the P–Q plane is also increased. When the controllable range extends to all four quadrants, bidirectional and independent control of active and reactive power flows are achieved. For the desired amount of active and reactive power flows (P_r^*, Q_r^*), within the allowable control range, the magnitude ($V_{s's}$) of the compensating voltage is related to the distance, $\sqrt{[(P_r^* - P_{rn})^2 + (Q_r^* - Q_{rn})^2]}$, between ($P_{rn}$, Q_{rn}) and (P_r^*, Q_r^*). The relative phase angle (β) of the compensating voltage ($\mathbf{V}_{s's}$) is the phase angle between the voltage regulator line (i.e., $\beta = 0°$ line) and the line that passes through (P_{rn}, Q_{rn}) and (P_r^*, Q_r^*).

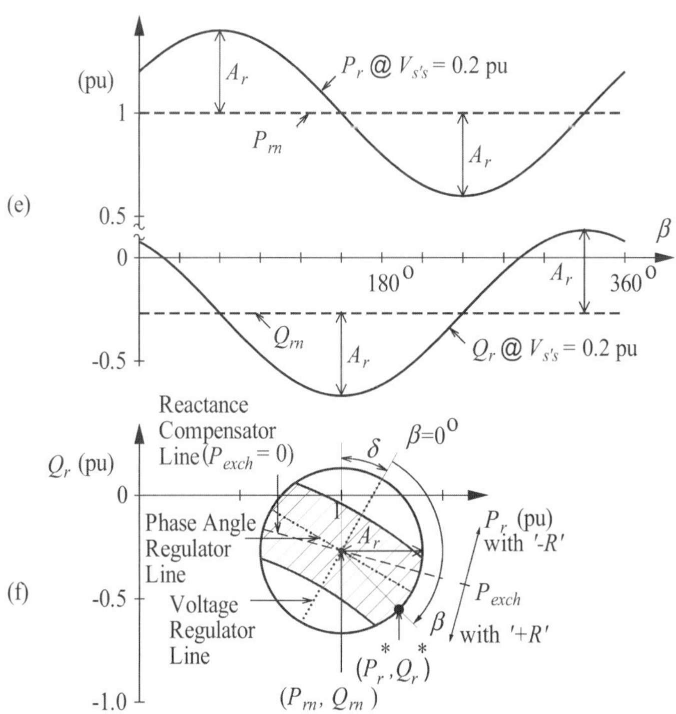

Figure 9-5 (e, f). Effect of a series-connected compensating voltage on power flow in a transmission line. (e) Variation of the receiving-end active and reactive power flows (P_r and Q_r) as a function of the relative phase angle (β). (f) Receiving-end Q_r versus P_r.

When a series-connected compensating voltage is used to modify the transmission line voltage, the following occur:

- The compensating voltage can be either in phase or out of phase with the phase-to-neutral voltage of the transmission line to implement the function of the VRT.
- The compensating voltage can be in quadrature with the phase-to-neutral voltage of the transmission line to implement the function of the PAR.
- The compensating voltage can be such that it provides a series reactance compensation because it is in quadrature with the prevailing line current. If the circular controllable area is equally divided by the reactance compensator line ($V_d = 0$ or $P_{exch} = 0$), the upper and lower halves represent P_{exch} due to $-R$ and $+R$, respectively.

Any compensator that provides compensation for one power flow control parameter (voltage magnitude, its phase angle, or line reactance) operates on a set of linear operating points inside the *P*–*Q* circle as defined above. Since these operating characteristic lines are neither horizontal nor vertical in the *P*–*Q* plane at the receiving end, controlling one parameter with the use of these compensators affects both the active and reactive power flows in the transmission line simultaneously, as shown in Figure 9-5f. The same conclusion can be drawn in regard to the active and reactive power flows at the modified sending end.

9.2.1 ST as a New Voltage Regulator

Figure 9-6a shows a new voltage regulator, which is a three-phase transformer with Y-connected primary windings (A, B, and C), called the exciter unit, and three secondary windings (a_1 on the core of the A phase, b_2 on the core of the B phase, and c_3 on the core of the C phase), called the compensating voltage unit. The three-phase sending-end voltages ($\mathbf{V}_{sA}$, $\mathbf{V}_{sB}$, and $\mathbf{V}_{sC}$) are applied to the exciter unit. The in-phase component of the compensating voltage for any phase is induced in a winding that is placed on the corresponding phase of the transformer core. By choosing the number of turns and, therefore, the magnitude of the compensating voltage ($\mathbf{V}_{s's}$), the magnitude of the modified compensating voltage is increased. Note that the three compensating windings (a_1, b_2, and c_3) have the same number of turns. Figure 9-6b shows the related phasor diagram.

Figure 9-7a shows a new voltage regulator, which is a three-phase transformer with Y-connected primary windings (A, B, and C), called the exciter unit, and six secondary windings (a_2 and a_3 on the core of the A phase; b_1 and b_3 on the core of the B phase; and c_1 and c_2 on the core of the C phase), called the compensating voltage unit. The three-phase sending-end voltages ($\mathbf{V}_{sA}$, $\mathbf{V}_{sB}$, and $\mathbf{V}_{sC}$) are applied to the exciter unit. The out-of-phase component of the compensating voltage for any phase is derived from the phasor sum of the voltages induced in two equal-turn windings that are placed on the remaining two phases of the transformer core. By choosing the number of turns

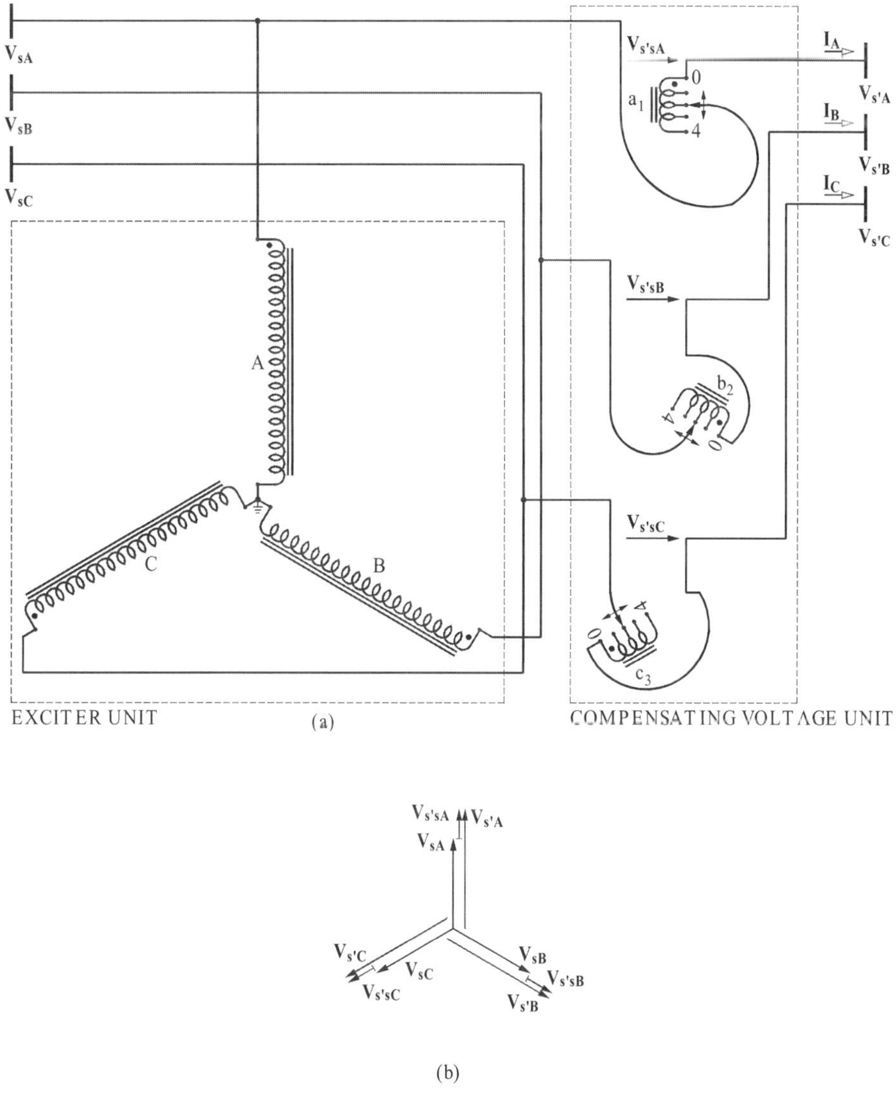

Figure 9-6. (a) Voltage regulating transformer for increasing line voltage. (b) Phasor diagram.

and, therefore, the magnitude of the compensating voltage ($\mathbf{V}_{s's}$), the magnitude of the modified sending-end voltage is decreased. Note that the six compensating windings (a_2, a_3, b_1, b_3, c_1, and c_2) have the same number of turns. Figure 9-7b shows the related phasor diagram.

Figure 9-8a shows the ST, which is a three-phase transformer with Y-connected primary windings (A, B, and C), called the exciter unit, and nine secondary windings (a_1,

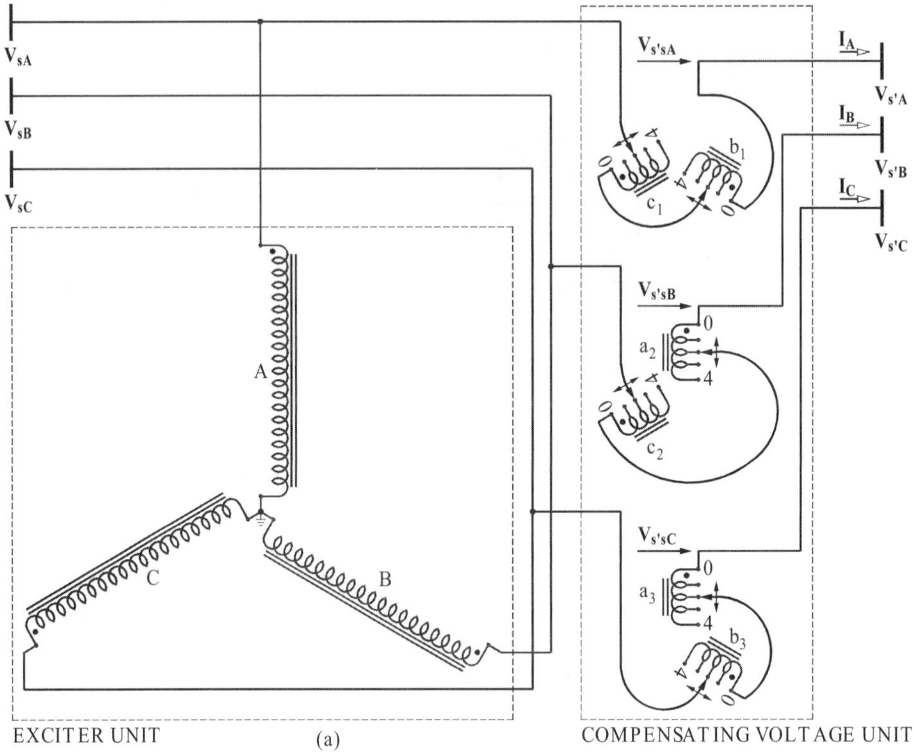

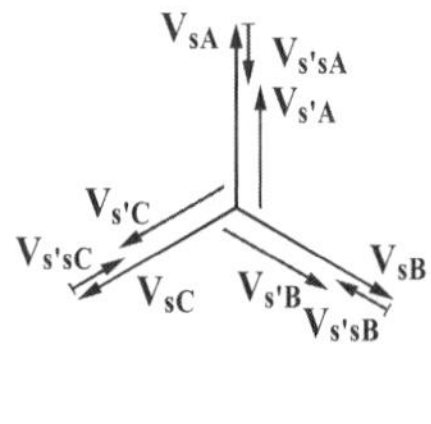

Figure 9-7. (a) Voltage regulating transformer for decreasing line voltage. (b) Phasor diagram.

a_2, and a_3 on the core of the A phase; b_1, b_2, and b_3 on the core of the B phase; and c_1, c_2, and c_3 on the core of the C phase), called the compensating voltage unit. Figure 9-8b shows the phasor diagram when the ST is used as a voltage regulator.

At this point, one might ask why one should use the ST as a voltage regulator when the VRT, as shown in Figure 9-2a, offers the same functionality with a simpler configuration. This is described in the next section.

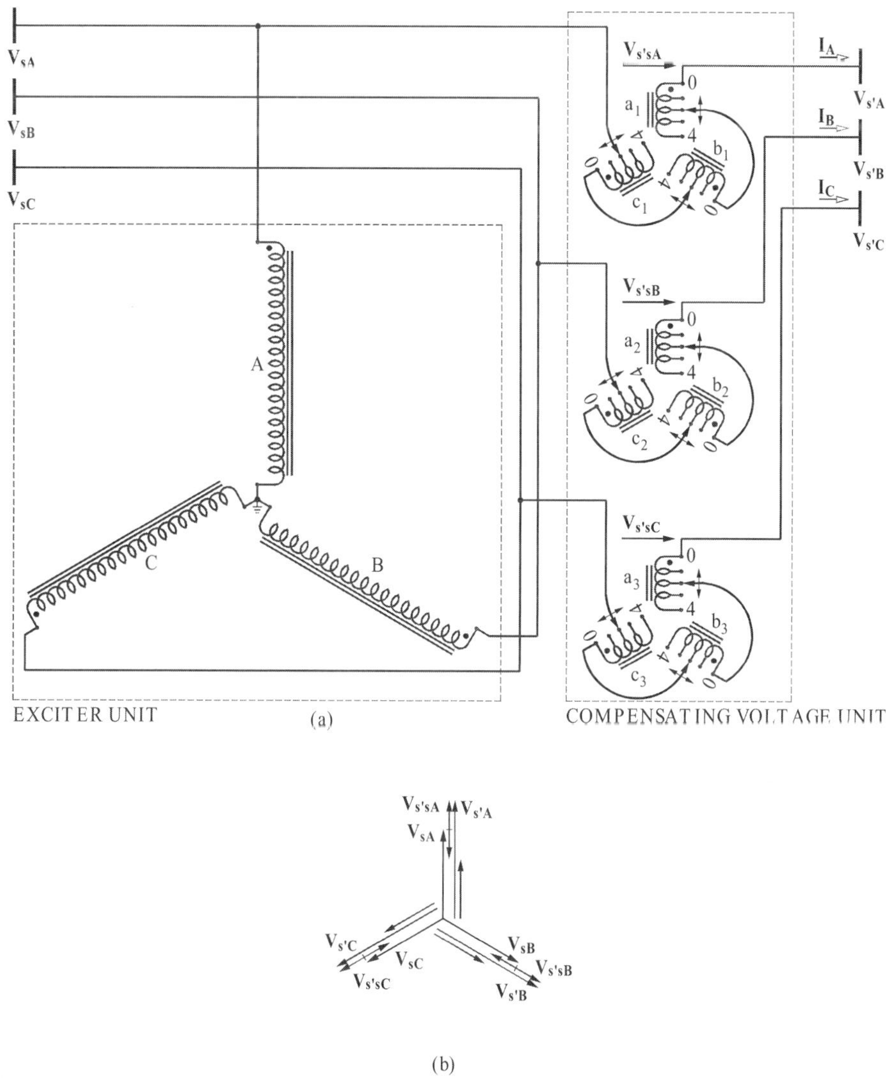

Figure 9-8. (a) ST for voltage regulation. (b) Phasor diagram.

9.2.2 ST as an Independent PFC

The circuit configuration, shown in Figure 9-8a, can also be used to form a series-connected compensating voltage that is at any phase angle with respect to the transmission line voltage, as shown in Figure 9-9a. By choosing the number of turns from each of the three windings and, therefore, the magnitudes of the components of the three 120° phase-shifted induced voltages, the compensating voltage ($\mathbf{V}_{s's}$) in any phase is derived from the phasor sum of the voltages induced in a three-phase winding set (a_1, b_1, and c_1 for compensation in the A phase; a_2, b_2, and c_2 for compensation in the B phase; and a_3, b_3, and c_3 for compensation in the C phase). The compensating voltage is also at any phase angle with respect to the prevailing line current. The active or direct component of the compensating voltage provides the series resistance emulation, which can be a positive resistor that absorbs active power from the line or a negative resistor

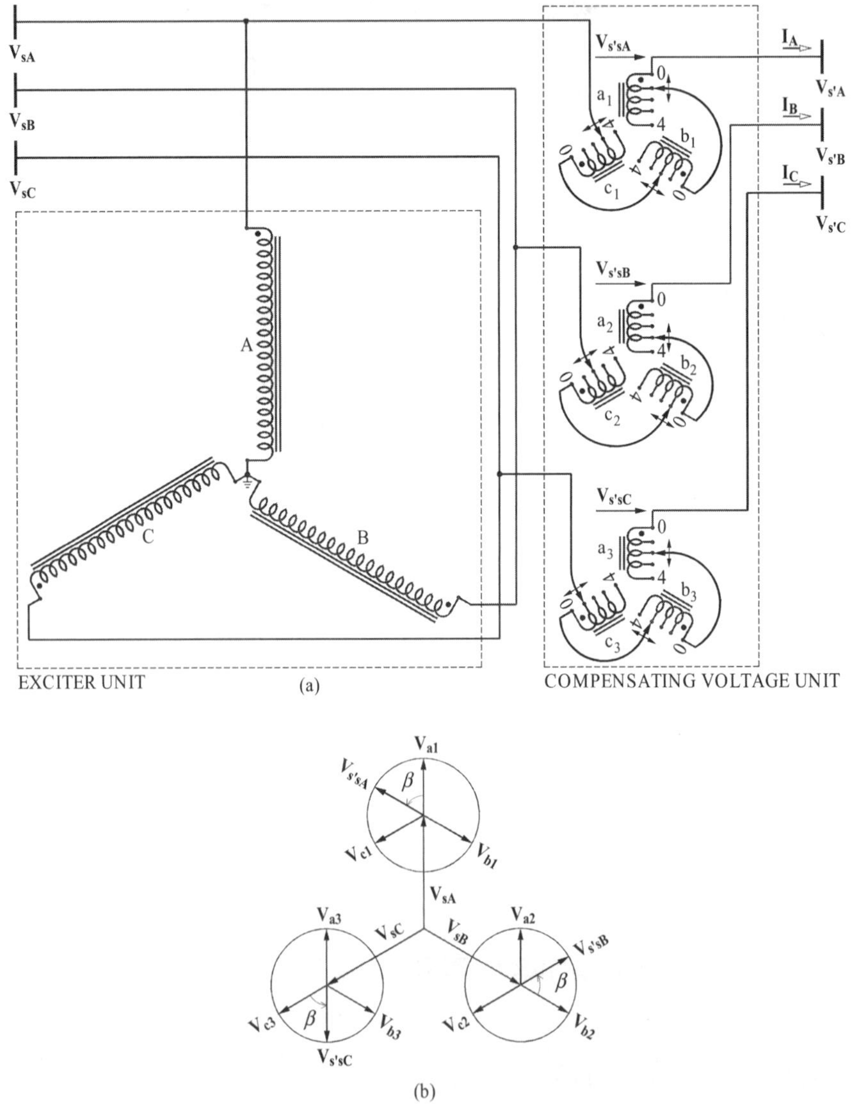

Figure 9-9. (a) ST for voltage compensation in the entire control range of 0° through 360°. (b) Phasor diagram.

that delivers active power to the line. The reactive or quadrature component of the compensating voltage provides the series reactance emulation, which can be a capacitor that delivers reactive power to the line and, in the process, increases the power flow of the line or an inductor that absorbs reactive power from the line and decreases the power flow of the line. Therefore, the desired compensating voltage is actually an impedance emulator. The effect of impedance emulation is such that the active and reactive power flows in a transmission line can be regulated independently. The power circuit is identical for both the voltage regulator and the impedance regulator. Both functions of voltage regulation and independent control of active and reactive power flows can be implemented in just one unit by proper programming of the tap control unit, as shown in Figure 9-11. Figure 9-9b shows the related phasor diagram.

It should be noted that each of a_1, b_2, and c_3 is tapped at the same number of turns; each of b_1, c_2, and a_3 is tapped at the same number of turns; and each of c_1, a_2, and b_3 is tapped at the same number of turns. However, the number of turns in the a_1–b_2–c_3 set, b_1–c_2–a_3 set, and c_1–a_2–b_3 set can be different from each other with one exception—when the ST is used as a voltage regulator to decrease the modified sending-end voltage, there are same number of turns in two windings that are connected to each phase in the case when the relative phase angle (β) is 180°.

Since the LTCs that are used in the ST change their positions in steps, the compensating points in the ST are discrete in the allowable control range. If the allowable magnitude ($V_{s's}$) of the compensating voltage ($\mathbf{V}_{s's}$) is assumed to be 0.2 pu and the change of voltage in each step of a LTC is 0.05 pu, there are $N = 4$ steps associated with each secondary winding. The total number of possible compensating points (cp) in that case is defined by

$$cp = 3N(N+1) \tag{9-1}$$

and it is 60. A comprehensive study of the power system network integrated with the PFC can only show how many steps is the right number of steps. Figure 9-10 shows that the theoretical circular control area is actually a hexagon. Shorter steps in the LTCs make the compensating points closer to each other and vice versa. Figure 9-10a shows the theoretically possible compensating points of the modified sending-end voltage with the use of the ST and Figure 9-10b shows the corresponding active and reactive power flows at the receiving end within the control range of 0° and 360°. Figure 9-10, parts c and d, show the practical control range of the ST. The filled dots, within the allowable voltage range in Figure 9-10c, show a total of 36 compensating points of the ST lie within the 10% range of the nominal voltage. Figure 9-10d shows that the maximum active power flow enhancement at the receiving end, within the allowable voltage range, is actually 0.35 pu instead of the theoretically possible 0.40 pu, as shown in Figure 9-10b.

9.2.3 Control of ST

The series-connected compensating voltage ($\mathbf{V}_{s's}$) in the transmission line emulates a positive or negative resistance (R_d) and a capacitive or inductive reactance (X_q) in se-

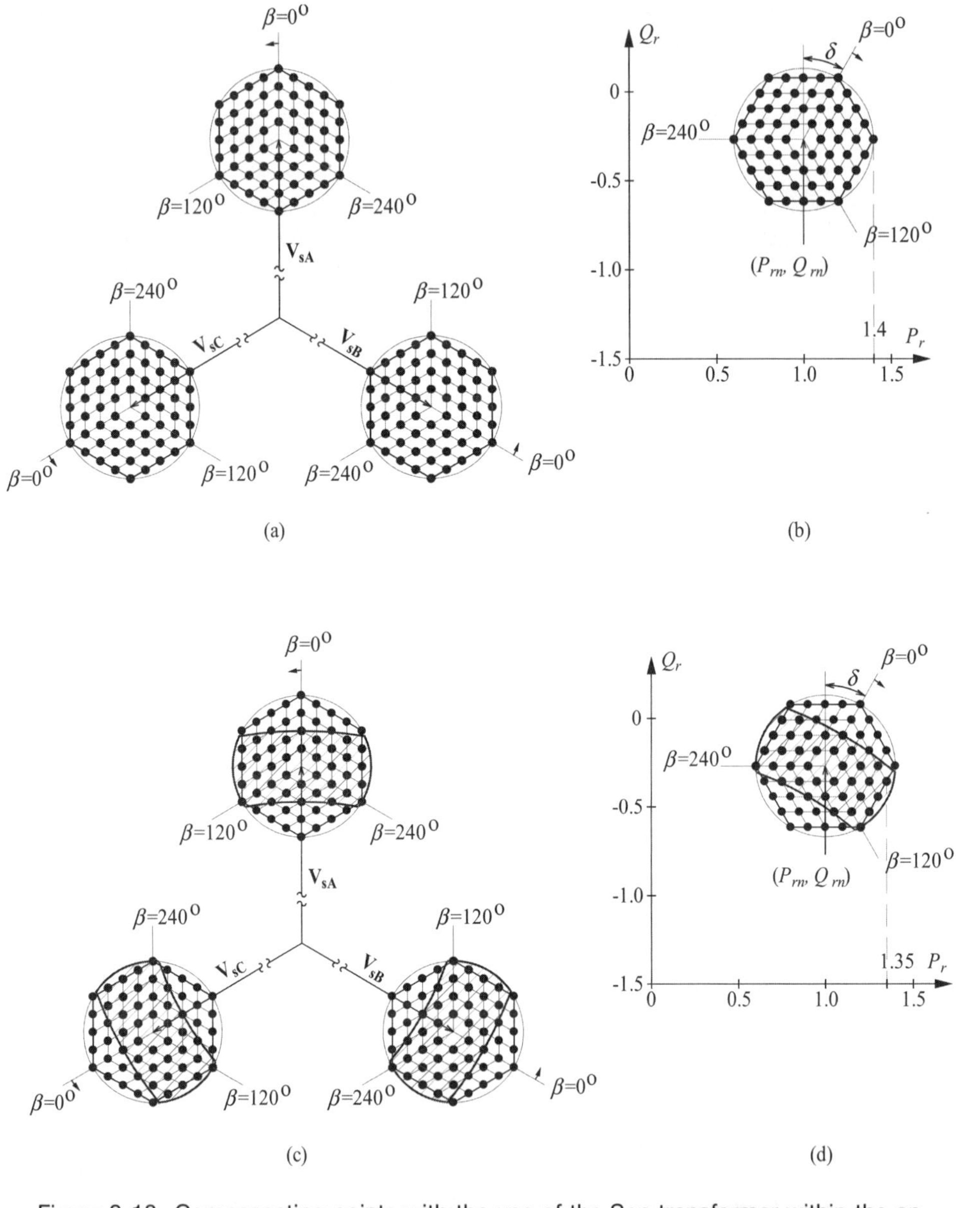

Figure 9-10. Compensating points with the use of the Sen transformer within the entire control range of 0° and 360°. Theoretically possible (a) modified sending-end voltage and (b) active and reactive power flows at the receiving end. Practically possible (c) modified sending-end voltage and (d) active and reactive power flows at the receiving end.

ries with the line without any discrete circuit components, but just with a transformer and LTCs. The control of the ST can be in several categories as described below. In each category, first the compensating voltage and its relative phase angle (β) with respect to the exciter voltage ($\mathbf{V_s}$) are calculated. Depending on the operating range of β ($0° < \beta < 120°$, $120° < \beta < 240°$, or $240° < \beta < 360°$), the two neighboring active compensating windings in each phase are selected. At $\beta = 0°$, 120°, and 240°, only one of the three secondary windings in each phase is active. Then, from the measured magnitude of the exciter voltage (V_s), the tap control unit determines the number of turns necessary on each winding of the compensating voltage unit. Based on this calculation, the appropriate taps are switched on to connect the required number of turns in series with the line. A special control algorithm must be implemented to avoid the jittering effect between the neighboring taps.

9.2.3.1 Impedance Emulation. Figure 9-11 shows the general control block diagram of the series impedance emulation using the ST. An instantaneous three-phase set of line voltages (v_s) is used to calculate the PLL angle (θ), which is phase locked to the phase a of the line voltage (v_{sa}). From an instantaneous three-phase set of measured line currents (i), the magnitude (I) and its relative phase angle (θ_{ir}) with respect to the PLL angle (θ) are calculated. From the compensating resistance demand (R_d^*) and the compensating reactance demand (X_q^*), the magnitude (Z_{dq}^*) and phase angle (θ_z) of the compensating impedance demand are calculated. The magnitude (I) of the line current multiplied by the magnitude of the compensating impedance (Z_{dq}^*) demand is the magnitude of the compensating voltage ($V_{s's}^*$) demand. The relative phase angle (β) of the compensating voltage demand is $\theta_{ir} + \theta_z + \pi$. Once the desired magnitude of the com-

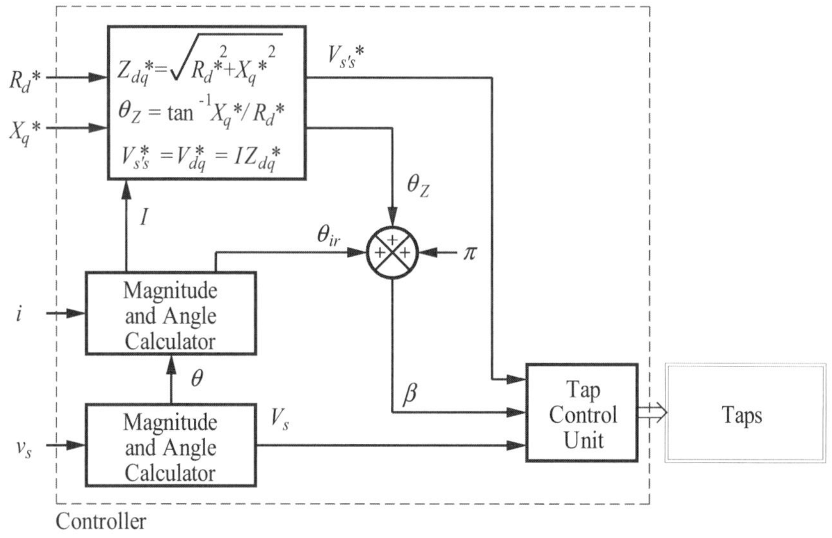

Figure 9-11. Series impedance emulation control block diagram of the ST.

pensating voltage ($V^*_{s's}$) demand and its relative phase angle (β) are defined, the tap control unit figures out the contribution from each winding of a three-phase set (a_1, b_1, and c_1 for compensation in the A phase; a_2, b_2, and c_2 for compensation in the B phase; and a_3, b_3, and c_3 for compensation in the C phase) to constitute ($V^*_{s's}$).

9.2.3.2 Resistance Emulation. As a special case, the control block diagram for the series resistance emulation (positive or negative) is shown in Figure 9-12. The calculated magnitude (I) of the line current multiplied by the compensating resistance (R^*_d) demand is the compensating voltage magnitude (V^*_d) demand. The compensating resistance (R^*_d) is defined to be positive when it is a positive resistor and negative when it is a negative resistor. The relative phase angle (β) of this compensating voltage demand is either $\theta_{ir} + \pi$ if the demanding compensating resistance is positive or $\theta_{ir} + 2\pi$ if the demanding compensating resistance is negative. Once the desired compensating voltage ($V^*_{s's} = |V^*_d|$) demand and its relative phase angle (β) are defined, the tap control unit figures out the contribution from each winding of a three-phase set (a_1, b_1 and c_1 for compensation in the A phase; a_2, b_2 and c_2 for compensation in the B phase; and a_3, b_3 and c_3 for compensation in the C phase) to constitute ($V^*_{s's}$).

9.2.3.3 Reactance Emulation. A simple modification to the control block diagram, shown in Figure 9-12, allows the emulation of a series reactance. The desired block diagram is shown in Figure 9-13. The calculated magnitude (I) of the line current multiplied by the compensating reactance demand (X^*_q) is the compensating voltage magnitude (V^*_q) demand. The compensating reactance (X^*_q) is defined to be positive when it is a capacitor and negative when it is an inductor. The relative phase angle

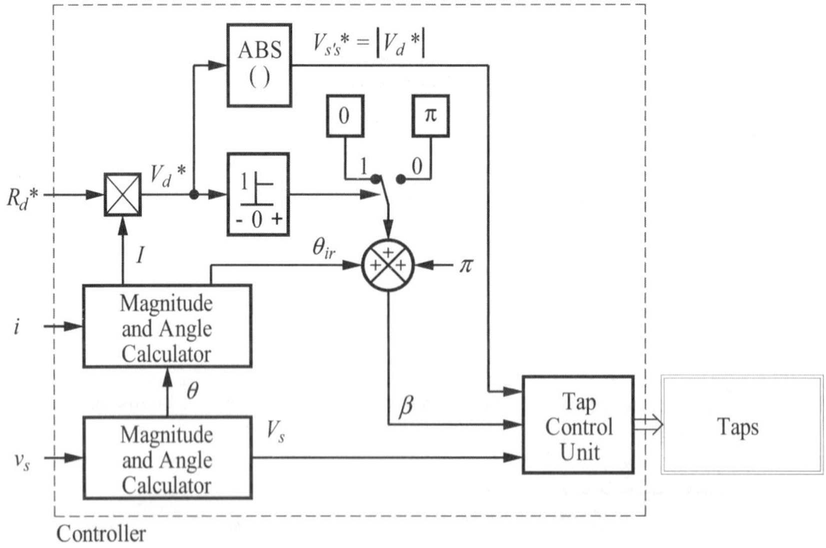

Figure 9-12. Series resistance emulation control block diagram of the ST.

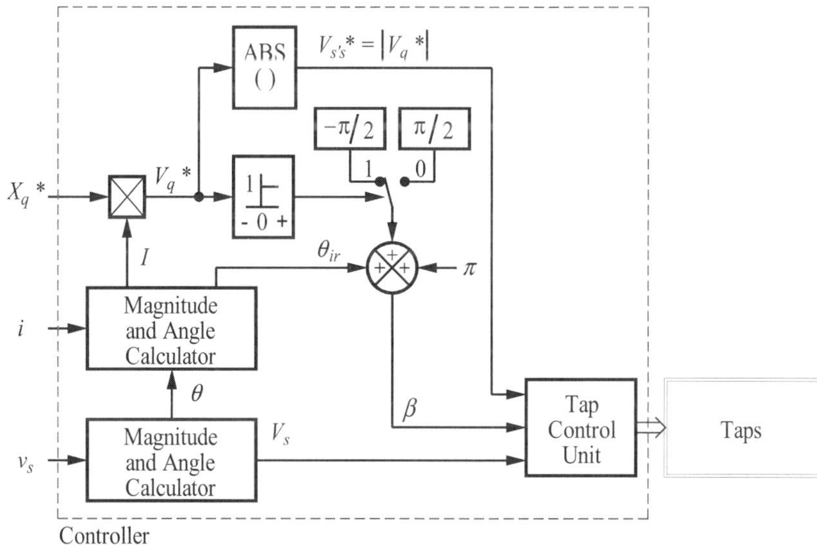

Figure 9-13. Series reactance emulation control block diagram of the ST.

(β) of this compensating voltage demand is either $\theta_{ir} - \pi/2 + \pi$ if the demanding compensating reactance is capacitive or $\theta_{ir} + \pi/2 + \pi$ if the demanding compensating reactance is inductive. Once the desired compensating voltage ($V^*_{s's} = |V^*_q|$) demand and its relative phase angle (β) are defined, the tap control unit figures out the contribution from each winding of a three-phase set (a_1, b_1 and c_1 for compensation in the A phase; a_2, b_2 and c_2 for compensation in the B phase; and a_3, b_3 and c_3 for compensation in the C phase) to constitute ($V^*_{s's}$).

9.2.3.4 Closed Loop Power Flow Control. The closed loop automatic power flow control block diagram is shown in Figure 9-14. The automatic power flow control algorithm works in such a way that the desired active and reactive power flows ($P^*_{s'}$ and $Q^*_{s'}$) are compared with the measured active and reactive power flows ($P_{s'}$ and $Q_{s'}$) and the errors are passed through an error amplifier, which produces, with respect to the exciter voltage ($\mathbf{V_s}$), the required direct and quadrature components (e_d and e_q) of the series-connected compensating voltage. Next, the magnitude ($V_{s's}$) of the compensating voltage, $v_{s's}$ (i.e., $-e_{dq}$), and its relative phase angle (β) with respect to the phase angle of the exciter voltage are calculated. Note that for a particular active and reactive power flows ($P^*_{s'}$ and $Q^*_{s'}$) demand, if the magnitude ($V_{s's}$) of the compensating voltage exceeds the rated voltage of the compensating voltage unit, then $V_{s's}$ is limited to its design rating and the active and reactive power flows are regulated to an intermediate value.

9.2.3.5 Open Loop Power Flow Control. The controllable range of the active and reactive power flows can easily be determined with an open loop control by

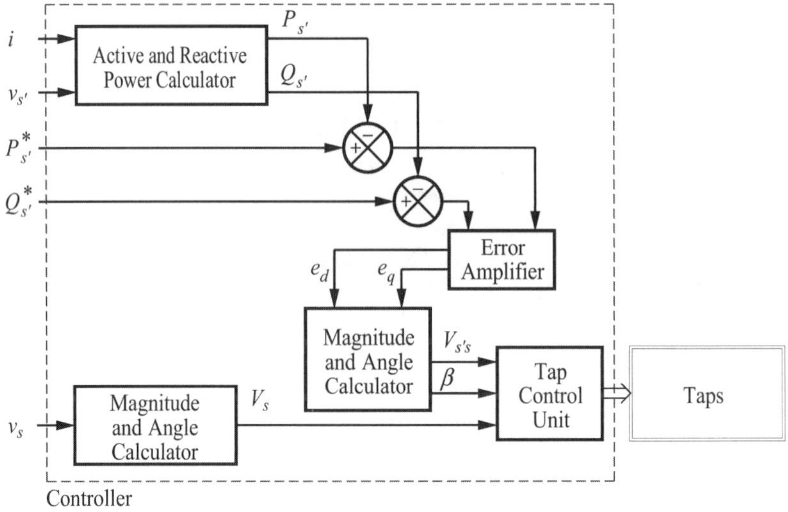

Figure 9-14. Closed loop automatic power flow control block diagram of the ST.

specifying to the tap control unit the rated compensating voltage ($V^*_{s's}$) and varying the relative phase angle (β) within its entire 360° range, as shown in Figure 9-15.

Once the compensating voltage ($\mathbf{V_{s's}}$) and its relative phase angle (β) are determined, the tip of phasor ($\mathbf{V_{s's}}$) falls within four possible operating points, as shown in Figure 9-16.

The selected operating point is based on producing the least error given by the following equation:

$$\varepsilon = \|r_k\|_2 = \sqrt{\left(V_k^x - V_{s's}^x\right)^2 + \left(V_k^y - V_{s's}^y\right)^2} \tag{9-2}$$

where ε is the least error and r_k is the error at a possible operating point for $k = 1, 2, 3,$ and 4. V_k^x and $V_{s's}^x$ are the projections of voltage of a possible operating point and the compensating voltage on the a phase (x-axis). V_k^y and $V_{s's}^y$ are the projections of volt-

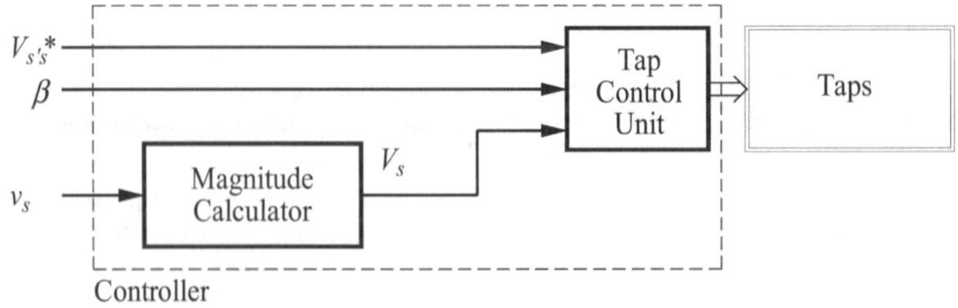

Figure 9-15. Open loop compensating voltage unit control block diagram of the ST.

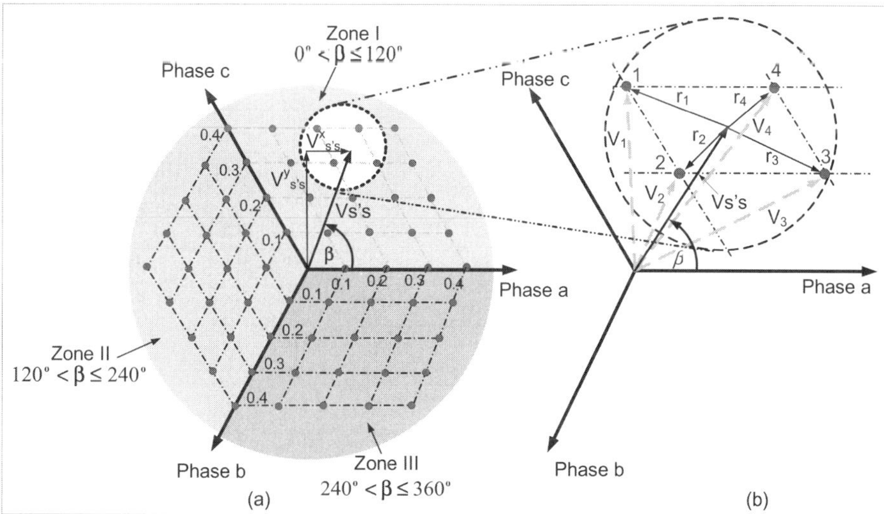

Figure 9-16. Selection of tap positions (Faruque and Dinavahi, 2007).

age of a possible operating point and the compensating voltage on the perpendicular to *a* phase (*y* axis). A hysteresis loop must be implemented to eliminate jittering effect among the four neighboring taps.

9.2.4 Simulation Results

A basic simulation of the ST was carried out and the results are shown in this section. Consider the power system network of Chapter 3. The PFC was replaced with the ST, as shown in Figure 9-9. The ST consists of three primary windings and nine secondary windings (three for series connection in each of the three phases). Each primary and secondary winding are considered to have one pu and 0.2 pu voltage ratings, respectively. Considering a LTC contact located at every 5% mark, each secondary winding consists of four LTC taps. The leakage reactance of the ST was set to zero to eliminate its secondary effect. The natural active and reactive power flows (P_{rn} and Q_{rn}) at the receiving end are 209.12 MW (1.31 pu) and –68.54 Mvar (–0.43 pu), respectively.

The control was implemented in an open loop voltage injection mode traveling from one operating point to the next in one particular hexagon in the entire control range of 0° to 360°. The various contributions from each of the three windings in phase A for operation in each hexagon (related to operation of 1, 2, 3, and 4 taps in each winding) and the corresponding magnitude ($V_{s's}$) and phase angle (β) of the compensating voltage ($\mathbf{V}_{s's}$) are listed in Table 9-1.

The actual magnitude of the compensating voltage ($V_{s's}$) in pu (considering the primary voltage to be one pu) and its relative phase angle (β) are shown in Figure 9-17. Figure 9-18 shows the active and reactive power flows (P_r and Q_r) at the receiving end

Table 9-1. Various contributions from each of the three windings in phase A for operation in each hexagon (related to operation of 1, 2, 3, and 4 taps on each winding) and the corresponding magnitude ($V_{s's}$) and relative phase angle (β) of the compensating voltage ($\mathbf{V_{s's}}$)

a_1	b_1	c_1	$V_{s's}$ (pu)	β (degree)
0.05	0.00	0.00	0.05	0.00
0.05	0.00	0.05	0.05	60.00
0.00	0.00	0.05	0.05	120.00
0.00	0.05	0.05	0.05	180.00
0.00	0.05	0.00	0.05	240.00
0.05	0.05	0.00	0.05	300.00
0.05	0.00	0.00	0.05	360.00

a_1	b_1	c_1	$V_{s's}$ (pu)	β (degree)
0.20	0.00	0.00	0.20	0.00
0.20	0.00	0.05	0.18	13.90
0.20	0.00	0.10	0.17	30.00
0.20	0.00	0.15	0.18	46.10
0.20	0.00	0.20	0.20	60.00
0.15	0.00	0.20	0.18	73.90
0.10	0.00	0.20	0.17	90.00
0.05	0.00	0.20	0.18	106.10
0.00	0.00	0.20	0.20	120.00
0.00	0.05	0.20	0.18	133.90
0.00	0.10	0.20	0.17	150.00
0.00	0.15	0.20	0.18	166.10
0.00	0.20	0.20	0.20	180.00
0.00	0.20	0.15	0.18	193.90
0.00	0.20	0.10	0.17	210.00
0.00	0.20	0.05	0.18	226.10
0.00	0.20	0.00	0.20	240.00
0.05	0.20	0.00	0.18	253.90
0.10	0.20	0.00	0.17	270.00
0.15	0.20	0.00	0.18	286.10
0.20	0.20	0.00	0.20	300.00
0.20	0.15	0.00	0.18	313.90
0.20	0.10	0.00	0.17	330.00
0.20	0.05	0.00	0.18	346.10
0.20	0.00	0.00	0.20	360.00

a_1	b_1	c_1	$V_{s's}$ (pu)	β (degree)
0.10	0.00	0.00	0.10	0.00
0.10	0.00	0.05	0.09	30.00
0.10	0.00	0.10	0.10	60.00
0.05	0.00	0.10	0.09	90.00
0.00	0.00	0.10	0.10	120.00
0.00	0.05	0.10	0.09	150.00
0.00	0.10	0.10	0.10	180.00
0.00	0.10	0.05	0.09	210.00
0.00	0.10	0.00	0.10	240.00
0.05	0.10	0.00	0.09	270.00
0.10	0.10	0.00	0.10	300.00
0.10	0.05	0.00	0.09	330.00
0.10	0.00	0.00	0.10	360.00

a_1	b_1	c_1	$V_{s's}$ (pu)	β (degree)
0.15	0.00	0.00	0.15	0.00
0.15	0.00	0.05	0.13	19.11
0.15	0.00	0.10	0.13	40.89
0.15	0.00	0.15	0.15	60.00
0.10	0.00	0.15	0.13	79.11
0.05	0.00	0.15	0.13	100.89
0.00	0.00	0.15	0.15	120.00
0.00	0.05	0.15	0.13	139.11
0.00	0.10	0.15	0.13	160.89
0.00	0.15	0.15	0.15	180.00
0.00	0.15	0.10	0.13	199.11
0.00	0.15	0.05	0.13	220.89
0.00	0.15	0.00	0.15	240.00
0.05	0.15	0.00	0.13	259.11
0.10	0.15	0.00	0.13	280.89
0.15	0.15	0.00	0.15	300.00
0.15	0.10	0.00	0.13	319.11
0.15	0.05	0.00	0.13	340.89
0.15	0.00	0.00	0.15	360.00

during the entire stepped control range of the relative phase angle (β) from 0° and 360°. Figure 9-19 shows the active power (P_r) versus reactive power (Q_r) flow at the receiving end during the entire control range of the relative phase angle (β) from 0° and 360°. The *P–Q* plot is somewhat different from an ideal hexagon due to the fact that the primary voltage ($\mathbf{V_s}$) keeps moving from its original (uncompensated) location because of the change of load in the line during compensation and the amount of

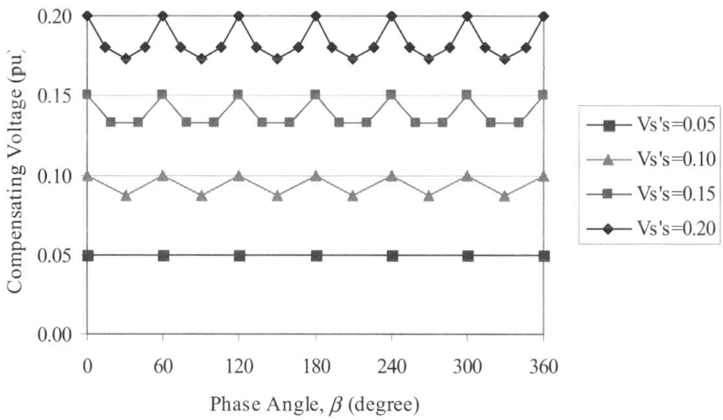

Figure 9-17. Magnitude of the series-connected compensating voltage ($V_{s's}$) in pu during the entire control range of its relative phase angle (β) from 0° to 360°.

Thèvenin impedance at the point of compensation and the shunt loading due to exciter unit of the ST. A more ideal hexagon-like characteristic results if the point of compensation is closer to a stiff bus.

9.2.5 Limited Angle Operation of ST

The ST requires the use of a three-phase transformer with three primary windings and nine secondary windings (or three single-phase transformers, each with one primary winding and three secondary windings). The compensating voltage ($\mathbf{V}_{s's}$) is of variable magnitude and at a variable phase angle. However, in many instances, the capability of connecting a compensating voltage in series with a line within its entire range of 360° is not needed. The active power flow can be increased to the maximum possible level within the first 120° of the possible 0° to 360° range of the relative phase angle (β). The active power flow can be decreased to the minimum possible level within the next 120° of the possible 0° to 360° range of β. In this case, the circuit configuration for the full 360° range of operation can be simplified to the "limited angle" configuration.

At $\beta = 0°$, 120°, and 360°, only one of the three secondary windings in each phase is active. This active winding in each phase carries the full line current at the rated winding voltage that is a fraction of the line voltage. Therefore, during the 360° range of operation, each of the nine secondary windings takes its turn to provide solely the full compensating voltage and carry the line current. Therefore, each of the nine secondary windings must be designed to carry the rated line current at rated winding voltage. This requires the magnetic rating of the 12 windings (three primary and nine secondary) of the ST to be 12/6 = 2 pu, considering a transformer with six windings (three primary and three secondary) to be of one pu rating.

The ST, with 360° range of voltage compensation capability, uses only three primary windings and at the most six secondary windings at any given operating point. The

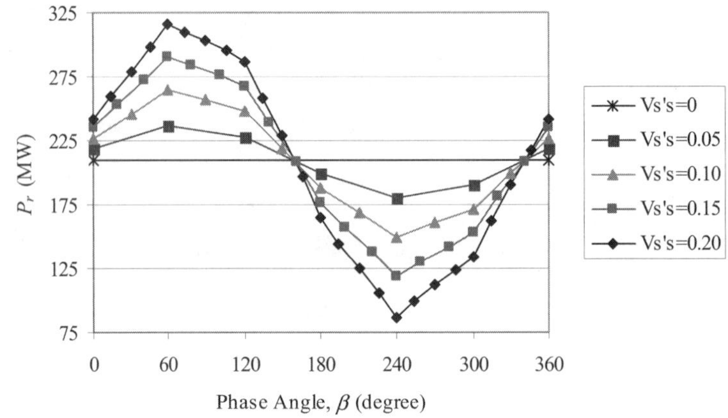

(a)

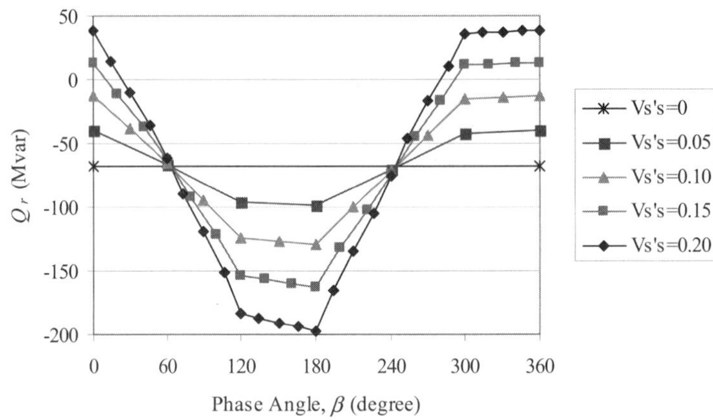

(b)

Figure 9-18. (a) Active power (P_r) and (b) reactive power (Q_r) flows at the receiving end during the entire control range of the relative phase angle (β) from 0° to 360°.

remaining three secondary windings stay inactive. Therefore, it is possible to achieve a 360° operating range just by using six secondary windings instead of nine secondary windings with hardware reconfiguration in every 120° range of operation. In this way, the number of windings required in the ST is reduced at the expense of additional disconnect switches. Considering the three primary windings, the total number of windings required is 3 + 6 = 9 instead of 3 + 9 = 12. Therefore, the practical magnetic rating of the ST is only 9/6 = 1.5 pu for 120° range of operation, instead of two pu for 360° range of operation. Note that by using six secondary windings, one of the six operating regions (0° to 120°, 120° to 240°, 240° to 360°, 300° to 60°, 60° to 180°, and 180° to 300°) can be selected.

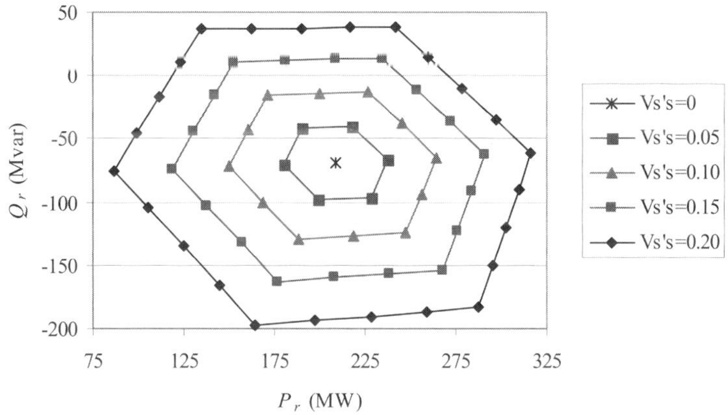

Figure 9-19. Active power (P_r) versus reactive power (Q_r) flows at the receiving end during the entire control range of the relative phase angle (β) from 0° to 360°.

Figures 9-20 through 9-25 show, through autotransformer action, the generation of a compensating voltage ($\mathbf{V}_{s's}$), which is of variable magnitude and operated within a limited angle. When this compensating voltage is connected in series with the line, the sending-end voltage is modified to ($\mathbf{V}_{s'}$).

Figure 9-20a shows that for the limited angle operation between 0° and 120°, the line voltage is applied to a shunt-connected, three-phase transformer's primary windings in the exciter unit. A total of six secondary windings (a_1 and a_2 on the core of the A phase, b_2 and b_3 on the core of the B phase, and c_3 and c_1 on the core of the C phase) constitute the limited angle compensating voltage unit. By choosing the number of turns from each of the two windings, the compensating voltage in any phase is derived from the phasor sum of the voltages induced in a two-phase winding set (a_1 and c_1 for connection in the A phase, a_2 and b_2 for connection in the B phase, and b_3 and c_3 for connection in the C phase). The relative phase angle (β) of the series-connected compensating voltage ($\mathbf{V}_{s's}$) with respect to the line voltage ($\mathbf{V}_s$) can vary between 0° and 120°. Figure 9-20b shows the related phasor diagram.

For $N = 4$ in each secondary winding, the total number of possible compensating points (cp) in the 120° control range of operation is given by

$$cp = N(N+2) \tag{9-3}$$

and it is 24. The filled dots, within the allowable voltage range between $\beta = 0°$ and $\beta = 120°$, in Figure 9-20c show a total of 16 compensating points in the ST. Figure 9-20d shows the corresponding P–Q diagram for the power flow at the receiving end.

Figure 9-20a can be further simplified to accommodate the local voltage regulation and independent control of the active and reactive power flows with minimum transformer rating, as shown in Figure 9-20e. The compensating windings that provide the in-phase voltages can be constructed to have either more or fewer turns than

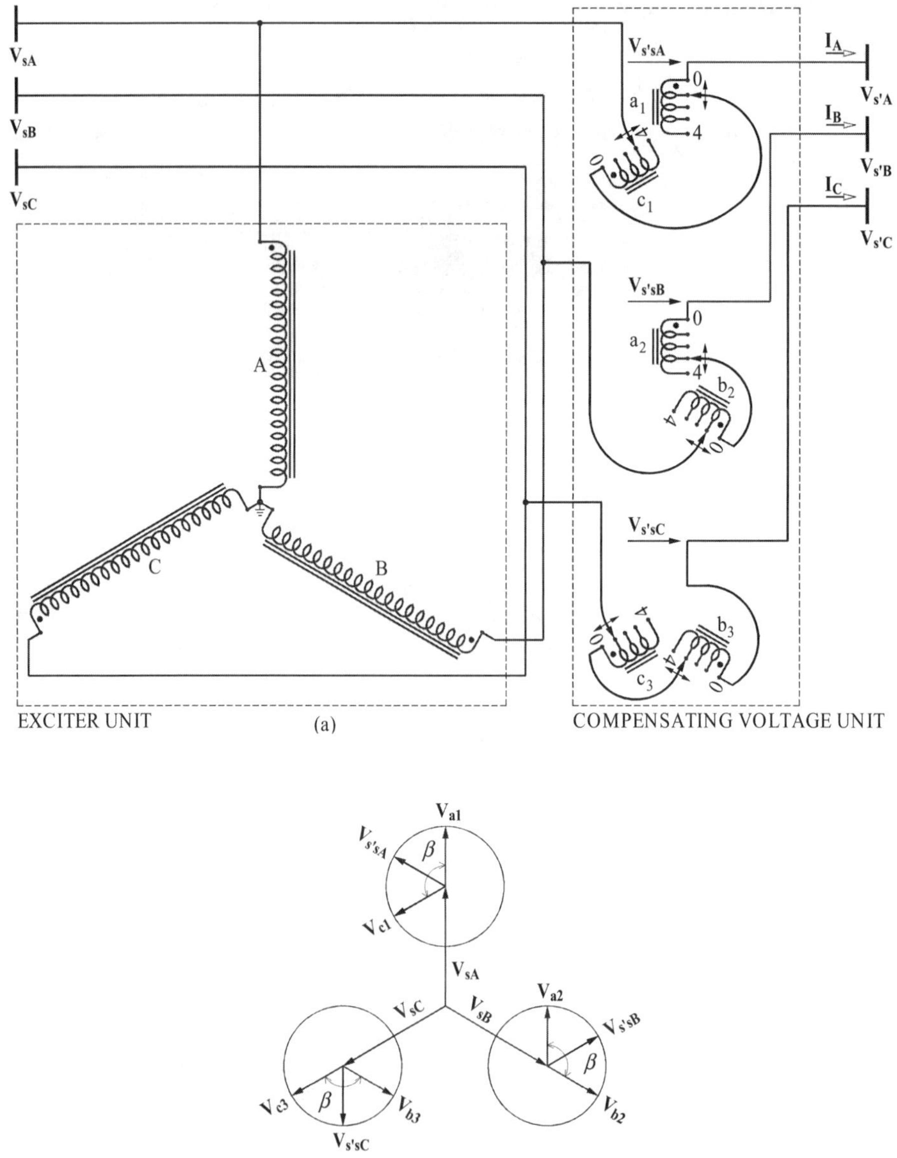

Figure 9-20(a, b). (a) ST for voltage compensation in the range of 0° through 120°. (b) Phasor diagram.

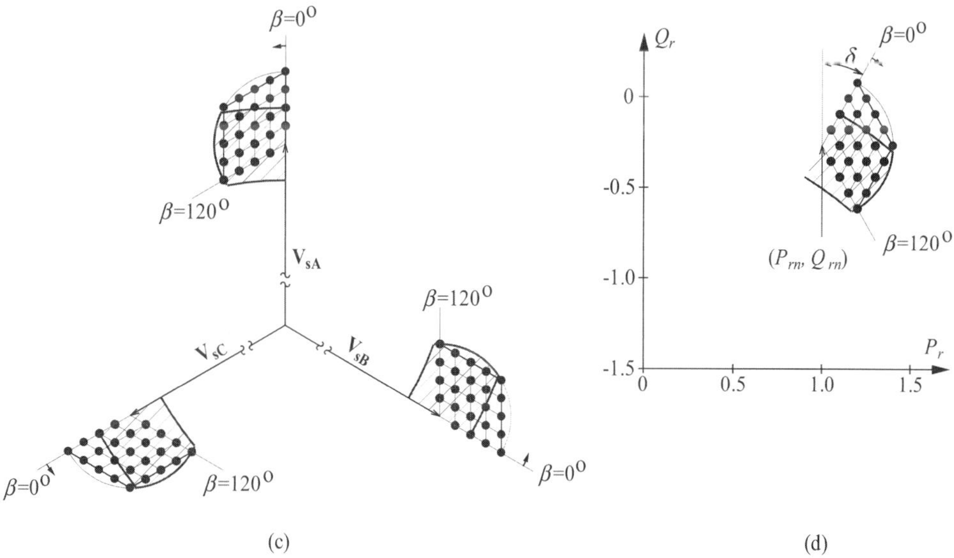

Figure 9-20(c, d). (c) Modified sending-end voltage points. (d) Corresponding receiving-end active and reactive power points for operation within the control range of 0° to 120°.

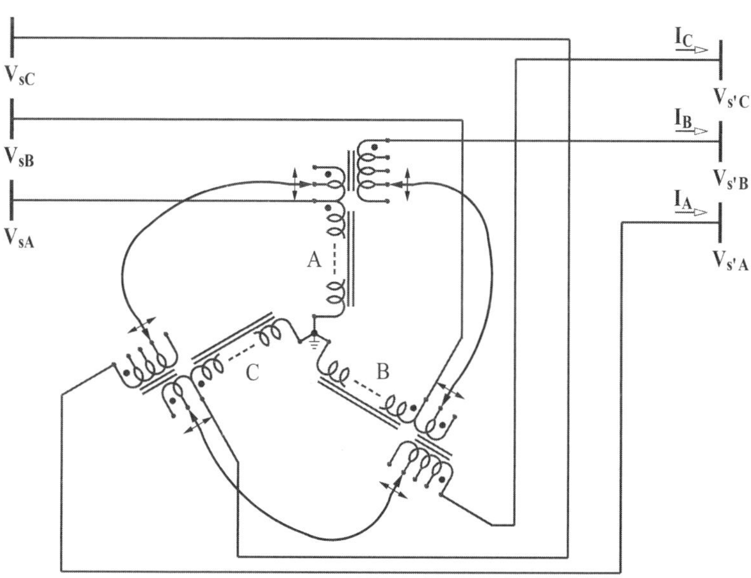

Figure 9-20(e). Most simplified configuration of the ST for operation in the range of 0° through 120°.

the rest of the compensating windings. This is to comply with the required local voltage limit. The number of isolated, compensating windings can be reduced by the following action. The compensating winding that provides the in-phase voltage and the corresponding exciting winding can be replaced with a partially tapped autotransformer winding as shown in the figure. As a result, this configuration requires only six windings: three for the exciter unit and partially for the compensating voltage unit and the remaining three for the dedicated compensating voltage unit.

Similarly, in an application in which there is a need to vary the relative phase angle (β) of the series-connected compensating voltage ($\mathbf{V}_{s's}$) between 120° and 240°, a compensating voltage unit with only six windings as shown in Figure 9-21a, is needed. This is achieved by constructing the compensating voltage from a combination of two voltages, each of which is induced in a separate winding of a two-phase set (b_1 and c_1 for connection in the A phase, a_2 and c_2 for connection in the B phase, and a_3 and b_3 for connection in the C phase). Figure 9-21b shows the related phasor diagram.

For $N = 4$ in each secondary winding, the total number of possible compensating points (cp) in the 120° control range of operation is 24 from equation 9-3. The filled dots, within the allowable voltage range between $\beta = 120°$ and $\beta = 240°$, in Figure 9-21c show a total of 14 compensating points in the ST. Figure 9-21d shows the corresponding P–Q diagram for the power flow at the receiving end.

Lastly, in an application in which there is a need to vary the relative phase angle (β) of the series-connected compensating voltage ($\mathbf{V}_{s's}$) between 240° and 360°, a compensating voltage unit with only six windings, as shown in Figure 9-22a, is needed. This is achieved by constructing the compensating voltage from a combination of two voltages, each of which is induced in a separate winding of a two-phase set (a_1 and b_1 for connection in the A phase, b_2 and c_2 for connection in the B phase, and a_3 and c_3 for connection in the C phase). Figure 9-22b shows the related phasor diagram.

For $N = 4$ in each secondary winding, the total number of possible compensating points (cp) in the 120° control range of operation is 24 from equation 9-3. The filled dots, within the allowable voltage range between $\beta = 240°$ and $\beta = 360°$, in Figure 9-22c show a total of 16 compensating points in the ST. Figure 9-22d shows the corresponding P–Q diagram for the power flow at the receiving end.

Extending the concept just presented, if the polarities of the windings in the compensating voltage unit in Figure 9-20a are reversed, the relative phase angle (β) of the series-connected compensating voltage ($\mathbf{V}_{s's}$) can vary between 180° and 300°. The schematic for this configuration is shown in Figure 9-23a. In the same way, if the polarities of the windings in the compensating voltage units, presented in Figures 9-21a and 9-22a, are reversed then the relative phase angle (β) of the compensating voltage ($\mathbf{V}_{s's}$) can vary between 300° and 60° and 60° and 180°, respectively. The schematics for these configurations are shown in Figures 9-24a and 9-25a. The related phasor diagrams are shown in Figures 9-23b, 9-24b, and 9-25b.

In all of these cases, the compensating voltage can be at any phase angle with the prevailing line current, which emulates, in series with the line, a capacitor or an induc-

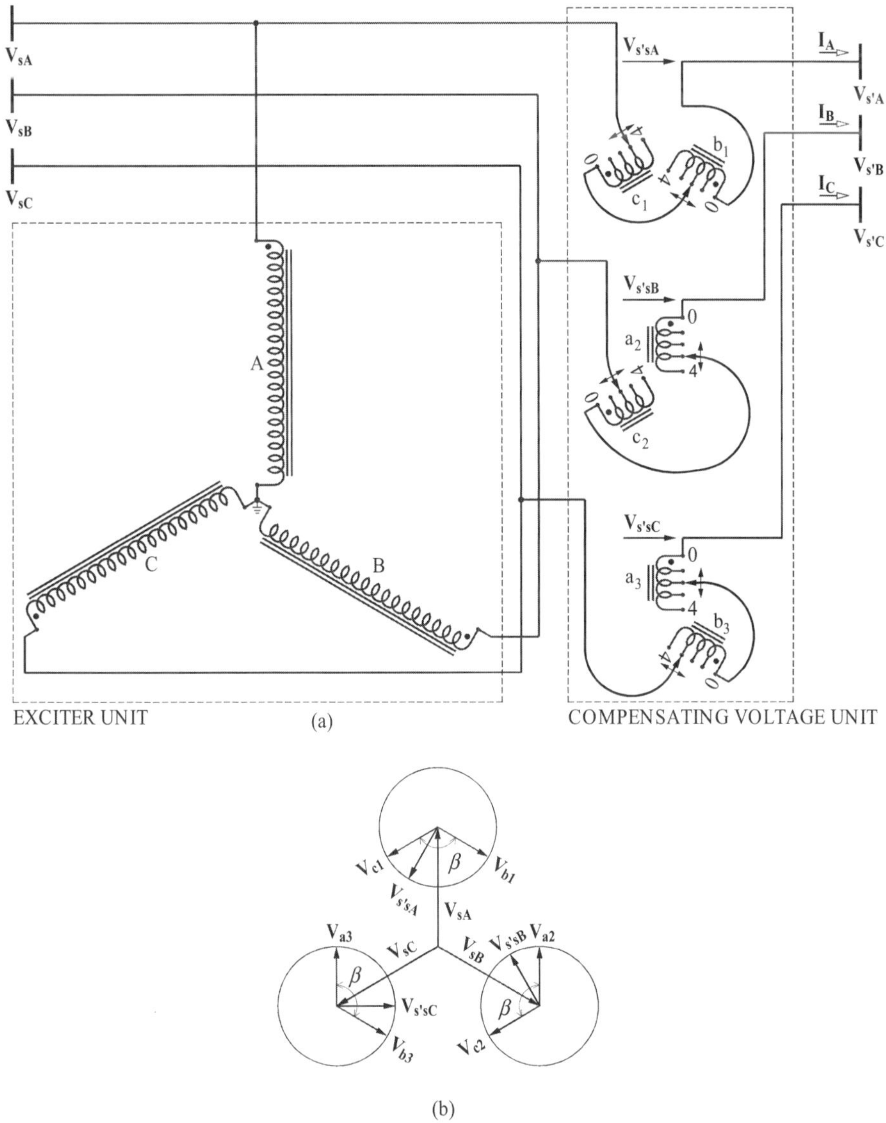

Figure 9-21(a, b). (a) ST for compensating voltage in the range of 120° through 240°. (b) Phasor diagram.

tor and a positive or a negative resistor. The effect is such that the active and reactive power flows in a transmission line can be regulated independently. In addition, the function of a voltage regulator can also be implemented, provided that the final variable magnitude and variable phase angle of the compensating voltage are within the controllable operating range.

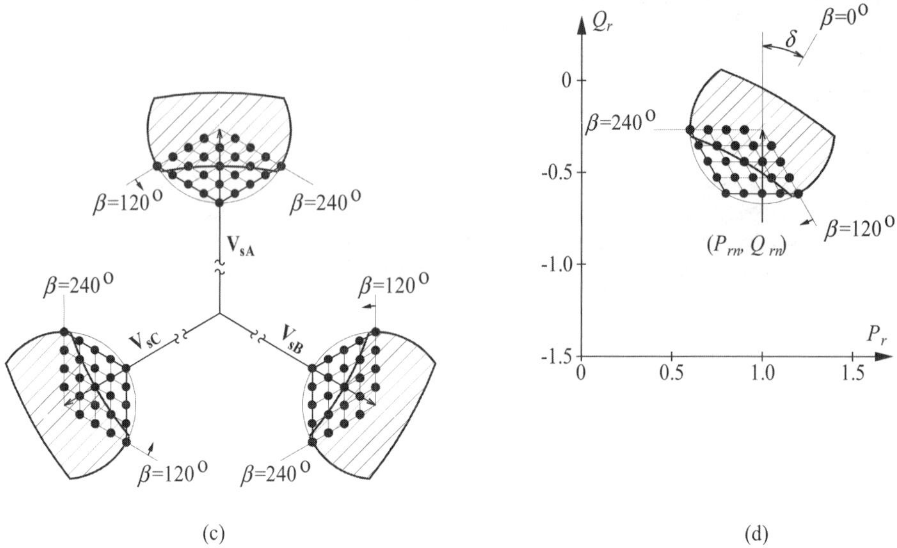

Figure 9-21(c, d). (c) Modified sending-end voltage points. (d) Corresponding receiving-end active and reactive power points for operation within the control range of 120° to 240°.

9.2.6 ST Using LTCs with Lower Current Rating

Figure 9-26 shows the generation of a line frequency compensating voltage ($\mathbf{V}_{s's}$), through autotransformer action, that is of variable magnitude and operated at variable phase angle. When this compensating voltage is connected in series with the line, the modified sending-end voltage becomes $\mathbf{V}_{s'}$. The result is that the active and reactive power flows in a transmission line are regulated independently.

Figure 9-26a shows that the three single-phase voltages ($\mathbf{V}_{sA}$, $\mathbf{V}_{sB}$, and $\mathbf{V}_{sC}$) of a three-phase transmission line voltage ($\mathbf{V}_s$) are applied across three three-winding sets. The voltage ($\mathbf{V}_{sA}$) is applied across the three windings a_1, a_2, and a_3. The voltage ($\mathbf{V}_{sB}$) is applied across the three windings b_1, b_2, and b_3. The voltage ($\mathbf{V}_{sC}$) is applied across the three windings c_1, c_2, and c_3. Each of a_1, b_1, and c_1 is tapped at the same number of turns; each of a_2, b_2, and c_2 is tapped at the same number of turns; each of a_3, b_3, and c_3 is tapped at the same number of turns. However, the number of turns in the a_1–b_1–c_1 set, a_2–b_2–c_2 set, and a_3–b_3–c_3 set can be different from each other. The composite voltage from the three windings (a_1, c_2, and b_3) on the primary side (exciter unit) is reflected in a single winding on the secondary side (compensating voltage unit) of the A phase. The composite voltage from the three windings (b_1, a_2, and c_3) on the primary side (exciter unit) is reflected in a single winding on the secondary side (compensating voltage unit) of the B phase. The composite voltage from the three windings (c_1, b_2, and a_3) on the primary side (exciter unit) is reflected in a single winding on the secondary side (compensating voltage unit) of the C phase. By choosing the number of turns from each of the three

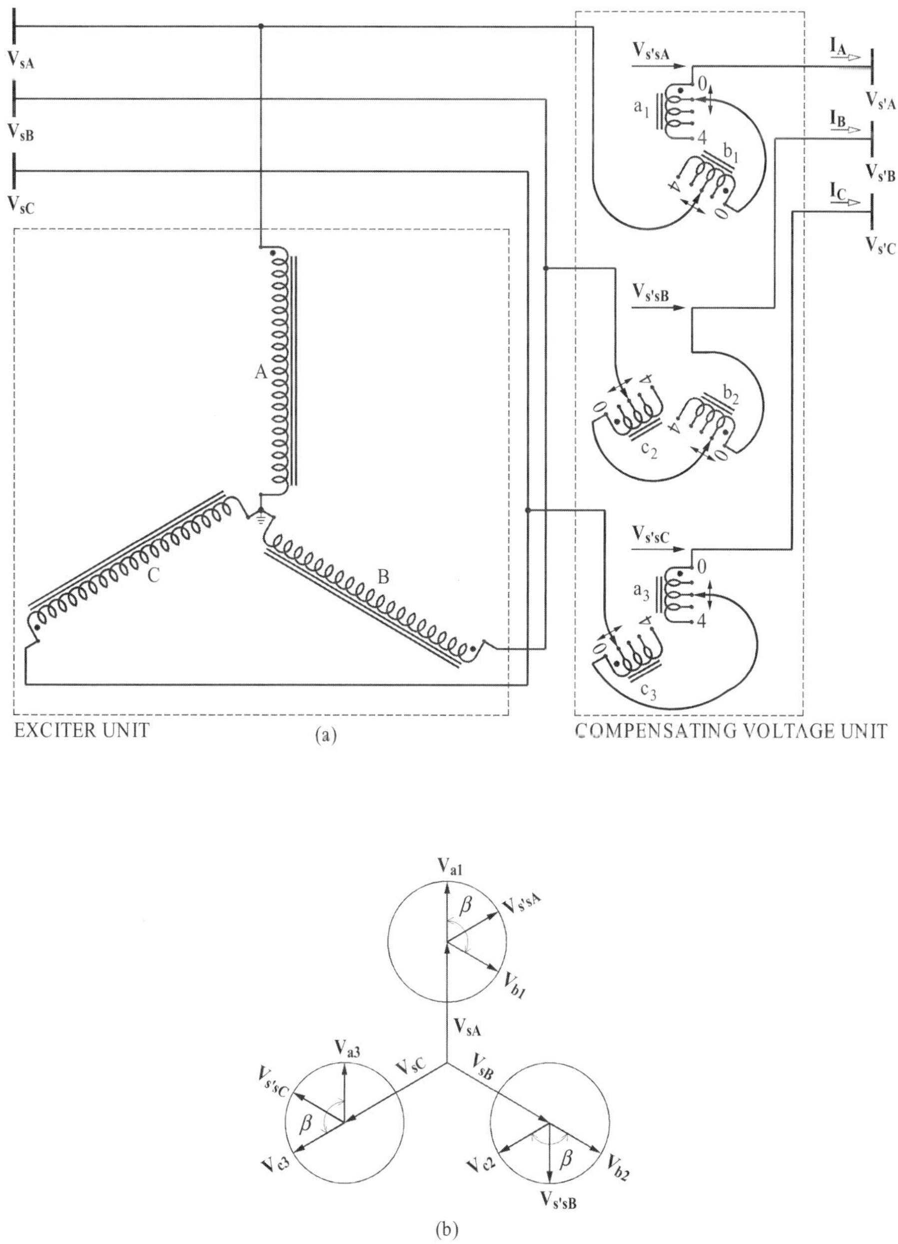

Figure 9-22(a, b). (a) ST for compensating voltage in the range of 240° through 360°. (b) Phasor diagram.

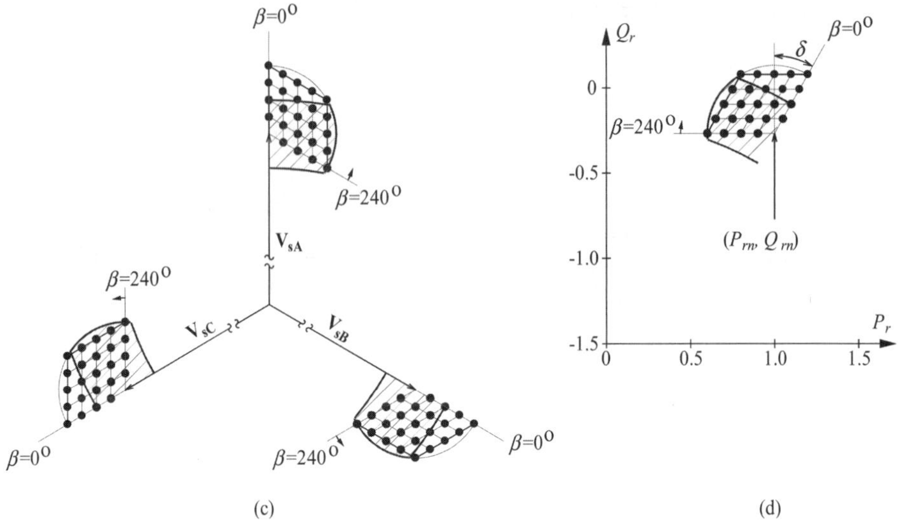

Figure 9-22(c, d). (c) Modified sending-end voltage points. (d) Corresponding receiving-end active and reactive power points for operation within the control range of 240° to 360°.

windings (a_1, c_2, and b_3), for example, the compensating voltage ($\mathbf{V}_{s's}$) can be of variable magnitude and at variable phase angle (β) as shown in Figure 9-26b.

Figure 9-26c shows the phasor diagram when the exciter voltage is applied across three windings (a_1, a_2, and a_3 in the A phase; b_1, b_2, and b_3 in the B phase; and c_1, c_2, and c_3 in the C phase), each with the same number of turns. Each of the nine primary windings is tapped at the same number of turns. The compensating voltage, which is the phasor sum of three equal voltages with 120° phase difference from one another, is zero. If equal voltages that are induced from three windings (a_1, c_2, and b_3 on the core of the A phase) are combined, the composite voltage is always zero. This property makes it possible to vary the compensating voltage from zero to its rated value in a configuration when the LTCs are on the exciter unit and the compensating voltage unit is in its simplest form with only one series insertion winding in each phase. The LTCs operate at a low current and ride through a higher fault current.

Figure 9-26d shows the phasor diagram when the exciter voltage is applied only across one winding in each phase. For example, $\mathbf{V}_{sA}$ is applied only across a_1, $\mathbf{V}_{sB}$ is applied only across b_1, and $\mathbf{V}_{sC}$ is applied only across c_1. In this case, the series-connected compensating voltage in each phase is in phase with the corresponding phase voltage of the transmission line and its magnitude is dependent on the number of turns in the compensating winding and the exciter winding.

Figure 9-26e shows that the three-phase voltages ($\mathbf{V}_{sA}$, $\mathbf{V}_{sB}$, and $\mathbf{V}_{sC}$) are applied across a_2, b_2 and c_2, respectively. In this case, the series-connected compensating voltage in each phase leads the corresponding phase voltage of the transmission line by 120°.

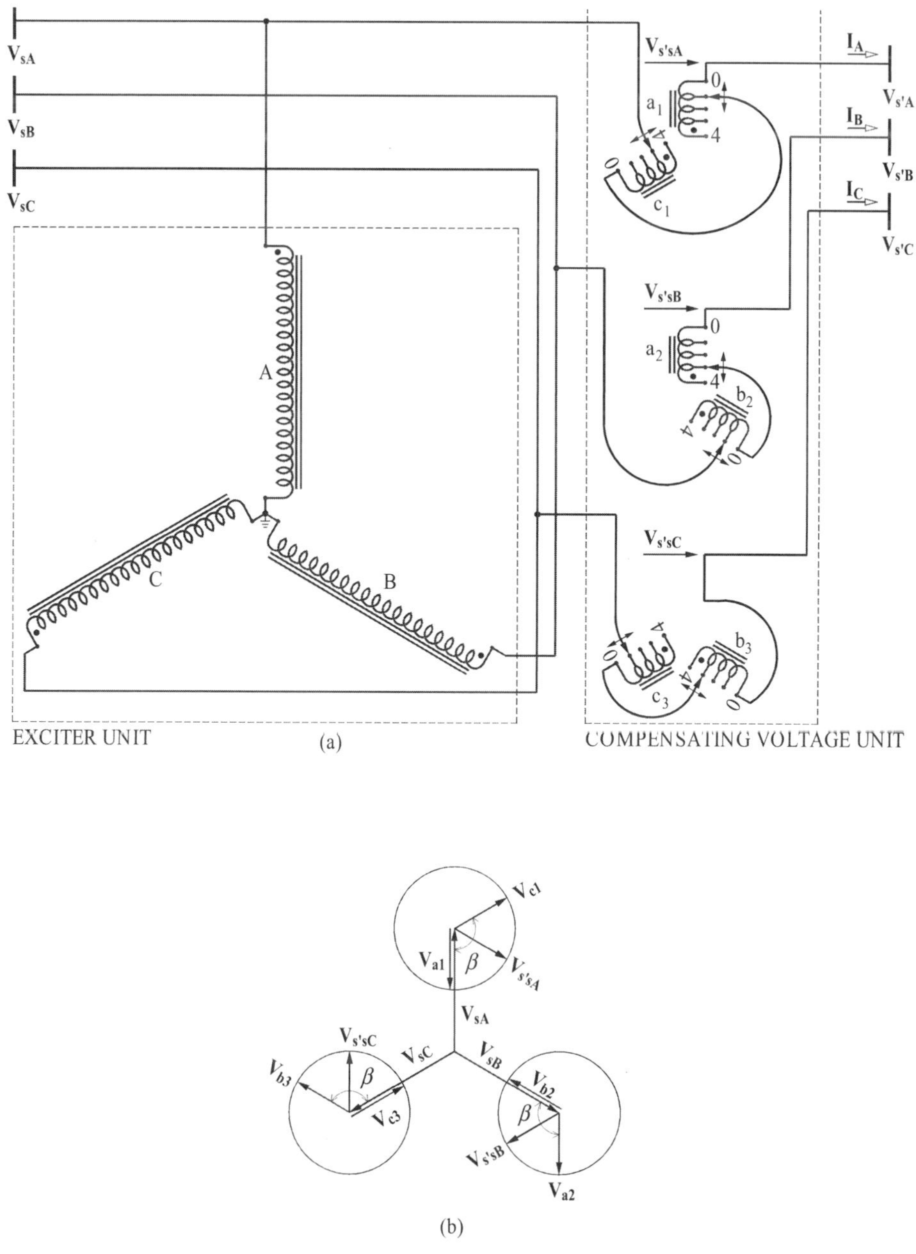

Figure 9-23. (a) ST for compensating voltage in the range of 180° through 300°. (b) Phasor diagram.

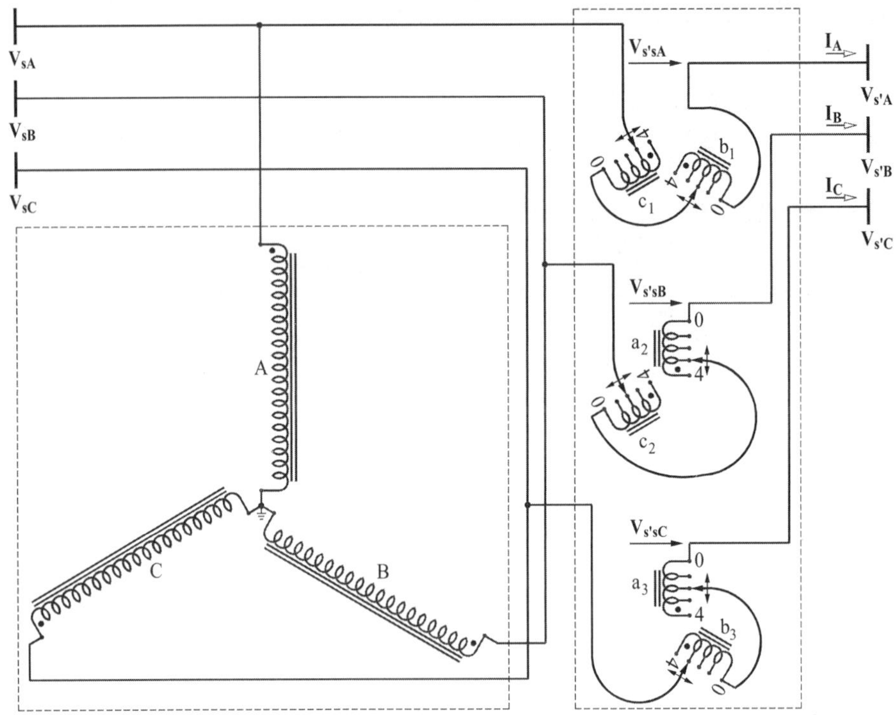

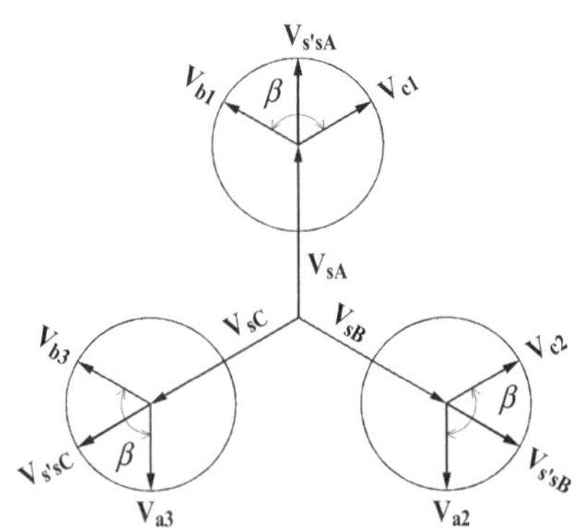

Figure 9-24. (a) ST for compensating voltage in the range of 300° through 60°. (b) Phasor diagram.

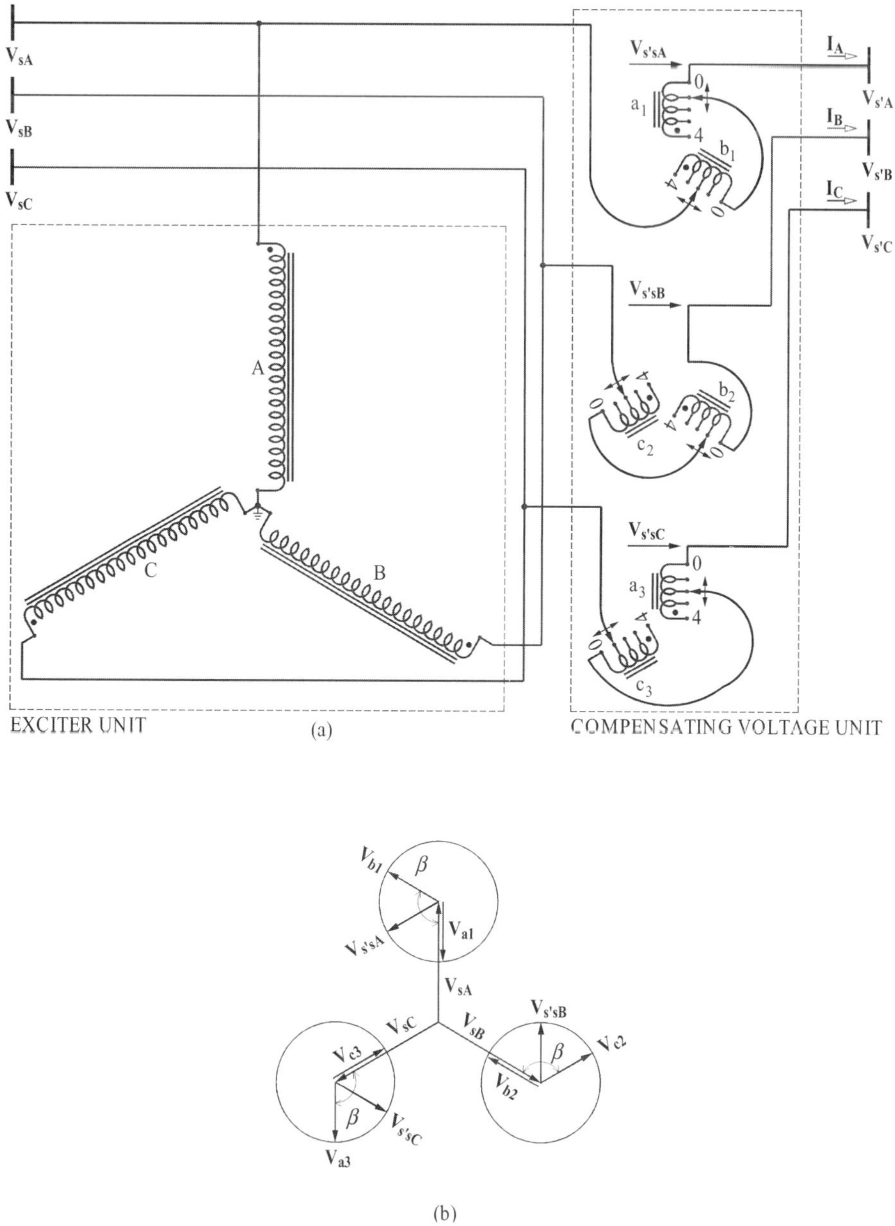

Figure 9-25. (a) ST for compensating voltage in the range of 60° through 180°. (b) Phasor diagram.

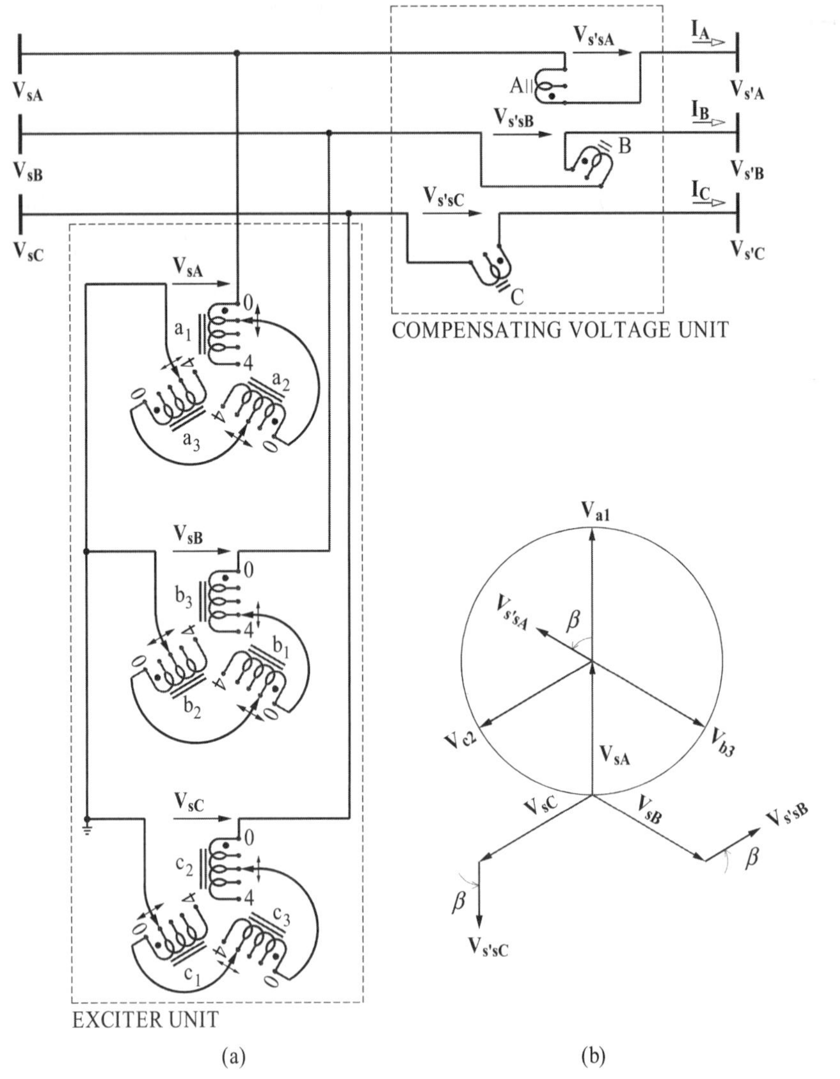

Figure 9-26(a, b). (a) ST configuration using LTCs with lower current rating. (b) Phasor diagram for compensating voltage with variable magnitude and phase angle.

Figure 9-26f shows that the three-phase voltages ($\mathbf{V}_{sA}$, $\mathbf{V}_{sB}$, and $\mathbf{V}_{sC}$) are applied across a_3, b_3 and c_3, respectively, and the compensating voltage in each phase now lags the corresponding phase voltage of the transmission line by 120°.

In summary, the compensating voltage can be at any phase angle with the prevailing line current, which emulates, in series with the line, a capacitor or an inductor and a positive or a negative resistor. The effect is such that the active and reactive power

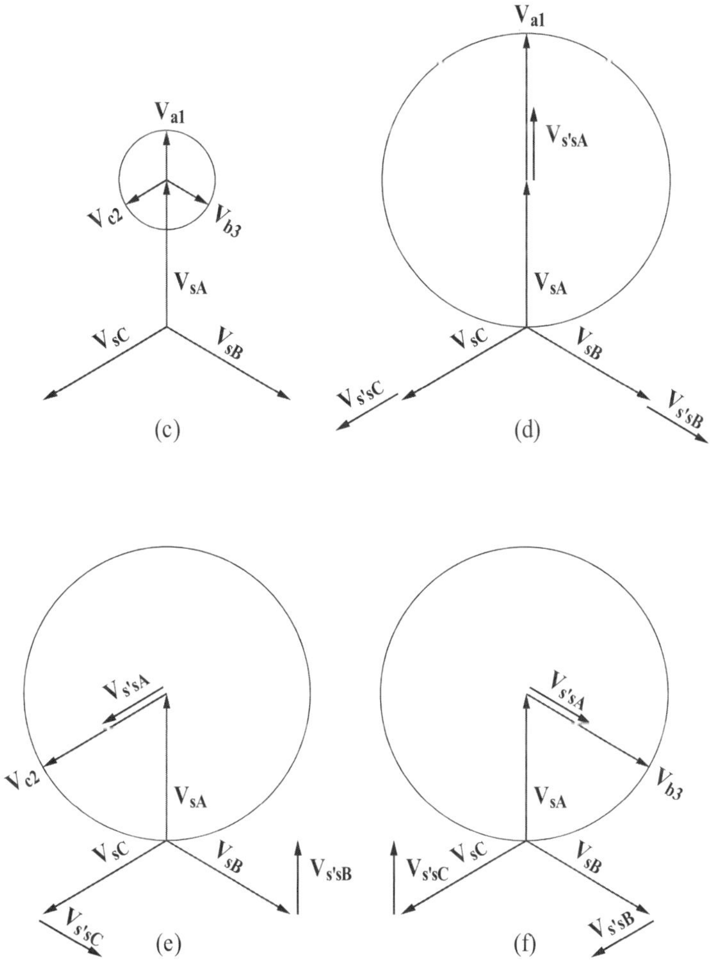

Figure 9-26(c, d, e, f). Phasor diagrams for (c) zero compensating voltage, (d) in-phase compensating voltage with variable magnitude, (e) 120° leading compensating voltage with variable magnitude, and (f) 120° lagging compensating voltage with variable magnitude.

flows in a transmission line can be regulated independently. In addition, the function of a voltage regulator can also be realized.

9.2.7 ST using LTCs with Lower Voltage and Current Ratings

In many applications, the series-connected compensating voltage is only a few percent of the line voltage. The series winding as well as the LTCs in a compensating voltage unit carry high line current and even higher fault current. Sometimes, the LTCs may be required to operate at a low current, so that they can ride through a high fault current. In such cases, three secondary voltages of the compensating voltage unit, each from a three-winding set, need to be high-voltage and low-current rated. Therefore, these secondary voltages may be used to excite three additional single-phase step-down transformers, whose secondary voltages are the compensating voltages that are connected in series with the line, as shown in Figure 9-27a. Figure 9-27b shows the related phasor diagram.

It would be advantageous to use such a scheme, which allows the generation of the required compensating voltage of variable magnitude and variable phase angle by using a three-phase transformer and LTCs and connection of the compensating voltage through three separate single-phase transformers (or one three-phase transformer with three isolated secondary windings), thus isolating the taps from any electrical connection with the high-voltage transmission line.

9.3 COMPARISON AMONG THE VRT, PAR, UPFC, AND ST

The comparison between the VSC-based UPFC and the transformer and LTCs-based ST can be done on several key points. The transformer and LTCs-based VRT and PAR are also included for reference.

The ST generates a compensating voltage ($\mathbf{V}_{s's}$) of variable magnitude and variable phase angle through autotransformer action between the shunt-connected primary windings and the series-connected secondary windings. When connected in series with the transmission line, the compensating voltage modifies the magnitude and phase angle of the uncompensated sending-end voltage ($\mathbf{V}_s$) to be the modified sending-end voltage ($\mathbf{V}_{s'}$) and, therefore, independently control the active and reactive power flows in the transmission line.

The ST offers an area of controllability in the *P–Q* plane just like the UPFC. Note that the VRT and PAR offer only linear characteristics in the *P–Q* plane and, therefore, do not regulate the active and reactive power flows independently.

9.3.1 Power Flow Enhancement

The simulation results for the network example considered in Chapter 3 shows that a 30 MVA-rated series unit (compensating voltage unit in the ST) increases the power flow in the line by 57 MW in the case of the ST and by 58 MW in the case of the UPFC. This increased capacity of the line may be used to redirect power from an overloaded line to an underloaded line that may be very useful during the peak demand period.

In the case of the ST, both active and reactive powers exchanged by the series-connected compensating voltage unit flow through the shunt-connected exciter unit. However in the case of the UPFC, only the exchanged active power flows through the shared DC link between the series-connected SSSC and the shunt-connected STATCOM. Each VSC is capable of generating or absorbing reactive power at its

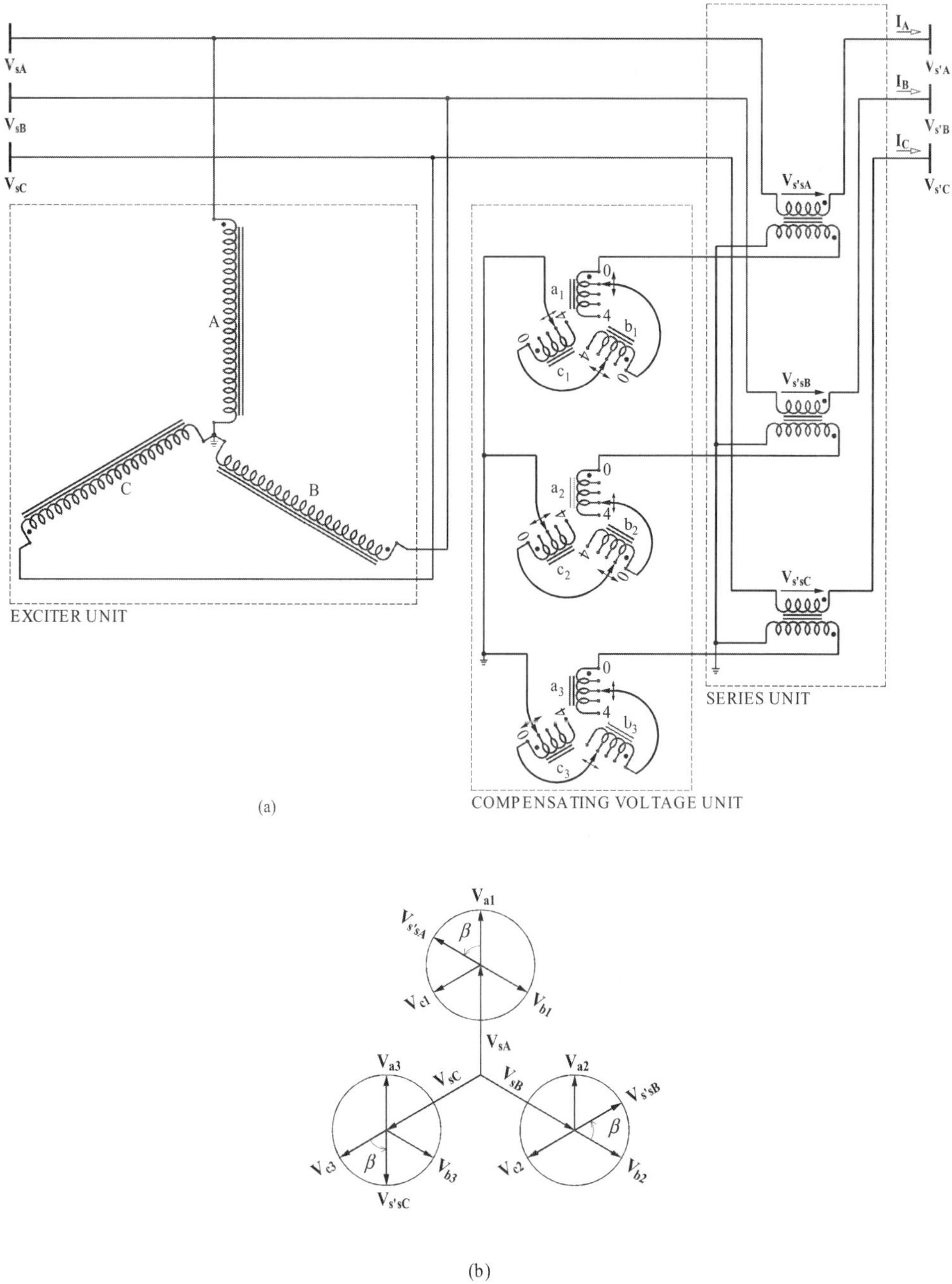

Figure 9-27. (a) ST configuration using taps with lower voltage and current ratings. (b) Phasor diagram.

AC terminal. The series VSC in the UPFC must be designed for the apparent power, which carries the exchanged active and reactive powers with the transmission line. The rating of the shunt VSC must be at least equal to the active power rating of the series VSC. Two VSCs in the UPFC are of equal ratings for simpler housekeeping and practical reasons, such as storing spare parts. Therefore, the only remaining rating of the STATCOM can be utilized to regulate the bus voltage. In the UPFC configuration, the STATCOM has to let the active power of the series VSC go through it first, thereby reducing its full reactive power carrying and voltage regulating capabilities.

Since the shunt and the series VSCs of the UPFC, considered in Chapter 3, are designed to be of the same rating, a 30 MVA-rated shunt unit is able to provide 21 Mvar of reactive power while delivering the maximum 21 MW of active power. If this capability of supplying shunt reactive power is really needed, particularly when the short-circuit impedance at the point of compensation is high, a separate capacitor bank as a shunt reactive compensator may be used along with the ST. Ultimately, the life cycle cost of a PFC must be the deciding factor for its installation and operation.

9.3.2 Speed of Operation

Figure 9-28a shows the test results from the world's first UPFC that consists of a shunt-connected ±160 MVA, 138 kV-rated VSC (STATCOM) and a series-connected ±160 MVA, 13.33 kV-rated VSC (SSSC). The STATCOM is operated in "voltage control" mode to hold the bus voltage ($\mathbf{V}_s$) at one pu. While maintaining unity power factor load on the line and the bus voltage at one pu level, the active power flow in the line is varied. The active power flow at the modified sending end is regulated at –145 MW, –65 MW, –240 MW, and –145 MW levels while holding the reactive power flow at the same point at zero (i.e., unity power factor). The STATCOM exchanges an appropriate amount of reactive power with the line to hold the BUS01 voltage at one pu level. The response time of the UPFC is several seconds.

The simulated transient responses of P_r and Q_r of the ST and UPFC are shown in Figure 9-28b. The natural power flows until t = 5 s. A series-connected compensating voltage ($\mathbf{V}_{s's}$) of magnitude 0.2 pu is applied at a relative phase angle (β) of 300° at t = 5 s and 240° at t = 14 s. While keeping the relative phase angle at 240°, the magnitude of the compensating voltage is increased to 0.4 pu at t = 23 s. The line power due to the application of the UPFC is changed in a linear fashion and that due to the application of the ST is changed in multiple steps. In both cases, the line power flow reversal is demonstrated.

The merit of the ST is that it allows independent control of active and reactive power flows in the line, similar to the UPFC. The UPFC can offer faster response (in ms at a given constant line impedance), but this capability cannot be utilized, as shown in Figure 9-28a, in order to assure continued operation under contingencies. In most utility applications in which the fast response is not required, the ST presents an economically attractive solution for power flow enhancement when compared with the UPFC, which has higher installation and operating costs.

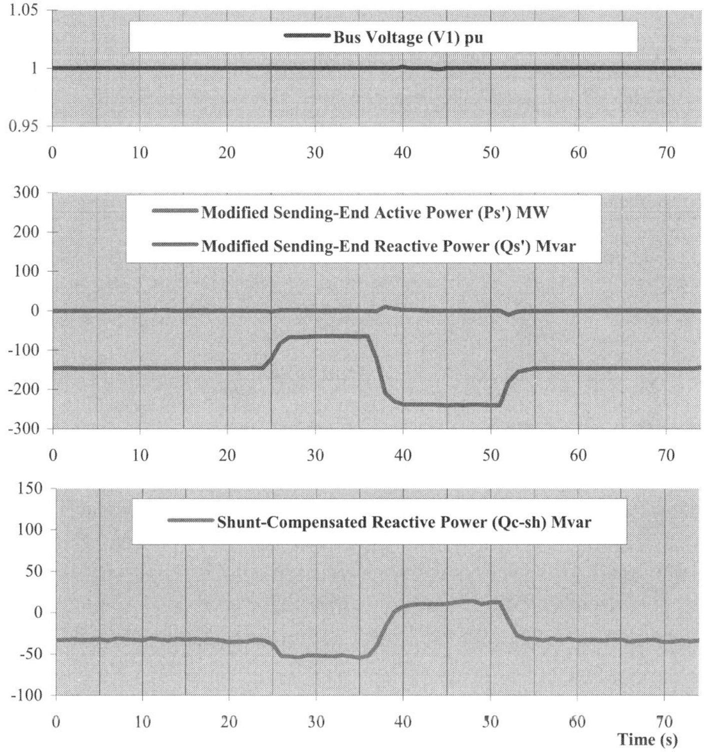

(a)

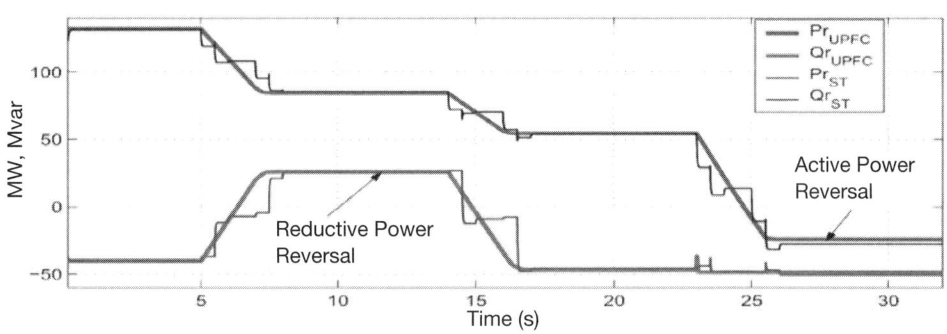

(b)

Figure 9-28. (a) UPFC test results for automatic power flow control mode of operation with maintaining unity power factor and one pu bus voltage while regulating the line power (Sen and Keri, 2002). (b) ST simulation results for open loop power flow control mode of operation (Faruque and Dinavahi, 2007).

9.3.3 Losses

The power loss in a VSC is defined as the total power consumed by various components of a VSC while carrying the rated current at rated voltage. A HN-VSC with GTO devices switching once per cycle has about 1.5% power loss under rated condition. That means for the UPFC that consists of two VSCs, the power loss is about 3% of the VSC rating. The power loss in a single PWM-operated VSC is about 4%. Therefore, for the UPFC consisting of two PWM-operated VSCs, the power loss is about 8%. This makes the operating cost of the VSC-based UPFC the highest among the available power flow solutions. A topology with a resonant-type switching scheme can bring the overall losses for a single VSC down from 4% to 3%, which is still a 6% power loss for the UPFC.

9.3.4 Switch Rating

Each VSC that carries its nominal current at its nominal voltage is said to have one-pu-rated semiconductor devices. In practice, the allowable current through a GTO is rated at 0.5 pu of the controllable current of the device. Considering that there are a number of GTOs connected in series and each one of these devices experiences a voltage overshoot at turn-off from the stored energy in the snubber and stray inductances, and the continuous operation is required even with a number of failed short GTOs in a switch, a conservative design dictates that each device be allowed to operate normally at 0.25 pu of the rated voltage. This means that the installed VA rating of the semiconductor devices is eight times the nominal rating in each VSC. In addition, there are two electronic switches: one for clamping the DC link voltage and another one for the electronic bypass. The semiconductor and snubber losses in the VSC require deionized-water cooling and heat exchangers to get rid of the heat.

The installed UPFCs were built with the use of relatively slow semiconductor devices, such as gate-turn-off (GTO) thyristors that are switched at fundamental frequency. This arrangement requires the use of a harmonically neutralized VSC (HN-VSC). A VSC that is built with relatively faster semiconductor devices, such as the insulated gate bipolar transistors (IGBTs) that are operated with pulse width modulation (PWM) techniques, limits the need for the magnetic circuits to just the coupling transformer. However, this arrangement requires the use of a filter at the output of each VSC in order to limit the need for a higher switching frequency with excessive switching loss. A VSC operated with the PWM techniques may also need shielding from electromagnetic interference (EMI).

9.3.5 Magnetic Circuit Design

Each coupling transformer in the UPFC carries its rated (one pu) power under steady-state condition. The intermediate magnetic circuit, associated with each VSC, is rated at 0.5 pu. A one-pu-rated series coupling transformer permits the compensating voltage to start at its peak value, so that a deenergized transformer can start operating from its zero flux condition and not be saturated. However, this restriction can delay the response of the VSC up to a quarter cycle. If the controller demands the compensating voltage to be applied at its zero value instead of its peak value, the flux in the transformer may ideally build up to the two pu level. Therefore, the series coupling transformer needs to be designed to be at the two pu level in order to avoid saturation. In this case, the magnetic circuit associated with the series VSC has to be rated at one pu instead of 0.5 pu. Therefore, the theoretical rating of the magnetic circuit of the HN-VSC-based UPFC is 4.5 pu (shunt coupling transformer: one pu; shunt intermediate transformer: 0.5 pu; series coupling transformer: two pu; and series intermediate transformer: one pu). In practice, the magnetic circuit for each VSC is typically designed to be rated at 15% to 25% more than the theoretical rating.

In most utility applications, it is adequate to apply the series-connected compensating voltage, calculated by the controller, over the following 10 to 20 cycles. This relaxes the requirement of the rating of the series coupling transformer. A series coupling transformer of 1.5 pu rating is considered to be adequate in a utility application. Then the question arises: Why should one invest in a fast-acting VSC when its output voltage cannot be delivered at the same speed of the electronic VSC because the series coupling transformer is not designed with full rating?

Each of the three primary windings in the exciter unit is designed for full line voltage and the rated single-phase power of the ST. Each of the nine secondary windings in the compensating voltage unit is designed to carry full line current and the rated single-phase power of the ST. This requires the magnetic rating of the 12 windings (three primary and nine secondary) of the ST to be 12/6 = 2 pu. Considering the three primary and three secondary windings, a total of six windings make a one pu transformer.

9.3.6 Optimization of Transformer Rating

The UPFC's rated compensating voltage in series with a transmission line in its possible 360° range has never been used. If the rated compensating voltage ($\mathbf{V}_{s's}$) is added to the sending-end voltage ($\mathbf{V}_s$) at 0° or 180°, the modified sending-end voltage, $\mathbf{V}_{s'}$ (i.e., $\mathbf{V}_s + \mathbf{V}_{s's}$) goes out of the allowable range, which may be within the ±10% of the nominal value. Moreover, in the 360° range of operation of the UPFC, the maximum power flow enhancement occurs within the first 120° and, therefore, the rest of the operating region is practically not necessary. The ST, in its simpler, customized configuration with a reduced amount of hardware, provides a limited angle (120°) operation, which is adequate in most utility applications.

Whether it is utilized or not, the UPFC can operate in the whole 360° range. In a UPFC with HN-VSCs, the required magnetic rating alone is at least 4.5 pu, whereas, in the ST, the magnetic rating is only two pu for the same 360° range of operation. In a practical implementation, only 120° range of operation is required. Consequently, this requirement is satisfied by constructing the ST as shown in Figure 9-20a. This configuration requires only three primary windings and six secondary windings. This results in a 1.5-pu-rated transformer. In comparison, the autotransformer and the PAR, each with one pu rating, can jointly perform in the 90° range of operation, requiring a total of two pu rating. Therefore, it can be concluded that the functions of the autotransformer and PAR are combined in a single unit of ST that results in a reduced amount of hardware from what is required for separate autotransformer and PAR. The number of isolated, compensating windings in the ST can be reduced by the following action. The compensating winding that provides the in-phase voltage and the corresponding exciting winding can be replaced with a partially tapped autotransformer winding, as shown in Figure 9-20e.

Figure 9-29 shows the practical control range of the VRT, PAR, UPFC, and ST. The filled dots, within the allowable voltage range in part a, show a total of 36 com-

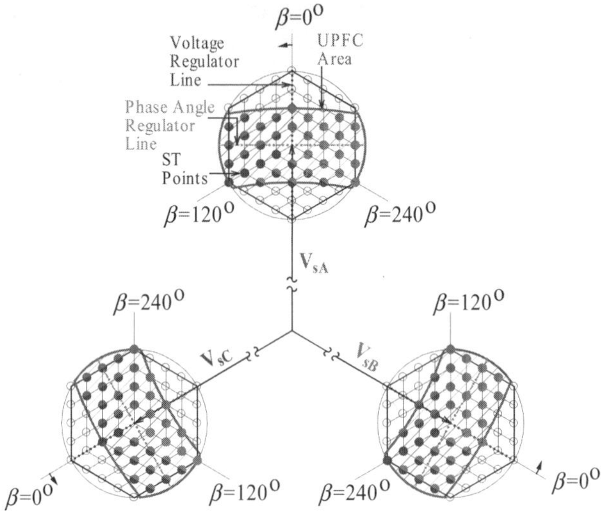

(a)

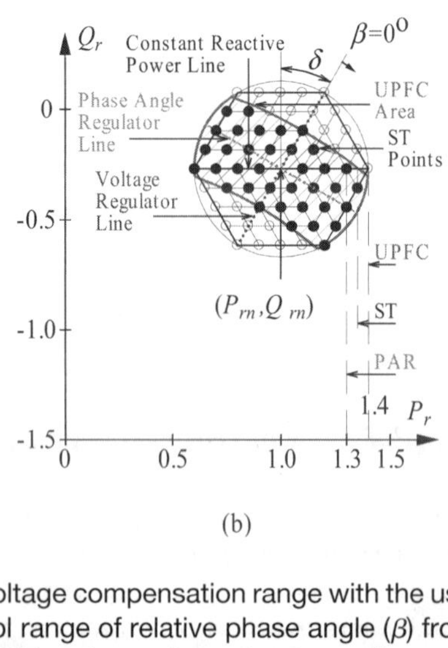

(b)

Figure 9-29. Practical voltage compensation range with the use of the VRT, PAR, UPFC, and ST within the control range of relative phase angle (β) from 0° to 360°: (a) modified sending-end voltage and (b) active and reactive power flows at the receiving end.

pensating points of the ST lie within the 10% range of the nominal voltage. Part b shows that the maximum active power flow enhancement at the receiving end from the natural flow of one pu, within the allowable voltage range, is 0.40 pu with the UPFC, 0.35 pu with the ST, and 0.30 pu with the PAR. Moreover, the ST can provide an active power flow enhancement of 0.3 pu as does the PAR, but along a constant reactive power line (without altering the natural flow of reactive power, Q_{rn}), whereas the PAR increases the reactive power from the natural flow of 0.268 pu to the simultaneously regulated flow of 0.441 pu. The VRT also regulates the active and reactive power flows simultaneously.

9.3.7 Harmonic Injection into the Power System Network

The ST always generates a sinusoidal compensating voltage of line frequency. Even though the compensating voltage emulates an impedance, this virtual impedance is actually a finite voltage and, therefore, cannot cause resonance with the line inductance or capacitance.

The UPFC always injects harmonic components (voltage and current) into the power system network. The magnitudes of the harmonic components are reduced with a higher leakage reactance in the coupling transformers, but the leakage reactance of the series coupling transformer increases the overall undesirable inductive compensation range that reduces power flow in the line and reduces the overall desirable capacitive compensation range that increases power flow in the line.

9.3.8 Operation During Line Faults

Generally, to limit a phase-to-neutral fault current, additional inductive impedance in the path of the fault is needed. If, for example, the additional emulated impedance is equal to the equivalent line impedance, the steady-state fault current is reduced to half of the natural fault current. The steady-state faulted voltage across the emulated impedance is half of the phase-to-neutral voltage. In the worst case, the emulated impedance must withstand twice the volt seconds that are needed to support half of the steady-state phase-to-neutral voltage in order to avoid saturation. The relatively insignificant (typically 0.05–0.15 pu) voltage rating of the series core of the ST would saturate early in the fault cycle. The saturated windings do not provide significant current-limiting effect in the ST, so the ST rides through a fault with little effect.

During a fault in the transmission line, the saturated core of the series coupling transformer does not provide significant current-limiting effect in the UPFC. The current through the series VSC will exceed its rated limit. As a protective measure, the semiconductor switches of a shunt-connected VSC are turned off to ensure zero current flow through it, but the semiconductor switches of the series-connected VSC have to be all turned on to ensure zero compensating voltage so that the VSC side of the series coupling transformer is shorted and the line is not open circuited. Therefore, to ride through the line faults, the series-connected VSC must be designed to carry the fault current. This is an expensive proposition. The alternate solution is to have a nominally rated VSC as a series compensator and an additional electronic bypass switch that is specifically designed to carry the fault current. Then the question arises as to whether the electronic bypass switch should be located on the high-voltage line side or on the low-voltage VSC side. If the electronic bypass switch is located on the high-voltage side, the VSC "sees" an additional inductance from the leakage reactance of the series coupling transformer in its forward path. Designers prefer to have the electronic bypass switch on the line side because when shorted, all of the leakage reactances of the coupling transformer and intermediate transformer appear across the output terminal of the VSC and a short circuit at its output terminal is prevented. However, it is quite expensive to have electronic switches mounted on an isolated platform and connected to the high-voltage transmission line. It is certainly less expensive to have the electronic bypass switch located on the low-voltage side (VSC side) of the series coupling transformer. In this case, when the electronic bypass switch is operated, the output voltage of the VSC is applied only to the leakage reactance of the intermediate transformer. Then the question arises as to whether the VSC is adequately designed to carry this resulting current until the fault current is transferred to the mechanical bypass, which takes a few line cycles to operate. At that time, the series VSC gating can be stopped without opening the line. If the electronic bypass switch is located on the low-voltage VSC side, the VSC should be protected in some way in order to avoid a high current through it. The obvious solution is to put an additional inductor on the low-voltage VSC side. This additional inductor always stays connected in the forward path of the VSC and, as a result, increases the overall undesirable inductive compensation range and reduces the overall desirable capacitive compensation range.

Note that a VSC that operates at fundamental switching frequency provides a positive-sequence fundamental voltage. In the case of a line fault, the negative-sequence component of the unbalanced line voltage causes the negative-sequence current to flow freely into the VSC, which is not desirable. A VSC with PWM operation provides a desirable voltage, not just a positive-sequence fundamental voltage, which is effective in operations during unbalanced condition and faults.

The basic differences between the traditional technology of transformer and LTCs versus the VSC are given in Table 9-2.

9.4 MULTILINE SEN TRANSFORMER

A multiline sen transformer (MST) connects a compensating voltage in series with each transmission line in a multiline transmission system and transfers active and reactive powers selectively from one or more transmission lines, referred to as "master" lines, to the others, referred to as "slave" lines. Each compensating voltage in the "master" lines is of any magnitude within its allowable limit and at any phase angle with respect to the transmission line voltage as well as the prevailing line current, so that the active and reactive power flows in that line are controlled independently. Each series-connected compensating voltage in the "slave" lines is of specific magnitude and phase angle with respect to the prevailing line current, so that the active and reac-

Table 9-2. Comparison among the voltage regulating transformer (VRT), phase angle regulator (PAR), unified power flow controller (UPFC), and Sen transformer (ST)

	Inter-dependent active and reactive power flow control		Shunt–series interconnected voltage injection with independent local voltage, active and reactive power flow control		
	VRT	PAR	UPFC with HN-VSCs	UPFC with PWM-VSCs	ST
In-phase voltage injection	Yes	No	Yes	Yes	Yes
Quadrature voltage injection	No	Yes	Yes	Yes	Yes
Independent control of active and reactive power flows in the line	No	No	Yes	Yes	Yes
Number of components/complexity	Low	Low	High	High	Low
Reliability and high availability	Yes	Yes	No	No	Yes
Compensating voltage at line frequency	Yes	Yes	No	No	Yes
Loss of useful capacitive rating due to high series leakage reactance in the coupling transformer	No	No	Yes	Yes	No
Fast fault current bypass switch needed	No	No	Yes	Yes	No
Adequate dynamic response for utility power flow regulation	Yes	Yes	Yes	Yes	Yes
Compensating voltage steps depend on number of taps installed	Yes	Yes	No	No	Yes
Capability of independent local reactive power generation and absorption	No	No	Yes	Yes	No
VSC output filter needed	No	No	No	Yes	No
VSC EMI shields may be needed	No	No	Yes	Yes	No
Estimated losses at rated power	<1%	<1%	3%	8%	<1%
Required VA rating of magnetic components (pu)	1	1	4.5	3	1.5
Estimated equipment cost normalized for a unit rated 50 MVA or more	1	1	7.5	6	1.5
Estimated losses/operating cost normalized for a unit rated 50 MVA or more	1	1	5	10	1

tive powers from the "master" transmission lines are transferred bidirectionally to the "slave" transmission lines. This technique provides a desirable power flow management for a multiline transmission system by decreasing the power flow in the overloaded lines and increasing it in the underloaded lines with little impact on the other uncompensated lines.

In the MST configuration, shown in Figure 9-30, the line voltage ($\mathbf{V}_s$) is applied to a shunt-connected three-phase transformer's Y-connected primary windings in the

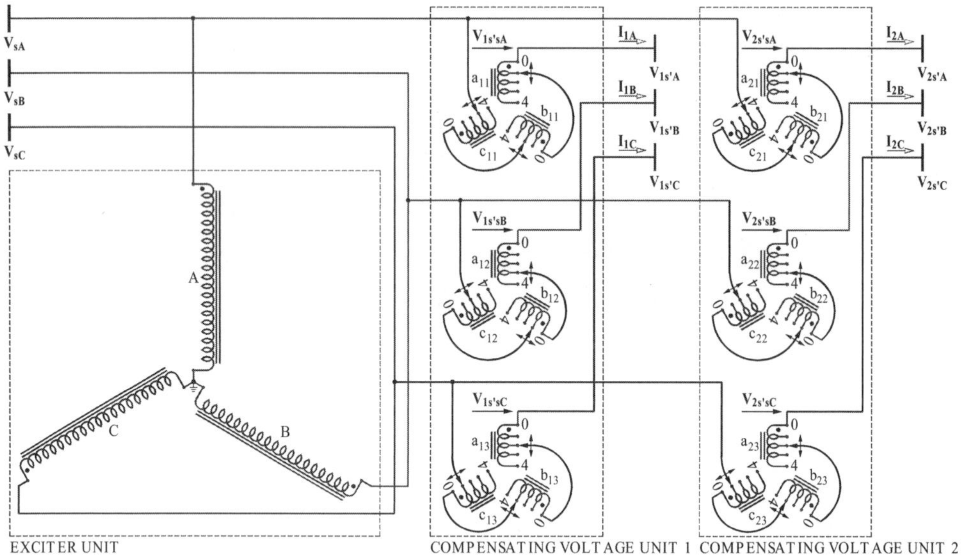

Figure 9-30. Multiline Sen transformer.

exciter unit. In the compensating voltage unit 1, a total of nine secondary windings (a_{11}, a_{12}, and a_{13} on the core of the A phase; b_{11}, b_{12}, and b_{13} on the core of the B phase; and c_{11}, c_{12}, and c_{13} on the core of the C phase) derive a three-phase compensating voltage, each phase of which is the phasor sum of the voltages induced in a three-winding set (a_{11}, b_{11}, and c_{11} for compensating voltage in the A phase; a_{12}, b_{12}, and c_{12} for compensating voltage in the B phase; and a_{13}, b_{13}, and c_{13} for compensating voltage in the C phase). In the compensating voltage unit 2, a total of nine secondary windings (a_{21}, a_{22}, and a_{23} on the core of the A phase; b_{21}, b_{22}, and b_{23} on the core of the B phase; and c_{21}, c_{22}, and c_{23} on the core of the C phase) derive a three-phase compensating voltage, each phase of which is the phasor sum of the voltages induced in a three-winding set (a_{21}, b_{21}, and c_{21} for compensating voltage in the A phase; a_{22}, b_{22}, and c_{22} for compensating voltage in the B phase; and a_{23}, b_{23}, and c_{23} for compensating voltage in the C phase). By choosing the number of turns from each of the three windings, and, therefore, the magnitudes of the components of the three 120° phase-shifted induced voltages, the composite compensating voltage magnitude and relative phase angle with respect to the transmission line voltage can be selected.

It should be noted that in the compensating voltage unit 1, each of a_{11}, b_{12}, and c_{13} is tapped at the same number of turns; each of b_{11}, c_{12}, and a_{13} is tapped at the same number of turns; and each of c_{11}, a_{12}, and b_{13} is tapped at the same number of turns. However, the number of turns in the a_{11}–b_{12}–c_{13} set, b_{11}–c_{12}–a_{13} set, and c_{11}–a_{12}–b_{13} set can be different from each other. Also, in the compensating voltage unit 2, each of a_{21}, b_{22}, and c_{23} is tapped at the same number of turns; each of b_{21}, c_{22}, and a_{23} is tapped at the same number of turns; and each of c_{21}, a_{22}, and b_{23} is tapped at the same number of turns. However, the number of turns in the a_{21}–b_{22}–c_{23} set, b_{21}–c_{22}–a_{23} set, and c_{21}–a_{22}–b_{23} set can be different from each other.

The compensating voltage ($\mathbf{V}_{1s's}$) in the compensating voltage unit 1 can be at any phase angle with the prevailing line current ($\mathbf{I}_1$), which emulates an impedance in series with the line 1. The compensating voltage ($\mathbf{V}_{2s's}$) in the compensating voltage unit 2 can be at any phase angle with the prevailing line current ($\mathbf{I}_2$), which emulates an impedance in series with line 2. An emulated impedance can be a capacitor or an inductor and a positive resistor or a negative resistor.

The series-connected compensating voltage ($\mathbf{V}_{1s's}$) controls the sending-end voltage ($\mathbf{V}_s$) to $\mathbf{V}_{1s'}$ in order to independently control the active and reactive power flows in line 1. The series-connected compensating voltage ($\mathbf{V}_{2s's}$) controls the sending-end voltage ($\mathbf{V}_s$) to $\mathbf{V}_{2s'}$ in order to provide the same function for line 2. The compensating voltage in each line is also at any phase angle with the prevailing line current. The active or direct component of the compensating voltage provides the series resistance emulation, whereas the reactive or quadrature component provides the series reactance emulation.

The operations of the compensating voltage units are such that the compensating voltage in the "master" line is at any phase angle with the line voltage as well as the prevailing line current, and that in the "slave" line is of specific magnitude and phase angle, so that the active and reactive powers from the "master" line are transferred bidirectionally to the "slave" line. The details of this control scheme are discussed in the next section. As discussed earlier, the 120° range of operation of the PFC is often sufficient. The MST, in its simpler customized configuration with a reduced amount of hardware, provides this limited angle (120°) operation.

In the "master" line (line 1), the controllable range of the active and reactive power flows can easily be determined by applying the series-connected, fully rated compensating voltage operating within its entire 360° range. As a consequence, the compensating voltage emulates an impedance in series with the line and, in turn, exchanges active and reactive powers with the line. For a certain compensating voltage ($\mathbf{V}_{1s's}$) in line 1, the exchanged active and reactive powers are P_{1exch} (i.e., $-\mathbf{V}_{1s's} \bullet \mathbf{I}_1$) and Q_{1exch} (i.e., $|-\mathbf{V}_{1s's} \times \mathbf{I}_1|$), respectively. To transfer these exchanged active and reactive powers to the "slave" line, the following relationships must be obeyed:

$$P_{1exch} + P_{2exch} = 0 \tag{9-4a}$$

$$Q_{1exch} + Q_{2exch} = 0 \tag{9-4b}$$

Therefore, the exchanged active and reactive powers in the "slave" line are

$$P_{2exch} = -P_{1exch} \tag{9-5a}$$

$$Q_{2exch} = -Q_{1exch} \tag{9-5b}$$

but the exchanged active and reactive powers in the "slave" line are

$$P_{2exch} = I_2^2 R_{2d}^* \tag{9-6a}$$

$$Q_{2exch} = I_2^2 X_{2q}^* \tag{9-6b}$$

where R_{2d}^* and X_{2q}^* are the emulated resistance and reactance commands, respectively.

From equations 9-5 and 9-6, the emulated resistance and reactance commands (R_{2d}^* and X_{2q}^*) for the "slave" line (line 2) are calculated as

$$R_{2d}^* = -P_{1exch} / I_2^2 \tag{9-7a}$$

$$X_{2q}^* = -Q_{1exch} / I_2^2 \tag{9-7b}$$

The required compensating voltage ($\mathbf{V}_{2s's}$) command can be given to the controller in terms of the emulated resistance and reactance commands (R_{2d}^* and X_{2q}^*). Substituting i_2 for i, R_{2d}^* for R_d, and X_{2q}^* for X_q in Figure 9-11, the same control block diagram can be executed to operate the taps for the compensating voltage unit 2.

The above concept can be extended to an n-line MST provided that the following relationships for the power balance are maintained:

$$P_{1exch} + P_{2exch} + \cdots + P_{nexch} = 0 \tag{9-8a}$$

$$Q_{1exch} + Q_{2exch} + \cdots + Q_{nexch} = 0 \tag{9-8b}$$

Any difference in power transfer between the transmission lines ("master" and "slave") will flow through the exciter winding. The exciter winding is rated to carry the differential power between the "master" and the "slave" lines. Note that in the implementation of an BTB-SSSC (also referred to as IPFC) for power flow management in a multiline transmission system, a balance of zero net active power at the DC link must be observed.

9.4.1 Basic Differences Between the MST and BTB-SSSC

A VSC-based BTB-SSSC is capable of transferring active and reactive powers selectively from one particular line to another. So is the transformer and the LTC-based MST. Note that in reality, a BTB-SSSC only transfers the active power from one line to another and, in addition, generates or absorbs the reactive power in each line independently.

Although a properly designed VSC is capable of applying a compensating voltage to the transmission line in subcycle time, it is not advisable to do so. System instability may occur during contingencies. The BTB-SSSC always needs an overrated series coupling transformer to avoid saturation. The response time of the MST is adequate in most utility applications, but it can be improved, when needed, by replacing the present LTCs with faster switches.

The MST with mechanical LTCs will ride through a line fault. To protect the VSC in a BTB-SSSC from the line fault current, a fast bypass switch is required.

A BTB-SSSC is capable of connecting its rated compensating voltage in series with a transmission line for a possible 360° range of operation. However, in order to maintain the line voltage within an allowable range, the rated compensating voltage is used only over a narrow range of the possible 360° phase angle. The MST offers the required operating range with less hardware.

Unlike a VSC, which always injects harmonic components into the power system network, the MST only injects a voltage of line frequency. A VSC may need additional filter circuits to reduce the injected harmonic components into the power system network.

The installation cost of the BTB-SSSC is in the order of five times that of the MST. Moreover, depending on the topology of the VSC, the BTB-SSSC with two VSCs has an efficiency in the range of 92% to 97%, whereas the MST with mechanical LTCs is more than 99% efficient. Therefore, the operating cost of the BTB-SSSC is an order of magnitude higher than that of the MST.

The PFC in a transmission system application is expected to operate reliably 99.9% of the time, except for scheduled maintenance. The reliability of the VSC-based BTB-SSSC concept has not yet been proven, but the components of the MST are transformers and LTCs, which are time-tested and proven to be reliable.

9.5 FLEXIBLE OPERATION OF THE ST

The basic scheme of the ST can be modified for special applications, such as limited angle operation with a reduced amount of hardware and interconnection of transmission lines of different voltages or phase angles while providing independent control of active and reactive power flows in those lines. Figure 9-31a shows one such scheme in which two transmission lines, one at 345 kV and the other at 138 kV, can be interfaced with the ST when the compensating voltage unit is connected at the 138 kV line side. The primary windings of the ST are energized with the voltage from the 345 kV line. The magnitude and phase angle of the stepped-down 138 kV voltages are controlled with the use of the compensating voltage unit. Figure 9-31b shows the related phasor diagram.

The above scheme can be extended, as shown in Figure 9-32a, to interconnect the transmission lines of different voltage levels while providing independent control of active and reactive power flows in the higher voltage lines. The primary windings of the ST are energized from the lower voltage side. The magnitude and phase angle of the stepped-up voltages are modified with the use of the compensating voltage unit. Figure 9-32b shows the related phasor diagram.

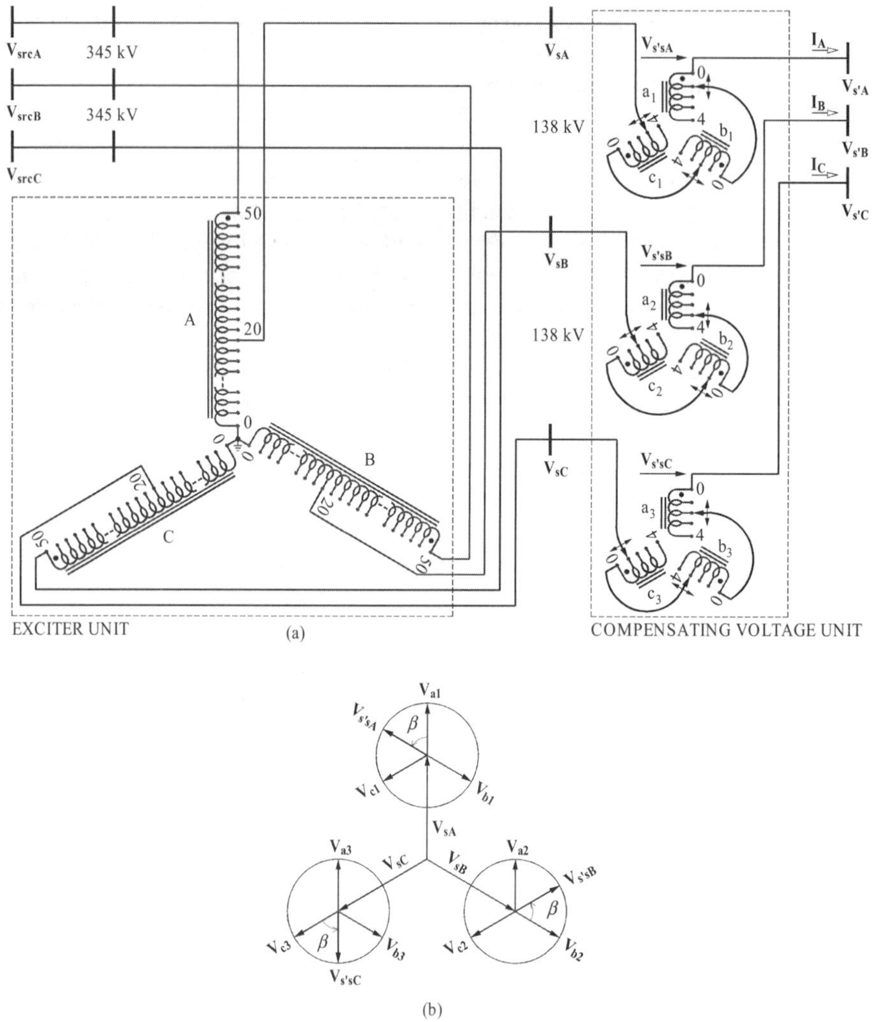

Figure 9-31. (a) ST's compensating voltage unit is connected to the stepped-down voltage of a transmission line. (b) Phasor diagram.

9.6 ST WITH SHUNT-CONNECTED COMPENSATING VOLTAGES

The ST, as explained so far, consists of a shunt-connected exciter unit and a series-connected compensating voltage unit that operates as an autotransformer. The basic concept of a series-connected compensating voltage in the ST is shown in Figure 9-33a. In an alternate approach, the compensating voltage can be shunt connected to the line, as shown in Figure 9-33b, providing an electrical isolation between the exciter unit's primary voltage and the compensating voltage unit's secondary voltages. Note that there is no

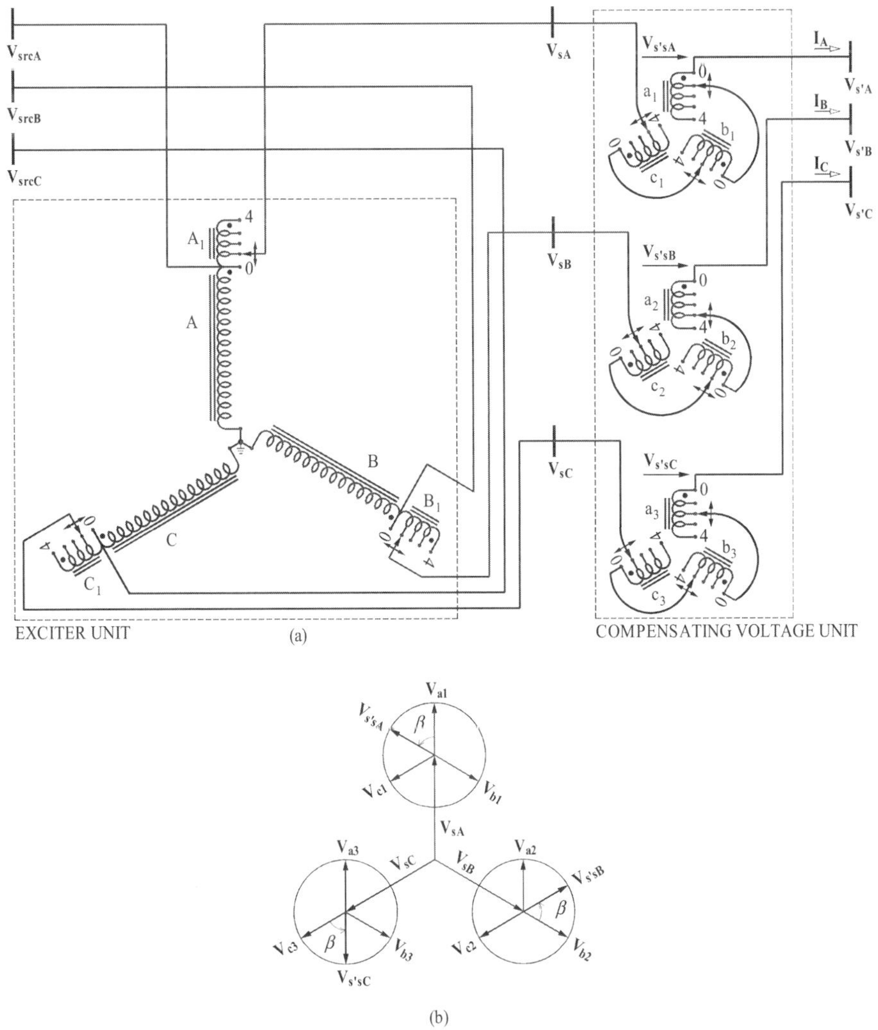

Figure 9-32. (a) ST's compensating voltage unit is connected to the stepped-up voltage of a transmission line. (b) Phasor diagram.

such electrical isolation between the series-connected compensating voltages and the shunt-connected exciting voltages. Also, the series-connected compensating voltage unit carries only a fraction of the transmitted power, whereas the shunt-connected compensating voltage unit carries the full transmitted power. The practical application of a shunt-connected compensating voltage unit might be in a case when two nearby transmission lines with different voltages or phase angles need to be interconnected.

The circuit configuration, shown in Figure 9-34a, can be used to form a shunt-connected compensating voltage ($\mathbf{V}_{s'}$) that is at any phase angle with respect to the trans-

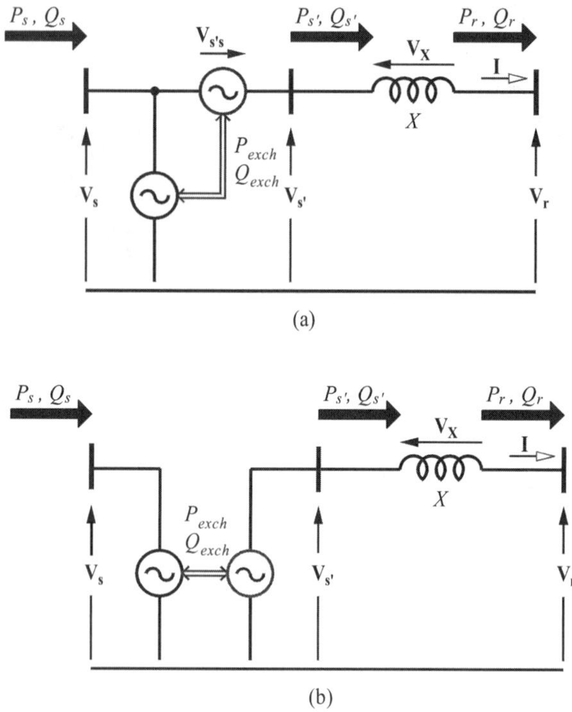

Figure 9-33. Two methods of modifying a transmission line voltage: (a) with a series-connected compensating voltage, (b) with a shunt-connected compensating voltage.

mission line voltage ($\mathbf{V}_s$). By choosing the number of turns from each of the three windings and, therefore, the magnitudes of the components of the three 120° phase-shifted induced voltages, the compensating voltage ($\mathbf{V}_{s'}$) in any phase is derived from the phasor sum of the voltages induced in a three-phase windings set (a_1, b_1, and c_1 for compensation in the A phase; a_2, b_2, and c_2 for compensation in the B phase; and a_3, b_3, and c_3 for compensation in the C phase). The compensating voltage can be at any phase angle with respect to the prevailing line current. The active or direct component of the compensating voltage provides the shunt resistance emulation, which can be a positive resistor or a negative resistor. The reactive or quadrature component of the compensating voltage provides the shunt reactance emulation, which can be a capacitor or an inductor. Therefore, the desired compensating voltage is actually an impedance emulator. The effect of impedance emulation is such that the transmission line voltage can be modified to a specific magnitude and phase angle. The power circuit is identical for both the series- and shunt-connected compensating voltage units, except that they are connected to the line in series and in shunt, respectively. Figure 9-34b shows the related phasor diagram. It should be noted that each of a_1, b_2, and c_3 is tapped at the same number of turns; each of b_1, c_2, and a_3 is tapped at the same number

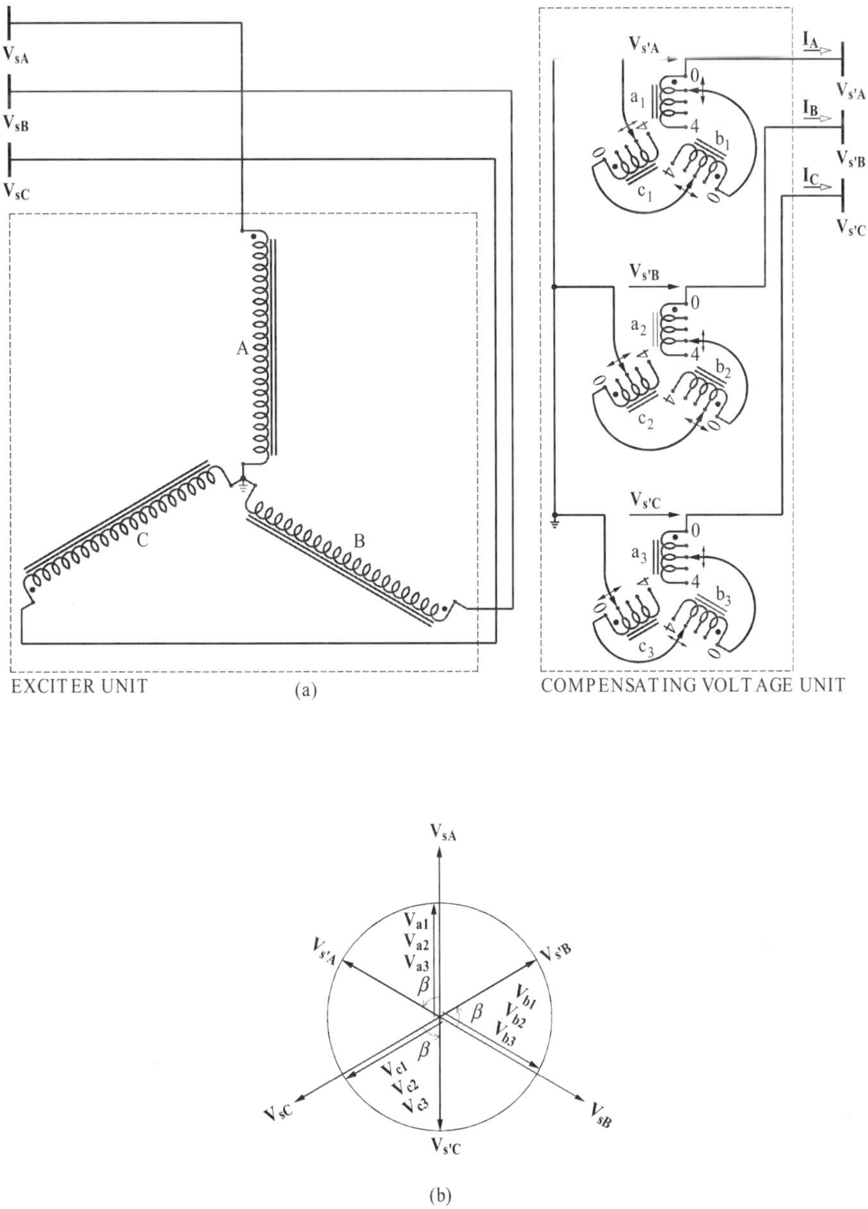

Figure 9-34. (a) Shunt-connected compensating voltage in the ST. (b) Phasor diagram.

of turns; and each of c_1, a_2, and b_3 is tapped at the same number of turns. However, the number of turns in the a_1-b_2-c_3 set, b_1-c_2-a_3 set, and c_1-a_2-b_3 set can be different from each other.

9.7 LIMITED ANGLE OPERATION OF THE ST WITH SHUNT-CONNECTED COMPENSATING VOLTAGES

The ST requires the use of a three-phase transformer with three primary windings and nine secondary windings. The shunt-connected compensating voltage ($\mathbf{V}_{s'}$) is of variable magnitude and variable phase angle. However, in many instances, the capability of connecting a voltage in shunt with a line within its entire range of 360° is not needed. In this case, the circuit configuration for the full 360° range of operation can be simplified to the "limited angle" configuration.

Figures 9-35 through 9-40 show, through transformer action, the generation of a line frequency, shunt-connected compensating voltage ($\mathbf{V}_{s'}$), which is of variable magnitude and operated within a limited angle. Figure 9-35a shows that for the limited angle operation between 0° and 120°, the voltage at any point in the electrical system is applied to a shunt-connected three-phase transformer's primary windings. A total of six secondary windings (a_1 and a_2 on the core of the A phase, b_2 and b_3 on the core of the B phase, and c_3 and c_1 on the core of the C phase) constitute the limited angle compensating voltage unit. By choosing the number of turns from each of the two windings, the compensating voltage in any phase is derived from the phasor sum of the voltages induced in a two-phase winding set (a_1 and c_1 for connection in the A phase, a_2 and b_2 for connection in the B phase, and b_3 and c_3 for connection in the C phase). The phase angle of the shunt-connected compensating voltage ($\mathbf{V}_{s'}$) with respect to the line voltage ($\mathbf{V}_s$) can vary between 0° and 120°. Figure 9-35b shows the related phasor diagram.

Similarly, in an application in which there is a need to vary the phase angle of the shunt-connected compensating voltage ($\mathbf{V}_{s'}$) between 120° and 240°, with respect to the exciter voltage ($\mathbf{V}_s$) a compensating voltage unit with only six windings, as shown in Figure 9-36a, is needed. This is achieved by constructing the compensating voltage from a combination of two voltages, each of which is induced in a separate winding of a two-phase set (b_1 and c_1 for connection in the A phase, a_2 and c_2 for connection in the B phase, and a_3 and b_3 for connection in the C phase). Figure 9-36b shows the related phasor diagram.

Finally, in an application in which there is a need to vary the phase angle of the shunt-connected compensating voltage ($\mathbf{V}_{s'}$) between 240° and 360°, with respect to the exciter voltage ($\mathbf{V}_s$) a compensating voltage unit with only six windings, as shown in Figure 9-37a, is needed. This is achieved by constructing the compensating voltage from a combination of two voltages, each of which is induced in a separate winding of a two-phase set (a_1 and b_1 for connection in the A phase, b_2 and c_2 for connection in the B phase, and a_3 and c_3 for connection in the C phase). Figure 9-37b shows the related phasor diagram.

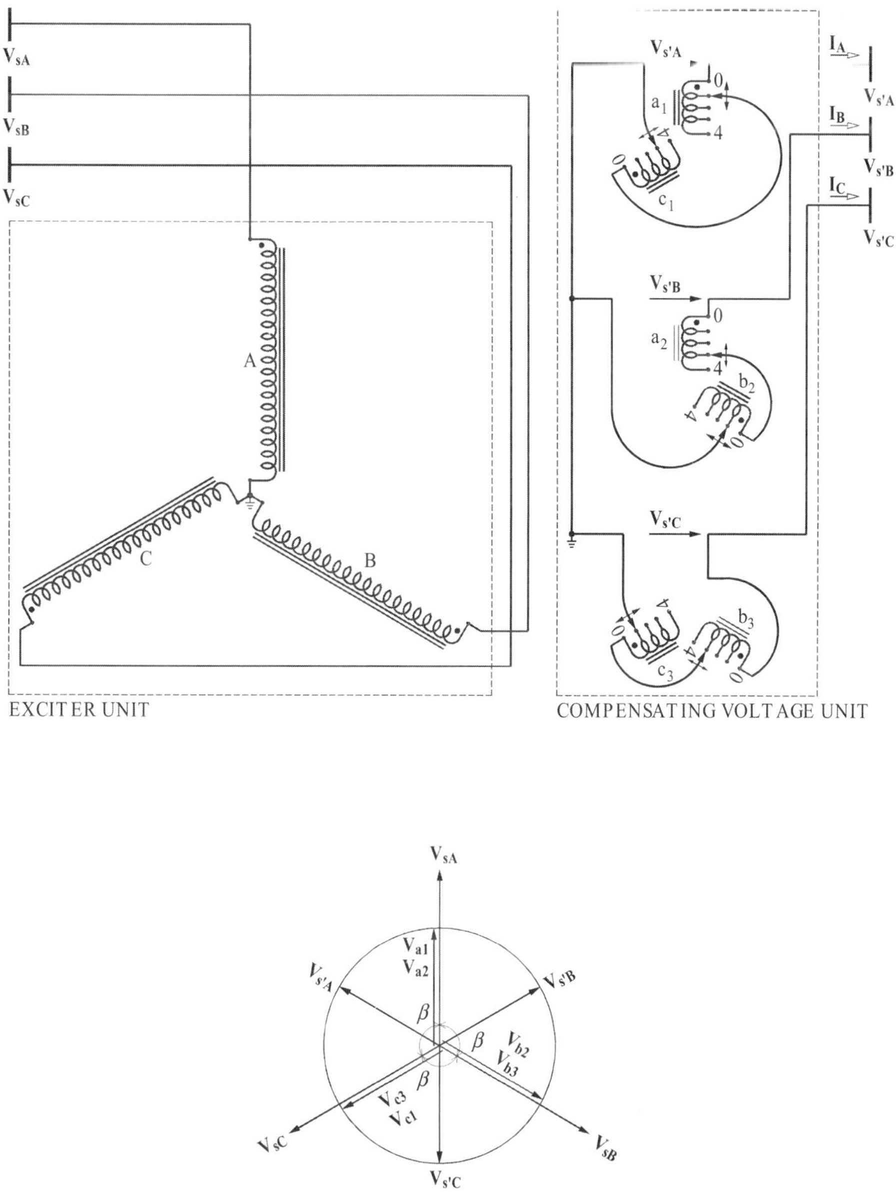

Figure 9-35. (a) ST with shunt-connected compensating voltage unit operating in the range of 0° through 120°. (b) Phasor diagram.

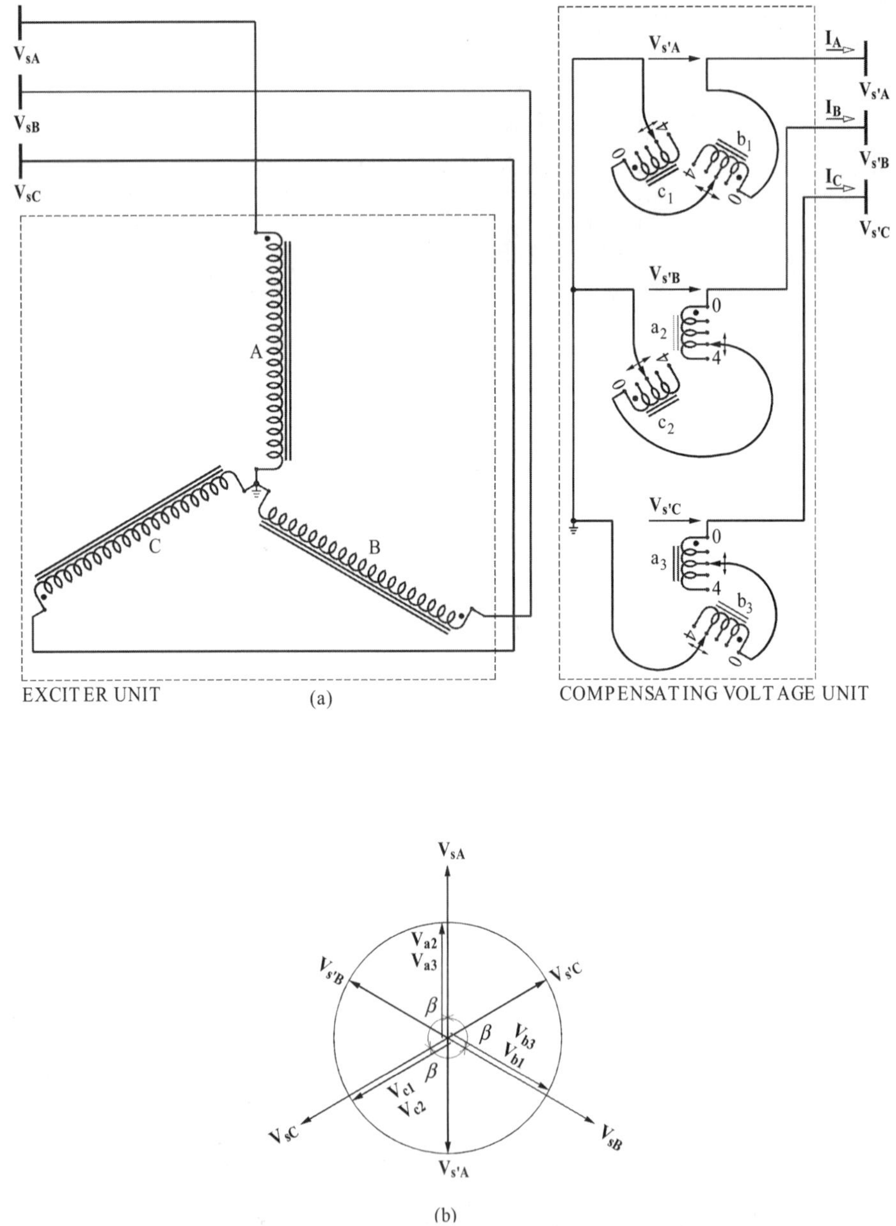

Figure 9-36. (a) ST with shunt-connected compensating voltage unit operating in the range of 120° through 240°. (b) Phasor diagram.

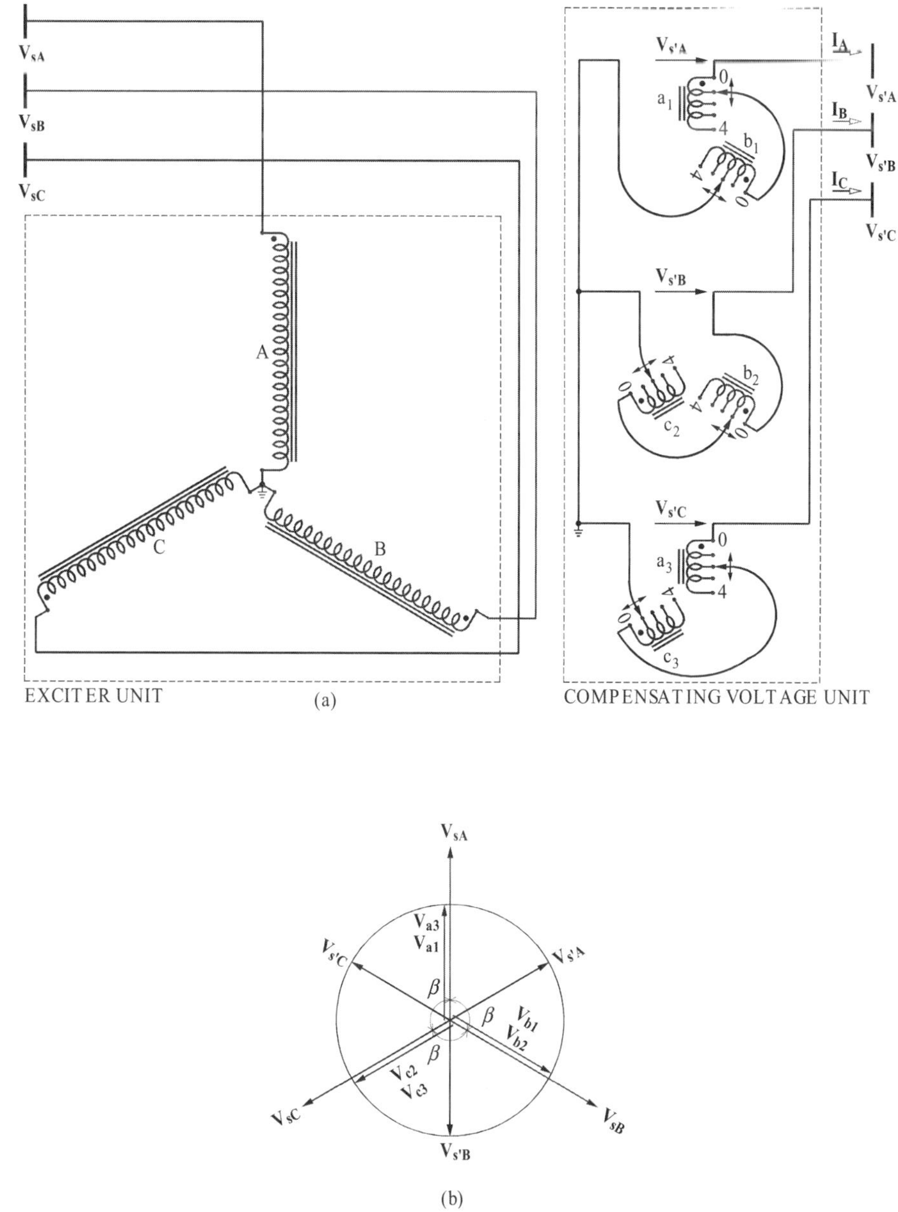

Figure 9-37. (a) ST with shunt-connected compensating voltage unit operating in the range of 240° through 360°. (b) Phasor diagram.

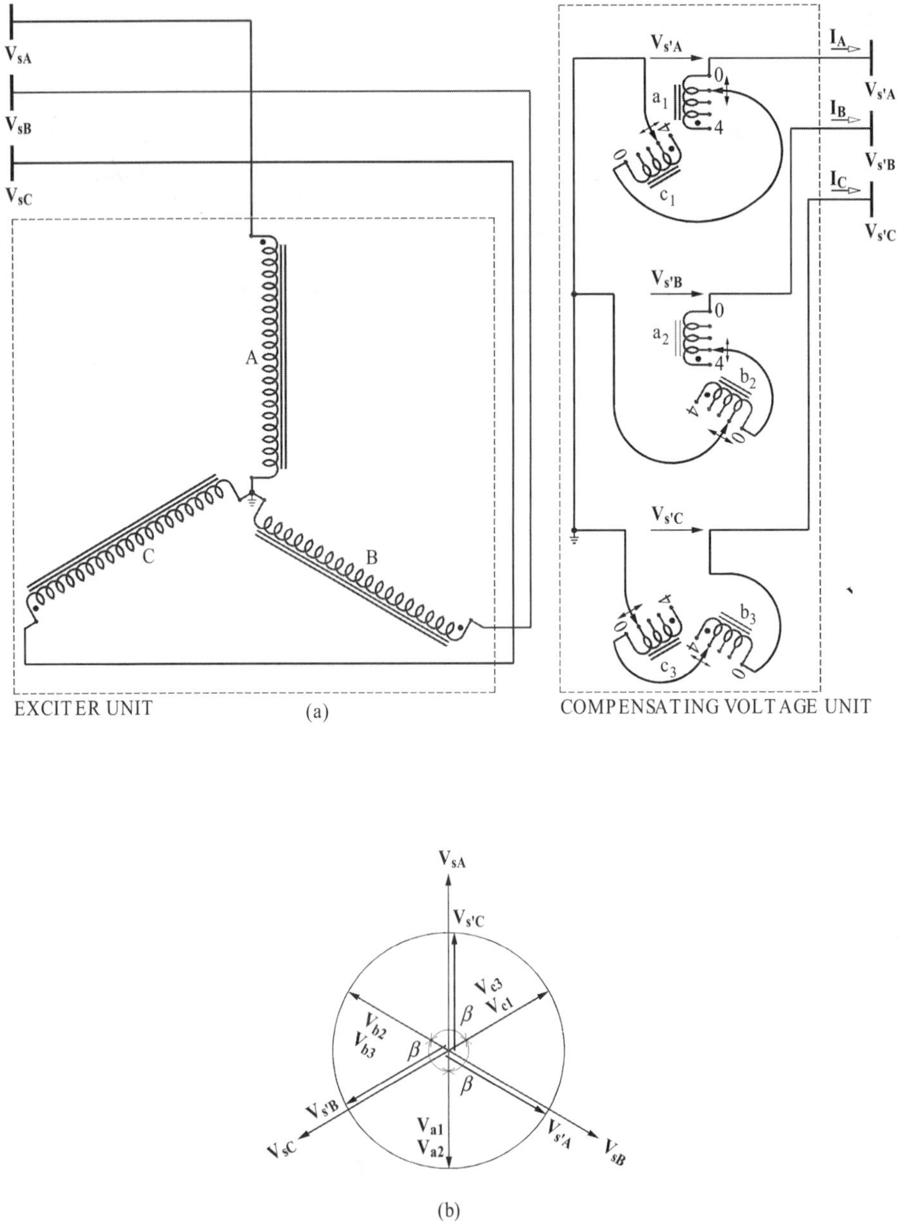

Figure 9-38. (a) ST with shunt-connected compensating voltage unit operating in the range of 180° through 300°. (b) Phasor diagram.

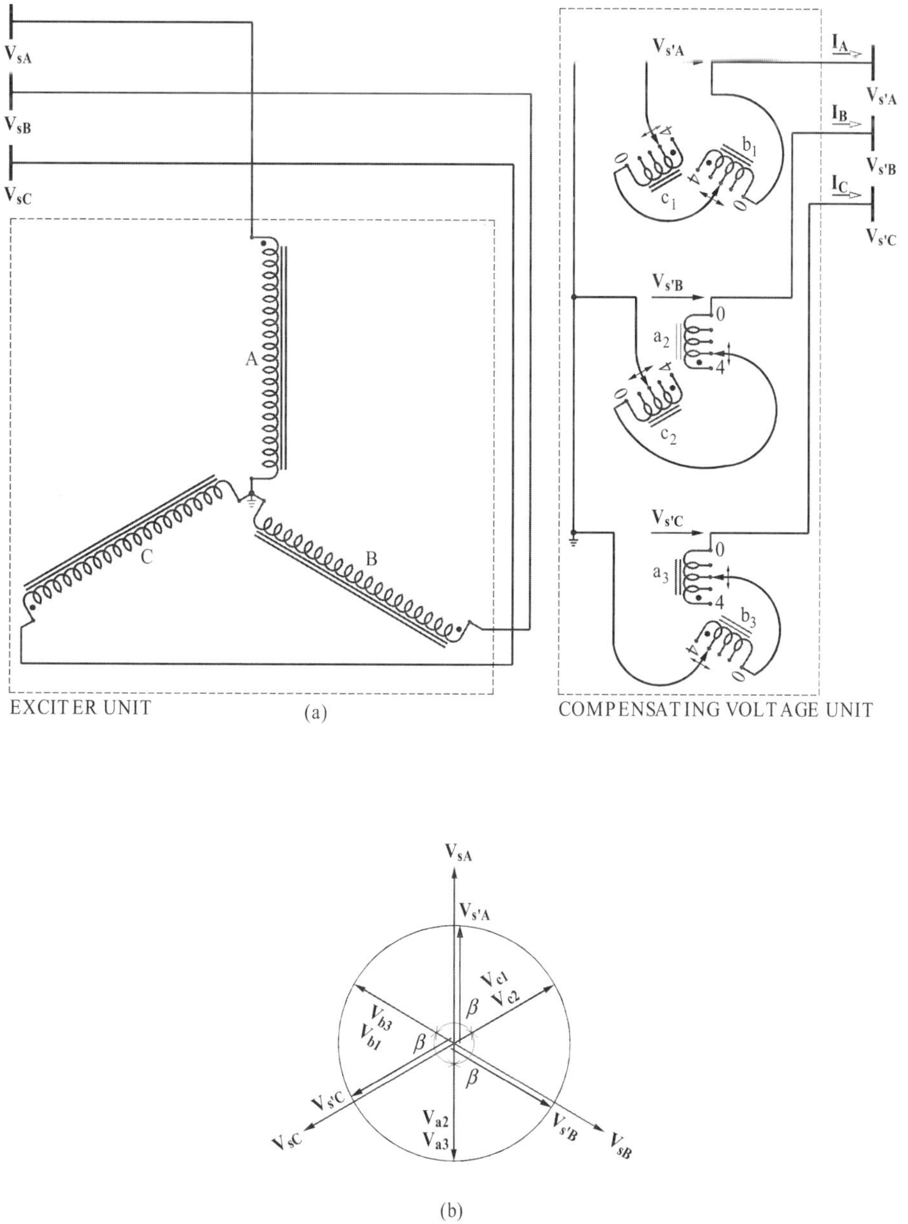

Figure 9-39. (a) ST with shunt-connected compensating voltage unit operating in the range of 300° through 60°. (b) Phasor diagram.

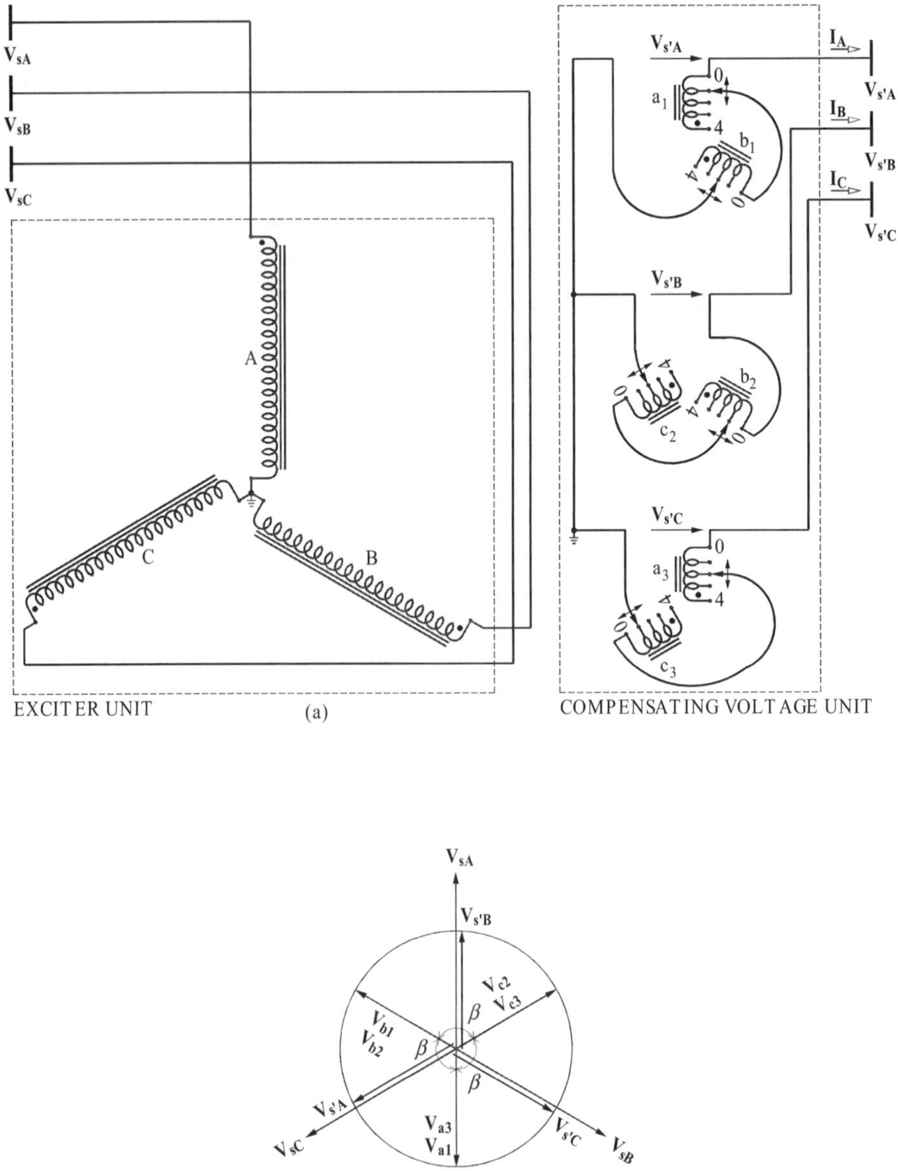

Figure 9-40. (a) ST with shunt-connected compensating voltage unit operating in the range of 60° through 180°. (b) Phasor diagram.

Extending the concept just presented, if the polarities of the windings in the shunt-connected compensating voltage unit in Figure 9-35a are reversed, the phase angle of the shunt-connected compensating voltage($\mathbf{V}_{s'}$) can vary between 180° and 300°. The schematic for this configuration is shown in Figure 9-38a. In the same way, if the polarities of the windings in the compensating voltage units, presented in Figures 9-36a and 9-37a, are reversed, then the phase angle of the compensating voltage ($\mathbf{V}_{s'}$) can vary from 300° to 60° and 60° to 180°, respectively. The schematics for these configurations are shown in Figures 9-39a and 9-40a. The related phasor diagrams are shown in Figures 9-38b, 9-39b, and 9-40b.

9.8 MST WITH SHUNT-CONNECTED COMPENSATING VOLTAGES

An MST connects a compensating voltage in shunt with each transmission line in a multiline transmission system. In certain special cases for point-to-point transfer of power between two isolated networks with different voltages or phase angles, the use of the shunt–shunt power conversion remains the preferred choice. In the MST configuration, shown in Figure 9-41, the line voltage ($\mathbf{V}_s$) is applied to a shunt-connected, three-phase transformer's Y-connected primary windings in the exciter unit. In the compensating voltage unit 1, a total of nine secondary windings (a_{11}, a_{12}, and a_{13} on the core of the A phase; b_{11}, b_{12}, and b_{13} on the core of the B phase; and c_{11}, c_{12}, and c_{13} on the core of the C phase) derive a three-phase compensating voltage, each phase of which is the phasor sum of the voltages induced in a three-winding set (a_{11}, b_{11}, and c_{11} for compensating voltage in the A phase; a_{12}, b_{12}, and c_{12} for compensating voltage in the B phase; and a_{13}, b_{13}, and c_{13} for compensating voltage in the C phase). In the compensating voltage unit 2, the same arrangement is used. By choosing the number of turns from each of the three windings and, therefore, the magnitudes of the components of the three 120° phase-shifted induced voltages, the composite compensating voltage's magnitude and its relative phase angle with respect to the transmission line voltage ($\mathbf{V}_s$) can be selected.

It should be noted that in the compensating voltage unit 1, each of a_{11}, b_{12}, and c_{13} is tapped at the same number of turns; each of b_{11}, c_{12}, and a_{13} is tapped at the same number of turns; and each of c_{11}, a_{12}, and b_{13} is tapped at the same number of turns. However, the number of turns in each set can be different from the other. The same is true for the compensating voltage unit 2.

The compensating voltage ($\mathbf{V}_{1s'}$) in the compensating voltage unit 1 can be at any phase angle with the prevailing line current ($\mathbf{I}_1$), which emulates an impedance in shunt with line 1. The compensating voltage ($\mathbf{V}_{2s'}$) in the compensating voltage unit 2 can be at any phase angle with the prevailing line current ($\mathbf{I}_2$), which emulates an impedance in shunt with line 2. An emulated impedance can be a capacitor or an inductor and a positive resistor or a negative resistor.

The operations of the compensating voltage units are such that the compensating voltage in the "master" line is at any phase angle with the line voltage as well as the prevailing line current and that in the "slave" line is of specific magnitude and phase

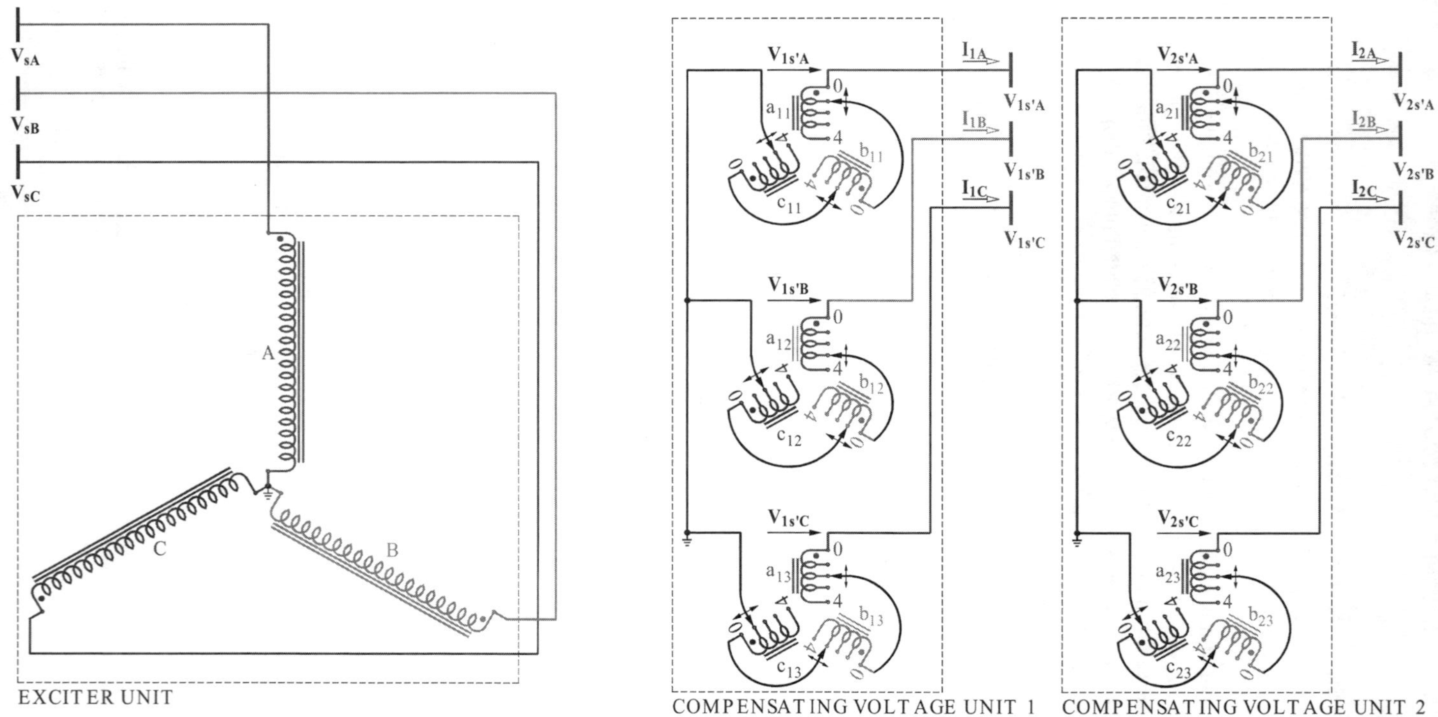

Figure 9-41. MST with shunt-connected compensating voltage units.

angle, so that the active and reactive powers from the "master" line are transferred bidirectionally to the "slave" line. As discussed earlier, the 120° range of operation of the PFC is often sufficient. The MST, in its simpler customized configuration with a reduced amount of hardware, provides a limited angle (120°) operation that may be sufficient in case of a point-to-point power transfer between two isolated networks.

9.9 GENERALIZED SEN TRANSFORMER

The generalized Sen transformer (GST), shown in Figure 9-42, can be used as an extension of the MST that generates both the shunt- and series-connected compensating voltages in a single unit. The series-connected compensating voltages serve as the PFCs that control the active and reactive power flows in each transmission line independently and transfer the active and reactive powers from one or more "master" transmission lines to the "slave" lines. The shunt compensating voltages connect the isolated networks with different voltages and phase angles. Each of the compensating voltages is induced from an exciter voltage through transformer action. Therefore, any mismatch of active and reactive powers between various compensating voltages flows to the line that supplies the exciting unit of the GST.

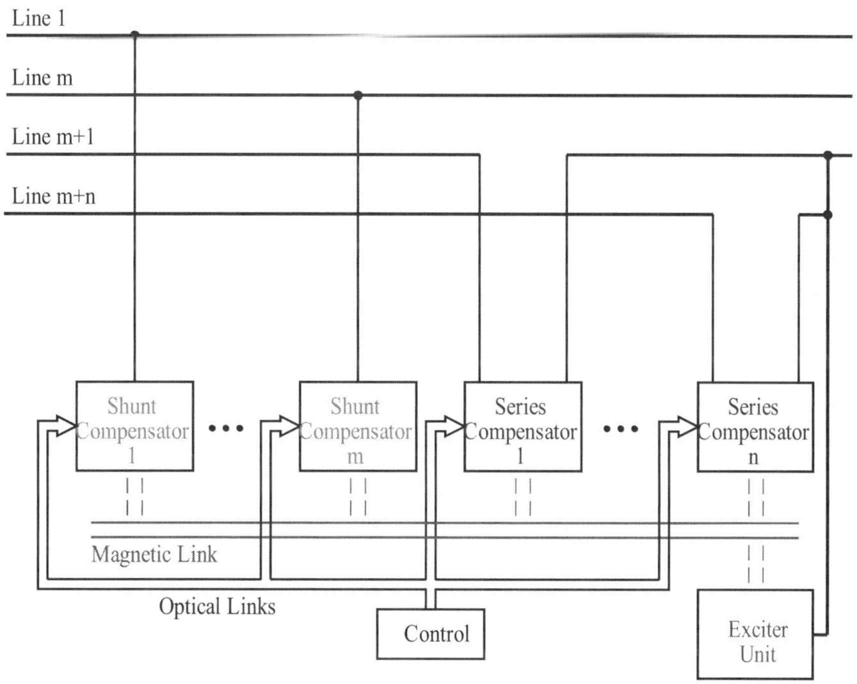

Figure 9-42. Generalized Sen transformer.

Thus, the GST can

1. Transfer the active and reactive powers between a number of transmission lines.
2. Control the active and reactive power flows in each line selectively.
3. Interconnect transmission lines with different voltages and phase angles.

The dynamic performance of this technology is limited by the operational speed of the LTCs. The mechanical LTCs are quite adequate for most utility applications. The dynamic performance can be improved, when needed, by replacing the present LTCs with faster switches.

9.10 SUMMARY

The Sen transformer (ST) is an alternate low-cost and much more efficient PFC that modifies the magnitude and phase angle of the transmission line voltage, which results in the same independent control of active and reactive power flows as the UPFC, albeit at a reduced dynamic rate. The ST uses the newest reliable, cost-effective, and proven transformer and LTC-based technology. Since the speed of operation is not essential in most utility applications, the low cost and high efficiency of the ST becomes attractive when compared with that of the VSC-based UPFC solution for the same amount of power flow enhancement. The series-connected compensating voltage in both the ST and the UPFC exchanges active and reactive powers with the line. In the case of the ST, these exchanged active and reactive powers flow back and forth through the shunt unit to the line. However, in the case of the UPFC, only the exchanged active power flows back and forth through the shunt unit to the line. Each of the series- and shunt-connected VSCs in the UPFC is capable of generating or absorbing reactive power. If this capability of supplying shunt reactive power is really needed, particularly when the short-circuit impedance at the point of connection is high, a separate capacitor bank as a shunt reactive compensator may be used along with the ST.

APPENDIX A

MISCELLANEOUS

A.I. THREE-PHASE BALANCED VOLTAGE, CURRENT, AND POWER

Consider a three-phase balanced voltage,

$$v_A = V_p \cos(\theta + \varphi_v) \tag{A-1a}$$

$$v_B = V_p \cos(\theta + \varphi_v - 120^\circ) \tag{A-1b}$$

$$v_C = V_p \cos(\theta + \varphi_v + 120^\circ) \tag{A-1c}$$

supplying a three-phase balanced current,

$$i_A = I_p \cos(\theta + \varphi_i) \tag{A-2a}$$

$$i_B = I_p \cos(\theta + \varphi_i - 120^\circ) \tag{A-2b}$$

$$i_C = I_p \cos(\theta + \varphi_i + 120^\circ) \tag{A-2c}$$

through a three-phase balanced impedance (**Z**), shown in Figure A-1, where the amplitudes of voltage and current in each phase are V_p and I_p, respectively, and $|\mathbf{Z}| = V_p/I_p$. The system angle, angular frequency, and time are denoted by θ, ω, and t, respectively. The phase angles of voltage and current from the reference are ϕ_v and ϕ_i, respectively. The timing diagrams for one cycle of balanced three-phase voltages and currents and the related phasor diagrams at the end of the cycle are shown in Figures A-2 and A-3, respectively. Note that the vertical projections of the rotating phasors are their instantaneous values.

The instantaneous power in the A, B, and C phases are

$$p_A = v_A i_A \tag{A-3a}$$

$$p_B = v_B i_B \tag{A-3b}$$

$$p_C = v_C i_C \tag{A-3c}$$

From equations A-1a, A-2a, and A-3a, the A phase instantaneous power is obtained as

$$p_A = V_p I_p \cos(\theta + \varphi_v)\cos(\theta + \varphi_i) = \frac{V_p I_p}{2}\left[\cos(2\theta + \varphi_v + \varphi_i) + \cos(\varphi_v - \varphi_i)\right] \tag{A-4a}$$

Similarly, the B phase instantaneous power is obtained from equations A-1b, A-2b, and A-3b as

$$\begin{aligned} p_B &= V_p I_p \cos(\theta + \varphi_v - 120^\circ)\cos(\theta + \varphi_i - 120^\circ) \\ &= \frac{V_p I_p}{2}\left[\cos(2\theta + \varphi_v + \varphi_i + 120^\circ) + \cos(\varphi_v - \varphi_i)\right] \end{aligned} \tag{A-4b}$$

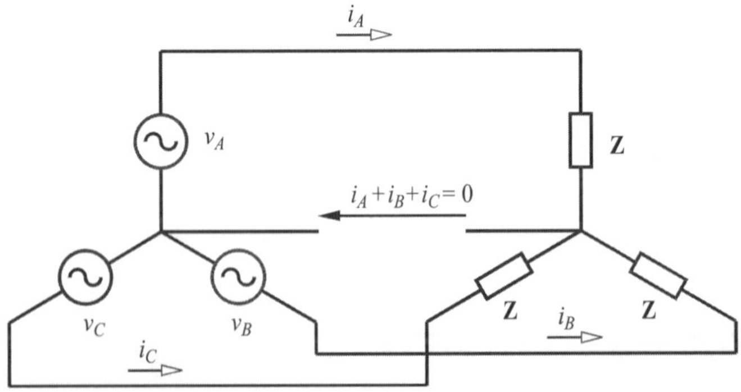

Figure A-1. Three-phase voltage source supplying currents to a three-phase load.

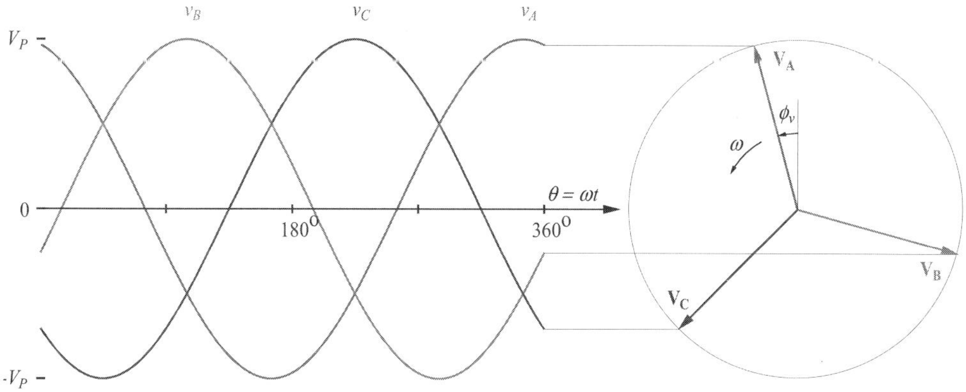

Figure A-2. Balanced three-phase voltages in a timing diagram and the corresponding phasor diagram at the end of the cycle.

Finally, the C phase instantaneous power is obtained from equations A-1c, A-2c, and A-3c as

$$p_C = V_p I_p \cos(\theta + \varphi_v + 120^\circ)\cos(\theta + \varphi_i + 120^\circ)$$

$$= \frac{V_p I_p}{2}\left[\cos(2\theta + \varphi_v + \varphi_i - 120^\circ) + \cos(\varphi_v - \varphi_i)\right] \qquad \text{(A-4c)}$$

In each of the three preceding expressions, there exists a double frequency cosine term with an amplitude of $0.5V_pI_p$. There also exists a fixed term $0.5V_pI_p\cos(\phi_v - \phi_i)$.

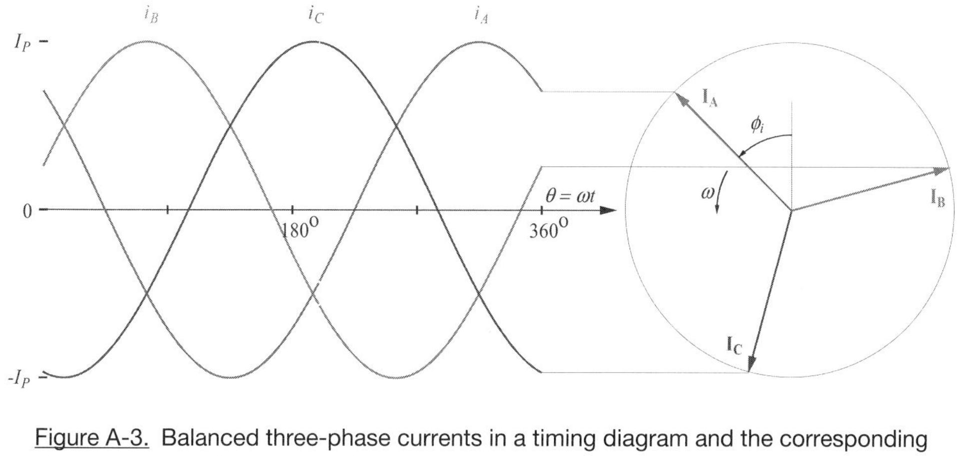

Figure A-3. Balanced three-phase currents in a timing diagram and the corresponding phasor diagram at the end of the cycle.

Since the integral of the cosine term over its one period is zero, the average power in each phase is $0.5V_pI_p \cos(\phi_v - \phi_i)$ and the three-phase average power is

$$P = \frac{3}{2}V_pI_p\cos(\varphi_v - \varphi_i) = \frac{3}{2}V_pI_p\cos\varphi \tag{A-5}$$

This is also referred to as active power whose maximum possible value is $1.5V_pI_p$ and this maximum value is referred to as apparent power (S). The ratio, P/S (i.e., cos ϕ), is referred to as the power factor. The amount of power that can be taken out of the apparent power without altering the active power is the reactive power and it is

$$Q = \sqrt{S^2 - P^2} = \frac{3}{2}V_pI_p\sqrt{1-\cos^2\varphi} = \frac{3}{2}V_pI_p\sin\varphi \tag{A-6}$$

Also,

$$P = \frac{3}{2}V_pI_p\cos\varphi = 3\frac{V_p}{\sqrt{2}}\frac{I_p}{\sqrt{2}}\cos\varphi = 3V_{LNrms}I_{rms}\cos\varphi = \sqrt{3}(\sqrt{3}V_{LNrms})I_{rms}\cos\varphi$$

$$= \sqrt{3}V_{LLrms}I_{rms}\cos\varphi \tag{A-7}$$

where V_{LNrms} and V_{LLrms} are the line-to-neutral and line-to-line rms voltages, respectively, and I_{rms} is the line rms current.

Also, the expression for the three-phase active power can be written as

$$P = \frac{3}{2}V_pI_p\cos(\varphi_v - \varphi_i) = \frac{3}{2}V_pI_p\,\mathrm{Re}\left[e^{j(\varphi_v-\varphi_i)}\right] = \frac{3}{2}\mathrm{Re}\left[V_pe^{j\varphi_v}I_pe^{-j\varphi_i}\right] = \frac{3}{2}\mathrm{Re}\left[\mathbf{V}\mathbf{I}^*\right] \tag{A-8}$$

The expression for the three-phase reactive power can be written as

$$Q = \frac{3}{2}V_pI_p\sin(\varphi_v - \varphi_i) = \frac{3}{2}V_pI_p\,\mathrm{Im}\left[e^{j(\varphi_v-\varphi_i)}\right] = \frac{3}{2}\mathrm{Im}\left[V_pe^{j\varphi_v}I_pe^{-j\varphi_i}\right] = \frac{3}{2}\mathrm{Im}\left[\mathbf{V}\mathbf{I}^*\right] \tag{A-9}$$

The expression for apparent power in its complex form can then be

$$\mathbf{S} = \frac{3}{2}\mathbf{V}\mathbf{I}^* = P + jQ \tag{A-10}$$

The three-phase instantaneous power that is defined as

$$p = p_A + p_B + p_C \tag{A-11}$$

is also $p = 1.5V_pI_p\cos\phi$ since the double frequency cosine terms in the three instantaneous phase power expressions are 120° apart and their sum is zero.

The three-phase instantaneous active and reactive powers are defined as

$$p = \frac{3}{2} V_p I_p \cos\varphi = p_A + p_B + p_C$$

from equation A-11,

$$p = v_A i_A + v_B i_B + v_C i_C \tag{A-12a}$$

from equation A-3(a, b, c), and

$$q = \frac{3}{2} V_p I_p \sin\varphi \tag{A-12b}$$

(more derivation is given later).

Note that when $\phi = 0$, the load is at the unity power factor. In this case, the three-phase active power is the same as the three-phase apparent power and the three-phase reactive power is zero. When $\phi = \pm\pi/2$, the load is at the zero power factor. In this case, the three-phase reactive power is the same as the three-phase apparent power and the three-phase active power is zero.

The analysis of balanced three-phase circuits is straightforward. The analysis of unbalanced three-phase circuits that occur due to unbalanced loads, unbalanced faults, or short circuits is, however, more difficult and can be simplified with the concept of symmetrical components.

A.II. SYMMETRICAL COMPONENTS

Any unbalanced three-phase variables, as illustrated in Figure A-4, can be resolved into two sets of balanced three-phase variables and a set of single-phase variables. The three-phase variables in the first set reach their respective peaks in the A-phase, B-phase, and C-phase sequence; therefore, this set is said to be in positive sequence. The three-phase variables in the second set reach their peaks in the A-phase, C-phase, and B-phase sequence; therefore, this set is said to be in negative sequence. The variables in the third set overlap with each other; therefore this set is said to be in zero sequence.

Each phase variable consists of positive, negative, and zero sequence components, such as

$$f_A = f_{A+} + f_{A-} + f_0 \tag{A-13a}$$

$$f_B = f_{B+} + f_{B-} + f_0 \tag{A-13b}$$

$$f_C = f_{C+} + f_{C-} + f_0 \tag{A-13c}$$

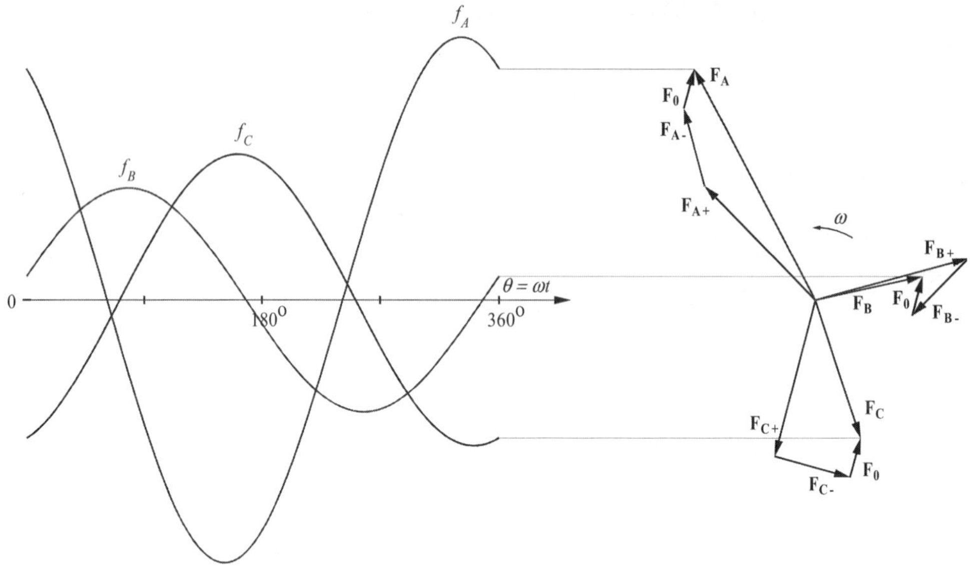

Figure A-4. Unbalanced three-phase variables in a timing diagram and the corresponding phasor diagram at the end of the cycle.

where f_{A+}, f_{B+}, and f_{C+} are the positive sequence components; f_{A-}, f_{B-}, and f_{C-} are the negative sequence components; and f_0 is the zero sequence component of the three-phase variables A, B, and C. The phase variable (f) can be voltage, current, flux, and so on.

In phasor form, the above three-phase variables A-13a to A-13c become

$$\mathbf{F_A} = \mathbf{F_{A+}} + \mathbf{F_{A-}} + \mathbf{F_0} \tag{A-14a}$$

$$\mathbf{F_B} = \mathbf{F_{B+}} + \mathbf{F_{B-}} + \mathbf{F_0} \tag{A-14b}$$

$$\mathbf{F_C} = \mathbf{F_{C+}} + \mathbf{F_{C-}} + \mathbf{F_0} \tag{A-14c}$$

Figure A-5 shows the balanced, three-phase, positive sequence components in a timing diagram for one cycle and the corresponding phasor diagram at the end of the cycle. The vertical projections of the three rotating phasors are the instantaneous values of the three-phase variables.

The positive sequence components can be related to each other with the help of a unit vector operator (a), where $a = e^{j120°}$, $a^2 = e^{j240°} = e^{-j120°}$, and $a^3 = e^{j360°} = 1$. The B- and C-phase variables are expressed in terms of the A-phase variable as

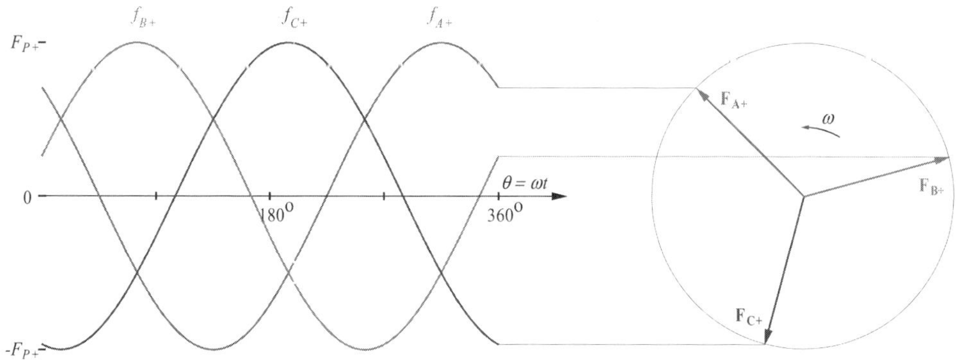

Figure A-5. Balanced three-phase, positive sequence components in a timing diagram and the corresponding phasor diagram at the end of the cycle.

$$\mathbf{F_{B+}} = a^2\mathbf{F_{A+}} \quad \text{(A-15a)}$$

$$\mathbf{F_{C+}} = a\mathbf{F_{A+}} \quad \text{(A-15b)}$$

Figure A-6 shows the balanced, three-phase, negative sequence components in a timing diagram for one cycle and the corresponding phasor diagram at the end of the cycle. The vertical projections of the three rotating phasors are the instantaneous values of the three-phase variables.

The negative sequence B- and C-phase components can be expressed in terms of the negative sequence A-phase component as

$$\mathbf{F_{B-}} = a\mathbf{F_{A-}} \quad \text{(A-16a)}$$

$$\mathbf{F_{C-}} = a^2\mathbf{F_{A-}} \quad \text{(A-16b)}$$

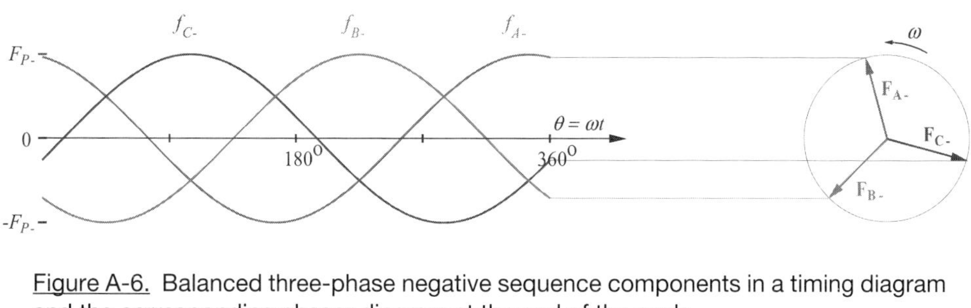

Figure A-6. Balanced three-phase negative sequence components in a timing diagram and the corresponding phasor diagram at the end of the cycle.

Figure A-7 shows the zero sequence component in a timing diagram for one cycle and the corresponding phasor diagram at the end of the cycle. Three zero sequence variables from the three phases overlap with each other. The vertical projection of the rotating phasor is the instantaneous value of the zero sequence component.

Therefore, the three-phase variables defined earlier can be rewritten as

$$\mathbf{F}_{\mathbf{A}} = \mathbf{F}_{\mathbf{A}+} + \mathbf{F}_{\mathbf{A}-} + \mathbf{F}_{0} \qquad \text{(A-17a)}$$

$$\mathbf{F}_{\mathbf{B}} = a^2\mathbf{F}_{\mathbf{A}+} + a\mathbf{F}_{\mathbf{A}-} + \mathbf{F}_{0} \qquad \text{(A-17b)}$$

$$\mathbf{F}_{\mathbf{C}} = a\mathbf{F}_{\mathbf{A}+} + a^2\mathbf{F}_{\mathbf{A}-} + \mathbf{F}_{0} \qquad \text{(A-17c)}$$

Keeping the expression for $\mathbf{F}_{\mathbf{A}}$ as it is and multiplying the expressions for $\mathbf{F}_{\mathbf{B}}$ and $\mathbf{F}_{\mathbf{C}}$ with a and a^2, respectively, equations A-17a to A-17c become

$$\mathbf{F}_{\mathbf{A}} = \mathbf{F}_{\mathbf{A}+} + \mathbf{F}_{\mathbf{A}-} + \mathbf{F}_{0} \qquad \text{(A-18a)}$$

$$a\mathbf{F}_{\mathbf{B}} = \mathbf{F}_{\mathbf{A}+} + a^2\mathbf{F}_{\mathbf{A}-} + a\mathbf{F}_{0} \qquad \text{(A-18b)}$$

$$a^2\mathbf{F}_{\mathbf{C}} = \mathbf{F}_{\mathbf{A}+} + a\mathbf{F}_{\mathbf{A}-} + a^2\mathbf{F}_{0} \qquad \text{(A-18c)}$$

The addition of the above three equations results in

$$\mathbf{F}_{\mathbf{A}} + a\mathbf{F}_{\mathbf{B}} + a^2\mathbf{F}_{\mathbf{C}} = 3\mathbf{F}_{\mathbf{A}+} + (\mathbf{F}_{\mathbf{A}-} + a^2\mathbf{F}_{\mathbf{A}-} + a\mathbf{F}_{\mathbf{A}-}) + (\mathbf{F}_{0} + a\mathbf{F}_{0} + a^2\mathbf{F}_{0}) = 3\,\mathbf{F}_{\mathbf{A}+} \qquad \text{(A-19)}$$

Therefore, the positive sequence component is found to be

$$\mathbf{F}_{\mathbf{A}+} = \frac{1}{3}(\mathbf{F}_{\mathbf{A}} + a\mathbf{F}_{\mathbf{B}} + a^2\mathbf{F}_{\mathbf{C}}) \qquad \text{(A-20)}$$

Keeping the expression for $\mathbf{F}_{\mathbf{A}}$ as it is and multiplying the expressions for $\mathbf{F}_{\mathbf{B}}$ and $\mathbf{F}_{\mathbf{C}}$ with a^2 and a, respectively, equations A-17a to A-17c become

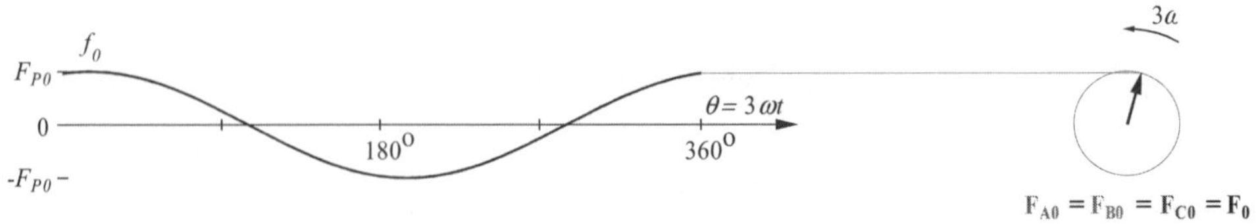

Figure A-7. Zero sequence component in a timing diagram and the corresponding phasor diagram at the end of the cycle.

$$\mathbf{F}_A = \mathbf{F}_{A+} + \mathbf{F}_{A-} + \mathbf{F}_0 \tag{A-21a}$$

$$a^2\mathbf{F}_B = a\mathbf{F}_{A+} + \mathbf{F}_{A-} + a^2\mathbf{F}_0 \tag{A-21b}$$

$$a\mathbf{F}_C = a^2\mathbf{F}_{A+} + \mathbf{F}_{A-} + a\mathbf{F}_0 \tag{A-21c}$$

The addition of the above three equations results in

$$\mathbf{F}_A + a^2\mathbf{F}_B + a\mathbf{F}_C = (\mathbf{F}_{A+} + a\mathbf{F}_{A+} + a^2\mathbf{F}_{A+}) + 3\mathbf{F}_{A-} + (\mathbf{F}_0 + a^2\mathbf{F}_0 + a\mathbf{F}_0) = 3\,\mathbf{F}_{A-} \tag{A-22}$$

Therefore, the negative sequence component is found to be

$$\mathbf{F}_{A-} = \frac{1}{3}(\mathbf{F}_A + a^2\mathbf{F}_B + a\mathbf{F}_C) \tag{A-23}$$

Simply by adding the expressions for $\mathbf{F}_A$, $\mathbf{F}_B$, and $\mathbf{F}_C$, equations A-17a to A-17c become

$$\mathbf{F}_A + \mathbf{F}_B + \mathbf{F}_C = (\mathbf{F}_{A+} + a^2\mathbf{F}_{A+} + a\mathbf{F}_{A+}) + (\mathbf{F}_{A-} + a\mathbf{F}_{A-} + a^2\mathbf{F}_{A-}) + 3\mathbf{F}_0 = 3\mathbf{F}_0 \tag{A-24}$$

Therefore, the zero sequence component is found to be

$$\mathbf{F}_0 = \frac{1}{3}(\mathbf{F}_A + \mathbf{F}_B + \mathbf{F}_C) \tag{A-25}$$

The following is an example of decomposing an unbalanced set of phasors, shown in Figure A-4, into three balanced sets of phasors, as shown in Figures A-5, A-6, and A-7.

The unbalanced set of phasors, as shown in Figure A-4, are

$$\mathbf{F}_A = 1.625\angle 28.35^\circ = 1.43 + j0.7716$$

$$\mathbf{F}_B = 0.6928\angle -77.77^\circ = 0.14676 - j0.6771$$

$$\mathbf{F}_C = 0.9013\angle -161.32^\circ = -0.85382 - j0.2888$$

From equation A-20, the A-phase positive sequence component is

$$\mathbf{F}_{A+} = \frac{1}{3}(\mathbf{F}_A + a\mathbf{F}_B + a^2\mathbf{F}_C) = 1\angle 45^\circ = 0.7071 + j0.7071$$

as shown in Figures A-5 and A-8.

From equations A-15a and A-15b, the B- and C-phase positive sequence components are

$$\mathbf{F}_{B+} = a^2\mathbf{F}_{A+} = 1\angle 240^\circ \times 1\angle 45^\circ = 1\angle 285^\circ = 0.2588 - j0.9659$$

as shown in Figures A-5 and A-9.

$$\mathbf{F}_{C+} = a\mathbf{F}_{A+} = 1\angle 120^\circ \times 1\angle 45^\circ = 1\angle 165^\circ = -0.9659 + j0.2588$$

as shown in Figures A-5 and A-10.

The three phasors are of the same magnitude and 120° apart from each other. Since their sum is zero, they are balanced. Since the A phase leads the B phase and the B phase leads the C phase, the phasors are in positive sequence.

From equation A-23, the A-phase negative sequence component is

$$\mathbf{F}_{A-} = \frac{1}{3}(\mathbf{F}_A + a^2\mathbf{F}_B + a\mathbf{F}_C) = 0.5\angle 15^\circ = 0.48296 + j0.1294$$

as shown in Figures A-6 and A-8.

From equations A-16a and A-16b, the B- and C-phase negative sequence components are

$$\mathbf{F}_{B-} = a\mathbf{F}_{A-} = 1\angle 120^\circ \times 0.5\angle 15^\circ = 0.5\angle 135^\circ = -0.3535 + j0.3535$$

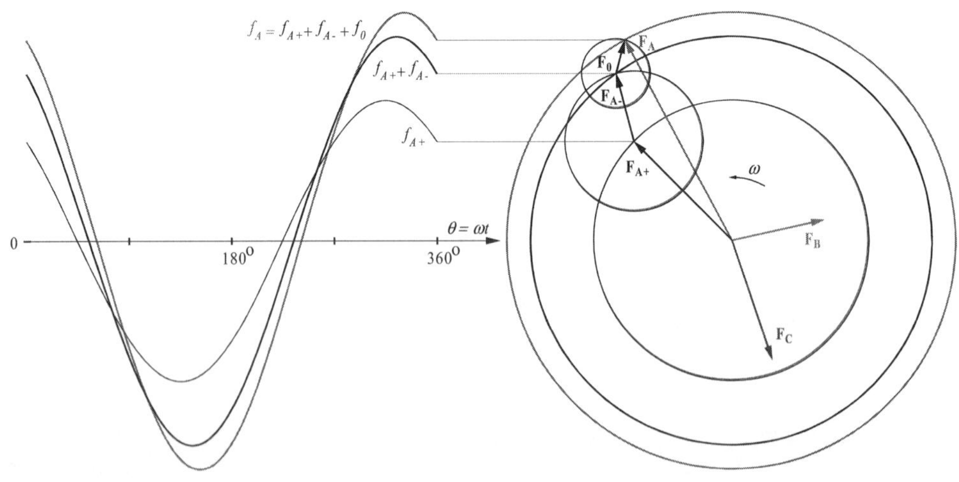

Figure A-8. A-phase positive, negative, and zero sequence components in a timing diagram and the corresponding phasor diagram at the end of the cycle.

as shown in Figures A-6 and A-9,

$$\mathbf{F}_{C-} = a^2\mathbf{F}_{A-} = 1\angle 240^\circ \times 0.5\angle 15^\circ = 0.5\angle 255^\circ = -0.1294 - j0.4829$$

as shown in Figures A-6 and A-10.

The three phasors are of the same magnitude and 120° apart from each other. Since their sum is zero, they are balanced. Since the A phase leads the C phase and the C phase leads the B phase, the phasors are in negative sequence.

From equation A-25, the A-phase zero sequence component is

$$\mathbf{F}_0 = \frac{1}{3}(\mathbf{F}_A + \mathbf{F}_B + \mathbf{F}_C) = \frac{1}{3}(1.43 + j0.7716 + 0.14676 - j0.6771 - 0.85382 - j0.2888)$$

$$= 0.241 - j0.064 = 0.25\angle -15^\circ$$

as shown in Figures A-7, A-8, A-9, and A-10.

The zero sequence phasors in the three phases are of the same magnitude and they overlap.

A.III. SEPARATION OF POSITIVE, NEGATIVE, AND ZERO SEQUENCE COMPONENTS IN A MULTIPLE FREQUENCY COMPOSITE VARIABLE

The various control algorithms in this book are implemented in a positive sequence, synchronously rotating reference frame that converts the positive sequence fundamental variable into DC. Any other component, if present, must be filtered out at the input of the control block. Consider the previous example with the same frequency positive

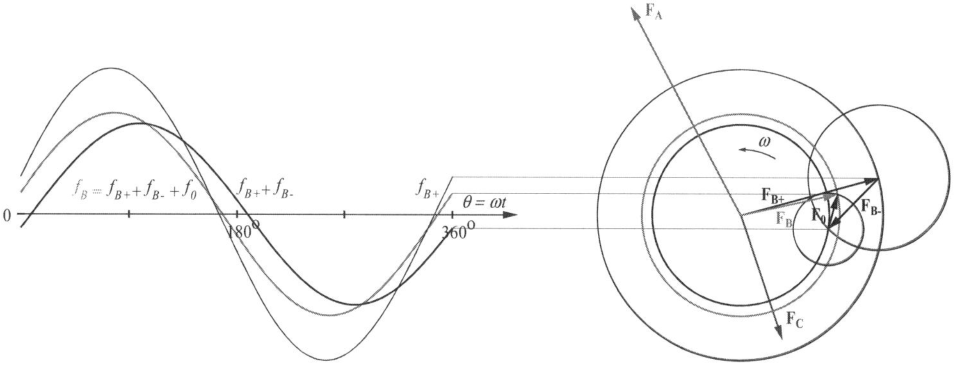

Figure A-9. B-phase positive, negative, and zero sequence components in a timing diagram and the corresponding phasor diagram at the end of the cycle.

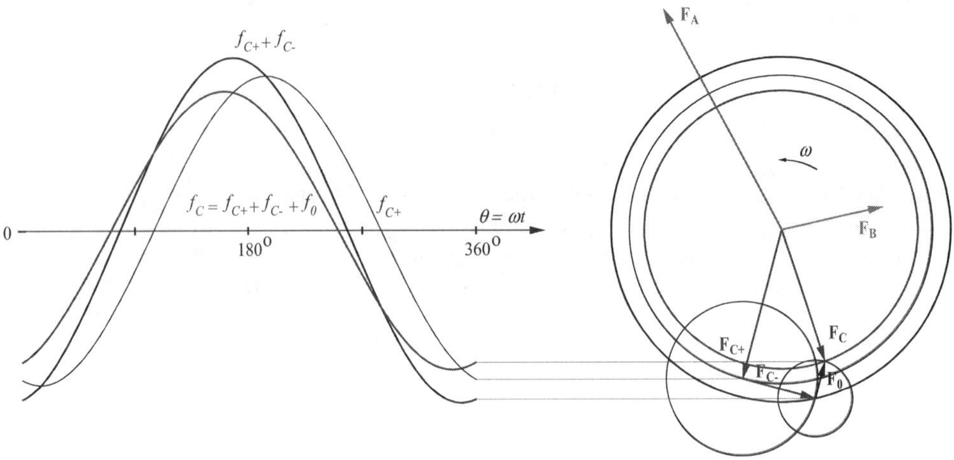

Figure A-10. C-phase positive, negative, and zero sequence components in a timing diagram and the corresponding phasor diagram at the end of the cycle.

and negative sequence components, but with a triple frequency zero sequence component, as shown in Figure A-11.

The timing diagram for one cycle of a set of unbalanced three-phase variables, consisting of positive, negative, and triple frequency zero sequence components and the corresponding phasor diagram at the end of the cycle are shown in Figure A-12.

The A–B–C three-phase variables positive, negative, and zero sequence components in three phases are shown in Figures A-13 through A-15, respectively.

The zero sequence component (f_0) is

$$f_0 = \frac{1}{3}(f_A + f_B + f_C) \tag{A-26}$$

Therefore, the three-phase variables that consist of positive and negative sequence components only are

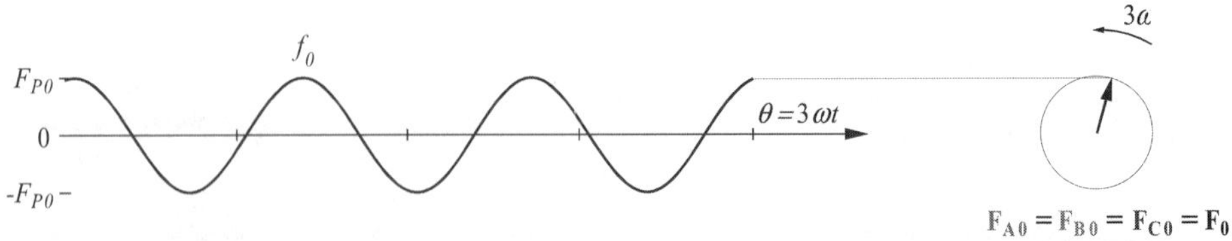

Figure A-11. Triple frequency zero sequence component in a timing diagram and the corresponding phasor diagram at the end of the cycle.

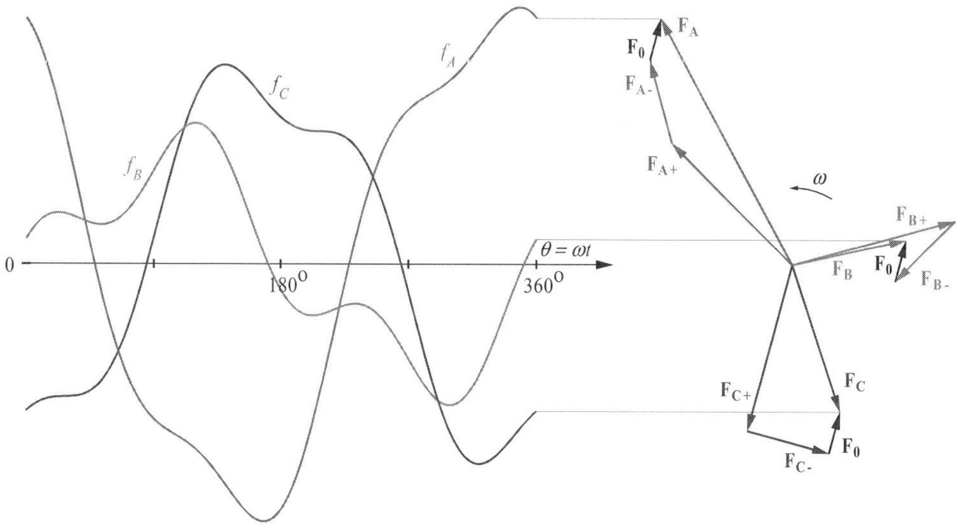

Figure A-12. Unbalanced three-phase variables in a timing diagram and the corresponding phasor diagram at the end of the cycle.

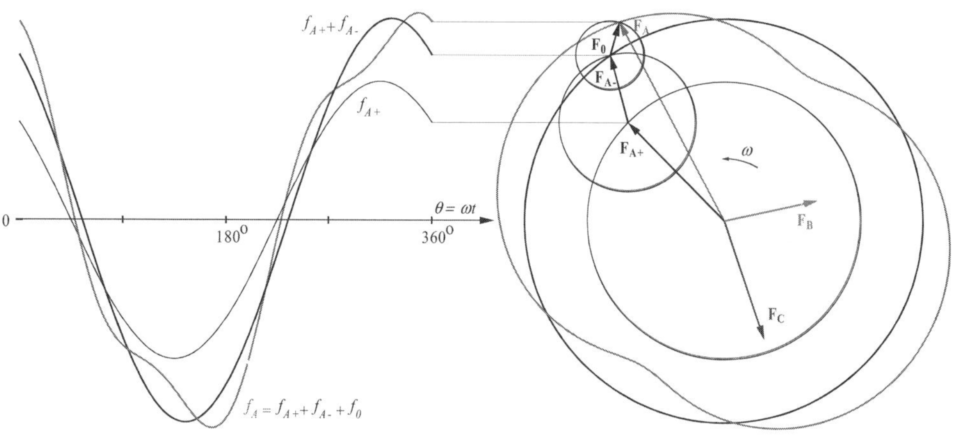

Figure A-13. A-phase positive, negative, and zero sequence components in a timing diagram and the corresponding phasor diagram at the end of the cycle.

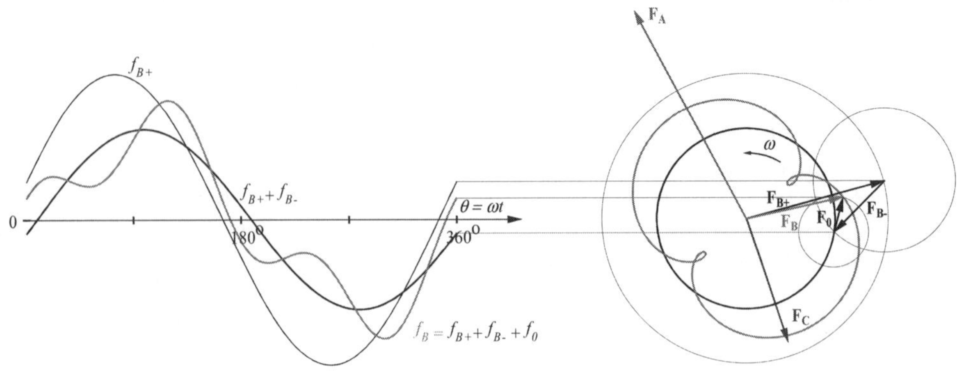

Figure A-14. B-phase positive, negative, and zero sequence components in a timing diagram and the corresponding phasor diagram at the end of the cycle.

$$f_A' = f_A - f_0 \tag{A-27a}$$

$$f_B' = f_B - f_0 \tag{A-27b}$$

$$f_C' = f_C - f_0 \tag{A-27c}$$

The three-phase variables ($f_{A'}$, $f_{B'}$, and $f_{C'}$) can be used in conjunction with a vector PLL, as discussed in Chapter 3, to calculate the PLL angle (θ) that is the angle of the positive sequence component (f_{A+}). The amplitude (F) consists of two compo-

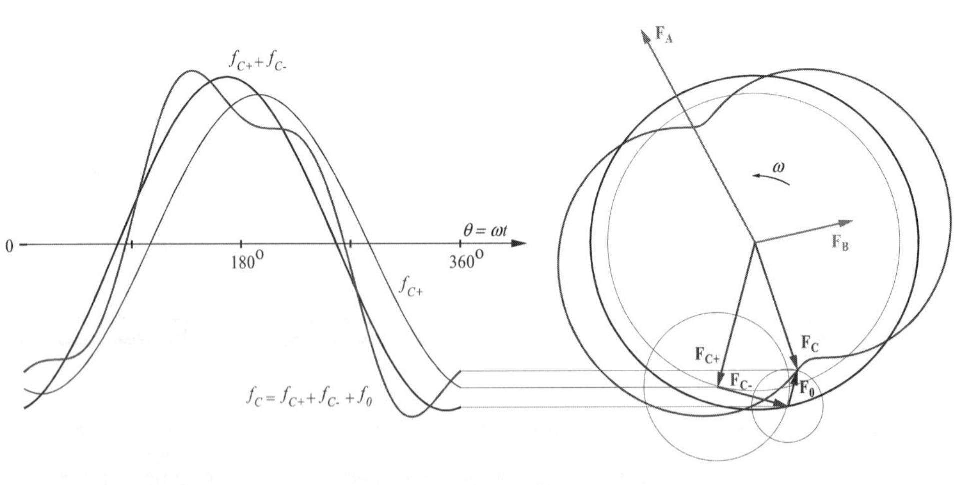

Figure A-15. C-phase positive, negative, and zero sequence components in a timing diagram and the corresponding phasor diagram at the end of the cycle.

nents: a DC component for the positive sequence amplitude (F_+) and a double frequency component for the negative sequence amplitude (F_-). The amplitude of the positive sequence component can be segregated using a filter (further discussed later). Once the positive sequence component is determined from its amplitude and angle, it can be subtracted from the three-phase variables ($f_{A'}$, $f_{B'}$, and $f_{C'}$) to calculate the negative sequence component. A composite variable consisting of positive and negative sequence components with multiple frequencies can be analyzed in various rotating reference frames to convert the amplitude of a particular frequency component into DC. All other frequency components can be filtered to isolate the DC component of interest. Each rotating reference frame can isolate one particular frequency component.

A.IV. THREE-PHASE UNBALANCED VOLTAGE, CURRENT, AND POWER

Consider a three-phase unbalanced voltage:

$$v_A = v_{A+} + v_{A-} + v_0 = V_{p+}\cos(\theta + \varphi_{v+}) + V_{p-}\cos(\theta + \varphi_{v-}) + v_0 \quad \text{(A-28a)}$$

$$v_B = v_{B+} + v_{B-} + v_0 = V_{p+}\cos(\theta + \varphi_{v+} - 120^\circ) + V_{p-}\cos(\theta + \varphi_{v-} + 120^\circ) + v_0 \quad \text{(A-28b)}$$

$$v_C = v_{C+} + v_{C-} + v_0 = V_{p+}\cos(\theta + \varphi_{v+} + 120^\circ) + V_{p-}\cos(\theta + \varphi_{v-} - 120^\circ) + v_0 \quad \text{(A-28c)}$$

is supplying a three-phase unbalanced current

$$i_A = i_{A+} + i_{A-} + i_0 = I_{p+}\cos(\theta + \varphi_{i+}) + I_{p-}\cos(\theta + \varphi_{i-}) + i_0 \quad \text{(A-29a)}$$

$$i_B = i_{B+} + i_{B-} + i_0 = I_{p+}\cos(\theta + \varphi_{i+} - 120^\circ) + I_{p-}\cos(\theta + \varphi_{i-} + 120^\circ) + i_0 \quad \text{(A-29b)}$$

$$i_C = i_{C+} + i_{C-} + i_0 = I_{p+}\cos(\theta + \varphi_{i+} + 120^\circ) + I_{p-}\cos(\theta + \varphi_{i-} - 120^\circ) + i_0 \quad \text{(A-29c)}$$

through a three-phase impedances ($\mathbf{Z_A}$, $\mathbf{Z_B}$, and $\mathbf{Z_C}$), as shown in Figure A-16. The amplitudes of the positive sequence voltage and current in each phase are V_{p+} and I_{p+}, respectively. The amplitudes of the negative sequence voltage and current in each phase are V_{p-} and I_{p-}, respectively. The zero sequence voltage and current in each phase are v_0 and i_0, respectively.

In the case of a three-phase, four-wire system, a zero sequence current (i_0) can be expressed from equations A-29a to A-29c as

$$i_0 = \frac{1}{3}(i_A + i_B + i_C) \quad \text{(A-30)}$$

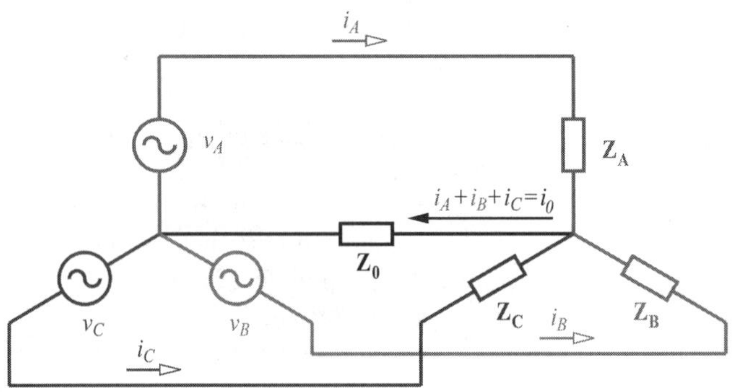

Figure A-16. Three-phase voltage source supplying three-phase unbalanced load.

This current flows through a zero sequence impedance to the neutral of the source. In the case of a three-phase, three-wire system, there is no path for zero sequence current (i_0) to flow, as shown in Figure A-1. Therefore,

$$i_A + i_B + i_C = 0 \tag{A-31}$$

A zero sequence voltage builds up between the neutral of the load and that of the source. The zero sequence voltage can be expressed from equations A-28a to A-28c as

$$v_0 = \frac{1}{3}(v_A + v_B + v_C) \tag{A-32}$$

The three-phase instantaneous power is defined as

$$p = p_A + p_B + p_C \tag{A-11}$$

where

$$\begin{aligned} p_A = v_A i_A &= (v_{A+} + v_{A-} + v_0)(i_{A+} + i_{A-} + i_0) \\ &= \left(V_{p+}\cos(\theta+\varphi_{v+}) + V_{p-}\cos(\theta+\varphi_{v-}) + v_0\right)\left(I_{p+}\cos(\theta+\varphi_{i+}) + I_{p-}\cos(\theta+\varphi_{i-}) + i_0\right) \end{aligned} \tag{A-33a}$$

from equations A-28a and A-29a,

$$\begin{aligned} p_B = v_B i_B = (v_{B+} + v_{B-} + v_0)(i_{B+} + i_{B-} + i_0) &= \left(V_{p+}\cos(\theta+\varphi_{v+} - 120^\circ)\right. \\ + V_{p-}\cos(\theta+\varphi_{v-} + 120^\circ) + v_0\Big)&\left(I_{p+}\cos(\theta+\varphi_{i+} - 120^\circ) + I_{p-}\cos(\theta+\varphi_{i-} + 120^\circ) + i_0\right) \end{aligned} \tag{A-33b}$$

from equations A-28b and A-29b,

$$p_C = v_C i_C = (v_{C+} + v_{C-} + v_0)(i_{C+} + i_{C-} + i_0) = \big(V_{p+}\cos(\theta + \varphi_{v+} + 120^\circ)$$
$$+ V_{p-}\cos(\theta + \varphi_{v-} - 120^\circ) + v_0\big)\big(I_{p+}\cos(\theta + \varphi_{i+} + 120^\circ) + I_{p-}\cos(\theta + \varphi_{i-} - 120^\circ) + i_0\big) \tag{A-33c}$$

from equations A-28c and A-29c.

The instantaneous powers in the A phase due to positive sequence voltage and positive, negative, and zero sequence currents are

$$v_{A+}i_{A+} = \frac{V_{p+}I_{p+}}{2}\left[\cos(2\theta + \varphi_{v+} + \varphi_{i+}) + \cos(\varphi_{v+} - \varphi_{i+})\right] \tag{A-34a}$$

$$v_{A+}i_{A-} = \frac{V_{p+}I_{p-}}{2}\left[\cos(2\theta + \varphi_{v+} + \varphi_{i-}) + \cos(\varphi_{v+} - \varphi_{i-})\right] \tag{A-34b}$$

$$v_{A+}i_0 = V_{p+}\cos(\theta + \varphi_{v+})i_0 \tag{A-34c}$$

The instantaneous powers in the A phase due to negative sequence voltage and positive, negative, and zero sequence currents are

$$v_{A-}i_{A+} = \frac{V_{p-}I_{p+}}{2}\left[\cos(2\theta + \varphi_{v-} + \varphi_{i+}) + \cos(\varphi_{v-} - \varphi_{i+})\right] \tag{A-35a}$$

$$v_{A-}i_{A-} = \frac{V_{p-}I_{p-}}{2}\left[\cos(2\theta + \varphi_{v-} + \varphi_{i-}) + \cos(\varphi_{v-} - \varphi_{i-})\right] \tag{A-35b}$$

$$v_{A-}i_0 = V_{p-}\cos(\theta + \varphi_{v-})i_0 \tag{A-35c}$$

The instantaneous powers in the A phase due to zero sequence voltage and positive, negative, and zero sequence currents are

$$v_0 i_{A+} = v_0 I_{p+}\cos(\theta + \varphi_{i+}) \tag{A-36a}$$

$$v_0 i_{A-} = v_0 I_{p-}\cos(\theta + \varphi_{i-}) \tag{A-36b}$$

$$v_0 i_0 = v_0 i_0 \tag{A-36c}$$

The instantaneous powers in the B phase due to positive sequence voltage and positive, negative, and zero sequence currents are

$$v_{B+}i_{B+} = \frac{V_{p+}I_{p+}}{2}\left[\cos(2\theta + \varphi_{v+} + \varphi_{i+} + 120^\circ) + \cos(\varphi_{v+} - \varphi_{i+})\right] \tag{A-37a}$$

$$v_{B+}i_{B-} = \frac{V_{p+}I_{p-}}{2}\left[\cos(2\theta + \varphi_{v+} + \varphi_{i-}) + \cos(\varphi_{v+} - \varphi_{i-} + 120^\circ)\right] \quad \text{(A-37b)}$$

$$v_{B+}i_0 = V_{p+}\cos(\theta + \varphi_{v+} - 120^\circ)i_0 \quad \text{(A-37c)}$$

The instantaneous powers in the B phase due to negative sequence voltage and positive, negative, and zero sequence currents are

$$v_{B-}i_{B+} = \frac{V_{p-}I_{p+}}{2}\left[\cos(2\theta + \varphi_{v-} + \varphi_{i+}) + \cos(\varphi_{v-} - \varphi_{i+} - 120^\circ)\right] \quad \text{(A-38a)}$$

$$v_{B-}i_{B-} = \frac{V_{p-}I_{p-}}{2}\left[\cos(2\theta + \varphi_{v-} + \varphi_{i-} - 120^\circ) + \cos(\varphi_{v-} - \varphi_{i-})\right] \quad \text{(A-38b)}$$

$$v_{B-}i_0 = V_{p-}\cos(\theta + \varphi_{v-} + 120^\circ)i_0 \quad \text{(A-38c)}$$

The instantaneous powers in the B phase due to zero sequence voltage and positive, negative, and zero sequence currents are

$$v_0 i_{B+} = v_0 I_{p+}\cos(\theta + \varphi_{i+} - 120^\circ) \quad \text{(A-39a)}$$

$$v_0 i_{B-} = v_0 I_{p-}\cos(\theta + \varphi_{i-} + 120^\circ) \quad \text{(A-39b)}$$

$$v_0 i_0 = v_0 i_0 \quad \text{(A-39c)}$$

The instantaneous powers in the C phase due to positive sequence voltage and positive, negative, and zero sequence currents are

$$v_{C+}i_{C+} = \frac{V_{p+}I_{p+}}{2}\left[\cos(2\theta + \varphi_{v+} + \varphi_{i+} - 120^\circ) + \cos(\varphi_{v+} - \varphi_{i+})\right] \quad \text{(A-40a)}$$

$$v_{C+}i_{C-} = \frac{V_{p+}I_{p-}}{2}\left[\cos(2\theta + \varphi_{v+} + \varphi_{i-}) + \cos(\varphi_{v+} - \varphi_{i-} - 120^\circ)\right] \quad \text{(A-40b)}$$

$$v_{C+}i_0 = V_{p+}\cos(\theta + \varphi_{v+} + 120^\circ)i_0 \quad \text{(A-40c)}$$

The instantaneous powers in the C phase due to negative sequence voltage and positive, negative, and zero sequence currents are

$$v_{C-}i_{C+} = \frac{V_{p-}I_{p+}}{2}\left[\cos(2\theta + \varphi_{v-} + \varphi_{i+}) + \cos(\varphi_{v-} - \varphi_{i+} + 120^\circ)\right] \quad \text{(A-41a)}$$

$$v_{C-}i_{C-} = \frac{V_{p-}I_{p-}}{2}\left[\cos(2\theta + \varphi_{v-} + \varphi_{i-} + 120^\circ) + \cos(\varphi_{v-} - \varphi_{i-})\right] \quad \text{(A-41b)}$$

$$v_{C-}i_0 = V_{p-}\cos(\theta + \varphi_{v-} - 120^\circ)\, i_0 \qquad \text{(A-41c)}$$

The instantaneous powers in the C phase due to zero sequence voltage and positive, negative, and zero sequence currents are

$$v_0 i_{C+} = v_0 I_{p+}\cos(\theta + \varphi_{i+} + 120^\circ) \qquad \text{(A-42a)}$$

$$v_0 i_{C-} = v_0 I_{p-}\cos(\theta + \varphi_{i-} - 120^\circ) \qquad \text{(A-42b)}$$

$$v_0 i_0 = v_0 i_0 \qquad \text{(A-42c)}$$

The three-phase power due to positive sequence voltage and positive, negative, and zero sequence currents are

$$p_{+} = v_{A+}i_{A+} + v_{B+}i_{B+} + v_{C+}i_{C+} = \frac{3}{2}V_{p+}I_{p+}\cos(\varphi_{v+} - \varphi_{i+}) \qquad \text{(A-43a)}$$

$$p_{+-} = v_{A+}i_{A-} + v_{B+}i_{B-} + v_{C+}i_{C-} = \frac{3}{2}V_{p+}I_{p-}\cos(2\theta + \varphi_{v+} + \varphi_{i-}) \qquad \text{(A-43b)}$$

$$p_{+0} = v_{A+}i_0 + v_{B+}i_0 + v_{C+}i_0 = 0 \qquad \text{(A-43c)}$$

The three-phase power due to negative sequence voltage and positive, negative, and zero sequence currents are

$$p_{-+} = v_{A-}i_{A+} + v_{B-}i_{B+} + v_{C-}i_{C+} = \frac{3}{2}V_{p-}I_{p+}\cos(2\theta + \varphi_{v-} + \varphi_{i+}) \qquad \text{(A-43d)}$$

$$p_{-} = v_{A-}i_{A-} + v_{B-}i_{B-} + v_{C-}i_{C-} = \frac{3}{2}V_{p-}I_{p-}\cos(\varphi_{v-} - \varphi_{i-}) \qquad \text{(A-43e)}$$

$$p_{-0} = v_{A-}i_0 + v_{B-}i_0 + v_{C-}i_0 = 0 \qquad \text{(A-43f)}$$

The three-phase power due to zero sequence voltage and positive, negative, and zero sequence currents are

$$p_{0+} = v_0 i_{A+} + v_0 i_{B+} + v_0 i_{C+} = 0 \qquad \text{(A-43g)}$$

$$p_{0-} = v_0 i_{A-} + v_0 i_{B-} + v_0 i_{C-} = 0 \qquad \text{(A-43h)}$$

$$p_0 = v_0 i_0 + v_0 i_0 + v_0 i_0 = 3 v_0 i_0 \qquad \text{(A-43i)}$$

Therefore, the three-phase instantaneous power is

$$\begin{aligned} p &= p_+ + p_- + p_{+-} + p_{-+} + p_0 \\ &= \frac{3}{2} V_{p+} I_{p+} \cos(\varphi_{v+} - \varphi_{i+}) + \frac{3}{2} V_{p-} I_{p-} \cos(\varphi_{v-} - \varphi_{i-}) \\ &\quad + \frac{3}{2} V_{p+} I_{p-} \cos(2\theta + \varphi_{v+} + \varphi_{i-}) + \frac{3}{2} V_{p-} I_{p+} \cos(2\theta + \varphi_{v-} + \varphi_{i+}) + 3\, v_0\, i_0 \end{aligned} \quad \text{(A-44)}$$

The first and second terms in the above expression are time-invariant fixed terms. The third and fourth terms are time-variant, double frequency, cosine terms. The fifth term is time-variant and is contributed by the zero sequence voltage and current. It is important to note that the sufficient condition for a time-variant, double frequency, cosine term to exist is the existence of the negative sequence term in voltage or current.

A.V. d–q TRANSFORMATION

A set of three instantaneous phase variables that sum to zero are referred to as a balanced three-phase set of either positive sequence or negative sequence as described before. A positive sequence set can be represented as the projections of three equal-length rotating phasors that are 120° apart (shown in Figure A-17a) on a stationary reference frame, as explained earlier. These instantaneous values can also be represented as the projections of a rotating vector upon three stationary axes that are 120° apart, as shown in Figure A-17b. Figure A-17b shows **F+** as a positive sequence

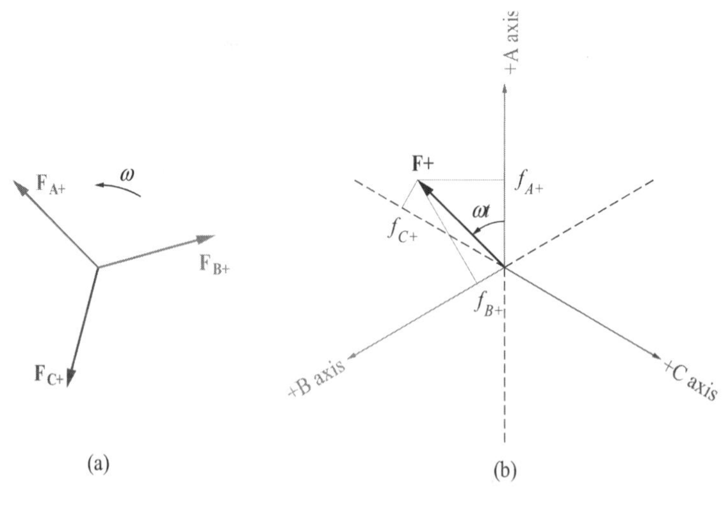

Figure A-17. Representation of three-phase positive sequence variables in (a) a phasor diagram and (b) a vector diagram.

spatial vector rotating in the counterclockwise direction. The projections of **F+** on the stationary A, B, and C axes are the positive sequence instantaneous phase variables.

A negative sequence set can be represented as the projections of three equal length rotating phasors that are 120° apart (shown in Figure A-18a) on a stationary reference frame. These instantaneous values can also be represented as the projections of a rotating vector upon three stationary axes that are 120° apart, as shown in Figure A-18b. Figure A-18b shows **F–** as a negative sequence spatial vector rotating in the clockwise direction. The projections of **F–** on the stationary A, B, and C axes are the negative sequence instantaneous phase variables.

The conversion of the A–B–C stationary variables into the d–q rotating frames is a two-step process. First, the positive and negative sequence variables are converted into their equivalent d–q stationary variables, denoted by subscript *s*, assuming that the d axis is aligned with the A axis, as shown in Figure A-19a. Second, these d–q stationary variables are rotated counterclockwise from their original position with an arbitrary angular speed of ω_a rad/s, as shown in Figure A-19b.

The three-phase set of variables

$$f_A = f_{A+} + f_{A-} + f_0 \tag{A-13a}$$

$$f_B = f_{B+} + f_{B-} + f_0 \tag{A-13b}$$

$$f_C = f_{C+} + f_{C-} + f_0 \tag{A-13c}$$

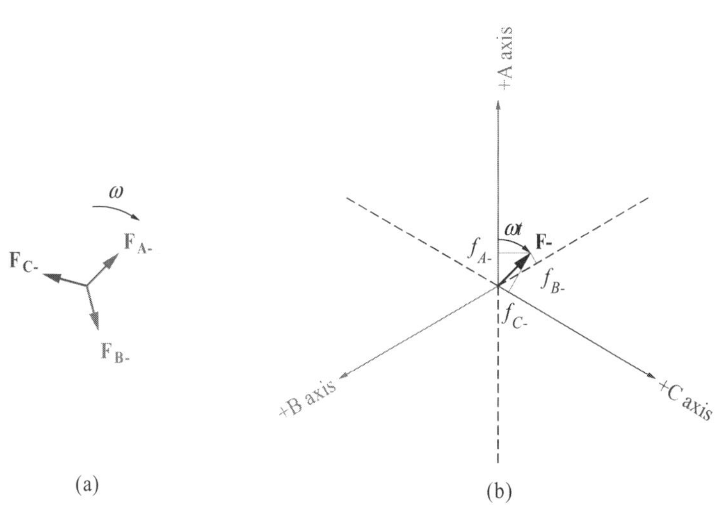

Figure A-18. Representation of three-phase negative sequence variables in (a) a phasor diagram and (b) a vector diagram.

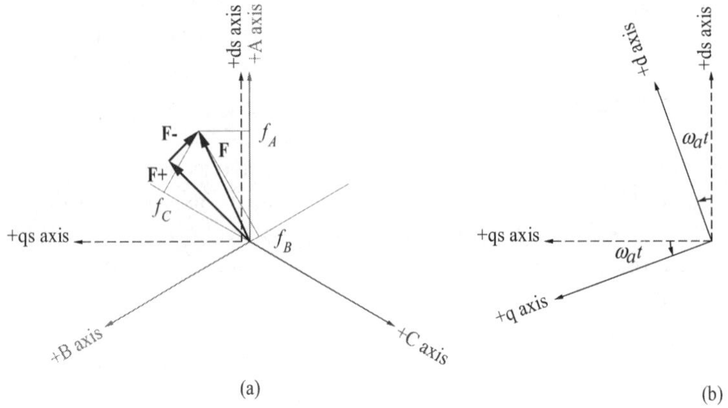

Figure A-19. Conversion of A, B, and C stationary variables into (a) d–q stationary and (b) d–q rotating frames.

can be defined as projections of f_{ds} and f_{qs} on the A–B–C axes as

$$f_A = f_{ds} + f_0 \tag{A-45a}$$

$$f_B = \cos(120^\circ) f_{ds} + \cos(30^\circ) f_{qs} + f_0 = -\frac{1}{2} f_{ds} + \frac{\sqrt{3}}{2} f_{qs} + f_0 \tag{A-45b}$$

$$f_C = \cos(120^\circ) f_{ds} + \cos(150^\circ) f_{qs} + f_0 = -\frac{1}{2} f_{ds} - \frac{\sqrt{3}}{2} f_{qs} + f_0 \tag{A-45c}$$

In matrix form,

$$\begin{bmatrix} f_A \\ f_B \\ f_C \end{bmatrix} = \begin{bmatrix} 1 & 0 & 1 \\ -\frac{1}{2} & \frac{\sqrt{3}}{2} & 1 \\ -\frac{1}{2} & -\frac{\sqrt{3}}{2} & 1 \end{bmatrix} \begin{bmatrix} f_{ds} \\ f_{qs} \\ f_0 \end{bmatrix} \tag{A-46a}$$

and

$$\begin{bmatrix} f_{ds} \\ f_{qs} \\ f_0 \end{bmatrix} = \frac{2}{3} \begin{bmatrix} 1 & -\frac{1}{2} & -\frac{1}{2} \\ 0 & \frac{\sqrt{3}}{2} & -\frac{\sqrt{3}}{2} \\ \frac{1}{2} & \frac{1}{2} & \frac{1}{2} \end{bmatrix} \begin{bmatrix} f_A \\ f_B \\ f_C \end{bmatrix} \tag{A-46b}$$

The relationship between the arbitrarily rotating d–q reference frame, denoted by subscript a, and the stationary d–q reference frame, as shown in Figure A-18b, is defined as

$$f_{ds} = f_{da}\cos(\omega_a t) - f_{qa}\sin(\omega_a t) \tag{A-47a}$$

and

$$f_{qs} = f_{da}\sin(\omega_a t) + f_{qa}\cos(\omega_a t) \tag{A-47b}$$

This relationship in matrix form is

$$\begin{bmatrix} f_{ds} \\ f_{qs} \end{bmatrix} = \begin{bmatrix} \cos(\omega_a t) & -\sin(\omega_a t) \\ \sin(\omega_a t) & \cos(\omega_a t) \end{bmatrix} \begin{bmatrix} f_{da} \\ f_{qa} \end{bmatrix} \tag{A-48a}$$

and

$$\begin{bmatrix} f_{da} \\ f_{qa} \end{bmatrix} = \begin{bmatrix} \cos(\omega_a t) & \sin(\omega_a t) \\ -\sin(\omega_a t) & \cos(\omega_a t) \end{bmatrix} \begin{bmatrix} f_{ds} \\ f_{qs} \end{bmatrix} \tag{A-48b}$$

The complete relationship including the zero sequence component is

$$\begin{bmatrix} f_{da} \\ f_{qa} \\ f_0 \end{bmatrix} = \begin{bmatrix} \cos(\omega_a t) & \sin(\omega_a t) & 0 \\ -\sin(\omega_a t) & \cos(\omega_a t) & 0 \\ 0 & 0 & 1 \end{bmatrix} \begin{bmatrix} f_{ds} \\ f_{qs} \\ f_0 \end{bmatrix} \tag{A-49}$$

From equations A-49 and A-46, the A-B-C phase variables can be converted into their equivalent d–q variables in an arbitrarily rotating reference frame as

$$\begin{bmatrix} f_{da} \\ f_{qa} \\ f_0 \end{bmatrix} = \begin{bmatrix} \cos(\omega_a t) & \sin(\omega_a t) & 0 \\ -\sin(\omega_a t) & \cos(\omega_a t) & 0 \\ 0 & 0 & 1 \end{bmatrix} \frac{2}{3} \begin{bmatrix} 1 & -\frac{1}{2} & -\frac{1}{2} \\ 0 & \frac{\sqrt{3}}{2} & -\frac{\sqrt{3}}{2} \\ \frac{1}{2} & \frac{1}{2} & \frac{1}{2} \end{bmatrix} \begin{bmatrix} f_A \\ f_B \\ f_C \end{bmatrix} \tag{A-50}$$

The above equation can be written in another form as

$$\begin{bmatrix} f_{da} \\ f_{qa} \\ f_0 \end{bmatrix} = \frac{2}{3} \begin{bmatrix} \cos(\omega_a t) & \cos(\omega_a t - 120^\circ) & \cos(\omega_a t + 120^\circ) \\ -\sin(\omega_a t) & -\sin(\omega_a t - 120^\circ) & -\sin(\omega_a t + 120^\circ) \\ \frac{1}{2} & \frac{1}{2} & \frac{1}{2} \end{bmatrix} \begin{bmatrix} f_A \\ f_B \\ f_C \end{bmatrix} \tag{A-51}$$

The following matrix is called Park's original transformation matrix. It represents the projections of the stationary A-B-C axes onto the rotating d–q axes:

$$[\mathbf{P}]=\frac{2}{3}\begin{bmatrix}\cos(\omega_a t) & \cos(\omega_a t-120^\circ) & \cos(\omega_a t+120^\circ)\\ -\sin(\omega_a t) & -\sin(\omega_a t-120^\circ) & -\sin(\omega_a t+120^\circ)\\ \frac{1}{2} & \frac{1}{2} & \frac{1}{2}\end{bmatrix} \tag{A-52}$$

The inverse of Park's original transformation matrix is

$$[\mathbf{P}]^{-1}=\begin{bmatrix}\cos(\omega_a t) & -\sin(\omega_a t) & 1\\ \cos(\omega_a t-120^\circ) & -\sin(\omega_a t-120^\circ) & 1\\ \cos(\omega_a t+120^\circ) & -\sin(\omega_a t+120^\circ) & 1\end{bmatrix} \tag{A-53}$$

A.V.1. Conversion of a Variable Containing Positive, Negative, and Zero Sequence Components into the d–q Frame

Consider a three-phase set of variables consisting of positive, negative, and zero sequence components as

$$f_A = f_{A+} + f_{A-} + f_0 = F_{p+}\cos(\theta+\varphi_{f+}) + F_{p-}\cos(\theta+\varphi_{f-}) + f_0 \tag{A-54a}$$

$$f_B = f_{B+} + f_{B-} + f_0 = F_{p+}\cos(\theta+\varphi_{f+}-120^\circ) + F_{p-}\cos(\theta+\varphi_{f-}+120^\circ) + f_0 \tag{A-54b}$$

$$f_C = f_{C+} + f_{C-} + f_0 = F_{p+}\cos(\theta+\varphi_{f+}+120^\circ) + F_{p-}\cos(\theta+\varphi_{f-}-120^\circ) + f_0 \tag{A-54c}$$

Then, the relationship between the d–q-0 variables in an arbitrarily rotating reference frame and A–B–C phase variables is

$$\begin{bmatrix}f_{da}\\ f_{qa}\\ f_0\end{bmatrix}=[\mathbf{P}]\begin{bmatrix}f_A\\ f_B\\ f_C\end{bmatrix}=[\mathbf{P}]\begin{bmatrix}f_{A+}+f_{A-}+f_0\\ f_{B+}+f_{B-}+f_0\\ f_{C+}+f_{C-}+f_0\end{bmatrix}=[\mathbf{P}]\begin{bmatrix}f_{A+}\\ f_{B+}\\ f_{C+}\end{bmatrix}+[\mathbf{P}]\begin{bmatrix}f_{A-}\\ f_{B-}\\ f_{C-}\end{bmatrix}+[\mathbf{P}]\begin{bmatrix}f_0\\ f_0\\ f_0\end{bmatrix} \tag{A-55}$$

But,

$$[\mathbf{P}]\begin{bmatrix}f_{A+}\\ f_{B+}\\ f_{C+}\end{bmatrix}=\frac{2}{3}\begin{bmatrix}\cos(\omega_a t) & \cos(\omega_a t-120^\circ) & \cos(\omega_a t+120^\circ)\\ -\sin(\omega_a t) & -\sin(\omega_a t-120^\circ) & -\sin(\omega_a t+120^\circ)\\ \frac{1}{2} & \frac{1}{2} & \frac{1}{2}\end{bmatrix}\begin{bmatrix}F_{p+}\cos(\theta+\varphi_{f+})\\ F_{p+}\cos(\theta+\varphi_{f+}-120^\circ)\\ F_{p+}\cos(\theta+\varphi_{f+}+120^\circ)\end{bmatrix}$$

(A-56)

Substituting $\theta = \omega t$, it is

$$[\mathbf{P}]\begin{bmatrix} f_{A+} \\ f_{B+} \\ f_{C+} \end{bmatrix} = \frac{2}{3}F_{p+}\begin{bmatrix} \cos(\omega_a t)\cos(\omega t+\varphi_{f+}) + \cos(\omega_a t-120^\circ)\cos(\omega t+\varphi_{f+}-120^\circ) \\ +\cos(\omega_a t+120^\circ)\cos(\omega t+\varphi_{f+}+120^\circ) \\ -\sin(\omega_a t)\cos(\omega t+\varphi_{f+}) - \sin(\omega_a t-120^\circ)\cos(\omega t+\varphi_{f+}-120^\circ) \\ -\sin(\omega_a t+120^\circ)\cos(\omega t+\varphi_{f+}+120^\circ) \\ \frac{1}{2}\left\{\cos(\omega t+\varphi_{f+}) + \cos(\omega t+\varphi_{f+}-120^\circ) + \cos(\omega t+\varphi_{f+}+120^\circ)\right\} \end{bmatrix}$$

$$= \frac{1}{3}F_{p+}\begin{bmatrix} \cos\left\{(\omega+\omega_a)t+\varphi_{f+}\right\} + \cos\left\{(\omega-\omega_a)t+\varphi_{f+}\right\} + \cos\left\{(\omega+\omega_a)t+\varphi_{f+}-240^\circ\right\} + \\ \cos\left\{(\omega-\omega_a)t+\varphi_{f+}\right\} + \cos\left\{(\omega+\omega_a)t+\varphi_{f+}+240^\circ\right\} + \cos\left\{(\omega-\omega_a)t+\varphi_{f+}\right\} \\ -\sin\left\{(\omega+\omega_a)t+\varphi_{f+}\right\} + \sin\left\{(\omega-\omega_a)t+\varphi_{f+}\right\} - \sin\left\{(\omega+\omega_a)t+\varphi_{f+}-240^\circ\right\} + \\ \sin\left\{(\omega-\omega_a)t+\varphi_{f+}\right\} - \sin\left\{(\omega+\omega_a)t+\varphi_{f+}+240^\circ\right\} + \sin\left\{(\omega-\omega_a)t+\varphi_{f+}\right\} \\ 0 \end{bmatrix}$$

$$= F_{p+}\begin{bmatrix} \cos\left\{(\omega-\omega_a)t+\varphi_{f+}\right\} \\ \sin\left\{(\omega-\omega_a)t+\varphi_{f+}\right\} \\ 0 \end{bmatrix} \tag{A-57}$$

$$[\mathbf{P}]\begin{bmatrix} f_{A-} \\ f_{B-} \\ f_{C-} \end{bmatrix} = \frac{2}{3}\begin{bmatrix} \cos(\omega_a t) & \cos(\omega_a t-120^\circ) & \cos(\omega_a t+120^\circ) \\ -\sin(\omega_a t) & -\sin(\omega_a t-120^\circ) & -\sin(\omega_a t+120^\circ) \\ \frac{1}{2} & \frac{1}{2} & \frac{1}{2} \end{bmatrix}\begin{bmatrix} F_{p-}\cos(\theta+\varphi_{f-}) \\ F_{p-}\cos(\theta+\varphi_{f-}+120^\circ) \\ F_{p-}\cos(\theta+\varphi_{f-}-120^\circ) \end{bmatrix} \tag{A-58}$$

Substituting $\theta = \omega t$, it is

$$[\mathbf{P}]\begin{bmatrix} f_{A-} \\ f_{B-} \\ f_{C-} \end{bmatrix} = \frac{2}{3}F_{p-}\begin{bmatrix} \cos(\omega_a t)\cos(\omega t+\varphi_{f-}) + \cos(\omega_a t-120^\circ)\cos(\omega t+\varphi_{f-}+120^\circ) + \\ \cos(\omega_a t+120^\circ)\cos(\omega t+\varphi_{f-}-120^\circ) \\ -\sin(\omega_a t)\cos(\omega t+\varphi_{f-}) - \sin(\omega_a t-120^\circ)\cos(\omega t+\varphi_{f-}+120^\circ) - \\ \sin(\omega_a t+120^\circ)\cos(\omega t+\varphi_{f-}-120^\circ) \\ \frac{1}{2}\left\{\cos(\omega t+\varphi_{f-}) + \cos(\omega t+\varphi_{f-}+120^\circ) + \cos(\omega t+\varphi_{f-}-120^\circ)\right\} \end{bmatrix}$$

$$= \frac{1}{3}F_{p-}\begin{bmatrix} \cos\{(\omega+\omega_a)t+\varphi_{f-}\}+\cos\{(\omega-\omega_a)t+\varphi_{f-}\}+\cos\{(\omega+\omega_a)t+\varphi_{f-}\}+ \\ \cos\{(\omega-\omega_a)t+\varphi_{f-}+240^\circ\}+\cos\{(\omega+\omega_a)t+\varphi_{f-}\}+\cos\{(\omega-\omega_a)t+\varphi_{f-}-240^\circ\} \\ -\sin\{(\omega+\omega_a)t+\varphi_{f-}\}+\sin\{(\omega-\omega_a)t+\varphi_{f-}\}-\sin\{(\omega+\omega_a)t+\varphi_{f-}\}+ \\ \sin\{(\omega-\omega_a)t+\varphi_{f-}+240^\circ\}-\sin\{(\omega+\omega_a)t+\varphi_{f-}\}+\sin\{(\omega-\omega_a)t+\varphi_{f-}-240^\circ\} \\ 0 \end{bmatrix}$$

$$= F_{p-}\begin{bmatrix} \cos\{(\omega+\omega_a)t+\varphi_{f-}\} \\ -\sin\{(\omega+\omega_a)t+\varphi_{f-}\} \\ 0 \end{bmatrix} \tag{A-59}$$

$$[\mathbf{P}]\begin{bmatrix} f_0 \\ f_0 \\ f_0 \end{bmatrix} = \frac{2}{3}\begin{bmatrix} \{\cos(\omega_a t)+\cos(\omega_a t-120^\circ)+\cos(\omega_a t+120^\circ)\}f_0 \\ -\{\sin(\omega_a t)+\sin(\omega_a t-120^\circ)+\sin(\omega_a t+120^\circ)\}f_0 \\ \frac{1}{2}3f_0 \end{bmatrix} = \begin{bmatrix} 0 \\ 0 \\ f_0 \end{bmatrix} \tag{A-60}$$

Then, it can be written from equation A-52 that

$$\begin{bmatrix} f_{da} \\ f_{qa} \\ f_0 \end{bmatrix} = [\mathbf{P}]\begin{bmatrix} f_{A+} \\ f_{B+} \\ f_{C+} \end{bmatrix} + [\mathbf{P}]\begin{bmatrix} f_{A-} \\ f_{B-} \\ f_{C-} \end{bmatrix} + [\mathbf{P}]\begin{bmatrix} f_0 \\ f_0 \\ f_0 \end{bmatrix}$$

$$= \begin{bmatrix} F_{p+}\cos\{(\omega-\omega_a)t+\varphi_{f+}\}+F_{p-}\cos\{(\omega+\omega_a)t+\varphi_{f-}\} \\ F_{p+}\sin\{(\omega-\omega_a)t+\varphi_{f+}\}-F_{p-}\sin\{(\omega+\omega_a)t+\varphi_{f-}\} \\ f_0 \end{bmatrix} \tag{A-61}$$

with substitutions from equations A-57, A-59, and A-60.

When $\omega_a = \omega$, meaning that the rotating reference frame is synchronous with the system angular frequency, equation A-62 becomes

$$\begin{bmatrix} f_d \\ f_q \\ f_0 \end{bmatrix} = \begin{bmatrix} F_{p+}\cos\varphi_{f+} + F_{p-}\cos(2\omega t+\varphi_{f-}) \\ F_{p+}\sin\varphi_{f+} - F_{p-}\sin(2\omega t+\varphi_{f-}) \\ f_0 \end{bmatrix} \tag{A-62}$$

Throughout the book, the phase-locked loop is derived from the line voltage with cosine reference. That means that the A phase voltage is at its peak when $\phi_{v+} = 0$. If the A–B–C line voltages are converted into their equivalent d–q variables in the synchronously rotating reference frame, equation A-62 can be rewritten as

$$\begin{bmatrix} v_d \\ v_q \\ v_0 \end{bmatrix} = \begin{bmatrix} V_{p+} + V_{p-} \cos(2\omega t + \varphi_{v-}) \\ -V_{p-} \sin(2\omega t + \varphi_{v-}) \\ v_0 \end{bmatrix} \tag{A-63}$$

If the current is at an angle (φ_{i+}) with respect to the line voltage, equation A-62 can be rewritten as

$$\begin{bmatrix} i_d \\ i_q \\ i_0 \end{bmatrix} = \begin{bmatrix} I_{p+} \cos\varphi_{i+} + I_{p-} \cos(2\omega t + \varphi_{i-}) \\ I_{p+} \sin\varphi_{i+} - I_{p-} \sin(2\omega t + \varphi_{i-}) \\ i_0 \end{bmatrix} \tag{A-64}$$

In the synchronously rotating reference frame, the positive sequence fundamental component turns into a DC value. The d–q components of the fundamental line voltage are V_{p+} and 0, respectively. The rotating d axis voltage contains a DC component that is the amplitude of the original A–B–C variables. The rotating q axis voltage contains no DC component. The d q components of the fundamental line current are I_{p+} $\cos(\phi_{i+})$ and I_{p+} $\sin(\phi_{i+})$. In addition, the rotating d and q axes components contain a twice-frequency component whose amplitude is the same as that of the negative sequence component in the original variables. The reason for the existence of the twice-frequency negative sequence component in the d–q frame is because the negative sequence components rotate at twice the angular speed in the opposite direction from the synchronously rotating reference frame. It is quite easy to eliminate these negative sequence terms from the d–q components using a half-cycle running averager. This simply uses the fact that if a variable contains an oscillatory component that is superimposed on a DC component, then averaging the variable over the period of the oscillatory component will result in only the DC component at the output of the averager. In this case, the averager acts as a filter. Once the oscillatory twice-frequency component is removed and the DC component is recovered from the d–q variables, the relative angle (ϕ_{i+}) of the current is then simply

$$\varphi_{i+} = \tan^{-1} \frac{i_q}{i_d} \tag{A-65}$$

A.V.2. Calculation of Instantaneous Power into the d–q Frame

The instantaneous active power in the three-phase system with phase voltages (v_A, v_B, and v_C) and phase currents (i_A, i_B, and i_C) is defined as

$$p = v_A i_A + v_B i_B + v_C i_C = \begin{bmatrix} v_A & v_B & v_C \end{bmatrix} \begin{bmatrix} i_A \\ i_B \\ i_C \end{bmatrix} = \begin{bmatrix} v_A \\ v_B \\ v_C \end{bmatrix}^T \begin{bmatrix} i_A \\ i_B \\ i_C \end{bmatrix} = \left[[\mathbf{P}]^{-1} \begin{bmatrix} v_d \\ v_q \\ v_0 \end{bmatrix} \right]^T [\mathbf{P}]^{-1} \begin{bmatrix} i_d \\ i_q \\ i_0 \end{bmatrix}$$

$$p = \begin{bmatrix} v_d \\ v_q \\ v_0 \end{bmatrix}^T \left[[\mathbf{P}]^{-1} \right]^T [\mathbf{P}]^{-1} \begin{bmatrix} i_d \\ i_q \\ i_0 \end{bmatrix} = \begin{bmatrix} v_d & v_q & v_0 \end{bmatrix} \frac{3}{2} \begin{bmatrix} 1 & 0 & 0 \\ 0 & 1 & 0 \\ 0 & 0 & 2 \end{bmatrix} \begin{bmatrix} i_d \\ i_q \\ i_0 \end{bmatrix}$$

$$p = \frac{3}{2}\left(v_d i_d + v_q i_q\right) + 3 v_0 i_0 \tag{A-66}$$

The instantaneous three-phase power consists of two terms: the first term is due to the positive and negative sequence voltage and current, and the second term is due to zero sequence voltage and current.

A.V.3. Calculation of Instantaneous Power into d–q Frame for a Three-Phase, Three-Wire System

Substituting $i_B = -(i_A + i_C)$ from equation A-31, equation A-46b can be written for current as

$$\begin{bmatrix} i_{ds} \\ i_{qs} \end{bmatrix} = \frac{2}{3} \begin{bmatrix} 1 & -\frac{1}{2} & -\frac{1}{2} \\ 0 & \frac{\sqrt{3}}{2} & -\frac{\sqrt{3}}{2} \end{bmatrix} \begin{bmatrix} i_A \\ i_B \\ i_C \end{bmatrix}$$

$$= \begin{bmatrix} \frac{2}{3} & -\frac{1}{3} & -\frac{1}{3} \\ 0 & \frac{1}{\sqrt{3}} & -\frac{1}{\sqrt{3}} \end{bmatrix} \begin{bmatrix} i_A \\ -i_A - i_C \\ i_C \end{bmatrix} = \begin{bmatrix} \frac{2}{3} + \frac{1}{3} & \frac{1}{3} - \frac{1}{3} \\ -\frac{1}{\sqrt{3}} & -\frac{1}{\sqrt{3}} - \frac{1}{\sqrt{3}} \end{bmatrix} \begin{bmatrix} i_A \\ i_C \end{bmatrix}$$

$$\begin{bmatrix} i_{ds} \\ i_{qs} \end{bmatrix} = \begin{bmatrix} 1 & 0 \\ -\frac{1}{\sqrt{3}} & -\frac{2}{\sqrt{3}} \end{bmatrix} \begin{bmatrix} i_A \\ i_C \end{bmatrix} \tag{A-67a}$$

and

$$\begin{bmatrix} i_A \\ i_C \end{bmatrix} = \begin{bmatrix} 1 & 0 \\ -\frac{1}{2} & -\frac{\sqrt{3}}{2} \end{bmatrix} \begin{bmatrix} i_{ds} \\ i_{qs} \end{bmatrix} \tag{A-67b}$$

Equation A-48b can then be written for current in the synchronously rotating reference frame as

$$\begin{bmatrix} i_d \\ i_q \end{bmatrix} = \begin{bmatrix} \cos(\omega t) & \sin(\omega t) \\ -\sin(\omega t) & \cos(\omega t) \end{bmatrix} \begin{bmatrix} i_{ds} \\ i_{qs} \end{bmatrix} = \begin{bmatrix} \cos(\omega t) & \sin(\omega t) \\ -\sin(\omega t) & \cos(\omega t) \end{bmatrix} \begin{bmatrix} 1 & 0 \\ -\frac{1}{\sqrt{3}} & -\frac{2}{\sqrt{3}} \end{bmatrix} \begin{bmatrix} i_A \\ i_C \end{bmatrix}$$

with a substitution from equation A-67a

$$= \begin{bmatrix} \cos(\omega t) - \frac{1}{\sqrt{3}}\sin(\omega t) & -\frac{2}{\sqrt{3}}\sin(\omega t) \\ -\sin(\omega t) - \frac{1}{\sqrt{3}}\cos(\omega t) & -\frac{2}{\sqrt{3}}\cos(\omega t) \end{bmatrix} \begin{bmatrix} i_A \\ i_C \end{bmatrix}$$

$$\begin{bmatrix} i_d \\ i_q \end{bmatrix} = \frac{2}{\sqrt{3}} \begin{bmatrix} \sin(\omega t + 120^\circ) & -\sin(\omega t) \\ \cos(\omega t + 120^\circ) & -\cos(\omega t) \end{bmatrix} \begin{bmatrix} i_A \\ i_C \end{bmatrix} \tag{A-68}$$

Equation A-68 can be rearranged as

$$\begin{bmatrix} i_A \\ i_C \end{bmatrix} = \begin{bmatrix} \cos(\omega t) & -\sin(\omega t) \\ \cos(\omega t + 120^\circ) & -\sin(\omega t + 120^\circ) \end{bmatrix} \begin{bmatrix} i_d \\ i_q \end{bmatrix} \tag{A-69}$$

Therefore, Park's original transformation matrix for three-phase three-wire system is

$$[\mathbf{p}] = \frac{2}{\sqrt{3}} \begin{bmatrix} \sin(\omega t + 120^\circ) & -\sin(\omega t) \\ \cos(\omega t + 120^\circ) & -\cos(\omega t) \end{bmatrix} \tag{A-70a}$$

and the inverse of Park's original transformation matrix for three-phase three-wire system is

$$[\mathbf{p}]^{-1} = \begin{bmatrix} \cos(\omega t) & -\sin(\omega t) \\ \cos(\omega t + 120^\circ) & -\sin(\omega t + 120^\circ) \end{bmatrix} \tag{A-70b}$$

This transformation matrix can be used for voltage, current, and flux as well, provided that the zero sequence component does not exist.

Similarly, in case of voltage,

$$\begin{bmatrix} v_d \\ v_q \end{bmatrix} = \frac{2}{\sqrt{3}} \begin{bmatrix} \sin(\omega t + 120^\circ) & -\sin(\omega t) \\ \cos(\omega t + 120^\circ) & -\cos(\omega t) \end{bmatrix} \begin{bmatrix} v_A \\ v_C \end{bmatrix} \tag{A-71}$$

In some applications, it may be easier to use the available phase-to-phase voltage. The phase-to-phase voltage is related to the phase-to-neutral voltage as

$$\begin{bmatrix} v_{AB} \\ v_{CB} \end{bmatrix} = \begin{bmatrix} 1 & -1 & 0 \\ 0 & -1 & 1 \end{bmatrix} \begin{bmatrix} v_A \\ v_B \\ v_C \end{bmatrix} = \begin{bmatrix} 1 & -1 & 0 \\ 0 & -1 & 1 \end{bmatrix} \begin{bmatrix} v_A \\ -v_A - v_C \\ v_C \end{bmatrix}$$

when $v_A + v_B + v_C = 0$. Then,

$$\begin{bmatrix} v_{AB} \\ v_{CB} \end{bmatrix} = \begin{bmatrix} 2 & 1 \\ 1 & 2 \end{bmatrix} \begin{bmatrix} v_A \\ v_C \end{bmatrix} \qquad \text{(A-72a)}$$

and

$$\begin{bmatrix} v_A \\ v_C \end{bmatrix} = \frac{1}{3} \begin{bmatrix} 2 & -1 \\ -1 & 2 \end{bmatrix} \begin{bmatrix} v_{AB} \\ v_{CB} \end{bmatrix} \qquad \text{(A-72b)}$$

Equations A-71 and A-72b can be combined as

$$\begin{bmatrix} v_d \\ v_q \end{bmatrix} = \frac{2}{\sqrt{3}} \begin{bmatrix} \sin(\omega t + 120^\circ) & -\sin(\omega t) \\ \cos(\omega t + 120^\circ) & -\cos(\omega t) \end{bmatrix} \begin{bmatrix} v_A \\ v_C \end{bmatrix}$$

$$= \frac{1}{3} \frac{2}{\sqrt{3}} \begin{bmatrix} \sin(\omega t + 120^\circ) & -\sin(\omega t) \\ \cos(\omega t + 120^\circ) & -\cos(\omega t) \end{bmatrix} \begin{bmatrix} 2 & -1 \\ -1 & 2 \end{bmatrix} \begin{bmatrix} v_{AB} \\ v_{CB} \end{bmatrix}$$

$$\begin{bmatrix} v_d \\ v_q \end{bmatrix} = \frac{2}{3} \begin{bmatrix} \cos(\omega t) & \cos(\omega t + 120^\circ) \\ -\sin(\omega t) & -\sin(\omega t + 120^\circ) \end{bmatrix} \begin{bmatrix} v_{AB} \\ v_{CB} \end{bmatrix} \qquad \text{(A-73)}$$

Equation A-73 can be rearranged as

$$\begin{bmatrix} v_{AB} \\ v_{CB} \end{bmatrix} = \sqrt{3} \begin{bmatrix} \sin(\omega t + 120^\circ) & \cos(\omega t + 120^\circ) \\ -\sin(\omega t) & -\cos(\omega t) \end{bmatrix} \begin{bmatrix} v_d \\ v_q \end{bmatrix} \qquad \text{(A-74)}$$

Substituting $i_B = -(i_A + i_C)$ from equation A-31, equation A-12a for instantaneous active power can be rewritten as

$$p = v_A i_A + v_B i_B + v_C i_C = v_A i_A - v_B (i_A + i_C) + v_C i_C = (v_A - v_B) i_A + (v_C - v_B) i_C$$

$$p = v_{AB} i_A + v_{CB} i_C = \begin{bmatrix} v_{AB} \\ v_{CB} \end{bmatrix}^T \begin{bmatrix} i_A \\ i_C \end{bmatrix} \qquad \text{(A-75)}$$

Substituting equations A-69 and A-73a into equation A-75, the expression for instantaneous active power can be derived as

$$p = \begin{bmatrix} v_{AB} \\ v_{CB} \end{bmatrix}^T \begin{bmatrix} i_A \\ i_C \end{bmatrix}$$

$$= \begin{bmatrix} \sqrt{3}\left\{\sin(\omega t + 120^\circ) v_d + \cos(\omega t + 120^\circ) v_q\right\} \\ \sqrt{3}\left\{-\sin(\omega t) v_d - \cos(\omega t) v_q\right\} \end{bmatrix}^T \begin{bmatrix} \cos(\omega t) i_d - \sin(\omega t) i_q \\ \cos(\omega t + 120^\circ) i_d - \sin(\omega t + 120^\circ) i_q \end{bmatrix}$$

$$p = \sqrt{3}\left\{\sin(\omega t + 120^\circ) v_d + \cos(\omega t + 120^\circ) v_q\right\}\left\{\cos(\omega t)\, i_d - \sin(\omega t)\, i_q\right\} + \sqrt{3}\left\{-\sin(\omega t) v_d - \cos(\omega t) v_q\right\}\left\{\cos(\omega t + 120^\circ)\, i_d - \sin(\omega t + 120^\circ)\, i_q\right\}$$

$$p = \frac{3}{2}\left(v_d i_d + v_q i_q\right) \tag{A-76}$$

Also, if the voltage and current rotating vectors are converted into two-phase rotating quantities as

$$\mathbf{v} = v_d + jv_q \tag{A-77a}$$

and

$$\mathbf{i} = i_d + ji_q \tag{A-77b}$$

the instantaneous power is

$$\mathbf{s} = \frac{3}{2}\mathbf{vi}^* = \frac{3}{2}(v_d + jv_q)(i_d - ji_q) = \frac{3}{2}(v_d i_d + v_q i_q) + j\frac{3}{2}(v_q i_d - v_d i_q) = p + jq \tag{A-78}$$

Then, the instantaneous active and reactive powers are

$$p = \mathrm{Re}\left[\mathbf{s}\right] = \frac{3}{2}(v_d i_d + v_q i_q) \tag{A-76}$$

$$q = \mathrm{Im}\left[\mathbf{s}\right] = \frac{3}{2}(v_q i_d - v_d i_q) \tag{A-79}$$

From equation A-68,

$$i_d = \frac{2}{\sqrt{3}}\sin(\omega t + 120^\circ)\, i_A - \frac{2}{\sqrt{3}}\sin(\omega t)\, i_C \tag{A-80a}$$

and

$$i_q = \frac{2}{\sqrt{3}}\cos(\omega t + 120^\circ)\, i_A - \frac{2}{\sqrt{3}}\cos(\omega t)\, i_C \tag{A-80b}$$

From equation A-73,

$$v_d = \frac{2}{3}\cos(\omega t)v_{AB} + \frac{2}{3}\cos(\omega t + 120^\circ)\, v_{CB} \quad \text{(A-81a)}$$

and

$$v_q = -\frac{2}{3}\sin(\omega t)v_{AB} - \frac{2}{3}\sin(\omega t + 120^\circ)\, v_{CB} \quad \text{(A-81b)}$$

Substituting equations A-80a and A-80b and A-81a and A-81b into equation A-79, the expression for instantaneous reactive power becomes

$$q = \frac{3}{2}(v_q i_d - v_d i_q) = \frac{3}{2}\frac{2}{3}\frac{2}{\sqrt{3}}\begin{bmatrix}\{-\sin(\omega t)v_{AB} - \sin(\omega t+120^\circ)v_{CB}\}\{\sin(\omega t+120^\circ)\, i_A - \sin(\omega t)\, i_C\} - \\ \{\cos(\omega t)v_{AB} + \cos(\omega t+120^\circ)v_{CB}\}\{\cos(\omega t+120^\circ)\, i_A - \cos(\omega t)\, i_C\}\end{bmatrix}$$

$$q = \frac{2}{\sqrt{3}}\begin{bmatrix} v_{AB} i_A \{-\sin(\omega t)\sin(\omega t+120^\circ) - \cos(\omega t)\cos(\omega t+120^\circ)\} + v_{AB}\, i_C \{\sin^2(\omega t) + \cos^2(\omega t)\} + \\ v_{CB} i_A \{-\sin^2(\omega t+120^\circ) - \cos^2(\omega t+120^\circ)\} + v_{CB}\, i_C \{\sin(\omega t+120^\circ)\sin(\omega t) + \cos(\omega t+120^\circ)\cos(\omega t)\}\end{bmatrix}$$

$$q = \frac{2}{\sqrt{3}}\left[-v_{AB} i_A \cos(120^\circ) + v_{AB} i_C - v_{CB} i_A + v_{CB}\, i_C \cos(120^\circ)\right] = \frac{2}{\sqrt{3}}\left[\frac{1}{2} v_{AB}\, i_A + v_{AB}\, i_C - v_{CB}\, i_A - \frac{1}{2} v_{CB}\, i_C\right]$$

$$q = \frac{2}{\sqrt{3}}\left[\frac{1}{2} v_{AB} i_A + \frac{1}{2} v_{AB} i_C - \frac{1}{2} v_{CB} i_A - \frac{1}{2} v_{CB}\, i_C\right] + \frac{2}{\sqrt{3}}\left[\frac{1}{2} v_{AB}\, i_C - \frac{1}{2} v_{CB}\, i_A\right]$$

$$q = \frac{1}{\sqrt{3}}\left[v_{AB} i_A + v_{AB} i_C - v_{CB} i_A - v_{CB} i_C\right] + \frac{1}{\sqrt{3}}\left[v_{AB}\, i_C - v_{CB}\, i_A\right]$$

$$q = \frac{1}{\sqrt{3}}\left[v_{AB}(i_A + i_C) - v_{CB}(i_A + i_C)\right] + \frac{1}{\sqrt{3}}\left[v_{AB}\, i_C + v_{BC}\, i_A\right]$$

$$= \frac{1}{\sqrt{3}}\left[(v_{AB} - v_{CB})(i_A + i_C)\right] + \frac{1}{\sqrt{3}}\left[v_{AB}\, i_C + v_{BC}\, i_A\right]$$

$$q = \frac{1}{\sqrt{3}}\left[v_{AC}(-i_B)\right] + \frac{1}{\sqrt{3}}\left[v_{AB} i_C + v_{BC} i_A\right] = \frac{1}{\sqrt{3}}\left[v_{CA} i_B\right] + \frac{1}{\sqrt{3}}\left[v_{AB} i_C + v_{BC} i_A\right]$$

Then,

$$q = \frac{1}{\sqrt{3}}\left[v_{BC} i_A + v_{CA} i_B + v_{AB} i_C\right] \quad \text{(A-82)}$$

Alternately from equation A-82, the instantaneous reactive power is

$$q = \frac{1}{\sqrt{3}}\left[v_{BC} i_A - v_{CA}(i_A + i_C) + v_{AB} i_C\right] = \frac{1}{\sqrt{3}}\left[(v_{BC} - v_{CA})\, i_A + (v_{AB} - v_{CA})\, i_C\right]$$

$$q = \frac{1}{\sqrt{3}}\left[(-2v_C + v_B + v_A) i_A + (2v_A - v_B - v_C)\, i_C\right] \quad \text{(A-83)}$$

In a three-phase, balanced network, $v_A + v_B + v_C = 0$. Substituting $v_A + v_B = -v_C$ and $-v_B - v_C - v_A$ and rewriting equation A-83,

$$q = \frac{1}{\sqrt{3}}\left[(-2v_C - v_C)i_A + (2v_A + v_A)i_C\right] = \frac{1}{\sqrt{3}}\left[-3\,v_C\,i_A + 3\,v_A\,i_C\right]$$

$$q = \sqrt{3}\left[v_A i_C - v_C i_A\right] \tag{A-84}$$

Check

Consider a three-phase voltage, $v_A = \cos\theta$, $v_B = \cos(\theta - 120°)$, and $v_C = \cos(\theta + 120°)$, driving a three-phase current $i_A = \cos(\theta + \varphi)$, $i_B = \cos(\theta + \varphi - 120°)$, and $i_C = \cos(\theta + \varphi + 120°)$. The instantaneous active and reactive powers are

$$\begin{aligned}
p &= v_A i_A + v_B i_B + v_C i_C = \cos\theta\cos(\theta+\varphi) + \cos(\theta - 120^\circ)\cos(\theta+\varphi-120^\circ) + \cos(\theta+120^\circ)\cos(\theta+\varphi+120^\circ) \\
&= \frac{1}{2}\left\{\cos 2(\theta + \frac{\varphi}{2}) + \cos\varphi + \cos 2\left(\theta - 120^\circ + \frac{\varphi}{2}\right) + \cos\varphi + \cos 2\left(\theta + 120^\circ + \frac{\varphi}{2}\right) + \cos\varphi\right\} \\
&= \frac{3}{2}\cos\varphi + \frac{1}{2}\left\{\cos(2\theta+\varphi) + \cos(2\theta + 120^\circ + \varphi) + \cos(2\theta - 120^\circ + \varphi)\right\} \\
&= \frac{3}{2}\cos\varphi
\end{aligned}$$

$$\begin{aligned}
q &= \sqrt{3}\left[v_A i_C - v_C i_A\right] = \sqrt{3}\left[\cos\theta\cos(\theta+\varphi+120^\circ) - \cos(\theta+120^\circ)\cos(\theta+\varphi)\right] \\
&= \frac{\sqrt{3}}{2}\left[\cos(2\theta+\varphi+120^\circ) + \cos(\varphi+120^\circ) - \cos(2\theta+\varphi+120^\circ) - \cos(\varphi-120^\circ)\right] \\
&= \frac{\sqrt{3}}{2}\left[\cos(\varphi+120^\circ) - \cos(\varphi-120^\circ)\right] = -\frac{\sqrt{3}}{2}\left[2\sin\varphi\sin 120^\circ\right] = -\frac{3}{2}\sin\varphi
\end{aligned}$$

At unity power factor, $\phi = 0$ and

$$p = 3/2,\ q = 0 \tag{A-85a}$$

At zero power factor lagging, $\phi = -90°$ and

$$p = 0,\ q = 3/2 \tag{A-85b}$$

At zero power factor leading, $\phi = 90°$ and

$$p = 0,\ q = -3/2 \tag{A-85c}$$

A.VI. FOURIER ANALYSIS

Any periodic variable (f) that is a function of angle (θ) can be represented as a Fourier series given by

$$f(\theta) = a_0 + a_1 \cos\theta + a_2 \cos(2\theta) + a_3 \cos(3\theta) + \ldots + b_1 \sin\theta + b_2 \sin(2\theta)$$

$$+ b_3 \sin(3\theta) + \ldots \qquad \text{(A-86)}$$

$$= a_0 + \sum_{n=1}^{\infty} \{a_n \cos(n\theta) + b_n \sin(n\theta)\} \qquad \text{(A-87)}$$

$$= a_0 + \sum_{n=1}^{\infty} A_n \sin(n\theta + \alpha_n) \qquad \text{(A-88)}$$

where

$$a_n = A_n \sin\alpha_n \qquad \text{(A-89a)}$$

$$b_n = A_n \cos\alpha_n \qquad \text{(A-89b)}$$

$$A_n = \sqrt{a_n^2 + b_n^2} \qquad \text{(A-89c)}$$

$$\alpha_n = \tan^{-1}\frac{a_n}{b_n} \qquad \text{(A-89d)}$$

Integrating both sides of equation A-87 results in

$$\int_0^{2\pi} f(\theta)d\theta = a_0 \int_0^{2\pi} d\theta + a_1 \int_0^{2\pi} \cos\theta\, d\theta + a_2 \int_0^{2\pi} \cos(2\theta)\, d\theta + \ldots + b_1 \int_0^{2\pi} \sin\theta\, d\theta$$

$$+ b_2 \int_0^{2\pi} \sin(2\theta)\, d\theta + \ldots = a_0 \theta\Big|_0^{2\pi} = 2a_0\pi$$

since the integral of a sine or a cosine variable over one period is zero. Then,

$$a_0 = \frac{1}{2\pi}\int_0^{2\pi} f(\theta)d\theta \qquad \text{(A-90)}$$

is the DC component in the original variable, $f(\theta)$.

The amplitudes (a_n) for $n = 1, 2, 3, \ldots$ of the cosine components can be expressed as follows. Multiplying both sides of equation A-86 with $\cos(n\theta)$ and integrating over one period results in

$$\int_0^{2\pi} f(\theta)\cos(n\theta)d\theta = a_0 \int_0^{2\pi} \cos(n\theta)\, d\theta + a_1 \int_0^{2\pi} \cos\theta \cos(n\theta)\, d\theta + \ldots$$

$$+ a_m \int_0^{2\pi} \cos(m\theta)\cos(n\theta)\, d\theta + \ldots$$

$$+ b_1 \int_0^{2\pi} \sin\theta \cos(n\theta)\, d\theta + \ldots + b_m \int_0^{2\pi} \sin(m\theta)\cos(n\theta)\, d\theta + \ldots$$

Of all the terms on the right-hand side, the only term where $m = n$ results in a nonzero coefficient of a_n and the rest of the terms are zero. Then,

$$\int_0^{2\pi} f(\theta)\cos(n\theta)\,d\theta = a_n \int_0^{2\pi} \cos^2(n\theta)\,d\theta = \frac{a_n}{2}\int_0^{2\pi} \{1 + \cos(2n\theta)\}d\theta$$

$$= \frac{a_n}{2}\int_0^{2\pi} d\theta + \frac{a_n}{2}\int_0^{2\pi} \cos(2n\theta)\,d\theta$$

$$= \frac{a_n}{2}\left[\theta\Big|_0^{2\pi} + \frac{\sin(2n\theta)}{2n}\Big|_0^{2\pi}\right] = \frac{a_n}{2}(2\pi) = a_n\pi$$

and

$$a_n = \frac{1}{\pi}\int_0^{2\pi} f(\theta)\cos(n\theta)\,d\theta \tag{A-91}$$

The amplitudes (b_n) for $n = 1, 2, 3, \ldots$ of the sine components can be expressed following a similar approach. Multiplying both sides of A-86 by $\sin(n\theta)$ and integrating over one period results in

$$\int_0^{2\pi} f(\theta)\sin(n\theta)\,d\theta = a_0\int_0^{2\pi} \sin(n\theta)\,d\theta + a_1\int_0^{2\pi} \cos\theta\sin(n\theta)\,d\theta + \ldots$$

$$+ a_m\int_0^{2\pi} \cos(m\theta)\sin(n\theta)\,d\theta + \ldots$$

$$+ b_1\int_0^{2\pi} \sin\theta\sin(n\theta)\,d\theta + \ldots + b_m\int_0^{2\pi} \sin(m\theta)\sin(n\theta)\,d\theta + \ldots$$

Of all the terms on the right-hand side, the only term where $m = n$ results in a nonzero coefficient of b_n and the rest of the terms are zero. Then,

$$\int_0^{2\pi} f(\theta)\sin(n\theta)\,d\theta = b_n\int_0^{2\pi} \sin^2(n\theta)\,d\theta = \frac{b_n}{2}\int_0^{2\pi} \{1 - \cos(2n\theta)\}d\theta$$

$$= \frac{b_n}{2}\left[\theta\Big|_0^{2\pi} - \frac{\sin(2n\theta)}{2n}\Big|_0^{2\pi}\right] = \frac{b_n 2\pi}{2} = b_n\pi$$

and

$$b_n = \frac{1}{\pi}\int_0^{2\pi} f(\theta)\sin(n\theta)\,d\theta \tag{A-92}$$

Check

$$f(\theta) = \frac{1}{2} + \sin\theta + \frac{1}{3}\sin(3\theta) + \frac{1}{5}\sin(5\theta)$$

From equation A-90, the DC component is

$$a_0 = \frac{1}{2\pi}\int_0^{2\pi}\left[\frac{1}{2} + \sin\theta + \frac{1}{3}\sin(3\theta) + \frac{1}{5}\sin(5\theta)\right]d\theta$$

$$= \frac{1}{2\pi}\left[\frac{1}{2}\int_0^{2\pi} d\theta + \int_0^{2\pi}\sin\theta\, d\theta + \frac{1}{3}\int_0^{2\pi}\sin(3\theta)\,d\theta + \frac{1}{5}\int_0^{2\pi}\sin(5\theta)\,d\theta\right]$$

$$= \frac{1}{2\pi}\left[\frac{1}{2}\theta\Big|_0^{2\pi} - \cos\theta\Big|_0^{2\pi} - \frac{1}{9}\cos(3\theta)\Big|_0^{2\pi} - \frac{1}{25}\cos(5\theta)\Big|_0^{2\pi}\right] = \frac{1}{2\pi}(\pi) = \frac{1}{2}$$

From equation A-91, the cosine component coefficient is

$$a_n = \frac{1}{\pi}\left[\frac{1}{2}\int_0^{2\pi}\cos(n\theta)\,d\theta + \int_0^{2\pi}\sin\theta\cos(n\theta)\,d\theta\right.$$
$$\left.+\frac{1}{3}\int_0^{2\pi}\sin(3\theta)\cos(n\theta)\,d\theta + \frac{1}{5}\int_0^{2\pi}\sin(5\theta)\cos(n\theta)\,d\theta\right]$$

For $n = 1$:

$$a_1 = \frac{1}{\pi}\left[\frac{1}{2}\int_0^{2\pi}\cos\theta\,d\theta + \int_0^{2\pi}\sin\theta\cos\theta\,d\theta + \frac{1}{3}\int_0^{2\pi}\sin(3\theta)\cos\theta\,d\theta + \frac{1}{5}\int_0^{2\pi}\sin(5\theta)\cos\theta\,d\theta\right]$$

$$= \frac{1}{\pi}\left[\frac{1}{2}\int_0^{2\pi}\cos\theta\,d\theta + \frac{1}{2}\int_0^{2\pi}\sin(2\theta)\,d\theta + \frac{1}{6}\int_0^{2\pi}\{\sin(4\theta) + \sin(2\theta)\}\,d\theta\right.$$
$$\left.+\frac{1}{10}\int_0^{2\pi}\{\sin(6\theta) + \sin(4\theta)\}\,d\theta\right] = 0$$

For $n = 3$:

$$a_3 = \frac{1}{\pi}\left[\frac{1}{2}\int_0^{2\pi}\cos(3\theta)\,d\theta + \int_0^{2\pi}\sin\theta\cos(3\theta)\,d\theta + \frac{1}{3}\int_0^{2\pi}\sin(3\theta)\cos(3\theta)\,d\theta\right.$$
$$\left.+\frac{1}{5}\int_0^{2\pi}\sin(5\theta)\cos(3\theta)\,d\theta\right] = 0$$

For $n = 5$:

$$a_5 = \frac{1}{\pi}\left[\frac{1}{2}\int_0^{2\pi}\cos(5\theta)\,d\theta + \int_0^{2\pi}\sin\theta\cos(5\theta)\,d\theta + \frac{1}{3}\int_0^{2\pi}\sin(3\theta)\cos(5\theta)\,d\theta\right.$$

$$\left. + \frac{1}{5}\int_0^{2\pi}\sin(5\theta)\cos(5\theta)\,d\theta\right] = 0$$

From equation A-92, the sine component coefficient is

$$b_n = \frac{1}{\pi}\left[\frac{1}{2}\int_0^{2\pi}\sin(n\theta)\,d\theta + \int_0^{2\pi}\sin\theta\,\sin(n\theta)\,d\theta + \frac{1}{3}\int_0^{2\pi}\sin(3\theta)\,\sin(n\theta)\,d\theta\right.$$

$$\left. + \frac{1}{5}\int_0^{2\pi}\sin(5\theta)\cos(n\theta)\,d\theta\right]$$

For $n = 1$:

$$b_1 = \frac{1}{\pi}\left[\frac{1}{2}\int_0^{2\pi}\sin\theta\,d\theta + \int_0^{2\pi}\sin\theta\sin\theta\,d\theta + \frac{1}{3}\int_0^{2\pi}\sin(3\theta)\sin\theta\,d\theta + \frac{1}{5}\int_0^{2\pi}\sin(5\theta)\sin\theta\,d\theta\right]$$

$$= \frac{1}{\pi}\left[0 + \int_0^{2\pi}\left\{\frac{1}{2} - \frac{1}{2}\cos(2\theta)\right\}d\theta + 0 + 0\right] = \frac{1}{\pi}\left[\frac{1}{2}\theta\Big|_0^{2\pi} - \frac{1}{4}\sin(2\theta)\Big|_0^{2\pi}\right]$$

$$= \frac{1}{\pi}\left[\frac{1}{2}(2\pi - 0) - 0\right] = 1$$

For $n = 3$:

$$b_3 = \frac{1}{\pi}\left[\frac{1}{2}\int_0^{2\pi}\sin(3\theta)\,d\theta + \int_0^{2\pi}\sin\theta\sin(3\theta)\,d\theta + \frac{1}{3}\int_0^{2\pi}\sin(3\theta)\sin(3\theta)\,d\theta\right.$$

$$\left. + \frac{1}{5}\int_0^{2\pi}\sin(5\theta)\sin(3\theta)\,d\theta\right] = \frac{1}{\pi}\left[0 + 0 + \frac{1}{6}\int_0^{2\pi}\{1 - \cos(6\theta)\}d\theta + 0\right]$$

$$= \frac{1}{\pi}\frac{1}{6}\left[\theta\Big|_0^{2\pi} - \frac{1}{6}\sin(6\theta)\Big|_0^{2\pi}\right] = \frac{1}{\pi}\frac{1}{6}\left[(2\pi - 0) - 0\right] = \frac{1}{3}$$

For $n = 5$:

$$b_5 = \frac{1}{\pi}\left[\frac{1}{2}\int_0^{2\pi} \sin(5\theta)\,d\theta + \int_0^{2\pi} \sin\theta \sin(5\theta)\,d\theta + \frac{1}{3}\int_0^{2\pi} \sin(3\theta)\sin(5\theta)\,d\theta\right.$$

$$\left. + \frac{1}{5}\int_0^{2\pi} \sin(5\theta)\sin(5\theta)\,d\theta\right] = \frac{1}{\pi}\left[0+0+0+\frac{1}{10}\int_0^{2\pi}\{1-\cos(10\theta)\}d\theta\right]$$

$$= \frac{1}{\pi}\frac{1}{10}\left[\theta\Big|_0^{2\pi} - \frac{1}{10}\sin(10\theta)\Big|_0^{2\pi}\right] = \frac{1}{\pi}\frac{1}{10}\left[(2\pi - 0) - 0\right] = \frac{1}{5}$$

Therefore, the composite variable is

$$f(\theta) = a_0 + b_1 \sin\theta + b_3 \sin(3\theta) + b_5 \sin(5\theta) = \frac{1}{2} + \sin\theta + \frac{1}{3}\sin(3\theta) + \frac{1}{5}\sin(5\theta)$$

A.VII. ADAMS–BASHFORTH NUMERICAL INTEGRATION FORMULA

The exact value of a function (f) at any time (t_{n+1}) can be written using Taylor's series as

$$f(t_{n+1}) = f(t_n) + hDf(t_n) + \frac{h^2}{2}D^2 f(t_n) + \frac{h^3}{6}D^3 f(t_n) + \frac{h^4}{24}D^4 f(t_n) + \varphi(h^5) \qquad \text{(A-93)}$$

where $f(t_n)$ is the exact value of the function (f) at time (t_n), $Df(t_n)$, $D^2f(t_n)$, $D^3f(t_n)$, and $D^4f(t_n)$ are the exact derivatives of the function (f) at time (t_n), $D \equiv d/dt$, and the time step for numerical integration is h (i.e., $t_{n+1} - t_n$). The total time (T) is divided into N equidistant segments where $T = Nh$.

The assumed difference equation for numerical integration is of the form

$$f_{n+1} = \alpha_1 f_n + h\beta_1 Df_n + h\beta_2 Df_{n-1} + h\beta_3 Df_{n-2} + h\beta_4 Df_{n-3} \qquad \text{(A-94)}$$

where f_{n+1} is the final value of the function (f) at the end of the time step n, f_n is the initial value of the function (f) at the beginning of the time step n, and Df_n, Df_{n-1}, Df_{n-2}, and Df_{n-3} are the derivatives of the function (f) at the beginning of the steps n, $n-1$, $n-2$, and $n-3$, respectively and α and β are constants. The initial value of the state variable (f) at the beginning of the ($n+1$)th step is the final value of the same state variable (f) at the end of the nth step. It is desired to choose the constants (α and β) to match the highest possible order.

The exact function given in equation A-93 is being truncated in equation A-94, thereby creating a total truncation error at every step of integration in the order of $\phi(h^5)$. Since, there are N steps in the total duration of integration (T), and $N \sim 1/h$, the global truncation error is in the order of $\phi(h^4)$.

Expanding each term on the both sides of equation A-94 using Taylor's series results in

$$\begin{aligned}
&f_n + hDf_n + \frac{h^2}{2}D^2 f_n + \frac{h^3}{6}D^3 f_n + \frac{h^4}{24}D^4 f_n + \ldots \\
&= \alpha_1 f_n \\
&+ h\beta_1 Df_n \\
&+ h\beta_2\left(Df_n - hD^2 f_n + \frac{h^2}{2}D^3 f_n - \frac{h^3}{6}D^4 f_n + \ldots\right) \\
&+ h\beta_3\left(Df_n - 2hD^2 f_n + 4\frac{h^2}{2}D^3 f_n - 8\frac{h^3}{6}D^4 f_n + \ldots\right) \\
&+ h\beta_4\left(Df_n - 3hD^2 f_n + 9\frac{h^2}{2}D^3 f_n - 27\frac{h^3}{6}D^4 f_n + \ldots\right)
\end{aligned} \tag{A-95}$$

Equating term by term,

$$\begin{aligned}
&f_n: \quad 1 = \alpha_1 \\
&Df_n: \quad h = h(\beta_1 + \beta_2 + \beta_3 + \beta_4) \Rightarrow \beta_1 + \beta_2 + \beta_3 + \beta_4 = 1 \\
&D^2 f_n: \quad \frac{h^2}{2} = -h^2(\beta_2 + 2\beta_3 + 3\beta_4) \Rightarrow \beta_2 + 2\beta_3 + 3\beta_4 = -\frac{1}{2} \\
&D^3 f_n: \quad \frac{h^3}{6} = \frac{h^3}{2}(\beta_2 + 4\beta_3 + 9\beta_4) \Rightarrow \beta_2 + 4\beta_3 + 9\beta_4 = \frac{1}{3} \\
&D^4 f_n: \quad \frac{h^4}{24} = -\frac{h^4}{6}(\beta_2 + 8\beta_3 + 27\beta_4) \Rightarrow \beta_2 + 8\beta_3 + 27\beta_4 = -\frac{1}{4}
\end{aligned} \tag{A-96}$$

From equation A-96, it can be found that $\alpha_1 = 1$, and $\beta_1 = 55/24$, $\beta_2 = -59/24$, $\beta_3 = 37/24$, and $\beta_4 = -9/24$.

Then, the difference equation A-94 becomes

$$f_{n+1} = f_n + \left(\frac{55}{24}Df_n - \frac{59}{24}Df_{n-1} + \frac{37}{24}Df_{n-2} - \frac{9}{24}Df_{n-3}\right)h \tag{A-97}$$

which is known as Adams–Bashforth's fourth-order integration method.

APPENDIX B

POWER FLOW CONTROL EQUATIONS IN A LOSSY TRANSMISSION LINE

Consider an AC transmission line with a resistance (R), an inductive reactance (X), and $X/R = \cot\phi = 3.7$ carrying power from the sending end with a voltage, $\mathbf{V_s}$ (i.e., $V_s \angle\delta_s$), to the receiving end with a voltage, $\mathbf{V_r}$ (i.e., $V_r \angle\delta_r$). The voltage ($\mathbf{V_t}$) across the transmission line is the difference between the sending- and receiving-end voltages. This voltage causes a current, $\mathbf{I}$ (i.e., $I \angle\theta_I$), to flow through the transmission line impedance (R, X), as shown in Figure B-1.

The current in the transmission line is

$$\mathrm{I} = \frac{V_S\angle\delta_S - V_r\angle\delta_r}{R + jX} = \frac{V_S(\cos\delta_S + j\sin\delta_S) - V_r(\cos\delta_r + j\sin\delta_r)}{R + jX}$$

$$= \frac{(V_S\cos\delta_S - V_r\cos\delta_r) + j(V_S\sin\delta_S - V_r\sin\delta_r)}{(R + jX)(R - jX)}(R - jX)$$

$$= \frac{C_1 + jC_2}{(R + jX)(R - jX)}(R - jX)$$

$$= \frac{RC_1 + XC_2}{R^2 + X^2} + j\frac{RC_2 - XC_1}{R^2 + X^2} = I\angle\theta_I = I(\cos\theta_I + j\sin\theta_I) \quad \text{(B-1)}$$

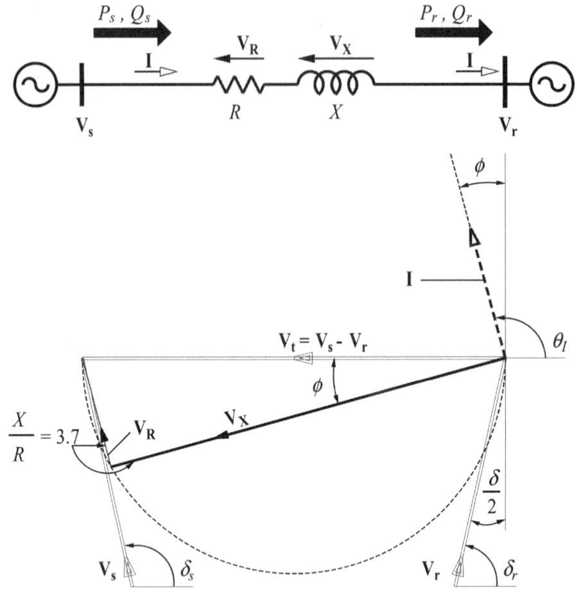

Figure B-1. Single line power transmission system and its phasor diagram.

where

$$C_1 = V_S \cos\delta_S - V_r \cos\delta_r \tag{B-2a}$$

and

$$C_2 = V_S \sin\delta_S - V_r \sin\delta_r \tag{B-2b}$$

Then,

$$I\cos\theta_I = \frac{RC_1 + XC_2}{R^2 + X^2} \tag{B-3a}$$

and

$$I\sin\theta_I = \frac{RC_2 - XC_1}{R^2 + X^2} \tag{B-3b}$$

From equations B-3a and B-3b, the magnitude of the current is

$$I = \sqrt{I^2\left(\cos^2\theta_I + \sin^2\theta_I\right)} = \sqrt{\left(I\cos\theta_I\right)^2 + \left(I\sin\theta_I\right)^2}$$

$$= \frac{1}{R^2 + X^2}\sqrt{\left(RC_1 + XC_2\right)^2 + \left(RC_2 - XC_1\right)^2}$$

$$= \frac{\sqrt{\left(R^2+X^2\right)\left(C_1^2+C_2^2\right)}}{R^2+X^2}$$

$$= \sqrt{\frac{C_1^2+C_2^2}{R^2+X^2}}$$

or

$$I = \sqrt{\frac{V_s^2 - 2V_sV_r\cos(\delta_s-\delta_r)+V_r^2}{R^2+X^2}} \qquad \text{(B-4a)}$$

after substituting for C_1 and C_2 from equations B-2a and B-2b.

From equations B-3a and B-3b, the phase angle of the current is

$$\theta_I = \tan^{-1}\frac{RC_2 - XC_1}{RC_1 + XC_2}$$

or

$$\theta_I = \tan^{-1}\frac{R(V_s\sin\delta_s - V_r\sin\delta_r) - X(V_s\cos\delta_s - V_r\cos\delta_r)}{R(V_s\cos\delta_s - V_r\cos\delta_r) + X(V_s\sin\delta_s - V_r\sin\delta_r)} \qquad \text{(B-4b)}$$

after substituting for C_1 and C_2 as before.

B.I. POWER FLOW EQUATIONS AT THE SENDING END OF AN UNCOMPENSATED TRANSMISSION LINE

The active power flow at the sending end of the transmission line is

$$P_s = V_sI\cos\left(\delta_s - \theta_I\right) = V_s\cos\delta_s I\cos\theta_I + V_s\sin\delta_s I\sin\theta_I$$

$$= \frac{V_s\cos\delta_s}{R^2+X^2}\left\{RC_1 + XC_2\right\} + \frac{V_s\sin\delta_s}{R^2+X^2}\left\{RC_2 - XC_1\right\}$$

using equations B-3a and B-3b,

$$= \frac{V_s}{R^2+X^2}\left[R\left\{C_1\cos\delta_s + C_2\sin\delta_s\right\} + X\left\{C_2\cos\delta_s - C_1\sin\delta_s\right\}\right]$$

$$= \frac{V_s}{R^2+X^2}\begin{bmatrix} R\left\{(V_s\cos\delta_s - V_r\cos\delta_r)\cos\delta_s + (V_s\sin\delta_s - V_r\sin\delta_r)\sin\delta_s\right\} \\ +X\left\{(V_s\sin\delta_s - V_r\sin\delta_r)\cos\delta_s - (V_s\cos\delta_s - V_r\cos\delta_r)\sin\delta_s\right\} \end{bmatrix}$$

after substituting for C_1 and C_2 from equations B-2a and B-2b,

$$= \frac{V_S}{R^2 + X^2}\left[R\left\{V_S - V_r \cos(\delta_S - \delta_r)\right\} + XV_r \sin(\delta_S - \delta_r)\right] \quad \text{(B-5a)}$$

$$= \frac{V_S}{R^2 + X^2}\left[R\left\{V_S - C_3\right\} + XC_4\right] \quad \text{(B-5b)}$$

or

$$P_S - \frac{RV_S^2}{R^2 + X^2} = \frac{V_S}{R^2 + X^2}\left[-RC_3 + XC_4\right] \quad \text{(B-5c)}$$

where

$$C_3 = V_r \cos(\delta_S - \delta_r) \quad \text{(B-6a)}$$

and

$$C_4 = V_r \sin(\delta_S - \delta_r) \quad \text{(B-6b)}$$

The reactive power flow at the sending end of the transmission line is

$$Q_S = V_S I \sin\left(\delta_S - \theta_I\right) = V_S \sin\delta_S I \cos\theta_I - V_S \cos\delta_S I \sin\theta_I$$

$$= \frac{V_S \sin\delta_S}{R^2 + X^2}\left\{RC_1 + XC_2\right\} - \frac{V_S \cos\delta_S}{R^2 + X^2}\left\{RC_2 - XC_1\right\}$$

using equations B-3a and B-3b,

$$= \frac{V_S}{R^2 + X^2}\left[R\left\{C_1 \sin\delta_S - C_2 \cos\delta_S\right\} + X\left\{C_2 \sin\delta_S + C_1 \cos\delta_S\right\}\right]$$

$$= \frac{V_S}{R^2 + X^2}\left[\begin{matrix} R\left\{(V_S \cos\delta_S - V_r \cos\delta_r)\sin\delta_S - (V_S \sin\delta_S - V_r \sin\delta_r)\cos\delta_S\right\} \\ +X\left\{(V_S \sin\delta_S - V_r \sin\delta_r)\sin\delta_S + (V_S \cos\delta_S - V_r \cos\delta_r)\cos\delta_S\right\} \end{matrix}\right]$$

after substituting for C_1 and C_2 as before,

$$= \frac{V_S}{R^2 + X^2}\left[-RV_r \sin(\delta_S - \delta_r) + X\left\{V_S - V_r \cos(\delta_S - \delta_r)\right\}\right] \quad \text{(B-7a)}$$

$$= \frac{V_S}{R^2 + X^2}\left[-RC_4 + X\left\{V_S - C_3\right\}\right] \quad \text{(B-7b)}$$

or

$$Q_s - \frac{XV_s^2}{R^2+X^2} = \frac{V_s}{R^2+X^2}\left[-RC_4 - XC_3\right] \tag{B-7c}$$

Equations B-5b and B-7b can be combined as

$$P_s^2 + Q_s^2 = \left(\frac{V_s}{R^2+X^2}\right)^2 \left[\left\{R(V_s - C_3) + XC_4\right\}^2 + \left\{-RC_4 + X(V_s - C_3)\right\}^2\right]$$

$$= \left(\frac{V_s}{R^2+X^2}\right)^2 \left(R^2+X^2\right)\left\{(V_s - C_3)^2 + C_4^2\right\}$$

$$= \frac{V_s^2}{R^2+X^2}\left\{V_s^2 - 2V_sV_r\cos(\delta_s - \delta_r) + V_r^2\right\}$$

after substituting for C_3 and C_4 from equations B-6a and B-6b,

$$= (V_s I)^2 \tag{B-8}$$

after substituting for I from equation B-4a.

From equations B-5c and B-7c, the active and reactive power flows at the sending end can be related for the entire range of power angle δ (i.e., $\delta_s - \delta_r$) as

$$\left(P_s - \frac{RV_s^2}{R^2+X^2}\right)^2 + \left(Q_s - \frac{XV_s^2}{R^2+X^2}\right)^2$$

$$= \left(\frac{V_s}{R^2+X^2}\right)^2 \left\{\left(-RC_3 + XC_4\right)^2 + \left(-RC_4 - XC_3\right)^2\right\}$$

$$= \left(\frac{V_s}{R^2+X^2}\right)^2 \left(R^2+X^2\right)\left(C_3^2 + C_4^2\right)$$

or

$$\left(P_s - \frac{RV_s^2}{R^2+X^2}\right)^2 + \left(Q_s - \frac{XV_s^2}{R^2+X^2}\right)^2 = \left(\frac{V_sV_r}{\sqrt{R^2+X^2}}\right)^2 \tag{B-9}$$

after substituting for C_3 and C_4 from equations B-6a and B-6b.

The above equation is the locus of a circle centered at

$$\left[\frac{RV_s^2}{R^2+X^2}, \frac{XV_s^2}{R^2+X^2}\right]$$

with a radius of

$$\frac{V_s V_r}{\sqrt{R^2 + X^2}}$$

From equations B-5c and B-7c, the relationship between the active and reactive power flows at the sending end is

$$\frac{Q_s - \dfrac{X V_s^2}{R^2 + X^2}}{P_s - \dfrac{R V_s^2}{R^2 + X^2}} = \frac{-RC_4 - XC_3}{-RC_3 + XC_4}$$

or

$$Q_s = \frac{-RC_4 - XC_3}{-RC_3 + XC_4}\left(P_s - \frac{R V_s^2}{R^2 + X^2}\right) + \frac{X V_s^2}{R^2 + X^2}$$

or

$$Q_s = \frac{-RV_r \sin(\delta_s - \delta_r) - XV_r \cos(\delta_s - \delta_r)}{-RV_r \cos(\delta_s - \delta_r) + XV_r \sin(\delta_s - \delta_r)}\left(P_s - \frac{R V_s^2}{R^2 + X^2}\right) + \frac{X V_s^2}{R^2 + X^2} \quad \text{(B-10)}$$

after substituting for C_3 and C_4 from equations B-6a and B-6b.

B.II. POWER FLOW EQUATIONS AT THE RECEIVING END OF AN UNCOMPENSATED TRANSMISSION LINE

The active power flow at the receiving end is

$$P_r = V_r I \cos\left(\delta_r - \theta_I\right) = V_r \cos\delta_r I \cos\theta_I + V_r \sin\delta_r I \sin\theta_I$$

$$= \frac{V_r \cos\delta_r}{R^2 + X^2}\left\{RC_1 + XC_2\right\} + \frac{V_r \sin\delta_r}{R^2 + X^2}\left\{RC_2 - XC_1\right\}$$

using equations B-3a and B-3b,

$$= \frac{V_r}{R^2 + X^2}\left[R\left\{C_1 \cos\delta_r + C_2 \sin\delta_r\right\} + X\left\{C_2 \cos\delta_r - C_1 \sin\delta_r\right\}\right]$$

$$= \frac{V_r}{R^2 + X^2}\left[\begin{array}{l} R\left\{(V_s \cos\delta_s - V_r \cos\delta_r)\cos\delta_r + (V_s \sin\delta_s - V_r \sin\delta_r)\sin\delta_r\right\} \\ +X\left\{(V_s \sin\delta_s - V_r \sin\delta_r)\cos\delta_r - (V_s \cos\delta_s - V_r \cos\delta_r)\sin\delta_r\right\} \end{array}\right]$$

after substituting for C_1 and C_2 from equations B-2a and B-2b,

$$= \frac{V_r}{R^2 + X^2}\left[R\left\{V_s \cos(\delta_s - \delta_r) - V_r\right\} + XV_s \sin(\delta_s - \delta_r)\right] \tag{B-11a}$$

$$= \frac{V_r}{R^2 + X^2}\left[R\left\{C_5 - V_r\right\} + XC_6\right] \tag{B-11b}$$

or

$$P_r + \frac{RV_r^2}{R^2 + X^2} = \frac{V_r}{R^2 + X^2}\left[RC_5 + XC_6\right] \tag{B-11c}$$

where

$$C_5 = V_s \cos(\delta_s - \delta_r) \tag{B-12a}$$

and

$$C_6 = V_s \sin(\delta_s - \delta_r) \tag{B-12b}$$

The reactive power flow at the sending end is

$$Q_r = V_r I \sin\left(\delta_r - \theta_I\right) = V_r \sin\delta_r I \cos\theta_I - V_r \cos\delta_r I \sin\theta_I$$

$$= \frac{V_r \sin\delta_r}{R^2 + X^2}\left\{RC_1 + XC_2\right\} - \frac{V_r \cos\delta_r}{R^2 + X^2}\left\{RC_2 - XC_1\right\}$$

using equation B-3a and B-3b,

$$= \frac{V_r}{R^2 + X^2}\left[R\left\{C_1 \sin\delta_r - C_2 \cos\delta_r\right\} + X\left\{C_2 \sin\delta_r + C_1 \cos\delta_r\right\}\right]$$

$$= \frac{V_r}{R^2 + X^2}\left[\begin{array}{l} R\left\{(V_s \cos\delta_s - V_r \cos\delta_r)\sin\delta_r - (V_s \sin\delta_s - V_r \sin\delta_r)\cos\delta_r\right\} \\ +X\left\{(V_s \sin\delta_s - V_r \sin\delta_r)\sin\delta_r + (V_s \cos\delta_s - V_r \cos\delta_r)\cos\delta_r\right\} \end{array}\right]$$

after substituting for C_1 and C_2 from equations B-2a and B-2b,

$$= \frac{V_r}{R^2 + X^2}\left[-RV_s \sin(\delta_s - \delta_r) + X\left\{V_s \cos(\delta_s - \delta_r) - V_r\right\}\right] \tag{B-13a}$$

$$= \frac{V_r}{R^2 + X^2}\left[-RC_6 + X\left\{C_5 - V_r\right\}\right] \tag{B-13b}$$

or

$$Q_r + \frac{X V_r^2}{R^2 + X^2} = \frac{V_r}{R^2 + X^2}\left[-RC_6 + XC_5\right] \tag{B-13c}$$

Equations B-11b and B-13b can be combined as

$$P_r^2 + Q_r^2 = \left(\frac{V_r}{R^2 + X^2}\right)^2 \left[\left\{R\left(C_5 - V_r\right) + XC_6\right\}^2 + \left\{-RC_6 + X\left(C_5 - V_r\right)\right\}^2\right]$$

$$= \left(\frac{V_r}{R^2 + X^2}\right)^2 \left(R^2 + X^2\right)\left\{\left(C_5 - V_r\right)^2 + C_6^2\right\}$$

$$= \frac{V_r^2}{R^2 + X^2}\left\{V_s^2 - 2V_s V_r \cos(\delta_s - \delta_r) + V_r^2\right\}$$

after substituting for C_5 and C_6 from equations B-12a and B-12b,

$$= (V_r I)^2 \tag{B-14}$$

after substituting for I from equation B-4a.

From equations B-11c and B-13c, the active and reactive power flows at the receiving end can be related for the entire range of power angle δ (i.e., $\delta_s - \delta_r$) as

$$\left(P_r + \frac{R V_r^2}{R^2 + X^2}\right)^2 + \left(Q_r + \frac{X V_r^2}{R^2 + X^2}\right)^2$$

$$= \left(\frac{V_r}{R^2 + X^2}\right)^2 \left\{\left(RC_5 + XC_6\right)^2 + \left(-RC_6 + XC_5\right)^2\right\}$$

$$= \left(\frac{V_r}{R^2 + X^2}\right)^2 \left(R^2 + X^2\right)\left(C_5^2 + C_6^2\right)$$

or

$$\left(P_r + \frac{R V_r^2}{R^2 + X^2}\right)^2 + \left(Q_r + \frac{X V_r^2}{R^2 + X^2}\right)^2 = \left(\frac{V_s V_r}{\sqrt{R^2 + X^2}}\right)^2 \tag{B-15}$$

after substituting for C_5 and C_6 from equations B-12a and B-12b.

The above equation is the locus of a circle centered at

$$\left[\frac{-R V_r^2}{R^2 + X^2}, \frac{-X V_r^2}{R^2 + X^2}\right]$$

with a radius of

$$\frac{V_s V_r}{\sqrt{R^2 + X^2}}$$

From equations B-11c and B-13c, the relationship between the active and reactive power flows at the receiving end is

$$\frac{Q_r + \dfrac{XV_r^2}{R^2 + X^2}}{P_r + \dfrac{RV_r^2}{R^2 + X^2}} = \frac{-RC_6 + XC_5}{RC_5 + XC_6}$$

or

$$Q_r = \frac{-RC_6 + XC_5}{RC_5 + XC_6}\left(P_r + \frac{RV_r^2}{R^2 + X^2}\right) - \frac{XV_r^2}{R^2 + X^2}$$

or

$$Q_r = \frac{-RV_s\sin(\delta_s - \delta_r) + XV_s\cos(\delta_s - \delta_r)}{RV_s\cos(\delta_s - \delta_r) + XV_s\sin(\delta_s - \delta_r)}\left(P_r + \frac{RV_r^2}{R^2 + X^2}\right) - \frac{XV_r^2}{R^2 + X^2} \qquad \text{(B-16)}$$

after substituting for C_5 and C_6 from equations B-12a and B-12b.

B.III. VERIFICATION OF POWER FLOW EQUATIONS AT THE SENDING AND RECEIVING ENDS OF AN UNCOMPENSATED TRANSMISSION LINE

The difference between the active power at the sending and receiving ends is

$$P_s - P_r = \frac{V_s}{R^2 + X^2}\left[R\{V_s - C_3\} + XC_4\right] - \frac{V_r}{R^2 + X^2}\left[R\{C_5 - V_r\} + XC_6\right]$$

using equations B-5b and B-11b,

$$= \frac{R}{R^2 + X^2}\left(V_s^2 + V_r^2\right) - \frac{R}{R^2 + X^2}\left(V_s C_3 + V_r C_5\right) + \frac{X}{R^2 + X^2}\left(V_s C_4 - V_r C_6\right)$$

$$= \frac{R}{R^2 + X^2}\left(V_s^2 + V_r^2\right) - \frac{RV_sV_r}{R^2 + X^2}(\cos(\delta_s - \delta_r) + \cos(\delta_s - \delta_r))$$

$$+ \frac{XV_sV_r}{R^2 + X^2}(\sin(\delta_s - \delta_r) - \sin(\delta_s - \delta_r))$$

after substituting for C_3, C_4, C_5, and C_6 from equations B-6a and B-6b, and B-12a and B-12b,

$$= \frac{R}{R^2 + X^2}\left(V_s^2 - 2V_sV_r\cos(\delta_s - \delta_r) + V_r^2\right)$$

$$P_s - P_r = I^2 R \tag{B-17}$$

after substituting for I from equation B-4a. It is the active power absorbed by the transmission line resistance (R).

Similarly, the difference between the reactive power at the sending and receiving ends is

$$Q_s - Q_r = \frac{V_s}{R^2 + X^2}\left[-RC_4 + X\{V_s - C_3\}\right] - \frac{V_r}{R^2 + X^2}\left[-RC_6 + X\{C_5 - V_r\}\right]$$

using equations B-7b and B-13b,

$$= \frac{X}{R^2 + X^2}\left(V_s^2 + V_r^2\right) - \frac{X}{R^2 + X^2}\left(V_sC_3 + V_rC_5\right) - \frac{R}{R^2 + X^2}\left(V_sC_4 - V_rC_6\right)$$

$$= \frac{X}{R^2 + X^2}\left(V_s^2 + V_r^2\right) - \frac{XV_sV_r}{R^2 + X^2}\left(\cos(\delta_s - \delta_r) + \cos(\delta_s - \delta_r)\right)$$

$$- \frac{RV_sV_r}{R^2 + X^2}\left(\sin(\delta_s - \delta_r) - \sin(\delta_s - \delta_r)\right)$$

after substituting for C_3, C_4, C_5, and C_6 from equations B-6a and B-6b, and B-12a and B-12b,

$$= \frac{X}{R^2 + X^2}\left(V_s^2 - 2V_sV_r\cos(\delta_s - \delta_r) + V_r^2\right)$$

$$Q_s - Q_r = I^2 X \tag{B-18}$$

after substituting for I from equation B-4a. It is the reactive power absorbed by the transmission line reactance (X).

B.IV. NATURAL POWER FLOW EQUATIONS IN AN UNCOMPENSATED TRANSMISSION LINE

The electric power flow through an AC transmission line is a function of the line impedance (R, X), the magnitudes of the sending end voltage ($\mathbf{V_s}$) and the receiving end voltage (**V**$_r$), and the difference of phase angles ($\delta_s - \delta_r$) between these voltages. For a given transmission line with impedance (R, X), sending end voltage, **V**$_s$ (i.e., $V_s \angle \delta_s$), and receiving end voltage, **V**$_r$ (i.e., $V_r \angle \delta_r$), the magnitude and angle of the natural line current are

$$I_n = \sqrt{\frac{V_s^2 - 2V_sV_r\cos(\delta_s - \delta_r) + V_r^2}{R^2 + X^2}} \tag{B-19a}$$

and

$$\theta_{In} = \tan^{-1}\frac{R(V_s\sin\delta_s - V_r\sin\delta_r) - X(V_s\cos\delta_s - V_r\cos\delta_r)}{R(V_s\cos\delta_s - V_r\cos\delta_r) + X(V_s\sin\delta_s - V_r\sin\delta_r)} \tag{B-19b}$$

Recall that for $R = 0$,

$$I_n = \frac{V_{Xn}}{X} = \frac{\sqrt{V_s^2 - 2V_sV_r\cos\delta + V_r^2}}{X} \tag{2-79}$$

where $\delta = \delta_s - \delta_r$.

Figure B-2 shows the magnitude (I_n) of the natural current (**I**$_n$) for the range of power angle δ (i.e., $\delta_s - \delta_r$) from 0 to 180° when $V_s = V_r = 1$, $X = 0.5$, and $R = 0$ ($X/R = \infty$), 0.2 ($X/R = 2.5$), and 0.5 ($X/R = 1$).

At first glance, it seems that if the transmission line is operated at a larger power angle, more current and more power will flow through the line. Let us examine the active and reactive power flows in the transmission line. From equation B-5a, the natural active power flow at the sending end of the transmission line is

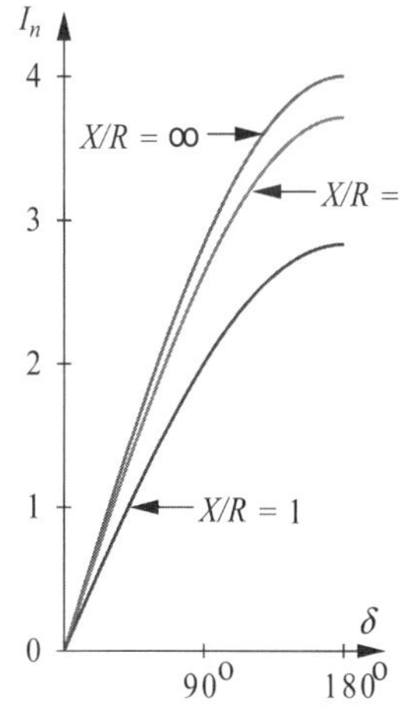

Figure B-2. (a) Natural current (I_n) as a function of power angle δ (i.e., $\delta_s - \delta_r$).

$$P_{sn} = \frac{V_s}{R^2 + X^2}\left[R\{V_s - V_r \cos(\delta_s - \delta_r)\} + XV_r \sin(\delta_s - \delta_r)\right] \quad \text{(B-20)}$$

From equation B-7a, the natural reactive power flow at the sending end of the transmission line is

$$Q_{sn} = \frac{V_s}{R^2 + X^2}\left[-RV_r \sin(\delta_s - \delta_r) + X\{V_s - V_r \cos(\delta_s - \delta_r)\}\right] \quad \text{(B-21)}$$

Recall that for $R = 0$,

$$P_{sn} = \frac{V_s V_r}{X}\left[\sin(\delta_s - \delta_r)\right] \quad \text{(2-15a)}$$

and

$$Q_{sn} = \frac{V_s V_r}{X}\left[\frac{V_s}{V_r} - \cos(\delta_s - \delta_r)\right] \quad \text{(2-15b)}$$

From equations B-20 and B-21, the natural active and reactive power flows at the sending end can be related for the entire range of power angle δ (i.e., $= \delta_s - \delta_r$) as

$$\left(P_{sn} - \frac{RV_s^2}{R^2 + X^2}\right)^2 + \left(Q_{sn} - \frac{XV_s^2}{R^2 + X^2}\right)^2 = \left(\frac{V_s V_r}{\sqrt{R^2 + X^2}}\right)^2 \quad \text{(B-22)}$$

From equations B-20 and B-21, it can be verified that

$$P_{sn}^2 + Q_{sn}^2 = \frac{V_s^2}{R^2 + X^2}\left\{V_s^2 - 2V_s V_r \cos(\delta_s - \delta_r) + V_r^2\right\} = (V_s I_n)^2 \quad \text{(B-23)}$$

after substituting for I_n from equation B-19a.

From equations B-20 and B-21, the relationship between the natural active and reactive power flows at the sending end is

$$Q_{sn} = \frac{-RV_r \sin(\delta_s - \delta_r) - XV_r \cos(\delta_s - \delta_r)}{-RV_r \cos(\delta_s - \delta_r) + XV_r \sin(\delta_s - \delta_r)}\left(P_{sn} - \frac{RV_s^2}{R^2 + X^2}\right) + \frac{XV_s^2}{R^2 + X^2} \quad \text{(B-24)}$$

Figure B-3 shows the natural active power (P_{sn}), natural reactive power (Q_{sn}), and natural apparent power (VA_{sn}) at the sending end, and Q_{sn} versus P_{sn} for the range of power angle δ (i.e., $\delta_s - \delta_r$) from 0 to 180° when $V_s = V_r = 1$, and $X = 0.5$, and $R = 0$ ($X/R = \infty$), 0.2 ($X/R = 2.5$), and 0.5 ($X/R = 1$).

From equation B-11a, the natural active power flow at the receiving end of the transmission line is

$$P_{rn} = \frac{V_r}{R^2 + X^2}\left[R\{V_s \cos(\delta_s - \delta_r) - V_r\} + XV_s \sin(\delta_s - \delta_r) \right] \quad \text{(B-25)}$$

And from equation B-13a, the natural reactive power flow at the receiving end of the transmission line is

$$Q_{rn} = \frac{V_r}{R^2 + X^2}\left[-RV_s \sin(\delta_s - \delta_r) + X\{V_s \cos(\delta_s - \delta_r) - V_r\} \right] \quad \text{(B-26)}$$

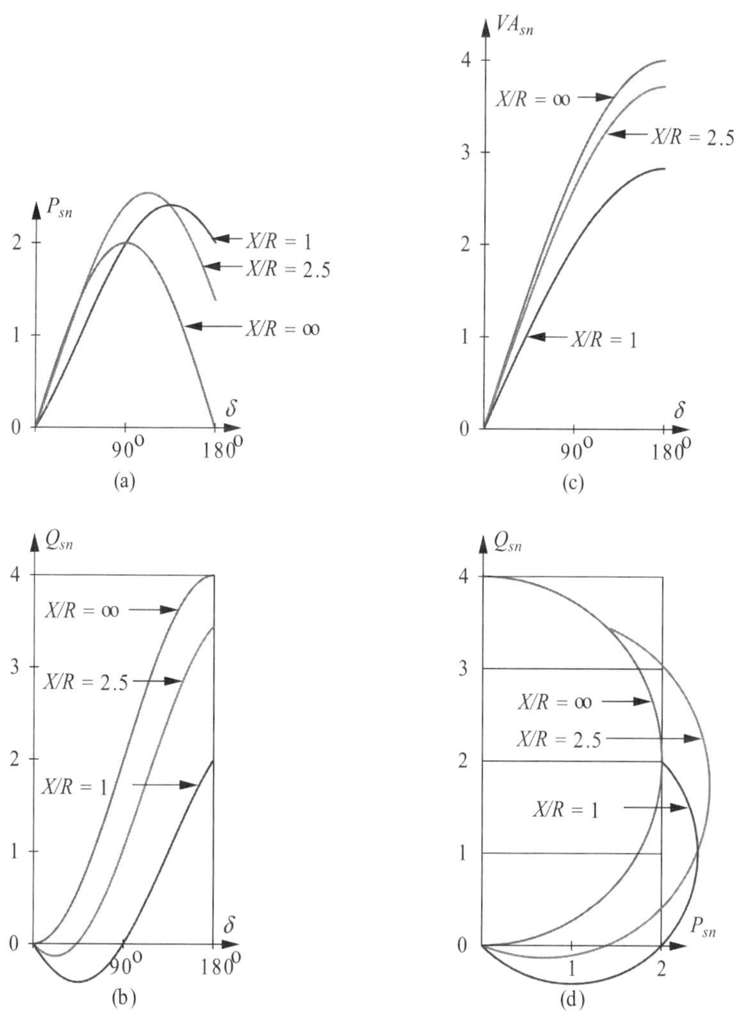

Figure B-3. (a) Natural active power (P_{sn}), (b) natural reactive power (Q_{sn}), and (c) natural apparent power (VA_{sn}) as a function of power angle (δ), and (d) Q_{sn} versus P_{sn} at the sending end of the transmission line.

Recall that for $R = 0$,

$$P_{rn} = \frac{V_s V_r}{X}\left[\sin(\delta_s - \delta_r)\right] \tag{2-16a}$$

and

$$Q_{rn} = \frac{V_s V_r}{X}\left[\cos(\delta_s - \delta_r) - \frac{V_r}{V_s}\right] \tag{2-16b}$$

From equations B-25 and B-26, the natural active and reactive power flows at the receiving end can be related for the entire range of power angle δ (i.e., $\delta_s - \delta_r$) as

$$\left(P_{rn} + \frac{RV_r^2}{R^2 + X^2}\right)^2 + \left(Q_{rn} + \frac{XV_r^2}{R^2 + X^2}\right)^2 = \left(\frac{V_s V_r}{\sqrt{R^2 + X^2}}\right)^2 \tag{B-27}$$

From equations B-25 and B-26, it can be verified that

$$P_{rn}^2 + Q_{rn}^2 = \frac{V_r^2}{R^2 + X^2}\left\{V_s^2 - 2V_s V_r \cos(\delta_s - \delta_r) + V_r^2\right\} = (V_r I_n)^2 \tag{B-28}$$

after substituting for I_n from equation B-19a.

From equations B-25 and B-26, the relationship between the natural active and reactive power flows at the receiving end is

$$Q_{rn} = \frac{-RV_s \sin(\delta_s - \delta_r) + XV_s \cos(\delta_s - \delta_r)}{RV_s \cos(\delta_s - \delta_r) + XV_s \sin(\delta_s - \delta_r)}\left(P_{rn} + \frac{R V_r^2}{R^2 + X^2}\right) - \frac{X V_r^2}{R^2 + X^2} \tag{B-29}$$

Figure B-4 shows the natural active power (P_{rn}), natural reactive power (Q_{rn}), and natural apparent power (VA_{rn}) at the receiving end, and Q_{rn} versus P_{rn} for the range of power angle δ (i.e., $\delta_s - \delta_r$) from 0 to 180° when $V_s = V_r = 1$, and $X = 0.5$, and $R = 0$ ($X/R = \infty$), 0.2 ($X/R = 2.5$), and 0.5 ($X/R = 1$).

Figure B-5 shows that the active power at the sending end is the same as that at the receiving end since there is no active power absorbed by the line reactance (X) when $R = 0$ ($X/R = \infty$). However, the reactive power at the sending end is inductive and that at the receiving end is capacitive. Figure B-3 shows that when the line becomes more and more resistive, the value of δ at which the maximum active power at the sending end takes place is higher. Figure B-4 shows that when the line becomes more and more resistive, the value of δ at which the maximum active power at the receiving end takes place is lower. In the case when $X/R = \infty$, the active power at the sending end as well as at the receiving end is maximum when δ (i.e., $\delta_s - \delta_r$) is 90°. However, for stable operation of the electric power system, δ is maintained at much lower value. The simulations throughout this book are set at $\delta = 20°$ and $X/R = 7.4$.

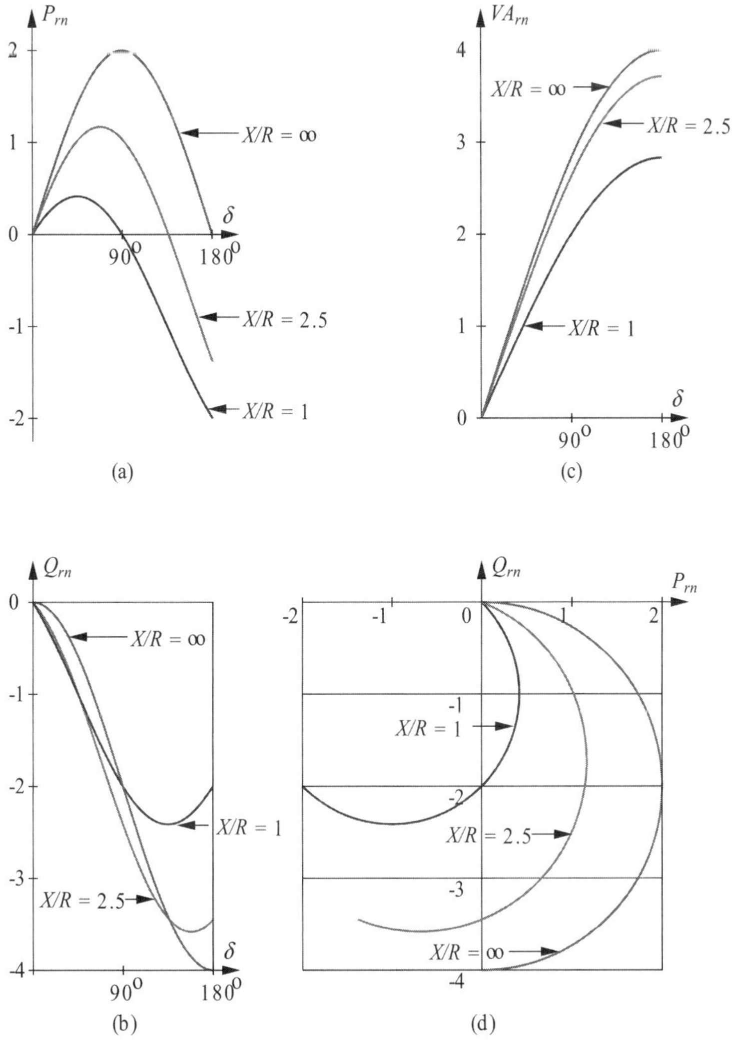

Figure B-4. (a) Natural active power (P_{rn}), (b) natural reactive power (Q_{rn}), and (c) natural apparent power (VA_{rn}) as a function of power angle (δ), and (d) Q_{rn} versus P_{rn} at the receiving end of the transmission line.

B.V. MOST IMPORTANT POWER FLOW CONTROL PARAMETERS

Assuming that the impedance of the transmission line does not change in a short period of time and there is little control over the distant bus voltage (e.g., receiving-end voltage), the only parameter that is able to vary the power flow in a transmission line is the modified sending-end voltage, $\mathbf{V}_{s'}$ (i.e., $V_{s'} \angle \delta_{s'}$). When the sending-end voltage is modified to $\mathbf{V}_{s'}$, the modified transmission line voltage ($\mathbf{V}_{t'}$), which is the difference

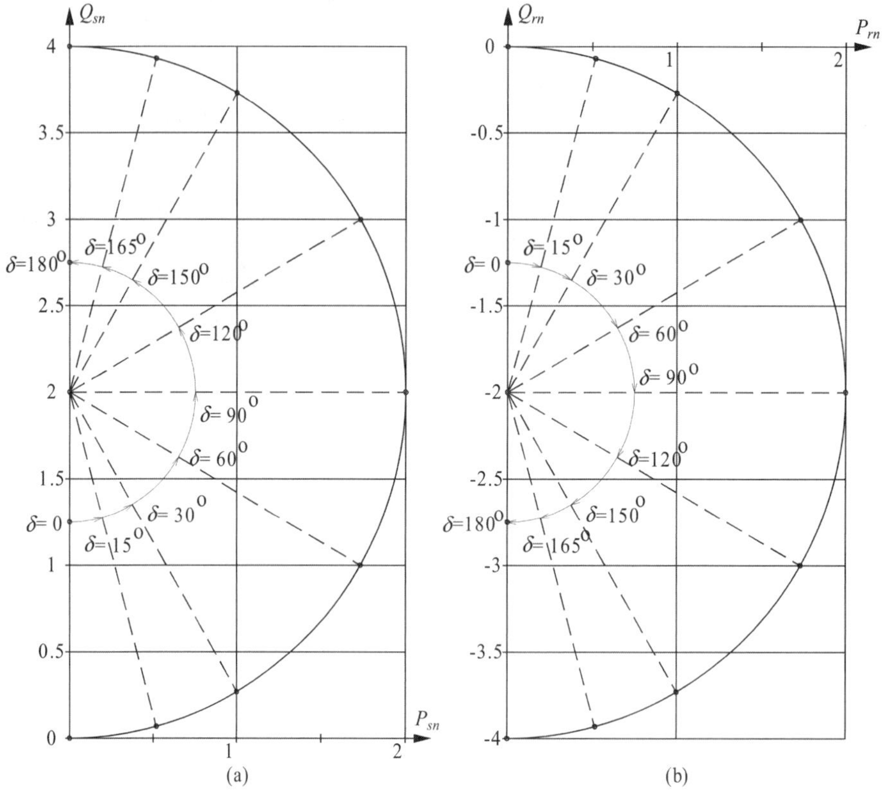

Figure B-5. (a) Q_{sn} versus P_{sn} at the sending end and (b) Q_{rn} versus V_{rn} at the receiving end of the transmission line for the range of power angle δ (i.e., $\delta_s - \delta_r$) from 0 to 180° when $V_s = V_r = 1$, $X = 0.5$, and $R = 0$ ($X/R = \infty$).

between the modified sending- and receiving-end voltages, causes a prevailing current, **I** (i.e., $I \angle \delta_I$), to flow through the transmission line impedance (R, X). Substituting the subscript s' for s in the previously derived equations, the following equations apply.

From equations B-4a amd B-4b, the magnitude and angle of the prevailing current can be written as

$$I = \sqrt{\frac{V_{s'}^2 - 2V_{s'}V_r \cos(\delta_{s'} - \delta_r) + V_r^2}{R^2 + X^2}} \tag{B-30a}$$

and

$$\theta_I = \tan^{-1} \frac{R(V_{s'} \sin\delta_{s'} - V_r \sin\delta_r) - X(V_{s'} \cos\delta_{s'} - V_r \cos\delta_r)}{R(V_{s'} \cos\delta_{s'} - V_r \cos\delta_r) + X(V_{s'} \sin\delta_{s'} - V_r \sin\delta_r)} \tag{B-30b}$$

From equation B-5a, the active power flow at the modified sending end of the transmission line is

$$P_{s'} = \frac{V_{s'}}{R^2 + X^2}\left[R\left\{V_{s'} - V_r \cos(\delta_{s'} - \delta_r)\right\} + XV_r \sin(\delta_{s'} - \delta_r)\right] \quad \text{(B-31)}$$

From equation B-7a, the reactive power flow at the modified sending end of the transmission line is

$$Q_{s'} = \frac{V_{s'}}{R^2 + X^2}\left[-RV_r \sin(\delta_{s'} - \delta_r) + X\left\{V_{s'} - V_r \cos(\delta_{s'} - \delta_r)\right\}\right] \quad \text{(B-32)}$$

From equation B-9, the active and reactive power flows at the modified sending end can be related for the entire range of power angle δ' (i.e., $\delta_{s'} - \delta_r$) as

$$\left(P_{s'} - \frac{RV_{s'}^2}{R^2 + X^2}\right)^2 + \left(Q_{s'} - \frac{XV_{s'}^2}{R^2 + X^2}\right)^2 = \left(\frac{V_{s'}V_r}{\sqrt{R^2 + X^2}}\right)^2 \quad \text{(B-33)}$$

The above equation is the locus of a circle centered at

$$\left[\frac{RV_{s'}^2}{R^2 + X^2}, \frac{XV_{s'}^2}{R^2 + X^2}\right]$$

with a radius of

$$\frac{V_{s'}V_r}{\sqrt{R^2 + X^2}}$$

From equations B-8, it can be verified that

$$P_{s'}^2 + Q_{s'}^2 = \frac{V_{s'}^2}{R^2 + X^2}\left\{V_{s'}^2 - 2V_{s'}V_r \cos(\delta_{s'} - \delta_r) + V_r^2\right\} = (V_{s'}I)^2 \quad \text{(B-34)}$$

after substituting for I from equation B-30a.

From equation B-10, the relationship between the active and reactive power flows at the modified sending end is

$$Q_{s'} = \frac{-RV_r \sin(\delta_{s'} - \delta_r) - XV_r \cos(\delta_{s'} - \delta_r)}{-RV_r \cos(\delta_{s'} - \delta_r) + XV_r \sin(\delta_{s'} - \delta_r)}\left(P_{s'} - \frac{RV_{s'}^2}{R^2 + X^2}\right) + \frac{XV_{s'}^2}{R^2 + X^2} \quad \text{(B-35)}$$

From equation B-11a, the active power flow at the receiving end of the transmission line is

$$P_r = \frac{V_r}{R^2 + X^2}\left[R\left\{V_{s'}\cos(\delta_{s'} - \delta_r) - V_r\right\} + XV_{s'}\sin(\delta_{s'} - \delta_r)\right] \qquad \text{(B-36)}$$

From equation B-13a, the reactive power flow at the receiving end of the transmission line is

$$Q_r = \frac{V_r}{R^2 + X^2}\left[-RV_{s'}\sin(\delta_{s'} - \delta_r) + X\left\{V_{s'}\cos(\delta_{s'} - \delta_r) - V_r\right\}\right] \qquad \text{(B-37)}$$

From equation B-15, the active and reactive power flows at the receiving end can be related for the entire range of power angle δ' (i.e., $\delta_{s'} - \delta_r$) as

$$\left(P_r + \frac{RV_r^2}{R^2 + X^2}\right)^2 + \left(Q_r + \frac{XV_r^2}{R^2 + X^2}\right)^2 = \left(\frac{V_{s'}V_r}{\sqrt{R^2 + X^2}}\right)^2 \qquad \text{(B-38)}$$

The above equation is the locus of a circle centered at

$$\left[\frac{-RV_r^2}{R^2 + X^2}, \frac{-XV_r^2}{R^2 + X^2}\right]$$

with a radius of

$$\frac{V_{s'}V_r}{\sqrt{R^2 + X^2}}$$

From equation B-14, it can be verified that

$$P_r^2 + Q_r^2 = \frac{V_r^2}{R^2 + X^2}\left\{V_{s'}^2 - 2V_{s'}V_r\cos(\delta_{s'} - \delta_r) + V_r^2\right\} = (V_r I)^2 \qquad \text{(B-39)}$$

after substituting for I from equation B-30a.

From equation B-16, the relationship between the active and reactive power flows at the receiving end is

$$Q_r = \frac{-RV_{s'}\sin(\delta_{s'} - \delta_r) + XV_{s'}\cos(\delta_{s'} - \delta_r)}{RV_{s'}\cos(\delta_{s'} - \delta_r) + XV_{s'}\sin(\delta_{s'} - \delta_r)}\left(P_r + \frac{RV_r^2}{R^2 + X^2}\right) - \frac{XV_r^2}{R^2 + X^2} \qquad \text{(B-40)}$$

For a given transmission line, with a fixed R, X, and $\mathbf{V_r}$, the loci of ($Q_{s'}$ versus $P_{s'}$) and (Q_r versus P_r) depend on the modified sending-end voltage, $\mathbf{V_{s'}}$ (i.e., $V_{s'}\angle\delta_{s'}$). There are basically two ways to modify the sending-end voltage, as follows.

B.V.1. Modifying Transmission Line Voltage with a Shunt-Connected Compensating Voltage

The transmission line voltage can be modified by connecting a compensating voltage, $\mathbf{V_{s'}}$ (i.e., $V_{s'} \angle \delta_{s'}$), with desired magnitude and angle in shunt with the modified sending end of the transmission line, as shown in Figure B-6. As an example, the sending-end voltage ($\mathbf{V_s}$) can be used to excite the primary windings of a transformer. The induced secondary voltage ($\mathbf{V_{s'}}$) can be shunt-connected as the modified sending-end voltage. For a given amount of transmitted power, the shunt–shunt topology is rated for the full amount of transmitted power. In certain special cases for point-to-point transfer of power between two isolated networks with different voltages or phase angles, the use of the shunt–shunt power converter remains the preferred topology. If a VSC-based shunt–shunt power converter is used for this purpose, the frequencies of both AC lines need not be the same.

B.V.2. Modifying Transmission Line Voltage with a Series-Connected Compensating Voltage

The transmission line voltage can be modified by connecting a compensating voltage, $\mathbf{V_{s's}}$ (i.e., $V_{s's} \angle \delta_s + \beta$), in series with the transmission line voltage, as shown in Figure B-7. As an example, the sending-end voltage ($\mathbf{V_s}$) can be used to excite the primary windings of a transformer. The induced secondary (compensating) voltage modifies the sending-end voltage to be the modified sending-end voltage ($\mathbf{V_{s'}}$). For a given amount of transmitted power, the shunt–series topology needs to be rated for only a fractional amount of transmitted power. Therefore, it is the preferred topology of a PFC.

The modified sending-end voltage is

$$\begin{aligned} \mathbf{V_{s'}} &= V_s \angle \delta_s + V_{s's} \angle(\delta_s + \beta) = V_s \cos \delta_s + V_{s's} \cos(\delta_s + \beta) \\ &\quad + j\{V_s \sin \delta_s + V_{s's} \sin(\delta_s + \beta)\} \\ &= V_{s'} \angle \delta_{s'} = V_{s'} (\cos \delta_{s'} + j \sin \delta_{s'}) \end{aligned} \tag{B-41}$$

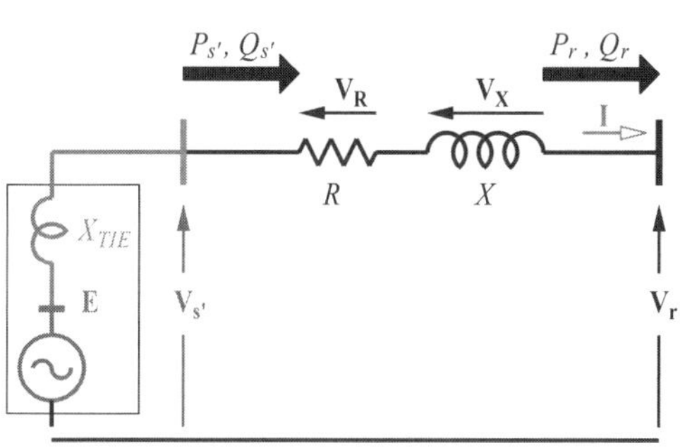

Figure B-6. Modified sending-end voltage ($\mathbf{V_{s'}}$) with a shunt-connected compensating voltage.

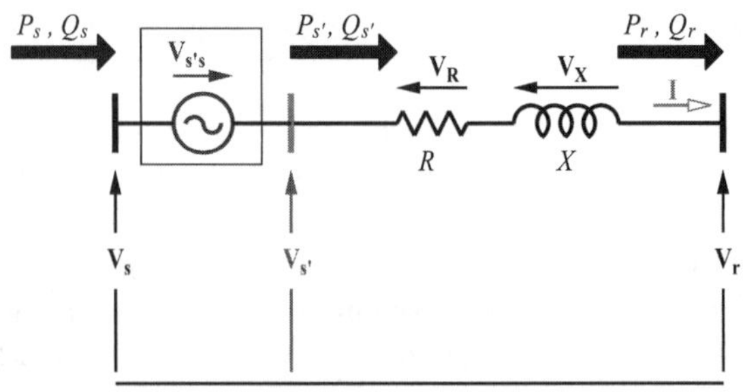

Figure B-7. Modified sending-end voltage ($\mathbf{V_{s'}}$) with a series-connected compensating voltage.

Separating the real and imaginary parts of the both sides of equation B-41,

$$V_{s'} \cos \delta_{s'} = V_s \cos \delta_s + V_{s's} \cos(\delta_s + \beta) \tag{B-42a}$$

and

$$V_{s'} \sin \delta_{s'} = V_s \sin \delta_s + V_{s's} \sin(\delta_s + \beta) \tag{B-42b}$$

Since

$$V_{s'}^2 = V_{s'}^2 \left(\cos^2 \delta_{s'} + \sin^2 \delta_{s'}\right) = \left(V_{s'} \cos \delta_{s'}\right)^2 + \left(V_{s'} \sin \delta_{s'}\right)^2$$

$$= V_s^2 + 2V_s V_{s's} \cos \beta + V_{s's}^2 \tag{B-43}$$

after substituting equations B-42a and B-42b, the magnitude of the modified sending-end voltage ($\mathbf{V_{s'}}$) is

$$V_{s'} = \sqrt{V_s^2 + 2V_s V_{s's} \cos \beta + V_{s's}^2} \tag{B-44}$$

The angle of the modified sending-end voltage ($\mathbf{V_{s'}}$) is

$$\delta_{s'} = \delta_s + \psi = \tan^{-1} \frac{V_{s'} \sin \delta_{s'}}{V_{s'} \cos \delta_{s'}} \tag{B-45a}$$

from equation B-41,

$$\delta_{s'} = \delta_s + \psi = \tan^{-1} \frac{V_s \sin \delta_s + V_{s's} \sin(\delta_s + \beta)}{V_s \cos \delta_s + V_{s's} \cos(\delta_s + \beta)} \tag{B-45b}$$

after substituting equations B-42a and B-42b.

The current flow in the transmission line is

$$\mathbf{I} = \frac{V_{s'}\angle\delta_{s'} - V_r\angle\delta_r}{R + jX}$$

$$= \frac{R(V_{s'}\cos\delta_{s'} - V_r\cos\delta_r) + X(V_{s'}\sin\delta_{s'} - V_r\sin\delta_r)}{R^2 + X^2}$$

$$+ j\frac{R(V_{s'}\sin\delta_{s'} - V_r\sin\delta_r) - X(V_{s'}\cos\delta_{s'} - V_r\cos\delta_r)}{R^2 + X^2}$$

$$= \frac{R\{V_s\cos\delta_s + V_{s's}\cos(\delta_s + \beta) - V_r\cos\delta_r\} + X\{V_s\sin\delta_s + V_{s's}\sin(\delta_s + \beta) - V_r\sin\delta_r\}}{R^2 + X^2}$$

$$+ j\frac{R\{V_s\sin\delta_s + V_{s's}\sin(\delta_s + \beta) - V_r\sin\delta_r\} - X\{V_s\cos\delta_s + V_{s's}\cos(\delta_s + \beta) - V_r\cos\delta_r\}}{R^2 + X^2}$$

with substitutions from equations B-42a and B-42b,

$$= I\angle\theta_I = I(\cos\theta_I + j\sin\theta_I) \qquad \text{(B-46)}$$

The magnitude of the current from equation B-30a is

$$I = \sqrt{\frac{V_{s'}^2 \quad 2V_{s'}V_r\cos(\delta_{s'} \quad \delta_r) + V_r^2}{R^2 + X^2}}$$

$$= \sqrt{\frac{V_{s'}^2 - 2V_r\cos\delta_r V_{s'}\cos\delta_{s'} - 2V_r\sin\delta_r V_{s'}\sin\delta_{s'} + V_r^2}{R^2 + X^2}}$$

$$= \sqrt{\frac{V_s^2 + 2V_sV_{s's}\cos\beta + V_{s's}^2 - 2V_r\cos\delta_r\{V_s\cos\delta_s + V_{s's}\cos(\delta_s + \beta)\} - 2V_r\sin\delta_r\{V_s\sin\delta_s + V_{s's}\sin(\delta_s + \beta)\} + V_r^2}{R^2 + X^2}}$$

with substitutions from equations B-44, B-42a, and B-42b,

$$= \sqrt{\frac{V_s^2 - 2V_sV_r\cos(\delta_s - \delta_r) + V_r^2 + V_{s's}\left[V_{s's} + 2\{V_s\cos\beta - V_r\cos(\delta_s - \delta_r + \beta)\}\right]}{R^2 + X^2}}$$

Then,

$$I = \sqrt{I_n^2 + \frac{V_{s's}\left[V_{s's} + 2\{V_s\cos\beta - V_r\cos(\delta_s - \delta_r + \beta)\}\right]}{R^2 + X^2}} \qquad \text{(B-47)}$$

after substituting for I_n from equation B-19a.

The angle of the current from equation B-30b is

$$\theta_I = \tan^{-1} \frac{R(V_{s'} \sin \delta_{s'} - V_r \sin \delta_r) - X(V_{s'} \cos \delta_{s'} - V_r \cos \delta_r)}{R(V_{s'} \cos \delta_{s'} - V_r \cos \delta_r) + X(V_{s'} \sin \delta_{s'} - V_r \sin \delta_r)}$$

$$\theta_I = \tan^{-1} \frac{R\{V_s \sin \delta_s + V_{s's} \sin(\delta_s + \beta) - V_r \sin \delta_r\} - X\{V_s \cos \delta_s + V_{s's} \cos(\delta_s + \beta) - V_r \cos \delta_r\}}{R\{V_s \cos \delta_s + V_{s's} \cos(\delta_s + \beta) - V_r \cos \delta_r\} + X\{V_s \sin \delta_s + V_{s's} \sin(\delta_s + \beta) - V_r \sin \delta_r\}} \tag{B-48}$$

after substituting equations B-42a and B-42b.

For a voltage regulator, $\beta = 0$ or π, depending on whether the voltage of the line is increasing or decreasing. Therefore, $\psi = 0$ and $\delta_{s'} = \delta_s$. The expressions for the magnitude and the angle of the current can be reduced to

$$I_{\beta=0,\,\pi} = \sqrt{\frac{V_{s'}^2 - 2V_{s'}V_r \cos(\delta_s - \delta_r) + V_r^2}{R^2 + X^2}} \tag{B-49a}$$

$$I_{\beta=0,\,\pi} = \sqrt{I_n^2 + \frac{V_{s's}\left[V_{s's} \pm 2\{V_s - V_r \cos(\delta_s - \delta_r)\}\right]}{R^2 + X^2}} \tag{B-49b}$$

and

$$\theta_{I\beta=0,\,\pi} = \tan^{-1} \frac{R(V_{s'} \sin \delta_s - V_r \sin \delta_r) - X(V_{s'} \cos \delta_s - V_r \cos \delta_r)}{R(V_{s'} \cos \delta_s - V_r \cos \delta_r) + X(V_{s'} \sin \delta_s - V_r \sin \delta_r)} \tag{B-50a}$$

$$\theta_{I\beta=0,\,\pi} = \tan^{-1} \frac{R\{(V_s \pm V_{s's})\sin \delta_s - V_r \sin \delta_r\} - X\{(V_s \pm V_{s's})\cos \delta_s - V_r \cos \delta_r\}}{R\{(V_s \pm V_{s's})\cos \delta_s - V_r \cos \delta_r\} + X\{(V_s \pm V_{s's})\sin \delta_s - V_r \sin \delta_r\}} \tag{B-50b}$$

For a phase angle regulator, $\beta = \pm\pi/2$, depending on whether the power flow in the line is increasing or decreasing. The expressions for the magnitude and the angle of the current can be reduced to

$$I_{\beta=\pm\pi/2} = \sqrt{\frac{V_s^2 - 2V_sV_r \cos(\delta_{s'} - \delta_r) + V_r^2}{R^2 + X^2}} \tag{B-51a}$$

$$I_{\beta=\pm\pi/2} = \sqrt{I_n^2 + \frac{V_{s's}[V_{s's} \pm 2V_r \sin(\delta_s - \delta_r)]}{R^2 + X^2}} \tag{B-51b}$$

and

$$\theta_{I\pm\pi/2} = \tan^{-1}\frac{R(V_s \sin\delta_{s'} - V_r \sin\delta_r) - X(V_s \cos\delta_{s'} - V_r \cos\delta_r)}{R(V_s \cos\delta_{s'} - V_r \cos\delta_r) + X(V_s \sin\delta_{s'} - V_r \sin\delta_r)} \quad \text{(B-52a)}$$

$$\theta_{I\pm\pi/2} = \tan^{-1}\frac{R\{V_s \sin\delta_s \pm V_{s's} \cos\delta_s - V_r \sin\delta_r\} - X\{V_s \cos\delta_s \mp V_{s's} \sin\delta_s - V_r \cos\delta_r\}}{R\{V_s \cos\delta_s \mp V_{s's} \sin\delta_s - V_r \cos\delta_r\} + X\{V_s \sin\delta_s \pm V_{s's} \cos\delta_s - V_r \sin\delta_r\}} \quad \text{(B-52b)}$$

B.VI. POWER FLOW AT THE SENDING END

The active power flow at the sending end of the transmission line is

$$P_s = V_s I \cos(\delta_s - \theta_I) = V_s \cos\delta_s \, I\cos\theta_I + V_s \sin\delta_s \, I\sin\theta_I$$

$$= \frac{V_s \cos\delta_s}{R^2 + X^2}\{R(V_{s'} \cos\delta_{s'} - V_r \cos\delta_r) + X(V_{s'} \sin\delta_{s'} - V_r \sin\delta_r)\}$$
$$+ \frac{V_s \sin\delta_s}{R^2 + X^2}\{R(V_{s'} \sin\delta_{s'} - V_r \sin\delta_r) - X(V_{s'} \cos\delta_{s'} - V_r \cos\delta_r)\}$$

with a substitution from equation B-46,

$$= \frac{V_s}{R^2 + X^2}\Big[R\{V_{s'} \cos\delta_{s'} \cos\delta_s + V_{s'} \sin\delta_{s'} \sin\delta_s - V_r \cos(\delta_s - \delta_r)\}$$
$$+ X\{V_{s'} \sin\delta_{s'} \cos\delta_s - V_{s'} \cos\delta_{s'} \sin\delta_s + V_r \sin(\delta_s - \delta_r)\}\Big] \quad \text{(B-53)}$$

$$= \frac{V_s}{R^2 + X^2}\Big[R\{(V_s \cos\delta_s + V_{s's} \cos(\delta_s + \beta))\cos\delta_s$$
$$+ (V_s \sin\delta_s + V_{s's} \sin(\delta_s + \beta))\sin\delta_s - V_r \cos(\delta_s - \delta_r)\}$$
$$+ X\{(V_s \sin\delta_s + V_{s's} \sin(\delta_s + \beta))\cos\delta_s$$
$$- (V_s \cos\delta_s + V_{s's} \cos(\delta_s + \beta))\sin\delta_s + V_r \sin(\delta_s - \delta_r)\}\Big]$$

with substitutions from equations B-42a and B-42b,

$$= \frac{V_s}{R^2 + X^2}\Big[R\{V_s + V_{s's} \cos\beta - V_r \cos(\delta_s - \delta_r)\} + X\{V_{s's} \sin\beta + V_r \sin(\delta_s - \delta_r)\}\Big]$$

$$P_s = P_{sn} + \frac{V_s V_{s's}}{R^2 + X^2}\left[R\cos\beta + X\sin\beta\right] \qquad \text{(B-54)}$$

with a substitution from equation B-20. Then,

$$P_s - P_{sn} = \frac{V_s V_{s's}}{R^2 + X^2}\left[R\cos\beta + X\sin\beta\right] \qquad \text{(B-55)}$$

The reactive power flow at the sending end of the transmission line is

$$Q_s = V_s I\sin(\delta_s - \theta_I) = V_s \sin\delta_s\, I\cos\theta_I - V_s\cos\delta_s\, I\sin\theta_I$$

$$= \frac{V_s\sin\delta_s}{R^2 + X^2}\{R(V_{s'}\cos\delta_{s'} - V_r\cos\delta_r) + X(V_{s'}\sin\delta_{s'} - V_r\sin\delta_r)\}$$
$$- \frac{V_s\cos\delta_s}{R^2 + X^2}\{R(V_{s'}\sin\delta_{s'} - V_r\sin\delta_r) - X(V_{s'}\cos\delta_{s'} - V_r\cos\delta_r)\}$$

with a substitution from equation B-46,

$$= \frac{V_s}{R^2 + X^2}\Big[R\{V_{s'}\cos\delta_{s'}\sin\delta_s - V_{s'}\sin\delta_{s'}\cos\delta_s - V_r\sin(\delta_s - \delta_r)\}$$
$$+ X\{V_{s'}\sin\delta_{s'}\sin\delta_s + V_{s'}\cos\delta_{s'}\cos\delta_s - V_r\cos(\delta_s - \delta_r)\}\Big] \qquad \text{(B-56)}$$

$$= \frac{V_s}{R^2 + X^2}\Big[R\{(V_s\cos\delta_s + V_{s's}\cos(\delta_s + \beta))\sin\delta_s$$
$$-(V_s\sin\delta_s + V_{s's}\sin(\delta_s + \beta))\cos\delta_s - V_r\sin(\delta_s - \delta_r)\}$$
$$+ X\{(V_s\sin\delta_s + V_{s's}\sin(\delta_s + \beta))\sin\delta_s$$
$$+(V_s\cos\delta_s + V_{s's}\cos(\delta_s + \beta))\cos\delta_s - V_r\cos(\delta_s - \delta_r)\}\Big]$$

with substitutions from equations B-42a and B-42b,

$$= \frac{V_s}{R^2 + X^2}\Big[-R\{V_r\sin(\delta_s - \delta_r) + V_{s's}\sin\beta\} + X\{V_s + V_{s's}\cos\beta - V_r\cos(\delta_s - \delta_r)\}\Big]$$

$$Q_s = Q_{sn} + \frac{V_{s's} V_s}{R^2 + X^2}\left[-R\sin\beta + X\cos\beta\right] \qquad \text{(B-57)}$$

with a substitution from equation B-21. Then,

$$Q_s - Q_{sn} = \frac{V_s V_{s's}}{R^2 + X^2}\left[-R\sin\beta + X\cos\beta\right] \tag{B-58}$$

Recall that for $R = 0$, the active and reactive power flows at the sending end of the transmission line are

$$P_s = P_{sn} + \frac{V_s V_{s's}}{X}\sin\beta \tag{2-21}$$

from equation B-54,

$$Q_s = Q_{sn} + \frac{V_s V_{s's}}{X}\cos\beta \tag{2-24}$$

from equation B-57.

From equations B-55 and B-58, the active and reactive power flows at the sending end can be related for the entire range of relative phase angle (β) as

$$P_s - P_{sn})^2 + (Q_s - Q_{sn})^2 = \left(\frac{V_s V_{s's}}{R^2 + X^2}\right)^2 \left[\{R\cos\beta + X\sin\beta\}^2 + \{-R\sin\beta + X\cos\beta\}^2\right]$$

$$(P_s - P_{sn})^2 + (Q_s - Q_{sn})^2 = \left(\frac{V_s V_{s's}}{R^2 + X^2}\right)^2 \left[\left(R^2 + X^2\right)\{\cos^2\beta + \sin^2\beta\}\right]$$

$$(P_s - P_{sn})^2 + (Q_s - Q_{sn})^2 = \left(\frac{V_s V_{s's}}{\sqrt{R^2 + X^2}}\right)^2 \tag{B-59}$$

The above equation is the locus of a circle, centered at (P_{sn}, Q_{sn}), with a radius of $(V_s V_{s's})/\sqrt{(R^2 + X^2)}$.

Recall that for $R = 0$, the active and reactive power flows at the sending end, independent of the relative phase angle (β), are related as

$$(P_s - P_{sn})^2 + (Q_s - Q_{sn})^2 = \left(\frac{V_s V_{s's}}{X}\right)^2 \tag{2-25}$$

The above equation is the locus of a circle, centered at (P_{sn}, Q_{sn}), with a radius of $(V_s V_{s's}/X)$.

From equations B-55 and B-58, the active and reactive power flows at the sending end are related as

$$\frac{Q_s - Q_{sn}}{P_s - P_{sn}} = \frac{-R\sin\beta + X\cos\beta}{R\cos\beta + X\sin\beta}$$

or

$$Q_s = \frac{-R\sin\beta + X\cos\beta}{R\cos\beta + X\sin\beta}(P_s - P_{sn}) + Q_{sn} \tag{B-60}$$

The relationship derived in B-60 for the active and reactive power flows at the sending end can be reduced to

$$Q_{s\,\beta=0,\,\pi} = \frac{X}{R}(P_{s\,\beta=0,\,\pi} - P_{sn}) + Q_{sn} \tag{B-61}$$

for $\beta = 0,\ \pi$,

$$Q_{s\,\beta=\pm\pi/2} = -\frac{R}{X}(P_{s\,\beta=\pm\pi/2} - P_{sn}) + Q_{sn} \tag{B-62}$$

for $\beta = \pm\pi/2$.

B.VII. POWER FLOW AT THE RECEIVING END

The active power flow at the receiving end of the transmission line is

$$P_r = V_r I \cos(\delta_r - \theta_I) = V_r \cos\delta_r\, I\cos\theta_I + V_r \sin\delta_r\, I\sin\theta_I$$

$$= \frac{V_r}{R^2 + X^2}\left[R\{V_{s'}\cos(\delta_{s'} - \delta_r) - V_r\} + XV_{s'}\sin(\delta_{s'} - \delta_r)\right] \tag{B-36}$$

$$\begin{aligned} &= \frac{V_r}{R^2 + X^2}\Big[R\{V_{s'}\cos\delta_{s'}\cos\delta_r + V_{s'}\sin\delta_{s'}\sin\delta_r - V_r\} \\ &\qquad + X\{V_{s'}\sin\delta_{s'}\cos\delta_r - V_{s'}\cos\delta_{s'}\sin\delta_r\}\Big] \end{aligned}$$

$$\begin{aligned} &= \frac{V_r}{R^2 + X^2}\Big[R\big\{(V_s\cos\delta_s + V_{s's}\cos(\delta_s + \beta))\cos\delta_r \\ &\qquad + (V_s\sin\delta_s + V_{s's}\sin(\delta_s + \beta))\sin\delta_r - V_r\big\} \\ &\qquad + X\big\{(V_s\sin\delta_s + V_{s's}\sin(\delta_s + \beta))\cos\delta_r \\ &\qquad - (V_s\cos\delta_s + V_{s's}\cos(\delta_s + \beta))\sin\delta_r\big\}\Big] \end{aligned}$$

with substitutions from equations B-42a and B-42b,

$$\begin{aligned} &= \frac{V_r}{R^2 + X^2}\Big[R\{V_s\cos(\delta_s - \delta_r) - V_r + V_{s's}\cos(\delta_s - \delta_r + \beta)\} \\ &\qquad + X\{V_s\sin(\delta_s - \delta_r) + V_{s's}\sin(\delta_s - \delta_r + \beta)\}\Big] \end{aligned}$$

$$P_r = P_{rn} + \frac{V_r V_{s's}}{R^2 + X^2}\left[R\cos(\delta_s - \delta_r + \beta) + X\sin(\delta_s - \delta_r + \beta)\right] \quad \text{(B-63)}$$

with a substitution from equation B-25. Then,

$$P_r - P_{rn} = \frac{V_r V_{s's}}{R^2 + X^2}\left[R\cos(\delta_s - \delta_r + \beta) + X\sin(\delta_s - \delta_r + \beta)\right] \quad \text{(B-64)}$$

The reactive power flow at the receiving end of the transmission line is

$$Q_r = V_r I\sin(\delta_r - \theta) = V_r \sin\delta_r\, I\cos\theta - V_r\cos\delta_r\, I\sin\theta$$

$$= \frac{V_r}{R^2 + X^2}\left[-RV_{s'}\sin(\delta_{s'} - \delta_r) + X\left\{V_{s'}\cos(\delta_{s'} - \delta_r) - V_r\right\}\right] \quad \text{(B-37)}$$

$$= \frac{V_r}{R^2 + X^2}\left[-R\left\{V_{s'}\sin\delta_{s'}\cos\delta_r - V_{s'}\cos\delta_{s'}\sin\delta_r\right\}\right.$$

$$\left. + X\left\{V_{s'}\cos\delta_{s'}\cos\delta_r + V_{s'}\sin\delta_{s'}\sin\delta_r - V_r\right\}\right]$$

$$= \frac{V_r}{R^2 + X^2}\left[-R\left\{\left(V_s\sin\delta_s + V_{s's}\sin(\delta_s + \beta)\right)\cos\delta_r\right.\right.$$

$$\left. - \left(V_s\cos\delta_s + V_{s's}\cos(\delta_s + \beta)\right)\sin\delta_r\right\}$$

$$+ X\left\{\left(V_s\cos\delta_s + V_{s's}\cos(\delta_s + \beta)\right)\cos\delta_r\right.$$

$$\left.\left. + \left(V_s\sin\delta_s + V_{s's}\sin(\delta_s + \beta)\right)\sin\delta_r - V_r\right\}\right]$$

with substitutions from equations B-42a and B-42b,

$$= \frac{V_r}{R^2 + X^2}\left[-R\left\{V_s\sin(\delta_s - \delta_r) + V_{s's}\sin(\delta_s - \delta_r + \beta)\right\}\right.$$

$$\left. + X\left\{V_s\cos(\delta_s - \delta_r) + V_{s's}\cos(\delta_s - \delta_r + \beta) - V_r\right\}\right]$$

$$Q_r = Q_{rn} + \frac{V_{s's} V_r}{R^2 + X^2}\left[-R\sin(\delta_s - \delta_r + \beta) + X\cos(\delta_s - \delta_r + \beta)\right] \quad \text{(B-65)}$$

with a substitution from equation B-26. Then,

$$Q_r - Q_{rn} = \frac{V_r V_{s's}}{R^2 + X^2}\left[-R\sin(\delta_s - \delta_r + \beta) + X\cos(\delta_s - \delta_r + \beta)\right] \quad \text{(B-66)}$$

Recall that for $R = 0$, the active and reactive power flows at the receiving end of the transmission line are

$$P_r = P_{rn} + \frac{V_r V_{s's}}{X} \sin(\delta + \beta) \tag{2-36}$$

from equation B-63,

$$Q_r = Q_{rn} + \frac{V_r V_{s's}}{X} \cos(\delta + \beta) \tag{2-39}$$

from equation B-65, where $\delta = \delta_s - \delta_r$.

From equations B-64 and B-66, the active and reactive power flows at the receiving end can be related for the entire range of relative phase angle (β) as

$$(P_r - P_{rn})^2 + (Q_r - Q_{rn})^2 = \left(\frac{V_r V_{s's}}{R^2 + X^2}\right)^2 \begin{bmatrix} \{R\cos(\delta_s - \delta_r + \beta) + X\sin(\delta_s - \delta_r + \beta)\}^2 \\ + \{-R\sin(\delta_s - \delta_r + \beta) + X\cos(\delta_s - \delta_r + \beta)\}^2 \end{bmatrix}$$

$$(P_r - P_{rn})^2 + (Q_r - Q_{rn})^2 = \left(\frac{V_r V_{s's}}{R^2 + X^2}\right)^2 \left[(R^2 + X^2)\{\cos^2(\delta_s - \delta_r + \beta) + \sin^2(\delta_s - \delta_r + \beta)\}\right]$$

$$(P_r - P_{rn})^2 + (Q_r - Q_{rn})^2 = \left(\frac{V_r V_{s's}}{\sqrt{R^2 + X^2}}\right)^2 \tag{B-67}$$

The above equation is the locus of a circle, centered at (P_{rn}, Q_{rn}), with a radius of $(V_r V_{s's})/\sqrt{(R^2 + X^2)}$.

Recall that for $R = 0$, the active and reactive power flows at the receiving end, independent of the relative phase angle (β), are related as

$$(P_r - P_{rn})^2 + (Q_r - Q_{rn})^2 = \left(\frac{V_r V_{s's}}{X}\right)^2 \tag{2-40}$$

The above equation is the locus of a circle, centered at (P_{rn}, Q_{rn}), with a radius of $(V_r V_{s's}/X)$.

From equations B-64 and B-66, the active and reactive powers flow at the receiving end are related as

$$\frac{Q_r - Q_{rn}}{P_r - P_{rn}} = \frac{-R\sin(\delta_s - \delta_r + \beta) + X\cos(\delta_s - \delta_r + \beta)}{R\cos(\delta_s - \delta_r + \beta) + X\sin(\delta_s - \delta_r + \beta)}$$

$$Q_r = \frac{-R\sin(\delta_s - \delta_r + \beta) + X\cos(\delta_s - \delta_r + \beta)}{R\cos(\delta_s - \delta_r + \beta) + X\sin(\delta_s - \delta_r + \beta)} (P_r - P_{rn}) + Q_{rn} \tag{B-68}$$

The relationship derived in B-68 for the active and reactive power flows at the receiving end can be reduced to

$$Q_{r\beta=0,\,\pi} = \frac{\mp R\sin(\delta_s - \delta_r) \pm X\cos(\delta_s - \delta_r)}{\pm R\cos(\delta_s - \delta_r) \pm X\sin(\delta_s - \delta_r)}(P_{r\beta=0,\,\pi} - P_{rn}) + Q_{rn} \qquad \text{(B-69)}$$

for $\beta = 0,\ \pi$,

$$Q_{r\beta=\pm\pi/2} = \frac{\mp R\cos(\delta_s - \delta_r) \mp X\sin(\delta_s - \delta_r)}{\mp R\sin(\delta_s - \delta_r) \pm X\cos(\delta_s - \delta_r)}(P_{r\beta=\pm\pi/2} - P_{rn}) + Q_{rn} \qquad \text{(B-70)}$$

for $\beta = \pm\pi/2$.

B.VIII. POWER FLOW AT THE MODIFIED SENDING END

The active power flow at the modified sending end of the transmission line is

$$P_{s'} = V_{s'} I\cos(\delta_{s'} - \theta_I) = V_{s'}\cos\delta_{s'} I\cos\theta_I + V_{s'}\sin\delta_{s'} I\sin\theta_I$$

$$= \frac{V_{s'}}{R^2 + X^2}\left[R\{V_{s'} - V_r\cos(\delta_{s'} - \delta_r)\} + XV_r\sin(\delta_{s'} - \delta_r)\right] \qquad \text{(B-31)}$$

$$= \frac{1}{R^2 + X^2}\left[R\left\{V_{s'}^2 - V_{s'}\cos\delta_{s'} V_r\cos\delta_r - V_{s'}\sin\delta_{s'} V_r\sin\delta_r\right\}\right.$$

$$\left. + X\{V_{s'}\sin\delta_{s'} V_r\cos\delta_r - V_{s'}\cos\delta_{s'} V_r\sin\delta_r\}\right]$$

$$= \frac{1}{R^2 + X^2}\begin{bmatrix} R\left\{V_{s'}^2 - (V_s\cos\delta_s + V_{s's}\cos(\delta_s + \beta))V_r\cos\delta_r - (V_s\sin\delta_s + V_{s's}\sin(\delta_s + \beta))V_r\sin\delta_r\right\} \\ + X\left\{(V_s\sin\delta_s + V_{s's}\sin(\delta_s + \beta))V_r\cos\delta_r - (V_s\cos\delta_s + V_{s's}\cos(\delta_s + \beta))V_r\sin\delta_r\right\} \end{bmatrix}$$

with substitutions from equations B-42a and B-42b,

$$= \frac{1}{R^2 + X^2}\left[R\left\{V_s^2 + 2V_s V_{s's}\cos\beta + V_{s's}^2 - V_s V_r\cos(\delta_s - \delta_r) - V_{s's} V_r\cos(\delta_s - \delta_r + \beta)\right\}\right.$$

$$\left. + X\{V_s V_r\sin(\delta_s - \delta_r) + V_{s's} V_r\sin(\delta_s - \delta_r + \beta)\}\right]$$

with a substitution from equation B-43,

$$= \frac{1}{R^2 + X^2}\left[R\left\{V_s^2 - V_s V_r\cos(\delta_s - \delta_r) + V_{s's}\left(V_{s's} + 2V_s\cos\beta - V_r\cos(\delta_s - \delta_r + \beta)\right)\right\}\right.$$

$$\left. + X\{V_s V_r\sin(\delta_s - \delta_r) + V_{s's} V_r\sin(\delta_s - \delta_r + \beta)\}\right]$$

$$P_{s'} = P_{sn} + \frac{V_{s's}}{R^2 + X^2}\left[R\{V_{s's} + 2V_s \cos\beta - V_r \cos(\delta_s - \delta_r + \beta)\} + XV_r \sin(\delta_s - \delta_r + \beta) \right] \tag{B-71}$$

with a substitution from equation B-20. Then,

$$P_{s'} - P_{sn} - \frac{RV_{s's}^2}{R^2 + X^2} = \frac{V_{s's}}{R^2 + X^2}\{RC_7 + XC_8\} \tag{B-72}$$

where

$$C_7 = 2V_s \cos\beta - V_r \cos(\delta_s - \delta_r + \beta) \tag{B-73a}$$

and

$$C_8 = V_r \sin(\delta_s - \delta_r + \beta) \tag{B-73b}$$

The reactive power flow at the modified sending end of the transmission line is

$$Q_{s'} = V_{s'} I \sin(\delta_{s'} - \theta_I) = V_{s'} \sin\delta_{s'} I \cos\theta_I - V_{s'} \cos\delta_{s'} I \sin\theta_I$$

$$= \frac{V_{s'}}{R^2 + X^2}\left[-R\, V_r \sin(\delta_{s'} - \delta_r) + X\{V_{s'} - V_r \cos(\delta_{s'} - \delta_r)\}\right] \tag{B-32}$$

$$= \frac{1}{R^2 + X^2}\Big[-R\,\{V_{s'} \sin\delta_{s'} V_r \cos\delta_r - V_{s'} \cos\delta_{s'} V_r \sin\delta_r\}$$
$$+X\{V_{s'}^2 - V_{s'} \cos\delta_{s'} V_r \cos\delta_r - V_{s'} \sin\delta_{s'} V_r \sin\delta_r\}\Big]$$

$$= \frac{1}{R^2 + X^2}\begin{bmatrix} -R\{(V_s \sin\delta_s + V_{s's} \sin(\delta_s + \beta))\, V_r \cos\delta_r - (V_s \cos\delta_s + V_{s's} \cos(\delta_s + \beta))\, V_r \sin\delta_r\} \\ +X\{V_{s'}^2 - (V_s \cos\delta_s + V_{s's} \cos(\delta_s + \beta)) V_r \cos\delta_r - (V_s \sin\delta_s + V_{s's} \sin(\delta_s + \beta))\, V_r \sin\delta_r\} \end{bmatrix}$$

with substitutions from equations B-42a and B-42b,

$$= \frac{1}{R^2 + X^2}\Big[-R\{V_s V_r \sin(\delta_s - \delta_r) + V_{s's} V_r \sin(\delta_s - \delta_r + \beta)\}$$
$$+X\{V_s^2 + 2V_s V_{s's} \cos\beta + V_{s's}^2 - V_s V_r \cos(\delta_s - \delta_r) - V_{s's} V_r \cos(\delta_s - \delta_r + \beta)\}\Big]$$

with a substitution from equation B-43,

$$= \frac{1}{R^2 + X^2}\Big[-R\{V_s V_r \sin(\delta_s - \delta_r) + V_{s's} V_r \sin(\delta_s - \delta_r + \beta)$$
$$+X\{V_s^2 - V_s V_r \cos(\delta_s - \delta_r) + V_{s's}\,(V_{s's} + 2V_s \cos\beta - V_r \cos(\delta_s - \delta_r + \beta))\}\Big]$$

$$Q_{s'} = Q_{sn} + \frac{V_{s's}}{R^2 + X^2}\left[-RV_r \sin(\delta_s - \delta_r + \beta) + X\{V_{s's} + 2V_s \cos\beta - V_r \cos(\delta_s - \delta_r + \beta)\}\right] \tag{B-74}$$

with a substitution from equation B-21. Then,

$$Q_{s'} - Q_{sn} - \frac{XV_{s's}^2}{R^2 + X^2} = \frac{V_{s's}}{R^2 + X^2}\{-RC_8 + XC_7\} \tag{B-75}$$

Recall that for $R = 0$, the active and reactive power flows at the modified sending end of the transmission line are

$$P_{s'} = P_{sn} + \frac{V_r V_{s's}}{X}\sin(\delta + \beta) \tag{2-55}$$

$$Q_{s'} = Q_{sn} + \frac{V_{s's}}{X}\left[V_{s's} + 2V_s \cos\beta - V_r \cos(\delta + \beta)\right] \tag{2-57}$$

where $\delta = \delta_s - \delta_r$.

From equations B-72 and B-75, the active and reactive power flows at the modified sending end can be related for the entire range of relative phase angle (β) as

$$\left(P_{s'} - P_{sn} - \frac{RV_{s's}^2}{R^2 + X^2}\right)^2 + \left(Q_{s'} - Q_{sn} - \frac{XV_{s's}^2}{R^2 + X^2}\right)^2$$

$$= \left(\frac{V_{s's}}{R^2 + X^2}\right)^2 \left\{(RC_7 + XC_8)^2 + (-RC_8 + XC_7)^2\right\}$$

$$= \left(\frac{V_{s's}}{R^2 + X^2}\right)^2 (R^2 + X^2)(C_7^2 + C_8^2)$$

$$\left(P_{s'} - P_{sn} - \frac{RV_{s's}^2}{R^2 + X^2}\right)^2 + \left(Q_{s'} - Q_{sn} - \frac{XV_{s's}^2}{R^2 + X^2}\right)^2$$

$$= \frac{V_{s's}^2}{R^2 + X^2}\left(4V_s^2 \cos^2\beta - 4V_s V_r \cos(\delta_s - \delta_r + \beta)\cos\beta + V_r^2\right) \tag{B-76}$$

with substitutions for C_7 and C_8 from equations B-73a and B-73b.

The above equation is a locus centered at

$$\left[P_{sn} + \frac{RV_{s's}^2}{R^2 + X^2},\ Q_{sn} + \frac{XV_{s's}^2}{R^2 + X^2}\right]$$

with a radius of

$$\frac{V_{s's}}{\sqrt{R^2+X^2}}\sqrt{4V_s^2\cos^2\beta-4V_sV_r\cos(\delta_s-\delta_r+\beta)\cos\beta+V_r^2}$$

Note that for a given series-connected compensating voltage ($V_{s'}$), the center of the locus is fixed, but the radius is a function of the relative phase angle (β).

From equations B-72 and B-75, the active and reactive power flows at the modified sending end are related as

$$\frac{Q_{s'}-Q_{sn}-\dfrac{XV_{s's}^2}{R^2+X^2}}{P_{s'}-P_{sn}-\dfrac{RV_{s's}^2}{R^2+X^2}}=\frac{-RC_8+XC_7}{RC_7+XC_8}$$

or

$$Q_{s'}=\frac{-RC_8+XC_7}{RC_7+XC_8}\left(P_{s'}-P_{sn}-\frac{RV_{s's}^2}{R^2+X^2}\right)+Q_{sn}+\frac{XV_{s's}^2}{R^2+X^2}$$

$$Q_{s'}=\frac{-RV_r\sin(\delta_s-\delta_r+\beta)+X\{2V_s\cos\beta-V_r\cos(\delta_s-\delta_r+\beta)\}}{R\{2V_s\cos\beta-V_r\cos(\delta_s-\delta_r+\beta)\}+XV_r\sin(\delta_s-\delta_r+\beta)}$$

$$\left(P_{s'}-P_{sn}-\frac{RV_{s's}^2}{R^2+X^2}\right)+Q_{sn}+\frac{XV_{s's}^2}{R^2+X^2} \qquad \text{(B-77)}$$

after substitutions from equations B-73a and B-73b.

For a voltage regulator, $\beta = 0$ or π, depending on whether the voltage of the line is increasing or decreasing. For $\beta = 0$, $\sin\beta = 0$, $\cos\beta = 1$, $\sin(\delta_s - \delta_r + \beta) = \sin(\delta_s - \delta_r)$, and $\cos(\delta_s - \delta_r + \beta) = \cos(\delta_s - \delta_r)$. For $\beta = \pi$, $\sin\beta = 0$, $\cos\beta = -1$, $\sin(\delta_s - \delta_r + \beta) = -\sin(\delta_s - \delta_r)$, and $\cos(\delta_s - \delta_r + \beta) = -\cos(\delta_s - \delta_r)$.

The relationship derived in equation B-77 for the active and reactive power flows at the modified sending end can be reduced to

$$Q_{s'\beta=0,\,\pi}=\frac{\mp RV_r\sin(\delta_s-\delta_r)+X\{\pm 2V_s\mp V_r\cos(\delta_s-\delta_r)\}}{R\{\pm 2V_s\mp V_r\cos(\delta_s-\delta_r)\}\pm XV_r\sin(\delta_s-\delta_r)}$$

$$\left(P_{s'\beta=0,\,\pi}-P_{sn}-\frac{RV_{s's}^2}{R^2+X^2}\right)+Q_{sn}+\frac{XV_{s's}^2}{R^2+X^2} \qquad \text{(B-78)}$$

For a phase angle regulator, $\beta = \pm\pi/2$, depending on whether the power flow in the line is increasing or decreasing. For $\beta = \pi/2$, $\sin\beta = 1$, $\cos\beta = 0$, $\sin(\delta_s - \delta_r + \beta) = \cos(\delta_s - \delta_r)$, and $\cos(\delta_s - \delta_r + \beta) = -\sin(\delta_s - \delta_r)$. For $\beta = -\pi/2$, $\sin\beta = -1$, $\cos\beta = 0$, $\sin(\delta_s - \delta_r + \beta) = -\cos(\delta_s - \delta_r)$, and $\cos(\delta_s - \delta_r + \beta) = \sin(\delta_s - \delta_r)$.

The relationship derived in equation B-77 for the active and reactive power flows at the modified sending end can be reduced to

$$Q_{s'\beta=\pm\pi/2} = \frac{\mp RV_r \cos(\delta_s - \delta_r) \pm XV_r \sin(\delta_s - \delta_r)}{\pm RV_r \sin(\delta_s - \delta_r) \pm XV_r \cos(\delta_s - \delta_r)}$$

$$\left(P_{s'\beta=\pm\pi/2} - P_{sn} - \frac{RV_{s's}^2}{R^2 + X^2} \right) + Q_{sn} + \frac{XV_{s's}^2}{R^2 + X^2} \qquad \text{(B-79)}$$

B.IX. EXCHANGED POWER BY THE COMPENSATING VOLTAGE

Regardless of whether the sending-end voltage is modified with the use of a shunt-connected or series-connected compensating voltage, the compensating voltage is an AC voltage source that carries a current that is at any angle with its terminal voltage, thereby emulating, in shunt or in series with the line, a capacitor or an inductor and a positive resistor or a negative resistor. The compensating voltage, essentially, emulates an impedance at its terminals. The active and reactive powers that are exchanged by the compensating voltage source are P_{exch} and Q_{exch}, respectively.

In the case of a shunt-connected compensating voltage,

$$P_{exch} = -P_{s'} \qquad \text{(B-80a)}$$

$$Q_{exch} = -Q_{s'} \qquad \text{(B-80b)}$$

where

$$P_{s'} = \frac{V_{s'}}{R^2 + X^2}\left[R\{V_{s'} - V_r \cos(\delta_{s'} - \delta_r)\} + XV_r \sin(\delta_{s'} - \delta_r) \right] \qquad \text{(B-31)}$$

$$Q_{s'} = \frac{V_{s'}}{R^2 + X^2}\left[-RV_r \sin(\delta_{s'} - \delta_r) + X\{V_{s'} - V_r \cos(\delta_{s'} - \delta_r)\} \right] \qquad \text{(B-32)}$$

In the case of a series-connected compensating voltage,

$$P_{exch} = P_s - P_{s'} \qquad \text{(B-81a)}$$

$$Q_{exch} = Q_s - Q_{s'} \qquad \text{(B-81b)}$$

where

$$P_s = P_{sn} + \frac{V_s V_{s's}}{R^2 + X^2}\left[R\cos\beta + X\sin\beta \right] \qquad \text{(B-54)}$$

$$Q_s = Q_{sn} + \frac{V_s V_{s's}}{R^2 + X^2}\left[-R\sin\beta + X\cos\beta\right] \tag{B-57}$$

$$P_{s'} = P_{sn} + \frac{V_{s's}}{R^2 + X^2}\left[R\{V_{s's} + 2V_s\cos\beta - V_r\cos(\delta_s - \delta_r + \beta)\} + XV_r\sin(\delta_s - \delta_r + \beta)\right] \tag{B-71}$$

$$Q_{s'} = Q_{sn} + \frac{V_{s's}}{R^2 + X^2}\left[-RV_r\sin(\delta_s - \delta_r + \beta) + X\{V_{s's} + 2V_s\cos\beta - V_r\cos(\delta_s - \delta_r + \beta)\}\right] \tag{B-74}$$

Then,

$$P_{exch} = P_s - P_{s'} = \frac{V_{s's}}{R^2 + X^2}\left[-R\{V_{s's} + V_s\cos\beta - V_r\cos(\delta_s - \delta_r + \beta)\}\right.$$
$$\left. + X\{V_s\sin\beta - V_r\sin(\delta_s - \delta_r + \beta)\}\right] \tag{B-82}$$

$$P_{exch} + \frac{RV_{s's}^2}{R^2 + X^2} = \frac{V_{s's}}{R^2 + X^2}\left(-RC_9 + XC_{10}\right) \tag{B-83}$$

where

$$C_9 = V_s\cos\beta - V_r\cos(\delta_s - \delta_r + \beta) \tag{B-84a}$$

and

$$C_{10} = V_s\sin\beta - V_r\sin(\delta_s - \delta_r + \beta) \tag{B-84b}$$

$$Q_{exch} = Q_s - Q_{s'} = \frac{V_{s's}}{R^2 + X^2}\left[-R\{V_s\sin\beta - V_r\sin(\delta_s - \delta_r + \beta)\}\right.$$
$$\left. - X\{V_{s's} + V_s\cos\beta - V_r\cos(\delta_s - \delta_r + \beta)\}\right] \tag{B-85}$$

$$Q_{exch} + \frac{XV_{s's}^2}{R^2 + X^2} = \frac{V_{s's}}{R^2 + X^2}\left(-RC_{10} - XC_9\right) \tag{B-86}$$

Recall that for $R = 0$,

$$P_{exch} = \frac{V_{s's}}{X}\left[V_s\sin\beta - V_r\sin(\delta + \beta)\right] \tag{2-74}$$

$$Q_{exch} = \frac{V_{s's}}{X}\left[V_r \cos(\delta + \beta) - V_{s's} - V_s \cos\beta\right] \qquad (2\text{-}83)$$

where $\delta = \delta_s - \delta_r$.

From equations B-83 and B-86, the active and reactive powers exchanged by the series-connected compensating voltage can be related as

$$\left(P_{exch} + \frac{RV_{s's}^2}{R^2 + X^2}\right)^2 + \left(Q_{exch} + \frac{XV_{s's}^2}{R^2 + X^2}\right)^2$$

$$= \left(\frac{V_{s's}}{R^2 + X^2}\right)^2 \left\{(-RC_9 + XC_{10})^2 + (-RC_{10} - XC_9)^2\right\}$$

$$= \left(\frac{V_{s's}}{R^2 + X^2}\right)^2 (R^2 + X^2)(C_9^2 + C_{10}^2) = \frac{V_{s's}^2}{R^2 + X^2}(C_9^2 + C_{10}^2)$$

$$\left(P_{exch} + \frac{RV_{s's}^2}{R^2 + X^2}\right)^2 + \left(Q_{exch} + \frac{XV_{s's}^2}{R^2 + X^2}\right)^2$$

$$= \frac{V_{s's}^2}{R^2 + X^2}\left\{V_s^2 + V_r^2 - 2V_s V_r \cos(\delta_s - \delta_r)\right\} = V_{s's}^2 I_n^2 \qquad (\text{B-87})$$

after substituting for C_9 and C_{10} from equations B-84a and B-84b, and for I_n from equation B-19a.

The above equation is the locus of a circle centered at

$$\left[\frac{-RV_{s's}^2}{R^2 + X^2}, \frac{-XV_{s's}^2}{R^2 + X^2}\right]$$

with a radius of $V_{s's}I_n$.

Recall that for $R = 0$, the exchanged active and reactive powers, independent of the relative phase angle (β), are related as

$$P_{exch}^2 + \left(Q_{exch} + \frac{V_{s's}^2}{X}\right)^2 = \left(\frac{V_{s's}}{X}\sqrt{V_s^2 + V_r^2 - 2V_s V_r \cos(\delta_s - \delta_r)}\right)^2$$

$$= \left(\frac{V_{s's}}{X} V_{Xn}\right)^2 = (V_{s's} I_n)^2 \qquad (2\text{-}87)$$

The above equation is the locus of a circle, centered at $(0, -V_{s's}^2/X)$, with a radius of $V_{s's}I_n$.

From equations B-83 and B-86, the exchanged active and reactive powers by the compensating voltage are related as

$$\frac{Q_{exch}+\dfrac{XV_{s's}^{2}}{R^{2}+X^{2}}}{P_{exch}+\dfrac{RV_{s's}^{2}}{R^{2}+X^{2}}}=\frac{-RC_{10}-XC_{9}}{-RC_{9}+XC_{10}}$$

or

$$Q_{exch}=\frac{-RC_{10}-XC_{9}}{-RC_{9}+XC_{10}}\left(P_{exch}+\frac{RV_{s's}^{2}}{R^{2}+X^{2}}\right)-\frac{XV_{s's}^{2}}{R^{2}+X^{2}}$$

or

$$Q_{exch}=\frac{-R\{V_s\sin\beta-V_r\sin(\delta_s-\delta_r+\beta)\}-X\{V_s\cos\beta-V_r\cos(\delta_s-\delta_r+\beta)\}}{-R\{V_s\cos\beta-V_r\cos(\delta_s-\delta_r+\beta)\}+X\{V_s\sin\beta-V_r\sin(\delta_s-\delta_r+\beta)\}}$$

$$\left(P_{exch}+\frac{RV_{s's}^{2}}{R^{2}+X^{2}}\right)-\frac{XV_{s's}^{2}}{R^{2}+X^{2}} \tag{B-88}$$

substituting for C_9 and C_{10} from equations B-84a and B-84b.

The relationship derived in equation B-88 for the exchanged active and reactive powers by the compensating voltage can be reduced to

$$Q_{exch\,\beta=0,\,\pi}=\frac{-R\{\mp V_r\sin(\delta_s-\delta_r)\}-X\{\pm V_s\mp V_r\cos(\delta_s-\delta_r)\}}{-R\{\pm V_s\mp V_r\cos(\delta_s-\delta_r)\}+X\{\mp V_r\sin(\delta_s-\delta_r)\}}$$

$$\left(P_{exch\,\beta=0,\,\pi}+\frac{RV_{s's}^{2}}{R^{2}+X^{2}}\right)-\frac{XV_{s's}^{2}}{R^{2}+X^{2}} \tag{B-89}$$

$$Q_{exch\,\beta=\pm\pi/2}=\frac{-R\{\pm V_s\mp V_r\cos(\delta_s-\delta_r)\}-X\{\pm V_r\sin(\delta_s-\delta_r)\}}{-R\{\pm V_r\sin(\delta_s-\delta_r)\}+X\{\pm V_s\mp V_r\cos(\delta_s-\delta_r)\}}$$

$$\left(P_{exch\,\beta=\pm\pi/2}+\frac{RV_{s's}^{2}}{R^{2}+X^{2}}\right)-\frac{XV_{s's}^{2}}{R^{2}+X^{2}} \tag{B-90}$$

Changing one parameter (voltage, $V_{s'}$, or phase angle, $\delta_{s'}$, of the modified sending-end voltage) using a PFC affects both the active and the reactive power flows in the transmission line simultaneously. The key to be able to independently control the active and reactive power flows (P and Q) in a transmission line is to be able to control the magnitude and phase angle of the line voltage independently. A PFC modifies the line voltage through a series-connected compensating voltage of a shunt–series power converter.

The UPFC can provide independent controls of the active and reactive power flows in the line through its shunt–series configuration. However, the faster (but, never close to subcycle) response of the UPFC is not needed in most utility applications in which the ST brings the right technology for independent power flow control at an estimated 20% of the installation cost and 10% of the operational cost of the UPFC.

APPENDIX C

EMTP FILES

The sample models in EMTP are listed in the following text A more comprehensive set of models is available at the following ftp address: ftp://ftp.wiley.com/public/sci_tech_med/facts_controllers.

1. 301NTWK1.DAT

```
BEGIN NEW DATA CASE
C File Name: 301NTWK1.DAT
C A SIMPLE TWO-GENERATOR, SINGLE LINE POWER STSYEM NETWORK
C 00000001111111111222222222233333333334444444444555555555566666666667777777778
C 34567890123456789012345678901234567890123456789012345678901234567890123456789
C ----dt<---tmax<---Xopt<---Copt<--Toler<        ><        >
16.666-6150.00-3
C 00000001111111111222222222233333333334444444444555555555566666666667777777778
C 34567890123456789012345678901234567890123456789012345678901234567890123456789
C -Iprnt<--Iplot<-Idoubl<-KssOut<-MaxOut<        ><        ><---Icat<-Nenerg
      20      10                                                       1
TACS HYBRID
99TWOPI    = 2.0 * PI
C BASE VOLTAGE, CURRENT, AND POWER
99Vbase    = 112676.528
99Ibase    = 946.662704
99Zbase    = Vbase / Ibase
99Pbase    = Vbase * Ibase
$INCLUDE 302MEAS1.SWT
$INCLUDE 303PTCT.SCL
$INCLUDE 304OPLL.PLL
$INCLUDE 305LINER.CMP
BLANK RECORD ENDING TACS
C 00000001111111111222222222233333333334444444444555555555566666666667777777778
C 34567890123456789012345678901234567890123456789012345678901234567890123456789
C <-Bus1<-Bus2<-Bus3<-Bus4<----R<----L<----C                                  V
  BUS01ABUS02A            0.001
  BUS01BBUS02B            0.001
  BUS01CBUS02C            0.001
$INCLUDE 306NTWK1.BRN
BLANK RECORD ENDING BRANCHES
C 00000001111111111222222222233333333334444444444555555555566666666667777777778
C 34567890123456789012345678901234567890123456789012345678901234567890123456789
C <-Bus1<-Bus2<---Tclose<----Topen<-------Ie<----Flash<--Request<-----Target<--O
$INCLUDE 307MEAS2.SWT
BLANK RECORD ENDING SWITCHES
C 00000001111111111222222222233333333334444444444555555555566666666667777777778
C 34567890123456789012345678901234567890123456789012345678901234567890123456789
$INCLUDE 308NTWK1.SRC
BLANK RECORD ENDING SOURCES
BLANK RECORD ENDING NODE VOLTAGE OUTPUT
BLANK RECORD ENDING PLOT
BEGIN NEW DATA CASE
BLANK RECORD ENDING ALL DATA CASES
```

2. 302MEAS1.SWT

```
C File Name: 302MEAS1.SWT
C INPUTS FROM THE MEASURING SWITCHES
C 0000000111111111122222222223333333333444444444455555555556666666666777777777778
C 3456789012345678901234567890123456789012345678901234567890123456789012345678901234567890
90BUS01A
90BUS01B
90BUS01C
90BUS02A
90BUS02B
90BUS02C
90VRA
90VRB
90VRC
91VSPA
91VSPB
91VSPC
```

3. 303PTCT.SCL

```
C File Name: 303PTCT.SCL
C NORMALIZATION OF MEASURED VARIABLES
C 0000000111111111122222222223333333333444444444455555555556666666666777777777778
C 3456789012345678901234567890123456789012345678901234567890123456789012345678901234567890
C BUS01 VOLTAGE COMPUTATION
99v1apu    = BUS01A / Vbase
99v1bpu    = BUS01B / Vbase
99v1cpu    = BUS01C / Vbase
99v1pu     = SQRT((v1apu * v1apu + v1bpu * v1bpu + v1cpu * v1cpu) * 2 / 3)
C BUS02 VOLTAGE COMPUTATION
99v2apu    = BUS02A / Vbase
99v2bpu    = BUS02B / Vbase
99v2cpu    = BUS02C / Vbase
99v2pu     = SQRT((v2apu * v2apu + v2bpu * v2bpu + v2cpu * v2cpu) * 2 / 3)
C RECEIVING-END VOLTAGE COMPUTATION
99vrapu    = VRA / Vbase
99vrbpu    = VRB / Vbase
99vrcpu    = VRC / Vbase
99vrpu     = SQRT((vrapu * vrapu + vrbpu * vrbpu + vrcpu * vrcpu) * 2 / 3)
C LINE CURRENT COMPUTATION AT Vs' BUS
99iapu     = VSPA / Ibase
99ibpu     = VSPB / Ibase
99icpu     = VSPC / Ibase
99ipu      = SQRT((iapu * iapu + ibpu * ibpu + icpu * icpu) * 2/3)
C SERIES-CONNECTED VOLTAGE COMPUTATION
99v12apu   = v1apu - v2apu
99v12bpu   = v1bpu - v2bpu
99v12cpu   = v1cpu - v2cpu
99v12pu    = SQRT((v12apu * v12apu + v12bpu * v12bpu + v12cpu * v12cpu) * 2 / 3)
C TACS OUTPUT
33v1pu  v12apuv12pu iapu
```

4. 304OPLL.PLL

```
C File Name: 304OPLL.PLL
C ORDINARY PHASE-LOCKED LOOP
C 0000000111111111122222222223333333333444444444455555555556666666666777777777778
```

```
C 34567890123456789012345678901234567890123456789012345678901234567890123456789O
C CONVERT FROM 3-PHASE TO 2-PHASE
99v1ds     = v1apu
99v1qs     = -(v1apu + 2.0 * v1cpu) / SQRT(3)
C PHASE-LOCKED LOOP (LOCKED TO PHASE A VOLTAGE OF BUS01)
99th       = ATAN(v1qs / (v1ds+1.0E-8))
99the      = th + PI * (v1ds .LT. 0)
99theta    = the + TWOPI * (the .LT. 0)
99COSPLL   = COS(theta)
99SINPLL   = SIN(theta)
C BUS01 VOLTAGE MAGNITUDE COMPUTATION
99v1d      =  v1ds * COSPLL + v1qs * SINPLL
99v1q      = -v1ds * SINPLL + v1qs * COSPLL
99v1dq     = SQRT(v1d * v1d + v1q * v1q)
 1v1dqf   +v1dq
       1.0
       1.0    0.0040
C LINE CURRENT MAGNITUDE COMPUTATION
99ids      =  iapu
99iqs      = -(iapu + 2.0 * icpu) / SQRT(3)
99id       =  ids * COSPLL + iqs * SINPLL
99iq       = -ids * SINPLL + iqs * COSPLL
99idq      = SQRT(id * id + iq * iq)
99thetaa   = ATAN( iq / (id + 1.0E-8) )
99thetab   = thetaa + PI * (id .LT. 0.0)
99thetir   = thetab + TWOPI * (thetab .LT. 0.0) - TWOPI * (thetab .GE. TWOPI)
C TACS OUTPUT
C 33theta
```

5. 305LINER.CMP

```
C File Name: 305LINER.CMP
C LINE RESISTANCE COMPUTATION
C 00000001111111111222222222233333333334444444444555555555566666666667777777777 8
C 34567890123456789012345678901234567890123456789012345678901234567890123456789O
C MODIFIED SENDING-END POWER COMPUTATION
99Psppu    = (v2apu*iapu + v2bpu*ibpu + v2cpu*icpu) / 1.5
99Qsppu    = (v2apu*icpu - v2cpu*iapu) * SQRT(3) / 1.5
C RECEIVING-END POWER COMPUTATION
99Prpu     = (vrapu*iapu + vrbpu*ibpu + vrcpu*icpu) / 1.5
99Qrpu     = (vrapu*icpu - vrcpu*iapu) * SQRT(3) / 1.5
99Rline    = (Psppu - Prpu) / ipu / ipu * Zbase
 1Rlinef  +Rline
       1.0
       1.0    0.0083
C TACS OUTPUT
C 33Psppu Qsppu Rline Rlinef
33Prpu  Qrpu
```

6. 306NTWK1.BRN

```
C File name: 306NTWK1.BRN
C SOURCE IMPEDANCE & LINE IMPEDANCE
C 000000011111111112222222222333333333344444444445555555555666666666677777777778
C 345678901234567890123456789012345678901234567890123456789012345678901234567890
C <-Bus1<-Bus2<-Bus3<-Bus4<----R<----L<----C                                   V
  VSRCA BUS01A            1.0053 19.73
  VSRCB BUS01BVSRCA BUS01A
```

```
VSRCC BUS01CVSRCA BUS01A
VSPA  VRA                3.0159 59.19
VSPB  VRB   VSPA  VRA
VSPC  VRC   VSPA  VRA
```

7. 307MEAS2.SWT

```
C File Name: 307MEAS2.SWT
C CURRENT MEASUREMENTS
C 000000011111111112222222222333333333344444444445555555555666666666677777777778
C 34567890123456789012345678901234567890123456789012345678901234567890123456789 0
C <-Bus1<-Bus2<---Tclose<----Topen<-------Ie<----Flash<--Request<-----Target<--O
  BUS02AVSPA                                              MEASURING
  BUS02BVSPB                                              MEASURING
  BUS02CVSPC                                              MEASURING
```

8. 308NTWK1.SRC

```
C File name: 308NTWK1.SRC
C SOURCE VOLTAGES AND RECEIVING-END VOLTAGES
C 000000011111111112222222222333333333344444444445555555555666666666677777777778
C 345678901234567890123456789012345678901234567890123456789012345678901234567890
C <--Bus<I<-----Ampl<-----Freq<----Phase<-------A1<------T1><---Tstart<----Tstop
14VSRCA   112676.528     60.00     0.00                          -1.0
14VSRCB   112676.528     60.00  -120.00                          -1.0
14VSRCC   112676.528     60.00   120.00                          -1.0
14VRA     112676.528     60.00   -20.00                          -1.0
14VRB     112676.528     60.00  -140.00                          -1.0
14VRC     112676.528     60.00   100.00                          -1.0
```

9. 309NTWK2.DAT

```
BEGIN NEW DATA CASE
C File Name: 309NTWK2.DAT
C A SIMPLE TWO-GENERATOR, SINGLE LINE POWER STSYEM NETWORK WITH
C LINE TO NEUTRAL FAULT
C 000000011111111112222222222333333333344444444445555555555666666666677777777778
C 345678901234567890123456789012345678901234567890123456789012345678901234567890
C ----dt<---tmax<---Xopt<---Copt<--Toler<        ><        >
16.666-6150.00-3
C 000000011111111112222222222333333333344444444445555555555666666666677777777778
C 345678901234567890123456789012345678901234567890123456789012345678901234567890
C -Iprnt<--Iplot<-Idoubl<-KssOut<-MaxOut<        ><        ><---Icat<-Nenerg
      20      10                                                  1
TACS HYBRID
99TWOPI   = 2.0 * PI
C BASE VOLTAGE, CURRENT, AND POWER
99Vbase   = 112676.528
99Ibase   = 946.662704
99Zbase   = Vbase / Ibase
99Pbase   = Vbase * Ibase
$INCLUDE 302MEAS1.SWT
$INCLUDE 303PTCT.SCL
C TACS OUTPUT
33v1apu
$INCLUDE 310VPLL.PLL
$INCLUDE 305LINER.CMP
BLANK RECORD ENDING TACS
```

```
C 000000011111111112222222222333333333344444444445555555555666666666677777777778
C 345678901234567890123456789012345678901234567890123456789012345678901234567890
C <-Bus1<-Bus2<-Bus3<-Bus4<----R<-----L<----C                                  V
  BUS01ABUS02A              0.001
  BUS01BBUS02B              0.001
  BUS01CBUS02C              0.001
$INCLUDE 311NTWK2.BRN
BLANK RECORD ENDING BRANCHES
C 000000011111111112222222222333333333344444444445555555555666666666677777777778
C 345678901234567890123456789012345678901234567890123456789012345678901234567890
C <-Bus1<-Bus2<---Tclose<----Topen<-------Ie<----Flash<--Request<-----Target<--O
$INCLUDE 307MEAS2.SWT
C FAULT STARTS
  BUS01A             .050      0.100       5.0                                 1
C FAULT ENDS
BLANK RECORD ENDING SWITCHES
C 000000011111111112222222222333333333344444444445555555555666666666677777777778
C 345678901234567890123456789012345678901234567890123456789012345678901234567890
$INCLUDE 308NTWK1.SRC
BLANK RECORD ENDING SOURCES
BLANK RECORD ENDING NODE VOLTAGE OUTPUT
BLANK RECORD ENDING PLOT
BEGIN NEW DATA CASE
BLANK RECORD ENDING ALL DATA CASES
```

10. 310VPLL.PLL

```
C File Name: 310VPLL.PLL
C VECTOR PHASE-LOCKED LOOP
C 000000011111111112222222222333333333344444444445555555555666666666677777777778
C 345678901234567890123456789012345678901234567890123456789012345678901234567890
C CONVERT FROM 3-PHASE TO 2-PHASE
99v1ds     = v1apu
99v1qs     = -(v1apu + 2.0 * v1cpu) / SQRT(3)
99KPLLp    = 100.0
99KPLLi    = 250.0
C PHASE-LOCKED LOOP (LOCKED TO PHASE A VOLTAGE OF BUS01)
99PLLerr   = v1qs * COS(PLLi) - v1ds * SIN(PLLi)
99PLLint   = PLLini + PLLerr * DELTAT
99PLLdot   = KPLLi * PLLint + KPLLp * PLLerr
  PLLini   +PLLint
99PLL      = PLLi + PLLdot * DELTAT
  PLLi     +PLL
99theta    = PLL - TWOPI * TRUNC(PLL/TWOPI)
99COSPLL   = COS(theta)
99SINPLL   = SIN(theta)
C BUS01 VOLTAGE MAGNITUDE COMPUTATION
99v1d      =  v1ds * COSPLL + v1qs * SINPLL
99v1q      = -v1ds * SINPLL + v1qs * COSPLL
99v1dq     = SQRT(v1d * v1d + v1q * v1q)
 1v1dqf    +v1dq
       1.0
       1.0    0.0040
C LINE CURRENT MAGNITUDE AND PHASE ANGLE COMPUTATION
99ids      =   iapu
99iqs      = -(iapu + 2.0 * icpu) / SQRT(3)
99id       =  ids * COSPLL + iqs * SINPLL
99iq       = -ids * SINPLL + iqs * COSPLL
99idq      = SQRT(id * id + iq * iq)
```

```
99thetaa  = ATAN( iq / (id + 1.0E-8) )
99thetab  = thetaa + PI * (id .LT. 0.0)
99thetir  = thetab + TWOPI * (thetab .LT. 0.0) - TWOPI * (thetab .GE. TWOPI)
C ------TACS INITIAL CONDITIONS------
C -name>xx<--------|
77PLLint  1.5080
C TACS OUTPUT
C 33theta
```

11. 311NTWK2.BRN

```
C File name: 311NTWK2.BRN
C SOURCE IMPEDANCE & LINE IMPEDANCE WITH SNUBBER RESISTORS ACROSS ALL INDUCTORS
C 000000011111111112222222222333333333344444444445555555555666666666677777777778
C 345678901234567890123456789012345678901234567890123456789012345678901234567890
C <-Bus1<-Bus2<-Bus3<-Bus4<----R<----L<----C                                   V
  VSRCA VSRCPA              1.0053
  VSRCPAVSA                         19.73
  VSRCB VSRCPB              1.0053
  VSRCPBVSB                         19.73
  VSRCC VSRCPC              1.0053
  VSRCPCVSC                         19.73
  VSPA  VRPA                3.0159
  VRPA  VRA                         59.19
  VSPB  VRPB                3.0159
  VRPB  VRB                         59.19
  VSPC  VRPC                3.0159
  VRPC  VRC                         59.19
  VSA   BUS01A               0.001
  VSB   BUS01B               0.001
  VSC   BUS01C               0.001
  VSRCPAVSA                 5.00+2
  VSRCPBVSB                 5.00+2
  VSRCPCVSC                 5.00+2
  VRPA  VRA                 4.00+3
  VRPB  VRB                 4.00+3
  VRPC  VRC                 4.00+3
```

12. 312TRAN1.TRN

```
C File Name: 312TRAN1.TRN
C 3-PHASE SHUNT COMPENSATING TRANSFORMER
C 000000011111111112222222222333333333344444444445555555555666666666677777777778
C 345678901234567890123456789012345678901234567890123456789012345678901234567890
C ------SHUNT COMPENSATING TRANSFORMER FOR PHASE A------
C TRANSFORMER <--Refxxxxxx<--Iss<--Phi<-Name<-Rmag<---------------------------O
  TRANSFORMER                           SHNTRA
C NEXT LINE (9999 IN COLUMNS 13-16) INDICATES A LINEAR TRANSFORMER
            9999
C <---Nodes--><----------><-Ohms<---mH<---kV
C <-Bus1<-Bus2<----------><---Rk<---Lk<Volt|---------------------------------->O
 1e1a                       0.892723.679   1.0
 2e11a                      0.892723.679   1.0
C ------SHUNT COMPENSATING TRANSFORMER FOR PHASE B------
  TRANSFORMER                           SHNTRB
            9999
 1e1b                       0.892723.679   1.0
 2e11b                      0.892723.679   1.0
```

```
C ------SHUNT COMPENSATING TRANSFORMER FOR PHASE C------
  TRANSFORMER                                SHNTRC
            9999
 1e1c                     0.892723.679   1.0
 2e11c                    0.892723.679   1.0
```

13. 313TRAN2.TRN

```
C File Name: 313TRAN2.TRN
C 3-PHASE SERIES COMPENSATING TRANSFORMER
C 00000001111111111222222222233333333334444444444555555555566666666667777777778
C 34567890123456789012345678901234567890123456789012345678901234567890123456789O
C ------SERIES COMPENSATING TRANSFORMER FOR PHASE A------
C TRANSFORMER <--Refxxxxxx<--Iss<--Phi<-Name<-Rmag<---------------------------O
  TRANSFORMER                                SRSTRA
C NEXT LINE (9999 IN COLUMNS 13-16) INDICATES A LINEAR TRANSFORMER
            9999
C <---Nodes--><----------><-Ohms<---mH<---kV
C <-Bus1<-Bus2<----------><---Rk<---Lk<Volt|---------------------------------->O
 1CNV02A                  0.892723.679   1.0
 2BUS01ABUS02A            0.892723.679   1.0
C ------SERIES COMPENSATING TRANSFORMER FOR PHASE B------
  TRANSFORMER                                SRSTRB
            9999
 1CNV02B                  0.892723.679   1.0
 2BUS01BBUS02B            0.892723.679   1.0
C ------SERIES COMPENSATING TRANSFORMER FOR PHASE C------
  TRANSFORMER                                SRSTRC
            9999
 1CNV02C                  0.892723.679   1.0
 2BUS01CBUS02C            0.892723.679   1.0
```

14. 314IDLT2.TRN

```
C File Name: 314IDLT2.TRN
C 3-PHASE IDEAL SERIES COMPENSATING TRANSFORMER WITH NO LEAKAGE IMPEDANCE
C 00000001111111111222222222233333333334444444444555555555566666666667777777778
C 34567890123456789012345678901234567890123456789012345678901234567890123456789O
C ------SERIES COMPENSATING TRANSFORMER FOR PHASE A------
C TRANSFORMER <--Refxxxxxx<--Iss<--Phi<-Name<-Rmag<---------------------------O
  TRANSFORMER                                SRSTRA
C NEXT LINE (9999 IN COLUMNS 13-16) INDICATES A LINEAR TRANSFORMER
            9999
C <---Nodes--><----------><-Ohms<---mH<---kV
C <-Bus1<-Bus2<----------><---Rk<---Lk<Volt|---------------------------------->O
 1CNV02A                    0.001 0.001   1.0
 2BUS01ABUS02A              0.001 0.001   1.0
C ------SERIES COMPENSATING TRANSFORMER FOR PHASE B------
  TRANSFORMER                                SRSTRB
            9999
 1CNV02B                    0.001 0.001   1.0
 2BUS01BBUS02B              0.001 0.001   1.0
C ------SERIES COMPENSATING TRANSFORMER FOR PHASE C------
  TRANSFORMER                                SRSTRC
            9999
 1CNV02C                    0.001 0.001   1.0
 2BUS01CBUS02C              0.001 0.001   1.0
```

15. 315PQPFC.DAT

```
BEGIN NEW DATA CASE
C File Name: 315PQPFC.DAT
C PQ POWER FLOW CONTROLLER WITH 1 INFINITE-PULSE VSC AND 1 INFINITE-PULSE CSC
C 000000011111111112222222222333333333344444444445555555555666666666677777777778
C 345678901234567890123456789012345678901234567890123456789012345678901234567890
C ----dt<---tmax<---Xopt<---Copt<--Toler<       ><       >
8.3333-6150.00-3
C 000000011111111112222222222333333333344444444445555555555666666666677777777778
C 345678901234567890123456789012345678901234567890123456789012345678901234567890
C -Iprnt<--Iplot<-Idoubl<-KssOut<-MaxOut<       ><       ><---Icat<-Nenerg
      30      10                                                1
TACS HYBRID
99CRD      = 180 / PI
99CDR      = 1 / CRD
99TWOPI    = 2.0 * PI
C BASE VOLTAGE, CURRENT, AND POWER
99Vbase    = 112676.528
99Ibase    = 946.662704
99Zbase    = Vbase / Ibase
99Pbase    = Vbase * Ibase
$INCLUDE 302MEAS1.SWT
91CNV01A
91CNV01B
91CNV01C
91CNV02A
91CNV02B
91CNV02C
$INCLUDE 303PTCT.SCL
$INCLUDE 310VPLL.PLL
$INCLUDE 316PQPFC.USR
$INCLUDE 317PQPFC.CON
$INCLUDE 305LINER.CMP
BLANK RECORD ENDING TACS
C 000000011111111112222222222333333333344444444445555555555666666666677777777778
C 345678901234567890123456789012345678901234567890123456789012345678901234567890
C <-Bus1<-Bus2<-Bus3<-Bus4<----R<----L<----C                                   V
$INCLUDE 311NTWK2.BRN
  BUS01ACNV01A            1.00+8
  BUS01BCNV01B            1.00+8
  BUS01CCNV01C            1.00+8
  e2a   e22a              1.00+8
  e2b   e22b              1.00+8
  e2c   e22c              1.00+8
  e22a                    1.00+8
  e22b                    1.00+8
  e22c                    1.00+8
  i1a                     1.00+8
  i1b                     1.00+8
  i1c                     1.00+8
$INCLUDE 313TRAN2.TRN
BLANK RECORD ENDING BRANCHES
C 000000011111111112222222222333333333344444444445555555555666666666677777777778
C 345678901234567890123456789012345678901234567890123456789012345678901234567890
C <-Bus1<-Bus2<---Tclose<----Topen<-------Ie<----Flash<--Request<-----Target<--O
13BUS01ACNV01A                                                        SHNBRK
13BUS01BCNV01B                                                        SHNBRK
13BUS01CCNV01C                                                        SHNBRK
```

```
13e2a   e22a                                                          SRSDS
13e2b   e22b                                                          SRSDS
13e2c   e22c                                                          SRSDS
13e22a                                                CLOSED          GNDSW
13e22b                                                CLOSED          GNDSW
13e22c                                                CLOSED          GNDSW
$INCLUDE 307MEAS2.SWT
  CNV01Ai1a                                           MEASURING
  CNV01Bi1b                                           MEASURING
  CNV01Ci1c                                           MEASURING
  CNV02Ae22a                                          MEASURING
  CNV02Be22b                                          MEASURING
  CNV02Ce22c                                          MEASURING
BLANK RECORD ENDING SWITCHES
C 00000001111111111222222222233333333334444444444555555555566666666667777777777778
C 34567890123456789012345678901234567890123456789012345678901234567890123456789012
60i1a   -1
60i1b   -1
60i1c   -1
60e2a
60e2b
60e2c
$INCLUDE 308NTWK1.SRC
BLANK RECORD ENDING SOURCES
BLANK RECORD ENDING NODE VOLTAGE OUTPUT
BLANK RECORD ENDING PLOT
BEGIN NEW DATA CASE
BLANK RECORD ENDING ALL DATA CASES
```

16. 316PQPFC.USR

```
C File Name: 316PQPFC.USR
C USER INPUT FILE FOR PQPFC
C 00000001111111111222222222233333333334444444444555555555566666666667777777777778
C 34567890123456789012345678901234567890123456789012345678901234567890123456789012
C A SHUNT BREAKER IS CLOSED.
99SHNBRK   = TIMEX .GE. 0.01
C SET ST = 1.0 FOR ST OR ST = 0.0 FOR UPFC
99ST       = 1.0
C SETTING THE COMPENSATING VOLTAGE
99e2cm1    = 0.0 * (TIMEX .GE. 0.020)
99e2dqcm   = e2cm1
 1e2dqrf  +e2dqcm
       1.0
       1.0    0.0160
99betcm1   = 0 * (TIMEX .GE. 0.020)
99betcmd   = betcm1 * CDR
 1beta     +betcmd
       1.0
       1.0    0.0020
C TACS OUTPUT
33v2pu
C 33betcm1e2cm1 Vs    Vsp   PrMW  QrMV  Pex2MWQex2MVSex2MV
```

17. 317PQPFC.CON

```
C File Name: 317PQPFC.CON
C PQPFC - VSC2 OPEN LOOP CONTROL
```

```
C 0000000011111111112222222222333333333344444444445555555555666666666677777777778
C 3456789012345678901234567890123456789012345678901234567890123456789012345678901234567890
C CSC1 CURRENT COMPUTATION
99Denom    = v1d * v1d + v1q * v1q
99i1d      = (Pex2pu * v1d + Qex2pu * ST * v1q) / Denom
99i1q      = (Pex2pu * v1q - Qex2pu * ST * v1d) / Denom
99i1dq     = SQRT(i1d * i1d + i1q * i1q)
99i1ds     = i1d * COSPLL - i1q * SINPLL
99i1qs     = i1d * SINPLL + i1q * COSPLL
99i1apu    = i1ds
99i1cpu    = -0.5 * (i1ds + i1qs * SQRT(3))
99i1bpu    = -(i1apu + i1cpu)
99i1pu     = SQRT((i1apu * i1apu + i1bpu * i1bpu + i1cpu * i1cpu) * 2/3)
99i1a      = i1apu * Ibase
99i1b      = i1bpu * Ibase
99i1c      = i1cpu * Ibase
99i1aapu   = CNV01A / Ibase
99i1bbpu   = CNV01B / Ibase
99i1ccpu   = CNV01C / Ibase
C VSC2 CURRENT COMPUTATION
99i2apu    = CNV02A / Ibase
99i2bpu    = CNV02B / Ibase
99i2cpu    = CNV02C / Ibase
99i2pu     = SQRT((i2apu * i2apu + i2bpu * i2bpu + i2cpu * i2cpu) * 2/3)
99i2ds     = i2apu
99i2qs     = -(i2apu + 2.0 * i2cpu) / SQRT(3)
99i2d      =  i2ds * COSPLL + i2qs * SINPLL
99i2q      = -i2ds * SINPLL + i2qs * COSPLL
99i2dq     = SQRT(i2d * i2d + i2q * i2q)
99tha2     = theta + PI / 2 + PI + beta
C This addition of PI/2 is because the gating signals are sine reference and
C the PLL signals are cosine reference.
C The PI is added to reverse the polarity of the series coupling transformer.
99theta2   = tha2 + TWOPI * (tha2 .LT. 0.0) - TWOPI * (tha2 .GE. TWOPI)
C VSC2 VOLTAGE
99vsrspk   = e2dqrf * Vbase
99e2a      = vsrspk * SIN(theta2)
99e2b      = vsrspk * SIN(theta2 - 120 * CDR)
99e2c      = vsrspk * SIN(theta2 + 120 * CDR)
99e2apu    = e2a / Vbase
99e2bpu    = e2b / Vbase
99e2cpu    = e2c / Vbase
99e2pu     = SQRT((e2apu * e2apu + e2bpu * e2bpu + e2cpu * e2cpu) * 2 /3)
99e2ds     =  e2apu
99e2qs     = -(e2apu + 2.0 * e2cpu) / SQRT(3)
99e2d      =  e2ds * COSPLL + e2qs * SINPLL
99e2q      = -e2ds * SINPLL + e2qs * COSPLL
99e2dq     = SQRT(e2d * e2d + e2q * e2q)
C VSC2 EXCHANGED POWER COMPUTATION
99Pex2pu   = (e2apu * i2apu + e2bpu * i2bpu + e2cpu * i2cpu) / 1.5
99Qex2pu   = (e2apu * i2cpu - e2cpu * i2apu) * SQRT(3) / 1.5
99Pex2MW   = Pex2pu * 1.5 * Pbase / 1e6
99Qex2MV   = Qex2pu * 1.5 * Pbase / 1e6
99Sex2MV   = SQRT(Pex2MW * Pex2MW + Qex2MV * Qex2MV)
C SHUNT-EXCHANGED POWER AT THE POINT OF COMPENSATION
99Pcshpu   = (v1apu * i1aapu + v1bpu * i1bbpu + v1cpu * i1ccpu) / 1.5
99Qcshpu   = (v1apu * i1ccpu - v1cpu * i1aapu) * SQRT(3) / 1.5
99PcshMW   = Pcshpu * 1.5 * Pbase / 1e6
99QcshMV   = Qcshpu * 1.5 * Pbase / 1e6
C SERIES-EXCHANGED POWER AT THE POINT OF COMPENSATION
```

```
99Pcsepu  = (v12apu * iapu + v12bpu * ibpu + v12cpu * icpu) / 1.5
99Qcsepu  = (v12apu * icpu - v12cpu * iapu) * SQRT(3) / 1.5
99PcseMW  = Pcsepu * 1.5 * Pbase / 1e6
99QcseMV  = Qcsepu * 1.5 * Pbase / 1e6
C MODIFIED SENDING-END POWER COMPUTATION IN MEGA UNITS
99PspMW   = Psppu * 1.5 * Pbase / 1e6
99QspMV   = Qsppu * 1.5 * Pbase / 1e6
C RECEIVING-END POWER COMPUTATION IN MEGA UNITS
99PrMW    = Prpu * 1.5 * Pbase / 1e6
99QrMV    = Qrpu * 1.5 * Pbase / 1e6
C BUS01 LINE-TO-LINE VOLTAGE COMPUTATION IN KILO UNITS
99Vs      = v1pu * Vbase * SQRT(3 / 2) / 1e3
C BUS02 LINE-TO-LINE VOLTAGE COMPUTATION IN KILO UNITS
99Vsp     = v2pu * Vbase * SQRT(3 / 2) / 1e3
C Setting up the disconnect switches
99GNDSW   = 0
99SRSDS   = .NOT. GNDSW
```

18. 320PQPFC.USR

```
C File Name: 320PQPFC.USR
C USER INPUT FILE FOR PQPFC
C 000000011111111112222222222333333333344444444445555555555666666666677777777778
C 345678901234567890123456789012345678901234567890123456789012345678901234567890
C A SHUNT BREAKER IS CLOSED.
99SHNBRK  = TIMEX .GE. 0.01
C SET ST = 1.0 FOR ST OR ST = 0.0 FOR UPFC
99ST      = 1.0
C SETTING THE COMPENSATING VOLTAGE
99e2cm1   = 0.15 * (TIMEX .GE. 0.020)
99e2dqcm  = e2cm1 * v1pu
 1e2dqrf  +e2dqcm
       1.0
       1.0     0.0160
99betcm1  = 0 * (TIMEX .GE. 0.020)
99betcmd  = betcm1 * CDR
 1beta    +betcmd
       1.0
       1.0     0.0020
C TACS OUTPUT
33v2pu
C 33betcm1e2cm1 Vs     Vsp    PrMW   QrMV   Pex2MWQex2MVSex2MV
```

19. 321SHRE1.BRN

```
C File name: 321SHRE1.BRN
C SHUNT CAPACITANCE TO DRAW 1 PU CURRENT
C 000000011111111112222222222333333333344444444445555555555666666666677777777778
C 345678901234567890123456789012345678901234567890123456789012345678901234567890
C <-Bus1<-Bus2<-Bus3<-Bus4<----R<----L<----C                                   V
  e11a                                21.532
  e11b                                21.532
  e11c                                21.532
```

20. 322SHRE2.BRN

```
C File name: 322SHRE2.BRN
C SHUNT CAPACITANCE (WITH SERIES INDUCTANCE) TO DRAW 1 PU CURRENT
```

```
C 0000000111111111122222222223333333333444444444455555555556666666666777777777778
C 3456789012345678901234567890123456789012345678901234567890123456789012345678901234567890
C <-Bus1<-Bus2<-Bus3<-Bus4<----R<----L<----C                                  V
  e11a  e1a                2.463865.355
  e11b  e1b                2.463865.355
  e11c  e1c                2.463865.355
  e1a                                  21.532
  e1b                                  21.532
  e1c                                  21.532
```

21. 323SHIN1.BRN

```
C File name: 323SHIN1.BRN
C SHUNT INDUCTOR TO DRAW 1 PU CURRENT
C 0000000111111111122222222223333333333444444444455555555556666666666777777777778
C 3456789012345678901234567890123456789012345678901234567890123456789012345678901234567890
C <-Bus1<-Bus2<-Bus3<-Bus4<----R<----L<----C                                  V
  e11a                     11.077293.84
  e11b                     11.077293.84
  e11c                     11.077293.84
```

22. 324SERE1.BRN

```
C File Name: 324SERE1.BRN
C 3-PHASE SERIES COMPENSATING INDUCTOR
C 0000000111111111122222222223333333333444444444455555555556666666666777777777778
C 3456789012345678901234567890123456789012345678901234567890123456789012345678901234567890
C <-Bus1<-Bus2<-Bus3<-Bus4<----R<----L<----C                                  V
  BUS01ABUS11A             1.785447.358
  BUS01BBUS11B             1.785447.358
  BUS01CBUS11C             1.785447.358
  BUS11ABUS12A                   47.358
  BUS11BBUS12B                   47.358
  BUS11CBUS12C                   47.358
  BUS12ABUS02A                   47.358
  BUS12BBUS02B                   47.358
  BUS12CBUS02C                   47.358
```

23. 325SERE2.BRN

```
C File Name: 325SERE2.BRN
C 3-PHASE SERIES COMPENSATING INDUCTOR/CAPACITOR
C 0000000111111111122222222223333333333444444444455555555556666666666777777777778
C 3456789012345678901234567890123456789012345678901234567890123456789012345678901234567890
C <-Bus1<-Bus2<-Bus3<-Bus4<----R<----L<----C                                  V
  BUS01ABUS11A             1.785447.358
  BUS01BBUS11B             1.785447.358
  BUS01CBUS11C             1.785447.358
  BUS11ABUS12A                         148.57
  BUS11BBUS12B                         148.57
  BUS11CBUS12C                         148.57
  BUS12ABUS02A              0.001
  BUS12BBUS02B              0.001
  BUS12CBUS02C              0.001
```

24. 401IDLAT.DAT

```
BEGIN NEW DATA CASE
C File Name: 401IDLAT.DAT
C IDEAL AUTOTRANSFORMER WITH NO LEAKAGE IMPEDANCE
C 00000001111111111222222222233333333334444444444555555555566666666667777777778
C 3456789012345678901234567890123456789012345678901234567890123456789012345678 90
C ----dt<---tmax<---Xopt<---Copt<--Toler<        ><        >
8.3333-6150.00-3
C 00000001111111111222222222233333333334444444444555555555566666666667777777778
C 3456789012345678901234567890123456789012345678901234567890123456789012345678 90
C -Iprnt<--Iplot<-Idoubl<-KssOut<-MaxOut<        ><        ><---Icat<-Nenerg
      30      10                                                   1
TACS HYBRID
99CRD      = 180 / PI
99CDR      = 1 / CRD
99TWOPI    = 2.0 * PI
C BASE VOLTAGE, CURRENT, AND POWER
99Vbase    = 112676.528
99Ibase    = 946.662704
99Zbase    = Vbase / Ibase
99Pbase    = Vbase * Ibase
$INCLUDE 302MEAS1.SWT
$INCLUDE 303PTCT.SCL
C SERIES-CONNECTED VOLTAGE COMPUTATION
99v21pu    = v2pu - v1pu
C TACS OUTPUT
33v2pu
$INCLUDE 305LINER.CMP
BLANK RECORD ENDING TACS
C 00000001111111111222222222233333333334444444444555555555566666666667777777778
C 3456789012345678901234567890123456789012345678901234567890123456789012345678 90
C <-Bus1<-Bus2<-Bus3<-Bus4<----R<----L<----C                                    V
$INCLUDE 311NTWK2.BRN
$INCLUDE 402IDLAT.TRN
BLANK RECORD ENDING BRANCHES
C 00000001111111111222222222233333333334444444444555555555566666666667777777778
C 3456789012345678901234567890123456789012345678901234567890123456789012345678 90
C <-Bus1<-Bus2<---Tclose<----Topen<-------Ie<----Flash<--Request<-----Target<--O
$INCLUDE 307MEAS2.SWT
BLANK RECORD ENDING SWITCHES
C 00000001111111111222222222233333333334444444444555555555566666666667777777778
C 3456789012345678901234567890123456789012345678901234567890123456789012345678 90
$INCLUDE 308NTWK1.SRC
BLANK RECORD ENDING SOURCES
BLANK RECORD ENDING NODE VOLTAGE OUTPUT
BLANK RECORD ENDING PLOT
BEGIN NEW DATA CASE
BLANK RECORD ENDING ALL DATA CASES
```

25. 402IDLAT.TRN

```
C File Name: 402IDLAT.TRN
C 3-PHASE IDEAL AUTOTRANSFORMER WITH NO LEAKAGE IMPEDANCE
C 00000001111111111222222222233333333334444444444555555555566666666667777777778
C 3456789012345678901234567890123456789012345678901234567890123456789012345678 90
C ------AUTOTRANSFORMER FOR PHASE A------
C TRANSFORMER <--Refxxxxxx<--Iss<--Phi<-Name<-Rmag<---------------------------O
  TRANSFORMER                         AUTOTA
C NEXT LINE (9999 IN COLUMNS 13-16) INDICATES A LINEAR TRANSFORMER
            9999
```

```
C <---Nodes--><---------- ><-Ohms<---mH<---kV
C <-Bus1<-Bus2<---------- ><---Rk<---Lk<Volt|----------------------------------->O
 1BUS01A                   0.001 0.001 1.000
 2BUS02ABUS01A             0.001 0.001 0.150
C ------AUTOTRANSFORMER FOR PHASE B------
  TRANSFORMER                          AUTOTB
           9999
 1BUS01B                   0.001 0.001 1.000
 2BUS02BBUS01B             0.001 0.001 0.150
C ------AUTOTRANSFORMER FOR PHASE C------
  TRANSFORMER                          AUTOTC
           9999
 1BUS01C                   0.001 0.001 1.000
 2BUS02CBUS01C             0.001 0.001 0.150
```

26. 404IDLTT.TRN

```
C File Name: 404IDLTT.TRN
C 3-PHASE IDEAL TWO-WINDING TRANSFORMER WITH NO LEAKAGE IMPEDANCE
C 000000011111111112222222222333333333344444444445555555555666666666677777777778
C 345678901234567890123456789012345678901234567890123456789012345678901234567890
C ------TWO-WINDING TRANSFORMER FOR PHASE A------
C TRANSFORMER <--Refxxxxxx<--Iss<--Phi<-Name<-Rmag<---------------------------O
  TRANSFORMER                          TWDGTA
C NEXT LINE (9999 IN COLUMNS 13-16) INDICATES A LINEAR TRANSFORMER
           9999
C <---Nodes--><---------- ><-Ohms<---mH<---kV
C <-Bus1<-Bus2<---------- ><---Rk<---Lk<Volt|----------------------------------->O
 1BUS01A                   0.001 0.001 1.000
 2BUS02A                   0.001 0.001 1.150
C ------TWO-WINDING TRANSFORMER FOR PHASE B------
  TRANSFORMER                          TWDGTB
           9999
 1BUS01B                   0.001 0.001 1.000
 2BUS02B                   0.001 0.001 1.150
C ------TWO-WINDING TRANSFORMER FOR PHASE C------
  TRANSFORMER                          TWDGTC
           9999
 1BUS01C                   0.001 0.001 1.000
 2BUS02C                   0.001 0.001 1.150
```

27. 406PAR1.TRN

```
C File Name: 406PAR1.TRN
C 3-PHASE, SINGLE-CORE PHASE ANGLE REGULATOR WITH NO LEAKAGE IMPEDANCE
C 000000011111111112222222222333333333344444444445555555555666666666677777777778
C 345678901234567890123456789012345678901234567890123456789012345678901234567890
C ------PHASE ANGLE REGULATOR FOR PHASE A------
C TRANSFORMER <--Refxxxxxx<--Iss<--Phi<-Name<-Rmag<---------------------------O
  TRANSFORMER                          PARA
C NEXT LINE (9999 IN COLUMNS 13-16) INDICATES A LINEAR TRANSFORMER
           9999
C <---Nodes--><---------- ><-Ohms<---mH<---kV
C <-Bus1<-Bus2<---------- ><---Rk<---Lk<Volt|----------------------------------->O
 1BUS01CBUS01B             0.001 0.001 1.732
 2BUS02ABUS01A             0.001 0.001 0.150
C ------PHASE ANGLE REGULATOR FOR PHASE B------
  TRANSFORMER                          PARB
```

```
         9999
 1BUS01ABUS01C              0.001 0.001 1.732
 2BUS02BBUS01B              0.001 0.001 0.150
C ------PHASE ANGLE REGULATOR FOR PHASE C------
  TRANSFORMER                                PARC
         9999
 1BUS01BBUS01A              0.001 0.001 1.732
 2BUS02CBUS01C              0.001 0.001 0.150
```

28. 408PAR2.TRN

```
C File Name: 408PAR2.TRN
C 3-PHASE, SINGLE-CORE PHASE ANGLE REGULATOR WITH NO LEAKAGE IMPEDANCE
C TRUE PAR WITH NO DEVIATION IN VOLTAGE
C 00000001111111111222222222233333333334444444444555555555566666666667777777778
C 34567890123456789012345678901234567890123456789012345678901234567890123456789 0
C ------PHASE ANGLE REGULATOR FOR PHASE A------
C TRANSFORMER <--Refxxxxxx<--Iss<--Phi<-Name<-Rmag<----------------------------0
  TRANSFORMER                                PARA
C NEXT LINE (9999 IN COLUMNS 13-16) INDICATES A LINEAR TRANSFORMER
         9999
C <---Nodes--><----------><-Ohms<---mH<---kV
C <-Bus1<-Bus2<----------><---Rk<---Lk<Volt|---------------------------------->0
 1BUSC  BUSB                0.001 0.001 1.732
 2BUS02ABUSA                0.001 0.001 0.075
 3BUSA  BUS01A              0.001 0.001 0.075
C ------PHASE ANGLE REGULATOR FOR PHASE B------
  TRANSFORMER                                PARB
         9999
 1BUSA  BUSC                0.001 0.001 1.732
 2BUS02BBUSB                0.001 0.001 0.075
 3BUSB  BUS01B              0.001 0.001 0.075
C ------PHASE ANGLE REGULATOR FOR PHASE C------
  TRANSFORMER                                PARC
         9999
 1BUSB  BUSA                0.001 0.001 1.732
 2BUS02CBUSC                0.001 0.001 0.075
 3BUSC  BUS01C              0.001 0.001 0.075
```

29. 501SHREA.DAT

```
EGIN NEW DATA CASE
C File Name: 501SHREA.DAT
C OPERATION OF A SHUNT-CONNECTED REACTANCE (CAPACITOR/INDUCTOR)
C 00000001111111111222222222233333333334444444444555555555566666666667777777778
C 34567890123456789012345678901234567890123456789012345678901234567890123456789 0
C ----dt<---tmax<---Xopt<---Copt<--Toler<        ><        >
16.666-6300.00-3
C 00000001111111111222222222233333333334444444444555555555566666666667777777778
C 34567890123456789012345678901234567890123456789012345678901234567890123456789 0
C -Iprnt<--Iplot<-Idoubl<-KssOut<-MaxOut<        ><        ><---Icat<-Nenerg
       5      10                                                 1
TACS HYBRID
99TWOPI   = 2.0 * PI
C BASE VOLTAGE, CURRENT, AND POWER
99Vbase   = 112676.528
99Ibase   = 946.662704
99Zbase   = Vbase / Ibase
```

```
99Pbase    = Vbase * Ibase
$INCLUDE 302MEAS1.SWT
91CNV01A
91CNV01B
91CNV01C
$INCLUDE 303PTCT.SCL
$INCLUDE 310VPLL.PLL
$INCLUDE 502SHREA.USR
$INCLUDE 503SHREA.CON
$INCLUDE 305LINER.CMP
BLANK RECORD ENDING TACS
C 00000001111111111222222222233333333334444444444555555555566666666667777777778
C 34567890123456789012345678901234567890123456789012345678901234567890123456789 0
C <-Bus1<-Bus2<-Bus3<-Bus4<----R<----L<----C                                   V
$INCLUDE 311NTWK2.BRN
  BUS01ABUS02A             0.001
  BUS01BBUS02B             0.001
  BUS01CBUS02C             0.001
  BUS01ACNV01A            1.00+8
  BUS01BCNV01B            1.00+8
  BUS01CCNV01C            1.00+8
$INCLUDE 321SHRE1.BRN
BLANK RECORD ENDING BRANCHES
C <-Bus1<-Bus2<---Tclose<----Topen<-------Ie<----Flash<--Request<-----Target<--O
13BUS01ACNV01A                                                        SHNBRK
13BUS01BCNV01B                                                        SHNBRK
13BUS01CCNV01C                                                        SHNBRK
$INCLUDE 307MEAS2.SWT
  CNV01Ae11a                                           MEASURING
  CNV01Be11b                                           MEASURING
  CNV01Ce11c                                           MEASURING
BLANK RECORD ENDING SWITCHES
C 00000001111111111222222222233333333334444444444555555555566666666667777777778
C 34567890123456789012345678901234567890123456789012345678901234567890123456789 0
$INCLUDE 308NTWK1.SRC
BLANK RECORD ENDING SOURCES
BLANK RECORD ENDING NODE VOLTAGE OUTPUT
BLANK RECORD ENDING PLOT
BEGIN NEW DATA CASE
BLANK RECORD ENDING ALL DATA CASES
```

30. 502SHREA.USR

```
C File Name: 502SHREA.USR
C USER INPUT FILE FOR SHUNT-CONNECTED REACTANCE (CAPACITOR/INDUCTOR)
C 00000001111111111222222222233333333334444444444555555555566666666667777777778
C 345678901234567890123456789012345678901234567890123456789012345678901234567890
C A SHUNT BREAKER IS CLOSED.
99SHNBRK  = (TIMEX .GE. 0.05) .AND. (TIMEX .LE. 0.25)
C TACS OUTPUT
33v1apu i1apu i1q   Qcshpu
```

31. 503SHREA.CON

```
C File Name: 503SHREA.CON
C SHUNT-CONNECTED REACTANCE
C 00000001111111111222222222233333333334444444444555555555566666666667777777778
C 345678901234567890123456789012345678901234567890123456789012345678901234567890
```

```
C SHUNT CURRENT COMPUTATION
99i1apu    = CNV01A / Ibase
99i1bpu    = CNV01B / Ibase
99i1cpu    = CNV01C / Ibase
99i1pu     = SQRT((i1apu * i1apu + i1bpu * i1bpu + i1cpu * i1cpu) * 2/3)
99i1ds     = i1apu
99i1qs     = -(i1apu + 2.0 * i1cpu) / SQRT(3)
99i1d      =  i1ds * COSPLL + i1qs * SINPLL
99i1q      = -i1ds * SINPLL + i1qs * COSPLL
99i1dq     = SQRT(i1d * i1d + i1q * i1q)
C SHUNT-EXCHANGED POWER AT THE POINT OF COMPENSATION
99Pcshpu   = (v1apu * i1apu + v1bpu * i1bpu + v1cpu * i1cpu) / 1.5
99Qcshpu   = (v1apu * i1cpu - v1cpu * i1apu) * SQRT(3) / 1.5
```

32. 505SHCAP.USR

```
C File Name: 505SHCAP.USR
C USER INPUT FILE FOR SHUNT-CONNECTED REACTANCE (CAPACITOR/INDUCTOR)
C 0000000111111111122222222223333333333444444444455555555556666666666777777777778
C 3456789012345678901234567890123456789012345678901234567890123456789012345678901234567890
C A SHUNT BREAKER IS CLOSED.
99SHNBRK   = (TIMEX .GE. 0.05) .AND. (TIMEX .LE. 0.25)
99BYPBRK   = (TIMEX .GE. 0.15)
C TACS OUTPUT
33v1apu i1apu i1q   Qcshpu
```

33. 506SEREA.DAT

```
BEGIN NEW DATA CASE
C File Name: 506SEREA.DAT
C OPERATION OF A SERIES-CONNECTED REACTANCE (INDUCTOR)
C 0000000111111111122222222223333333333444444444455555555556666666666777777777778
C 3456789012345678901234567890123456789012345678901234567890123456789012345678901234567890
C ----dt<---tmax<---Xopt<---Copt<--Toler<       ><       >
16.666-6600.00-3
C 0000000111111111122222222223333333333444444444455555555556666666666777777777778
C 3456789012345678901234567890123456789012345678901234567890123456789012345678901234567890
C -Iprnt<--Iplot<-Idoubl<-KssOut<-MaxOut<       ><       ><---Icat<-Nenerg
      40      10                                                 1
TACS HYBRID
99TWOPI    = 2.0 * PI
C BASE VOLTAGE, CURRENT, AND POWER
99Vbase    = 112676.528
99Ibase    = 946.662704
99Zbase    = Vbase / Ibase
99Pbase    = Vbase * Ibase
$INCLUDE 302MEAS1.SWT
$INCLUDE 303PTCT.SCL
$INCLUDE 310VPLL.PLL
$INCLUDE 507SEREA.USR
$INCLUDE 508SEREA.CON
$INCLUDE 305LINER.CMP
BLANK RECORD ENDING TACS
C 0000000111111111122222222223333333333444444444455555555556666666666777777777778
C 3456789012345678901234567890123456789012345678901234567890123456789012345678901234567890
C <-Bus1<-Bus2<-Bus3<-Bus4<----R<----L<----C                                    V
$INCLUDE 311NTWK2.BRN
  BUS01ABUS02A             1.00+8
```

```
  BUS01BBUS02B              1.00+8
  BUS01CBUS02C              1.00+8
$INCLUDE 324SERE1.BRN
BLANK RECORD ENDING BRANCHES
C <-Bus1<-Bus2<--Vig---|<-Ihold--|<-Tdeion-|xxxxxxxxxxCLOSEDSAMEGRID->SIGNAL  oo
C <-Bus1<-Bus2xxxxxxxxxxxxxxxxxxxxxxxxxxxxxxxxxxxxxxxxCLOSEDxxxxxxxxxxSIGNAL  oo
13BUS01ABUS02A                                        CLOSED          SRSBK1
13BUS01BBUS02B                                        CLOSED          SRSBK1
13BUS01CBUS02C                                        CLOSED          SRSBK1
13BUS11ABUS12A                                        CLOSED          SRSBK2
13BUS11BBUS12B                                        CLOSED          SRSBK2
13BUS11CBUS12C                                        CLOSED          SRSBK2
13BUS12ABUS02A                                        CLOSED          SRSBK3
13BUS12BBUS02B                                        CLOSED          SRSBK3
13BUS12CBUS02C                                        CLOSED          SRSBK3
$INCLUDE 307MEAS2.SWT
BLANK RECORD ENDING SWITCHES
C 00000001111111111222222222233333333334444444444555555555566666666667777777778
C 34567890123456789012345678901234567890123456789012345678901234567890123456789 0
$INCLUDE 308NTWK1.SRC
BLANK RECORD ENDING SOURCES
BLANK RECORD ENDING NODE VOLTAGE OUTPUT
BLANK RECORD ENDING PLOT
BEGIN NEW DATA CASE
BLANK RECORD ENDING ALL DATA CASES
```

34. 507SEREA.USR

```
C File Name: 507SEREA.USR
C USER INPUT FILE FOR SERIES-CONNECTED REACTANCE (CAPACITOR/INDUCTOR)
C 00000001111111111222222222233333333334444444444555555555566666666667777777778
C 345678901234567890123456789012345678901234567890123456789012345678901234567890
C A SERIES BREAKER IS OPENED.
99SRSBK1  = 1 - ((TIMEX .GE. 0.025) .AND. (TIMEX .LE. 0.30))
99SRSBK2  = 1 - ((TIMEX .GE. 0.050) .AND. (TIMEX .LE. 0.30))
99SRSBK3  = 1 - ((TIMEX .GE. 0.175) .AND. (TIMEX .LE. 0.30))
C TACS OUTPUT
33Qcsepu
```

35. 508SEREA.CON

```
C File Name: 508SEREA.CON
C SERIES-CONNECTED REACTANCE
C 00000001111111111222222222233333333334444444444555555555566666666667777777778
C 345678901234567890123456789012345678901234567890123456789012345678901234567890
C SERIES-EXCHANGED POWER AT THE POINT OF COMPENSATION
99Pcsepu  = (v12apu * iapu + v12bpu * ibpu + v12cpu * icpu) / 1.5
99Qcsepu  = (v12apu * icpu - v12cpu * iapu) * SQRT(3) / 1.5
```

36. 601IDL21.DAT

```
BEGIN NEW DATA CASE
C File Name: 601IDL21.DAT
C INFINITE-PULSE, 2-LEVEL VSC #1
C 00000001111111111222222222233333333334444444444555555555566666666667777777778
C 345678901234567890123456789012345678901234567890123456789012345678901234567890
C ----dt<---tmax<---Xopt<---Copt<--Toler<        ><        >
8.3333-625.000-3
```

```
C 00000001111111111222222222233333333334444444444555555555566666666667777777777778
C 34567890123456789012345678901234567890123456789012345678901234567890123456789012
C -Iprnt<--Iplot<-Idoubl<-KssOut<-MaxOut<       ><       ><---Icat<-Nenerg
       4       1                                                       1
TACS HYBRID
99TWOPI   = 2 * PI
99freq    = 60
C PHASE-LOCKED LOOP
99theta   = TWOPI * freq * TIMEX
99theta1  = theta - TWOPI * TRUNC(theta/TWOPI)
C DC VOLTAGE
99vDC     = PI / 2
$INCLUDE 602IDL21.VSC
C TACS output
33theta theta1e1a   e1b   e1c
BLANK RECORD ENDING TACS
C 00000001111111111222222222233333333334444444444555555555566666666667777777777778
C 34567890123456789012345678901234567890123456789012345678901234567890123456789012
C <-Bus1<-Bus2<-Bus3<-Bus4<----R<----L<----C                                    V
  VSRCA                     1.0000
BLANK RECORD ENDING BRANCHES
C 00000001111111111222222222233333333334444444444555555555566666666667777777777778
C 34567890123456789012345678901234567890123456789012345678901234567890123456789012
C <-Bus1<-Bus2<---Tclose<----Topen<-------Ie<----Flash<--Request<-----Target<--O
BLANK RECORD ENDING SWITCHES
14VSRCA    112676.528      60.00       0.00                          -1.0
BLANK RECORD ENDING SOURCES
BLANK RECORD ENDING NODE VOLTAGE OUTPUT
BLANK RECORD ENDING PLOT
BEGIN NEW DATA CASE
BLANK RECORD ENDING ALL DATA CASES
```

37. 602IDL21.VSC

```
C File Name: 602IDL21.VSC
C INFINITE-PULSE, 2-LEVEL VSC1 OUTPUT VOLTAGE
C 00000001111111111222222222233333333334444444444555555555566666666667777777777778
C 34567890123456789012345678901234567890123456789012345678901234567890123456789012
99Kvsc1   = 2 / PI
99Vvsc1   = vDC * Kvsc1
99e1a     = Vvsc1 * SIN(theta1)
99e1b     = Vvsc1 * SIN(theta1 - TWOPI / 3)
99e1c     = Vvsc1 * SIN(theta1 + TWOPI / 3)
```

38. 6072POLE.DAT

```
BEGIN NEW DATA CASE
C File Name: 6072POLE.DAT
C 2-LEVEL POLE VSC
C 00000001111111111222222222233333333334444444444555555555566666666667777777777778
C 34567890123456789012345678901234567890123456789012345678901234567890123456789012
C ----dt<---tmax<---Xopt<---Copt<--Toler<       ><       >
8.3333-625.000-3
C 00000001111111111222222222233333333334444444444555555555566666666667777777777778
C 34567890123456789012345678901234567890123456789012345678901234567890123456789012
C -Iprnt<--Iplot<-Idoubl<-KssOut<-MaxOut<       ><       ><---Icat<-Nenerg
       4       1                                                       1
TACS HYBRID
```

```
99CRD      = 180/PI
99CDR      = 1/CRD
99TWOPI    = 2 * PI
99freq     = 60
C INPUTS FROM THE MEASURING SWITCHES
91VSC01A
C PHASE-LOCKED LOOP
99theta    = TWOPI * freq * TIMEX
99theta1   = theta - TWOPI * TRUNC(theta/TWOPI)
C DC VOLTAGE
99vDC      = PI / 2
C VSC OUTPUT CURRENT
99i1a      = VSC01A
$INCLUDE 6082POLE.GAT
$INCLUDE 6092POLE.POL
$INCLUDE 6102POLE.MAG
C TACS OUTPUT
33theta theta1e1a   i1a
BLANK RECORD ENDING TACS
C 00000001111111111222222222233333333334444444444555555555566666666667777777777778
C 34567890123456789012345678901234567890123456789012345678901234567890123456789O
C <-Bus1<-Bus2<-Bus3<-Bus4<----R<----L<----C                                   V
  VSRCA                     1.0000
  VSC01A                    1.00002.6525
BLANK RECORD ENDING BRANCHES
C 00000001111111111222222222233333333334444444444555555555566666666667777777777778
C 34567890123456789012345678901234567890123456789012345678901234567890123456789O
C <-Bus1<-Bus2<---Tclose<----Topen<-------Ie<----Flash<--Request<-----Target<--O
  e1a   VSC01A                                               MEASURING
BLANK RECORD ENDING SWITCHES
60e1a
14VSRCA   112676.528      60.00      0.00                               -1.0
BLANK RECORD ENDING SOURCES
BLANK RECORD ENDING NODE VOLTAGE OUTPUT
BLANK RECORD ENDING PLOT
BEGIN NEW DATA CASE
BLANK RECORD ENDING ALL DATA CASES
```

39. 6082POLE.GAT

```
C FILE NAME: 6082POLE.GAT (2-Level VSC as Converter #1)
C 2-LEVEL POLE VSC GATING SIGNALS
C 00000001111111111222222222233333333334444444444555555555566666666667777777777778
C 34567890123456789012345678901234567890123456789012345678901234567890123456789O
99Kvsc1    = 2 / PI
99an1A1    = theta1
99ang1A1   = an1A1 + TWOPI*(an1A1 .LT. 0.0) - TWOPI*(an1A1 .GE. TWOPI)
99pos1A1   = (ang1A1 .GE. 0) .AND. ((ang1A1 - PI) .LT. 0)
99neg1A1   = ((ang1A1 - PI) .GE. 0) .AND. ((ang1A1 - TWOPI) .LT. 0)
```

40. 6092POLE.POL

```
C File name: 6092POLE.POL
C TWO-LEVEL POLE VSC OUTPUT VOLTAGE
C 00000001111111111222222222233333333334444444444555555555566666666667777777777778
C 34567890123456789012345678901234567890123456789012345678901234567890123456789O
99v1A1     = vDC / 2 * pos1A1 - vDC / 2 * neg1A1
```

41. 6102POLE.MAG

```
C File name: 6102POLE.MAG
C MAGNETIC CIRCUIT
C 0000000111111111122222222223333333333444444444455555555556666666666777777777778
C 3456789012345678901234567890123456789012345678901234567890123456789012345678901234567890
99e1a1    = v1A1
C VSC1 OUTPUT VOLTAGE
99e1a     = e1a1
```

42. 6112POLE.DAT

```
BEGIN NEW DATA CASE
C File Name: 6112POLE.DAT
C 2-LEVEL POLE VSC OPERATING WITH ACTUAL SWITCHES
C 0000000111111111122222222223333333333444444444455555555556666666666777777777778
C 3456789012345678901234567890123456789012345678901234567890123456789012345678901234567890
C ----dt<---tmax<---Xopt<---Copt<--Toler<      ><      >
8.3333-625.000-3
C 0000000111111111122222222223333333333444444444455555555556666666666777777777778
C 3456789012345678901234567890123456789012345678901234567890123456789012345678901234567890
C -Iprnt<--Iplot<-Idoubl<-KssOut<-MaxOut<      ><      ><---Icat<-Nenerg
       4       1                                               1
TACS HYBRID
99TWOPI   = 2 * PI
99freq    = 60
C INPUTS FROM THE MEASURING SWITCHES
90e1a
90POS
90NEG
91VSC01A
C PHASE-LOCKED LOOP
99theta   = TWOPI * freq * TIMEX
99theta1  = theta - TWOPI * TRUNC(theta/TWOPI)
C DC VOLTAGE
99vDCp    = PI / 4
99vDCn    = -PI / 4
99vDC     = POS - NEG
C VSC OUTPUT CURRENT
99i1a     = VSC01A
$INCLUDE 6082POLE.GAT
C TACS OUTPUT
33theta theta1e1a   i1a   vDC
BLANK RECORD ENDING TACS
C 0000000111111111122222222223333333333444444444455555555556666666666777777777778
C 3456789012345678901234567890123456789012345678901234567890123456789012345678901234567890
C <-Bus1<-Bus2<-Bus3<-Bus4<----R<----L<----C                                   V
  VSRCA                   1.0000
  VSC01A                  1.00002.6525
  POS                        0.1       5.0+0
  NEG                        0.1       5.0+0
  e1a                     1.00+8
  vDCp  POS                0.001
  vDCp                    1.00+8
  vDCn  NEG                0.001
  vDCn                    1.00+8
BLANK RECORD ENDING BRANCHES
C 0000000111111111122222222223333333333444444444455555555556666666666777777777778
C 3456789012345678901234567890123456789012345678901234567890123456789012345678901234567890
```

```
C <-Bus1<-Bus2<---Tclose<----Topen<-------Ie<----Flash<--Request<-----Target<--O
  e1a   VSC01A                                            MEASURING
13POS   e1a                                                               pos1A1
13e1a   NEG                                                               neg1A1
BLANK RECORD ENDING SWITCHES
60vDCp
60vDCn
14VSRCA   112676.528     60.00       0.00                             -1.0
BLANK RECORD ENDING SOURCES
BLANK RECORD ENDING NODE VOLTAGE OUTPUT
BLANK RECORD ENDING PLOT
BEGIN NEW DATA CASE
BLANK RECORD ENDING ALL DATA CASES
```

43. 61206P21.DAT

```
BEGIN NEW DATA CASE
C File Name: 61206P21.DAT
C 6-PULSE, 2-LEVEL VSC #1
C 00000001111111111222222222233333333334444444444555555555566666666667777777778
C 345678901234567890123456789012345678901234567890123456789012345678901234567890
C ----dt<---tmax<---Xopt<---Copt<--Toler<       ><       >
8.3333-625.000-3
C 00000001111111111222222222233333333334444444444555555555566666666667777777778
C 345678901234567890123456789012345678901234567890123456789012345678901234567890
C -Iprnt<--Iplot<-Idoubl<-KssOut<-MaxOut<       ><       ><---Icat<-Nenerg
       4       1                                                      1
TACS HYBRID
99CRD      = 180/PI
99CDR      = 1/CRD
99TWOPI    = 2 * PI
99freq     = 60
C INPUTS FROM THE MEASURING SWITCHES
91VSC01A
91VSC01B
91VSC01C
C PHASE-LOCKED LOOP
99theta    = TWOPI * freq * TIMEX
99theta1   = theta - TWOPI * TRUNC(theta/TWOPI)
C DC VOLTAGE
99vDC      = PI / 2
C VSC OUTPUT CURRENT
99i1a      = VSC01A
99i1b      = VSC01B
99i1c      = VSC01C
$INCLUDE 61306P21.GAT
$INCLUDE 61406P21.PCL
$INCLUDE 61506P21.MAG
C TACS OUTPUT
33theta theta1e1a   e1b   e1c   i1a   i1b   i1c
BLANK RECORD ENDING TACS
C 00000001111111111222222222233333333334444444444555555555566666666667777777778
C 345678901234567890123456789012345678901234567890123456789012345678901234567890
C <-Bus1<-Bus2<-Bus3<-Bus4<----R<----L<----C                                   V
  VSRCA                  1.0000
  VSC01A                 1.00002.6525
  VSC01B                 1.00002.6525
  VSC01C                 1.00002.6525
BLANK RECORD ENDING BRANCHES
```

```
C 00000001111111111222222222233333333334444444444555555555566666666667777777778
C 345678901234567890123456789012345678901234567890123456789012345678901234567890
C <-Bus1<-Bus2<---Tclose<----Topen<--      Ie<----Flash<--Request<-----Target<--O
  e1a   VSC01A                                            MEASURING
  e1b   VSC01B                                            MEASURING
  e1c   VSC01C                                            MEASURING
BLANK RECORD ENDING SWITCHES
60e1a
60e1b
60e1c
14VSRCA   112676.528     60.00      0.00                             -1.0
BLANK RECORD ENDING SOURCES
BLANK RECORD ENDING NODE VOLTAGE OUTPUT
BLANK RECORD ENDING PLOT
BEGIN NEW DATA CASE
BLANK RECORD ENDING ALL DATA CASES
```

44. 61306P21.GAT

```
C FILE NAME: 61306P21.GAT
C 6-PULSE, 2-LEVEL VSC1 GATING SIGNALS
C 00000001111111111222222222233333333334444444444555555555566666666667777777778
C 345678901234567890123456789012345678901234567890123456789012345678901234567890
99Kvsc1    = 2 / PI
99an1A1    = theta1
99ang1A1   = an1A1 + TWOPI*(an1A1 .LT. 0.0) - TWOPI*(an1A1 .GE. TWOPI)
99pos1A1   = (ang1A1 .GE. 0) .AND. ((ang1A1 - PI) .LT. 0)
99neg1A1   = ((ang1A1 - PI) .GE. 0) .AND. ((ang1A1 - TWOPI) .LT. 0)
99an1B1    = theta1 - 120.0 * CDR
99ang1B1   = an1B1 + TWOPI*(an1B1 .LT. 0.0) - TWOPI*(an1B1 .GE. TWOPI)
99pos1B1   = (ang1B1 .GE. 0) .AND. ((ang1B1 - PI) .LT. 0)
99neg1B1   = ((ang1B1 - PI) .GE. 0) .AND. ((ang1B1 - TWOPI) .LT. 0)
99an1C1    = theta1 - 240.0 * CDR
99ang1C1   = an1C1 + TWOPI*(an1C1 .LT. 0.0) - TWOPI*(an1C1 .GE. TWOPI)
99pos1C1   = (ang1C1 .GE. 0) .AND. ((ang1C1 - PI) .LT. 0)
99neg1C1   = ((ang1C1 - PI) .GE. 0) .AND. ((ang1C1 - TWOPI) .LT. 0)
```

45. 61406P21.POL

```
C File name: 61406P21.POL
C 00000001111111111222222222233333333334444444444555555555566666666667777777778
C 345678901234567890123456789012345678901234567890123456789012345678901234567890
C 6-PULSE, 2-LEVEL VSC1 POLE VOLTAGES
99v1A1     = vDC / 2 * pos1A1 - vDC / 2 * neg1A1
99v1B1     = vDC / 2 * pos1B1 - vDC / 2 * neg1B1
99v1C1     = vDC / 2 * pos1C1 - vDC / 2 * neg1C1
```

46. 61506P21.MAG

```
C File name: 61506P21.MAG
C 6-PULSE, 2-LEVEL VSC1 MAGNETIC CIRCUIT
C 00000001111111111222222222233333333334444444444555555555566666666667777777778
C 345678901234567890123456789012345678901234567890123456789012345678901234567890
99v1N1     = (v1A1 + v1B1 + v1C1) / 3
99v1A1N    = v1A1 - v1N1
99v1B1N    = v1B1 - v1N1
99v1C1N    = v1C1 - v1N1
99e1a1     = v1A1N
```

```
99e1b1     = v1B1N
99e1c1     = v1C1N
C VSC1 OUTPUT VOLTAGE
99e1a      = e1a1
99e1b      = e1b1
99e1c      = e1c1
```

47. 62512P21.GAT

```
C FILE NAME: 62512P21.GAT
C 12-PULSE, 2-LEVEL VSC1 GATING SIGNALS
C 00000001111111111222222222233333333334444444444555555555566666666667777777778
C 34567890123456789012345678901234567890123456789012345678901234567890123456789 0
99Kvsc1    = 2 / PI
99an1A1    = theta1
99ang1A1   = an1A1 + TWOPI*(an1A1 .LT. 0.0) - TWOPI*(an1A1 .GE. TWOPI)
99pos1A1   = (ang1A1 .GE. 0) .AND. ((ang1A1 - PI) .LT. 0)
99neg1A1   = ((ang1A1 - PI) .GE. 0) .AND. ((ang1A1 - TWOPI) .LT. 0)
99an1B1    = theta1 - 120.0 * CDR
99ang1B1   = an1B1 + TWOPI*(an1B1 .LT. 0.0) - TWOPI*(an1B1 .GE. TWOPI)
99pos1B1   = (ang1B1 .GE. 0) .AND. ((ang1B1 - PI) .LT. 0)
99neg1B1   = ((ang1B1 - PI) .GE. 0) .AND. ((ang1B1 - TWOPI) .LT. 0)
99an1C1    = theta1 - 240.0 * CDR
99ang1C1   = an1C1 + TWOPI*(an1C1 .LT. 0.0) - TWOPI*(an1C1 .GE. TWOPI)
99pos1C1   = (ang1C1 .GE. 0) .AND. ((ang1C1 - PI) .LT. 0)
99neg1C1   = ((ang1C1 - PI) .GE. 0) .AND. ((ang1C1 - TWOPI) .LT. 0)
99an1D1    = theta1 - 30.0 * CDR
99ang1D1   = an1D1 + TWOPI*(an1D1 .LT. 0.0) - TWOPI*(an1D1 .GE. TWOPI)
99pos1D1   = (ang1D1 .GE. 0) .AND. ((ang1D1 - PI) .LT. 0)
99neg1D1   = ((ang1D1 - PI) .GE. 0) .AND. ((ang1D1 - TWOPI) .LT. 0)
99an1E1    = theta1 - 150.0 * CDR
99ang1E1   = an1E1 + TWOPI*(an1E1 .LT. 0.0) - TWOPI*(an1E1 .GE. TWOPI)
99pos1E1   = (ang1E1 .GE. 0) .AND. ((ang1E1 - PI) .LT. 0)
99neg1E1   = ((ang1E1 - PI) .GE. 0) .AND. ((ang1E1 - TWOPI) .LT. 0)
99an1F1    = theta1 - 270.0 * CDR
99ang1F1   = an1F1 + TWOPI*(an1F1 .LT. 0.0) - TWOPI*(an1F1 .GE. TWOPI)
99pos1F1   = (ang1F1 .GE. 0) .AND. ((ang1F1 - PI) .LT. 0)
99neg1F1   = ((ang1F1 - PI) .GE. 0) .AND. ((ang1F1 - TWOPI) .LT. 0)
```

48. 62612P21.POL

```
C File name: 62612P21.POL
C 00000001111111111222222222233333333334444444444555555555566666666667777777778
C 345678901234567890123456789012345678901234567890123456789012345678901234567890
C 12-PULSE, 2-LEVEL VSC1 POLE VOLTAGES
99v1A1     = vDC / 2 * pos1A1 - vDC / 2 * neg1A1
99v1B1     = vDC / 2 * pos1B1 - vDC / 2 * neg1B1
99v1C1     = vDC / 2 * pos1C1 - vDC / 2 * neg1C1
99v1D1     = vDC / 2 * pos1D1 - vDC / 2 * neg1D1
99v1E1     = vDC / 2 * pos1E1 - vDC / 2 * neg1E1
99v1F1     = vDC / 2 * pos1F1 - vDC / 2 * neg1F1
```

49. 62712P21.MAG

```
C File name: 62712P21.MAG
C 12-PULSE, 2-LEVEL VSC1 MAGNETIC CIRCUIT
C 00000001111111111222222222233333333334444444444555555555566666666667777777778
C 345678901234567890123456789012345678901234567890123456789012345678901234567890
```

```
99v1N1    = (v1A1 + v1B1 + v1C1) / 3
99v1A1N   = v1A1 - v1N1
99v1B1N   = v1B1 - v1N1
99v1C1N   = v1C1 - v1N1
99v1D1E1  = v1D1 - v1E1
99v1E1F1  = v1E1 - v1F1
99v1F1D1  = v1F1 - v1D1
99e1a1    = (v1A1N + v1D1E1 / SQRT(3)) / 2
99e1b1    = (v1B1N + v1E1F1 / SQRT(3)) / 2
99e1c1    = (v1C1N + v1F1D1 / SQRT(3)) / 2
C VSC1 OUTPUT VOLTAGE
99e1a     = e1a1
99e1b     = e1b1
99e1c     = e1c1
```

50. 63724P21.GAT

```
C FILE NAME: 63724P21.GAT
C 24-PULSE, 2-LEVEL VSC1 GATING SIGNALS
C 000000011111111112222222222333333333344444444445555555555666666666677777777778
C 345678901234567890123456789012345678901234567890123456789012345678901234567890
99Kvsc1   = 2 / PI
99an1A1   = theta1 + 22.5 * CDR
99ang1A1  = an1A1 + TWOPI * (an1A1 .LT. 0.0) - TWOPI * (an1A1 .GE. TWOPI)
99pos1A1  = (ang1A1 .GE. 0) .AND. ((ang1A1 - PI) .LT. 0)
99neg1A1  = ((ang1A1 - PI) .GE. 0) .AND. ((ang1A1 - TWOPI) .LT. 0)
99an1B1   = theta1 + (-120.0 + 22.5) * CDR
99ang1B1  = an1B1 + TWOPI * (an1B1 .LT. 0.0) - TWOPI * (an1B1 .GE. TWOPI)
99pos1B1  = (ang1B1 .GE. 0) .AND. ((ang1B1 - PI) .LT. 0)
99neg1B1  = ((ang1B1 - PI) .GE. 0) .AND. ((ang1B1 - TWOPI) .LT. 0)
99an1C1   = theta1 + (-240.0 + 22.5) * CDR
99ang1C1  = an1C1 + TWOPI * (an1C1 .LT. 0.0) - TWOPI * (an1C1 .GE. TWOPI)
99pos1C1  = (ang1C1 .GE. 0) .AND. ((ang1C1 - PI) .LT. 0)
99neg1C1  = ((ang1C1 - PI) .GE. 0) .AND. ((ang1C1 - TWOPI) .LT. 0)
99an1A2   = theta1 + 7.5 * CDR
99ang1A2  = an1A2 + TWOPI * (an1A2 .LT. 0.0) - TWOPI * (an1A2 .GE. TWOPI)
99pos1A2  = (ang1A2 .GE. 0) .AND. ((ang1A2 - PI) .LT. 0)
99neg1A2  = ((ang1A2 - PI) .GE. 0) .AND. ((ang1A2 - TWOPI) .LT. 0)
99an1B2   = theta1 + (-120.0 + 7.5) * CDR
99ang1B2  = an1B2 + TWOPI * (an1B2 .LT. 0.0) - TWOPI * (an1B2 .GE. TWOPI)
99pos1B2  = (ang1B2 .GE. 0) .AND. ((ang1B2 - PI) .LT. 0)
99neg1B2  = ((ang1B2 - PI) .GE. 0) .AND. ((ang1B2 - TWOPI) .LT. 0)
99an1C2   = theta1 + (-240.0 + 7.5) * CDR
99ang1C2  = an1C2 + TWOPI * (an1C2 .LT. 0.0) - TWOPI * (an1C2 .GE. TWOPI)
99pos1C2  = (ang1C2 .GE. 0) .AND. ((ang1C2 - PI) .LT. 0)
99neg1C2  = ((ang1C2 - PI) .GE. 0) .AND. ((ang1C2 - TWOPI) .LT. 0)
99an1D1   = theta1 + (-30.0 + 22.5) * CDR
99ang1D1  = an1D1 + TWOPI * (an1D1 .LT. 0.0) - TWOPI * (an1D1 .GE. TWOPI)
99pos1D1  = (ang1D1 .GE. 0) .AND. ((ang1D1 - PI) .LT. 0)
99neg1D1  = ((ang1D1 - PI) .GE. 0) .AND. ((ang1D1 - TWOPI) .LT. 0)
99an1E1   = theta1 + (-150.0 + 22.5) * CDR
99ang1E1  = an1E1 + TWOPI * (an1E1 .LT. 0.0) - TWOPI * (an1E1 .GE. TWOPI)
99pos1E1  = (ang1E1 .GE. 0) .AND. ((ang1E1 - PI) .LT. 0)
99neg1E1  = ((ang1E1 - PI) .GE. 0) .AND. ((ang1E1 - TWOPI) .LT. 0)
99an1F1   = theta1 + (-270.0 + 22.5) * CDR
99ang1F1  = an1F1 + TWOPI * (an1F1 .LT. 0.0) - TWOPI * (an1F1 .GE. TWOPI)
99pos1F1  = (ang1F1 .GE. 0) .AND. ((ang1F1 - PI) .LT. 0)
99neg1F1  = ((ang1F1 - PI) .GE. 0) .AND. ((ang1F1 - TWOPI) .LT. 0)
99an1D2   = theta1 + (-30.0 + 7.5) * CDR
```

```
99ang1D2  = an1D2 + TWOPI * (an1D2 .LT. 0.0) - TWOPI * (an1D2 .GE. TWOPI)
99pos1D2  = (ang1D2 .GE. 0) .AND. ((ang1D2 - PI) .LT. 0)
99neg1D2  = ((ang1D2 - PI) .GE. 0) .AND. ((ang1D2 - TWOPI) .LT. 0)
99an1E2   = theta1 + (-150.0 + 7.5) * CDR
99ang1E2  = an1E2 + TWOPI * (an1E2 .LT. 0.0) - TWOPI * (an1E2 .GE. TWOPI)
99pos1E2  = (ang1E2 .GE. 0) .AND. ((ang1E2 - PI) .LT. 0)
99neg1E2  = ((ang1E2 - PI) .GE. 0) .AND. ((ang1E2 - TWOPI) .LT. 0)
99an1F2   = theta1 + (-270.0 + 7.5) * CDR
99ang1F2  = an1F2 + TWOPI * (an1F2 .LT. 0.0) - TWOPI * (an1F2 .GE. TWOPI)
99pos1F2  = (ang1F2 .GE. 0) .AND. ((ang1F2 - PI) .LT. 0)
99neg1F2  = ((ang1F2 - PI) .GE. 0) .AND. ((ang1F2 - TWOPI) .LT. 0)
```

51. 63824P21.POL

```
C File name: 63824P21.POL
C 00000000111111111122222222223333333333444444444455555555556666666666777777777778
C 34567890123456789012345678901234567890123456789012345678901234567890123456789 0
C 24-PULSE, 2-LEVEL VSC1 POLE VOLTAGES
99v1A1     = vDC / 2 * pos1A1 - vDC / 2 * neg1A1
99v1B1     = vDC / 2 * pos1B1 - vDC / 2 * neg1B1
99v1C1     = vDC / 2 * pos1C1 - vDC / 2 * neg1C1
99v1A2     = vDC / 2 * pos1A2 - vDC / 2 * neg1A2
99v1B2     = vDC / 2 * pos1B2 - vDC / 2 * neg1B2
99v1C2     = vDC / 2 * pos1C2 - vDC / 2 * neg1C2
99v1D1     = vDC / 2 * pos1D1 - vDC / 2 * neg1D1
99v1E1     = vDC / 2 * pos1E1 - vDC / 2 * neg1E1
99v1F1     = vDC / 2 * pos1F1 - vDC / 2 * neg1F1
99v1D2     = vDC / 2 * pos1D2 - vDC / 2 * neg1D2
99v1E2     = vDC / 2 * pos1E2 - vDC / 2 * neg1E2
99v1F2     = vDC / 2 * pos1F2 - vDC / 2 * neg1F2
```

52. 63924P21.MAG

```
C File name: 63924P21.MAG
C 24-PULSE, 2-LEVEL VSC1 MAGNETIC CIRCUIT
C 00000000111111111122222222223333333333444444444455555555556666666666777777777778
C 345678901234567890123456789012345678901234567890123456789012345678901234567890
99K111     = 2 * SIN(97.5 * CDR) / SQRT(3)
99K112     = 2 * SIN(22.5 * CDR) / SQRT(3)
99K121     = 2 * SIN(112.5 * CDR) / SQRT(3)
99K122     = 2 * SIN(7.5 * CDR) / SQRT(3)
99v1N11    = (v1A1 + v1B1 + v1C1) / 3
99v1A1N    = v1A1 - v1N11
99v1B1N    = v1B1 - v1N11
99v1C1N    = v1C1 - v1N11
99v1N12    = (v1A2 + v1B2 + v1C2) / 3
99v1A2N    = v1A2 - v1N12
99v1B2N    = v1B2 - v1N12
99v1C2N    = v1C2 - v1N12
99v1N21    = (v1D1 + v1E1 + v1F1) / 3
99v1D1N    = v1D1 - v1N21
99v1E1N    = v1E1 - v1N21
99v1F1N    = v1F1 - v1N21
99v1N22    = (v1D2 + v1E2 + v1F2) / 3
99v1D2N    = v1D2 - v1N22
99v1E2N    = v1E2 - v1N22
99v1F2N    = v1F2 - v1N22
99e1a1     = (K111 * v1A1N + K112 * v1B1N + K121 * v1D1N + K122 * v1F1N) / 2
```

```
99e1b1    = (K111 * v1B1N + K112 * v1C1N + K121 * v1E1N + K122 * v1D1N) / 2
99e1c1    = (K111 * v1C1N + K112 * v1A1N + K121 * v1F1N + K122 * v1E1N) / 2
99e1a2    = (K121 * v1A2N + K122 * v1D2N + K111 * v1D2N + K112 * v1F2N) / 2
99e1b2    = (K121 * v1B2N + K122 * v1C2N + K111 * v1E2N + K112 * v1D2N) / 2
99e1c2    = (K121 * v1C2N + K122 * v1A2N + K111 * v1F2N + K112 * v1E2N) / 2
C VSC1 OUTPUT VOLTAGE
99e1a     = (e1a1 + e1a2) / 2
99e1b     = (e1b1 + e1b2) / 2
99e1c     = (e1c1 + e1c2) / 2
```

53. 64924Q21.GAT

```
C FILE NAME: 64924Q21.GAT
C 24-PULSE, QHN, 2-LEVEL VSC1 GATING SIGNALS
C 0000000111111111122222222223333333333444444444455555555556666666666777777777778
C 3456789012345678901234567890123456789012345678901234567890123456789012345678901234567890
99Kvsc1   = (2/PI) * COS(PI/24)
99an1A1   = theta1 + 7.5 * CDR
99ang1A1  = an1A1 + TWOPI * (an1A1 .LT. 0.0) - TWOPI * (an1A1 .GE. TWOPI)
99pos1A1  = (ang1A1 .GE. 0) .AND. ((ang1A1 - PI) .LT. 0)
99neg1A1  = ((ang1A1 - PI) .GE. 0) .AND. ((ang1A1 - TWOPI) .LT. 0)
99an1B1   = theta1 + (-120.0 + 7.5) * CDR
99ang1B1  = an1B1 + TWOPI * (an1B1 .LT. 0.0) - TWOPI * (an1B1 .GE. TWOPI)
99pos1B1  = (ang1B1 .GE. 0) .AND. ((ang1B1 - PI) .LT. 0)
99neg1B1  = ((ang1B1 - PI) .GE. 0) .AND. ((ang1B1 - TWOPI) .LT. 0)
99an1C1   = theta1 + (-240.0 + 7.5) * CDR
99ang1C1  = an1C1 + TWOPI * (an1C1 .LT. 0.0) - TWOPI * (an1C1 .GE. TWOPI)
99pos1C1  = (ang1C1 .GE. 0) .AND. ((ang1C1 - PI) .LT. 0)
99neg1C1  = ((ang1C1 - PI) .GE. 0) .AND. ((ang1C1 - TWOPI) .LT. 0)
99an1A2   = theta1 - 7.5 * CDR
99ang1A2  = an1A2 + TWOPI * (an1A2 .LT. 0.0) - TWOPI * (an1A2 .GE. TWOPI)
99pos1A2  = (ang1A2 .GE. 0) .AND. ((ang1A2 - PI) .LT. 0)
99neg1A2  = ((ang1A2 - PI) .GE. 0) .AND. ((ang1A2 - TWOPI) .LT. 0)
99an1B2   = theta1 + (-120.0 - 7.5) * CDR
99ang1B2  = an1B2 + TWOPI * (an1B2 .LT. 0.0) - TWOPI * (an1B2 .GE. TWOPI)
99pos1B2  = (ang1B2 .GE. 0) .AND. ((ang1B2 - PI) .LT. 0)
99neg1B2  = ((ang1B2 - PI) .GE. 0) .AND. ((ang1B2 - TWOPI) .LT. 0)
99an1C2   = theta1 + (-240.0 - 7.5) * CDR
99ang1C2  = an1C2 + TWOPI * (an1C2 .LT. 0.0) - TWOPI * (an1C2 .GE. TWOPI)
99pos1C2  = (ang1C2 .GE. 0) .AND. ((ang1C2 - PI) .LT. 0)
99neg1C2  = ((ang1C2 - PI) .GE. 0) .AND. ((ang1C2 - TWOPI) .LT. 0)
99an1D1   = theta1 + (-30.0 + 7.5) * CDR
99ang1D1  = an1D1 + TWOPI * (an1D1 .LT. 0.0) - TWOPI * (an1D1 .GE. TWOPI)
99pos1D1  = (ang1D1 .GE. 0) .AND. ((ang1D1 - PI) .LT. 0)
99neg1D1  = ((ang1D1 - PI) .GE. 0) .AND. ((ang1D1 - TWOPI) .LT. 0)
99an1E1   = theta1 + (-150.0 + 7.5) * CDR
99ang1E1  = an1E1 + TWOPI * (an1E1 .LT. 0.0) - TWOPI * (an1E1 .GE. TWOPI)
99pos1E1  = (ang1E1 .GE. 0) .AND. ((ang1E1 - PI) .LT. 0)
99neg1E1  = ((ang1E1 - PI) .GE. 0) .AND. ((ang1E1 - TWOPI) .LT. 0)
99an1F1   = theta1 + (-270.0 + 7.5) * CDR
99ang1F1  = an1F1 + TWOPI * (an1F1 .LT. 0.0) - TWOPI * (an1F1 .GE. TWOPI)
99pos1F1  = (ang1F1 .GE. 0) .AND. ((ang1F1 - PI) .LT. 0)
99neg1F1  = ((ang1F1 - PI) .GE. 0) .AND. ((ang1F1 - TWOPI) .LT. 0)
99an1D2   = theta1 + (-30.0 - 7.5) * CDR
99ang1D2  = an1D2 + TWOPI * (an1D2 .LT. 0.0) - TWOPI * (an1D2 .GE. TWOPI)
99pos1D2  = (ang1D2 .GE. 0) .AND. ((ang1D2 - PI) .LT. 0)
99neg1D2  = ((ang1D2 - PI) .GE. 0) .AND. ((ang1D2 - TWOPI) .LT. 0)
99an1E2   = theta1 + (-150.0 - 7.5) * CDR
99ang1E2  = an1E2 + TWOPI * (an1E2 .LT. 0.0) - TWOPI * (an1E2 .GE. TWOPI)
```

```
99pos1E2   = (ang1E2 .GE. 0) .AND. ((ang1E2 - PI) .LT. 0)
99neg1E2   = ((ang1E2 - PI) .GE. 0) .AND. ((ang1E2 - TWOPI) .LT. 0)
99an1F2    = theta1 + (-270.0 - 7.5) * CDR
99ang1F2   = an1F2 + TWOPI * (an1F2 .LT. 0.0) - TWOPI * (an1F2 .GE. TWOPI)
99pos1F2   = (ang1F2 .GE. 0) .AND. ((ang1F2 - PI) .LT. 0)
99neg1F2   = ((ang1F2 - PI) .GE. 0) .AND. ((ang1F2 - TWOPI) .LT. 0)
```

54. 65024Q21.MAG

```
C File name: 65024Q21.MAG
C 24-PULSE QUASI HARMONIC NEUTRALIZED, 2-LEVEL VSC1 MAGNETIC CIRCUIT
C 0000000111111111122222222223333333333444444444455555555556666666666777777777778
C 3456789012345678901234567890123456789012345678901234567890123456789012345678901234567890
99v1N1     = (v1A1 + v1B1 + v1C1) / 3
99v1A1N    = v1A1 - v1N1
99v1B1N    = v1B1 - v1N1
99v1C1N    = v1C1 - v1N1
99v1D1E1   = v1D1 - v1E1
99v1E1F1   = v1E1 - v1F1
99v1F1D1   = v1F1 - v1D1
99e1a1     = (v1A1N + v1D1E1 / SQRT(3)) / 2
99e1b1     = (v1B1N + v1E1F1 / SQRT(3)) / 2
99e1c1     = (v1C1N + v1F1D1 / SQRT(3)) / 2
99v1N2     = (v1A2 + v1B2 + v1C2) / 3
99v1A2N    = v1A2 - v1N2
99v1B2N    = v1B2 - v1N2
99v1C2N    = v1C2 - v1N2
99v1D2E2   = v1D2 - v1E2
99v1E2F2   = v1E2 - v1F2
99v1F2D2   = v1F2 - v1D2
99e1a2     = (v1A2N + v1D2E2 / SQRT(3)) / 2
99e1b2     = (v1B2N + v1E2F2 / SQRT(3)) / 2
99e1c2     = (v1C2N + v1F2D2 / SQRT(3)) / 2
C VSC1 OUTPUT VOLTAGE
99e1a      = (e1a1 + e1a2) / 2
99e1b      = (e1b1 + e1b2) / 2
99e1c      = (e1c1 + e1c2) / 2
```

55. 65848Q21.GAT

```
C FILE NAME: 65848Q21.GAT
C 48-PULSE, QHN, 2-LEVEL VSC1 GATING SIGNALS
C 0000000111111111122222222223333333333444444444455555555556666666666777777777778
C 3456789012345678901234567890123456789012345678901234567890123456789012345678901234567890
99Kvsc1    = (2/PI) * COS(PI/24) * COS(PI/48)
99an1A1    = theta1 + 11.25 * CDR
99ang1A1   = an1A1 + TWOPI * (an1A1 .LT. 0.0) - TWOPI * (an1A1 .GE. TWOPI)
99pos1A1   = (ang1A1 .GE. 0) .AND. ((ang1A1 - PI) .LT. 0)
99neg1A1   = ((ang1A1 - PI) .GE. 0) .AND. ((ang1A1 - TWOPI) .LT. 0)
99an1B1    = theta1 + (-120.0 + 11.25) * CDR
99ang1B1   = an1B1 + TWOPI * (an1B1 .LT. 0.0) - TWOPI * (an1B1 .GE. TWOPI)
99pos1B1   = (ang1B1 .GE. 0) .AND. ((ang1B1 - PI) .LT. 0)
99neg1B1   = ((ang1B1 - PI) .GE. 0) .AND. ((ang1B1 - TWOPI) .LT. 0)
99an1C1    = theta1 + (-240.0 + 11.25) * CDR
99ang1C1   = an1C1 + TWOPI * (an1C1 .LT. 0.0) - TWOPI * (an1C1 .GE. TWOPI)
99pos1C1   = (ang1C1 .GE. 0) .AND. ((ang1C1 - PI) .LT. 0)
99neg1C1   = ((ang1C1 - PI) .GE. 0) .AND. ((ang1C1 - TWOPI) .LT. 0)
99an1A2    = theta1 + 3.75 * CDR
```

```
99ang1A2   = an1A2 + TWOPI * (an1A2 .LT. 0.0) - TWOPI * (an1A2 .GE. TWOPI)
99pos1A2   = (ang1A2 .GE. 0) .AND. ((ang1A2 - PI) .LT. 0)
99neg1A2   = ((ang1A2 - PI) .GE. 0) .AND. ((ang1A2 - TWOPI) .LT. 0)
99an1B2    = theta1 + (-120.0 + 3.75) * CDR
99ang1B2   = an1B2 + TWOPI * (an1B2 .LT. 0.0) - TWOPI * (an1B2 .GE. TWOPI)
99pos1B2   = (ang1B2 .GE. 0) .AND. ((ang1B2 - PI) .LT. 0)
99neg1B2   = ((ang1B2 - PI) .GE. 0) .AND. ((ang1B2 - TWOPI) .LT. 0)
99an1C2    = theta1 + (-240.0 + 3.75) * CDR
99ang1C2   = an1C2 + TWOPI * (an1C2 .LT. 0.0) - TWOPI * (an1C2 .GE. TWOPI)
99pos1C2   = (ang1C2 .GE. 0) .AND. ((ang1C2 - PI) .LT. 0)
99neg1C2   = ((ang1C2 - PI) .GE. 0) .AND. ((ang1C2 - TWOPI) .LT. 0)
99an1A3    = theta1 - 3.75 * CDR
99ang1A3   = an1A3 + TWOPI * (an1A3 .LT. 0.0) - TWOPI * (an1A3 .GE. TWOPI)
99pos1A3   = (ang1A3 .GE. 0) .AND. ((ang1A3 - PI) .LT. 0)
99neg1A3   = ((ang1A3 - PI) .GE. 0) .AND. ((ang1A3 - TWOPI) .LT. 0)
99an1B3    = theta1 + (-120.0 - 3.75) * CDR
99ang1B3   = an1B3 + TWOPI * (an1B3 .LT. 0.0) - TWOPI * (an1B3 .GE. TWOPI)
99pos1B3   = (ang1B3 .GE. 0) .AND. ((ang1B3 - PI) .LT. 0)
99neg1B3   = ((ang1B3 - PI) .GE. 0) .AND. ((ang1B3 - TWOPI) .LT. 0)
99an1C3    = theta1 + (-240.0 - 3.75) * CDR
99ang1C3   = an1C3 + TWOPI * (an1C3 .LT. 0.0) - TWOPI * (an1C3 .GE. TWOPI)
99pos1C3   = (ang1C3 .GE. 0) .AND. ((ang1C3 - PI) .LT. 0)
99neg1C3   = ((ang1C3 - PI) .GE. 0) .AND. ((ang1C3 - TWOPI) .LT. 0)
99an1A4    = theta1 - 11.25 * CDR
99ang1A4   = an1A4 + TWOPI * (an1A4 .LT. 0.0) - TWOPI * (an1A4 .GE. TWOPI)
99pos1A4   = (ang1A4 .GE. 0) .AND. ((ang1A4 - PI) .LT. 0)
99neg1A4   = ((ang1A4 - PI) .GE. 0) .AND. ((ang1A4 - TWOPI) .LT. 0)
99an1B4    = theta1 + (-120.0 - 11.25) * CDR
99ang1B4   = an1B4 + TWOPI * (an1B4 .LT. 0.0) - TWOPI * (an1B4 .GE. TWOPI)
99pos1B4   = (ang1B4 .GE. 0) .AND. ((ang1B4 - PI) .LT. 0)
99neg1B4   = ((ang1B4 - PI) .GE. 0) .AND. ((ang1B4 - TWOPI) .LT. 0)
99an1C4    = theta1 + (-240.0 - 11.25) * CDR
99ang1C4   = an1C4 + TWOPI * (an1C4 .LT. 0.0) - TWOPI * (an1C4 .GE. TWOPI)
99pos1C4   = (ang1C4 .GE. 0) .AND. ((ang1C4 - PI) .LT. 0)
99neg1C4   = ((ang1C4 - PI) .GE. 0) .AND. ((ang1C4 - TWOPI) .LT. 0)
99an1D1    = theta1 + (-30.0 + 11.25) * CDR
99ang1D1   = an1D1 + TWOPI * (an1D1 .LT. 0.0) - TWOPI * (an1D1 .GE. TWOPI)
99pos1D1   = (ang1D1 .GE. 0) .AND. ((ang1D1 - PI) .LT. 0)
99neg1D1   = ((ang1D1 - PI) .GE. 0) .AND. ((ang1D1 - TWOPI) .LT. 0)
99an1E1    = theta1 + (-150.0 + 11.25) * CDR
99ang1E1   = an1E1 + TWOPI * (an1E1 .LT. 0.0) - TWOPI * (an1E1 .GE. TWOPI)
99pos1E1   = (ang1E1 .GE. 0) .AND. ((ang1E1 - PI) .LT. 0)
99neg1E1   = ((ang1E1 - PI) .GE. 0) .AND. ((ang1E1 - TWOPI) .LT. 0)
99an1F1    = theta1 + (-270.0 + 11.25) * CDR
99ang1F1   = an1F1 + TWOPI * (an1F1 .LT. 0.0) - TWOPI * (an1F1 .GE. TWOPI)
99pos1F1   = (ang1F1 .GE. 0) .AND. ((ang1F1 - PI) .LT. 0)
99neg1F1   = ((ang1F1 - PI) .GE. 0) .AND. ((ang1F1 - TWOPI) .LT. 0)
99an1D2    = theta1 + (-30.0 + 3.75) * CDR
99ang1D2   = an1D2 + TWOPI * (an1D2 .LT. 0.0) - TWOPI * (an1D2 .GE. TWOPI)
99pos1D2   = (ang1D2 .GE. 0) .AND. ((ang1D2 - PI) .LT. 0)
99neg1D2   = ((ang1D2 - PI) .GE. 0) .AND. ((ang1D2 - TWOPI) .LT. 0)
99an1E2    = theta1 + (-150.0 + 3.75) * CDR
99ang1E2   = an1E2 + TWOPI * (an1E2 .LT. 0.0) - TWOPI * (an1E2 .GE. TWOPI)
99pos1E2   = (ang1E2 .GE. 0) .AND. ((ang1E2 - PI) .LT. 0)
99neg1E2   = ((ang1E2 - PI) .GE. 0) .AND. ((ang1E2 - TWOPI) .LT. 0)
99an1F2    = theta1 + (-270.0 + 3.75) * CDR
99ang1F2   = an1F2 + TWOPI * (an1F2 .LT. 0.0) - TWOPI * (an1F2 .GE. TWOPI)
99pos1F2   = (ang1F2 .GE. 0) .AND. ((ang1F2 - PI) .LT. 0)
99neg1F2   = ((ang1F2 - PI) .GE. 0) .AND. ((ang1F2 - TWOPI) .LT. 0)
99an1D3    = theta1 + (-30.0 - 3.75) * CDR
```

```
99ang1D3   = an1D3 + TWOPI * (an1D3 .LT. 0.0) - TWOPI * (an1D3 .GE. TWOPI)
99pos1D3   = (ang1D3 .GE. 0) .AND. ((ang1D3 - PI) .LT. 0)
99neg1D3   = ((ang1D3 - PI) .GE. 0) .AND. ((ang1D3 - TWOPI) .LT. 0)
99an1E3    = theta1 + (-150.0 - 3.75) * CDR
99ang1E3   = an1E3 + TWOPI * (an1E3 .LT. 0.0) - TWOPI * (an1E3 .GE. TWOPI)
99pos1E3   = (ang1E3 .GE. 0) .AND. ((ang1E3 - PI) .LT. 0)
99neg1E3   = ((ang1E3 - PI) .GE. 0) .AND. ((ang1E3 - TWOPI) .LT. 0)
99an1F3    = theta1 + (-270.0 - 3.75) * CDR
99ang1F3   = an1F3 + TWOPI * (an1F3 .LT. 0.0) - TWOPI * (an1F3 .GE. TWOPI)
99pos1F3   = (ang1F3 .GE. 0) .AND. ((ang1F3 - PI) .LT. 0)
99neg1F3   = ((ang1F3 - PI) .GE. 0) .AND. ((ang1F3 - TWOPI) .LT. 0)
99an1D4    = theta1 + (-30.0 - 11.25) * CDR
99ang1D4   = an1D4 + TWOPI * (an1D4 .LT. 0.0) - TWOPI * (an1D4 .GE. TWOPI)
99pos1D4   = (ang1D4 .GE. 0) .AND. ((ang1D4 - PI) .LT. 0)
99neg1D4   = ((ang1D4 - PI) .GE. 0) .AND. ((ang1D4 - TWOPI) .LT. 0)
99an1E4    = theta1 + (-150.0 - 11.25) * CDR
99ang1E4   = an1E4 + TWOPI * (an1E4 .LT. 0.0) - TWOPI * (an1E4 .GE. TWOPI)
99pos1E4   = (ang1E4 .GE. 0) .AND. ((ang1E4 - PI) .LT. 0)
99neg1E4   = ((ang1E4 - PI) .GE. 0) .AND. ((ang1E4 - TWOPI) .LT. 0)
99an1F4    = theta1 + (-270.0 - 11.25) * CDR
99ang1F4   = an1F4 + TWOPI * (an1F4 .LT. 0.0) - TWOPI * (an1F4 .GE. TWOPI)
99pos1F4   = (ang1F4 .GE. 0) .AND. ((ang1F4 - PI) .LT. 0)
99neg1F4   = ((ang1F4 - PI) .GE. 0) .AND. ((ang1F4 - TWOPI) .LT. 0)
```

56. 65948P21.POL

```
C File name: 65948P21.POL
C 000000011111111112222222222333333333344444444445555555555666666666677777777778
C 345678901234567890123456789012345678901234567890123456789012345678901234567890
C 48-PULSE, 2-LEVEL VSC1 POLE VOLTAGES
99v1A1     = vDC / 2 * pos1A1 - vDC / 2 * neg1A1
99v1B1     = vDC / 2 * pos1B1 - vDC / 2 * neg1B1
99v1C1     = vDC / 2 * pos1C1 - vDC / 2 * neg1C1
99v1A2     = vDC / 2 * pos1A2 - vDC / 2 * neg1A2
99v1B2     = vDC / 2 * pos1B2 - vDC / 2 * neg1B2
99v1C2     = vDC / 2 * pos1C2 - vDC / 2 * neg1C2
99v1A3     = vDC / 2 * pos1A3 - vDC / 2 * neg1A3
99v1B3     = vDC / 2 * pos1B3 - vDC / 2 * neg1B3
99v1C3     = vDC / 2 * pos1C3 - vDC / 2 * neg1C3
99v1A4     = vDC / 2 * pos1A4 - vDC / 2 * neg1A4
99v1B4     = vDC / 2 * pos1B4 - vDC / 2 * neg1B4
99v1C4     = vDC / 2 * pos1C4 - vDC / 2 * neg1C4
99v1D1     = vDC / 2 * pos1D1 - vDC / 2 * neg1D1
99v1E1     = vDC / 2 * pos1E1 - vDC / 2 * neg1E1
99v1F1     = vDC / 2 * pos1F1 - vDC / 2 * neg1F1
99v1D2     = vDC / 2 * pos1D2 - vDC / 2 * neg1D2
99v1E2     = vDC / 2 * pos1E2 - vDC / 2 * neg1E2
99v1F2     = vDC / 2 * pos1F2 - vDC / 2 * neg1F2
99v1D3     = vDC / 2 * pos1D3 - vDC / 2 * neg1D3
99v1E3     = vDC / 2 * pos1E3 - vDC / 2 * neg1E3
99v1F3     = vDC / 2 * pos1F3 - vDC / 2 * neg1F3
99v1D4     = vDC / 2 * pos1D4 - vDC / 2 * neg1D4
99v1E4     = vDC / 2 * pos1E4 - vDC / 2 * neg1E4
99v1F4     = vDC / 2 * pos1F4 - vDC / 2 * neg1F4
```

57. 66048Q21.MAG

```
C File name: 66048Q21.MAG
```

```
C 48-PULSE QUASI HARMONIC NEUTRALIZED, 2-LEVEL VSC1 MAGNETIC CIRCUIT
C 00000001111111111222222222233333333334444444444555555555566666666667777777777778
C 34567890123456789012345678901234567890123456789012345678901234567890123456789O
99v1N1     = (v1A1 + v1B1 + v1C1) / 3
99v1A1N    = v1A1 - v1N1
99v1B1N    = v1B1 - v1N1
99v1C1N    = v1C1 - v1N1
99v1D1E1   = v1D1 - v1E1
99v1E1F1   = v1E1 - v1F1
99v1F1D1   = v1F1 - v1D1
99e1a1     = (v1A1N + v1D1E1 / SQRT(3)) / 2
99e1b1     = (v1B1N + v1E1F1 / SQRT(3)) / 2
99e1c1     = (v1C1N + v1F1D1 / SQRT(3)) / 2
99v1N2     = (v1A2 + v1B2 + v1C2) / 3
99v1A2N    = v1A2 - v1N2
99v1B2N    = v1B2 - v1N2
99v1C2N    = v1C2 - v1N2
99v1D2E2   = v1D2 - v1E2
99v1E2F2   = v1E2 - v1F2
99v1F2D2   = v1F2 - v1D2
99e1a2     = (v1A2N + v1D2E2 / SQRT(3)) / 2
99e1b2     = (v1B2N + v1E2F2 / SQRT(3)) / 2
99e1c2     = (v1C2N + v1F2D2 / SQRT(3)) / 2
99v1N3     = (v1A3 + v1B3 + v1C3) / 3
99v1A3N    = v1A3 - v1N3
99v1B3N    = v1B3 - v1N3
99v1C3N    = v1C3 - v1N3
99v1D3E3   = v1D3 - v1E3
99v1E3F3   = v1E3 - v1F3
99v1F3D3   = v1F3 - v1D3
99e1a3     = (v1A3N + v1D3E3 / SQRT(3)) / 2
99e1b3     = (v1B3N + v1E3F3 / SQRT(3)) / 2
99e1c3     = (v1C3N + v1F3D3 / SQRT(3)) / 2
99v1N4     = (v1A4 + v1B4 + v1C4) / 3
99v1A4N    = v1A4 - v1N4
99v1B4N    = v1B4 - v1N4
99v1C4N    = v1C4 - v1N4
99v1D4E4   = v1D4 - v1E4
99v1E4F4   = v1E4 - v1F4
99v1F4D4   = v1F4 - v1D4
99e1a4     = (v1A4N + v1D4E4 / SQRT(3)) / 2
99e1b4     = (v1B4N + v1E4F4 / SQRT(3)) / 2
99e1c4     = (v1C4N + v1F4D4 / SQRT(3)) / 2
C VSC1 OUTPUT VOLTAGE
99v1a1     = (e1a1 + e1a3) / 2
99v1b1     = (e1b1 + e1b3) / 2
99v1c1     = (e1c1 + e1c3) / 2
C
99v1a2     = (e1a2 + e1a4) / 2
99v1b2     = (e1b2 + e1b4) / 2
99v1c2     = (e1c2 + e1c4) / 2
C
99e1a      = (v1a1 + v1a2) / 2
99e1b      = (v1b1 + v1b2) / 2
99e1c      = (v1c1 + v1c2) / 2
```

58. 67512P31.DAT

```
BEGIN NEW DATA CASE
```

```
C File Name: 67512P31.DAT
C 12-PULSE, 3-LEVEL VSC #1
C 000000011111111112222222222333333333344444444445555555555666666666677777777778
C 345678901234567890123456789012345678901234567890123456789012345678901234567890
C ----dt<---tmax<---Xopt<---Copt<--Toler<      ><      >
8.3333-625.000-3
C 000000011111111112222222222333333333344444444445555555555666666666677777777778
C 345678901234567890123456789012345678901234567890123456789012345678901234567890
C -Iprnt<--Iplot<-Idoubl<-KssOut<-MaxOut<      ><      ><---Icat<-Nenerg
       4       1                                                      1
TACS HYBRID
99CRD      = 180/PI
99CDR      = 1/CRD
99TWOPI    = 2 * PI
99freq     = 60
C INPUTS FROM THE MEASURING SWITCHES
91VSC01A
91VSC01B
91VSC01C
C PHASE-LOCKED LOOP
99theta    = TWOPI * freq * TIMEX
99theta1   = theta - TWOPI * TRUNC(theta/TWOPI)
C DC VOLTAGE
99vDC      = PI / 2
C VSC1 DEAD ANGLE
99gamma1   = 7.5 * CDR
C VSC OUTPUT CURRENT
99i1a      = VSC01A
99i1b      = VSC01B
99i1c      = VSC01C
$INCLUDE 67612P31.GAT
$INCLUDE 62612P21.POL
$INCLUDE 62712P21.MAG
C TACS OUTPUT
33theta theta1e1a   e1b   e1c   i1a   i1b   i1c
BLANK RECORD ENDING TACS
C 000000011111111112222222222333333333344444444445555555555666666666677777777778
C 345678901234567890123456789012345678901234567890123456789012345678901234567890
C <-Bus1<-Bus2<-Bus3<-Bus4<----R<----L<----C                                   V
  VSRCA                    1.0000
  VSC01A                   1.00002.6525
  VSC01B                   1.00002.6525
  VSC01C                   1.00002.6525
BLANK RECORD ENDING BRANCHES
C 000000011111111112222222222333333333344444444445555555555666666666677777777778
C 345678901234567890123456789012345678901234567890123456789012345678901234567890
C <-Bus1<-Bus2<---Tclose<----Topen<-------Ie<----Flash<--Request<-----Target<--O
  e1a   VSC01A                                         MEASURING
  e1b   VSC01B                                         MEASURING
  e1c   VSC01C                                         MEASURING
BLANK RECORD ENDING SWITCHES
60e1a
60e1b
60e1c
14VSRCA   112676.528     60.00      0.00                              -1.0
BLANK RECORD ENDING SOURCES
BLANK RECORD ENDING NODE VOLTAGE OUTPUT
BLANK RECORD ENDING PLOT
BEGIN NEW DATA CASE
BLANK RECORD ENDING ALL DATA CASES
```

59. 67612P31.GAT

```
C FILE NAME: 67612P31.GAT
C 12-PULSE, 3-LEVEL VSC1 GATING SIGNALS
C 00000001111111111222222222233333333334444444444555555555566666666667777777778
C 34567890123456789012345678901234567890123456789012345678901234567890123456789
99Kvsc1    = (2 / PI)
99an1A1    = theta1
99ang1A1   = an1A1 + TWOPI * (an1A1 .LT. 0.0) - TWOPI * (an1A1 .GE. TWOPI)
99pos1A1   = ((ang1A1 - gamma1) .GE. 0) .AND. ((ang1A1 - PI + gamma1) .LT. 0)
99neg1A1   = ((ang1A1 - PI-gamma1) .GE. 0) .AND. ((ang1A1 - TWOPI+gamma1) .LT. 0)
99an1B1    = theta1 - 120.0 * CDR
99ang1B1   = an1B1 + TWOPI * (an1B1 .LT. 0.0) - TWOPI * (an1B1 .GE. TWOPI)
99pos1B1   = ((ang1B1 - gamma1) .GE. 0) .AND. ((ang1B1 - PI + gamma1) .LT. 0)
99neg1B1   = ((ang1B1 - PI-gamma1) .GE. 0) .AND. ((ang1B1 - TWOPI+gamma1) .LT. 0)
99an1C1    = theta1 - 240.0 * CDR
99ang1C1   = an1C1 + TWOPI * (an1C1 .LT. 0.0) - TWOPI * (an1C1 .GE. TWOPI)
99pos1C1   = ((ang1C1 - gamma1) .GE. 0) .AND. ((ang1C1 - PI + gamma1) .LT. 0)
99neg1C1   = ((ang1C1 - PI-gamma1) .GE. 0) .AND. ((ang1C1 - TWOPI+gamma1) .LT. 0)
99an1D1    = theta1 - 30.0 * CDR
99ang1D1   = an1D1 + TWOPI * (an1D1 .LT. 0.0) - TWOPI * (an1D1 .GE. TWOPI)
99pos1D1   = ((ang1D1 - gamma1) .GE. 0) .AND. ((ang1D1 - PI + gamma1) .LT. 0)
99neg1D1   = ((ang1D1 - PI-gamma1) .GE. 0) .AND. ((ang1D1 - TWOPI+gamma1) .LT. 0)
99an1E1    = theta1 - 150.0 * CDR
99ang1E1   = an1E1 + TWOPI * (an1E1 .LT. 0.0) - TWOPI * (an1E1 .GE. TWOPI)
99pos1E1   = ((ang1E1 - gamma1) .GE. 0) .AND. ((ang1E1 - PI + gamma1) .LT. 0)
99neg1E1   = ((ang1E1 - PI-gamma1) .GE. 0) .AND. ((ang1E1 - TWOPI+gamma1) .LT. 0)
99an1F1    = theta1 - 270.0 * CDR
99ang1F1   = an1F1 + TWOPI * (an1F1 .LT. 0.0) - TWOPI * (an1F1 .GE. TWOPI)
99pos1F1   = ((ang1F1 - gamma1) .GE. 0) .AND. ((ang1F1 - PI + gamma1) .LT. 0)
99neg1F1   = ((ang1F1 - PI-gamma1) .GE. 0) .AND. ((ang1F1 - TWOPI+gamma1) .LT. 0)
```

60. 68124Q31.DAT

```
BEGIN NEW DATA CASE
C File Name: 68124Q31.DAT
C 24-PULSE, QHN, 3-LEVEL VSC #1
C 00000001111111111222222222233333333334444444444555555555566666666667777777778
C 34567890123456789012345678901234567890123456789012345678901234567890123456789
C ----dt<---tmax<---Xopt<---Copt<--Toler<        ><        >
8.3333-625.000-3
C 00000001111111111222222222233333333334444444444555555555566666666667777777778
C 34567890123456789012345678901234567890123456789012345678901234567890123456789
C -Iprnt<--Iplot<-Idoubl<-KssOut<-MaxOut<        ><        ><---Icat<-Nenerg
       4       1                                                        1
TACS HYBRID
99CRD      = 180/PI
99CDR      = 1/CRD
99TWOPI    = 2 * PI
99freq     = 60
C INPUTS FROM THE MEASURING SWITCHES
91VSC01A
91VSC01B
91VSC01C
C PHASE-LOCKED LOOP
99theta    = TWOPI * freq * TIMEX
99theta1   = theta - TWOPI * TRUNC(theta/TWOPI)
```

```
C DC VOLTAGE
99vDC      = PI / 2
C VSC1 DEAD ANGLE
99gamma1   = 3.75 * CDR
C VSC OUTPUT CURRENT
99i1a      = VSC01A
99i1b      = VSC01B
99i1c      = VSC01C
$INCLUDE 68224Q31.GAT
$INCLUDE 63824P21.POL
$INCLUDE 65024Q21.MAG
C TACS OUTPUT
33theta theta1e1a   e1b   e1c   i1a   i1b   i1c
BLANK RECORD ENDING TACS
C 00000001111111111222222222233333333334444444444555555555566666666667777777777778
C 34567890123456789012345678901234567890123456789012345678901234567890123456789890
C <-Bus1<-Bus2<-Bus3<-Bus4<----R<----L<----C                                     V
  VSRCA                   1.0000
  VSC01A                  1.00002.6525
  VSC01B                  1.00002.6525
  VSC01C                  1.00002.6525
BLANK RECORD ENDING BRANCHES
C 00000001111111111222222222233333333334444444444555555555566666666667777777777778
C 34567890123456789012345678901234567890123456789012345678901234567890123456789890
C <-Bus1<-Bus2<---Tclose<----Topen<-------Ie<----Flash<--Request<-----Target<--O
  e1a   VSC01A                                           MEASURING
  e1b   VSC01B                                           MEASURING
  e1c   VSC01C                                           MEASURING
BLANK RECORD ENDING SWITCHES
60e1a
60e1b
60e1c
14VSRCA   112676.528     60.00       0.00                              -1.0
BLANK RECORD ENDING SOURCES
BLANK RECORD ENDING NODE VOLTAGE OUTPUT
BLANK RECORD ENDING PLOT
BEGIN NEW DATA CASE
BLANK RECORD ENDING ALL DATA CASES
```

61. 68224Q31.GAT

```
C FILE NAME: 68224Q31.GAT
C 24-PULSE, QHN, 3-LEVEL VSC1 GATING SIGNALS
C 00000001111111111222222222233333333334444444444555555555566666666667777777777778
C 34567890123456789012345678901234567890123456789012345678901234567890123456789890
99Kvsc1    = (2 / PI) * COS(PI / 24)
99an1A1    = theta1 + 7.5 * CDR
99ang1A1   = an1A1 + TWOPI * (an1A1 .LT. 0.0) - TWOPI * (an1A1 .GE. TWOPI)
99pos1A1   = ((ang1A1 - gamma1) .GE. 0) .AND. ((ang1A1 - PI + gamma1) .LT. 0)
99neg1A1   = ((ang1A1 - PI-gamma1) .GE. 0) .AND. ((ang1A1 - TWOPI+gamma1) .LT. 0)
99an1B1    = theta1 + (-120.0 + 7.5) * CDR
99ang1B1   = an1B1 + TWOPI * (an1B1 .LT. 0.0) - TWOPI * (an1B1 .GE. TWOPI)
99pos1B1   = ((ang1B1 - gamma1) .GE. 0) .AND. ((ang1B1 - PI + gamma1) .LT. 0)
99neg1B1   = ((ang1B1 - PI-gamma1) .GE. 0) .AND. ((ang1B1 - TWOPI+gamma1) .LT. 0)
99an1C1    = theta1 + (-240.0 + 7.5) * CDR
99ang1C1   = an1C1 + TWOPI * (an1C1 .LT. 0.0) - TWOPI * (an1C1 .GE. TWOPI)
99pos1C1   = ((ang1C1 - gamma1) .GE. 0) .AND. ((ang1C1 - PI + gamma1) .LT. 0)
99neg1C1   = ((ang1C1 - PI-gamma1) .GE. 0) .AND. ((ang1C1 - TWOPI+gamma1) .LT. 0)
99an1A2    = theta1 - 7.5 * CDR
```

```
99ang1A2   = an1A2 + TWOPI * (an1A2 .LT. 0.0) - TWOPI * (an1A2 .GE. TWOPI)
99pos1A2   = ((ang1A2 - gamma1) .GE. 0) .AND. ((ang1A2 - PI + gamma1) .LT. 0)
99neg1A2   = ((ang1A2 - PI-gamma1) .GE. 0) .AND. ((ang1A2 - TWOPI+gamma1) .LT. 0)
99an1B2    = theta1 + (-120.0 - 7.5) * CDR
99ang1B2   = an1B2 + TWOPI * (an1B2 .LT. 0.0) - TWOPI * (an1B2 .GE. TWOPI)
99pos1B2   = ((ang1B2 - gamma1) .GE. 0) .AND. ((ang1B2 - PI + gamma1) .LT. 0)
99neg1B2   = ((ang1B2 - PI-gamma1) .GE. 0) .AND. ((ang1B2 - TWOPI+gamma1) .LT. 0)
99an1C2    = theta1 + (-240.0 - 7.5) * CDR
99ang1C2   = an1C2 + TWOPI * (an1C2 .LT. 0.0) - TWOPI * (an1C2 .GE. TWOPI)
99pos1C2   = ((ang1C2 - gamma1) .GE. 0) .AND. ((ang1C2 - PI + gamma1) .LT. 0)
99neg1C2   = ((ang1C2 - PI-gamma1) .GE. 0) .AND. ((ang1C2 - TWOPI+gamma1) .LT. 0)
99an1D1    = theta1 + (-30.0 + 7.5) * CDR
99ang1D1   = an1D1 + TWOPI * (an1D1 .LT. 0.0) - TWOPI * (an1D1 .GE. TWOPI)
99pos1D1   = ((ang1D1 - gamma1) .GE. 0) .AND. ((ang1D1 - PI + gamma1) .LT. 0)
99neg1D1   = ((ang1D1 - PI-gamma1) .GE. 0) .AND. ((ang1D1 - TWOPI+gamma1) .LT. 0)
99an1E1    = theta1 + (-150.0 + 7.5) * CDR
99ang1E1   = an1E1 + TWOPI * (an1E1 .LT. 0.0) - TWOPI * (an1E1 .GE. TWOPI)
99pos1E1   = ((ang1E1 - gamma1) .GE. 0) .AND. ((ang1E1 - PI + gamma1) .LT. 0)
99neg1E1   = ((ang1E1 - PI-gamma1) .GE. 0) .AND. ((ang1E1 - TWOPI+gamma1) .LT. 0)
99an1F1    = theta1 + (-270.0 + 7.5) * CDR
99ang1F1   = an1F1 + TWOPI * (an1F1 .LT. 0.0) - TWOPI * (an1F1 .GE. TWOPI)
99pos1F1   = ((ang1F1 - gamma1) .GE. 0) .AND. ((ang1F1 - PI + gamma1) .LT. 0)
99neg1F1   = ((ang1F1 - PI-gamma1) .GE. 0) .AND. ((ang1F1 - TWOPI+gamma1) .LT. 0)
99an1D2    = theta1 + (-30.0 - 7.5) * CDR
99ang1D2   = an1D2 + TWOPI * (an1D2 .LT. 0.0) - TWOPI * (an1D2 .GE. TWOPI)
99pos1D2   = ((ang1D2 - gamma1) .GE. 0) .AND. ((ang1D2 - PI + gamma1) .LT. 0)
99neg1D2   = ((ang1D2 - PI-gamma1) .GE. 0) .AND. ((ang1D2 - TWOPI+gamma1) .LT. 0)
99an1E2    = theta1 + (-150.0 - 7.5) * CDR
99ang1E2   = an1E2 + TWOPI * (an1E2 .LT. 0.0) - TWOPI * (an1E2 .GE. TWOPI)
99pos1E2   = ((ang1E2 - gamma1) .GE. 0) .AND. ((ang1E2 - PI + gamma1) .LT. 0)
99neg1E2   = ((ang1E2 - PI-gamma1) .GE. 0) .AND. ((ang1E2 - TWOPI+gamma1) .LT. 0)
99an1F2    = theta1 + (-270.0 - 7.5) * CDR
99ang1F2   = an1F2 + TWOPI * (an1F2 .LT. 0.0) - TWOPI * (an1F2 .GE. TWOPI)
99pos1F2   = ((ang1F2 - gamma1) .GE. 0) .AND. ((ang1F2 - PI + gamma1) .LT. 0)
99neg1F2   = ((ang1F2 - PI-gamma1) .GE. 0) .AND. ((ang1F2 - TWOPI+gamma1) .LT. 0)
```

62. 69424Q21.MAG

```
C File name: 69424Q21.MAG
C 24-PULSE QUASI HARMONIC NEUTRALIZED, 2-LEVEL VSC1 MAGNETIC CIRCUIT WITH IPT
C 00000001111111111222222222233333333334444444444555555555566666666667777777778
C 34567890123456789012345678901234567890123456789012345678901234567890123456789 0
99v1A12    = (v1A1 + v1A2) / 2
99v1B12    = (v1B1 + v1B2) / 2
99v1C12    = (v1C1 + v1C2) / 2
99v1D12    = (v1D1 + v1D2) / 2
99v1E12    = (v1E1 + v1E2) / 2
99v1F12    = (v1F1 + v1F2) / 2
99v1N1     = (v1A12 + v1B12 + v1C12) / 3
99v1A12N   = v1A12 - v1N1
99v1B12N   = v1B12 - v1N1
99v1C12N   = v1C12 - v1N1
99v1DE12   = v1D12 - v1E12
99v1EF12   = v1E12 - v1F12
99v1FD12   = v1F12 - v1D12
C VSC1 OUTPUT VOLTAGE
99e1a      = (v1A12N + v1DE12 / SQRT(3)) / 2
99e1b      = (v1B12N + v1EF12 / SQRT(3)) / 2
99e1c      = (v1C12N + v1FD12 / SQRT(3)) / 2
```

63. 801CSC1.DAT

```
BEGIN NEW DATA CASE
C File Name: 801CSC1.DAT
C OPEN LOOP OPERATION OF A SHUNT-CONNECTED CSC
C 00000001111111111222222222233333333334444444444555555555566666666667777777778
C 34567890123456789012345678901234567890123456789012345678901234567890123456789
C ----dt<---tmax<---Xopt<---Copt<--Toler<      ><      >
16.666-6300.00-3
C 00000001111111111222222222233333333334444444444555555555566666666667777777778
C 34567890123456789012345678901234567890123456789012345678901234567890123456789
C -Iprnt<--Iplot<-Idoubl<-KssOut<-MaxOut<      ><      ><---Icat<-Nenerg
      20      10                                                1
TACS HYBRID
99TWOPI   = 2.0 * PI
C BASE VOLTAGE AND CURRENT
99Vbase   = 112676.528
99Ibase   = 946.662704
99Zbase   = Vbase / Ibase
99Pbase   = Vbase * Ibase
$INCLUDE 302MEAS1.SWT
91CNV01A
91CNV01B
91CNV01C
$INCLUDE 303PTCT.SCL
$INCLUDE 310VPLL.PLL
$INCLUDE 802CSC1.USR
$INCLUDE 803CSC1.CON
$INCLUDE 305LINER.CMP
BLANK RECORD ENDING TACS
C 00000001111111111222222222233333333334444444444555555555566666666667777777778
C 34567890123456789012345678901234567890123456789012345678901234567890123456789
C <-Bus1<-Bus2<-Bus3<-Bus4<----R<----L<----C                                    V
$INCLUDE 311NTWK2.BRN
  BUS01ABUS02A             0.001
  BUS01BBUS02B             0.001
  BUS01CBUS02C             0.001
  BUS01ACNV01A            1.00+8
  BUS01BCNV01B            1.00+8
  BUS01CCNV01C            1.00+8
  i1qa                    1.00+8
  i1qb                    1.00+8
  i1qc                    1.00+8
BLANK RECORD ENDING BRANCHES
13BUS01ACNV01A                                                         SHNBRK
13BUS01BCNV01B                                                         SHNBRK
13BUS01CCNV01C                                                         SHNBRK
$INCLUDE 307MEAS2.SWT
  CNV01Ai1qa                                           MEASURING
  CNV01Bi1qb                                           MEASURING
  CNV01Ci1qc                                           MEASURING
BLANK RECORD ENDING SWITCHES
C 00000001111111111222222222233333333334444444444555555555566666666667777777778
C 34567890123456789012345678901234567890123456789012345678901234567890123456789
60i1qa  -1
60i1qb  -1
60i1qc  -1
$INCLUDE 308NTWK1.SRC
```

```
BLANK RECORD ENDING SOURCES
BLANK RECORD ENDING NODE VOLTAGE OUTPUT
BLANK RECORD ENDING PLOT
BEGIN NEW DATA CASE
BLANK RECORD ENDING ALL DATA CASES
```

64. 802CSC1.USR

```
C File Name: 802CSC1.USR
C USER INPUT FILE FOR SHUNT-CONNECTED CURRENT-SOURCED CONVERTER
C 0000000111111111122222222223333333333444444444455555555556666666666777777777778
C 3456789012345678901234567890123456789012345678901234567890123456789012345678901234567890
C A SHUNT BREAKER IS CLOSED.
99SHNBRK  = TIMEX .GE. 0.05
C SETTING THE STATCOM REFERENCE REACTIVE CURRENT
99i1qcm1  = TIMEX .GE. 0.125
99i1qcm2  = 2 * (TIMEX .GE. 0.175)
99i1qcm3  = 2 * (TIMEX .GE. 0.25)
99i1qcmd  = (i1qcm1 - i1qcm2 + i1qcm3)
 1i1qref  +i1qcmd
       1.0
       1.0     0.0040
C TACS OUTPUT
33v1apu i1apu i1q   Qcshpu
```

65. 803CSC1.CON

```
C File Name: 803CSC1.CON
C SHUNT-CONNECTED CURRENT-SOURCED CONVERTER
C 0000000111111111122222222223333333333444444444455555555556666666666777777777778
C 3456789012345678901234567890123456789012345678901234567890123456789012345678901234567890
99i1qa     = -i1qref * Ibase * COS(theta + PI/2)
99i1qb     = -i1qref * Ibase * COS(theta + PI/2 - TWOPI/3)
99i1qc     = -i1qref * Ibase * COS(theta + PI/2 + TWOPI/3)
C CSC1 CURRENT COMPUTATION
99i1apu    = CNV01A / Ibase
99i1bpu    = CNV01B / Ibase
99i1cpu    = CNV01C / Ibase
99i1pu     = SQRT((i1apu * i1apu + i1bpu * i1bpu + i1cpu * i1cpu) * 2/3)
99i1ds     = i1apu
99i1qs     = -(i1apu + 2.0 * i1cpu) / SQRT(3)
99i1d      =  i1ds * COSPLL + i1qs * SINPLL
99i1q      = -i1ds * SINPLL + i1qs * COSPLL
99i1dq     = SQRT(i1d * i1d + i1q * i1q)
C SHUNT-EXCHANGED POWER AT THE POINT OF COMPENSATION
99Pcshpu   = (v1apu * i1apu + v1bpu * i1bpu + v1cpu * i1cpu) / 1.5
99Qcshpu   = (v1apu * i1cpu - v1cpu * i1apu) * SQRT(3) / 1.5
```

66. 804VSC1.DAT

```
BEGIN NEW DATA CASE
C File Name: 804VSC1.DAT
C OPEN LOOP OPERATION OF A VSC SHUNT-CONNECTED THROUGH A TIE IMPEDANCE
C 0000000111111111122222222223333333333444444444455555555556666666666777777777778
C 3456789012345678901234567890123456789012345678901234567890123456789012345678901234567890
C ----dt<---tmax<---Xopt<---Copt<--Toler<       ><       >
16.666-6750.00-3
C 0000000111111111122222222223333333333444444444455555555556666666666777777777778
```

```
C 34567890123456789012345678901234567890123456789012345678901234567890123456789012345678901234567890
C -Iprnt<--Iplot<-Idoubl<-KssOut<-MaxOut<      ><      ><---Icat<-Nenerg
      50      10                                                  1
TACS HYBRID
99TWOPI   = 2.0 * PI
C BASE VOLTAGE AND CURRENT
99Vbase   = 112676.528
99Ibase   = 946.662704
99Zbase   = Vbase / Ibase
99Pbase   = Vbase * Ibase
$INCLUDE 302MEAS1.SWT
91CNV01A
91CNV01B
91CNV01C
$INCLUDE 303PTCT.SCL
$INCLUDE 310VPLL.PLL
$INCLUDE 805VSC1.USR
$INCLUDE 806VSC1.CON
$INCLUDE 305LINER.CMP
BLANK RECORD ENDING TACS
C 00000001111111111222222222233333333334444444444555555555566666666667777777777778
C 34567890123456789012345678901234567890123456789012345678901234567890123456789012345678901234567890
C <-Bus1<-Bus2<-Bus3<-Bus4<----R<----L<----C                                    V
$INCLUDE 306NTWK1.BRN
  BUS01ABUS02A              0.001
  BUS01BBUS02B              0.001
  BUS01CBUS02C              0.001
  BUS01ACNV01A             1.00+8
  BUS01BCNV01B             1.00+8
  BUS01CCNV01C             1.00+8
  e11a  e1a                1.785447.358
  e11b  e1b                1.785447.358
  e11c  e1c                1.785447.358
BLANK RECORD ENDING BRANCHES
13BUS01ACNV01A                                                          SHNBRK
13BUS01BCNV01B                                                          SHNBRK
13BUS01CCNV01C                                                          SHNBRK
$INCLUDE 307MEAS2.SWT
  CNV01Ae11a                                            MEASURING
  CNV01Be11b                                            MEASURING
  CNV01Ce11c                                            MEASURING
BLANK RECORD ENDING SWITCHES
C 00000001111111111222222222233333333334444444444555555555566666666667777777777778
C 34567890123456789012345678901234567890123456789012345678901234567890123456789012345678901234567890
60e1a
60e1b
60e1c
$INCLUDE 308NTWK1.SRC
BLANK RECORD ENDING SOURCES
BLANK RECORD ENDING NODE VOLTAGE OUTPUT
BLANK RECORD ENDING PLOT
BEGIN NEW DATA CASE
BLANK RECORD ENDING ALL DATA CASES
```

67. 805VSC1.USR

```
C File Name: 805VSC1.USR
C USER INPUT FILE FOR SHUNT-CONNECTED VOLTAGE-SOURCED CONVERTER
C 00000001111111111222222222233333333334444444444555555555566666666667777777777778
```

```
C 345678901234567890123456789012345678901234567890123456789012345678901234567890
C A SHUNT BREAKER IS CLOSED.
99SHNBRK  = TIMEX .GE. 0.05
C SETTING THE STATCOM REFERENCE VOLTAGE
99e1rcm1  = 0.2 * (TIMEX .GE. 0.3)
99e1rcm2  = 0.4 * (TIMEX .GE. 0.5)
99e1cmd   = (0.987 + e1rcm1 - e1rcm2) * SHNBRK + 1.068 * (.NOT. SHNBRK)
 1e1ref   +e1cmd
       1.0
       1.0      0.0320
C TACS OUTPUT
33Qcshpui1q   i1apu e1apu e1pu
C ------TACS INITIAL CONDITIONS------
C -name>xx<--------|
77e1ref   1.068000
```

68. 806VSC1.CON

```
C File Name: 806VSC1.CON
C SHUNT-CONNECTED VOLTAGE-SOURCED CONVERTER
C 000000011111111112222222222333333333344444444445555555555666666666677777777778
C 345678901234567890123456789012345678901234567890123456789012345678901234567890
99e1a     = e1ref * Vbase * COS(theta)
99e1b     = e1ref * Vbase * COS(theta-TWOPI/3)
99e1c     = e1ref * Vbase * COS(theta+TWOPI/3)
99e1apu   = e1a / Vbase
99e1bpu   = e1b / Vbase
99e1cpu   = e1c / Vbase
99e1pu    = SQRT((e1apu * e1apu + e1bpu * e1bpu + e1cpu * e1cpu) * 2 / 3)
C VSC1 current computation
99i1apu   = CNV01A / Ibase
99i1bpu   = CNV01B / Ibase
99i1cpu   = CNV01C / Ibase
99i1pu    = SQRT((i1apu * i1apu + i1bpu * i1bpu + i1cpu * i1cpu) * 2 / 3)
99i1ds    = i1apu
99i1qs    = -(i1apu + 2.0 * i1cpu) / SQRT(3)
99i1d     =  i1ds * COSPLL + i1qs * SINPLL
99i1q     = -i1ds * SINPLL + i1qs * COSPLL
99i1dq    = SQRT(i1d * i1d + i1q * i1q)
C SHUNT-EXCHANGED POWER AT THE POINT OF COMPENSATION
99Pcshpu  = (v1apu * i1apu + v1bpu * i1bpu + v1cpu * i1cpu) / 1.5
99Qcshpu  = (v1apu * i1cpu - v1cpu * i1apu) * SQRT(3) / 1.5
```

69. 808VSCDC.DAT

```
BEGIN NEW DATA CASE
C File Name: 808VSCDC.DAT
C OPEN LOOP OPERATION OF A SHUNT-CONNECTED VSC WITH A DC LINK CAPACITOR
C 000000011111111112222222222333333333344444444445555555555666666666677777777778
C 345678901234567890123456789012345678901234567890123456789012345678901234567890
C ----dt<---tmax<---Xopt<---Copt<--Toler<        ><        >
16.666-6750.00-3
C 000000011111111112222222222333333333344444444445555555555666666666677777777778
C 345678901234567890123456789012345678901234567890123456789012345678901234567890
C -Iprnt<--Iplot<-Idoubl<-KssOut<-MaxOut<        ><        ><---Icat<-Nenerg
      50        10                                                  1
TACS HYBRID
99TWOPI   = 2.0 * PI
```

```
C BASE VOLTAGE AND CURRENT
99Vbase    = 112676.528
99Ibase    = 946.662704
99Zbase    = Vbase / Ibase
99Pbase    = Vbase * Ibase
$INCLUDE 302MEAS1.SWT
91CNV01A
91CNV01B
91CNV01C
$INCLUDE 303PTCT.SCL
$INCLUDE 310VPLL.PLL
$INCLUDE 809VSCDC.USR
$INCLUDE 810VSCDC.CON
$INCLUDE 602IDL21.VSC
$INCLUDE 305LINER.CMP
BLANK RECORD ENDING TACS
C 00000001111111111222222222233333333334444444444555555555566666666667777777778
C 34567890123456789012345678901234567890123456789012345678901234567890123456789 0
C <-Bus1<-Bus2<-Bus3<-Bus4<----R<----L<----C                                 V
$INCLUDE 306NTWK1.BRN
  BUS01ABUS02A            0.001
  BUS01BBUS02B            0.001
  BUS01CBUS02C            0.001
  BUS01ACNV01A           1.00+8
  BUS01BCNV01B           1.00+8
  BUS01CCNV01C           1.00+8
$INCLUDE 312TRAN1.TRN
BLANK RECORD ENDING BRANCHES
13BUS01ACNV01A                                                          SHNBRK
13BUS01BCNV01B                                                          SHNBRK
13BUS01CCNV01C                                                          SHNBRK
$INCLUDE 307MEAS2.SWT
  CNV01Ae11a                                               MEASURING
  CNV01Be11b                                               MEASURING
  CNV01Ce11c                                               MEASURING
BLANK RECORD ENDING SWITCHES
C 00000001111111111222222222233333333334444444444555555555566666666667777777778
C 34567890123456789012345678901234567890123456789012345678901234567890123456789 0
60e1a
60e1b
60e1c
$INCLUDE 308NTWK1.SRC
BLANK RECORD ENDING SOURCES
BLANK RECORD ENDING NODE VOLTAGE OUTPUT
BLANK RECORD ENDING PLOT
BEGIN NEW DATA CASE
BLANK RECORD ENDING ALL DATA CASES
```

70. 809VSCDC.USR

```
C File Name: 809VSCDC.USR
C USER INPUT FILE FOR SHUNT-CONNECTED VSC WITH A DC LINK CAPACITOR
C 00000001111111111222222222233333333334444444444555555555566666666667777777778
C 34567890123456789012345678901234567890123456789012345678901234567890123456789 0
C A SHUNT BREAKER IS CLOSED.
99SHNBRK  = TIMEX .GE. 0.05
C SETTING THE DC LINK REFERENCE VOLTAGE
99vDCcm1  = 0.314 * (TIMEX .GE. 0.3)
99vDCcm2  = 0.627 * (TIMEX .GE. 0.5)
```

```
99vDCcmd  = (1.551 + vDCcm1 - vDCcm2) * SHNBRK + 1.677 * (.NOT. SHNBRK)
 1vDCpu    +vDCcmd
       1.0
       1.0     0.0320
C TACS OUTPUT
33Qcshpui1q   i1apu e1apu e1pu  vDCpu
C ------TACS INITIAL CONDITIONS------
C -name>xx<--------|
77vDCpu    1.677000
```

71. 810VSCDC.CON

```
C File Name: 810VSCDC.CON
C OPEN LOOP OPERATION OF A SHUNT-CONNECTED VSC WITH A DC LINK CAPACITOR
C 000000011111111112222222222333333333344444444445555555555666666666677777777778
C 345678901234567890123456789012345678901234567890123456789012345678901234567890
C DC LINK CAPACITOR VOLTAGE COMPUTATION
99vDC      = vDCpu * Vbase
C VSC1 CONTROL ANGLE COMPUTATION
99tha1     = theta + PI / 2
C This addition of PI/2 is because the gating signals are sine reference and
C the PLL signals are cosine reference.
99theta1   = tha1 + TWOPI*(tha1.LT.0.0) - TWOPI*(tha1.GE.TWOPI)
99e1apu    = e1a / Vbase
99e1bpu    = e1b / Vbase
99e1cpu    = e1c / Vbase
99e1pu     = SQRT((e1apu * e1apu + e1bpu * e1bpu + e1cpu * e1cpu) * 2 /3)
C VSC1 CURRENT COMPUTATION
99i1apu    = CNV01A / Ibase
99i1bpu    = CNV01B / Ibase
99i1cpu    = CNV01C / Ibase
99i1pu     = SQRT((i1apu * i1apu + i1bpu * i1bpu + i1cpu * i1cpu) * 2/3)
99i1ds     = i1apu
99i1qs     = -(i1apu + 2.0 * i1cpu) / SQRT(3)
99i1d      =  i1ds * COSPLL + i1qs * SINPLL
99i1q      = -i1ds * SINPLL + i1qs * COSPLL
99i1dq     = SQRT(i1d * i1d + i1q * i1q)
C SHUNT-EXCHANGED POWER AT THE POINT OF COMPENSATION
99Pcshpu   = (v1apu * i1apu + v1bpu * i1bpu + v1cpu * i1cpu) / 1.5
99Qcshpu   = (v1apu * i1cpu - v1cpu * i1apu) * SQRT(3) / 1.5
```

72. 812STATI.USR

```
C File Name: 812STATI.USR
C USER INPUT FILE FOR STATCOM CURRENT CONTROL
C 000000011111111112222222222333333333344444444445555555555666666666677777777778
C 345678901234567890123456789012345678901234567890123456789012345678901234567890
C A SHUNT BREAKER IS CLOSED.
99SHNBRK   = TIMEX .GE. 0.05
99KpshI    = .45
99KishI    = 40
99Cap      = 42.0E-6
C SETTING THE STATCOM REACTIVE REFERENCE CURRENT
99i1qcm1   = TIMEX .GE. 0.125
99i1qcm2   = 2 * (TIMEX .GE. 0.175)
99i1qcm3   = 2 * (TIMEX .GE. 0.25)
99i1qcmd   = i1qcm1 - i1qcm2 + i1qcm3
 1i1qrfI   +i1qcmd
```

```
        1.0
        1.0     0.0040
C TACS OUTPUT
33i1qrefi1q   vDCpu v1apu e1apu i1apu alpha
```

73. 813STATI.CON

```
C File Name: 813STATI.CON
C STATCOM CURRENT CONTROL
C 0000000111111111122222222223333333333444444444455555555556666666666777777777778
C 3456789012345678901234567890123456789012345678901234567890123456789012345678890
C INNER CURRENT CONTROL LOOP STARTS
99i1qref  = i1qrfI
99i1qerr  = (i1qref - i1q) * SHNBRK
99i1qint  = i1qini + i1qerr * DELTAT
  i1qini  +i1qint
99alpha   = -(KpshI * i1qerr + KishI * i1qint)
C INNER CURRENT CONTROL LOOP ENDS
99tha1    = theta + PI / 2 + alpha
C This addition of PI/2 is because the gating signals are sine reference and
C the PLL signals are cosine reference.
99theta1  = tha1 + TWOPI * (tha1.LT.0.0) - TWOPI * (tha1.GE.TWOPI)
99e1apu   = e1a / Vbase
99e1bpu   = e1b / Vbase
99e1cpu   = e1c / Vbase
99e1pu    = SQRT((e1apu * e1apu + e1bpu * e1bpu + e1cpu * e1cpu) * 2 / 3)
C VSC1 EXCHANGED POWER COMPUTATION
99Pex1pu  = (e1apu * i1apu + e1bpu * i1bpu + e1cpu * i1cpu) / 1.5
99Pexch1  = Pex1pu * 1.5 * Pbase
99Qex1pu  = (e1apu * i1cpu - e1cpu * i1apu) * SQRT(3) / 1.5
C DC LINK CAPACITOR VOLTAGE COMPUTATION
99iDC1    = (Pexch1 / vDC) * SHNBRK
99DvDC    = iDC1 / Cap
 1vDC     +DvDC
       1.0
       0.0       1.0
99vDCpu   = vDC / Vbase
C VSC1 CURRENT COMPUTATION
99i1apu   = CNV01A / Ibase
99i1bpu   = CNV01B / Ibase
99i1cpu   = CNV01C / Ibase
99i1pu    = SQRT((i1apu * i1apu + i1bpu * i1bpu + i1cpu * i1cpu) * 2 / 3)
99i1ds    = i1apu
99i1qs    = -(i1apu + 2.0 * i1cpu) / SQRT(3)
99i1d     =  i1ds * COSPLL + i1qs * SINPLL
99i1q     = -i1ds * SINPLL + i1qs * COSPLL
99i1dq    = SQRT(i1d * i1d + i1q * i1q)
C SHUNT-EXCHANGED POWER AT THE POINT OF COMPENSATION
99Pcshpu  = (v1apu * i1apu + v1bpu * i1bpu + v1cpu * i1cpu) / 1.5
99Qcshpu  = (v1apu * i1cpu - v1cpu * i1apu) * SQRT(3) / 1.5
C ------TACS INITIAL CONDITIONS------
C -name>xx<--------|
77vDC      189000.0
```

74. 815STATV.USR

```
C File Name: 815STATV.USR
C USER INPUT FILE FOR STATCOM VOLTAGE CONTROL
```

```
C 000000001111111111222222222233333333334444444444555555555566666666667777777778
C 345678901234567890123456789012345678901234567890123456789012345678901234567890
C A SHUNT BREAKER IS CLOSED.
99SHNBRK  = TIMEX .GE. 0.05
99KpshI   = .45
99KishI   = 40
99Cap     = 42.0E-6
99KpshV   = 0.0
99KishV   = 2500.0
99Kdroop  = 0.0
C SETTING THE STATCOM REFERENCE VOLTAGE
99v1rcm1  = 0.025 * (TIMEX .GE. 0.05)
99v1rcm2  = 0.085 * (TIMEX .GE. 0.125)
99v1rcm3  = 0.2 * (TIMEX .GE. 0.175)
99v1rcm4  = 0.06 * (TIMEX .GE. 0.25)
99v1ref   = 0.99 + v1rcm1 + v1rcm2 - v1rcm3 + v1rcm4
C TACS OUTPUT
33i1qrefi1q   vDCpu v1apu e1apu i1apu alpha v1ref i1qlmt
```

75. 816STATV.CON

```
C File Name: 816STATV.CON
C STATCOM VOLTAGE CONTROL
C 000000001111111111222222222233333333334444444444555555555566666666667777777778
C 345678901234567890123456789012345678901234567890123456789012345678901234567890
C OUTER VOLTAGE CONTROL LOOP STARTS
99v1derr  = (v1ref - v1dqf + i1qdly * Kdroop) * SHNBRK
99v1dfnc  = v1dini + v1derr * DELTAT
99v1diup  = v1dfnc .GE. (i1qlmt / KishV)
99v1dilo  = v1dfnc .LE. (-i1qlmt / KishV)
99v1dimd  = .NOT. (v1diup .OR. v1dilo)
99v1dint  = v1dfnc*v1dimd + (i1qlmt*v1diup-i1qlmt*v1dilo)/KishV
  v1dini  +v1dint
99v1dpi   = -(KpshV * v1derr + KishV * v1dint)
99i1qup   = v1dpi .GE. i1qlmt
99i1qlo   = v1dpi .LE. -i1qlmt
99i1qmd   = .NOT. (i1qup .OR. i1qlo)
99i1qrfV  = v1dpi * i1qmd + i1qlmt * i1qup - i1qlmt * i1qlo
  i1qdly  +i1qrfV
C OUTER VOLTAGE CONTROL LOOP ENDS
C INNER CURRENT CONTROL LOOP STARTS
99i1qref  = -i1qrfV
99i1qerr  = (i1qref - i1q) * SHNBRK
99i1qint  = i1qini + i1qerr * DELTAT
  i1qini  +i1qint
99alpha   = -(KpshI * i1qerr + KishI * i1qint)
C INNER CURRENT CONTROL LOOP ENDS
99tha1    = theta + PI / 2 + alpha
C This addition of PI/2 is because the gating signals are sine reference and
C the PLL signals are cosine reference.
99theta1  = tha1 + TWOPI*(tha1.LT.0.0) - TWOPI*(tha1.GE.TWOPI)
99e1apu   = e1a / Vbase
99e1bpu   = e1b / Vbase
99e1cpu   = e1c / Vbase
99e1pu    = SQRT((e1apu * e1apu + e1bpu * e1bpu + e1cpu * e1cpu) * 2 / 3)
C VSC1 EXCHANGED POWER COMPUTATION
99Pexch1  = e1a * CNV01A + e1b * CNV01B + e1c * CNV01C
99Pex1pu  = Pexch1 / 1.5 / Pbase
99Qexch1  = (e1a * CNV01C - e1c * CNV01A) * SQRT(3)
```

```
C DC LINK CAPACITOR VOLTAGE COMPUTATION
99iDC1     = (Pexch1 / vDC) * SHNBRK
99DvDC     = iDC1 / Cap
 1vDC      +DvDC
        1.0
        0.0       1.0
99vDCpu    = vDC / Vbase
C VSC1 CURRENT COMPUTATION
99i1apu    = CNV01A / Ibase
99i1bpu    = CNV01B / Ibase
99i1cpu    = CNV01C / Ibase
99i1pu     = SQRT((i1apu * i1apu + i1bpu * i1bpu + i1cpu * i1cpu) * 2 / 3)
99i1ds     = i1apu
99i1qs     = -(i1apu + 2.0 * i1cpu) / SQRT(3)
99i1d      =  i1ds * COSPLL + i1qs * SINPLL
99i1q      = -i1ds * SINPLL + i1qs * COSPLL
99i1dq     = SQRT(i1d * i1d + i1q * i1q)
99i1qlmt   = SQRT(1.0 - i1d * i1d)
C SHUNT-EXCHANGED POWER AT THE POINT OF COMPENSATION
99Pcshpu   = (v1apu * i1apu + v1bpu * i1bpu + v1cpu * i1cpu) / 1.5
99Qcshpu   = (v1apu * i1cpu - v1cpu * i1apu) * SQRT(3) / 1.5
C ------TACS Initial Conditions------
C -name>xx<--------|
77vDC      189000.0
77v1dqf    0.990000
```

76. 824SSSCO.USR

```
C File Name: 824SSSCO.USR
C USER INPUT FILE FOR SSSC OPEN LOOP CONTROL
C 0000000111111111122222222223333333333444444444455555555556666666666777777777778
C 3456789012345678901234567890123456789012345678901234567890123456789012345678890
C A SERIES BREAKER IS OPENED.
99SRSBRK   = 0
C Setting the SSSC reference voltage
99e2rcm1   = 0.09375 * (TIMEX .GE. 0.05)
99e2rcm2   = 0.05475 * (TIMEX .GE. 0.175)
99e2rcm3   = 0.2627 * (TIMEX .GE. 0.30)
99e2rcm4   = 0.0904 * (TIMEX .GE. 0.45)
99e2cmd    = -e2rcm1 - e2rcm2 + e2rcm3 + e2rcm4
 1e2ref    +e2cmd
        1.0
        1.0     0.0320
C TACS OUTPUT
33e2apu i2apu Qex2pue2ref
```

77. 825SSSCO.CON

```
C File Name: 825SSSCO.CON
C SSSC OPEN LOOP CONTROL
C 0000000111111111122222222223333333333444444444455555555556666666666777777777778
C 3456789012345678901234567890123456789012345678901234567890123456789012345678890
C VSC2 CONTROL ANGLE COMPUTATION
99tha2     = theta + thetir - PI / 2
C This addition of -PI/2 is because the capacitive compensating voltage
C lags the line current by PI/2.
99theta2   = tha2 + TWOPI * (tha2 .LT. 0.0) - TWOPI * (tha2 .GE. TWOPI)
99e2a      = e2ref * Vbase * COS(theta2)
```

```
99e2b      = e2ref * Vbase * COS(theta2 - TWOPI / 3)
99e2c      = e2ref * Vbase * COS(theta2 + TWOPI / 3)
99e2apu    = e2a / Vbase
99e2bpu    = e2b / Vbase
99e2cpu    = e2c / Vbase
99e2pu     = SQRT((e2apu * e2apu + e2bpu * e2bpu + e2cpu * e2cpu) * 2 / 3)
C VSC2 OUTPUT CURRENT COMPUTATION
99i2apu    = CNV02A / Ibase
99i2bpu    = CNV02B / Ibase
99i2cpu    = CNV02C / Ibase
99i2pu     = SQRT((i2apu * i2apu + i2bpu * i2bpu + i2cpu * i2cpu) * 2 / 3)
C VSC2 EXCHANGED POWER COMPUTATION
99Pex2pu   = (e2apu * i2apu + e2bpu * i2bpu + e2cpu * i2cpu) / 1.5
99Qex2pu   = (e2apu * i2cpu - e2cpu * i2apu) * SQRT(3) / 1.5
C SETTING UP THE DISCONNECT SWITCHES
99GNDSW    = 0
99SRSDS    = .NOT. GNDSW
```

78. 826SSSC1.DAT

```
BEGIN NEW DATA CASE
C File Name: 826SSSC1.DAT
C SSSC CLOSED LOOP OPERATION WITH AN INFINITE-PULSE VSC
C 00000001111111111222222222233333333334444444444555555555566666666667777777777778
C 34567890123456789012345678901234567890123456789012345678901234567890123456789 0
C ----dt<---tmax<---Xopt<---Copt<--Toler<        ><        >
16.666-6600.00-3
C 00000001111111111222222222233333333334444444444555555555566666666667777777777778
C 34567890123456789012345678901234567890123456789012345678901234567890123456789 0
C -Iprnt<--Iplot<-Idoubl<-KssOut<-MaxOut<        ><        ><---Icat<-Nenerg
      40      10                                                 1
TACS HYBRID
99TWOPI    = 2.0 * PI
99CRD      = 180 / PI
99CDR      = 1 / CRD
C BASE VOLTAGE, CURRENT, AND POWER
99Vbase    = 112676.528
99Ibase    = 946.662704
99Zbase    = Vbase / Ibase
99Pbase    = Vbase * Ibase
$INCLUDE 302MEAS1.SWT
91CNV02A
91CNV02B
91CNV02C
$INCLUDE 303PTCT.SCL
$INCLUDE 310VPLL.PLL
$INCLUDE 827SSSCC.USR
$INCLUDE 828SSSCC.CON
$INCLUDE 604IDL22.VSC
$INCLUDE 305LINER.CMP
BLANK RECORD ENDING TACS
C 00000001111111111222222222233333333334444444444555555555566666666667777777777778
C 34567890123456789012345678901234567890123456789012345678901234567890123456789 0
C <-Bus1<-Bus2<-Bus3<-Bus4<----R<----L<----C                                    V
$INCLUDE 306NTWK1.BRN
  BUS01ABUS02A            1.00+8
  BUS01BBUS02B            1.00+8
  BUS01CBUS02C            1.00+8
  e2a   e22a              1.00+8
```

```
  e2b   e22b              1.00+8
  e2c   e22c              1.00+8
  e22a                    1.00+8
  e22b                    1.00+8
  e22c                    1.00+8
$INCLUDE 313TRAN2.TRN
BLANK RECORD ENDING BRANCHES
C <-Bus1<-Bus2<--Vig---|<-Ihold--|<-Tdeion-|xxxxxxxxxxCLOSEDSAMEGRID->SIGNAL  oo
C <-Bus1<-Bus2xxxxxxxxxxxxxxxxxxxxxxxxxxxxxxxxxxxxxxxxxCLOSEDxxxxxxxxxxSIGNAL  oo
13BUS01ABUS02A                                                        SRSBRK
13BUS01BBUS02B                                                        SRSBRK
13BUS01CBUS02C                                                        SRSBRK
13e2a   e22a                                                          SRSDS
13e2b   e22b                                                          SRSDS
13e2c   e22c                                                          SRSDS
13e22a                                              CLOSED            GNDSW
13e22b                                              CLOSED            GNDSW
13e22c                                              CLOSED            GNDSW
$INCLUDE 307MEAS2.SWT
  CNV02Ae22a                                        MEASURING
  CNV02Be22b                                        MEASURING
  CNV02Ce22c                                        MEASURING
BLANK RECORD ENDING SWITCHES
C 0000000111111111122222222223333333333444444444455555555556666666666777777777778
C 3456789012345678901234567890123456789012345678901234567890123456789012345678901234567890
60e2a
60e2b
60e2c
$INCLUDE 308NTWK1.SRC
BLANK RECORD ENDING SOURCES
BLANK RECORD ENDING NODE VOLTAGE OUTPUT
BLANK RECORD ENDING PLOT
BEGIN NEW DATA CASE
BLANK RECORD ENDING ALL DATA CASES
```

79. 827SSSCC.USR

```
C File Name: 827SSSCC.USR
C USER INPUT FILE FOR SSSC CLOSED LOOP CONTROL
C 0000000111111111122222222223333333333444444444455555555556666666666777777777778
C 3456789012345678901234567890123456789012345678901234567890123456789012345678901234567890
C A SERIES BREAKER IS OPENED.
99SRSBRK  = 0
99KpseV   = 10
99KiseV   = 100
99Cap     = 42.0E-6
C SETTING THE SSSC REFERENCE REACTANCE
99Xccm1   = 0.15 * (TIMEX .GE. 0.05)
99Xccm2   = 0.15 * (TIMEX .GE. 0.175)
99Xccm3   = 0.4 * (TIMEX .GE. 0.30)
99Xccm4   = 0.05 * (TIMEX .GE. 0.45)
99Xccmd   = -Xccm1 - Xccm2 + Xccm3 + Xccm4
 1Xcref   +Xccmd
       1.0
       1.0    0.0020
C TACS OUTPUT
33vDCpu e2apu i2apu Xcref Qex2pu
```

80. 828SSSCC.CON

```
C File Name: 828SSSCC.CON
C SSSC CLOSED LOOP CONTROL
C 0000000111111111122222222223333333333444444444455555555556666666666777777777778
C 3456789012345678901234567890123456789012345678901234567890123456789012345678901234567890
C VSC2 CONTROL ANGLE COMPUTATION
99Vqcmd    = Xcref * idq
99VDCcmd   = ABS(Vqcmd / Kvsc2)
99VDCsgn   = SIGN(Vqcmd)
99VDCerr   = (VDCcmd - vDCpu) * SRSDS
99VDCint   = VDCini + VDCerr * DELTAT
  VDCini   +VDCint
99zeta     = VDCsgn * (KpseV * VDCerr + KiseV * VDCint)
99thetai   = theta + thetir
99thetav   = thetai - VDCsgn * PI / 2
99tha2     = thetav + PI / 2 + zeta
C This addition of PI/2 is because the gating signals are sine reference and
C the PLL signals are cosine reference.
99theta2   = tha2 + TWOPI * (tha2 .LT. 0.0) - TWOPI * (tha2 .GE. TWOPI)
99e2apu    = e2a / Vbase
99e2bpu    = e2b / Vbase
99e2cpu    = e2c / Vbase
99e2pu     = SQRT((e2apu * e2apu + e2bpu * e2bpu + e2cpu * e2cpu) * 2 / 3)
C VSC2 OUTPUT CURRENT COMPUTATION
99i2apu    = CNV02A / Ibase
99i2bpu    = CNV02B / Ibase
99i2cpu    = CNV02C / Ibase
99i2pu     = SQRT((i2apu * i2apu + i2bpu * i2bpu + i2cpu * i2cpu) * 2 / 3)
C VSC2 EXCHANGED POWER COMPUTATION
99Pex2pu   = (e2apu * i2apu + e2bpu * i2bpu + e2cpu * i2cpu) / 1.5
99Qex2pu   = (e2apu * i2cpu - e2cpu * i2apu) * SQRT(3) / 1.5
99Pexch2   = Pex2pu * 1.5 * Pbase
C DC LINK CAPACITOR VOLTAGE COMPUTATION
99iDC2     = (Pexch2 / (vDCpu + 1.0E-8) / Vbase) * SRSDS
99DvDCpu   = iDC2 / Cap / Vbase
C 0000000111111111122222222223333333333444444444455555555556666666666777777777778
C 3456789012345678901234567890123456789012345678901234567890123456789012345678901234567890
C <-Namexx+<In1--x+<In2--x+<In3--x+<In4--x+<In5--x<-GAIN<-FVLO<-FXHI<-NMLO<-NMHI
 1vDCpu    +DvDCpu                                        1.0   0.0001  1.00
C <---Num0<-----Num1<-----Num2<-----Num3<-----Num4<-----Num5<-----Num6<-----Num7
      1.0
C <---Den0<-----Den1<-----Den2<-----Den3<-----Den4<-----Den5<-----Den6<-----Den7
      0.0       1.0
99vDC      = vDCpu * Vbase
C SETTING UP THE DISCONNECT SWITCHES
99GNDSW    = 0
99SRSDS    = .NOT. GNDSW
```

81. 831SSSCI.USR

```
C File Name: 831SSSCI.USR
C USER INPUT FILE FOR SSSC OPEN LOOP CONTROL WITH VOLTAGE CONTROL METHOD
C 0000000111111111122222222223333333333444444444455555555556666666666777777777778
C 3456789012345678901234567890123456789012345678901234567890123456789012345678901234567890
C A SERIES BREAKER IS OPENED.
99SRSBRK   = 0
C SETTING THE COMPENSATING REACTANCE REFERENCE
99Xccm1    = 0.20 * (TIMEX .GE. 0.050)
99Xccm2    = 0.20 * (TIMEX .GE. 0.125)
```

```
99Xccm3    = 0.46 * (TIMEX .GE. 0.200)
99Xccm4    = 0.69 * (TIMEX .GE. 0.300)
99Xccm5    = 0.40 * (TIMEX .GE. 0.500)
99Xccmd    = -Xccm1 - Xccm2 + Xccm3 + Xccm4 + Xccm5
 1Xcref   +Xccmd
        1.0
        1.0      0.016
C TACS OUTPUT
33e2apu i2apu Xcref Qex2pu
```

82. 832SSSCI.CON

```
C File Name: 832SSSCI.CON
C SSSC OPEN LOOP CONTROL WITH CURRENT CONTROL METHOD
C 0000000111111111122222222223333333333444444444455555555556666666666777777777778
C 3456789012345678901234567890123456789012345678901234567890123456789012345678890
C VSC2 CONTROL ANGLE COMPUTATION
99Vqcmd    = Xcref * idq
99VDCsgn   = SIGN(Vqcmd)
99thetai   = theta + thetir
99thetav   = thetai - VDCsgn * PI / 2
99tha2     = thetav + PI / 2
C This addition of PI/2 is because the gating signals are sine reference and
C the PLL signals are cosine reference.
99theta2   = tha2 + TWOPI * (tha2 .LT. 0.0) - TWOPI * (tha2 .GE. TWOPI)
99e2apu    = e2a / Vbase
99e2bpu    = e2b / Vbase
99e2cpu    = e2c / Vbase
99e2pu     = SQRT((e2apu * e2apu + e2bpu * e2bpu + e2cpu * e2cpu) * 2 / 3)
C VSC2 OUTPUT CURRENT COMPUTATION
99i2apu    = CNV02A / Ibase
99i2bpu    = CNV02B / Ibase
99i2cpu    = CNV02C / Ibase
99i2pu     = SQRT((i2apu * i2apu + i2bpu * i2bpu + i2cpu * i2cpu) * 2 / 3)
C VSC2 EXCHANGED POWER COMPUTATION
99Pex2pu   = (e2apu * i2apu + e2bpu * i2bpu + e2cpu * i2cpu) / 1.5
99Qex2pu   = (e2apu * i2cpu - e2cpu * i2apu) * SQRT(3) / 1.5
C DC LINK CAPACITOR VOLTAGE COMPUTATION
99vDC      = ABS(Vqcmd) / Kvsc2 * Vbase
C SETTING UP THE DISCONNECT SWITCHES
99GNDSW    = 0
99SRSDS    = .NOT. GNDSW
```

83. 834SSSCV.USR

```
C File Name: 834SSSCV.USR
C USER INPUT FILE FOR SSSC OPEN LOOP CONTROL WITH VOLTAGE CONTROL METHOD
C 0000000111111111122222222223333333333444444444455555555556666666666777777777778
C 3456789012345678901234567890123456789012345678901234567890123456789012345678890
C A SERIES BREAKER IS OPENED.
99SRSBRK   = 0
C SETTING THE COMPENSATING VOLTAGE REFERENCE
99Vqcm1    = 0.060 * (TIMEX .GE. 0.050)
99Vqcm2    = 0.205 * (TIMEX .GE. 0.125)
99Vqcm3    = 0.200 * (TIMEX .GE. 0.250)
99Vqcm4    = 0.200 * (TIMEX .GE. 0.400)
99Vqcmd    = Vqcm1 - Vqcm2 - Vqcm3 - Vqcm4
 1Vqref   +Vqcmd
```

```
        1.0
        1.0     0.016
C TACS OUTPUT
33e2apu i2apu Vqref Qex2pu
```

84. 835SSSCV.CON

```
C File Name: 835SSSCV.CON
C SSSC OPEN LOOP CONTROL WITH VOLTAGE CONTROL METHOD
C 0000000111111111122222222223333333333444444444455555555556666666666777777777778
C 3456789012345678901234567890123456789012345678901234567890123456789012345678901234567890
C TRANSMISSION LINE VOLTAGE COMPUTATION
99vta      = v1apu - vrapu
99vtb      = v1bpu - vrbpu
99vtc      = v1cpu - vrcpu
99vtds     =  vta
99vtqs     = -(vta + 2.0 * vtc) / SQRT(3)
99vtd      =  vtds * COSPLL + vtqs * SINPLL
99vtq      = -vtds * SINPLL + vtqs * COSPLL
99vtdq     = SQRT(vtd * vtd + vtq * vtq)
99thata    = ATAN( vtq / (vtd + 1.0E-8) )
99thatb    = thata + PI * (vtd .LT. 0.0)
99thavt    = thatb + TWOPI * (vtd .LT. 0.0) - TWOPI * (vtd .GE. TWOPI)
C VSC2 CONTROL ANGLE COMPUTATION
C COUPLING TRANSFORMER RES = 1.7854 OHMS, TRANSMISSION LINE RES = 3.0159 OHMS
99VR       = idq * (1.7854 + 3.0159) / Zbase
99VXeff    = SQRT(vtdq * vtdq - VR * VR)
99phi      = ATAN(VR / (VXeff + 1.0E-8))
99cap1     = (Vqref .GT. 0.0)
99ind      = (Vqref .LT. 0.0) * (ABS(Vqref) .LT. vtdq)
99cap2     = (Vqref .LT. 0.0) * (ABS(Vqref) .GT. vtdq)
C capind   = -ind + (cap1 .OR. cap2)
99capind   = -(ind .OR. cap2) + cap1
99psi      = phi * (capind .EQ. 0.0) + (phi + PI) * cap1 + phi * ind - phi * cap2
99thetav   = theta + thavt + psi
99tha2     = thetav + PI / 2
C This addition of PI/2 is because the gating signals are sine reference and
C the PLL signals are cosine reference.
99theta2   = tha2 + TWOPI * (tha2 .LT. 0.0) - TWOPI * (tha2 .GE. TWOPI)
99e2apu    = e2a / Vbase
99e2bpu    = e2b / Vbase
99e2cpu    = e2c / Vbase
99e2pu     = SQRT((e2apu * e2apu + e2bpu * e2bpu + e2cpu * e2cpu) * 2 /3)
C VSC2 OUTPUT CURRENT COMPUTATION
99i2apu    = CNV02A / Ibase
99i2bpu    = CNV02B / Ibase
99i2cpu    = CNV02C / Ibase
99i2pu     = SQRT((i2apu * i2apu + i2bpu * i2bpu + i2cpu * i2cpu) * 2 /3)
C VSC2 EXCHANGED POWER COMPUTATION
99Pex2pu   = (e2apu * i2apu + e2bpu * i2bpu + e2cpu * i2cpu) / 1.5
99Qex2pu   = (e2apu * i2cpu - e2cpu * i2apu) * SQRT(3) / 1.5
C DC LINK CAPACITOR VOLTAGE COMPUTATION
99vDC      = ABS(Vqref) / Kvsc2 * Vbase
C SETTING UP THE DISCONNECT SWITCHES
99GNDSW    = 0
99SRSDS    = .NOT. GNDSW
```

85. 836UPFC1.DAT

```
BEGIN NEW DATA CASE
C File Name: 836UPFC1.DAT
C UPFC OPERATION WITH 2 INFINITE-PULSE VSCs
C 00000001111111111222222222233333333334444444444555555555566666666667777777778
C 34567890123456789012345678901234567890123456789012345678901234567890123456789 0
C ----dt<---tmax<---Xopt<---Copt<--Toler<       ><       >
16.666-6450.00-3
C 00000001111111111222222222233333333334444444444555555555566666666667777777778
C 34567890123456789012345678901234567890123456789012345678901234567890123456789 0
C -Iprnt<--Iplot<-Idoubl<-KssOut<-MaxOut<       ><       ><---Icat<-Nenerg
      30      10                                                 1
TACS HYBRID
99CRD     = 180 / PI
99CDR     = 1 / CRD
99TWOPI   = 2.0 * PI
C BASE VOLTAGE, CURRENT, AND POWER
99Vbase   = 112676.528
99Ibase   = 946.662704
99Zbase   = Vbase / Ibase
99Pbase   = Vbase * Ibase
$INCLUDE 302MEAS1.SWT
90POS
90NEG
91CNV01A
91CNV01B
91CNV01C
91CNV02A
91CNV02B
91CNV02C
$INCLUDE 303PTCT.SCL
$INCLUDE 310VPLL.PLL
$INCLUDE 837UPFC.USR
$INCLUDE 838UPFC.CON
$INCLUDE 670IDL31.VSC
$INCLUDE 672IDL32.VSC
$INCLUDE 305LINER.CMP
BLANK RECORD ENDING TACS
C 00000001111111111222222222233333333334444444444555555555566666666667777777778
C 34567890123456789012345678901234567890123456789012345678901234567890123456789 0
C <-Bus1<-Bus2<-Bus3<-Bus4<----R<----L<----C                                    V
$INCLUDE 306NTWK1.BRN
  BUS01ACNV01A             1.00+8
  BUS01BCNV01B             1.00+8
  BUS01CCNV01C             1.00+8
  BUS01ABUS02A             1.00+8
  BUS01BBUS02B             1.00+8
  BUS01CBUS02C             1.00+8
  e2a   e22a               1.00+8
  e2b   e22b               1.00+8
  e2c   e22c               1.00+8
  e22a                     1.00+8
  e22b                     1.00+8
  e22c                     1.00+8
 $INCLUDE 312TRAN1.TRN
 $INCLUDE 313TRAN2.TRN
   POS   NEG                            42.0+0
   iDCn1 POS                0.001
   iDCn1                   1.00+8
   iDCn2 NEG                0.001
```

```
  iDCn2                     1.00+5
BLANK RECORD ENDING BRANCHES
C <-Bus1<-Bus2<--Vig---|<-Ihold--|<-Tdeion-|xxxxxxxxxxCLOSEDSAMEGRID->SIGNAL  oo
C <-Bus1<-Bus2xxxxxxxxxxxxxxxxxxxxxxxxxxxxxxxxxxxxxxxxCLOSEDxxxxxxxxxxSIGNAL  oo
13BUS01ACNV01A                                                        SHNBRK
13BUS01BCNV01B                                                        SHNBRK
13BUS01CCNV01C                                                        SHNBRK
13BUS01ABUS02A                                                        SRSBRK
13BUS01BBUS02B                                                        SRSBRK
13BUS01CBUS02C                                                        SRSBRK
13e2a   e22a                                                          SRSDS
13e2b   e22b                                                          SRSDS
13e2c   e22c                                                          SRSDS
13e22a                                                CLOSED          GNDSW
13e22b                                                CLOSED          GNDSW
13e22c                                                CLOSED          GNDSW
$INCLUDE 307MEAS2.SWT
  CNV01Ae11a                                          MEASURING
  CNV01Be11b                                          MEASURING
  CNV01Ce11c                                          MEASURING
  CNV02Ae22a                                          MEASURING
  CNV02Be22b                                          MEASURING
  CNV02Ce22c                                          MEASURING
BLANK RECORD ENDING SWITCHES
60e1a
60e1b
60e1c
60e2a
60e2b
60e2c
60iDCn1 -1
60iDCn2 -1
$INCLUDE 308NTWK1.SRC
BLANK RECORD ENDING SOURCES
 2POS    94500.0
 2NEG    -94500.0
 3POS    NEG                      189000.0
BLANK RECORD ENDING NODE VOLTAGE OUTPUT
BLANK RECORD ENDING PLOT
BEGIN NEW DATA CASE
BLANK RECORD ENDING ALL DATA CASES
```

86. 837UPFC.USR

```
C File Name: 837UPFC.USR
C USER INPUT FILE FOR UPFC - VSC1 CLOSED LOOP CONTROL, VSC2 OPEN LOOP CONTROL
C 000000011111111112222222222333333333344444444445555555555666666666677777777778
C 345678901234567890123456789012345678901234567890123456789012345678901234567890
C A SHUNT BREAKER IS CLOSED.
99SHNBRK  = TIMEX .GE. 0.05
C A SERIES BREAKER IS OPENED.
99SRSBRK  = 0
99KpshI   = .45
99KishI   = 40
99KpshV   = 0.0
99KishV   = 2500.0
99Kdroop  = 0.0
C SET vctrl = 1.0 FOR VOLTAGE CONTROL OR vctrl = 0.0 FOR CURRENT CONTROL
99vctrl   = 0.0
```

```
C SETTING THE STATCOM REFERENCE VOLTAGE
99v1rcm1  = 0.025 * (TIMEX .GE. 0.05)
99v1rcm2  = 0.085 * (TIMEX .GE. 0.125)
99v1rcm3  = 0.2 * (TIMEX .GE. 0.175)
99v1rcm4  = 0.06 * (TIMEX .GE. 0.25)
99v1ref   = (0.99 + v1rcm1 + v1rcm2 - v1rcm3 + v1rcm4) * 0 + 1
C SETTING THE STATCOM REFERENCE REACTIVE CURRENT
99i1qcm1  = TIMEX .GE. 0.125
99i1qcm2  = 2 * (TIMEX .GE. 0.175)
99i1qcm3  = 2 * (TIMEX .GE. 0.25)
99i1qcmd  = (i1qcm1 - i1qcm2 + i1qcm3) * 0
 1i1qrfI  +i1qcmd
       1.0
       1.0    0.0040
C SETTING THE SSSC REFERENCE VOLTAGE
99e2cm1   = 0.2 * (TIMEX .GE. 0.100)
99e2cm2   = 0.2 * (TIMEX .GE. 0.300)
99e2dqcm  = (e2cm1 + e2cm2)
 1e2dqrf  +e2dqcm
       1.0
       1.0    0.0040
99beta1   = 300 * (TIMEX .GE. 0.100)
99beta2   = 60 * (TIMEX .GE. 0.200)
99beta    = (beta1 - beta2) * CDR
C TACS OUTPUT
33i1qrefi1q   vDCpu v1apu e1apu i1apu alpha e2apu i2apu PcsepuQcsepui1qlmt
```

87. 838UPFC.CON

```
C File Name: 838UPFC.CON
C UPFC - VSC1 CLOSED LOOP CONTROL, VSC2 OPEN LOOP CONTROL
C 00000001111111111222222222233333333334444444444555555555566666666667777777778
C 345678901234567890123456789012345678901234567890123456789012345678901234567890
C VSC1 CURRENT COMPUTATION
99i1apu   = CNV01A / Ibase
99i1bpu   = CNV01B / Ibase
99i1cpu   = CNV01C / Ibase
99i1pu    = SQRT((i1apu * i1apu + i1bpu * i1bpu + i1cpu * i1cpu) * 2 / 3)
99i1ds    = i1apu
99i1qs    = -(i1apu + 2.0 * i1cpu) / SQRT(3)
99i1d     =  i1ds * COSPLL + i1qs * SINPLL
99i1q     = -i1ds * SINPLL + i1qs * COSPLL
99i1dq    = SQRT(i1d * i1d + i1q * i1q)
99i1qlmt  = SQRT(1.0 - i1d * i1d)
C VSC2 CURRENT COMPUTATION
99i2apu   = CNV02A / Ibase
99i2bpu   = CNV02B / Ibase
99i2cpu   = CNV02C / Ibase
99i2pu    = SQRT((i2apu * i2apu + i2bpu * i2bpu + i2cpu * i2cpu) * 2 / 3)
99i2ds    = i2apu
99i2qs    = -(i2apu + 2.0 * i2cpu) / SQRT(3)
99i2d     =  i2ds * COSPLL + i2qs * SINPLL
99i2q     = -i2ds * SINPLL + i2qs * COSPLL
99i2dq    = SQRT(i2d * i2d + i2q * i2q)
C OUTER VOLTAGE CONTROL LOOP STARTS
99v1derr  = (v1ref - v1dqf + i1qdly * Kdroop) * SHNBRK
99v1dfnc  = v1dini + v1derr * DELTAT
99v1diup  = v1dfnc .GE. (i1qlmt / KishV)
99v1dilo  = v1dfnc .LE. (-i1qlmt / KishV)
```

```
99v1dimd  = .NOT. (v1diup .OR. v1dilo)
99v1dint  = v1dfnc*v1dimd + (i1qlmt*v1diup-i1qlmt*v1dilo)/KishV
  v1dini  +v1dint
99v1dpi   = -(KpshV * v1derr + KishV * v1dint)
99i1qup   = v1dpi .GE. i1qlmt
99i1qlo   = v1dpi .LE. -i1qlmt
99i1qmd   = .NOT. (i1qup .OR. i1qlo)
99i1qrfV  = v1dpi * i1qmd + i1qlmt * i1qup - i1qlmt * i1qlo
  i1qdly  +i1qrfV
C OUTER VOLTAGE CONTROL LOOP ENDS
C INNER CURRENT CONTROL LOOP STARTS
99i1qref  = -i1qrfV * vctrl + i1qrfI * (.NOT. vctrl)
99i1qerr  = (i1qref - i1q) * SHNBRK
99i1qint  = i1qini + i1qerr * DELTAT
  i1qini  +i1qint
99alpha   = -(KpshI * i1qerr + KishI * i1qint)
C INNER CURRENT CONTROL LOOP ENDS
99tha1    = theta + PI / 2 + alpha
C This addition of PI/2 is because the gating signals are sine reference
C and the Phase-Locked Loop signals are cosine reference.
99theta1  = tha1 + TWOPI * (tha1 .LT. 0.0) - TWOPI * (tha1 .GE. TWOPI)
99gamma1  = 3.75 * CDR
99e1apu   = e1a / Vbase
99e1bpu   = e1b / Vbase
99e1cpu   = e1c / Vbase
99e1pu    = SQRT((e1apu * e1apu + e1bpu * e1bpu + e1cpu * e1cpu) * 2 / 3)
99tha2    = theta + PI / 2 + PI + beta
C This addition of PI/2 is because the gating signals are sine reference and
C the PLL signals are cosine reference.
C The PI is added to reverse the polarity of the series coupling transformer.
99thet2   = tha2 + TWOPI * (tha2 .LT. 0.0) - TWOPI * (tha2 .GE. TWOPI)
99theta2  = thet2 + TWOPI * (thet2 .LT. 0.0) - TWOPI * (thet2 .GE. TWOPI)
99cosga2  = e2dqrf / vDCpu / Kvsc2
99ga2d    = cosga2 + .00000001
99ga2q    = SQRT(1 - cosga2 * cosga2)
99gamma2  = ATAN(ga2q / ga2d)
99e2apu   = e2a / Vbase
99e2bpu   = e2b / Vbase
99e2cpu   = e2c / Vbase
99e2pu    = SQRT((e2apu * e2apu + e2bpu * e2bpu + e2cpu * e2cpu) * 2 / 3)
C VSC1 EXCHANGED POWER COMPUTATION
99Pex1pu  = (e1apu * i1apu + e1bpu * i1bpu + e1cpu * i1cpu) / 1.5
99Pexch1  = Pex1pu * 1.5 * Pbase
99Qex1pu  = (e1apu * i1cpu - e1cpu * i1apu) * SQRT(3) / 1.5
C VSC2 EXCHANGED POWER COMPUTATION
99Pex2pu  = (e2apu * i2apu + e2bpu * i2bpu + e2cpu * i2cpu) / 1.5
99Pexch2  = Pex2pu * 1.5 * Pbase
99Qex2pu  = (e2apu * i2cpu - e2cpu * i2apu) * SQRT(3) / 1.5
C DC LINK CAPACITOR VOLTAGE COMPUTATION
99vDC     = POS - NEG
99vDCpu   = vDC / Vbase
99iDC1    = (Pexch1 / vDC) * SHNBRK
99iDC2    = (Pexch2 / vDC) * SHNBRK
99iDCnet  = iDC1 + iDC2
99iDCn1   = iDCnet
99iDCn2   = -iDCn1
C SERIES-EXCHANGED POWER AT THE POINT OF COMPENSATION
99Pcsepu  = (v12apu * iapu + v12bpu * ibpu + v12cpu * icpu) / 1.5
99Qcsepu  = (v12apu * icpu - v12cpu * iapu) * SQRT(3) / 1.5
C Setting up the disconnect switches
99GNDSW   = 0
99SRSDS   = .NOT. GNDSW
```

BIBLIOGRAPHY

I. BOOKS

1. IEEE Working Group (15.05.07) on HVDC Bibliography and Records, "An Annotated Bibliography of HVDC Transmission and FACTS Devices 1991–1993," in *Bonneville Power Administration, Western Area Power Administration,* 1993.
2. IEEE Working Group (15.05.07) on HVDC Bibliography and Records, "An Annotated Bibliography of HVDC Transmission and FACTS Devices 1994–1995," in *Bonneville Power Administration,* 1995.
3. IEEE Working Group (15.05.07) on HVDC Bibliography and Records, "An Annotated Bibliography of HVDC Transmission and FACTS Devices 1996–1997," in *Bonneville Power Administration,* Jun. 1998.
4. Y. H. Song and A. T. Johns, *Flexible AC Transmission Systems (FACTS).* London, UK: IEE, 1999.
5. N. G. Hingorani and L. Gyugyi, *Understanding FACTS: Concepts and Technology of Flexible AC Transmission Systems.* New York: IEEE Press, 2000.
6. R. M. Mathur and R. K. Varma, *Thyristor-Based FACTS Controllers for Electrical Transmission Systems.* New York: IEEE Press and Wiley, 2002.
7. E. Acha, C. R. Fuerte-Esquivel, H. Ambriz-Pérez and C. Angeles-Camacho, *FACTS: Modelling and Simulation in Power Networks,* Hoboken, NJ: Wiley, 2004.
8. K. H. Sueker, *Power Electronics Design: A Practitioner's Guide,* 1st Edition, Elsevier, 2005
9. K. R. Padiyar, *Facts: Controllers in Power Transmission and Distribution.* Bangalore, India: Anshan, 2007.

II. GENERAL

10. Westinghouse, "AC voltage regulator," in *Electrical Transmission and Distribution Reference Book,* East Pittsburgh, PA: Westinghouse Electric Corp., 1942.

Introduction to FACTS Controllers. By Kalyan K. Sen and Mey Ling Sen

11. L. Gyugyi and M. B. Brennen, "Method and control apparatus for static VAR generator and compensator," U.S. Patent no. 3,999,117, 1976.
12. W. S. Meyer, "EMTP data modularization and sorting by class: a foundation upon which EMTP data bases can be built," *EMTP Newsletter,* vol. 4, no. 2, pp. 28–40, Nov. 1983.
13. E. J. Stacey and M. B. Brennen, "Active power conditioner system," U.S. Patent 4,651,265, 1987.
14. T. T. Nguyen and W. D. Humpage, "TCR and SR compensators in composite electromagnetic transient simulation," *IEE Proceedings,* pt. C, vol. 136, no. 3, pp. 195–205, May 1989.
15. *ATP Rule Book,* Leuven EMTP Center, 1990.
16. L. J. Bohmann and R. H. Lasseter, "Stability and harmonics in thyristor controlled reactors," *IEEE Transactions Power Delivery,* vol. 5, no. 2, pp. 1175–1181, Apr. 1990.
17. A. M. Gole and V. K. Sood, "A static compensator model for use with electromagnetic transients simulation programs," *IEEE Transactions Power Delivery,* vol. 5, no. 3, pp. 1398–1407, Jul. 1990.
18. C. K. K. Tang and D. R. Turner, "Modeling of large nonlinear loads and thyristor controlled shunt compensators," *IEE Proceedings,* pt. C, vol. 138, no. 6, pp. 505–517, Nov. 1991.
19. A. N. Vasconcelos, A. J. P. Ramos, J. S. Monteiro, M. B. V. C. Lima, H. D. Silva, and L. R. Lins, "Detailed modeling of an actual static VAR compensator for electromagnetic transients studies," *IEEE Transactions Power Systems,* vol. 7, no. 1, pp. 11–19, Feb. 1992.
20. J. A. Martinez, "How to adopt the EMTP for classroom instruction," *IEEE Transactions Power Systems,* vol. 7, no. 1, pp. 351–358, Feb. 1992.
21. S. Y. Lee, S. Bhattacharya, T. Lejonberg, A. E. Hammad, and S. Lefebvre, "Detailed modeling of static VAR compensators using the Electromagnetic Transients Program (EMTP)," *IEEE Transactions Power Delivery,* vol. 7, no. 2, pp. 836–847, Apr. 1992.
22. S. Lefebvre and L. Gerin-Lajoie, "A static compensator model for the EMTP," *IEEE Transactions Power Systems,* vol. 7, no. 2, pp. 477–486, May 1992.
23. A. J. F. Keri, et. al., "Improving transmission system performance using controlled series capacitors," Paris (CIGRE), Paper 14/37/38-07, 1992.
24. B. Pilvelait, T. H. Ortmeyer, and D. Maratukulam, "Advanced series compensation for transmission system using a switched capacitor module," *IEEE Transactions Power Delivery,* vol. 8, no. 2, pp. 584–590, Apr. 1993.
25. A. T. Johns and Q. Y. Xuan, "Digital study of thyristor controlled series capacitor compensated EHV transmission systems," in *IEEE Proceedings 10th IEEE Regional Conference Computer, Communication, Control and Power Engineering, TENCON,* pp. 263–265.
26. L. Gyugyi, et al, "Apparatus and method for dynamic voltage restoration of utility distribution networks," U.S. Patent 5 329 222, 1994.
27. S. G. Helbing and G. G. Karady, "Investigations of an advanced form of series compensation," *IEEE Transactions Power Delivery,* vol. 9, no. 2, pp. 939–947, Apr. 1994.
28. M. R. Iravani, P. L. Dandeno, K. H. Nguyen, D. Zhu, and D. Maratukulam, "Applications of static phase shifters in power systems," *IEEE Transactions Power Delivery,* vol. 9, no. 3, pp. 1600–1608, Jul. 1994.
29. M. B. Brennen and B. Banerjee, "Low cost, high performance active power line condition-

ers," in *Third International Conference on Power Quality: End-Use Applications and Perspectives,* EPRI, Amsterdam, Oct. 1994.

30. M. R. Iravani and D. J. Maratukulam, "Review of semiconductor controlled (static) phase shifters for power system application," *IEEE Transactions Power Systems,* vol. 9, no. 4, Nov. 1994.
31. T. F. Godart, A. F. Imece, J. C. McIver, and E. A. Chebli, "Feasibility of thyristor controlled series capacitor for distribution substation enhancements," *IEEE Transactions Power Delivery,* vol. 10, no. 1, pp. 203–209, Jan. 1995.
32. M. Kezunovic, M. Aganagic, S. McKenna, and D. Hamai, "Computing responses of series compensation capacitors with MOV protection in real-time," *IEEE Transactions Power Delivery,* vol. 10, no. 1, pp. 244–251, Jan. 1995.
33. A. Ghosh and G. Ledwich, "Modeling and control of thyristor-controlled series capacitors," *IEE Proceedings Generation, Transmission and Distribution,* vol. 142, no. 3, pp. 297–304, May 1995.
34. J. A. Martinez, "EMTP simulation of a digitally-controlled static VAr system for optimal load compensation," *IEEE Transactions Power Delivery,* vol. 10, no. 3, pp. 1408–1415, Jul. 1995.
35. W. Zhu, R. Spee, R. R. Mohler, G. C. Alexander,W. A. Mittelstadt, and D. Maratukulam, "An EMTP study of SSR mitigation using the thyristor controlled series capacitor," *IEEE Transactions Power Delivery,* vol. 10, no. 3, pp. 1479–1485, Jul. 1995.
36. D. J. Nyati, S. Eitzmann, M. Kappenman, J. VanHouse, D. Mohan, and A. N. Edris, "Design issues for a single core transformer thyristor controlled phase-angle regulator," *IEEE Transactions Power Delivery,* vol. 10, no. 4, pp. 2013–2019, Oct. 1995.
37. S. Lefebvre, A. M. Gole, J. Reeve, L. Piliotto, N. Martins, and S. Bhattacharya, "Working group on dynamic performance and modeling of DC systems and power electronics for transmission systems: Report on test systems for AC/DC interaction studies," *IEEE Transactions Power Delivery,* vol. 10, no. 4, pp. 2027–2034, Oct. 1995.
38. G. E. Lee and D. L. Goldsworthy, "Bonneville Power Administration's Pacific AC intertie series capacitors: experience, equipment and protection," *IEEE Transactions Power Delivery,* vol. 11, no. 1, pp. 253–259, Jan. 1996.
39. M. Junya, I. Keiichi, T. Takeshi, and A. Minoru, "Dynamic performance of a thyristor controlled series capacitor system," in *Proceedings 31st Universities Power Engineering Conference,* vol. 3, pp. 961–964, Iraklio, Greece, 1996.
40. H. A. Othman and L. Angquist, "Analytical modeling of thyristor controlled series capacitors for SSR studies," *IEEE Transactions Power Systems,* vol. 11, no. 1, pp. 119–127, Feb. 1996.
41. S. G. Jalali, R. A. Hedin, M. Pereira, and K. Sadek, "Stability model for the advanced series compensator (ASC)," *IEEE Transactions Power Delivery,* vol. 11, no. 2, pp. 1128–1137, Apr. 1996.
42. A. R. M. Tenorio, J. B. Ekanayake, and N. Jenkins, "Modeling of FACTS devices," in *Proceedings IEE Conference AC and DC Power Transmission,* pp. 340–345, London, U.K., Apr./May 1996.
43. A. C. S. de Lima, S. S. Wanderley, and R. M. Stephan, "Modeling a static VAR compensator using EMTP," in *Proceedings 38th Midwest Symposium on Circuits and Systems,* vol. 2, pp. 219–222, Rio de Janeiro, Brazil, 1996.
44. J. Ndayizamba, S. Lefebvre, and D. D. Xuan, "Efficient and flexible thyristor controlled

series capacitor simulation in EMTP-TACS," in *Proceedings 12th Power Systems Computation Conference,* vol. 2, pp. 1122–1127, Dresden, Germany, 1996.

45. J. Ndayizamba, S. Lefebvre, and D. D. Xuan, "Torsional interaction in flexible AC transmission systems with a thyristor controlled series capacitor," in *Proceedings 5th Electrimacs, Modeling and Simulation of Electric Machines, Converters and Systems,* pp. 1187–1192, Saint Nazaire, France, 1996.
46. B. Das, A. Ghosh, and A. Sachchidanand, "A new control scheme for dynamic brake," in *Proceedings Ninth National Power Systems Conference, NPSC,* vol. 2, pp. 501–507, Kanpur, India, 1996.
47. A. M. Gole, A. Keri, C. Kwankpa, E. W. Gunther, H. W. Dommel, I. Hassan, J. R. Marti, J. A. Martinez, K. G. Fehrle, L. Tang, M. F. McGranaghan, O. B. Nayak, P. F. Ribeiro, R. Iravani, and R. Lasseter, "Guidelines for modeling power electronics in electric power engineering applications," *IEEE Transactions Power Delivery,* vol. 12, no. 1, pp. 505–514, Jan. 1997.
48. P. Parihar and G. G. Karady, "Characterization of a thyristor controlled reactor," *Electric Power Systems Research,* vol. 41, no. 2, pp. 141–149, May 1997.
49. L. Wang, "Simulations of prefiring NGH damping scheme on suppressing torsional oscillations using EMTP," *IEEE Transactions Power Systems,* vol. 12, no. 2, pp. 882–888, May 1997.
50. L. Prikler and G. Banfai, "EMTP models for simulation of shunt reactor switching transients," *International Journal Electrical Power Energy Systems,* vol. 19, no. 4, pp. 235–240, May 1997.
51. M. R. Sultan, J. Reeve, and R. Adapa, "Transient Behavior of Systems Containing FACTS Devices: Modeling of the Thyristor Controlled Series Capacitor," EPRI Final Report TR-108191, 2149-15, Jul. 1997.
52. P. Mattavelli, G. C. Verghese, and A. M. Stankovic, "Phasor dynamics of thyristor controlled series capacitor systems," *IEEE Transactions Power Systems,* vol. 12, no. 3, pp. 1259–1267, Aug. 1997.
53. E. A. Leonidaki, B. C. Papadias, N. D. Hatziargyriou, and S. Giorgas, "A physical laboratory model of a thyristor controlled series capacitor," in *Proceedings 33rd Universities Power Engineering Conference,* Edinburgh, U.K., 1998.
54. O. M. Neto, P. C. A. Leao, and A. E. A. Araujo, "Controlled series compensation: Dynamic performance under fault conditions," in *Proceedings 32nd Universities Power Engineering Conference, UPEC,* vol. 1, pp. 5–8, Manchester, U.K., 1997.
55. O. M. Neto, R. A. Meneguim, F. D. Oliveira, and J. M. Oliveira, "Modeling of a thyristor controlled series capacitor in matlab environment," in *Proceedings 32nd Univertsities Power Engineering Conference, UPEC,* vol. 1, pp. 21–14, Manchester, U.K., 1997.
56. J. A. Martinez and N. Mohan, "ATP simulation of power electronics systems using a data-module approach," in *Proceedings 32nd Universities Power Engineering Conference, UPEC,* vol. 1, pp. 495–498, Manchester, U.K., 1997.
57. E. A. Leonidaki, N. D. Hatziargyriou, and G. J. Georgantzis, "Simulation of thyristor controlled series capacitors and subsynchronous studies," in *Proceedings 32nd Universities Power Engineering Conference, UPEC,* vol. 2, pp. 883–886, Manchester, U.K., 1997.
58. H.-G. Han, J.-K. Park, and B.-H. Le, "Analysis of thyristor controlled series compensator dynamics using the state variable approach of a periodic system model," *IEEE Transactions Power Delivery,* vol. 12, no. 4, pp. 1744–1750, Oct. 1997.

59. B. K. Perkins and M. R. Iravani, "Dynamic modeling of a TCSC with application to SSR analysis," *IEEE Transactions Power Systems,* vol. 12, no. 4, pp. 1619–1625, Nov. 1997.
60. A. A. Girgis, A. A. Sallam, and A. K. El-Din, "An adaptive protection scheme for Advanced Series Compensated (ASC) transmission lines," *IEEE Transactions Power Delivery,* vol. 13, no. 2, pp. 414–420, Apr. 1998.
61. L. Palav and A. M. Gole, "On using the solid-state breaker in distribution systems," in *Proceedings IEEE Canadian Conference Electrical and Computer Engineering,* vol. 1, pp. 693–700, Waterloo, Ontario, 1998.
62. A. R. M. Tenorio and N. Jenkins, "Use of EMTP-ATP MODELS language for modeling thyristor controlled series capacitors," in *Proceedings 33rd Universities Power Engineering Conference, UPEC,* vol. 2, pp. 679–682, Edinburgh, U.K., 1998.
63. M. Sultan, J. Reeve, and R. Adapa, "Combined transient and dynamic analysis of HVDC and FACTS systems," *IEEE Transactions Power Delivery,* vol. 13, no. 4, pp. 1271–1277, Oct. 1998.
64. P. S. S. Holenarsipur, D. J. Christofersen, and N. Mohan, "Thyristor switched arrester for improved protection of power system equipment," in *Proceedings IEEE PES Winter Meeting,* vol. 2, pp. 1009–1014, New York, 1999.
65. H. Xueqiang-Zhao and H. Chen-Chen, "Circuit analysis of a thyristor controlled series compensation," in *Proceedings IEEE PES Winter Meeting,* vol. 2, pp. 1067–1072, New York, 1999.
66. Y. Tanaka, H. Taniguchi, M. Egawa, H. Fujita, M. Watanabe, and H. Konishi, "Using a miniature model and EMTP simulations to evaluate new methods to control and protect a thyristor controlled series compensator," in *Proceedings IEEE PES Winter Meeting,* vol. 2, pp. 1196–1201, New York, 1999.
67. P. Mattavelli, G. C. Verghese, and A. M. Stankovic, "SSR analysis with dynamic phasor model of thyristor controlled series capacitor," *IEEE Transactions Power Systems,* vol. 14, no. 1, pp. 200–208, Feb. 1999.
68. Z. Xu and C. Chen, "Damping subsynchronous resonance using an improved NGH SSR damping," in *Proceedings IEEE PES Summer Meeting,* vol. 2, pp. 780–785, Edmonton, AB, Canada, 1999.
69. D. D. Xueqiang-Zhao, "Study of TCSC model and prospective application in the power systems of China," in *Proceedings International Conference Power Electronics and Drive Systems, PEDS,* vol. 2, pp. 688–691, Kowloon, Hong Kong, 1999.
70. C. Gama and R. Tenorio, "Improvements for power system performance: modeling, analysis and benefits of TCSCs," in *Proceedings IEEE PES Winter Meeting,* vol. 2, pp. 1462–1467, 2000.
71. L. F. W. De Souza, E. H. Watanabe, and M. Aredes, "GTO controlled series capacitor," in *Proceedings IEEE PES Winter Meeting,* vol. 4, pp. 2520–2525, 2000.
72. L. F. W. De Souza, E. H. Watanabe, and M. Aredes, "GTO controlled series capacitors: multi-module and multi-pulse arrangements," *IEEE Transactions Power Delivery,* vol. 15, no. 2, pp. 725–731, Apr. 2000.
73. P. K. Muttik, P. Wang, and M. W. H. Minchin, "Detailed simulation of SVC transient performance using PSCAD," *Proceedings International Conference Power System Technology, PowerCon,* Perth, Australia, 2000.
74. R. J. Davalos and J. M. Ramirez, "A review of a quasistatic and a dynamic TCSC model," *IEEE Power Engineering Rev.,* vol. 20, no. 11, pp. 63–65, Nov. 2000.

75. EPRI, "Evaluate solid-state LTC options for medium power transformers: Project 41C3084/6658-6424," Electric Power Research Institute, Palo Alto, CA, EPRI Tech. Rep. 1,000,916, Dec. 2000.
76. B. K. Johnson, "Simulation of TSC and TSSC," in *Proceedings IEEE PES Winter Meeting,* vol. 2, pp. 637–638, 2001.
77. L. L. Grigsby, *The Electric Power Engineering Handbook,* vol. 3, pp. 191–200, Boca Raton, FL: CRC 2001.
78. PSCAD/EMTDC Manual, Manitoba HVDC Research Centre, 2001.
79. B. K. Johnson, "Benchmark system for the simulation of TCSC and SVC," in *Proceedings IEEE PES Winter Meeting,* pp. 484–487, New York, 2002.
80. S. Al-Senaidi, "An application of thyristor controlled series capacitor to selective pole switching," *Electric Power Components & Systems,* vol. 30, no. 8, pp. 14–20, Aug. 2002.
81. K. K. Sen and M. L. Sen, "Versatile power flow transformers for compensating power flow in a transmission line," U.S. Patents 6,335,613, 6,384,581, 6,396,248, and 6,420,856, 2002.
82. K. K. Sen and M. L. Sen, "Introducing the family of Sen transformers: a set of power flow controlling transformers," *IEEE Transactions Power Delivery,* vol. 18, no. 1, pp. 149–157, Jan. 2003.
83. K. K. Sen, M. L. Sen, "Comparison of the Sen transformer with the unified power flow controller," *IEEE Transactions Power Delivery,* vol. 18, no. 3, pp. 1523–1533, Oct. 2003.
84. K. K. Sen and M. L. Sen, "Multiline power flow transformer for compensating power flow among transmission lines," U.S. Patent 6,841,976, 2005.
85. M. O. Faruque, V. Dinavahi, S. Santoso, R. Adapa, "Review of electromagnetic transient models of non-VSC FACTS," *IEEE Transactions on Power Delivery,* vol. 20, no. 2, pp. 1065–1079, April 2005.
86. M. O. Faruque, V. Dinavahi, "A tap-changing algorithm for the implementation of Sen transformer," *IEEE Transactions on Power Delivery,* vol. 22, no. 3, pp. 1750–1757, July 2007.
87. B. Asghari, M. O. Faruque, V. Dinavahi, "Detailed real-time transient model of Sen transformer," *IEEE Transactions on Power Delivery,* vol. 23, no. 3, pp. 1513–1521, July 2008.
88. W. Geraldes, E. H. Rose, C. Machado, J. T. Honda, and P. M. Vendramini, "Static VAR compensator from Coxipo substation (Brazil), comparison between computer simulations and field tests," in *Proceedings 5th International Conference AC and DC Power Transmission,* pp. 301–306, London.

III. STATCOM

89. C. Schauder and H. Mehta, "Vector analysis and control of advanced static VAR compensators," *IEE Proceedings Generation, Transmission and Distribution,* vol. 140, no. 4, pp. 299–306, July 1993.
90. Y. Ni and L. Snider, "STATCOM power frequency model with VSC charging dynamics and its application in the power system stability analysis," in *Proceedings Fourth International Conference on Advances in Power System Control, Operation and Management, APSCOM,* vol. 1, pp. 119–124, Nov. 1997.

91. C. Hochgraf and R. H. Lasseter, "Statcom controls for operation with unbalanced voltages," *IEEE Transactions on Power Delivery,* vol. 13, pp. 538–544, no. 2, April 1998.
92. K. V. Patil, J. Senthil, J. Jiang, and R. M. Mathur, "Application of STATCOM for damping torsional oscillations in series compensated AC systems ," *IEEE Transactions on Energy Conversion,* vol. 13, no. 3, pp. 237–243, Sept. 1998.
93. P. W. Lehn and M. R. Iravani, "Experimental evaluation of STATCOM closed loop dynamics," *IEEE Transactions on Power Delivery,* vol. 13, no. 4, pp. 1378–1384, Oct. 1998.
94. H. Okuno, T. Tominaka, S. Fujishima, T. Mitsumoto, T. Kubo, T. Kawaguchi, J. Kim, K. Ikegami, N. Sakamoto, N. Yokouchi, T. Morikawa, T. Tanaka, A. Goto, and Y. Yano, "Physical experiments on a 10 kVAr STATCOM model," in *Proceedings IEEE PES Winter Meeting,* vol. 2, pp. 1075–1077, 1999.
95. K. K. Sen, "STATCOM-STATic synchronous COMpensator: Theory, Modeling, and Applications," in *Proceedings IEEE PES Winter Meeting,* vol. 2, pp. 1177–1183, 1999.
96. H. F. Wang, "Phillips-Heffron model of power systems installed with STATCOM and applications," *IEE Proceedings Generation, Transmission and Distribution,* vol. 146, no. 5, pp. 521–527, Sept. 1999.
97. H. Yonezawa, T. Shimato, M. Tsukada, K. Matsuno, I. Iyoda, J. J. Paserba, and G. F. Reed, "Study of a STATCOM application for voltage stability evaluated by dynamic PV curves and time simulations," in *Proceedings IEEE PES Winter Meeting,* vol. 2, pp. 1471–1476, 2000.
98. Y. Yu, C. Jianye, and H. Yingduo, "STATCOM modeling and analysis in damping power system oscillations," in *Proceedings 35th Intersociety Conference and Exhibition on Energy Conversion Engineering, IECEC,* vol. 2, pp. 756–762, July 2000.
99. P. Garica-Gonzalez and A. Garcia-Cerrada, "Control system for a PWM-based STATCOM," *IEEE Transactions on Power Delivery,* vol. 15, no. 4, pp. 1252–1257, Oct. 2000.
100. P. Rao, M. L. Crow, and Z. Yang, "STATCOM control for power system voltage control applications," *IEEE Transactions on Power Delivery,* vol. 15, no. 4, pp. 1311–1317, Oct. 2000.
101. J. A. Martinez, "Modeling of Custom Power equipment using electromagnetic transients programs," in *Proceedings Ninth International Conference on Harmonics and Quality of Power,* vol. 3, pp. 769–774, Oct. 2000.
102. L. Chun, J. Qirong, and W. Zhonghong, "Study of STATCOM control for power swings damping improvement," *Proceedings International Conference on Power System Technology, PowerCon,* vol. 1, pp. 535–540, Dec. 2000.
103. S. Chen, G. Joos, and L. T. Moran, "Dynamic performance of PWM STATCOMs operating under unbalance and fault conditions in distribution systems," in *Proceedings IEEE PES Winter Meeting,* vol. 2, pp. 950–955, 2001.
104. P. S. Sensarma, K. R. Padiyar, and V. Ramanarayanan, "Analysis and performance evaluation of a distribution STATCOM for compensating voltage fluctuations," *IEEE Transactions on Power Delivery,* vol. 16, no. 2, pp. 259–264, April 2001.
105. Z. Yang, C. Shen, L. Zhang, M. L. Crow, and S. Atcitty, "Integration of StatCom and battery energy storage," *IEEE Transactions on Power Systems,* vol. 16, no. 2, pp. 254–260, May 2001.
106. V. Dinavahi, R. Iravani, and R. Bonert, "Real-time digital simulation of power electronic apparatus interfaced with digital controllers," *IEEE Transactions on Power Delivery,* vol. 16, no. 4, pp. 775–781, Oct. 2001.
107. B. K. Keshavan and N. Prabhu, "Damping of subsynchronous oscillations using STAT-

COM—a FACTS device," in *Proceedings IEEE/PES Transmission and Distribution Conference and Exposition,* vol. 1, pp. 1–7, 2001.

108. P. W. Lehn, "Exact modeling of the voltage source converter," *IEEE Transactions on Power Delivery,* vol. 17, no. 1, pp. 217–222, Jan. 2002.
109. O. Anaya-Lara and E. Acha, "Modeling and analysis of custom power systems by PSCAD/EMTDC," *IEEE Transactions on Power Delivery,* vol. 17, no. 1, pp. 266–272, Jan. 2002.
110. L. Ran, L. Holdsworth, and G. A. Putrus, "Dynamic selective harmonic elimination of a three-level inverter used for static VAr compensation," *IEE Proceedings Generation, Transmission and Distribution,* vol. 149, no. 1, pp. 83–89, Jan. 2002.
111. P. W. Lehn, "A benchmark system for simulation of the DSTATCOM," in *Proceedings IEEE PES Winter Meeting,* vol. 1, pp. 496–498, 2002.
112. C. Qian and M. L. Crow, "A cascaded converter-based StatCom with energy storage," in *Proceedings IEEE PES Winter Meeting,* vol. 1, pp. 544–549, 2002.
113. C. K. Sao, P. W. Lehn, M. R. Iravani, and J. A. Martinez, "A benchmark system for digital time-domain simulation of a pulse-width modulated D-STATCOM," *IEEE Transactions on Power Delivery,* vol. 17, no. 4, pp. 1113–1120, Oct. 2002.
114. L. Cong and Y. Wang, "Coordinated control of generator excitation and STATCOM for rotor angle stability and voltage regulation enhancement of power systems," *IEE Proceedings Generation, Transmission and Distribution,* vol. 149, no. 6, pp. 659–666, Nov. 2002.
115. K. K. Sen and A. J. F. Keri, "Comparison of field results and digital simulation results of voltage-sourced converter-based FACTS controllers," *IEEE Transactions on Power Delivery,* vol. 18, no. 1, pp. 300–306, Jan 2003.
116. A. B. Arsoy, Y. Liu, P. F. Ribeiro, and F. Wang , "StatCom-SMES," *IEEE Industry Applications Magazine,* vol. 9, no. 2, pp. 21–28, March-April 2003.
117. V. Dinavahi, R. Iravani, and R. Bonert, "Design of a real-time digital simulator for a D-STATCOM system," *IEEE Transactions on Industrial Electronics,* vol. 51, no. 5, pp. 1001–1008, Oct. 2004.

IV. SSSC

118. L. Gyugyi, C. D. Schauder, and K. K. Sen, "Static synchronous series compensator: a solid-state approach to the series compensation of transmission lines," *IEEE Transactions on Power Delivery,* vol. 12, no. 1, pp. 406–417, Jan. 1997.
119. K. K. Sen, "SSSC-static synchronous series compensator: theory, modeling, and application," *IEEE Transactions on Power Delivery,* vol. 13, no. 1, pp. 241–246, Jan. 1998.
120. R. Mihalic, "Power flow control with controllable reactive series elements," *IEE Proceedings Generation, Transmission and Distribution,* vol. 145, no. 5, pp. 493–498, Sept. 1998.
121. H. Okuno, T. Tominaka, S. Fujishima, T. Mitsumoto, T. Kubo, T. Kawaguchi, J.-W. Kim, K. Ikegami, N. Sakamoto, S. Yokouchi, T. Morikawa, T. Tanaka, A. Goto, and Y. Yano, "Switching-level simulation model for SSSC with EMTP," in *Proceedings IEEE PES Winter Meeting,* vol. 2, pp. 1075–1077, 1999.
122. R. Mihalic and I. Papic, "Mathematical models and simulation of a static synchronous se-

ries compensator," in *Proceedings IEEE International Conference on Electric Power Engineering, PowerTech,* 1999.
123. L. Sunil Kumar and A. Ghosh, "Modeling and control design of a static synchronous series compensator," *IEEE Transactions on Power Delivery,* vol. 14, no. 4, pp. 1448–1453, Oct. 1999.
124. B. Han, S. Moon, J. Park, and G. Karady, "Dynamic characteristic analysis of multi-bridge PWM inverter for SSSC," in *Proceedings IEEE PES Summer Meeting,* vol. 3, pp. 1618–1623, 2000.
125. J. Seo, S. Moon, J. Park, and J. Choe, "Design of a robust SSSC supplementary controller to suppress the SSR in the series compensated system," in *Proceedings IEEE PES Winter Meeting,* vol. 3, pp. 1283–1288, 2001.
126. B. Han, S. Baek, H. Kim, and G. Karady, "Dynamic characteristic analysis of SSSC based on multi-bridge inverter," *IEEE Transactions on Power Delivery,* vol. 17, no. 2, pp. 623–629, April 2002.
127. G. N. Pillai, A. Ghosh, and A. Joshi, "Torsional interaction between an SSSC and a PSS in a series compensated power system," *IEE Proceedings Generation, Transmission and Distribution,* vol. 149, no. 6, pp. 653–658, Nov. 2002.

V. UPFC

128. L. Gyugyi, "A unified power flow control concept for flexible AC transmission systems," *Proceedings Institute Electrical Engineering C,* vol. 139, no. 4, July 1992
129. L. Gyugyi, C. D. Schauder, S. L. Williams, T. R. Rietman, D. R. Torgerson, and A. Edris, "The unified power flow controller: A new approach to power transmission control," *IEEE Transactions on Power Delivery,* vol. 10, no. 2, pp. 1085–1097, April 1995.
130. R. Mihalic, P. Zunko, and D. Povh, "Improvement of transient stability using unified power flow controller," *IEEE Transactions on Power Delivery,* vol. 11, no. 1, pp. 485–492, Jan. 1996.
131. A. Nabavi-Niaki and M. R. Iravani, "Steady-state and dynamic models of unified power flow controller (UPFC) for power system studies," *IEEE Transactions on Power Systems,* vol. 11, no. 4, pp. 1937–1943, Nov. 1996.
132. K.S. Smith, L. Ran, and J. Penman, "Dynamic modelling of a unified power flow controller," *IEE Proceedings Generation, Transmission and Distribution,* vol. 144, no. 1, pp. 7–12, Jan. 1997.
133. X. Lombard, and P. G. Therond, "Control of unified power flow controller: comparison of methods on the basis of a detailed numerical model," *IEEE Transactions on Power Systems,* vol. 12, no. 2, pp. 824–830, May 1997.
134. A. M. Kulkarni and K. R. Padiyar, "Performance evaluation of unified power flow controller using transient simulation," *Proceedings International Conference on Power Electronics and Drive Systems,* vol. 2, pp. 734–739, 1997.
135. I. Papic, P. Zunko, D. Povh, and M. Weinhold, "Basic control of unified power flow controller," *IEEE Transactions on Power Systems,* vol. 12, no. 4, pp. 1734–1739, Nov. 1997.
136. M. Rahman, M. Ahmed, R. Gutman, R.J. O'Keefe, R. J. Nelson, and J. Bian, "UPFC application on the AEP system: planning considerations," *IEEE Transactions on Power Systems,* vol. 12, no. 4, pp. 1695–1701, Nov. 1997.

137. C. D. Schauder, L. Gyugyi, M. R. Lund, D. M. Hamai, T. R. Rietman, D. R. Torgerson, and A. Edris, "Operation of the unified power flow controller (UPFC) under practical constraints" *IEEE Transactions on Power Delivery,* vol. 13, no. 2, pp. 630–639, April 1998.

138. A. S. Mehraban, A. Edris, C. D. Schauder, and J. H. Provanzana, "Installation, commissioning, and operation of the world's first UPFC on the AEP system," *Proceedings International Conference on Power System Technology, POWERCON,* vol. 1, pp. 323–327, Aug. 1998.

139. K.R. Padiyar and A. M. Kulkarni, "Control design and simulation of unified power flow controller," *IEEE Transactions on Power Delivery,* vol. 13, no. 4, pp. 1348–1354, Oct. 1998.

140. K. K. Sen and E. J. Stacey, "UPFC-unified power flow controller: theory, modeling, and applications," *IEEE Transactions on Power Delivery,* vol. 13, no. 4, pp. 1453–1460, Oct. 1998.

141. C. Schauder, E. Stacey, M. Lund, L. Gyugyi, L. Kovalsky, A. Keri, A. Mehraban, and A. Edris, "AEP UPFC project: installation, commissioning and operation of the 160 MVA STATCOM (phase I)," *IEEE Transactions on Power Delivery,* vol. 13, no. 4, pp. 1530–1535, Oct. 1998.

142. Y. Ni, Z. Huang, S. Chen, and B. Zhang, "Incorporating UPFC Model into the Power System Toolbox of the MATLAB for Transient Stability Study," in *Proceedings IEEE Region 10 International Conference on Global Connectivity in Energy, Computer, Communication and Control, TENCON,* vol. 2, pp. 506–509, Dec. 1998.

143. H. Okuno, T. Tominaka, S. Fujishima, T. Mitsumoto, T. Kubo, T. Kawaguchi, J. W. Kim, K. Ikegami, N. Sakamoto, S. Yokouchi, T. Morikawa, T. Tanaka, A. Goto, and Y. Yano, "Unified power flow controller on the AEP system: commissioning and operation," in *Proceedings IEEE PES Winter Meeting,* vol. 2, 31, pp. 1075–1077, 1999.

144. A. J. F. Keri, A. S. Mehraban, X. Lombard, A. Eiriachy, and A. A. Edris, "Unified power flow controller (UPFC): modeling and analysis," *IEEE Transactions on Power Delivery,* vol. 14, no. 2, pp. 648–654, April 1999.

145. Y. Morioka, M. Kato, Y. Mishima, Y. Nakachi, M. Asada, and K. Tokuhara, "Implementation of unified power flow controller and verification for transmission capability improvement," *IEEE Transactions on Power Systems,* vol. 14, no. 2, pp. 575–581, May 1999.

146. H. F. Wang, "Applications of modelling UPFC into multimachine power systems," *IEE Proceedings Generation, Transmission and Distribution,* vol. 146, no. 3, pp. 306–312, May 1999.

147. T. Makombe and N. Jenkins, "Investigation of a unified power flow controller," *IEE Proceedings Generation, Transmission and Distribution,* vol. 146, no. 4, pp. 400–408, Jul. 1999.

148. R. Caldon, A. Mari, A. Paolucci, and N. Turri, "An efficient modular model for static and dynamic analysis of FACTS performances on electric power systems," in *Proceedings International Conference on Electric Power Engineering, PowerTech,* 1999.

149. B. A. Renz, A. Keri, A. S. Mehraban, C. Schauder, E. Stacey, L. Kovalsky, L. Gyugyi, and A. Edris, "AEP unified power flow controller performance," *IEEE Transactions on Power Delivery,* vol. 14, no. 4, pp. 1374–1381, Oct. 1999.

150. H. Fujita, Y. Watanabe, and H. Akagi, "Control and analysis of a unified power flow controller," *IEEE Transactions on Power Electronics,* vol. 14, no. 6, pp. 1021–1027, Nov. 1999.

151. S. Zheng, Y. L. Tan, S. Cheng, and Y. Zhu, "Dynamic characteristic study of UPFC based on a detailed simulation model," in *Proceedings IEEE PES Winter Meeting,* vol. 2, pp. 1360–1364, 2000.
152. C. A. Canizares, "Power flow and transient stability models of FACTS controllers for voltage and angle stability studies," in *Proceedings IEEE PES Winter Meeting,* vol. 2, pp. 1447–1454, 2000.
153. Z. J. Meng and P. L. So, "A current injection UPFC model for enhancing power system dynamic performance," in *Proceedings IEEE PES Winter Meeting,* vol. 2, pp. 1544–1549, 2000.
154. S. Arabi, P. Kundur, and R. Adapa, "Innovative techniques in modeling UPFC for power system analysis," *IEEE Transactions on Power Systems,* vol. 15, no. 1, pp. 336–341, Feb. 2000.
155. Z. Huang, Y. Ni, C. M. Shen, F. F. Wu, S. Chen, and B. Zhang, "Application of unified power flow controller in interconnected power systems-modeling, interface, control strategy, and case study," *IEEE Transactions on Power Systems,* vol. 15 no. 2, pp. 817–824, May 2000.
156. H. Wang, "A unified model for the analysis of FACTS devices in damping power system oscillations. III. Unified power flow controller," *IEEE Transactions on Power Delivery,* vol. 15, no. 3, pp. 978–983, Jul. 2000.
157. D. J. Won and S. I. Moon, "A study on the dynamic responses of a PWM based UPFC and its application to power system," in *Proceedings IEEE PES Summer Meeting,* vol. 3, pp. 1590–1595, 2000.
158. K. Schoder, A. Hasanovic, and A. Feliachi, "Load-flow and dynamic model of the unified power flow controller (UPFC) within the Power System Toolbox (PST)," in *Proceedings 43rd IEEE Midwest Symposium on Circuits and Systems,* vol. 2, pp. 634–637, Aug. 2000.
159. K. Sen, "Simulation of UPFC," in *Proceedings IEEE PES Winter Meeting,* vol. 2, pp. 641–642, 2001.
160. J. Seo, S. Moon, J. Park, and J. Choe, "Design of a robust UPFC controller for enhancing the small signal stability in the multimachine power systems," in *Proceedings IEEE PES Winter Meeting,* vol. 3, pp. 1197–1202, 2001.
161. Y. L. Kang, G. B. Shrestha, and T. T. Lie, "Consideration of DC capacitor and line inductance in the design of UPFC to improve transient stability," in *Proceedings IEEE PES Winter Meeting,* vol. 3, pp. 1277–1282, 2001.
162. K. Sedraoui, K. Al-haddad, and G. Olivier, "A new approach for the dynamic control of unified power flow controller (UPFC)," in *Proceedings IEEE PES Summer Meeting,* vol. 2, pp. 955–960, 2001.
163. Y. L. Kang, G.B. Shresta, and T.T. Lie, "Application of an NLPID controller on a UPFC to improve transient stability of a power system," *IEE Proceedings Generation, Transmission and Distribution,* vol. 148, no. 6, pp. 523–529, 2001.
164. P. C. Stefanov and A. M. Stankovic, "Modeling of UPFC operation under unbalanced conditions with dynamic phasors," *IEEE Transactions on Power Systems,* vol. 17, no. 2, pp. 395–403, May 2002.
165. H. Fujita, H. Akagi and Y. Watanabe, "Dynamic performance of a unified power flow controller for stabilizing AC transmission systems," in *Proceedings 33rd Annual IEEE Power Electronics Specialists Conference,* vol. 1, pp. 81–87, 2002.

166. C. T. Chang and Y. Y. Hsu, "Design of UPFC controllers and supplementary damping controller for power transmission control and stability enhancement of a longitudinal power system," *IEE Proceedings Generation, Transmission and Distribution,* vol. 149, no. 4, pp. 463–471, Jul. 2002.
167. J. B. Choo, et al, "Development of FACTS operation technology to the KEPCO power network—installation & operation," in *Proceedings IEEE Power Engineering Society Transmission & Distribution Conference,* Yokohama, Japan, Paper no. 231, Oct. 2002.
168. P. Garcia-Gonzalez and A. Garcia-Cerrada, "Detailed analysis and experimental results of the control system of a UPFC," *IEE Proceedings Generation, Transmission and Distribution,* vol. 150, no. 2, pp. 147–154, Mar. 2003.

VI. IPFC

169. B. Fardanesh, et al., "Convertible Static Compensator application to the New York transmission system," Paris (CIGRE), Paper 14-103, 1998.
170. L. Gyugyi, K. K. Sen, and C. D. Schauder, "The interline power flow controller concept: a new approach to power flow management in transmission systems," *IEEE Transactions on Power Delivery,* vol. 14, no. 3, pp. 1115–1123, Jul. 1999.
171. B. Fardanesh, B. Shperling, E. Uzunovic, and S. Zelingher, "Multiconverter FACTS devices: the generalized unified power flow controller (GUPFC)," in *Proceedings IEEE PES Summer Meeting,* vol. 2, 16–20, pp. 1020–1025, 2000.
172. J. Chen, T. T. Lie, and D. M. Vilathgamuwa, "Basic control of interline power flow controller," in *Proceedings IEEE PES Winter Meeting,* vol. 1, pp. 521–525, 2002.
173. X.-P. Zhang, "Modelling of the interline power flow controller and the generalised unified power flow controller in Newton power flow," *IEE Proceedings Communications,* vol. 150, no. 3, pp. 268–274, May 2003.

INDEX

Pages followed by f and t indicate figures and tables.

ABOUT THE AUTHORS

Kalyan K. Sen was born in Bankura, West Bengal, India. He received B.E.E (first class honors), M.S.E.E, and Ph.D. degrees, all in Electrical Engineering, from Jadavpur University (India), Tuskegee University, and Worcester Polytechnic Institute, respectively. He spent 22 years in academia and industry and became a Westinghouse Fellow Engineer. He was a key member of the FACTS development team at Westinghouse Science and Technology Center in Pittsburgh. In the last 19 years, he worked mostly at Westinghouse and its successor companies, except during 1999–2001 when he worked at ABB in Sweden. He contributed to all aspects (concept development, simulation, design, and commissioning) of FACTS projects at Westinghouse. Dr. Sen conceived some of the basic concepts in FACTS technology. He has 25 patents and publications in the areas of FACTS and power electronics. His interests are in power converters, control, electrical machines, and power system simulations and studies. He is a licensed Professional Engineer in the Commonwealth of Pennsylvania.

Dr. Sen, a Senior Member of IEEE, has served the organization in many positions. In 2003, he reestablished the Pittsburgh Chapters of the Power and Energy Society and the Industry Applications Society. Both chapters received the "Outstanding Large Chapter" awards for their activities in 2004. Under his chairmanship, the Pittsburgh Section received the "Outstanding Large Section" award for its activities in 2005. He served as an Editor of the *IEEE Transactions on Power Delivery* from 2002 to 2007. He served as the Technical Program Chair of the 2008 Power and Energy Society General Meeting in Pittsburgh, and the Chapters and Sections Activities Track Chair at the 2008 IEEE Sections Congress in Quebec City, Canada. He has been serving as an IEEE Distinguished Lecturer since 2002.

Mey Ling Sen was born in Aruba, Dutch Caribbean. She received B.S.E.E. (high distinction) and M.E.E. degrees, all in Electrical Engineering, from Worcester Polytechnic Institute and Rice University, respectively. As an engineering consultant, she worked at Westinghouse and ABB. She is the President of Sen Engineering Solutions. She is the co-inventor of the Sen transformer, which is the most efficient, reliable, and cost-effective FACTS controller. Her interests are in electronics, electrical machines, and power engineering. As a member of IEEE, she has served the Pittsburgh Chapters of the Power and Energy Society and Industry Applications Society in various positions, including Chapter Chair.